W9-CLR-211

ANNUAL REVIEW OF PLANT PHYSIOLOGY AND PLANT MOLECULAR BIOLOGY

ANNUAL REVIEW OF PLANT PHYSIOLOGY AND PLANT MOLECULAR BIOLOGY

VOLUME 47, 1996

RUSSELL L. JONES, *Editor*
University of California, Berkeley

CHRISTOPHER R. SOMERVILLE, *Associate Editor*
Carnegie Institution of Washington, Stanford, California

VIRGINIA WALBOT, *Associate Editor*
Stanford University

http://annurev.org science@annurev.org 415-493-4400
ANNUAL REVIEWS INC. 4139 EL CAMINO WAY P.O. BOX 10139 PALO ALTO, CALIFORNIA 94303-0139

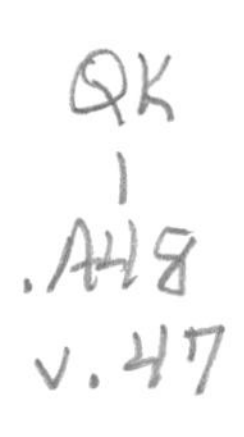

ANNUAL REVIEWS INC.
Palo Alto, California, USA

International Standard Serial Number: 1040-2519
International Standard Book Number: 0-8243-0647-3
Library of Congress Catalog Card Number: A-51-1660

The paper used in this publication meets the minimum requirements of American National Standards for Information Sciences—Permanence of Paper for Printed Library Materials, ANZI Z39.48-1984

Typesetting by Ruth McCue Saavedra and the Annual Reviews Inc. Editorial Staff

PRINTED AND BOUND IN THE UNITED STATES OF AMERICA

ANNUAL REVIEWS INC. is a nonprofit scientific publisher established to promote the advancement of the sciences. Beginning in 1932 with the *Annual Review of Biochemistry,* the Company has pursued as its principal function the publication of high-quality, reasonably priced *Annual Review* volumes. The volumes are organized by Editors and Editorial Committees who invite qualified authors to contribute critical articles reviewing significant developments within each major discipline. The Editor-in-Chief invites those interested in serving as future Editorial Committee members to communicate directly with him. Annual Reviews Inc. is administered by a Board of Directors, whose members serve without compensation.

AR Annual Review of Plant Physiology and Plant Molecular Biology
Volume 47 (1996)

CONTENTS

INDEXES

SOME RELATED ARTICLES IN OTHER *ANNUAL REVIEWS*

From the *Annual Review of Biochemistry,* Volume 65 (1996)

Relationships Between DNA Repair and Transcription, EC Friedberg
DNA Excision Repair, A Sancar
Selenocysteine, TC Stadtman
DNA Repair in Eukaryotes, RD Wood
Mechanisms of Helicase-Catalyzed DNA Unwinding, TM Lohman and KP Bjornson
Protein Prenylation: Molecular Mechanisms and Functional Consequences, FL Zhang and PJ Casey
Protein Transport Across the Eukaryotic Endoplasmic Reticulum and Bacterial Inner Membranes, TA Rapoport, B Jungnickel, and U Kutay
Molecular Biology of Mammalian Amino Acid Transporters, MS Malandro and MS Kilberg
Telomere Length Regulation, CW Greider
Molecular Genetics of Signal Transduction in Dictyostelium, CA Parent and PN Devreotes
Structural Basis of Lectin-Carbohydrate Recognition, WI Weis and K Drickamer
Rhizobium Lipo-Chitooligosaccharide Nodulation Factors: Signaling Molecules Mediating Recognition and Morphogenesis, J Dénarié, F Debellé, and J-C Promé
Electron Transfer in Proteins, HB Gray and JR Winkler
Crosstalk Between Nuclear and Mitochondrial Genomes, RO Poyton and JE McEwen
DNA Topoisomerases, JC Wang
Interrelationships of the Pathways of mRNA Decay and Translation in Eukaryotic Cells, A Jacobson and SW Peltz
Recoding: Dynamic Reprogramming of Translation, RF Gesteland and JF Atkins
Biochemistry and Structural Biology of Transcription Factor IID (TFIID), SK Burley and RG Roeder

From the *Annual Review of Biophysics and Biomolecular Structure,* Volume 24 (1995)

Membrane-Structure Studies Using X-Ray Standing Waves, M Caffrey and J Wang
Structure and Function of DNA Methyltransferases, X Cheng
NMR Spectroscopic Studies of Paramagnetic Proteins: Iron-Sulfur Proteins, H Cheng and JL Markley
Molecular and Structural Basis of Target Recognition by Calmodulin, A Crivici and M Ikura
Compact Intermediate States in Protein Folding, AL Fink
Mass Spectrometry of Nucleic Acids: The Promise of Matrix-Assisted Laser Desorption-Ionization (MALDI) Mass Spectrometry, MC Fitzgerald and LM Smith
Thermodynamics of Partly Folded Intermediates in Proteins, E Freire

Complexes of the Minor Groove of DNA, BH Geierstanger and DE Wemmer
Fluorescent Protein Biosensors: Measurement of Molecular Dynamics in Living Cells, KA Giuliano, PL Post, KM Hahn, and DL Taylor
Capillary Electrophoresis of Proteins and Nucleic Acids, BL Karger, Y-H Chu, and F Foret
Actin-Binding Protein Complexes at Atomic Resolution, PJ McLaughlin and AG Weeds
Site-Directed Mutagenesis with an Expanded Genetic Code, D Mendel, VW Cornish, and PG Schultz
Design of Molecular Function: Channels of Communication, M Montal
DNA Analogues with Nonphosphodiester Backbones, PE Nielsen
Applications of Parallel Computing to Biological Problems, B Ostrovsky, MA Smith, and Y Bar-Yam
Nucleic Acid Hybridization: Triplex Stability and Energetics, GE Plum, DS Pilch, SF Singleton, and KJ Breslauer
Lectin Structure, JM Rini
Site-Specific Dynamics in DNA: Theory, BH Robinson and GP Drobny
Flexible Docking and Design, R Rosenfeld, S Vajda, and C DeLisi
Exceptionally Stable Nucleic Acid Hairpins, G Varani
Structure and Mechanism of DNA Topoisomerases, DB Wigley

From the *Annual Review of Cell and Developmental Biology,* Volume 11 (1995)

Control of Actin Assembly at Filament Ends, A Schafer and JA Cooper
Integrins: Emerging Paradigms of Signal Transduction, M Ginsberg, MA Schwartz, and MD Schaller
Protein Import into the Nucleus: An Integrated View, GR Hicks and NV Raikhel
COPs Regulating Membrane Traffic, TE Kreis, M Lowe, and R Pepperkok
Silencing and Heritable Domains of Gene Expression, S Loo and J Rine
Biological Atomic Force Microscopy: From Microns to Nanometers and Beyond, Z Shao, J Yang, and AP Somlyo
The Nucleolus, PJ Shaw and EG Jordan
Receptor-Mediated Protein Sorting to the Vacuole in Yeast: Roles for Protein Kinase, Lipid Kinase, and GTP-Binding Proteins, H Stack, B Horazdovsky, and SD Emr
Heat Shock Transcsription Factors: Structure and Regulation, C Wu

From the *Annual Review of Genetics,* Volume 29 (1995)

Light-Harvesting Complexes in Oxygenic Photosynthesis: Diversity, Control, and Evolution: AR Grossman, D Bhaya, KE Apt, and DM Kehoe
Genetic Analysis of the Form and Function of Photosystem I and Photosystem II, HB Pakrasi
The Plant Response in Pathogenesis, Symbiosis, and Wounding: Variations on a Common Theme? C Baron and PC Zambryski
Membrane Protein Assembly: Genetic, Evolutionary, and Medical Perspectives, Colin Manoil and Beth Traxler

Conservation Genetics, R Frankham
The Genetics of Flower Development: From Floral Induction to Ovule Morphogenesis, D Weigel
Molecular Genetic Aspects of Human Mitochondrial Disorders: N-G Larsson and DA Clayton
Genetics of Ustilago maydis, *a Fungus Pathogen that Induces Tumors in Maize,* F Banuett
Chlamydomonas reinhardtii *as the Photosynthetic Yeast,* J-D Rochaix
Homologous Recombination Proteins in Prokaryotes and Eukaryotes, RD Camerini-Otero and P Hsieh
Meiotic Recombination Hotspots, M Lichten and ASH Goldman
DNA Repair in Humans, A Sancar

From the *Annual Review of Microbiology,* Volume 49 (1995)

Nitrogenase Structure and Function: A Biochemical-Genetic Perspective, JW Peters, K Fisher, and DR Dean

From the *Annual Review of Physiology,* Volume 57 (1995)

The Multifunctional Calcium/Calmodulin-Dependent Protein Kinase: From Form to Function: AP Braun and H Schulman
Review of Mechanosensitive Channels, H Sackin

From the *Annual Review of Phytopathology,* Volume 34 (1996)

Breeding Disease-Resistant Wheats for Tropical Highlands and Lowlands, HJ Dubin and S Rajaram
Microbial Elicitors and Their Receptors in Plants, MG Hahn
Molecular Biology of Rice Tungro Viruses, R Hull
Genetics of the Resistance to Wheat Leaf Rust, JA Kolmer
Avirulence Genes of Bacteria, JE Leach and FF White
Morphogenesis and Mechanisms of Penetration by Plant-Pathogenic Fungi, K Mendgen, M Hahn, and H Deising
Ozone and Plant Health, H Sandermann
Plant Virus Gene Vectors for Transient Expression of Foreign Proteins in Plants, HB Scholthof, K-BG Scholthof, and AO Jackson
QTL Mapping and Quantitative Disease Resistance in Plants, ND Young

J. MacMillan

Annu. Rev. Plant Physiol. Plant Mol. Biol. 1996. 47:1–21

REFLECTIONS OF A BIO-ORGANIC CHEMIST

Jake MacMillan

IACR-Long Ashton Research Station, Department of Agricultural Sciences, University of Bristol, Long Ashton, Bristol BS18 9AF, United Kingdom

KEY WORDS: A reprise of 50 years in plant sciences, from colchicine to gibberellins via fungal metabolites; interdisciplinary research

ABSTRACT

The chapter provides a personal and anecdotal account of the author's attempts to keep the horizons of advancing science in sight. It sketches his background to entering science and chronicles various episodes across fifty years in research. Milestones are noted on the author's journey from his structural studies on colchicine, griseofulvin, and gibberellic acid to the isolation, analysis, biosynthesis, and molecular biology of plant gibberellins. The author discusses his personal and professional interactions with plant physiologists and plant biochemists over the years. Philosophical observations are offered on some of the attributes important to conducting research and on changing attitudes toward research.

CONTENTS

1040-2519/96/0601-0001$08.00

Oats and beans and barley grow
Oats and beans and barley grow
Do you or I or anyone know
How oats and beans and barley grow?
Traditional childrens' singing game

INTRODUCTION

The invitation to provide the prefatory chapter to Volume 47 of the *Annual Review of Plant Physiology and Plant Molecular Biology* is, of course, a much appreciated compliment. However, my first reaction to the invitation was, "So the editors think my days as an active scientist are over!" My philosophy has always been to keep looking forward. My working hypothesis of old age has been that it is the time when an individual stops looking forward and is content to look back. Nevertheless, I must confess that I was recently stung into writing my childhood memories for my children when I discovered that the house, and school, of that early childhood had been bull-dozed. So I had already begun to reminisce. But in that case, at least, I have an assured and receptive readership! In mulling over whether to accept this invitation, thoughts crept into my mind that I should perhaps have spent more time looking back. There are certainly several scientific reviews, as well as research papers, that I should have taken time to write. For the present, I am pleased with the opportunity to undertake this account.

I have two caveats. First, I have never kept a diary. Second, I am first and foremost an organic chemist. I hope plant biologists who read this will be as attentive as I have had to be when reading their literature! For, if I were to convey one message in particular, it would be to underline the rewards of interdisciplinary research.

EARLY YEARS

I was born in Wishaw, Scotland, and grew up there for ten years. My mother came from farming stock. My father, like his father, was a signalman on the old London Midland and Scottish (LMS) railway line. I had one brother, who was four years younger. We lived in a two-room apartment, reached by an outside staircase that ran past a communal WC on the first landing. One room was reserved as the best room and was never used. We lived in the other room, sleeping in two double beds in alcoves. I remember the nights that I lay in bed thinking, and wondered about death. Do all young children? I recall the mornings when my father came home from night-shift, lit the coal fire, and made breakfast for us; such mornings were in contrast to the cold, dank mornings of his other shifts. I can also remember my first day at school, and how puzzled I was that the other children were crying. I was scolded when I returned home without my new schoolcap that had probably cost my parents an arm and a leg.

I have never worn headware since. Even in those early years soccer was a consuming passion, and I played in all hours of daylight in a quarry, using jackets as goalposts. There was little else to do. One diversion was to go to cinema matinees on Saturday afternoons where the price of admission was two jam jars or twopence and the fare was the Wild West. I have no memories of deprivation. Money was tight, and when my father brought home his meager pay packet each week my mother divided it out for rent, food, and pipe tobacco. My brother and I had no pocket money, but we did not seem to need any. The houses in the tenement were modest but clean, and our neighbours were proud, lower-working-class families. Yet these were the years of the Great Depression.

When I was ten years old the family moved to Lanark, about 15 miles south of Wishaw. The change of school in my last year of primary schooling was traumatic. I was placed in the top class, but I was very unhappy. I had been a top dog in my previous school and I was far from that status in my new form. In vain I begged to be allowed to move to the lower form, which seemed to have a more relaxed teacher than the tyrant who taught me. But she must have done a good job, because at the end of the year I did so well in the selection examination for Lanark Grammar School that I was placed in the top entry form. And competition was fierce because the school was made up of students from many outlying primary schools. In retrospect, I am greatly indebted to that strict primary school teacher. I relate this episode to illustrate the following point. In the current climate of assessing teachers it is not only the opinions of the current students that should be sought but also those of ex-students.

Our two-room cottage in Lanark had no kitchen or bathroom and it was over two miles from school. The school served lunches we could not afford. I could have taken sandwiches, but my mother insisted that I had a cooked midday meal. So I had to run home and back each lunchtime. Twenty minutes each way and twenty minutes to eat was the schedule. This was relatively easy to meet in reasonable weather, but winters could be severe and deep snow drifts made the journey difficult. Although the temptation must have been great, I do not remember a single occasion when I played truant. In school, a strict discipline was maintained by use of the tawse, a leather strap cut into strips at the end and applied with force to the outstretched hand. This weapon was frequently used, even for whispering in class, and it brought tears of pain, especially if the punishment was considered unjust.

The teaching regime was also tough. There were endless formal exams upon which progress to the next form depended. There was also endless homework in preparation for spelling tests, reciting verse, and standing before the class to parse passages in English, French, and Latin. However, an adequate knowledge of spelling and syntax was to remain with me for the rest of my life. I did not like Latin, especially translations from Virgil usually assigned as homework. I have never regretted the ordeal, because my basic knowledge of Latin has helped enormously in spelling and in deriving the meaning of words.

Two teachers had a great influence on me. One taught English in an imaginative and mind-opening way. He did not succeed in making me a scholar of language or literature, but he taught me how to think. The other was the chemistry teacher. He was not a good formal teacher, but he did allow us to do experiments on our own. Thereafter, chemistry was the only subject for me. Not that my headmaster agreed. I remember at the age of fifteen he called me to his office and proudly announced that he had found a good job for me with excellent prospects. The manager of a local bank had asked him to find a likely lad to fill a vacancy as a bank clerk, and I was the headmaster's choice. This was the first of several big decisions in my life. What should I do? I knew my family were keen that I should start earning. They were also taken by the idea that I would be entering the white-collar work force. It was clear that the headmaster wanted me to accept the offer. Did he therefore think that banking was the limit of my capabilities? Fortunately, my parents did not press me to accept the job, though they could have done with the additional wage packet, and I was allowed to make up my own mind, which I did on the following basis. My form at school contained a distillation of talent from many primary schools and many backgrounds, including children whose parents were professionals. Many of them had planned to continue their education at university. I soberly compared myself with my classmates and decided I was at least as intelligent as they were. So I refused the job as bank clerk and set my sights on qualifying for university entrance. I must have been very competitive at school. The episode also illustrates peer-group influence. I was to observe subsequently that successive years of university students were greatly influenced, for good or bad, by the attitudes of the perceived social leaders in the cohort. The study of the ethos of peer groups is much neglected.

The outbreak of World War II had little effect on me initially. I continued my studies and playing soccer at school and in the local adult league. I received several offers to become a professional player, but I stuck firmly to my goal of obtaining entry to a university and studying chemistry. When I did obtain the required qualification, there was the the problem of finance. I sat a University of Glasgow Scholarship Examination, which was a sobering revelation of my lack of general knowledge. I was placed in the middle of about 1000 examinees and failed to get a scholarship. As a result, I had to rely on my parents and working during vacations to pay my way.

These early preuniversity years were character building. I do not remember spending any time at home, and I had great freedom to roam and explore the world. The children of today miss out greatly in this respect. My parents must have had confidence in society, in sharp contrast to the present when parents are afraid to let their children out of their sight. There was also a cultural belief in education in Scotland, and this, together with my relative poverty, certainly fueled my ambition. Also, being raised in a Presbyterian background instilled in me the philosophy that there is always a price to pay for the good things in life and that you only get out of life what you put into it. Perhaps this philosophy is essential in scientific research, with its many ups and downs.

UNIVERSITY YEARS

In the first year, we were required to take three subjects at the ordinary level. I took chemistry, natural philosophy, and mathematics. In each subject we were examined each term to earn class certificates that allowed us to sit the examination at the end of the year. If the class certificates were "Excellent," we were exempted from the examination at the end of the year. This exemption was a wartime innovation introduced to save paper, but I suspect the real reason was to reduce the work load on staff. In the second year I chose botany as an elective one-year subject. All I remember of the course were the long tedious laboratory classes, peering down a microscope and trying to draw what I saw or was supposed to see, and can recall nothing of the course except my indignation when I did not get an "Excellent" class certificate. Many years later I revisited the botany department as a member of a panel to assess a research grant application. I took particular delight in putting the members of staff (some of whom I remembered) through *my* hoops!

My third and fourth years at university were devoted entirely to chemistry. They were particularly difficult years. Not only did our studies became more and more serious, but the war had a profound effect in two ways. First, as students we were exempted from conscription, but this exemption was limited. As a result, our last two years were shortened to six successive terms without summer vacations. This continuous study was a strain both on health and pocket. The second effect of the war was that we were obliged to join the University Senior Training Corps. In the evenings and at weekends we were trained as an elite squad that would be rushed to the beachhead of the first invasion by the Germans. Thank God there was no invasion; we would have been slaughtered. However, I did make time for soccer, still as a nonprofessional player, although I received a brown envelope after each match containing much-needed cash.

In the final year we had to do a research project. My objective was to prepare chiral derivatives of trivalent nitrogen using Troger's base as a model. We also had to give a short talk on a topic from the recent literature, chosen by ourselves. I chose papers on the preparation of oxazolines by Cornforth. I do not know why. I had no idea that the work was related to penicillin. Indeed, in 1945, I had never heard of this historic antibiotic, nor had I heard of Cornforth, whom I subsequently did get to know. I was simply struck by the exotic chemical structures described by Cornforth. Of the finals examination I have a vivid memory of the stir created by the inorganic paper. The questions were all on first- and third-year courses—not one on the final year's work. As we read the paper there was an audible groan that threatened to become a riot. I looked round and realized that most of my fellow students were very nervous and upset. I remember telling myself that the absolute marks on this paper were going to be very low and that, if I stayed calm, I could do well on a relative scale, compared with my panicking fellow students. My calculated assessment must have worked, because I was awarded a First Class Honours Degree. I was thrilled, but not quite so much as my parents, who got their first glimpse of a university when they attended my graduation.

After graduating there was the question of what to do. I started looking for employment in the chemical industry. I went for an interview at Imperial Chemical Industries Ltd., Dyestuffs Division, Manchester. However, I was rather half-hearted because I wanted to continue as a postgraduate student at Glasgow. Fortunately, I obtained a postgraduate studentship from the Department of Scientific and Industrial Research.

My PhD topic was on the structure of colchicine. I was not a particularly good student, but I did establish the structure of the central 7-membered ring, and my PhD thesis (100 typewritten pages and hand-drawn structures) led to four publications. My supervisor was Dr. JD Loudon, to whom I owe a great debt. He gave me a sound grounding in the practice and ethics of research as well as many memorable *bon mots.* I recall two examples that have served me well. The first arose from my comment to him that Windaus, a giant among organic chemists earlier this century, had made many mistakes in his early studies on colchicine. Loudon's immediate reply was that when I had made as many mistakes as Windaus, I could call myself a chemist. (Have I now made the requisite number of mistakes?) The second was in response to my question on what I should do eventually in research. His answer was that it mattered less what I chose than that I did it well.

ICI YEARS

After earning my PhD in 1948 I gave no thought to postdoctoral studies or to an academic career. I did not consider myself suitable. Besides, academic salaries were extremely low, and I felt it was time to earn real money. As luck would have it (and I have had my fair share of it) an ideal position with Imperial Chemical Industries (ICI) fell into my lap.

In the late 1940s ICI set up a basic research establishment called the Butterwick (later Akers) Research Laboratories in the grounds of a Victorian house, the Frythe. The main thrust of the small group of organic chemists, under JF Grove, was the isolation and determination of the chemical structures of fungal metabolites, produced by a group of microbiologists, under PW Brian. My first significant project was the determination of the structure of griseofulvin, a metabolite of *Penicillium griseofulvum* Dierckx and *Penicillium nigricans* (Banier) Thom. The interest in griseofulvin was its fungistatic effect, selective for those fungi, such as dermatophytes, that had chitinous cell walls. This property led to its commercial exploitation, and griseofulvin is still used in the systemic treatment of mammalian mycoses. My involvement in the determination of the structure of griseofulvin was an exciting one. The Pharmaceutical Division of ICI became very interested in the commercial possibilities of griseofulvin and arranged a discussion with their consultants, Professors Sir Robert Robinson, Oxford University, and Lord Alexander Todd, Cambridge University. This meeting required preparing a briefing document on our work, which was sent to the consultants. The act of preparing this report made me think long and hard on our results, and I deduced a structure for griseofulvin. Being a cautious young man I told no one of this, preferring to wait and hear the opinions of the two wise men. At the meeting, Sir Robert was impressed by the large optical rotation of griseofulvin and thought it must be a biphenyl derivative with restricted rotation around the central bond. On this basis he proposed a structure, but it did not account for an important transformation product of griseofulvin. I pointed out this difficulty and then drew my proposed structure on the blackboard. Sir Robert immediately closed his folder containing our briefing, jumped up, and declared the problem solved. He then pointed out that this correct structure was a spiran and therefore explained the high optical activity. Thereby, honor was satisfied, and I was left to bask in my success. After the meeting, which was held in Manchester, I traveled back to the South with Todd by train. I was very apprehensive at having to converse with the great man for several hours. But I need not have worried. He did all the talking and regaled me with anecdotes of the chemists of the day—all untouchables to an apprentice like me. One comment was that he considered Kogl's deduction of the structures of auxins A and B to be one of the outstanding chemical achievements of the 1930s. I would have liked to have been a fly on the wall when he read the paper by Vliegenthart & Vliegenthart (11) in 1966! Sam Wildman is still ferreting away in an attempt to unravel the mystery of auxins A and B.

In the 1950s the importance of the three-dimensionality of molecules was being emphasized. Thus, the completion of the structural studies on griseofulvin required the determination of the absolute stereochemistry. For this, a sample of (±)-methylsuccinic acid was required for optical resolution and comparision with the (±)-methylsuccinic acid, obtained by oxidation of griseofulvin. The required compound was ordered. After two months I was told that the order had been farmed out to a free-lance supplier who had committed suicide. (Can anyone beat that for the most macabre excuse for late delivery of an order?) Eventually, the (±)-methylsuccinic acid was delivered. It was resolved into its enantiomers by fractional crystallization of its alkaloid salt, and the problem was solved. I am happy to record that this was the only time I have had to perform a classical optical resolution. Other points of scientific satisfaction on the griseofulvin story were the coming to fruition of the ideas that leaving out the potassium chloride, or substituting potassium bromide, in the culture media should lead to dechlorogriseofulvin or bromogriseofulvin. The latter compound was the first bromine-containing fungal metabolite to be described. Unfortunately, the dechloro- and bromo- analogues were, respectively, two and one orders of magnitude less active than griseofulvin. In recalling this work on griseofulvin I came across the manuscript of a lecture that I gave at a Gordon Conference in 1962. It was a detailed review, prepared to obtain clearance from ICI to make the contents of the lecture public. I regret that it never appeared in print.

I was to continue my interest on the structures of fungal metabolites (for example, colletodiol, wortmannin, heveadride, xanthomegnin and related pigments, and colletotrichin) subsequently at the University of Bristol. However, my next research project on fungal metabolites at ICI was on the gibberellins.

GIBBERELLINS

At ICI, Akers Research Laboratories

The ICI interest in the gibberellins as fungal products was initiated by *Chemical Abstracts,* 1950, 44(22):10814–17, which contained abstracts of 12 papers published by Yabuta, Sumiki, and coworkers in the *Journal of the Agricultural and Chemical Society of Japan* between 1940 and 1944. The abstracts summarized the work by the Tokyo group on the production, isolation, chemistry, and biological activity of gibberellins A and B. These abstracts were noted by AW Sexton, Research Director of ICI Pharmaceutical Division, who suggested to PW Brian that he should produce the gibberellins and evaluate their physiological properties. So began the work on gibberellins at the Akers Research Labs at the Frythe. It soon became clear that we were not the first, outside Japan, to be attracted to the gibberellins. In the United States, JE Mitchell, at the Chemical Corps Biological Labs, Camp Detrick, and subsequently FH Stodola, at the USDA Labs, Peoria, had also made progress on the production and isolation of gibberellin. The history of the gibberellins up to the mid-1950s has been well chronicled in English by BB Stowe & T Yamaki (8), FH Stodola (7), and BO Phinney (6). A Japanese perspective is provided by S Tamura (9, 10). I have translations of Tamura's two Japanese articles, kindly provided by Bernie Phinney. Given the availability of these excellent and detailed accounts of the history of the gibberellins, I only add here personal recollections and comments on my involvement at ICI and subsequently at the University of Bristol.

In his 1977 article, Tamura (10) expressed his great disappointment that the rapid development of "gibberellinology" occurred outside Japan. He points out that the existence of gibberellins and their biological properties were known in Japan from 1935, yet it was outside Japan that the subject rapidly developed in the 1950s. Tamura suggested a number of possible reasons, including (*a*) the submerged fermentation technology developed for penicillin production during World War II was not readily available in Japan, (*b*) the obsession with auxin as the hormone in plants blinded the plant biologists to the importance of gibberellins, and (*c*) the attitude of Japanese culture was such that the importance of gibberellins was accepted only after the West became interested in them. I do not know if these factors played a part, but there is no doubt in my mind that the overriding reason for the rapid advances made at the ICI Akers Research Labs was the fact that we were a closely knit group of microbiologists, plant physiologists, and chemists, coordinated by an exceptionally gifted scientist, PW Brian. Thus, the research was focused and interdisciplinary from the beginning. It began with a thorough search for high-yielding strains of *Gibberella fujikuroi* and the best fermentation conditions for the production of gibberellins. Chemists were at hand to isolate, purify, and determine the chemical structure of the resultant gibberellic acid. Likewise, plant physiologists were available to study the details of the plant growth-promoting effects. The shadow of auxin (indole-3-acetic acid) was evident in an early paper by PW Brian. However, when he (1) showed that gibberellic acid reversed the dwarf habit of Meteor peas he was able to suggest that "the simplest physiological explanation...would be that tall plants normally produce a substance similar to gibberellic acid, which acts as an internal growth regulator *in addition to the auxin system*" (1; my emphasis).

Thus, by 1956–1957 a search had begun at the ICI Akers Research Labs for direct evidence that gibberellins were present in higher plants. Again the approach was interdisciplinary. First, Margaret Radley looked for gibberellic acid-like substances by extraction, paper chromatography, and bioassay. She found, inter alia, that immature seeds of runner beans (*Phaseolus multiflorus*) were a relatively rich source (ca 0.25 μg/g). It was therefore the job of a chemist to isolate and identify the gibberellin-like compounds, and I was that lucky chemist. About two metric tons of runner beans were obtained from local growers, and from these pods about 100 kg of immature seed were laboriously collected. By the autumn of 1957, milligram amounts of GA_1 had been isolated and identified, largely through the technical skills and patience of PJ Suter. Over the next two years, with the additional help of JC Seaton, three hitherto unknown gibberellins, GA_5, GA_6, and GA_8, were isolated, and their structures were determined. I still have specimens of these original isolates. I repeat that this achievement was the result of chemists and biologists working closely together—a lesson I have never forgotten. Of course, we were not alone in the search for gibberellins in higher plants. Advances in science are usually made when the required clues are available. Bernie Phinney and Charlie West had also deciphered the clues and were simultaneously investigating the gibberellin-like compounds (a phrase first used by Bernie Phinney) in seeds of *Phaseolus vulgaris*. These investigators had also picked up another clue that we at ICI had overlooked, namely, the paper in 1951 by JW Mitchell et al (4) showing that ether extracts of immature seeds of Black Valentine beans contained gibberellin-like biological activity. I know that PW Brian never forgave himself for missing this paper.

The year 1957 was an exciting one for me. The ICI gibberellin group received invitations to provide speakers at a Gordon Conference in New Hampshire on natural products and to a symposium on the gibberellins, organized by Frank Stodola at the 132nd Meeting of the American Chemical Society in New York. The natural choice was Brian Cross, who had done most of the early work on the chemistry of gibberellic acid. Because of ill health, he had to decline, and I was substituted. Thus, by default, I made my first visit to the United States in a propeller-driven plane that took 19 hours to cross the Atlantic. I was chauffeur-driven to the airport in a company car. I traveled first-class, seated next to Douglas Fairbanks, Jr, and slept in a hammock that was unrolled from the roof of the plane. Those were the days! I made the most of that trip. Before the Gordon Conference I visited RB Woodward, who bled me dry on gibberellin chemistry in his intimidating office. He then took me out to dinner, which was an amazing feast for someone from the food-rationed United Kingdom. At the Gordon Conference I met many of the big names of organic chemistry and was complimented on my presentation for a reason that still makes me smile, namely, that I told it as it was and did not try to make a detective story out of it. On reporting to the ICI office in New York, I was told that there were meetings in Stanford on the gibberellins that I should attend and that they would make the arrangements. This raised a problem. The previous day I had persuaded my wife, back in England, to fly out and share the wonderful time I was having. As I met her off the plane in New York, I kissed her good-bye and departed for the West Coast—and she forgave me! (We did have a very fine holiday together when I returned to New York.) At Stanford, I met Bernie Phinney for the first time. He took me to Muir Woods and told me about his and West's work on gibberellins from bean seeds. To my eternal shame I could not tell him that we were also working flat out on the same problem and had already isolated a compound that I had concluded must be GA_1. The final proof came as soon as I returned to the lab after the ACS meeting in New York when we obtained an IR spectrum of the acid and its methyl ester. In the course of this work we had isolated several other constituents of the seed of *P. multiflorus* (*coccineus*); one of them turned out to be a close relative of abscisic acid, and we called it phaseic acid.

My second visit to the United States was in 1960 to attend and give two lectures at a gibberellin symposium during the 138th Meeting of the American Chemical Society in New York. As a result of these two visits I met the main players in the development of the gibberellin story in the United States in the 1950s. I have kept in touch with most of them, both professionally and personally. One, Larry Rappaport, persuaded me to join him as coauthor of a book on the gibberellins. I still have copies of my chapters of this unfinished work! A positive outcome was a memorable summer with my family at the University of California, Davis, and a reciprocal sabbatical year at Bristol by the Rappaport family. On the way to Davis I made the first of many visits to the MSU-DOE Plant Research Laboratories at East Lansing, Michigan. At Davis, I met Russ Jones, and the meeting resulted in a joint study on in vitro hydroxylating systems and in an enduring friendship.

By the early 1960s the days of the Akers Research Labs were numbered. Like so many basic industrial research establishments where there is insufficient throughput of young, thrusting scientists, the lifetime of the organization is about one generation. There are many examples of this phenomenon. By this time I was the director of the labs and an associate research manager of ICI, Pharmaceuticals Division. I was responsible for the transfer of the site to the new owners, Unilever, who were moving in as we were moving out. I also had to organize the move of the Akers staff to the Pharmaceuticals Division. These tasks involved weekly trips, by train, between the Akers Labs in Hertfordshire and the Pharmaceuticals site in Cheshire. It was a stressful time. On one return journey I realized that I was looking forward to getting on the train and gulping down two double gin and tonics to return to a feeling of normality. It was then that I resolved to get out. There were also other reasons. For example, at the Akers Labs I sat in a huge office, remote from the labs, doing paperwork and trying to sort out the personal problems of staff. I decided to seek a post in academia. Again I was lucky, for the 1960s was a time of rapid expansion of the universities in the United Kingdom. After briefly considering a position at Aston University, I was offered a position at the University of Bristol, where the only obstacle to my appointment was the disbelief that I was willing to accept such a large cut in salary. However, I had the full support of my wife in this matter, as in all others (except one when she later refused to emigrate to the United States). I was somewhat unhappy at reneging on what I felt were my responsibilities to ICI and my colleagues there. Before I finally accepted the Bristol offer, I made an appointment to see the research director on the ICI main board. I was shown into his office and began, hesitantly, to explain my predicament in wanting to leave ICI and my unhappiness with leaving the Akers group in the lurch. Almost immediately he stood up, shook my hand,

and said, "No man is indispensible—the best of luck in your new job." He then showed me the door. I appreciated this straight talk and gave a whoop of delight on the other side of the door. However, I must not leave a wrong impression of my 15 years at the Akers Research Laboratories. They were extremely enjoyable and rewarding, both professionally and personally.

In the turmoil of moving I never took time to write up a small piece of work showing that GA_3 moves in the transpiration stream with little if any metabolism. In the experiment, 5 mg of GA_3 were applied to 100 seedlings of maize *dwarf-1* (50 μg in 100 μl water per seedling). The seedlings were grown in a water-saturated atmosphere in bell jars, and guttation drops were collected over three days from the tip of the leaves. From the guttation drops 4.75 mg of crystalline GA_3 were recovered and identified by infrared spectroscopy. Perhaps someone will complete this work.

At the University of Bristol

It was hard to get into first-hand research again. I had moved into a lab, bereft of state-of-the-art instrumentation that had burgeoned in the electronic revolution of the 1950s. But I quickly learned the art of grantsmanship and begging for funds from my industrial friends. I had almost decided to stop work on gibberellins and to concentrate on the chemistry of other natural products and on nuclear magnetic resonance (NMR) studies. The latter interest arose through the delivery of a new Varian A60 NMR Spectrometer, ordered by my predecessor, and therefore without an operator. So I attended a Varian course in Zurich where I ended up spending a night in a police cell! (But that is another story.) I then took charge of the Bristol NMR instrument. Through determining the NMR spectra of several dimers of substituted propenylbenzenes, I became very interested in the mechanism of the dimerization reaction and devoted much time to this study (most of it still unpublished). However, a friendship with another member of staff, Bob Binks, brought me back to the gibberellins. He had become interested in gas-liquid chromatography (GC) which, by that time, had become demystified as a physical chemist's plaything and had become available in a form that was simple enough for organic chemists to use. We were not the first to explore the use of GC for gibberellins; that had been done for the methyl esters of GA_1 through GA_9 by the group at the University of Tokyo. What was new was the use of the MeTMSi derivatives, which gave sharp peaks and separation of 17 of the gibberellins known at that time. Binks was also very interested in mass spectrometry because the department was about to take delivery of a high-resolution mass spectrometer, an MS 9. Our ultimate goal was the analysis of gibberellins by combined gas chromatography–mass spectrometry (GC-MS), a technique that was then in its infancy. In the event, we were never able to develop a GC interface to the MS 9; however, through the energy and enthusiasm of RJ

Pryce, who traveled back and forth to Glasgow University and to Stockholm to use the commercially available LKB 9000 instrument, we had shown by 1966 that gibberellins could be identified by GC-MS in derivatized crude acid extracts of immature seeds of *P. coccineus.* I still remember the excitement of those days—the first GC-MS spectra of the MeTMSi derivatives of GA_1, GA_5, GA_6, and GA_8 from crude acid extracts from seeds of *Phaseolus* and the mass separation of the methyl esters of GA_5 and GA_{20} by accelerating voltage alternation.

These early results were described at the Sixth International Conference on Plant Growth Substances at Carlton University, Ottawa, in 1967. There, my personal friendship with Nobutaka Takahashi began, and we sat down and thrashed out an agreement to allocate GA numbers to the naturally occurring gibberellins as they were isolated. Initially, Takahashi had used this procedure in naming the four gibberellins that he had isolated from the fungus as gibberellins A_1, A_2, A_3, and A_4. When we had first isolated gibberellins from runner bean seeds I was tempted to use a nomenclature that distinguished between plant and fungal gibberellins but had abandoned this idea on the grounds that many of the fungal gibberellins would also be found in plants. So, the Takahashi initiative had been continued in naming GA_5 through GA_{17}. However, by the time of the Ottawa meeting, complications had arisen through the use of the names, bamboo gibberellin and *Pharbitis* gibberellin (both also isolated from *P. multiflorus* seeds) as well as *Canavalia* gibberellins-I and -II and lupin gibberellin. Takahashi and I agreed to allocate GA-numbers to these and to gibberellins subsequently discovered. We described this proposal in *Nature* (3). The system has worked well and we have now reached GA_{108}.

At Bristol University we continued our GC-MS studies on the gibberellins and extended them to other plant hormones. Drawing line diagrams of spectra manually from traces on photosensitive paper was a real pain. Through the ingenuity of R Binks, RL Cleaver, and JS Littler, real-time processing of GC-MS had been developed by 1970. However, it was not until 1972 that we had an instrument of our own (an AEI MS 30) to use it on. In the meantime we ran spectra on instruments in other labs, often working overnight. I described our progress at the Seventh International Conference on Plant Growth Substances at Canberra in 1970. At this meeting Jan Graebe described the conversion of mevalonic acid into *ent*-kaurenoids using his pumpkin cell-free system, and he asked me to help identifiy his products from the incubation of these *ent*-kaurenoids. So began an enjoyable and fruitful collaboration that lasted until the Graebe group obtained their own GC-MS equipment. An early and exciting outcome was the discovery that the ^{14}C-content, and therefore the specific radioactivity, of the metabolites from ^{14}C-mevalonic acid could be determined directly by GC-MS. Other highlights of this collaboration were the identification of GA_{12}-aldehyde and GA_4 as metabolites of mevalonic acid and the elucidation of the stereochemical process in the conversion of *ent*-7α-hydroxykaurenoic acid to GA_{12}-aldehyde.

I had traveled to the conference in Canberra via Moscow and Tokyo. In 1970 it was still necessary to be briefed by the United Kingdom Foreign Office before visiting the (former) USSR. But those instructions were quickly forgotten in the friendly atmosphere of my hosts, principally MKh Chailakyan. He pressed me hard to undertake large-scale extractions of long- and short-day tobacco plants in the search for gibberellins and anthesins. It was difficult to decline, but I did! My most vivid memories of the visit are not scientific. They include a visit to the humble apartment of the Chailakyans, the privileges of an academician to summon an official car at any time of day or night and to walk straight into a Bolshoi Ballet performance ahead of a long queue, the thunder of the military parade rehearsing the October Revolution celebrations, and the cultural delights of Leningrad. Traveling from the early winter snows of Moscow to the autumn colours of Japan was an even greater cultural shock. I had previously met Y Sumiki at the Sixth International Congress for Microbiology in Rome in 1953, and I had invited him to the Akers Research Laboratories after the conference. I had subsequently heard that he was disappointed by the hospitality shown to him by ICI, and I soon understood how he must have felt when I experienced the boundless hospitality and generosity of the Japanese. I cemented my friendship with Nobutaka Takahashi and met the members of the group, who could not wait to tell me that I had once addressed Takahashi in a letter as "Dear Nobuta," which means "Dear wild pig." It is, perhaps, appropriate that I am writing this article in the Year of Nobuta! During this visit Sumiki had arranged a formal lab dinner in my honour. When we arrived at the dining room, Sumiki was sitting cross-legged at the table to welcome his guests. At the door, each of the guests threw themselves to the floor and slid up to the great man on the polished wooden floor with head down. I was astonished, amused, and worried. Should I do likewise? What would the reader have done? I confess I walked up to him and bowed from the waist. The dinner was a feast of Japanese food at its best, delicately offered by geishas and accompanied by saki which Sumiki drank from a very large tumbler. Afterward there was a sing-song of Scottish songs, played on Japanese guitars. I was very embarrassed that they knew the words of the songs much better than I did. At the end of the evening Sumiki took me aside and warned me to be careful in Japan because I was far too trusting. I still do not know what he meant. Perhaps he was revealing more about himself than about me. I also visited Kyoto and gave a talk at a meeting of the Japanese Society for the Chemical Regulation of Plants. My first visit to Japan was a marvelous experience, one that has been repeated many times since.

Until 1972 we used packed GC columns, prepared in-house with poor GC resolution, and mass spectrometers with low sensitivity. A distinct improvement in both respects came with the ability to use SCOT columns with the AEI MS 30. But a quantum leap in both GC resolution and MS sensitivity came in 1981 with the acquisition of a VG 7050, coupled to capillary GC columns and with a dedicated on-line data acquisition system. An added sophistication was the use of Kovats Retention Indices, which became essential to distinguish between GA epimers with similar MS but different GC retention times. The source of the *n*-alkanes was Parafilm, which had previously been banned from the lab because of the contamination it introduced into plant extracts! Two other serious problems of contamination of plant extracts were plasticizers in solvents and GAs in glassware, used in the chemical studies. The main source of plasticizer was from the tubing used in the lab. When I asked the suppliers for the identity of their plasticizer, they refused to reveal their "trade secrets." This was a challenge. We quickly showed that it was a C_{20}-compound and thought we had isolated a new diterpene, but it turned out to be tri-n-butyl citrate. The glassware problem was solved by physically separating the chemistry and the plant extract activities and forbidding interchange of glassware between the two labs. In all the GC-MS developments the leading light was Paul Gaskin, who became—and still is—a wizard in obtaining and interpretating GC-MS spectra. Our labors on GC-MS at Bristol are now summarized in a book (2). The technical improvements in the early 1980s enabled an expansion of our gibberellin studies, as detailed later.

The collaborative studies with the Graebe group were not our first entry into the biosynthesis of gibberellins. In 1969, I spent the summer with my family at Calgary with Dick Pharis. I took the opportunity to attend the 11th International Botanical Congress in Seattle. There, I was excited by a paper, "Amylase mutants in the fungus, *Gibberella fujikuroi*" by BO Phinney and M Fukuyama (Dilworth). After the talk I told Bernie that he needed a chemist. He asked if I was offering. I replied "Yes, why don't you come to Bristol?" He agreed and spent the following summer in my lab with Machi Fukuyama. Thus began a close personal and professional relationship with Bernie that still continues. After this first visit he was to spend two sabbatical years and alternate summers at Bristol, and I was to make many visits to UCLA. With the crucial contribution of John Bearder, metabolic studies with the fungal mutants progressed steadily throughout the 1970s, leading to the elucidation of the biosynthesis of GA_3 in the fungus and the demonstration of the nonspecificity of the enzymes to metabolize analogues of *ent*-kaurenoic acid to analogues of the fungal gibberellins. So too did the studies with seeds, both in vitro with Jan Graebe and in vivo in our own lab with Val Sponsel (Frydman). But metabolic studies with vegetative tissues only became possible in the early 1980s with the increased GC-MS sensitivity mentioned in the preceding paragraph. Thus the continuing collaboration with Bernie Phinney expanded to a detailed study of the gibberellin metabolic pathways in shoots of maize. These studies have provided the most complete in vivo metabolic evidence to date for gibberellin biosynthesis in the vegetative tissue of a higher plant. Thus, each step in the formation of GA_1 and GA_3 from *ent*-kaurene has been established. The determination of the position of the d1 mutation in this pathway enabled the elimination of GA_{20} and its precursors as bioactive promoters of stem elongation in maize. A bonus was the discovery that the biosynthesis of GA_3 in maize was different from that in the fungus. The latter discovery led to the detection of an enzyme preparation from Marah seeds that converts GA_5 to GA_3 and 2,3-dehydroGA_9 to GA_7, a subject that I am still actively studying. A collaboration with Jim Reid and his group in Tasmania using pea shoots was also made possible by the improved GC-MS sensitivity, and we showed, inter alia, that GA_1 levels and stem elongation were directly correlated.

It is important to remember that this interdisciplinary approach was taking place in a department of organic chemistry. There was always a core of enthusiastic chemists, graduate students, and postdocs in my group. Inter alia, these chemists were developing methods of synthesizing new gibberellins that were being detected by GC-MS, preparing isotopically labeled gibberellins for metabolic and enzyme mechanistic studies, and designing new derivatives of gibberellins for probing biological action. Specific details of the chemistry are not appropriate in the context of this account. In addition to the chemists, there were always one or more postdocs from other disciplines. This provided a special atmosphere that attracted many visiting plant scientists from different countries to the group. It also engendered internal interdiscplinary research: for example, structure/activity relationships, purification and properties of GA biosynthetic enzymes, the design and use of GA-antigen for the preparation of GA-epitope-specific monoclonal antibodies, and the preparation of GA derivatives to probe receptors. I am pleased that similar studies are continuing at Long Ashton Research Station (see later).

Eventually, and possibly inevitably, as at the Akers Research Labs, my administrative responsibilities increased. I wish I had known of the "Peter Principle" (5) earlier than I did! During the last decade at the University I became Head of the Department of Organic Chemistry and eventually Chairman of the School of Chemistry. As a result I was gradually drawn away from direct participation in the ongoing research of the laboratory. I regret this and the lost opportunity to become more directly involved in the molecular biology revolution in plant sciences. However, the arrival of mandatory retirement in 1990 had its compensations, and it gave me the opportunity to become the postdoc that I had never been.

At Long Ashton Research Station

I declined an offer of facilities within the School of Chemistry at Bristol University on principle. I wanted to start a new life without the temptation to interfere in my old one. I therefore made prior arrangements to move to what is now IACR-Long Ashton Research Station. I had had a long association with the Long Ashton Research Station (LARS) both through the University of Bristol, of which it is now the Department of Agricultural Sciences, and through the former Agricultural and Food Research Council (AFRC), which funded both me and LARS. My initial contact with LARS began with Leonard Luckwill's studies on the gibberellins of apple seeds in the 1960s and had continued, both through common research interests and administratively through membership of the Agricultural Committee of the University of Bristol and AFRC working groups.

My main concern was that the expertise of my group at Bristol University should not be lost. I therefore approached Ken Treharne, who was then the director of LARS, with the proposition that members of the group, including Paul Gaskin and the GC-MS instrumentation, would be moved to LARS, that they would be integrated into the research programs as individuals, and that I would have facilities to continue my own research. With Ken's help, and the active support of his staff, this arrangement was formally agreed to by the AFRC. The arrangement had been made in 1988 and, after Ken's untimely death in that year, it was honoured by his successor, Peter Shewry. I hope that LARS are as pleased with the outcome as I am.

For the first three years as a postdoc I spent six months of each year in Bernie Phinney's lab at UCLA and the other half at LARS. Since then I have worked full-time at LARS. I have been able to work at the bench, mainly continuing preretirement interests on the pathways and enzymology of gibberellin biosynthesis and learning much about molecular biology from colleagues. Perhaps I am not achieving much but at least I am experiencing firsthand my repeated advice to graduate students to be patient. At present, I am excited about current studies on the enzyme catalyzing the formation of GA_3 and GA_7 from their immediate precursors, proud of my ability still to synthesize the appropriate ^{14}C-labeled precursors, and amazed that I am still able to learn new techniques. I am equally excited at the advances that are now being made on the molecular biology of the biosynthesis and action of GAs that are taking place at Long Ashton and elsewhere.

COMMENTARY

I have referred to the phenomenon of independent discoveries that occur because the clues are in place. I have also been struck by the cyclic nature of scientific research. I give three examples from my own experience. First, colchicine, the topic of my PhD studies, has reared its head again in relation to the possible effects of gibberellins on microtubules. Second, griseofulvin became very useful in limiting the radial growth of *Gibberella fujikuroi* in the ha-ha (half seed-halo) assay, used by Fukuyama and Phinney in screening for fungal mutants. Third, current research on the formation of GA_3 from GA_5 in plants by a dioxygenase is a pleasing reminder of my early speculation that GA_3 may be formed from GA_5 by activated dioxygen. Will such connections be as easy to make with the increasing reliance on computerized retrieval of information? Browsing through the literature is a fast-disappearing luxury.

The physical and biological sciences have made unimagined advances in my lifetime, and the pace of change is still accelerating. I feel that I have been running all my life to avoid the horizons of science disappearing completely from sight. Would I do it again? The answer is Yes. Would I enter science now? The answer is Maybe. There are many reasons for each of these answers. I have enjoyed and continue to enjoy research. I feel privileged and fortunate to have entered science at a time of great optimism and expansion. I have had the luxury of being in a position to make career decisions that, for the most part, have either been good choices or made to be good choices. On the whole, I have been able to resist pressures to tailor my endeavors for reasons other than perceived scientific merit. I have always had time to plan and develop research interests without the need to publish or perish. The motivation and accountability have always been self-generated.

I am not so certain that such favorable conditions apply today. The shortage of research funding per researcher is driving the younger workers to a philosophy of quantity rather than quality and the established workers to a competitive back-biting that is a disservice to science. I feel that there is less and less room for scholarship and more and more emphasis on wealth creation and accountability. Wealth creation is a noble cause, but the problem with the concept is that there is little understanding between innovative discovery and wealth creation. There are more examples of the practical exploitation of scientific discoveries through serendipity than by foresight. It is therefore essential that curiosity-driven research be nurtured from the bottom up. For example, ICI obtained a rich return from the research on griseofulvin and the gibberellins at the Akers Reseach Labs. It is equally evident that highly skilled scientists are essential if wealth is to be created from fundamental discoveries wherever in the world they occur. But this factor is rarely included in deciding overall research budgets. With the shortage of funding and greater emphasis on short-term research contracts, from where are future scientific leaders to come? Accountability in the use of research funding cannot be disputed. The most important part of this process must surely be to inform the public, who foot the bill; scientists have a poor record of doing so. But accountability can rapidly become the excuse for bureaucracy, whose currency is numbers (statistics). Thus, there is a growing tendency, at least in the United Kingdom, to quantify *numerically* the *quality* of research. The use of a Journal Impact Value, generated by the Science Citation Index, to assess publications is particularly insidious. For example, it overemphasizes reviews and citations from them that are already used too much. Too many authors cite references to reviews rather than the original papers that, I suspect, may not always have been read by the citing authors. Thus, there is a danger of creating and perpetuating tentative conclusions as facts. I would urge authors, reviewers of manuscripts, and editors of journals to minimize the use of reference to general reviews and, when used, to make sure that they are clearly identified in the text.

Collaborative research is one way to eke out limited research funding. It can lead to excellent innovative science. It can also decrease direct competitiveness and unnecessary duplication in the literature, which is already too large. But it cannot be imposed by funding bodies. It must come from the coal-face, and it is as much a matter of personalities as of science. Interdisciplinary research is even more difficult. It is discouraged by reactionaries, who wish to keep their disciplines pure. For example, the ultimate put-down is "but so-and-so is not a straight organic chemist (or whatever)." Such attitudes do not help where funding is discipline driven and peer reviewed. Fortunately this attitude is less prevalent in the biological sciences than in the physical sciences. To be successful, each partner in a collaboration must bring something that the other partner(s) cannot provide. One of the delights of international collaboration is to enjoy, in depth, an international community with common interests. In this I have been particularly fortunate. In addition to the scientific rewards, I have made many lifelong, reciprocal open-house friendships all over the globe. Let this, at least, be a spur to aspiring scientists.

What other advice can be offered to young scientists? First, I refer again to the two *bon mots* of Professor JD Loudon. Do what you believe in and do it well. Do not be afraid of making mistakes; the only way to avoid them is to do nothing. I would add that AQ (politely defined as Application Quotient) is as important as IQ. I have known too many young scientists who have not had enough AQ to realize the potential of their high IQ, possibly because they found life too easy in the beginning. Let me again emphasize the virtue of patience. Remember that postgraduate studies provide, principally, a training in research. I have direct proof of this. In 1971 an extensive fire destroyed my labs in the School of Chemistry at Bristol University. It also destroyed all the records and research material of two graduate students in their second year of studies. When they repeated their work they did so in a fraction of the time and to a much higher standard. I am not advocating that all graduate students' research records be burned after two years. I only pass on the result of this unfortunate experiment.

My advice to senior scientists is to nurture the younger scientists but allow them to make their own mistakes. I also recommend that the size of a research group should be no greater than the number that allows you to enjoy the successes—and agonize at the failures—of each member of the group on a daily basis.

Finally, let me remind researchers, funding bodies, industrialists, and politicians that the limitations to the advance of knowledge are usually conceptual and not factual. Ideas plus AQ advance science. Ideas come in various ways, but my best ones have come during mind-emptying work, such as digging the garden, interior decorating, or soaking in a hot bathtub. AQ is self-generated.

ACKNOWLEDGMENTS

Recently a young researcher asked me if election to the Fellowship of the Royal Society of London was the greatest event in my life. My response was, "Thrilling though that event was, my wife was, and still is."

I am also grateful to many others who have influenced my development. I have referred to JD Loudon and PW Brian. Others include my collaborators, many of whom I have mentioned, and especially Bernie Phinney (Los Angeles has become a second home to me) and Paul Gaskin, who has been a stalwart in the development of the GC-MS technology. I also take this opportunity to acknowledge the debt I owe to all the Bristol graduate students and postdoctoral colleagues who have helped me to keep young in thought.

Thc graduate students were Maurice F. Barnes, Michael H. Beale, John R. Bearder, Lee J. Beeley, Gillian M. Boother, David Bowen, Barry D. Cavell, Simon J. Castellaro, Christopher Cloke, Richard I. Crane, Simon J. Dolan, Graeme J. Down, Richard C. Durley, Christopher D. Foulds, Andrew M. Fowles, Sarah J. Gilmour, Ian K. Hatton, Peter Hedden, David W. Holdup, Michael Hutchison, Paul S. Kirkwood, Paul Lewer, Norman J. Lewis, Martin W. Lunnon, Ian K. Makinson, Roger Metcalf, Nasser-ud-din Ahmad, Robert J. Pryce. Jill E. Readman, Tom J. Simpson, Arun K Singh, David A. Taylor, Edward R. H. Walker, A. John Weir, Colin M. Wels, and Samuel K. Yeboah.

The postdoctoral colleagues were Michael H. Beale, John R. Bearder, Sarah J. Gilmour, David M. Harrison, Timothy J. Ingram, Paul S. Kirkwood, J. Paul Knox, Joan E. Nester-Hudson, Frances M. Semenenko (one of my daughters), Valerie A. Smith, Valerie M. Sponsel, Clive R. Spray, Christine L. Willis, and Anthony E. Vanstone.

Finally, I am indebted to the Institute of Arable Crops (IACR)-Long Ashton for providing the facilities to prepare this account. The IACR receives grant-aided support from the Biotechnilogical and Biological Sciences Research Council of the United Kingdom.

Literature Cited

1. Brian PW. 1957. The effects of some microbial metabolic products on plant growth. *Symp. Soc. Exp. Biol.* 11:166–81
2. Gaskin P, MacMillan J. 1991. *GC-MS of the Gibberellins and Related Compounds: Methodology and a Library of Spectra.* Bristol: Cantocks Enterprises
3. MacMillan J, Takahashi N. 1968. Proposed procedure for the allocation of trivial names to the gibberellins. *Nature* 217: 170–71
4. Mitchell JW, Skaggs DP, Anderson WP. 1951. Plant growth stimulating hormones in immature bean seeds. *Science* 114: 159–61
5. Peter LJ, Hull R. 1969. *The Peter Principle.* New York: William Morrow
6. Phinney BO. 1983. The history of gibberellins. In *The Biochemistry and Physiology of Gibberellins,* ed. A Crozier, 1:19-52. New York: Praeger
7. Stodola FH. 1958. *Source Book on Gibberellins 1828–1957.* Washington, DC: Agric. Res. Serv., US Dep. Agric.
8. Stowe BB, Yamaki T. 1957. The history and physiological action of the gibberellins. *Annu. Rev. Plant Physiol.* 8:181–216
9. Tamura S. 1969. The history of research on gibberellin. In *Gibberellins, Chemistry, Biochemistry and Physiology,* ed. S Tamura, 1:3-27. Tokyo: Tokyo Taigakushuppankai. 372 pp. (In Japanese)
10. Tamura S. 1977. The history of plant hormone gibberellin. In *Plant Hormones* 2:18-50. Tokyo: Dai Nippon Tosho Co. Ltd. (In Japanese)
11. Vliegenthart JA, Vliegenthart JFG. 1966. Reinvestigation of authentic samples of auxins A and B, and related products by mass spectrometry. *Rec. Pays-Bas* 85: 1266–72

Annu. Rev. Plant Physiol. Plant Mol. Biol. 1996. 47:23–48

HOMOLOGY-DEPENDENT GENE SILENCING IN PLANTS

P. Meyer

Max-Delbrück-Laboratorium in der MPG, Carl-von-Linné Weg 10, D-50829 Köln, Germany, and University of Leeds, Centre for Plant Biochemistry & Biotechnology and Department of Genetics, Leeds LS2 9JT, United Kingdom

H. Saedler

Max-Planck-Institut für Züchtungsforschung, Carl-von-Linné Weg 10, D-50829 Köln, Germany

KEY WORDS: gene silencing, trans-inactivation, paramutation, cosuppression, DNA methylation

ABSTRACT

Homology-dependent gene silencing phenomena in plants have received considerable attention, especially when it was discovered that the presence of homologous sequences not only affected the stability of transgene expression, but that the activity of endogenous genes could be altered after insertion of homologous transgenes into the genome. Homology-mediated inactivation most likely comprises at least two different molecular mechanisms that induce gene silencing at the transcriptional or posttranscriptional level, respectively. In this review we discuss different mechanistic models for plant-specific inactivation mechanisms and their relationship with repeat-specific silencing phenomena in other species.

CONTENTS

1040-2519/96/0601-0023$08.00

INTRODUCTION

With the rapidly increasing application of transgene technology in plants, the control of transgene expression has become an important point of concern. A common aspect of many cases of inactivation of transgenes is the presence of duplicated homologous sequences. Apparently homology serves as a signal that can trigger gene inactivation at either the transcriptional or posttranscriptional level. Homology-dependent gene silencing is the basic feature for several phenomena that apparently each have distinct regulatory mechanisms. This includes, for example, the inactivation of tandem repeats, *trans*-inactivation of allelic or ectopic copies, and the coordinated silencing of a transgene and the endogenous homologous gene. In this review we summarize the recent data on homology-dependent gene silencing, focusing on the different models for the regulation of transcriptional and posttranscriptional silencing. Most likely, transgene research has uncovered the existence of an endogenous control mechanism for multiple sequences, which does not affect transgenes exclusively. On the basis of the mechanistic models we therefore discuss the biological function of homology-scanning systems and implications for genome organization and evolution.

HOMOLOGY-DEPENDENT SILENCING PHENOMENA

Inactivation of Homologous Transgenes

The importance of homologous sequences for the induction of gene silencing was discovered when transgenic tobacco plants were retransformed with constructs that were partly homologous with the integrated transgene. In the presence of the second construct, the primary transgene became inactivated and hypermethylated within the promoter region, the site of homology between the "suppressor" and the "target" locus (76). Since this remarkable discovery, numerous cases of homology-based silencing in transgenic plants have been reported. Although single transgene copies can become inactivated (104), the integration of multiple copies enhances silencing efficiencies, particularly if repeated sequences are inserted in concatameric arrangements at one locus (5, 88), but also when homologous transgenes are located at alleles of a locus (83) or are present at unlinked sites (75). Transgene inactivation can comprise both transcriptional (83) and posttranscriptional silencing (27, 57) of marker genes.

Silencing is influenced by the length of the homology and especially by the position of the interacting sequences. Linked copies are more efficiently silenced than unlinked copies, and unlinked loci show characteristic differences in silencing capacity (132) and susceptibility to being silenced (91). The most efficient example of *trans*-inactivation is a tobacco line carrying a transgene insert with two genes driven by the 19S and the 35S promoter of CaMV, respectively. Both genes linked to the two promoters are suppressed, and this locus *trans*-inactivates newly introduced constructs that provide at least 90 bp of common homology (132).

HERITABILITY AND REVERSION OF SILENCING When interacting loci are separated in genetic crosses, reversion of the silenced state occurs slowly over several generations (91). The silenced state of concatameric transgenes, which are always inherited as a block, is preferentially transmitted to the progeny (5, 63, 88). In some transformants, silencing is progressively enhanced in subsequent generations (5, 63). Other silenced transgenes show a defined resetting phase. A silenced *rolB* transgene is reactivated and remains active in young seedlings, while silencing occurs again erratically during further development of the seedling (27).

In concatemeric transgenes, a reduction of repeats enhances the probability of reversion of the silencing event. In *Arabidopsis* transformants in which transgene copy numbers had been reduced because of intrachromosomal recombination, transgene inactivation was observed at a lower frequency than in the parental line that still contained multiple repeats (4, 5). A similar observation was made by another laboratory for an *Arabidopsis* transformant that had also lost several transgene repeats via intrachromosomal recombination. When lines were selected from the parental plant and the deletion line that had reactivated the transgene, only the deletion lines conserved the active state of the transgene during meiosis (87).

THE ROLE OF DNA-METHYLATION A correlation between gene inactivation and DNA methylation has been shown for transgenes (2, 131), transposable elements (19, 116), and some endogenous genes (123). For other genes no such correlation was observed (93). With respect to a correlation between DNA methylation and homology-based silencing, we can group the different silencing events into three classes. Silencing events that show a direct correlation between transcriptional inactivation and DNA methylation within the promoter (76, 83) or coding region (57), silencing events that are not associated with detectable changes in DNA methylation (27, 44), and silencing phenomena where hypermethylation patterns build up over successive generations (86).

An interesting aspect of DNA methylation in plants is the presence of methylated C residues outside of CG or CNG sequences, the symmetrical target sequences for maintenance methylation in plants. To date, nonsymmetrical methylation patterns, which are probably not encoded in the sequence—but more likely in the secondary structure of a sequence—have only been detected in transgenes (57, 85) and not in endogenous genes (93). It is therefore unclear whether they are specifically imposed on transgenes or genes that have been transferred into new chromosomal environments. In the latter case, they should also occur in transposable elements.

Paramutation

An indication that homology-based silencing events are not specific for transgene DNA but reflect an endogenous mechanism comes from the analysis of paramutation. Paramutation was described more than 60 years ago (139) and has been examined in several species (11, 21, 48, 49). The term refers to the interaction of homologous plant alleles that leads to heritable epigenetic effects. A detailed review of paramutation, in the context of genetic imprinting, has recently been published (74). We therefore only summarize a few aspects relevant to the mechanistic models for gene silencing. A *paramutagenic* allele can cause a *paramutable* allele to undergo an epigenetic conversion to become a *paramutant* allele of lower function. The new *paramutant* state is metastable, because it can be somatically and germinally inherited in the absence of the *paramutagenic* allele, but it also reverts with different frequencies. Paramutation requires a metastable state of the *paramutable* allele, which is only amplified through its interaction with the *paramutagenic* allele. Paramutation and the frequencies of reversion are dependent on environmental and developmental factors.

Molecular studies in snapdragon (10, 22, 65), maize (101), and *Petunia* (83) provided insights into the mechanisms involved in paramutation. A correlation between the expression of the *paramutable* gene and its methylation state was observed for the *R* locus in maize (30) and for an *A1* maize transgene in *Petunia* (83). In contrast, no differences in cytosine methylation could be detected between *paramutagenic* and *paramutable* alleles at the *B* locus of maize, despite an extensive analysis over a distance of 12 kb (100). The latter study does not exclude the involvement of DNA modification in paramutation, as certain types of nucleotide modifications, such as hydroxy-methylcytosine, A-methylation, or methylcytosines located in nonsymmetrical positions would have gone undetected. Nevertheless, the analysis of *B* suggests either that methylation is not the cause but a secondary effect of paramutation, or that different classes and mechanisms of paramutation exist. The latter is a possibility because *B* and *R* paramutation differ in other characteristics (100). Paramutated alleles of *B* are extremely stable, whereas *R* paramutation shows frequent reversions. Furthermore, the *B* locus contains a single allele, whereas in most *R* alleles multiple homologous genes occur at the *R* locus. A similar complexity has been found for two semidominant alleles of the *nivea* locus in *Antirrhinum majus* that show structural rearrangements such as inverted duplications or concatamerization of truncated copies of the *nivea* gene (10, 22).

Mutual Inactivation of Transgenes and Endogenous Genes

The term cosuppression (59) was coined to describe the inhibition of gene expression of an endogenous gene after the introduction of a homologous transgene. This phenomenon was first described for the chalcone synthase (CHS) gene in *Petunia* (90, 129). Up to half of the transformants that contained a CHS sense copy produced white flowers or floral sectors because of the loss of CHS activity. Nuclear run-on analysis showed normal CHS transcription rates but a reduction in steady-state levels of CHS mRNA, apparently as a result of posttranscriptional effects (38, 128). Frequently, not all flowers showed the same cosuppression phenotype. Individual plants developed branches with purple, white, or sectored flowers. Among the flowers of individual branches, cosuppression patterns usually remained very similar, which suggests that cosuppression was somatically inherited and initiated during formation of the meristem of individual branches (38).

Cosuppression is not unique to CHS but appears to be a general phenomenon affecting many endogenous genes. Detection can be difficult if the inhibition of the gene does not produce a visible phenotype. Examples of genes that showed an unstable expression after the introduction of homologous sequences are dihydroflavanol reductase (129) and the homeotic *fbp2* gene in *Petunia* (3), tomato polygalacturonidase (120), phytoene synthase (40), pectinesterase (119), an *Arabidopsis* cab140 gene (13), phenylalanine ammonia-lyase (33), β-1,3-glucanase (25), chitinase (50), nitrate reductase (24), S-adenosyl-L-methionine synthetase (9), and glutamine synthetase in tobacco (G Coruzzi, unpublished data). The efficiency of cosuppression varies for individual transformants. The transfer of the same sense construct often generates cosuppressed transformants as well as transformants that overexpress the sense construct (9, 24). Individual genes show characteristic differences in their susceptibility to cosuppression. Two extreme examples are the *Petunia* chalcone isomerase gene, for which cosuppression has not been observed thus far (28), and the nuclear-encoded cytosolic tobacco GS2 gene that is cosuppressed, to variable extents, in all transformants tested so far (G Coruzzi, unpublished data).

It is important to recognize that there are multiple steps within the expression pathway that contribute to the production and activity of a gene product. This might also explain the partly contradictory features found for individual cosuppression events. Inhibition of gene expression at the posttranscriptional level has been confirmed for many types of cosuppression (25, 38, 128), but transcriptional suppression can also be found (13). It is possible that cosuppression is mediated by different mechanisms in different species or for individual genes. Alternatively, for certain genes posttranscriptional silencing could be the primary event that induces transcriptional inactivation as a second step of a common cosuppression mechanism.

REQUIREMENT FOR TRANSCRIPTION An important aspect of cosuppression is the question as to whether the transgene and the endogenous genes need to be transcribed. Evidence for the requirement of mutual transcription comes from studies on cosuppression of polygalacturonase (PG) in tomato induced by constitutive expression of a truncated PG transcript. In ripe fruits where the endogenous gene is active, expression of both genes is reduced, and transcript levels of the constitutively expressed transgene are significantly lower in ripe fruits compared with green fruits (120). On the other hand, a promoterless CHS construct induced cosuppression effects in 15% of transgenic *Petunia* plants (128), which suggests that expression is not required for cosuppression. However, traces of CHS antisense RNA are found in these transformants (128), which allow the speculation that an endogenous promoter reads into the transgene creating an antisense-mediated inhibition.

There is some evidence that critical levels of transcription are required for efficient induction of cosuppression. For several examples, silencing was found to be enhanced or even dependent on the homozygous state of a transgene (24, 25, 50). In contrast to meiotically transmittable examples of cosuppression (60), silencing in homozygous lines was not inherited by outcross progeny now containing only one transgene. One interpretation for this effect is that the primary transgene is transcribed at a relatively high rate, and that by duplication of these rates, transcript levels in homozygous plants reach a critical threshold. Alternatively, cosuppression might be stimulated by a DNA-based interaction between the transgene alleles in homozygous lines. At least for cosuppression of the β-1,3-glucanase in tobacco, the latter assumption could be excluded. In one line, inactivation of the β-1,3-glucanase genes occurred exclusively in plants homozygous for a homologous transgene. In haploid plants of this line, suppression was observed regardless of whether the transgene derived from homozygous or hemizygous transformants (25). This result excludes a function for allelic interactions and suggests a dose-dependent regulation for silencing that is determined by the ratio between the transgene transcripts and the copies of the endogenous genes or the entire genomes.

DEVELOPMENTAL AND ENVIRONMENTAL CONTROL OF COSUPPRESSION Several cases of cosuppression show developmental regulation and a dependence on environmental factors. Cosuppression of CHS genes in *Petunia* produces a variety of anthocyanin pigmentation patterns in the flower, among which highly ordered patterns can be found that are somatically heritable (90). These observations suggest a linkage between regulatory mechanisms of morphological differentiation and the induction of cosuppression (61). Frequently, silencing is triggered after a lag period, either stochastically at different stages during development (24, 50) or synchronously at a specific stage of development (25; H Vaucheret, unpublished data).

Various cases of environmental influences on silencing have been observed. Cosuppression of CHS genes in *Petunia* (129) and β-1,3-glucanase genes in tobacco (25) are stimulated by high light intensities. Silencing of chitinase genes in *Nicotiana sylvestris* (50) and nitrate reductase in tobacco (24) are dependent on germination and growth conditions.

Inactivation Mediated by RNA Viruses

An unexpected link between cosuppression and transgene-mediated viral resistance was observed in transgenic plants resistant to different members of the potyvirus group (69, 89). Untranslatable constructs of the viral coat protein gene or the RNA polymerase gene of potyviruses generated a strain-specific resistance against the virus, accompanied by very low steady-state levels of the transgenic RNA. In virus-resistant lines, homologous transgenes were also *trans*-inactivated, which suggests that viral resistance is mediated by a homology-based inactivation mechanism.

MODELS FOR HOMOLOGY-BASED SILENCING

The complexity of experimental details concerning homology-based silencing makes it difficult to allocate defined mechanistic models exclusively to certain silencing categories. We have therefore avoided linking detailed models to the description of different types of silencing presented above. In the following, we discuss several models for the molecular mechanisms involved in silencing, which are not mutually exclusive but which may apply individually or even synergistically for individual silencing types.

Silencing Mediated Via DNA-DNA Pairing

The interaction of homologous DNA copies has been proposed as a mechanistic model for certain types of cosuppression (59), *trans*-inactivation (73), and paramutation (83). It was proposed (61) that silencing reflects changes in the physical state of a transgene and that mutual silencing of a transgene and an endogenous homologue are caused by regular changes in the epigenetic states of the transgene. A DNA-DNA pairing model would provide an explanation for the differences in efficiency at which silencing occurs within individual transformants, because the interaction between two homologous sequences would be determined by the probability with which the two loci associate in interphase nuclei. This probability should be higher for tandemly linked copies, compared with unlinked, ectopic copies. The tandem arrangement of transgenes may not only enhance the efficiency of DNA-DNA pairing, but the formation of stemloop structures on single strands of a region carrying inverted repeats could mediate an efficient spread of de novo methylation patterns. Foldback DNA is specifically recognized by the human methyltransferase (121).

Individual transgenes differ significantly in their capacity to *trans*-inactivate homologous copies (132), which probably reflects their potential to scan other chromosomal locations for homology. The presence of very efficient *trans*-silencers close to the telomere suggests that telomeric regions are favorable sites for the interaction with homologous sequences (72).

RNA-Mediated Models for Silencing

THE ROLE OF RNA-DNA HYBRIDS The DNA pairing model suggests that epigenetic patterns, characterized by a specific state of DNA methylation or chromatin structure, are exchanged during a potential somatic hybridization (60, 75). The observation that RNA molecules can induce hypermethylation patterns within homologous DNA sequences suggested that changes in epigenetic states could also be mediated by DNA-RNA pairing. Evidence for the participation of RNA molecules in the induction of DNA methylation came from a study of tobacco transformants carrying the cDNA of potato spindle tuber viroid. Specific methylation of the viroid DNA was observed whenever viroid RNA replication had occurred (135). These data suggest that transcripts can induce methylation in the homologous DNA region, which might be especially relevant for transformants that accumulate large amounts of nuclear transcripts because of high transcription rates or imperfect RNA processing. The specific methylation of coding regions in certain posttranscriptional silencing events (57) might reflect such an RNA-mediated induction of DNA methylation.

DEGRADATION OF THRESHOLD LEVELS OF RNA The dosage-dependency of certain silencing effects (25, 27, 57) and the observation of a linkage between silencing and the onset of expression of the endogenous gene (120) suggest that silencing can be induced by the production of defined threshold levels. Such a model might especially apply for particular genes that carry target sequences for RNA degradation to control high expression levels generated by gene induction (92).

THE AUTOREGULATION MODEL Inspired by a model for cytokinin habituation (79), an autoregulation model for silencing was proposed (80). In this model, transcription of a target gene, susceptible to silencing, leads to the production of a diffusible activator that increases steady-state mRNA levels of the target gene. Expression of the target gene therefore depends on the concentration of the activator in a positive feedback loop. The model suggests that activator synthesis depends on the transcription rates of the target gene and that activator degradation is proportional to activator concentration. The activator will also stabilize mRNA levels of transgenes that are homologous to the endogenous target gene. Transcription of homologous transgenes will enhance activator concentrations. In this model, the activator therefore mediates the linked stabilization or repression of its target gene and a homologous transgene. The system is stable, when synthesis rates equal the degradation rates of the activator. Variations in transcription rates, however, will induce instabilities in the feedback control system. Development-dependent changes in transcription rates would increase the concentration of activator molecules, which would in turn enhance the activator degradation system. When transcription rates decline again, the high degradation rates would rapidly reduce the number of activator molecules, thus decreasing steady-state mRNA levels of the target gene and the homologous transgene. Steady-state mRNA levels could recover when low activator concentrations raise again. The postulation of activator molecules could explain the developmental modulation and resetting effect of certain cosuppression systems.

ANTISENSE-MEDIATED RNA DEGRADATION An obvious element to account for the sequence specificity of cosuppression is the production of antisense RNA. RNA duplexes would be targets for RNAseH-like endogenous enzymes. Antisense transcripts could be generated by promoters present on the transgene DNA or by endogenous plant promoters at the 3′ end of the transgene (47). Alternatively, they could be produced by a plant RNA-dependent RNA polymerase (39, 69). If antisense RNA is only produced at particular developmental stages or if promoters located 3′ to the silenced gene are regulated by environmental stimuli, this would explain the developmental and environmental dependence of certain silencing phenomena.

The production of antisense molecules by an RNA-dependent RNA polymerase may depend on the production of specific threshold levels of sense RNA or on the accumulation of RNA intermediates during a delay in RNA transport or processing (38, 61). This hypothesis assumes that RNA-dependent RNA polymerases recognize "aberrant" transcripts, which may derive from incorrect transcription, transport, or translation of the transgene. The production of aberrant RNA may be modulated by changes in epigenetic states of a gene that influence the mode or efficiency of RNA processing (38).

Certain transgenes only cosuppress sequences that contain a homologous 3′ end, whereas genes only homologous to the 5′ region are not affected (J English, unpublished data). This observation suggests that antisense transcripts are preferentially made against the 3′ end region of the transgene. On the other hand, constructs that contained the 5′ end of one gene and a second gene at the 3′ end, efficiently silenced both endogenous genes, which argues against a general function of the 3′ end (119). Nevertheless, specific modifications at the 3′ end have been detected, such as the accumulation of processing intermediates (J Kooter, unpublished data) and incorrect splicing within the 3′ end of silenced transcripts (D Flavell and M Metzlaff, unpublished data).

CELLULAR MECHANISMS INVOLVED IN HOMOLOGY-DEPENDENT SILENCING

Many aspects of the models listed above are still speculative, and in cases where particular molecular features, such as hypermethylation or high transcription rates, have been associated with silencing events, their general importance is still unclear. It would be premature and possibly detrimental to favor one common model for the many different silencing events, because we would narrow the scope of our investigations. From the experimental details published for various gene silencing systems we can draw two important conclusions. First, the growing number of reports on transgene silencing no longer correspond with the early assessment that we are dealing with a few rare events of minor importance. Second, gene silencing was not specifically developed for transgenes, but it reflects endogenous functions that most likely participate in the regulation of gene expression and plant development.

At present, we can formulate three major areas for future research activities: the function of chromatin in a dynamic regulatory system in plant development, the control of RNA turnover within RNA processing routes that are involved in the fine tuning of gene expression, and the importance of a homology-detection mechanism for gene expression and genome organization. We discuss the general importance of these three aspects for the regulation of gene expression in plants and other eukaryotes. We do not know whether and which of the mechanisms that have been found in other eukaryotes are also relevant for plants. Nevertheless, the examples shown below should be helpful to define primary models for mechanisms of homology-dependent silencing in plants, and they should explain why we think that an improvement of our knowledge in the three areas mentioned above might be necessary to understand the control of gene expression in plants.

The Regulatory Function of Chromatin States

It has been proposed that epigenetic patterns can be established not only by DNA methylation (54) but also by supramolecular chromatin structures (136). Our present knowledge about the formation and control of different states of chromatin in plants is still very limited. Assumptions that changes in expression patterns are based on modifications of chromatin conformation are mainly grounded in the indirect evidence that changes in DNA methylation occur and that dense methylation patterns induce the formation of highly packed chromatin (8).

The most advanced studies on the role of chromatin structure in heritable gene repression come from *Drosophila* and yeast, two species that lack C-methylation and that have proven especially suitable for chromatin studies. Their small genomes simplify the analysis of individual genomic regions and genetic analysis has identified modifiers of chromatin complexes. Yeast offers the advantages of rapid generation of mutants, easy physiological analysis, and gene replacement by homologous recombination. The polytene chromosomes in the salivary gland of *Drosophila* allow a precise localization of chromatin complexes.

POSITION-EFFECT VARIEGATION Position-effect variegation (PEV) is a partial inactivation of gene expression in *Drosophila* caused by a rearrangement that places a normally euchromatic gene near a heterochromatic region (51, 106). PEV at the *white* locus, involved in eye color, can be monitored in individual cells by the reduction of red eye pigment. Inactivation of the gene, which results from the spreading of the heterochromatic state into the euchromatic neighborhood, causes a mosaic phenotype of red and white cells that gave PEV its name (124). PEV demonstrates the differences between two kinds of chromatin: heterochromatin, which is located within the pericentric regions and which remains condensed throughout the cell cycle, and euchromatin, which is located in the chromosome arms and decondenses during interphase. Inactivation of a gene is accompanied by cytologically visible spreading of heterochromatin over 50–100 polytene bands, corresponding to hundreds of kilobases. The *Drosophila* genome consists of more than 100 loci that suppress or enhance PEV, some of which have been characterized to encode chromosomal proteins (113). A set of heterochromatin-associated genes, the products of the *Su(var)* genes, assemble cooperatively to form complexes in heterochromatin regions. These protein complexes can continue to expand. Complex formation requires the interaction of large amounts of different gene products, and insufficiency of one of the *Su(var)* genes reduces the spreading of heterochromatin (70).

REPRESSION MEDIATED BY MEMBERS OF THE POLYCOMB GROUP Another example for the control of gene activity by large chromosomal complexes is the regulation of homeotic gene expression. During early embryogenesis, the maternal and segmentation gene products catalyze the assembly of inhibitory proteins of the Polycomb group (Pc-G) and activating proteins of the trithorax group (trx-G) at homeotic gene loci. Patterns of chromatin conformation, mediated by Pc-G and trx-G proteins, are mitotically transmitted and provide the basis for differential expression of homeotic genes along the anterior-posterior axis (98). Pc-G proteins can be localized at specific regions on polytene chromosomes, which implies that the complexes recognize specific regions or secondary structures. Most likely, there exists a molecular relationship between *Pc-G* genes and *Su(var)* genes, because some mutations in *Pc-G* genes affect PEV, and certain *Su(var)* genes influence *Pc-G*-mediated effects (35). Moreover, the Pc protein shares an amino-acid sequence, called the chromodomain, with the Su(var)3-9 product, the heterochromatin protein HP1 (99).

CHROMATIN-MEDIATED REPRESSION IN YEAST In yeast, at least two modes of transcriptional repression are mediated by chromatin conformation: (*a*) repression by the global regulator complex Ssn6/Tup1 (111), and (*b*) silencing of the yeast mating type loci (107) and genes at telomeres (96). The gene products of the *TUP1* gene and the *SSN6* gene are physically associated in a large protein complex, required for repression of cell type-specific genes and genes repressed by glucose or oxygen, respectively. The complex does not bind to DNA directly but is targeted to particular promoters via protein-protein interactions with specific promoter binding proteins. In this interaction, Tup1 provides the repressor activity and Ssn6 the targeting function (127). Repression by Ssn6/Tup1 is mediated by organizing repressed chromatin domains, possibly through interactions with histone H4 (111). Silencing at the yeast mating type loci, HML and HMR, and at telomeres is also regulated by the creation of a defined chromatin structure. Formation of a silencing chromatin structure is mediated by an interaction of the Sir proteins and their interaction with histones H3 and H4 (125). Targeting of the complex to specific regions is mediated by the origin-recognition complex protein (ORC), the Rap1 protein, and the Abf1 protein (111).

In summary, we can define three important aspects for chromatin-mediated gene repression: the formation of heterochromatin-related protein complexes, the targeting of these complexes to particular locations, and the role of histones as modulators for the formation of certain complexes. Not all three aspects may be relevant for potential chromatin-mediated silencing effects in plants, but a search for modifier functions of silencing, either by mutagenesis (27) or by searching for plant proteins that share common domains with known regulators of silencing in *Drosophila* or yeast, might clarify whether chromatin-mediated repression is a universal feature in eukaryotes. Encouraging support for this assumption comes from a report about the identification of the murine *bmi-1* gene, a homologue of the *Pc-G* gene *Posterior sex combs* gene. Mice deficient for *bmi-1* show multiple posterior-directed homeotic transformations (130), caused by ectopic expression of genes of the HOX4 cluster (95).

A ROLE FOR CHROMATIN-MEDIATED REPRESSION IN HOMOLOGY-DEPENDENT SILENCING The formation of chromatin states can be relevant for two aspects of homology-dependent silencing mechanisms, the *trans*-inactivation of homologous sequences and the developmental regulation of silencing. Even a transient pairing of two homologous sequences could favor the exchange of chromatin components that have formed a repressed complex on one copy. Thus, silencing would be the result of the establishment of a repressed chromatin state in the transgene region, some part of which is transferred to other homologous transgenes or endogenes genes. Support for this assumption comes from the observations that multiple transgene copies are preferentially silenced and that transgenes can be specific targets for DNA methylation, which is associated with chromatin condensation (81). It has been proposed that DNA methylation acts as a defense mechanism against foreign DNA (7, 29). Plant genes have a relatively narrow range of AT-content and are embedded into 200-kb large chromosomal regions of a matching AT-content, termed isochores. Monocotyledonous and dicotyledonous species contain distinct isochore compositions (112). Therefore, transgenes with deviant base compositions may become specific targets for de novo methylation. A possible case of this occurs when a single copy of the GC-rich *A1*-gene from maize can become specifically methylated in transgenic *Petunia* (82), whereas its homologue, with a GC-content similar to *Petunia,* from *Gerbera* remains unmethylated (34). Because chromosomal integration sites differ somewhat in base composition, transgenes will become methylated with different efficiencies at different sites, and certain transgenes will not be inactivated at all (26). It is possible that the insertion of multiple copies will increase the probability of individual transgenes being methylated and condensed heterochromatin because the entire region is different from a typical isochore. Furthermore, multiple transgenes integrated in tandem or as inverted repeats might enhance the formation of condensed chromosomal complexes, if they provide target chromatin-associated factors, similar to repeat-induced heterochromatinization processes in *Drosophila.* Transient interactions of transgene sequences inserted at different chromosomal locations would then induce a spread of condensed chromatin states, which explains the preferential inactivation of multiple transgene copies, even at dispersed sites.

CHROMATIN-MEDIATED REGULATION OF CHROMATIN STATES The chromatin conformation may be regularly modified during development and could also be influenced by environmental conditions. Any change could influence transcriptional activity of the gene, the efficiency of RNA transport, or the competence of the locus for somatic pairing with homologous sequences. Any of these features could explain the developmental regulation and the environmental dependence of some homology-based silencing phenomena.

It has been suggested (61) that cosuppression of CHS genes in *Petunia* reflects developmental and physiological factors that impose heritable metastable changes in the plant genome. This idea was deduced from models developed from the analysis of transposable elements in plants (12, 77). Brink proposed that the genome possesses a *paragenetic* function that is distinct from its genetic function. Based on the concept of paramutation, he defines an *orthochromatin* that harbors the DNA that is subject to mutations and a *parachromatin* that is sensitive to the cellular environment and capable of receiving, recording, and mitotically transmitting information from outside the chromosome. During development, changes in *parachromatin* would therefore condition differential activities of genetic loci in the *orthochromatin* (12). A similar concept is found in Mc Clintock's interpretation of epiallelic states of transposable elements. She observed that individual epialleles of an element showed characteristic *phases* of activity, which could be influenced by the expression of other elements. This *presetting* effect was mitotically transmitted but erased in the next generation. She proposed that individual genes are embedded or dissociated from condensed chromatin clusters in a regular manner, which regulates their differential expression during development (77). If extracellular signals induce changes in paragenetic states, this could also modify the position and association of genes within such clusters of a cell and its somatic derivatives.

It is unknown which molecular factors regulate changes in chromatin structure, but it has been suggested that particular chromatin states are generated or conserved by changes in DNA methylation (81). Evidence that DNA methylation is involved in the determination and modulation of epigenetic states comes from the analysis of the *En/Spm* transposable element in maize (36). This autonomous element can exist in three distinct but interconvertible forms termed cryptic, programmable, and active. These forms can be distinguished by the methylation levels of GC-rich sequences in the downstream control region (DCR), near the promotor. Cryptic elements are almost stably inactive and exhibit somatic reversion frequencies to active states of 10^{-5}: They are highly methylated. Active elements are unmethylated. Programmable elements that revert more frequently than cryptic elements and that can be *trans*-activated by an active element have an intermediate methylation state. The modulation of epigenetic states is mediated by TpnA, an autoregulatory protein encoded by the element. TpnA has three distinct functions. It is required for transposition of the element (41), it activates methylated promoters of programmable elements, and it represses the unmethylated promoter of active elements (115).

Besides the determination of their phases of activity, transposable elements are regulated by a developmental control mechanism that determines the heritability of the phases and the activity of the elements during development. Therefore, changes in the epigenetic state influence the activity of the element in the next generation (37). A related observation has been made for the activity of an *A1* transgene in *Petunia* that was also correlated with changes in DNA methylation. In an F1 progeny from one transgenic line homozygous for *A1,* plants derived from pollination of the first flowers showed a significantly more stable expression of the marker gene, while in progeny from pollination of older flowers the transgene became inactivated and methylated at high frequencies (84). These data also suggest a developmental regulation of epigenetic patterns that can be transferred to the next generation.

Posttranscriptional Control of Gene Expression

Posttranscriptional regulatory mechanisms have an important function in the control of gene expression (43, 52). The efficiency at which a gene will be expressed depends on mRNA processing, transcript stability, nucleocytoplasmic transport, translation efficiency, and protein modification and half life. We do not discuss these items in detail but concentrate on a few examples that indicate how changes in RNA stability might be involved in gene silencing.

RNA TRANSPORT In recent years several partly contradictory models have emerged about the transport of RNA from the site of transcription to the cytoplasm. In analogy to the established model of an organized movement of newly synthesized polypeptides through the cytoplasmic secretion machinery (105), it was proposed that transcripts move to the cytoplasm in an ordered fashion, passing through localized spots that harbor individual steps of the processing machinery (16, 139). This model is supported by reports of the localization of splicing components in subnuclear domains, called speckles and foci, and by observations that intron-containing RNAs are targeted to speckles upon microinjection into mammalian nuclei (122).

In contrast with these reports, other studies argue against a compartmentalization of RNA processing, because splicing occurred at the sites of transcription. These sites were not coincident with intranuclear speckles that harbor components of the splicing machinery (142). Assuming that the position of certain genes within the interphase nucleus determines the entrance of the transcript into defined processing routes, transcripts of multiple transgenes localized at different positions may enter common or separate processing routes. Local concentrations of homologous transcripts would be enhanced significantly, if transcripts of the transgene and the homologous endogenous gene pass the same processing track. We therefore not only have to consider the general quantities of steady-state RNA levels within the nucleus but also have to account for the local concentrations of homologous RNA molecules within the processing track.

RNA STABILITY RNA stability is influenced by a number of factors. The 5′ cap structure and the 3′ poly(A) tail stabilize mRNA against degradation. The poly(A) tail also regulates the efficiency of translation, but only when the transcript is capped (42). Other posttranscriptional modifications have been proposed to serve as signals for degradation. Adenine residues can be methylated or converted into inosines that might serve as a tag for RNA degradation (64). Destabilizing and stabilizing sequence elements have been identified in specific mRNAs. These elements either provide target sequences for RNA degradation (92) or binding sites for stabilizing factors (17).

RNA stability is also influenced by the efficiency of translation. An interesting linkage between translation efficiency and mRNA metabolism has been detected for the human β-39 mRNA, a mutation of the β-globin gene that carries a stop codon at position 39 (6). The nontranslatability of the β-39 mRNA induced a significant reduction in mRNA accumulation, although transcription, splicing, and polyadenylation of the β-39 mRNA are not altered. This observation prompted the ideas that either a nuclear mechanism exists that is capable of sensing nonsense mutations or that there is a feedback communication from the cytoplasm to the nucleus. It has been suggested that this feedback interaction occurs at the nuclear membrane at points of contact with the rough endoplasmatic reticulum, where translation can occur (6).

PRODUCTION OF ANTISENSE RNA As mentioned earlier, it has been proposed that antisense transcripts are involved in posttranscriptional silencing. Antisense transcripts could be produced by promoters located in the 3′ region of a gene or by an RNA-dependent RNA polymerase. Several examples exist for the use of endogenous antisense transcripts for transcriptional (94) or posttranscriptional (53, 55) control of expression. In barley, a lack of alpha-amylase expression coincides with the appearance of a transcript complementary to the alpha-amylase mRNA (108). RNA-dependent RNA polymerases are widely distributed among plants, although there has been a dispute about the possible contamination of the material by viral RNA polymerases (39). These enzymes are usually present in low amounts, which can be significantly increased upon viral infection. The enzymes in different plants are clearly distinctive in size and template specificity. The host-specificity is conserved after induction of higher enzyme levels by infections with the same virus, and this supports the hypothesis that the enzymes are not derived from the virus but are encoded by the host plant. The biological role of RNA-dependent RNA polymerases has not been fully elucidated, but it is obvious that they can create antisense molecules against existing cellular transcripts.

To evaluate the importance of posttranscriptional control mechanisms for silencing we still need to answer several key questions. We do not know whether transcription, translation, or polysome-association is a prerequisite for silencing. We need to define what determines the "aberrant" state of RNA and whether this induces RNA degradation or the production of antisense transcripts. We also need to understand how transcripts are transported into the cytoplasm and whether and which factors exist that induce feedback responses at the DNA level when RNA processing or translation is disturbed.

Repeat-Specific Control Mechanisms

The participation of repeated sequences in gene inactivation phenomena is not limited to plants but can be found in several other eukaryotes. A comparison of homology-dependent silencing in plants with other eukaryotic silencing systems illuminates some interesting similarities that suggest common biological functions.

THE RIP- AND MIP-MECHANISMS OF FILAMENTOUS FUNGI In the filamentous fungi *Neurospora crassa* and *Ascobulus immersus,* the presence of DNA repeats triggers methylation and inactivation of the repeated regions. In *Neurospora* transformants that contain linked or unlinked duplicated sequences, a mechanism named Repeat Induced Point Mutation (RIP) induces methylation of C-residues followed by mutation of C to T, preferentially at CA dinucleotides (14, 117). In *Ascobolus,* gene duplication leads to de novo methylation and premeiotic inactivation because of a mechanism termed Methylation Induced Premeiotically (MIP) (45). The efficiency of MIP depends on a gene's location and on the length of homologous repeats. Clustered repeats of a critical length always become methylated, whereas efficiencies for ectopic homologues vary considerably (110). Methylation not only encompasses C residues within CpG dinucleotides, it also extends to C residues located in nonsymmetrical sequences, which implies a novel type of methyltransferase activity (46). Similar nonsymmetrical 5mC patterns have been observed for sequences that had undergone RIP in *Neurospora* (118) and for plant transgenes that had become transcriptionally (85) or posttranscriptionally silenced (57), respectively.

An inverse correlation between copy number and the expression of transgenes was also shown for the asexual cycle of *Neurospora* (97). After transformation with a resistance marker, vegetative and reversible inactivation of the marker occurred in multicopy transformants. Inactivation was accompanied by hypermethylation of marker genes. Treatment with 5-azacytidine induced a stable reactivation of the marker in some, but not all, transformants, which suggests a functional impact of DNA methylation on gene expression. Methylation might also be involved in a unidirectional silencing event in *Neurospora,* termed quelling (109). The expression of endogenous genes was impaired when several homologous copies were integrated at an ectopic site. Reversion of the quelling effect was correlated with a reduction in the number of ectopically integrated gene fragments. These data demonstrate that *Neurospora* and *Ascobolus* contain homology-searching mechanisms responsible for the specific methylation of repeated sequences. These show some similarities to silencing phenomena in plants. Because filamentous fungi are excellent subjects for mutation analysis, it is very likely that endogenous genes involved in the regulation of silencing will be first identified in *Neurospora* or *Ascobolus* before they are found in plants.

HOMOLOGY-MEDIATED REPRESSION IN *DROSOPHILA* In *Drosophila,* repeated sequences are involved in the initiation of position-effect variegation, in *trans*-effects of PEV, and in transvection. PEV-like effects were induced when tandemly linked copies of a P-transposon carrying a *white* transgene were integrated into the *Drosophila* genome (31), which suggests that pairing of repeats contributes to heterochromatin formation. Homologous chromosomes of nondividing nuclei are physically paired in somatic cells of *Drosophila,* providing the basis for the transmission of heterochromatin states to homologous alleles. One example for pairing-dependent *trans*-inactivation is dominant PEV at the *brown* locus. Heterochromatin that was imposed on a rearranged *brown* allele was transmitted to the unrearranged homologous copy, which also became inactivated (32).

Other pairing-dependent modulations of gene expression have been described under the term transvection (68). Like paramutation, transvection depends on the interaction of susceptible alleles, but the effect is not preserved after segregation of the interacting alleles during meiosis. A well-characterized example for allelic interaction is the regulation of the *white* gene by the product of the *zeste* gene (103). The *white* gene carries a set of *zeste* binding sites in its promotor, in the eye enhancer region. A lack of *zeste* function leads to a moderate decrease in *white* activity. The z^1 mutation of *zeste* is responsible for the *zeste* product forming hyperaggregates (20), and this causes severe repression in lines that contain two paired copies of *white.* Repression is not observed for single *white* copies, but if a homologous *white* gene, together with the *zeste* binding sites, is inserted at ectopic positions, about one third of these lines show repression in a z^1 background. This observation is interpreted as indicating that specific loci can interact with the *white* locus to form z^1-specific aggregates. *Zeste-white* interactions are further modified by several genetic loci, most of which are members of the *polycomb group*.

REPEAT-INDUCED DNA METHYLATION IN MAMMALS In contrast to the large number of repeat-induced inactivation events found in transgenic plants, it has been unclear whether a similar mechanism exists in mammals. Sense inactivation of an endogenous gene (15) and an inverse correlation between the number of transgene copies and their methylation state have been reported (78). Because such events are rare, it was doubtful whether these events reflected the presence of a homology-search mechanism as has been postulated for plants and fungi.

More compelling evidence that repeat-induced inactivation is also present in mammals came from the analysis of human genetic diseases associated with the amplification of triplet repeats (67). Most of these genes, which encode transcription factors, carry multimers of CAG- or CCG-triplets that are responsible for poly-glutamine or poly-proline stretches within the coding sequence. The triplet repeats are variable in length and can expand or contract in somatic cells or when passed to the next generation. As the repeat expansion is not uniform, individuals carry a mosaic pattern; individual cells harbor various repeat lengths in the affected gene (102). Excessive repeat amplification causes disease. The onset of disease is developmentally regulated; for most diseases there seems to be an inverse relation between the number of triplet repeats and the age at which the first symptoms of the disease become manifest (56, 126). Because the probability and the length of repeat extension significantly increase in cells that show deficiencies in DNA repair (1), it is unclear whether diseases are caused by a defective mismatch repair or whether the repeat elements are causally involved. For triplets located within protein-coding regions, repeat amplification results in a significant increase in the length of the single amino acid stretches. It has been shown that long poly-glutamine stretches reduce the activity of transcription factors (18), and the disease phenotype could reflect a lack of function of the protein. Amplified triplets are not restricted to protein-coding regions, however, but are found in the 5′ noncoding region (141) or in 3′ untranslated DNA of some genes (138). For these diseases, the repeat-amplification does not affect protein function but transcription. The expansion of triplet repeats in the *FMR1* gene (140) actually leads to a reduction of gene expression. The increase in CGG repeats makes the CpG island-type promoter of the FMR1 gene, which is normally unmethylated, susceptible to methylation. The level of methylation is not uniform. Individuals have different levels and the mental retardation phenotype correlates with the extent of methylation (71). In analogy to repeat-induced heterochromatinization in *Drosophila* (31), it is tempting to speculate that amplification of triplet repeats in humans induces heterochromatin formation, which is induced or followed by cytosine methylation, converting the gene into condensed and transcriptionally impaired chromatin.

Indirect evidence for the existence of a mechanism for homology-based DNA methylation came from a study of the distribution of CpG dinucleotides in the mammalian genome (66). Based on the criterion that CpG depletion and TpG overrepresentation for a particular genomic region indicates a prior history of high methylation, repetitive sequences were identified as preferred targets for methylation. These data suggest that sequence repeats are specifically recognized to become methylated. This would not only affect highly repetitive DNA but also homologous members of gene families and pseudogenes. It was proposed (66) that the parental sequences are protected from the methylation mechanism, because the insertion of introns into their sequence masks the sequence homology with their pseudogenes.

A ROLE FOR REPEAT-INDUCED METHYLATION IN GENOME EVOLUTION If we consider that many plant genomes carry a large proportion of duplicated loci (137), homology-dependent silencing mechanisms should influence the expression of endogenous duplicated genes. It is conceivable that duplicated sequences escape silencing if they are embedded in noninteracting chromosomal environments or if they contain a significant degree of sequence divergence. In this respect, the presence and efficiency of repeat-dependent silencing mechanisms should influence the potential of the plant genome to develop new allelic variations during evolution. In this context, repeat-induced methylation counterbalances DNA amplification processes generating heterogeneous epigenetic patterns in repeated sequences that can be further modulated by environmental stress. In mammalian tissue cultures, certain resistance genes can be amplified under selection pressure (114). In plants, the most dramatic examples for DNA amplification are the environmentally induced morphological changes in flax. Flax plants treated with high levels of fertilizers amplified specific genomic subsets (23). It has been proposed (58) that changes in heterochromatin content and DNA methylation are also associated with this phenomenon. Epigenetic states are also responsive to changing environmental conditions as demonstrated by the activation of transposable elements (133, 134) or the change of DNA methylation patterns in tissue culture (62) or in field-grown plants (84). A combination of environmentally regulated mechanisms of gene amplification and epimutation would provide the cell with an efficient system to adapt to changing external conditions. Because of the stochastic character of amplification and epigenetic modification mechanisms, the plant would be a chimera of distinctive somatic sectors subject to selection. Novel phenotypes of the genome could be manifested but also corrected in subsequent generations. In mammals the separation of germline and somatic cells, as well as the high degree of complexity of differentiated cells, puts severe constraints on the efficiency of such a mechanism. For plants cells, however, totipolence and the production of germ cells late in somatic development, a balanced activity of amplification and epimutation should have a considerable evolutionary advantage.

CONCLUSIONS AND OUTLOOK

The term homology-dependent silencing refers to inactivation events at the transcriptional or posttranscriptional level that differ in efficiency, resetting, and heritability. Besides the requirement for homology, a common feature of all homology-dependent inactivation events is the disturbance of the sequence context because of DNA rearrangements. This applies for the inactivation of transgenes that integrate randomly into the genome via illegitimate recombination and for nontransgenic silencing events such as transposon inactivation and paramutation. On the other hand, not every illegitimately recombined transgene is subject to silencing, which suggests that the chromosomal location plays an important role. The efficient silencing activity of certain loci may result from a particular sequence context, a specific DNA rearrangement, or a secondary structure. Alternatively, the position of a locus within the nucleus may determine the efficiency of homology pairing or RNA processing.

To clarify the underlying mechanisms of gene silencing and to understand their importance in the regulation of plant development, we must expand our very limited knowledge about the location of genes within the nucleus, the factors involved in chromatin conformation, and the regulation of transcript transport and processing. Finally, we have to understand how developmental programs and environmental conditions influence the formation of epigenetic patterns at the DNA level and whether there are feedback control signals in the RNA processing pathway. Most certainly, we can expect more than one mechanism of gene silencing to exist, and it is unlikely that the silencing phenomena reported so far already represent a complete selection. For the scientific community, it will be rewarding to note and pursue silencing events with greater emphasis than in previous years, when silenced genes were out of the scope of many scientists. Understanding the molecular basis of gene silencing will not only improve the control of the application of transgene technology, it will most likely unveil new endogenous control mechanisms involved in the regulation of plant development and genome evolution.

ACKNOWLEDGMENTS

We thank B. Davies for critical reading of the manuscript and the following colleagues for communicating unpublished results or supplying preprints: D. Baulcombe, G. Coruzzi, P. Elomaa, A. Depicker, J. English, R. Flavell, R. Jorgensen, M. Metzlaff, and H. Vaucheret. We also thank the members of the Human Capitol and Mobility network (grant number ERBCHRXCT940530) on gene silencing for continuous discussion.

Literature Cited

1. Aaltonen LA, Peltomäki P, Leach FS, Sistonen P, Pylkkanen L, et al. 1993. Clues to the pathogenesis of familial colorectal cancer. *Science* 260:812–16
2. Amasino RM, Powell ALT, Gordon MP. 1984. Changes in T-DNA methylation and expression are associated with phenotypic variation and plant regeneration in a crown gall tumor line. *Mol. Gen. Genet.* 197: 437–46
3. Angenent GC, Franken J, Busscher M, Weiss D, van Tunen AJ. 1994. Co-suppression of the petunia homeotic gene *fbp*2 affects the identity of the generative meristem. *Plant J.* 5:33–44
4. Assad FF, Signer ER. 1992. Somatic and germinal recombination of a direct repeat in *Arabidopsis. Genetics* 132:553–66
5. Assad FF, Tucker KL, Signer ER. 1993. Epigenetic repeat-induced gene silencing (RIGS) in *Arabidopsis. Plant Mol. Biol.* 22:1067–85
6. Baserga S, Benz JEJ. 1992. β-globin nonsense mutation: deficient accumulation of mRNA occurs despite normal cytoplasmic stability. *Proc. Natl. Acad. Sci. USA* 89:2935–39
7. Bestor TH. 1990. DNA methylation: evolution of a bacterial immune function into a regulator of gene expression and genome structure in higher eukaryotes. *Philos. Trans. R. Soc. London Ser. B* 326: 179–87
8. Bird A. 1992. The essentials of DNA methylation. *Cell* 70:5–8
9. Boerjan W, Bauw G, Van Montagu M, Inze D. 1994. Distinct phenotypes generated by overexpression and suppression of S-adenosyl-L-methionine synthetase reveal developmental patterns of gene silencing in tobacco. *Plant Cell* 6:1401–14
10. Bollmann J, Carpenter R, Coen ES. 1991. Allelic interactions at the nivea locus of Antirrhinum. *Plant Cell* 3:1327–36
11. Brink RA. 1956. A genetic change associated with the R locus in maize which is directed and potentially reversible. *Genetics* 41:872–89
12. Brink RA. 1960. Paramutation and chromosome organization. *Q. Rev. Biol.* 35: 120–37
13. Brusslan JA, Karlin-Neumann GA, Huang L, Tobin EM. 1993. An Arabidopsis mutant with a reduced level of cab140 RNA is a result of cosuppression. *Plant Cell* 5: 667–77
14. Cambareri EB, Jensen BC, Schabtach E, Selker EU. 1989. Repeat-induced G-C to A-T mutations in *Neurospora. Science* 244: 1571–775
15. Cameron FH, Jennings PA. 1991. Inhibition of gene expression by a short sense fragment. *Nucleic Acids Res.* 19:469–75
16. Carter KC, Bowman D, Carrington W, Fogarty K, McNeil JA, et al. 1993. A three-dimensional view of precursor messenger RNA metabolism within the mammalian nucleus. *Science* 259:1330–35
17. Casey JL, Koeller DM, Ramin VC, Klausner RD, Harford JB. 1989. Iron regulation of transferrin receptor mRNA levels requires iron-responsive elements and rapid turnover determinant in the 3′ untranslated region of the mRNA. *EMBO J.* 8:3693–99
18. Chamberlain NL, Driver ED, Miesfeld RL. 1994. The length and location of CAG trinucleotide repeats in the androgen receptor N-terminal domain affect transactivation function. *Nucleic Acids Res.* 22:3181–86
19. Chandler VL, Walbot V. 1986. DNA modification of a maize transposable element correlates with loss of activity. *Proc. Natl. Acad. Sci. USA* 83:1767–71
20. Chen JD, Pirrotta V. 1993. Stepwise assembly of hyperaggregated forms of *Drosophila zeste* mutant protein suppreses *white* gene expression in vivo. *EMBO J.* 12: 2061–73
21. Coe EH Jr. 1966. The properties, origin, and mechanism of conversion-type inheritance at the B locus in maize. *Genetics* 53: 1035–63
22. Coen ES, Carpenter R. 1988. A semidominant allele, niv-525, acts in trans to inhibit expression of its wild-type homologue in *Antirrhinum majus. EMBO J.* 7:877–83
23. Cullis CA. 1986. Phenotypic consequences of environmentally induced changes in plant DNA. *Trends Genet.* 2:307–9
24. Dorlhac de Borne FD, Vincentz M, Chupeau Y, Vaucheret H. 1994. Co-suppresion of nitrate reductase host genes and transgenes in transgenic plants. *Mol. Gen. Genet.* 243:613–21
25. De Carvalho F, Gheysen G, Kushnir S, Van Montagu M, Inze D, Castresana C. 1992. Suppression of beta-1,3-glucanase transgene expression in homozygous plants. *EMBO J.* 11:2595–602
26. Dehio C, Schell J. 1993. Stable expression of a single-copy rolA gene in transgenic *Arabidopsis thaliana* allows an exhaustive mutagenesis analysis of the transgene-associated phenotype. *Mol. Gen. Genet.* 241: 359–66
27. Dehio C, Schell J. 1994. Identification of plant genetic loci involved in posttranscriptional mechanism for meiotically reversible transgene silencing. *Proc. Natl. Acad. Sci. USA* 91:5538–42
28. De Lange P. 1994. *Inhibition of gene ex-*

pression by antisense genes in Petunia hybrida. PhD thesis. Vrije Universiteit, Amsterdam, Netherlands

29. Doerfler W. 1991. Patterns of DNA methylation: evolutionary vestiges of foreign DNA inactivation as a host defense mechanism. *Biol. Chem. Hoppe-Seyler* 372: 557–64
30. Dooner HK, Robbins TP, Jorgensen RA. 1991. Genetic and developmental control of anthocyanin biosynthesis. *Annu. Rev. Genet.* 25:173–99
31. Dorer RD, Henikoff S. 1994. Expansion of transgene repeats cause heterochromatin formation and gene silencing in *Drosophila. Cell* 77:993–1002
32. Dreesen TD, Henikoff S, Loughney K. 1991. A pairing-sensitive element that mediates trans-inactivation is associated with the Drosophila brown gene. *Genes Dev.* 5:331–40
33. Elkind Y, Edwards R, Mavandad M, Hedrick SA, Ribak O, et al. 1990. Abnormal plant development and down-regulation of phenylpropanoid biosythesis in transgenic tobacco containing a heterologous phenylalanine ammonia-lyase gene. *Proc. Natl. Acad. Sci. USA* 87:9057–61
34. Elomaa P, Helariutta Y, Griesbach RJ, Kotilainen M, Seppänen P, Teeri TH. 1996. Transgene inactivation in *Petunia hybrida* is influenced by the properties of the foreign gene. *Mol. Gen. Genet.* In press
35. Fauvarque MO, Dura JM. 1993. *Polyhomeotic* regulatory sequences induce developmental regulator-dependent variegation and targeted P-element insertions in *Drosophila. Genes Dev.* 7:1508–20
36. Fedoroff N, Masson P, Banks JA. 1989. Mutations, epimutations, and the developmental programming of the maize *suppressor-mutator* transposable element. *BioEssays* 10:139–44
37. Fedoroff NV, Banks JA. 1988. Is the suppressor-mutator element controlled by a basic developmental mechanism? *Genetics* 120:559–70
38. Flavell RB. 1994. Inactivation of gene expression in plants as a consequence of specific sequence duplication. *Proc. Natl. Acad. Sci. USA* 91:3490–96
39. Fraenkel-Conrad H. 1983. RNA-dependent RNA polymerases of plants. *Proc. Natl. Acad. Sci. USA* 80:422–24
40. Fray RG, Grierson D. 1993. Identification and genetic analysis of normal and mutant phytoene synthase genes of tomato by sequencing, complementation and co-suppression. *Plant Mol. Biol.* 22:589–602
41. Frey M, Reinicke J, Grant S, Saedler H. 1990. Excision of the *En/Spm* transposable element of *Zea Mays* requires two element-encoded proteins. *EMBO J.* 9:4037–44
42. Gallie DR. 1991. The cap and poly(A)tail function synergistically to regulate mRNA translation efficiency. *Genes Dev.* 5: 2108–16
43. Gallie DR. 1993. Posttranscriptional regulation of gene expression in plants. *Annu. Rev. Plant Physiol. Plant Mol. Biol.* 44: 77–105
44. Goring DR, Thomson L, Rothstein SJ. 1991. Transformation of a partial nopaline synthase gene into tobacco suppresses the expression of a resident wild-type gene. *Proc. Natl. Acad. Sci. USA* 88:1770–74
45. Goyon C, Faugeron G. 1989. Targeted transformation of *Ascobulus immersus* and de novo methylation of the resulting duplicated DNA sequences. *Mol. Cell. Biol.* 9: 2818–27
46. Goyon C, Nogueira TIV, Faugeron G. 1994. Perpetuation of cytosine methylation in *Ascobulus immersus* implies a novel type of maintenance methylase. *J. Mol. Biol.* 2:42–51
47. Grierson D, Fray RG, Hamilton AJ, Smith CJS, Watson CF. 1991. Does co-suppression of sense genes in transgenic plants involve antisense RNA? *Trends Biotechnol.* 9:122–23
48. Hagemann R. 1958. Somatische Konversion bei *Lycopersicon esculentum. Mill. Z. Vererbl.* 89:587–13
49. Harrison BJ, Carpenter R. 1973. A comparison of the instabilities at the nivea and pallida loci in *Antirrhinum majus. Heredity* 31:309–23
50. Hart CM, Fischer B, Neuhaus JM, Meins F. 1992. Regulated inactivation of homologous gene expression in transgenic *Nicotiana sylvestris* plants containing a defense-related tobacco chitinase gene. *Mol. Gen. Genet.* 235:179–88
51. Henikoff S. 1990. Position-effect variegation after 60 years. *Trends Genet.* 6:422–25
52. Hentze MW. 1991. Determinants and regulation of cytoplasmic mRNA stability in eukaryotic cells. *Biochim. Biophys. Acta* 1090:281–92
53. Hildebrandt M, Nellen W. 1992. Differential antisense transcription from the *Dictyostelium EB4* gene locus: implications on antisense-mediated regulation of mRNA stability. *Cell* 69:197–204
54. Holliday R, Pugh JE. 1975. DNA modification mechanisms and gene activity during development. *Science* 187:226–32
55. Hoopes BC, McClure WR. 1985. A cII-dependent promoter is located within the Q gene of bacteriophage lambda. *Proc. Natl. Acad. Sci. USA* 82:3134–38
56. Igarashi S, Tanno Y, Onodera O, Yamazaki M, Sato S, et al. 1992. Strong correlation between the number of CAG repeats in androgen receptor genes and the clinical onset of features of spinal and bulbar muscular atrophy. *Neurology* 42:2300–2

57. Ingelbrecht I, Van Houdt H, Van Montagu M, Depicker A. 1994. Posttranscriptional silencing of reporter transgenes in tobacco correlates with DNA methylation. *Proc. Natl. Acad. Sci. USA* 91:10502–6
58. Jablonka E, Lamb MJ. 1989. The inheritance of acquired epigenetic variations. *J. Theor. Biol.* 139:69–83
59. Jorgensen R. 1990. Altered gene expression in plants due to trans interactions between homologous genes. *Trends Biotechnol.* 8:340–44
60. Jorgensen R. 1993. The germinal inheritance of epigenetic information in plants. *Philos. Trans. R. Soc. London Ser. B* 339: 173–81
61. Jorgensen R. 1994. Developmental significance of epigenetic imposition on the plant genome: a paragenetic function for chromosomes. *Dev. Genet.* 15:523–32
62. Kaeppler SM, Phillips RL. 1993. DNA methylation and tissue culture-induced variation in plants. *In Vitro Cell. Dev. Biol.* 29:125–30
63. Kilby NJ, Ottoline Leyser HM, Furner IJ. 1992. Promoter methylation and progressive transgene inactivation in *Arabidopsis*. *Plant Mol. Biol.* 20:103–12
64. Kimelman D, Kirschner MW. 1989. An antisense mRNA directs the covalent modification of the transcript encoding fibroblast growth factor in Xenopus oocytes. *Cell* 59:687–96
65. Krebbers E, Hehl R, Piotrowiak R, Lönnig W, Sommer H, Saedler H. 1987. Molecular analysis of a paramutant plant of *Antirrhinum majus* and the involvement of transposable elements. *Mol. Gen. Genet.* 209: 499–507
66. Kricker MC, Drake JW, Radman M. 1992. Duplication-targeted DNA methylation and mutagenesis in the evolution of eukaryotic chromosomes. *Proc. Natl. Acad. Sci. USA* 89:1075–79
67. Künzler P, Matsuo K, Schaffner W. 1995. Pathological, physiological and evolutionary aspects of short unstable DNA repeats in the human genome. *Biol. Chem. Hoppe-Seyler* 376:201–11
68. Lewis EB. 1954. The theory and application of a new method of detecting chromosomal rearrangements in *Drosophila melanogaster*. *Am. Nat.* 88:225–39
69. Lindbo JA, Silva-Rosales L, Proebsting WM, Dougherty WG. 1993. Induction of a highly specific antiviral state in transgenic plants: implications for regulation of gene expression and virus resistance. *Plant Cell* 5:1749–59
70. Locke J, Kotarski A, Tartof KD. 1988. Dosage-dependent modifiers of position effect variegation in Drosophila and mass action model that explains their effect. *Genetics* 120:181–98
71. Loesch DZ, Huggins R, Hay DA, Gedeon AK, Mulley JC, Sutherland GR. 1993. Genotype-phenotype relationships in fragile X syndrome: a family study. *Am. J. Hum. Genet.* 53:1064–73
72. Matzke AJM, Neuhuber F, Park YD, Ambros PF, Matzke MA. 1994. Homology-dependent gene silencing in transgenic plants: epistatic silencing loci contain multiple copies of methylated transgenes. *Mol. Gen. Genet.* 244:219–29
73. Matzke MA, Matzke AJM. 1990. Gene interactions and epigenetic variation in transgenic plants. *Dev. Genet.* 11:214–23
74. Matzke MA, Matzke AJM. 1993. Genomic imprinting in plants: parental effects and *trans*-inactivation phenomena. *Annu. Rev. Plant Physiol. Plant Mol. Biol.* 44:53–76
75. Matzke MA, Neuhuber F, Matzke AJM. 1993. A variety of epistatic interactions can occur between partially homologous transgene loci brought together by sexual crossing. *Mol. Gen. Genet.* 236:379–86
76. Matzke MA, Priming M, Trnovsky J, Matzke AJM. 1989. Reversible methylation and inactivation of marker genes in sequentially transformed tobacco plants. *EMBO J.* 8:643–49
77. McClintock B. 1967. Regulation of pattern of gene expression by controlling elements in maize. *Carnegie Inst. Yearb.* 65:568–78
78. Mehtali M, LeMeur M, Lathe R. 1990. The methylation-free status of a housekeeping transgene is lost at high copy number. *Gene* 91:179–84
79. Meins F Jr. 1989. Habituation: heritable variation in the requirement of cultured plant cells for hormones. *Annu. Rev. Genet.* 23:395–408
80. Meins JF, Kunz C. 1994. Silencing of chitinase expression in transgenic plants: an autoregulatory model. In *Gene Inactivation and Homologous Recombination in Plants*, ed. J Paszkowski, pp. 335–48. Dordrecht: Kluwer
81. Meyer P. 1995. DNA methylation and transgene silencing in *Petunia hybrida*. See Ref. 81a, pp. 15–28
81a. Meyer P, ed. 1995. *Gene Silencing in Higher Plants and Related Phenomena in Other Eukaryotes*. Berlin: Springer-Verlag
82. Meyer P, Heidmann I. 1994. Epigenetic variants of a transgenic petunia line show hypermethylation in transgene DNA: an indication for specific recognition of foreign DNA in transgenic plants. *Mol. Gen. Genet.* 243:390–99
83. Meyer P, Heidmann I, Niedenhof I. 1993. Differences in DNA-methylation are associated with a paramutation phenomenon in transgenic petunia. *Plant J.* 4:86–100
84. Meyer P, Linn F, Heidmann I, Meyer AH, Niedenhof I, Saedler H. 1992. Endogenous and environmental factors influence 35S

promoter methylation of a maize A1 gene construct in transgenic petunia and its colour phenotype. *Mol. Gen. Genet.* 231: 345–52
85. Meyer P, Niedenhof I, Ten Lohuis M. 1994. Evidence for cytosine methylation of non-symmetrical sequences in transgenic *Petunia hybrida. EMBO J.* 13:2084–88
86. Mittlesten Scheid O. 1995. Transgene inactivation in *Arabidopsis thaliana.* See Ref. 81a, pp. 29–42
87. Mittlesten Scheid O, Afsar K, Paszkowski J. 1994. Gene inactivation in *Arabidopsis thaliana* is not accompanied by an accumulation of repeat-induced point mutations. *Mol. Gen. Genet.* 244:325–30
88. Mittlesten Scheid O, Paszkowski J, Potrykus I. 1991. Reversible inactivation of a transgene in *Arabidopsis thaliana. Mol. Gen. Genet.* 228:104–12
89. Mueller E, Gilbert J, Davenport G, Brigneti G, Baulcombe DC. 1995. Homology-dependent resistance: transgenic virus resistance in plants related to homology-dependent gene silencing. *Plant J.* 7:1001–13
90. Napoli C, Lemieux C, Jorgensen R. 1990. Introduction of a chimeric chalcone synthese gene into petunia results in reversible cosuppression of homologous genes in trans. *Plant Cell* 2:279–89
91. Neuhuber F, Park YD, Matzke AJM, Matzke MA. 1994. Susceptibility of transgene loci to homology-dependent gene silencing. *Mol. Gen. Genet.* 244:230–41
92. Newman TC, Ohme-Takagi MO, Taylor CB, Green PJ. 1993. DST sequences, highly conserved among plant SAUR genes, target reporter transcripts for rapid decay in tobacco. *Plant Cell* 5:701–14
93. Nick H, Bowen B, Ferl RJ, Gilbert W. 1986. Detection of cytosine methylation in the maize alcohol dehydrogenase gene by genomic sequencing. *Nature* 319:243–46
94. Okamoto K, Hara S, Ramaninder B, Freundlich M. 1988. Evidence in vivo for autogeneous control of the cyclic AMP receptor protein gene *(crp)* in *Eschericia coli* by divergent RNA. *J. Bacteriol.* 170:5076–79
95. Orlando V, Paro R. 1995. Chromatin multiprotein complexes involved in the maintenance of transcription patterns. *Curr. Opin. Genet. Dev.* 5:174–79
96. Palladino F, Gasser S. 1994. Telomeric maintenance and gene repression: a common end? *Curr. Opin. Cell Biol.* 6:373–79
97. Pandit NN, Russo VEA. 1992. Reversible inactivation of a foreign gene, *hph*, during the sexual cycle of *Neurospora crassa* transformants. *Mol. Gen. Genet.* 234: 412–22
98. Paro R. 1993. Mechanisms of heritable gene repression during development of Drosophila. *Curr. Opin. Cell Biol.* 5: 999–1005
99. Paro R, Hogness D. 1991. The polycomb protein shares a homologous domain with a heterochromatin-associated protein in *Drosophila. Proc. Natl. Acad. Sci. USA* 88: 263–67
100. Patterson GI, Chandler VL. 1995. Paramutation in maize and related allelic interactions. See Ref. 81a, pp. 121–41
101. Patterson GI, Thorpe CJ, Chandler VL. 1993. Paramutation, an allelic interaction, is associated with a stable and heritable reduction of transcription of the maize b regulatory gene. *Genetics* 135:881–94
102. Pieretti M, Zhang FP, Fu YH, Warren ST, Oostra BA, et al. 1991. Absence of expression of the FMR-1 gene in fragile x syndrome. *Cell* 66:817–22
103. Pirrotta V. 1991. The genetics and molecular biology of *zeste* in *Drosophila melanogaster. Adv. Genet.* 29:301–48
104. Pröls F, Meyer P. 1992. The methylation patterns of chromosomal integration regions influence gene activity of transferred DNA in *Petunia hybrida. Plant J.* 2:465–75
105. Pryer KN, Wuestehube LJ, Schekman R. 1992. Vesicle-mediated protein sorting. *Annu. Rev. Biochem.* 61:471–516
106. Reuter G, Spierer P. 1992. Position-effect variegation and chromatin proteins. *BioEssays* 14:605–12
107. Rivier DH, Pillus L. 1994. Silencing speaks up. *Cell* 76:963–66
108. Rogers JC. 1988. RNA complementary to alpha-amylase mRNA in barley. *Plant Mol. Biol.* 11:125–38
109. Romano N, Macino G. 1992. Quelling: transient inactivation of gene expression in *Neurospora crassa* by transformation with homologous sequences. *Mol. Microbiol.* 6: 3343–53
110. Rossignol JL, Faugeron G. 1995. MIP: an epigenetic gene silencing process in *Ascobulus immersus.* See Ref. 81a, pp. 179–91
111. Roth SY. 1995. Chromatin-mediated transcriptional repression in yeast. *Curr. Opin. Genet. Dev.* 5:168–73
112. Salinas J, Matassi G, Montero LM, Bernardi G. 1988. Compositional compartmentalization and compositional patterns in the nuclear genomes of plants. *Nucleic Acids Res.* 16:4269–85
113. Shaffer CD, Wallrath LL, Elgin SCR. 1993. Regulating genes by packaging domains: bits of heterochromatin in euchromatin? *Trends Genet.* 9:35–37
114. Schimke RT. 1984. Gene amplification in cultured animal cells. *Cell* 37:705–13
115. Schläppi M, Raina R, Fedoroff N. 1994. Epigenetic regulation of the maize *Spm* transposable element: novel activation of a methylated promoter by TnpA. *Cell* 77: 427–37
116. Schwartz D, Dennis E. 1986. Transposase

activity of the Ac controlling element in maize is regulated by its degree of methylation. *Mol. Gen. Genet.* 205:476–82

117. Selker EU, Cambareri EB, Jensen BC, Haack KR. 1987. Rearrangement of duplicated DNA in specialized cells of *Neurospora*. *Cell* 51:741–52
118. Selker EU, Fritz DY, Singer MJ. 1993. Dense non-symmetrical DNA methylation resulting from repeat-induced point mutation (RIP) in *Neurospora*. *Science* 262: 1724–28
119. Seymour GB, Fray RG, Hill P, Tucker GA. 1993. Down-regulation of two non-homologous endogenous tomato genes with a single chimeric sense gene construct. *Plant Mol. Biol.* 23:1–9
120. Smith CJS, Watson CF, Bird CR, Ray J, Schuch W, Grierson D. 1990. Expression of a truncated tomato polygalacturonase gene inhibits expression of the endogenous gene in transgenic plants. *Mol. Gen. Genet.* 244:447–81
121. Smith SS, Lingeman RG, Kaplan BE. 1992. Recognition of foldback DNA by the human DNA (cytosine-5-)-methyltransferase. *Biochemistry* 31:850–54
122. Spector DL, Fu XD, Maniatis T. 1991. Associations between distinct pre-mRNA splicing components and the cell nucleus. *EMBO J.* 10:3467–81
123. Spena A, Viotti A, Pirrotta V. 1983. Two adjacent genomic zein sequences: structure, organization and tissue-specific restriction pattern. *J. Mol. Biol.* 169:799–811
124. Spofford JB. 1976. Position-effect variegation in *Drosophila*. In *Genetics and Biology of Drosophila*, pp. 955–1019. London: Academic
125. Thompson JS, Ling XF, Grunstein M. 1993. Histone H3 amino-terminus is required for telomeric and silent mating locus repression in yeast. *Nature* 369:245–47
126. Trottier Y, Biancalana V, Mandel JL. 1994. Instability of CAG repeats in Huntington's disease: relation to parental transmission and age of onset. *J. Med. Genet.* 31:377–82
127. Tzamarias D, Struhl K. 1994. Functional dissection of the yeast Cyc8-Tup1 transcriptional co-repressor complex. *Nature* 369:758–61
128. Van Blokland R, Van der Geest N, Mol JNM, Kooter JM. 1994. Transgene-mediated suppression of chalcone synthase expression in *Petunia hybrida* results from an increase in RNA turnover. *Plant J.* 6: 861–77
129. Van der Krol AR, Mur LA, Beld M, Mol J, Stuitje AR. 1990. Flavonoid genes in petunia: addition of a limiting number of copies may lead to a suppression of gene expression. *Plant Cell* 2:291–99
130. Van der Lugt NMT, Domen J, Linders K, Van Roon M, Robanus-Maandag E, et al. 1994. Posterior transformation, neurological abnormalities, and severe hematopoetic defects in mice with a targeted deletion in the Bmi-1 protooncogene. *Genes Dev.* 8: 757–69
131. Van Slogteren GMS, Hooykaas PJJ, Schilperoot RA. 1984. Silent T-DNA genes in plant lines transformed by *Agrobacterium tumefaciens* are activated by grafting and 5-azacytidine treatment. *Plant Mol. Biol.* 3:333–36
132. Vaucheret H. 1993. Identification of a general silencer for 19S and 35S promoters in a transgenic tobacco plant: 90bp of homology in the promotersequence are sufficient for trans-inactivation. *C. R. Acad. Sci. III* 316:1471–83
133. Walbot V. 1988. Reactivation of the mutator transposable element system following gamma irradiation of seed. *Mol. Gen. Genet.* 212:259–64
134. Walbot V. 1992. Reactivation of mutator transposable elements of maize by ultraviolet light. *Mol. Gen. Genet.* 234:353–60
135. Wassenegger M, Heimes S, Riedel L, Sänger HL. 1994. RNA-directed de novo methylation of genomic sequences in plants. *Cell* 76:567–76
136. Weintraub H. 1985. Assembly and propagation of repressed and derepressed chromosomal states. *Cell* 42:705–11
137. Whitkus R, Doebley J, Lee M. 1992. Comparative genome mapping of sorghum and maize. *Genetics* 132:1119–30
138. Wieringa PJ. 1994. Myotonic dystrophy reviewed: back to the future? *Hum. Mol. Genet.* 3:1–7
139. Winkler H. 1930. *Die Konversion der Gene*. Jena: Gustav-Fischer-Verlag
140. Xing Y, Johnson CV, Dobner PR, Lawrence JB. 1993. Higher level organization of individual gene transcription and RNA splicing. *Science* 259:1326–30
141. Yu S, Pritchard M, Kremer E, Lynch M, Nancarrow J, et al. 1991. Fragile-X genotype characterized by an unstable region of DNA. *Science* 252:1179–81
142. Zhang G, Taneja KL, Singer RH, Green MR. 1994. Localization of pre-mRNA splicing in mammalian nuclei. *Science* 372: 809–12

Annu. Rev. Plant Physiol. Plant Mol. Biol. 1996. 47:49–73

14-3-3 PROTEINS AND SIGNAL TRANSDUCTION

Robert J. Ferl

Program in Plant Molecular and Cellular Biology, Department of Horticultural Sciences, University of Florida, Box 110690, Gainesville, Florida 32611-0690

KEY WORDS: protein interactions, signaling, phosphorylation

ABSTRACT

Perhaps in keeping with their enigmatic name, 14-3-3 proteins offer a seemingly bewildering array of opportunities for interaction with signal transduction pathways. In each organism there are many isoforms that can form both homo- and heterodimers, and many biochemical activities have been attributed to the 14-3-3 group. The potential for diversity—and also confusion—is high. The mammalian literature on 14-3-3 proteins provides an appropriate context to appreciate the potential roles of 14-3-3s in plant signal transduction pathways. In addition, functional and structural themes emerge when 14-3-3s are examined and compiled in ways that draw attention to their participation in protein phosphorylation and protein-protein interactions. These themes allow examination of plant 14-3-3s from two perspectives: the ways in which plant 14-3-3s contribute to and extend ideas already described in animals, and the ways that plant 14-3-3s present unique contributions to the field. The crystal structure of an animal 14-3-3 has been solved. When considered with the evolutionary stability of large segments of the 14-3-3 protein, the structure illuminates several aspects of 14-3-3 function. However, diversity in other regions of the 14-3-3s and their presence as multigene families offer many opportunities for cell-specific specialization of individual functions.

CONTENTS

1040-2519/96/0601-0049$08.00

INTRODUCTION

Although the processes involved in biological signal transduction have long captured the attention of biochemists and cellular biologists, the 14-3-3 proteins have long remained obscure, attracting little interest though they have been studied since 1967 (52). Only recently have the 14-3-3 proteins come to the forefront of biochemical science, owing largely to their appearance as players in the field of signal transduction. In recent years the literature on 14-3-3 proteins has entered a logarithmic phase of growth as roles for 14-3-3s are uncovered within well-characterized processes and systems or have been discovered as part of new signal pathways.

There is, however, no real consensus about just what these proteins actually do. This situation may be due at least in part to the fact that 14-3-3s may actually have a number of roles. As we better understand those diverse roles and the biochemical characteristics of the 14-3-3s, the fundamental properties of these proteins will reveal the underlying principles that guide 14-3-3 participation in diverse pathways (1, 2, 9, 26, 55). Therefore now is an appropriate time to gather information on 14-3-3s and to examine the emerging understanding of their roles in cellular signaling systems.

EARLY HISTORY

Discovery, Naming, and Initial Characterization

The 14-3-3 proteins initially were described as a part of an extensive characterization by Moore & Perez of the acidic, soluble proteins within the mammalian brain (52). The study involved the purification of brain proteins of unknown function; names were assigned based on fractionation on DEAE cellulose and electrophoretic mobility upon starch gel electrophoresis. The interesting observations at the time were that many of the proteins (including the 14-3-3 proteins) were abundant in brain [up to 1% of the total soluble protein (6) for the 14-3-3s], were found apparently only in brain, and were immunologically conserved among animals. This and several following studies over twenty years (6, 7, 16, 29) led to the conclusion that 14-3-3 proteins were uniquely important to brain and neuronal function. In fact, on through the late 1980s the 14-3-3 proteins were thought to be found exclusively within the neurons of the brain (7) [and perhaps in other organs at much lower concentration (29)], and it was thought that they could be axonally transported in retinal ganglion cells (16).

By the mid-1980s, and still as parts of larger, systematic studies of acidic brain proteins, the 14-3-3s were understood to be a heterogeneous family of proteins with subunit masses of 25–32 kDa and with various isoelectric points (7, 27, 52, 80). It was also known that they were found as dimers of 50–60 kDa in their native form and that their amino termini were blocked (27). Peptide maps were developed, and amino acid analyses were completed (28). However, no data on the possible biochemical functions of the 14-3-3s were developed in the twenty years following their initial discovery, until a single observation began what was to become a cascade of functional discoveries that continue to the present.

Discovery of a Biochemical Function

In 1987, Ichimura et al (28) made a pivotal observation, noticing that the 14-3-3 proteins in their continuing survey of acidic brain proteins were similar in amino acid composition to a protein known to activate tyrosine and tryptophan hydoxylases in the presence of calcium and calmodulin-dependent protein kinase II or cAMP-dependent kinase (16, 80). Later experiments confirmed that the biochemical characteristics of purified 14-3-3s were indistinguishable from those of purified activators of tyrosine and tryptophan hydroxylases. This observation is notable for several reasons. First, it seemed to cement the relationship between 14-3-3 and brain function. Tyrosine and tryptophan hydoxylases are the first enzymes in the metabolic pathways that lead to serotonin and dopamine, important neurotransmitters (16). The activation of the first enzymes of the pathway seems a logical and necessary component of neuronal function, thus explaining at least in part the specificity of 14-3-3 within the neurons of the brain and the need for 14-3-3s in abundance. Second, the observation began what has become one of the recurrent themes in 14-3-3 biology, their association and interaction with proteins and protein kinases. Third, the observation was a harbinger of much of what was to come in that many of the roles of 14-3-3 proteins were to be discovered by people not looking for 14-3-3s.

This rediscovery of 14-3-3 proteins was rapidly followed by the further characterization of 14-3-3s from bovine brain and by cloning of a representative cDNA (27). At this point, the several component members of the 14-3-3 protein family present in the bovine brain were given Greek letter designations (α, β, γ, δ, ε, ζ, and η) on the basis of their order of elution during reversed phase chromatography. [It should be noted that other isoforms subsequently isolated from other animal species also retain Greek letter designations, and the elution profile is apparently well conserved among animals (2). However, it is not yet clear that a similarly designated isoform is *always* directly or functionally homologous to the original bovine or sheep isoforms.] The η 14-3-3 isoform was purified to homogeneity and found to activate tryptophan hydroxylase. Cyanogen bromide cleavage fragments were sequenced to provide data for cloning via hybridization with representative oligonucleotides. The derived amino acid sequence of 14-3-3 η was unlike anything then in the databases, and other than the presence of an extremely acidic C-terminal domain, little in the sequence suggested features explaining its ability to activate the tyrosine and tryptophan hydoxylases.

PERIOD OF DISCOVERY

Since 1990, 14-3-3 proteins have been uncovered in many biological systems. The 14-3-3s are no longer considered to be brain-specific, nor is their presence limited to the animal kingdom. Now they are thought to be nearly ubiquitous in one form or another, in all cells of every organism. In fact, the consensus now building in the literature is that 14-3-3s are highly conserved and found in a broad range of tissues, that different isoforms have different functions, and that "no [eukaryotic] cell type has been identified which does not contain members of the 14-3-3 family" (62). It is likely that improved immunological detection methods are largely responsible for this insight. Despite detection of more and different 14-3-3 isoforms, many of the initial conclusions regarding the original brain isoforms remain largely intact because of cell or tissue-specific expression of individual isoforms.

Problems with nomenclature persist, owing at least in part to the rapid discovery of many functions and properties of the 14-3-3s (Table 1). Some 14-3-3s had already been given different names based on their activities or properties long before their sequence identified them as part of the family. Some yeast forms had been characterized genetically and hence bear their gene name instead of a 14-3-3 moniker (19). In addition, the different isoforms present in brains were given Greek letter designations based on operational criteria before any sequence information allowed assessment of evolutionary homologies. Therefore, not only is the naming of the protein family itself still enigmatic, but reference to the various subtypes present in an organism is without extended convention as well (Table 2).

In several cases rediscoveries of the same isoform have led to some understanding of the diverse activities that may be associated with certain individual isoforms (Table 2). In addition, further biochemical characterization has

Table 1 Key discoveries of 14-3-3 involving new species or forms

Year[a]	Organism/Organ	Activity/Expression[b]	Refs[c]
1967	Bovine/brain	14-3-3s initially described	52
1987	Bovine/brain	14-3-3s found to be activators of Tyrosine and Tryptophan hydroxylases	28, 80
1988	Bovine/brain	14-3-3 η cloned	27
1990	Sheep/brain	Purified Protein Kinase C Inhibitor Proteins (KCIPs) were found to be a family of proteins substantially identical in amino acid sequence to 14-3-3	71
1991	Bovine/brain	The β and γ isoforms are sequenced from partial cDNAs and from fragments of the proteins	33
1991	Rat/brain	The rat γ isoform is cloned by hybridization and shown to have tissue-specific expression within the rat brain and neurons of the spinal cord	76
1991	Human/T-cell	Fortuitous cloning of a human isoform, latter called τ, that is *not* found in the brain. At this time it was also realized that a cloning artifact had recovered a partial cDNA sequence from *Drosophila* (66)	56
1992	Human/brain, tumor cell lines	The human η isoform is cloned by hybridization with the bovine clone	30
1992	Chromaffin cells	Partial sequencing of the Exo1 protein reveals homology to 14-3-3s—Exo1 stimulates calcium-dependent exocytosis	53
1992	Sheep/brain	The ζ form is found to activate rather than inhibit PKC	32
1992	*Arabidopsis*	Physical interaction of a 14-3-3 homolog, called GF14, within the G-box protein–DNA complex	44
1992	Spinach, *Oenothera*	cDNAs are recovered by immunoscreen with antibodies to KCIP peptides. KCIP activity was also purified from pea leaves	25
1992	Maize	GF14–14-3-3s are found to be part of a protein–DNA complex	15
1992	*Xenopus*/ Pituitary	Although found in the pituitary, the sequence is also expressed in the brain	48
1992	Bovine, *Drosophila*, Sheep/brain	Multiple isoforms of the KCIPs are sequenced from the protein. Protein kinase C is shown to phosphorylate bovine β and γ and sheep ε and ζ	72
1992	Yeast	The first yeast homolog, BMH1, is cloned and found to inhibit growth when overexpressed	74
1992	Barley	A 14-3-3 is induced by pathogen attack	8
1992	*Drosophila*	Found as one of several head-specific clones	70
1992	Human	A 14-3-3 homolog is an epithelial cell marker	60

[a]Discoveries are presented by year of publication only; no order within years is indicated or implied. The table is complete through the 1994 publication year.

[b]Only new or previously undefined functions are indicated, and the list does not include further characterizations of existing forms or functions.

[c]In several cases, activities were found by separate references within one publication year, or several references allow a single inference of 14-3-3 structure or function.

Table 1 (*continued*)

Year[a]	Organism/Organ	Activity/Expression[b]	Refs[c]
1992	Human	A human 14-3-3 is cloned and shown to have phospholipase activity. This activity was largely discredited by later experiments	81
1993	Bovine/brain	Physical interaction of 14-3-3 with the phosphorylated form of tryptophan hydroxylase is part of its activation	23
1993	Bovine/brain	FAS, a eukaryotic factor absolutely required for the ADP ribosylation of Ras and other substrates by the virulence factor ExoS of *Pseudomonas aeruginosa*	21
1993	Rice	Random cloning of a rice homolog	36
1993	*Arabidopsis*	GF14ω is found to bind calcium and is phosphorylated by a membrane-bound kinase	46
1994		Evolutionary analysis reveals that, to date, all plant isoforms are evolutionarily distinct from the animal forms	18
1994		The ζ and β isoforms, as well as yeast BMH1, are shown to interact with the protein kinase Raf and affect its activation	20, 31
1994	*Arabidopsis*	A single plant isoform is shown to activate PKC, tryptophan hydroxylase, activate ExoS, and bind calcium	43
1994	*Arabidopsis,* tomato	DNAs are cloned by immunoscreening with anti-GF14	40, 45
1994	Yeast	In *S. pombe, rad24* and *rad25* are two 14-3-3 homologs that are genetically required for DNA damage checkpoint before mitosis	19
1994	Rat/brain	The ζ and θ isoforms are cloned, and in situ hybridization shows overlapping and unique distribution patterns of θ, γ, β, η, and ζ isoforms	75
1994	Rat	MSF, a mitochondrial import stimulation factor, is shown to be an apparent heterodimer of 14-3-3s	5, 24
1994	Tobacco	A 14-3-3 is found to be down-regulated by NaCl adaptation of culture cells and is found to inhibit PKC and activate ADP ribosylation	10
1994		14-3-3s interact with and activate Raf protein kinase and together activate downstream kinases	17, 22, 42, 68
1994	*Arabidopsis*	Two 14-3-3s, RCI1 and RCI2, are cloned as being rapidly induced by cold treatment	34
1994	Maize	First genomic clones of 14-3-3s reveal intron/exon structure	13
1994		Some 14-3-3s can bind phospholipids	65
1994	Chicken/brain	Avian form cloned	59
1994	Oat, maize	The fusicoccin receptor [Fusicoccin Binding Protein (FCBP)] is found to be a 14-3-3 homolog	38, 47, 57
1994		Physical association of 14-3-3s with the c-Bcr and Bcr-Abl protein kinases	61

Table 2 Names and cross index of 14-3-3 homologs

Name	Basis of name	Cross index[a]	Refs
14-3-3	Chromatography and electrophoretic mobility: Subtypes are α, β, δ, γ, ε, ζ, and η from brain protein extracts. Later a θ was identified as a cDNA, and the α and δ forms were found to be phosphorylated versions of β and ζ, respectively	α is phosphorylated β. δ is phosphorylated ζ. θ and τ are apparently the same	4, 27, 33, 49, 52, 75
BMH1	Brain modulosignalin homolog: BMH1 and BMH2 are required for growth in *S. cerevisiae*		74
BAP1	Bcr-associated protein; associates with the Bcr protein kinase	Same as τ, the human T-cell form, also called θ	61
Exo1	Exo(cytosis) stimulating proteins: stimulate calcium-dependent exocytosis		32, 53
FAS	Factor-activating ExoS: required for the ADP-ribosylation of Ras	Identical to human PLA2 and bovine ζ	21
FCBP	Fusicoccin binding protein: receptor for the plant phytotoxin	The maize FCBP is identical to maize GF14-12	38, 47
GF14	G-box factor 14-3-3 homolog: found within a protein–DNA complex in plants	Subtypes are ω, ψ, χ, ϕ, and υ in *Arabidopsis,* and GF14-12 and GF14-6 in maize	15, 44, 45
MSF	Mitochondrial import stimulation factor: two forms, MSFL and MSFS, together stimulate import of mitochondrial precursor proteins	MSFL is the rat ε isoform MSFS is a rat η isoform	5, 24
KCIP	Kinase C inhibitor proteins: as a class inhibit protein kinase C but also are found to activate protein kinase C under different assay conditions	KCIP-1 is the η isoform	32, 71
PLA2	Phospholipase A2: from human, but later evidence shows this form *not* to have PLA2 activity	PLA2 is identical to the bovine ζ isoform and FAS	17, 62, 81
stratifin	Epithelial marker proteins, HME1 and HS1: a group of proteins shown to be various members of the 14-3-3 family	Stratifin and HME1 is σ, HS1 is τ	41, 49, 56, 65
rad24, rad24	RADiation sensitive mutants in *S. pombe* that are required for DNA damage checkpoint		19
RCI	Rare cold-inducible: from *Arabidopsis*	RCI1 is GF14ψ	34

[a]Only those names that are correlated by sequence to the original Greek letter designations are given. These crossreferences are either given or derived from the sequences presented in the references.

helped simplify the system somewhat, as two isoforms are now known to be posttranslational modifications of a single gene protein product (4).

The current confusion over the name of the family and the complexities of naming the subtypes within the family will likely continue until a consensus emerges regarding function. In the interim, the use of the Greek letter designations for isotypes (members of the gene family) appears to be logical and holding up to evolutionary considerations (see below). Especially when the derived amino acid sequences are identical between or among animals, it seems appropriate to revert the name to the Greek letter subtype from the original sheep and bovine brain extracts. The original Greek letter designations for animals have no evolutionary basis for being applied to plants (18) (and see below), therefore a plant basis is necessary for at least interim naming. The largest well-characterized gene family in plants is in *Arabidopsis,* whose subtypes also bear Greek letter designations. Therefore new occurrences of 14-3-3s in plants should perhaps relate to those sequences where identities or homologies are evident. With the advent of genome initiatives in both animals and plants, a huge number of 14-3-3 homologous Expressed Sequence Tags are now becoming available, contributing both to the classification difficulties as well as to the resolution of the size and diversity of the gene family.

FUNCTIONAL THEMES

No single activity or function yet emerges from the literature as the archetypical role for the 14-3-3s. There are, however, recurrent themes apparent that are beginning to coalesce into an understanding of the fundamental functional potentials of the 14-3-3 proteins. These themes, presented here with no inferences as to priority, are participation in protein-protein interactions and involvement with protein kinase-related events. Other activities have been reported, but these two main themes are consistent in the 14-3-3 literature, from the first reported biochemical functions on through to the latest reports of participation in signal transduction pathways.

Involvement with Protein Kinase-Related Events

It seems fitting that the first reported functions of 14-3-3 proteins involved effects with protein kinases. The activation of tyrosine and tryptophan hydroxylases is dependent upon phosphorylation of their regulatory domains, most effectively by calcium-calmodulin-dependent protein kinase II (33). This activation process is essentially dependent upon 14-3-3 homologs (16, 28), with the α, β, γ, δ, ε, ζ, and η isoforms (33) as well the plant GF14ω isoform all demonstrating similar activation abilities (43). The relationship between the 14-3-3 and the kinase-dependent activation is not completely clear, and the activation of the hydroxylases by 14-3-3s appears to occur after the phosphorylation event. Nonetheless, the juncture involving 14-3-3, protein kinases, and the activation of tyrosine and tryptophan hydoxylases is a key point in the understanding of one role of 14-3-3s in relation to signal transduction within neuronal cells.

Protein kinase C is clearly regulated by 14-3-3s, yet the question of whether the regulation is positive or negative remains somewhat controversial (1). The KCIPs, or kinase C inhibitor proteins, were among the first 14-3-3s to be described (2, 71). The KCIPs have been described as potent inhibitors of PKC (2, 3, 71), yet the 14-3-3s have also been described as activators of PKC (32, 43). The discrepancy between activation and inhibition of PKC is yet to be fully resolved, but the answer may lie in the assay procedures and conditions, protein preparations, or particular isoforms employed by different laboratories (1).

Another activity associated with 14-3-3s, that of stimulating calcium-dependent exocytosis in adrenal chromaffin cells (53, 64), may also be through effects on protein kinase C. Exo1 is a soluble cytosolic protein of the 14-3-3 family (53), and the stimulation activity is present in at least two recombinant 14-3-3 isoforms (65) and is inhibited by anti-14-3-3 antibodies (77). Because the stimulation of exocytosis is mediated by calcium and PKC, the conclusion is that the 14-3-3s may directly activate PKC to start the series of events leading to exocytosis. Additional studies with isoform-specific antibodies (65) indicate that the β, γ, and ζ isoforms are present in the chromaffin cells and are likely responsible for the stimulation activity, whereas the σ, τ, ε, and η were either absent or present at very low levels and probably not involved in exocytosis. The conclusion that simple activation of PKC is responsible for stimulation of exocytosis is complicated by studies showing that Exo1 has no direct stimulatory effect on PKC (54) and by localization and lipid-binding assays (65). The β and ζ isoforms were found in the cytosol and cytoskeletal fractions, while the γ isoform was found in the cytosol and associated with chromaffin granule membrane fractions. Brain 14-3-3s were shown to aggregate phospholipid vesicles, and the γ isoform (as well as the ε isoform) was found to be able to bind directly to phospholipids (65). Thus the stimulation of exocytosis seems to involve synergy between PKC and certain 14-3-3s, as well as possible membrane associations.

The c-Raf-1 protein kinase is apparently activated by 14-3-3s (17, 20, 22, 31, 68), bringing 14-3-3s into the forefront of signal transduction events. In several experimental systems, activation of Raf-1 kinase, together with its attendant recruitment to the membrane and activation of downstream signaling events, can be influenced by 14-3-3s. In yeast, the BMH1 14-3-3 homolog is required for Ras-induced activation of c-Raf-1 (31). Expression of 14-3-3 ζ in *Xenopus* oocytes stimulated c-Raf-dependent maturation (17). Most of the compelling data linking 14-3-3s with c-Raf are derived from protein-protein interaction studies (see below). Direct activation of c-Raf by 14-3-3s is controversial because 14-3-3s interact with inactive Raf in the cytosol and direct activation of c-Raf activity in vitro has been difficult to detect (22, 31). Nonetheless, 14-3-3s accompany active c-Raf to the membrane and 14-3-3 (BMH1) deletion mutants fail to demonstrate Ras-dependent c-Raf activation (31). Thus 14-3-3s are clearly part of Ras-mediated signal transductions through their association with and potential regulation of c-Raf-1 kinase (1, 9, 26, 55).

Bcr-Abl is an oncogene kinase that can be regulated by association with 14-3-3s (61). The Bcr gene has kinase signaling functions of its own, but it is mostly known as part of the Bcr-Abl oncogene chimera resulting from the Philadelphia translocation chromosome of humans. Association between Bcr and 14-3-3s apparently contributes to the oncogenicity of Bcr-Abl and is required for full transforming ability. Therefore the 14-3-3s may participate in associating Bcr with its signal pathways and may connect Bcr-Abl to inappropriate signal pathways.

The 14-3-3s can themselves be phosphorylated, and phosphorylation may influence their activities. In brains, α is the phosphorylated form of β, and δ is phosphorylated ζ, with the phosphorylation in both forms at suspected proline-directed kinase sites (4). The human τ isoform is phosphorylated at a serine(s) by Bcr and (at low levels) at tyrosine by Bcr-Abl (61). Sheep brain isoforms are phosphorylated at serine(s) (72), and the plant 14-3-3, GF14ω, is phosphorylated at serine(s) by a membrane-bound kinase from *Arabidopsis* (46).

Protein-Protein Interactions

Many of the biochemical characteristics of the 14-3-3s, including most of their potential functions, appear to depend on extended protein-protein interactions. The 14-3-3s are often found associated with larger protein complexes. These complexes appear to be more than merely the aggregation of sticky proteins; rather, they appear to be specific associations within which the 14-3-3s have explicit roles.

PHYSICAL COMPLEXES WITH KINASES Many of the kinase–proto-oncogene-related activities ascribed to the 14-3-3s involve direct physical interactions between the 14-3-3 and the target protein. Several examples of interaction exist, all tying 14-3-3s to important roles early in several signal transduction pathways (26, 55).

The activation of c-Raf-1 kinase is apparently accomplished during its recruitment to the membrane, and this activation and recruitment is augmented by 14-3-3s that accompany c-Raf-1 to the membrane. The association of 14-3-3s and c-Raf has been demonstrated by immunoprecipitation, in vitro binding assays, and in vivo association methods in the yeast two-hybrid system (17, 20, 22, 31, 55). To date, the β and ζ mammalian isoforms (17, 20, 22, 31) and yeast BMH1 (31) have been identified as associating with Raf in a variety of systems including expression in yeast, immunoprecipitation from yeast and insect expression cells, and in *Xenopus* and mammalian cells. Isoforms of 14-3-3 have been recovered blindly from yeast two-hybrid screens with c-Raf as the "bait" (17, 20), as well as from immunoprecipitates of cells expressing recombinant c-Raf (20, 22). Thus the case for physical association between c-Raf and 14-3-3s is strong and based on diverse but complementary experimental data, clearly placing the 14-3-3s among the players within the mitogen-activated protein kinase (MAP kinase) cascade.

In addition, REKS [Ras-dependent Extracellular signal-regulated kinase (ERK)-mitogen-activated protein kinase Kinase] is a protein kinase kinase kinase that participates in the midstream events of the MAP kinase pathway that begins with Ras and Raf (39). REKS is a large complex composed of B-Raf kinase and (at least) two different 14-3-3s (79). Thus 14-3-3s may participate in the MAP kinase cascade in several key spots.

Polyomavirus middle tumor antigen association with 14-3-3s was discovered by immunoprecipitation. The middle T (MT) antigen also associates with other proteins, including protein kinases, that are implicated in signaling growth regulation (58).

The protein kinases c-Bcr and Bcr-Abl are also associated with 14-3-3s, initially called Bap-1 (61). This association was discovered by direct screening of an expression library with peptides from the Bcr kinase and then was backed up by immunoprecipitation experiments. The Bcr gene is noted for having many potential functions. The kinase activity is located in the N-terminus of the protein. Bcr is associated with the Abl tyrosine kinase in the translocations of the leukemia-inducing Philadelphia chromosome of humans (51). Apparently, the association of the kinase domain of Bcr with the Abl kinase activates the Abl tyrosine kinase in an oncogenic fashion. Immunological precipitation of Bcr and Bcr-Abl detected 14-3-3s as part of a stable protein-protein complex. In addition, the 14-3-3 was phosphorylated by both the serine-threonine kinase of Bcr and the tyrosine kinase of Bcr-Abl but not by Abl itself. The site for interaction was mapped to the N-terminus of Bcr. The physical association with 14-3-3 may allow for proper localization of the Bcr kinase in normal cells; in transformed cells improper targeting of Abl activity may occur through linkage with Bcr.

It is interesting to note that all of the 14-3-3–kinase associations were described within eukaryotic expression systems. This is not likely a coincidence but probably reflects the need for eukaryotic modification of the 14-3-3s or their targets in order for association to occur.

PARTICIPATION IN DNA BINDING COMPLEXES One of the properties of 14-3-3s that is (to date) limited to plants is physical association in complexes with transcription factors (15, 44). This was among the very first discoveries of 14-3-3s in plants and it takes the signal transduction participation of 14-3-3s into the nucleus.

The G-box is a well-characterized *cis*-acting element involved in stress and ABA activation of genes, and the G-box factors (GBFs) are a class of bZIP transcriptional activators. During early experiments designed to clone members of the GBF-DNA binding complex in *Arabidopsis,* partially purified GBF activity was used to produce monoclonal antibodies (44). Individual monoclonal antibodies were then selected based on their ability to "supershift" the protein-DNA complex observed in the bandshift (electrophoretic mobility shift) assay with G-box oligonucleotides. Several monoclonal antibodies were recovered that specifically supershifted the G-box–protein complex; these antibodies were used to clone genes encoding proteins that turned out to be 14-3-3 homologs (44). Similar experiments were conducted in maize (15), and diverse G-box- and GBF-related extracts were confirmed to be similarly affected by the anti-14-3-3 antibodies (10). The conclusion is that 14-3-3 proteins are found associated with transcription factors and that association does not interfere with DNA binding.

The monoclonal antibody used for cloning many of the plant isoforms does not recognize all 14-3-3s present in plants and therefore may be specific to a certain subset of isoforms. The GF14–14-3-3 isoforms ω, ψ, χ, ϕ, and υ were all cloned with the antibody, as were the maize and tomato isoforms (see Table 2). While several groups have examined plant 14-3-3s with the same antibody (see below), other plant isoforms have been recovered that apparently do not react with the antibody (RJ Ferl, unpublished observation).

The interaction of 14-3-3 homologs with transcription is of unknown biochemical function. However, the interaction has recently been confirmed using immunoprecipitation and the two-hybrid in vivo interaction system in yeast (PC Sehnke & RJ Ferl, personal communication, 1995). Because both the 14-3-3s (46) and the GBFs (37) are known to be phosphorylated and because phosphorylation of GBF increases its binding to DNA (37), it is tempting to speculate that the association of 14-3-3s and GBFs is a part of the kinase-regulated activation of the DNA binding complex.

PHYSICAL ASSOCIATION WITH TRYPTOPHAN HYDROXYLASE Phosphorylation is part of the activation process for tryptophan hydroxylase and is known to be affected by the 14-3-3s (16, 28). Using a column matrix with one of the cofactors for tryptophan hydroxylase as a specific substrate, an association between 14-3-3 ζ and the hydroxylase was detected only if the hydroxylase was phosphorylated (23). This physical association between the phosphorylated tryptophan hydroxylase enzyme and the 14-3-3 further relates the themes of protein-protein interactions and protein kinase-related activities.

CHAPERONE ACTIVITY Mitochondrial import Stimulation Factor (MSF) is a cytoplasmic chaperone from rat liver that appears to help target protein precursors to the mitochondrion (24). MSF is apparently a heterodimer of the rat ζ and ε isoforms that is capable of forming a stable, immunoprecipitation complex with adrenodoxin precursor (5). MSF–14-3-3 also recognizes import-incompetent precursor aggregates and (using an inherent ATPase activity) unfolds the aggregated precursors in a manner similar to hsp70s. Further, the 14-3-3–precursor complex binds to the outer membrane of the mitochondrion, implicating the 14-3-3 in a targeting role to mitochondrial receptors in addition to the ATPase-dependent, chaperone-like activity.

This chaperone function, together with the attendant ATPase activity, offers novel perspectives on the protein-protein interaction theme. Although the name MSF is new, the ε and ζ 14-3-3 components have been seen in other roles including association with and activation of Raf, PKC, and exocytosis (see Tables 1 and 2). Most importantly, it highlights the participation of 14-3-3s in membrane-associated functions, perhaps a parallel to accompaniment of Raf to the plasma membrane. It therefore seems likely that a single fundamental mechanism underlies the protein-protein interactions that characterize much of the 14-3-3 functional data.

Other Biochemical Activities

Several additional biochemical activities have been associated with the 14-3-3 proteins. Some but not all of these fit into the categories of protein-protein interactions or influences on kinases. Nonetheless, these activities are part of the data from which ultimate understanding must emerge.

The ATPase activity of 14-3-3s discovered in association with their chaperone-like capabilities (5, 24), is the only enzymatic activity yet associated with the 14-3-3s. All other "activities" are either enhancements of extrinsic activities or physical associations with enzymatic products. It remains to be seen whether ATPase activity is found in other isoforms or as part of the other associations indicated for the 14-3-3.

The ζ isoform was once thought to have phospholipase A2 activity (81). Though this activity was later discounted for both animal (62) and plant isoforms, the interactions with lipids and calcium that were part of the supposed phospholipase activity remain to be completely elucidated or discounted. The ζ isoform binds calcium in a blot assay (81), and the plant GF14ω isoform binds calcium on a mole-mole basis at 0.5 mM affinity (46). Later studies demonstrated that certain isoforms (most notably the γ and ε isoforms) can indeed bind phospholipids and aggregate phospholipid vesicles

(65). Therefore, while phospholipase A2 activity is an unlikely 14-3-3 activity, binding of calcium and phospholipids must be considered part of the 14-3-3 repertoire.

Another unique attribute of 14-3-3s is their ability to activate the ADP ribosylation activity of Exoenzyme S (ExoS), a virulence factor associated with infection by *Pseudomonas aeruginosa* (21). FAS (14-3-3 ζ) was found to be absolutely required for the activation of ExoS on its preferred substrate, Ras. Later studies showed that even plant 14-3-3 isoforms could substitute for ζ in the assay, making this activity a likely property of all 14-3-3s (10, 43). It is unclear just how this activation effect on a bacterial toxin protein is related to normal in vivo functions of 14-3-3s. However, because there is evidence for ADP ribosylation activity in normal eukaryotic cells, it is possible that this activity plays a role in the trafficking of proteins such as Ras. If so, the 14-3-3s could participate in the MAP kinase cascade at least twice—through effects on Ras and association with Raf (1).

In a similar situation, 14-3-3s have been shown to be the receptor molecules for the plant phytotoxin fusicoccin (38, 47, 57). Fusicoccin is a metabolite of the pathogen *Fusicoccum amygdali* and is the causative agent of pathological wilting in several species. The Fusicoccin Binding Protein (FCBP) is a conserved, plasma membrane bound protein, and its sequence analysis eventually placed it among the 14-3-3s. In maize, the FCBP (47) is identical to the GF14 form that had been identified as part of the G-box DNA binding complex (15). In oats, the FCBP is composed of at least two 14-3-3s that cross-react with antibodies to yeast BMH1 (38). In the dayflower (*Commelina communis*) the FCBP is made up of at least two isoforms similar to the maize and *Arabidopsis* forms (57).

There are several remarkable aspects of the FCBP as a 14-3-3. First, the FCBP is known to be membrane bound (12), whereas the largest portions of 14-3-3s are cytosolic. Thus, if the FCBP is a 14-3-3, it might have been a novel 14-3-3. In the case of the maize, however, the FCBP is a 14-3-3 known to be both cytoplasmic and nuclear (15, 47). Therefore it appears that the FCBP is simply a restricted subset of the cytoplasmic 14-3-3 forms; only those forms found on the plasma membrane are receptors for fusicoccin (38). This idea highlights the differential membrane associations of 14-3-3s and places the possible effects of fusicoccin and 14-3-3 at the plasma membrane, in a situation essentially parallel to that for the colocalization of Raf and 14-3-3. Second, the effect of fusicoccin is to activate the plasma membrane proton ATPase (which causes the stomata to freeze open and thereby induces wilting) (38). It is possible that the proton ATPases are regulated by kinases and that 14-3-3s are a factor in that process. In that regard, the possibility of a kinase copurifying with the FCBP has been noted (38). Finally, the FCBP presents a unique aspect of 14-3-3s, or at least a unique perspective on 14-3-3 function in plant cell signal transduction.

Yeast 14-3-3s

For several reasons, the yeasts offer perhaps the most straightforward presentation of 14-3-3 functions. First, yeasts appear to have fewer 14-3-3 isoforms, maybe only two (19, 73). Thus the problems of understanding the properties of 14-3-3s that may be complicated by cell-specific expression or specialization of function are largely eliminated. Second, mutants disrupting the 14-3-3 genes are at this time available only in *Saccharomyces cerevisiae* and *Schizosaccharomyces pombe,* allowing the only assignment of phenotypes to the 14-3-3 genes. Last, the yeasts are amenable to engineering, which allows us to more fully understand 14-3-3 functions and interactions.

In *Saccharomyces cerevisiae,* the 14-3-3 homolog BMH1 was discovered as an open reading frame near the PDA1 gene (74). Its derived amino acid sequence does not suggest direct homology with any animal or plant Greek letter subtypes or isoforms (18). A second yeast gene, BMH2, has been isolated by homology with BMH1 (73). Overexpression of BMH1 depressed growth (74), and the BMH1-BMH2 double disruptant is lethal (73). This likely indicates that the BMH1 and BMH2 genes encode most 14-3-3 protein in this yeast. The *Arabidopsis* GF14ω homolog can complement the yeast double disruption (73). Thus it appears that whatever the fundamental function or property of 14-3-3s that regulates growth in yeast, it is an attribute most likely shared by plant and animal homologs.

Additional insights on 14-3-3 function are provided by *Schizosaccharomyces pombe* mutants *rad24* and *rad25.* Genetic studies on *rad24* and *rad25* predate their assignment as 14-3-3s (19). These two genes serve as checkpoint controls to ensure that DNA damage has been repaired before mitosis begins. As with BMH1 and BMH2, the *rad24 rad25* double mutant is inviable. Mutants in *rad24* and *rad25* enter mitosis prematurely after radiation. Interestingly, *rad24* and *rad25* do not produce exactly the same phenotype. While *rad24* has drastic effects, *rad25* has only marginal effects on resistance to radiation-induced damage. In fact, *rad24* was the known checkpoint mutant, and *rad25* was isolated as a suppressor of the *rad24* mutation. These two 14-3-3s overlap in an essential function, but condition unique phenotypes. Thus, it is clear that even in this two-gene yeast system, specialization of individual 14-3-3 isoform functions exists.

Summary and Comments on Function

Figure 1 presents a summary of potential roles for 14-3-3s in cellular signal transduction. Much like the data set from which it is derived, the summary is ambiguous and incomplete. Some consistent themes are emerging from the

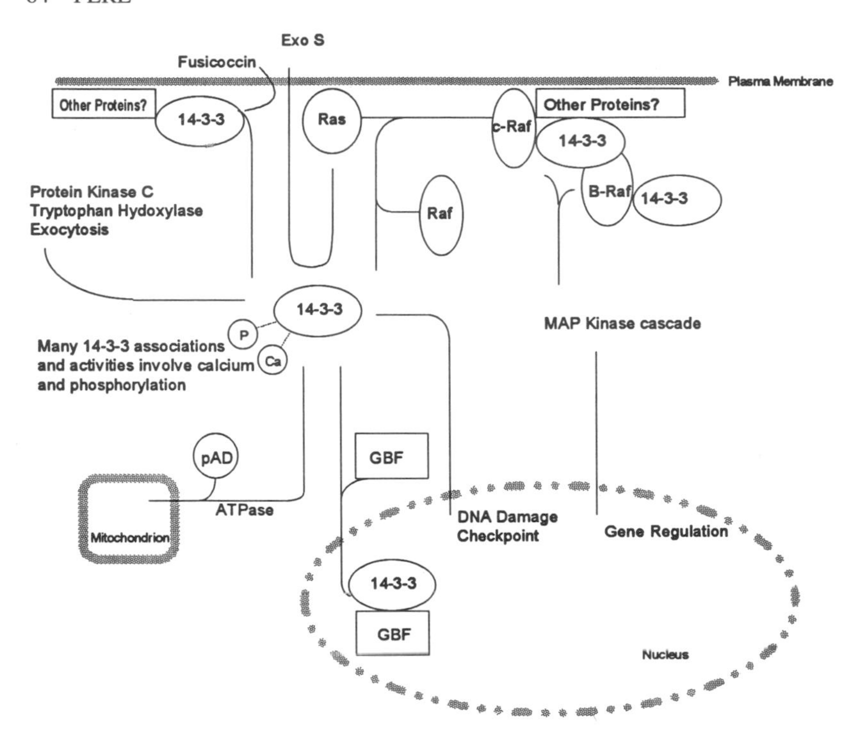

Figure 1 Potential roles for 14-3-3 proteins in cell signaling.

literature on 14-3-3 function. Clearly, protein-protein interactions and associations with protein kinases are fundamental properties that must be a part of any model of 14-3-3s participation in cellular signaling. Other properties, such as calcium binding, ATPase activity, fusicoccin binding, and phospholipid binding, may be understood as components of these central themes. On the other hand, each may be a peculiar property limited to certain isoforms; and the diversity of isoforms may yet account for the diversity of activities.

Returning to the main themes of protein-protein complexes and interaction with kinases, some of these interactions occur at the plasma membrane. The plasma membrane is where the fusicoccin receptor activity of 14-3-3 is localized. Could it be that fusicoccin binds only to the 14-3-3s that are complexed with signaling pathway components there? Membrane associations must also be involved in the chaperone activity for mitochondrial import, and protein receptors on that membrane may be similar but distinct from those associating with 14-3-3s at the plasma membrane. There are at least two direct associations of 14-3-3 with the nucleus (that of DNA damage checkpoint activity and the association with transcription factors) as well as the indirect effects by association with the MAP kinase signaling pathway.

In considering Figure 1, the fact that fusicoccin binds only to membrane-bound 14-3-3 is consistent with 14-3-3 being found associated with the plasma membrane with c-Raf, and the activation of ExoS may be related to the ADP ribosylation of Ras, which is part of the series of events that bring 14-3-3–Raf complex to the membrane as part of the MAP kinase cascade that included B-Raf–14-3-3 complexes. Thus the model serves as a framework within which to view themes in the present data and upon which to examine future information regarding 14-3-3 function.

CELL-SPECIFIC EXPRESSION THEMES

Why do animals and plants have multiple genes, each encoding a different isoform? At the present time, we simply do not know how restricted in function(s) each isoform is. Nonetheless, an appealing supposition is that different cell types have different, perhaps overlapping, needs for individual 14-3-3s and that the different isoforms present alternate ranges of activities.

Support for this idea can be drawn from studies demonstrating unique or different cell-specific expression profiles for particular isoforms. Certainly hints of specialization were present from the inception of 14-3-3 research, when the 14-3-3s were thought to be brain- and neuron-specific (52). Later, 14-3-3s were found in other cells and organs, but some of the nonbrain forms (such as the T-cell and epithelial markers τ and σ) do show cell-specific expression (41, 49, 50, 56, 65). Thus on the gross level, certain specializations in isoform expression do exist.

Evidence for specialization also exists at the cell and tissue levels within the brain. The rat η isoform is specifically and highly expressed in certain cell layers of the brain (76). In a larger study of isoform expression in developing rat embryos, the ζ, θ/τ, β, γ, and η isoforms had an overlapping pattern, with all but the θ/τ form showing distinct specific expression in neuronal tissues (75).

The characterized *Arabidopsis* isoforms, GF14 ω, ψ, χ, φ, and υ demonstrate some organ specificity at the RNA blot hybridization level (CJ Daugherty & RJ Ferl, personal communication, 1995; 44). Within the flower, the χ isoform is concentrated in developing pollen and in papillar cells of the stigma, as detected both by in situ hybridization and staining of transgenic plants carrying a GF14χ promoter gus reporter fusion construct. GF14χ transcripts are also very highly expressed in the endosperm of developing seeds (CJ Daugherty & RJ Ferl, personal communication, 1995). Thus, for the characterized members of the 14-3-3 family in *Arabidopsis,* there appears to be both cell- and organ-specific expression.

In plants we also find that external stimuli can influence 14-3-3 expression. The RCI (Rare Cold Inducible) 14-3-3s of *Arabidopsis* are named based on their low-temperature induction in stems, leaves, and flowers at 4°C (34). The maize 14-3-3, GF14-12, is induced about threefold by hypoxia (14). A tobacco homolog 14-3-3 is down-regulated during acclimation to high salt (10), and a rice isoform is regulated by both salt and low temperature (36). A barley 14-3-3 homolog is induced after infection with the powdery mildew fungus *Erysiphe graminis* (8). This observation potentially links pathogen attack to signal transduction pathways, paralleling identification of the fusicoccin receptor as a 14-3-3. These stimulus-directed examples are interesting and may be very useful for future dissection of 14-3-3 function and expression in plants.

EVOLUTIONARY THEMES

Can an evolutionary perspective provide a context to understand the relationships among isoforms and organisms? Many reports of 14-3-3s include discussion of conservation of amino acid sequence conservation among the many forms. Even among the most divergent plant and animal proteins, overall similarity is greater than 70%. There are large blocks of sequence that are nearly completely conserved among all isoforms (2, 18, 48). In the face of such conservation of sequence, is there any evolutionary evidence to suggest that specialized subfamilies have developed? Is there any evidence for homologous specialized functions shared between animal and plant isoforms?

The answer to these questions appears to be simply no. Analysis of the sequences available through early 1994 (18) suggested that the animal and plant 14-3-3 lineages have been on distinct and separate evolutionary tracts since their earliest divergences. In the interim, many new 14-3-3 sequences have been published that confirm this conclusion.

Figure 2 is an unrooted network describing the basic phylogeny of the published 14-3-3 isoforms. It is based on a parsimony analysis of 14-3-3

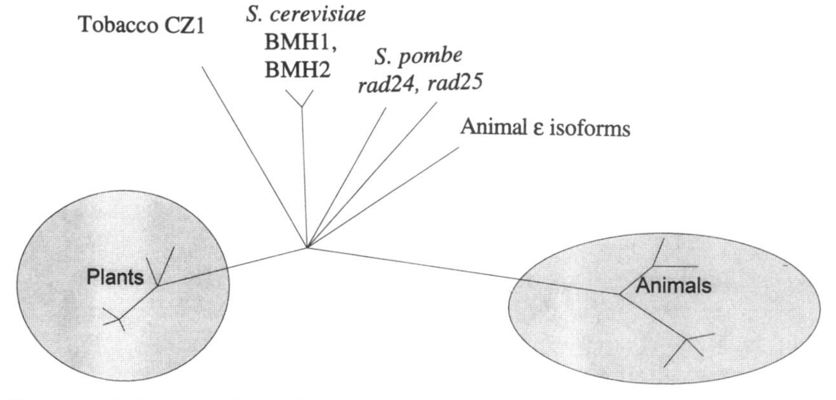

Figure 2 Evolutionary relationships among the published 14-3-3 isoforms

sequences truncated to remove the C-terminal regions, because these regions are so divergent that proper alignment of amino acid positions becomes problematic (18). The length of the lines roughly corresponds to the evolutionary divergence among the sequences, and only those branches that survive bootstrap analysis at 90% are indicated.

Most of the known plant and animal isoforms are late evolutionary branches from single ancient forms within their kingdom. Given these independent trajectories, there can be no *shared specializations* among the known animal and plant isoforms. Fundamental properties may be, and probably are, conserved over the full spectrum. As yet there are no shared specializations. It is also then entirely likely that some properties or functions could be entirely limited to either plants or animals.

The notable exceptions are the tobacco 14-3-3 (CZ1) (10) and the animal ε isoforms. Each of these isoforms, much like the yeast isoforms, have been on their own unique trajectories. From the present sequence data, no evolutionary homologies exist within either of these isoform lineages, except that the yeast BMH1 and BMH2 genes appear to be a relatively recent divergence. The ε isoform is present in several animals, and it is therefore likely to be found in all. So far, tobacco is the only plant representative of a similarly distinct lineage, though a safe prediction is that isoforms evolutionarily homologous to cz1 will be found in other plants.

Evolutionary considerations suggest that the fundamental themes of 14-3-3 function likely apply to many or all isoforms. For example, a main line plant isoform is known to activate tryptophan hydroxylase and regulate PKC much like the main line animal forms (43); therefore this property is likely a fundamental characteristic shared by all isoforms in those evolutionary lines. Because a main line plant isoform can rescue a yeast BMH disruption (73), it is likely that other plant forms will as well. And because both the tobacco and a main line plant isoform can activate ExoS ribosylation much like the animal ζ isoform (10, 43), this activity is likely shared by all 14-3-3s. There is also some homology between 14-3-3s and open reading frames in cyanobacteria (69), therefore the search for fundamental answers may extend to the prokaryotes.

STRUCTURAL THEMES

The crystal structure of much of the animal ζ and τ isoforms has recently been solved (42a, 78). This structure, though missing several residues and lacking fine resolution, consolidates earlier biochemical analysis of structural features of the 14-3-3 proteins and allows for an approach to understanding a structural basis for 14-3-3 function. Perhaps the most important aspect of the structure is that the residues that are variable in an evolutionary context do not alter the central character of the protein. Thus the fundamental features of this structure should be present in all 14-3-3s of plant, animal, and yeast origin.

The structure of the 14-3-3 dimer is a flattened and open-ended cup shape, forming a channel that, when viewed from the side, is similar to a carpenter's C-clamp (Figure 3). The two monomers interact at the base of the clamp through the overlapping and largely hydrophobic interaction of the first three antiparallel α-helices (35). The remainder of the floor of the clamp, as well as the sides, are made from six additional antiparallel α-helices. The interior of the clamp is largely negative in character and is thought to be the main face of contact with other proteins. The main site for phosphorylation in animals, Ser 186 in the β and ζ forms (4), lies on the outside face of the clamp, near an area of the protein thought to be annexin-like (2). A loop corresponding to the proposed calcium-binding EF-hand of the *Arabidopsis* GF14ω isoform (46) also lies on the outer surface of the clamp, near the open end and the C-terminus.

The size, shape, and charge distribution of the 14-3-3 structure immediately suggests that it actually serves as a docking clamp, holding the target protein or proteins nestled into its negatively charged base. Indeed, the peptides known to interact with PKC (3, 64) are on the inner surface of the base of the clamp. Additional interactions with kinases are proposed to occur between their basic zinc-finger regions and the acidic domains of 14-3-3s (1). It is possible, then, that the 14-3-3s' dimer facilitates interactions with proteins that fit within its grasp, perhaps even proteins that would fail to interact otherwise.

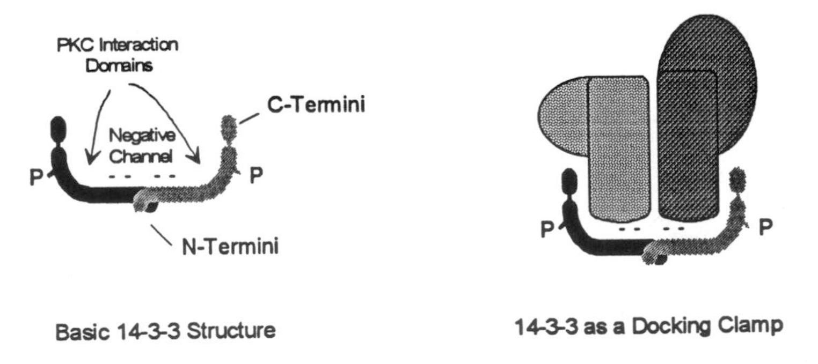

Figure 3 Basic structural features of the 14-3-3 proteins and their possible role as docking clamps that facilitate protein interactions. The overall shape was determined from the crystal structure. The 14-3-3s may be phosphorylated, and the crystal structure predicts that phosphorylation sites are on the external surface. In addition, 14-3-3s are often detected as heterodimers, indicated here by different shading of the monomers. Heterodimerization may allow the distinctive C-termini to bring heterologous paring partners into the grasp of the 14-3-3s, as indicated by the dissimilar proteins docked into the negative channel. The C-termini are presented here as extensions of the sides of the channel in order to indicate their possible role in specifically securing protein associations within the channel.

The extreme N- and C-termini are missing from the crystal structure, indicating that these regions are likely in a disordered state in the crystals of pure 14-3-3. Both the N- and C-termini are highly variable. It is quite possible that these termini may offer specialized activity to the fundamental core structure. For example, because parts of the N-terminus are indicated in potential membrane interactions, variations there may differentially influence potential membrane associations. In addition, because the C-termini are at the open end of the clamp, they may form specialized interactions with the proteins held within the clamp and therefore may direct the nature of the protein-protein interactions facilitated by 14-3-3s. If these termini do, in fact, interact with the proteins that are held in the clamp, their structures may not become ordered until the 14-3-3 is docked with the target proteins. Further understanding of 14-3-3 function and interactions with other proteins awaits the analysis of complexed cocrystals.

CONCLUDING REMARKS

In spite of the many activities associated with the 14-3-3 proteins, fundamental properties are indeed beginning to emerge. The 14-3-3s apparently play a major role in getting proteins together, especially proteins associated with phosphorylation events and signal transduction pathways. The 14-3-3 structure provides a context for understanding the properties of 14-3-3s that permit them to exercise those roles.

Much of our understanding of 14-3-3s is now being derived from protein-protein interaction studies, and it is from these that new and better understanding of 14-3-3s is likely to appear. There are specific priorities for progress in plant 14-3-3 research. First, it is very likely that additional isoforms have yet to be discovered. Second, certain benchmark organisms should be fully characterized with respect to all 14-3-3s present. This would allow not only a full characterization of all of the players on the stage but would also provide the opportunity for a coherent naming convention. Third, given that 14-3-3s serve as chaperones for mitochondrial protein import, it is likely that (some) plant isoforms have arisen with similar functions with the plastid. Fourth, the recovery of mutants, either by screening, direct knockout, selection based on interaction with fusicoccin, or other strategies, should be attempted. Defining phenotypes with mutants will clarify the pathways that individual 14-3-3s participate in. Finally, and by way of general conclusion, biochemical and gene expression work should continue with one eye focused on the evolutionary homology of the isoform under investigation. Conclusions could then be based on more than isoform-specific criteria, and the attendant data could be viewed within a context that allows greater understanding of the entire 14-3-3 family of proteins.

ACKNOWLEDGMENTS

I am extraordinarily indebted to members of my laboratory who have worked on 14-3-3 proteins in *Arabidopsis,* tomato, and maize. Gui Hua Lu, Nick deVetten, Alice DeLisle, Paul Sehnke, Brian Bowen, Beth Laughner, Phil Miller, Ke Wu, Mike Rooney, and Chris Daugherty have all contributed to our knowledge of plant 14-3-3s, and especially to my opportunity to present this review. I am truly grateful. Many of the observations and conclusions raised in this review are the result of extended discussions with these people, and I thank them and L. C. Hannah for reviewing the manuscript. However, any misrepresentations or errors are surely my own, and I apologize for them ahead of time. Our work on 14-3-3s is supported by USDA/NRI grant 93-37304-9608 for the structure and function of 14-3-3 proteins and genes, and by NIH grant GM40061 for 14-3-3 interaction with transcription factors. This is No. R-04809 journal series of the Florida Agricultural Experiment Station.

Literature Cited

1. Aitken A. 1995. 14-3-3 proteins on the MAP. *Trends Biol. Sci.* 20:95–97
2. Aitken A, Collinge DB, van Heusden GPH, Isobe T, Roseboom PH, et al. 1992. 14-3-3 proteins: a highly conserved, widespread family of eukaryotic proteins. *Trends Biol. Sci.* 17: 498–501
3. Aitken A, Ellis CA, Harris A, Sellers LA, Token A. 1990. Kinase and neurotransmitters. *Nature* 344:594
4. Aitken A, Howell S, Jones D, Madrazo J, Patel Y. 1995. 14-3-3 α and δ are the phosphorylated forms of Raf-activating 14-3-3 β and ζ. *J. Biol. Chem.* 270:5706–9
5. Alam R, Hachiya N, Sakaguchi M, Kawabata S, Iwanago S, et al. 1994. cDNA cloning and characterization of mitochondrial import stimulation factor (MSF) purified from rat liver cytosol. *Biochem. J.* 116:416–25
6. Boston PF, Jackson P, Kynoch PAM, Thompson RJ. 1982. Purification, properties, and immunohistochemical localization of human brain 14-3-3 protein. *J. Neurochem.* 38:1466–74
7. Boston PF, Jackson P, Thompson RJ. 1982. Human 14-3-3 protein radioimmunoassay, tissue distribution, and cerebrospinal fluid levels in patients with neurological disorders. *J. Neurochem.* 38:1475–82
8. Brandt J, Thordal-Christensen H, Vad K, Gregersen PL, Collinge DB. 1992. A pathogen-induced gene of barley encodes a protein showing high similarity to a protein kinase regulator. *Plant J.* 2:815–20
9. Burbelo PD, Hall A. 1995. Hot numbers in signal transduction. *Curr. Biol.* 5:95–96
10. Chen Z, Fu H, Liu D, Chang PL, Narasimhan M, et al. 1994. A NaCl-regulated plant gene encoding a brain protein homolog that activates ADP ribosyltransferase and inhibits protein kinase C. *Plant J.* 6:729–40
11. Deleted in proof
12. de Boer AH, Watson BE, Cleland RE. 1989. Purification and identification of the fusicoccin binding protein from oat root plasma membrane. *Plant Physiol.* 89:250–59
13. de Vetten NC, Ferl RJ. 1994. Two genes encoding GF14 (14-3-3) proteins in *Zea mays. Plant Physiol.* 106:1593–604
14. de Vetten NC, Ferl RJ. 1995. Characterization of a maize G-box binding factor that is induced by hypoxia. *Plant J.* 7: 589–601
15. de Vetten NC, Lu G, Ferl RJ. 1992. A maize protein associated with the G-Box binding complex has homology to brain regulatory proteins. *Plant Cell* 4:1295–307
16. Erikson OF, Moore BW. 1980. Investigation of the axonal transport of three acidic soluble proteins (14-3-2, 14-3-3, and S100) in the rabbit visual system. *J. Neurochem.* 35:232–41
17. Fantl WJ, Muslin AJ, Kikuchi A, Martin

JA, MacNicol AM, et al. 1994. Activation of Raf-1 by 14-3-3 proteins. *Nature* 371: 612–13
18. Ferl RJ, Lu G, Bowen BW. 1994. Evolutionary implications of the family of 14-3-3 brain protein homologs in *Arabidopsis thaliana*. *Genetica* 92:129–38
19. Ford JC, Al-Khodairy F, Fotou E, Sheldrick KS, Griffiths DJF, Carr AM. 1994. 14-3-3 protein homologs required for the DNA damage checkpoint in fission yeast. *Science* 265:533–37
20. Freed E, Symons M, Macdonald SG, McCormick F, Ruggieri R. 1994. Binding of 14-3-3 proteins to the protein kinase Raf and effects on its activation. *Science* 265: 1713–16
21. Fu H, Coburn J, Collier RJ. 1993. The eukaryotic host factor that activates exoenzyme S of *Pseudomonas aeruginosa* is a member of the 14-3-3 protein family. *Proc. Natl. Acad. Sci. USA* 90:2320–24
22. Fu H, Xia K, Pallas DC, Cui C, Conroy K et al. 1994. Interaction of the protein kinase Raf-1 with 14-3-3 proteins. *Science* 266: 126–33
23. Furukawa Y, Ikuta N, Omata S, Yamauchi T, Isobe T, Ichimura T. 1993. Demonstration of the phosphorylation-dependent interaction of tryptophan hydroxylase with the 14-3-3 protein. *Biochem. Biophys. Res. Commun.* 194:144–49
24. Hachiya N, Komiya T, Alam R, Iwahashi J, Sakaguchi M, et al. 1994. MSF, a novel cytoplasmic chaperone which functions in precursor targeting to mitochondria. *EMBO J.* 13:5146–54
25. Hirsch S, Aitken A, Bertsch U, Soll J. 1992. A plant homologue to mammalian brain 14-3-3 protein and protein kinase C inhibitor. *FEBS Lett.* 296:222–24
26. Hopkin K. 1994. Just past Ras: divergence in a signal-transduction pathway. *J. NIH Res.* 6:62–65
27. Ichimura T, Isobe T, Okuyama T, Takahashi N, Araki K, et al. 1988. Molecular cloning of cDNA coding for brain-specific 14-3-3 protein, a protein kinase-dependent activator of tyrosine and tryptophan hydroxylases. *Proc. Natl. Acad. Sci. USA* 85: 7084–88
28. Ichimura T, Isobe T, Yamauchi T, Fujisawa H. 1987. Brain 14-3-3 protein is an activator protein that activates tryptophan 5-monooxygenase in the presence of Ca^{2+}, calmodulin-dependent protein kinase II. *FEBS Lett.* 219:79–82
29. Ichimura T, Sugano H, Kuwano R, Sunaya T, Okuyama T, Isobe T. 1991. Widespread distribution of the 14-3-3 protein in vertebrate brains and bovine tissues: correlation with the distributions of calcium-dependent protein kinases. *J. Neurochem.* 1449–51
30. Ichimura-Ohshima Y, Morii K, Ichimura T, Araki K, Takahashi Y, et al. 1992. cDNA cloning and chromosome assignment of the gene for human brain 14-3-3 protein η chain. *J. Neurosci. Res.* 31:600–5
31. Irie K, Gotoh Y, Yashar BM, Errede B, Nishida E, Matsumoto K. 1994. Stimulatory effects of yeast and mammalian 14-3-3 proteins on the Raf protein kinase. *Science* 265:1716–19
32. Isobe T, Hiyane Y, Ichimura T, Okuyama T, Takahashi N, et al. 1992. Activation of protein kinase C by the 14-3-3 proteins homologous with Exo1 protein that stimulates calcium-dependent exocytosis. *FEBS Lett.* 308:121–24
33. Isobe T, Ichimura T, Sunaya T, Okuyama T, Takahashi N, et al. 1991. Distinct forms of the protein kinase-dependent activator of tyrosine and tryptophan hydroxylases. *J. Mol. Biol.* 217:125–32
34. Jarillo JA, Capel J, Leyva A, Martinez-Zapater JM, Salinas J. 1994. Two related low-temperature-inducible genes of *Arabidopsis* encode proteins showing high homology to 14-3-3 proteins, a family of putative kinase regulators. *Plant Mol. Biol.* 25: 693–704
35. Jones DHA, Martin H, Madrazo J, Robinson K, Nielsen PJ, et al. 1995. Expression and structural analysis of 14-3-3 proteins. *J. Mol. Biol.* 245:375–84
36. Kidou S, Umeda M, Kato A, Uchimiya H. 1993. Isolation and characterization of a rice cDNA similar to the bovine brain-specific 14-3-3 protein gene. *Plant Mol. Biol.* 21:191–94
37. Klimczak LI, Schindler U, Cashmore AR. 1992. DNA binding activity of the *Arabidopsis* G-box binding factor GBF1 is stimulated by phosphorylation by casein kinase II from broccoli. *Plant Cell* 4:87–98
38. Korthout HAAJ, de Boer AH. 1994. A fusicoccin binding protein belongs to the family of 14-3-3 brain protein homologs. *Plant Cell* 6:1681–92
39. Kuroda S, Shimizu K, Yamamori B, Matsuda S, Imazumi K, et al. 1995. Purification and characterization of REKS from *Xenopus* eggs: identification of REKS as a Ras-dependent mitogen-activated protein kinase kinase kinase. *J. Biol. Chem.* 270: 2460–65
40. Laughner B, Lawrence SD, Ferl RJ. 1994. Two tomato fruit homologs of 14-3-3 mammalian brain proteins. *Plant Physiol.* 105: 1457–58
41. Leffers H, Madsen P, Rasmussen HH, Honore B, Andersen AH, et al. 1993. Molecular cloning and expression of the transformation sensitive epithelial marker stratifin: a member of a protein family that has been involved in the protein kinase C signalling pathway. *J. Mol. Biol.* 231:982–98

42. Li S, Janosch P, Tanji M, Rosenfeld GC, Waymire JC, et al. 1995. Regulation of Raf-1 kinase activity by the 14-3-3 family of proteins. *EMBO J.* 14:685–96
42a. Liu D, Bienkowska J, Petssa C, Collier RJ, Fu H, Liddington R. 1995. Crystal structure of the zeta isoform of the 14-3-3 protein. *Nature* 376:191–94
43. Lu G, de Vetten NC, Sehnke PC, Isobe T, Ichimura T, et al. 1994. A single *Arabidopsis* GF14 isoform possesses biochemical characteristics of diverse 14-3-3 homologues. *Plant Mol. Biol.* 25:659–67
44. Lu G, DeLisle AJ, de Vetten NC, Ferl RJ. 1992. Brain proteins in plants: an *Arabidopsis* homolog to neurotransmitter pathway activators in part of a DNA binding complex. *Proc. Natl. Acad. Sci. USA* 89: 11490–94
45. Lu G, Rooney MF, Wu K, Ferl RJ. 1994. Five cDNAs encoding *Arabidopsis* GF14 proteins. *Plant Physiol.* 105:1459–60
46. Lu G, Sehnke PC, Ferl RJ. 1994. Phosphorylation and calcium binding properties of an *Arabidopsis* GF14 brain protein homolog. *Plant Cell* 6:501–10
47. Marra M, Fullone MR, Fogliano V, Pen J, Mattei M, et al. 1995. The 30-KD protein present in purified fusicoccin receptor preparations is a 14-3-3 like protein. *Plant Physiol.* 106:1497–501
48. Martens GJM, Piosik PA, Danen EHJ. 1992. Evolutionary conservation of the 14-3-3 protein. *Biochem. Biophys. Res. Com.* 184:1456–59
49. Martin H, Patel Y, Jones D, Howell S, Robinson K, Aitken A. 1993. Antibodies against the major brain isoforms of 14-3-3 protein. *FEBS Lett.* 331:296–303
50. Martin H, Rostas J, Patel Y, Aitken A. 1994. Subcellular localisation of 14-3-3 isoforms in rat brain using specific antibodies. *J. Neurochem.* 63:2259–65
51. Maru Y, Witte ON. 1991. The *BCR* gene encodes a novel serine/threonine kinase activity within a single exon. *Cell* 67: 459–68
52. Moore BW, Perez VJ. 1967. Specific acidic proteins of the nervous system. In *Physiological and Biochemical Aspects of Nervous Integration,* ed. FD Carlson, pp. 343–59 Woods Hole, MA: Prentice Hall
53. Morgan A, Burgoyne RD. 1992. Exo1 and Exo2 proteins stimulate calcium-dependent exocytosis in permeabilized adrenal chromaffin cells. *Nature* 355:833–36
54. Morgan A, Burgoyne RD. 1992. Interaction between protein kinase C and Exo1 (14-3-3 protein) and its relevance to exocytosis in permeabilized adrenal chromaffin cells. *Biochem. J.* 286:807–11
55. Morrison D. 1994. 14-3-3: modulators of signaling proteins. *Science* 266:56–57
56. Nielsen PJ. 1991. Primary structure of a human protein kinase regulator protein. *Biochem. Biophys. Acta* 1088:425–28
57. Oecking C, Eckerskorn C, Weiler EW. 1994. The fusicoccin receptor of plants is a member of the 14-3-3 superfamily of eukaryotic regulatory proteins. *FEBS Lett.* 352:163–66
58. Pallas DC, Fu H, Haehnel LC, Weller W, Collier RJ, Roberts TM. 1994. Association of polyomavirus middle tumor antigen with 14-3-3 proteins. *Science* 265:535–37
59. Patel Y, Martin H, Howell S, Jones D, Robinson K, Aitken A. 1994. Purification of 14-3-3 protein and analysis of isoforms in chicken brain. *Biochem. Biophys. Acta* 1222:405–9
60. Prasad GL, Valverius EM, McDuffie E, Cooper HL. 1992. Complementary DNA cloning of a novel epithelial cell marker protein, HME1, that may be down-regulated in neoplastic mammary cells. *Cell Growth Diff.* 3:507–13
61. Reuther GW, Fu H, Cripe LD, Collier RJ, Pendergast AM. 1994. Association of the protein kinases c-Bcr and Bcr-Abl with proteins of the 14-3-3 family. *Science* 266: 129–33
62. Robinson K, Jones D, Patel Y, Martin H, Madrazo J, et al. 1994. Mechanism of inhibition of protein kinase C by 14-3-3 isoforms. *Biochem. J.* 299:853–61
63. Roseboom PH, Weller JL, Babila T, Aitken A, Sellers LA, et al. 1994. Cloning and characterization of the ϵ and ζ isoforms of the 14-3-3 proteins. *DNA Cell Biol.* 13: 629–40
64. Roth D, Morgan A, Burgoyne RD. 1993. Identification of a key domain in annexin and 14-3-3 proteins that stimulate calcium-dependent exocytosis in permeabilized adrenal chromaffin cells. *FEBS Lett.* 320: 207–10
65. Roth D, Morgan A, Martin H, Jones D, Martens GJM, et al. 1994. Characterization of 14-3-3 proteins in adrenal chromaffin cells and demonstration of isoform-specific phospholipid binding. *Biochem. J.* 301: 305–10
66. Schejter ED, Segal D, Glazer L, Shilo B. 1986. Alternative 5' exons and tissue-specific expression of the Drosophila EGF receptor homolog transcripts. *Cell* 46: 1091–101
67. Deleted in proof
68. Shimizu K, Kuroda S, Yamamori B, Matsuda S, Kaibuchi K, et al. 1994. Synergistic activation by Ras and 14-3-3 protein of a mitogen-activated protein kinase kinase kinase named Ras-dependent extracellular signal-regulated kinase kinase stimulator. *J. Biol. Chem.* 269:22917–20
69. Simpson GG, Clark G, Brown JWS. 1994. Isolation of a maize cDNA encoding a protein with extensive similarity to an inhibitor

of protein kinase C and a cyanobacterial open reading frame. *Biochem. Biophys. Acta* 1222:306–8

70. Swanson KD, Ganguly R. 1992. Characterization of a *Drosophila melanogaster* gene similar to the mammalian genes encoding the tyrosine/tryptophan hydroxylase activator and protein kinase C inhibitor proteins. *Gene* 113:183–90
71. Token A, Ellis CA, Sellers LA, Aitken A. 1990. Protein kinase C inhibitor proteins. *FEBS Lett.* 191:421–29
72. Token A, Sellers LA, Amess B, Patel Y, Harris A, Aitken A. 1992. Multiple isoforms of a protein kinase C inhibitor (KCIP-1/14-3-3) from sheep brain. *FEBS Lett.* 206:453–61
73. van Heusden GPH, Griffiths DJF, Ford JC, Chin-A-Woeng TFC, Schrader PAT, et al. 1995. The 14-3-3 proteins encoded by the *BMH1* and *BMH2* genes are essential in the yeast *Saccharomyces cerevisiae* and can be replaced by a plant homologue. *Eur. J. Biochem.* 229:45–53
74. van Heusden GPH, Wenzel TJ, Lagendijk EL, de Steensma HY, van den Berg JA. 1992. Characterization of the yeast *BMH1* gene encoding a putative protein homologous to mammalian protein kinase II activators and protein kinase C inhibitors. *FEBS. Lett.* 302:145–50
75. Watanabe M, Isobe T, Ichimura T, Kuwano R, Takahashi Y, et al. 1994. Molecular cloning of rat cDNAs for the zeta and theta subtypes of 14-3-3 protein and differential distributions of their mRNAs in the brain. *Mol. Brain Res.* 25:113–21
76. Watanabe M, Isobe T, Okuyama T, Ichimura T, Kuwano R, et al. 1991. Molecular cloning of cDNA to rat 14-3-3 n chain polypeptide and the neuronal expression of the mRNA in central nervous system. *Mol. Brain Res.* 10:151–58
77. Wu YN, Vu N, Wagner PD. 1992. Anti-(14-3-3 protein) antibody inhibits stimulation of noradrenaline (norepinephrine) secretion by chromaffin-cell cytosolic proteins. *Biochem. J.* 285:697–700
78. Xiao B, Smerdon SJ, Jones DH, Dodson GG, Soneji Y, et al. 1995. Structure of a 14-3-3 protein and implications for coordination of multiple signalling pathways. *Nature* 376:188–91
79. Yamamori B, Kuroda S, Shimizu K, Fukui K, Ohtsuka T, Takai Y. 1995. Purification of a Ras-dependent mitogen-activated protein kinase kinase kinase from bovine brain cytosol and its identification as a complex of B-Raf and 14-3-3 proteins. *J. Biol. Chem.* 270:11723–26
80. Yamauchi T, Nagata H, Fujisawa H. 1981. A new activator protein that activates tryptophan 5-monooxygenase and tyrosine monooxygenase in the presence of Ca^{2+}-calmodulin-dependent protein kinase: purification and characterization. *J. Biol. Chem.* 256:5404–9
81. Zupan LA, Steffens DL, Berry CA, Landt M, Gross RW. 1992. Cloning and expression of a human 14-3-3 protein mediating phospholipolysis. *J. Biol. Chem.* 267: 8707–10

Annu. Rev. Plant Physiol. Plant Mol. Biol. 1996. 47:75–100

DNA DAMAGE AND REPAIR IN PLANTS

Anne B. Britt

Section of Plant Biology, University of California, Davis, California 95616

KEY WORDS: DNA damage, DNA repair, UV radiation, *Arabidopsis*, photoreactivation

ABSTRACT

The biological impact of any DNA damaging agent is a combined function of the chemical nature of the induced lesions and the efficiency and accuracy of their repair. Although much has been learned from microbes and mammals about both the repair of DNA damage and the biological effects of the persistence of these lesions, much remains to be learned about the mechanism and tissue-specificity of repair in plants. This review focuses on recent work on the induction and repair of DNA damage in higher plants, with special emphasis on UV-induced DNA damage products.

1040-2519/96/0601-0075$08.00

INTRODUCTION

DNA is constantly subject to chemical modification; even under the best of circumstances, purine bases are lost from the mammalian genome at a rate of several thousand bases per cell per day (62). Alkylating agents, essential for a variety of biosynthetic processes, can turn a legitimate base into either a mutagenic, miscoding deviant, or a lethal, noncoding lesion. Hydrolytic deamination can directly change one base into another. Fresh air (oxygen) and sunshine (UV) are undoubtedly the two major genotoxic agents for most organisms, and plants are obliged to be exposed to both of these mutagens. For this reason, plants, like all living things, have mechanisms that enable them to tolerate or repair the DNA damage they inevitably experience.

DNA damage has both genotoxic and cytotoxic effects. The study of the induction of DNA damage and its repair in humans has been of interest largely because of the demonstrable role of mutagenesis in carcinogenesis and a postulated role of DNA damage in aging. The contribution of DNA damaging agents to genetic load in animals has also been of interest, particularly in the study of radiation-induced mutagenesis (85). While carcinogenesis is not particularly relevant to most agronomically important plants, it is possible that DNA damage may play a significant role in the "aging" of seeds stocks and perennial crops. Although DNA damage is often thought of primarily in regard to its mutagenic effects, the persistence of damaged bases also has a significant growth-inhibitory influence. Many DNA damage products act as blocks to the progress of both DNA and RNA polymerases. Accumulated damage will not only preclude cell division but will eventually kill even terminally differentiated, nonreplicating tissues such as a mature leaf. For this reason DNA repair mechanisms are required in all living plant tissues to alleviate the toxic effects of the accumulation of DNA damage on plant metabolism.

The role of DNA damage and repair in the creation of genetic diversity is also of interest. Some DNA damage "tolerance" pathways, which enable the cell to replicate in spite of the persistence of damage, are actually responsible for the mutagenic effects of many DNA damaging agents. Because the genetic variation created in part through point mutation and recombination are prerequisites of both natural and artificial selection, understanding the mechanisms of genetic change is relevant to both theoretical evolution and genetic engineering. Although the two major damage tolerance mechanisms, lesion bypass and recombinational "repair," have been clearly established as mutagenic events in microbes, it remains to be determined whether these pathways exist in higher plants.

DNA damage can be broadly classified into three types of lesions: mismatched bases, double-strand breaks, and chemically modified bases. Each of these classes of lesions is corrected via distinct repair pathways. Reviews of

DNA repair often focus on the last class of lesions, but the repair pathways for the first two classes of damage are particularly interesting to the plant geneticist. Although no work has been published on any aspect of mismatch repair in plants, there have been some very interesting recent developments pertaining to the repair of double-strand breaks. Our knowledge of general DNA repair pathways in plants lags far behind our knowledge of these pathways in bacteria, yeast, and mammals. For an in-depth review of DNA damage products and their repair in these organisms, I refer the reader to the recently published textbook *DNA Repair and Mutagenesis* (31).

DNA DAMAGE PRODUCTS

Assessing the biological significance of any single type of lesion is complex, requiring knowledge of the frequency at which it occurs, the immediate effects of its persistence (whether the lesion can mispair directly, or whether it acts as a block to replication and/or transcription), and the number and efficiency of repair and tolerance pathways that pertain to the lesion. Repair and tolerance pathways are addressed below. In this section I review some of the most common naturally occurring DNA damage products and their immediate biological effects.

As is the case with many of the repair mechanisms I discuss, the information on the induction of damage was derived from the study of microbes and mammals. The relative importance of individual lesion types may be quite different in plants. It is possible that some genotoxic, unique plant metabolites [i.e. psoralens or metabolized xenobiotics (90a)] occasionally find their way into the plant nucleus or perhaps reach an organellar genome. It is unknown whether plant metabolites have a significant effect on the stability of plant DNA. In addition, the DNA present in seeds, like that present in bacterial or fungal spores, experiences a very different chemical environment from that enjoyed by the DNA in the nucleus of actively metabolizing cells (84). The type and extent of damage that occurs during seed storage is an important and still developing area of research. Finally, the distribution of naturally occurring DNA damage products changes not only from one organism to the next but also from one tissue to the next. To the epidermal cells of a plant, the damage induced by UV radiation is as inevitable as hydrolytic damage induced by the water present in the cell's nucleus. Even subtle changes produced under controlled laboratory conditions can substantially alter the type and frequency of spontaneous mutations (121). For this reason, the reader should understand that the relative importance of the various DNA damage products discussed below (and this is not an all-inclusive list) and, in fact, the repair pathways discussed later, should *always* be regarded within the context of the organism in question, the tissue studied, and the nature of its environment.

Hydrolytic Damage

For an excellent review of the chemical stability of DNA (both in vivo and, interestingly, from ancient sources) I direct the reader to a recent article (61). The DNA of living cells is subject to a variety of hydrolytic reactions, the most common being the hydrolysis of the glycosylic bond between purine bases and the DNA backbone. Although the overall rate of depurination is quite slow, in organisms with a large genome, such as humans or maize, spontaneous hydrolysis would be expected to induce the loss of several thousand purine bases per day per cell. If an abasic site were to persist, it would block DNA replication and transcription and it would be lethal in replicating cells. There is also some evidence that AP (apurinic) sites are potentially mutagenic as a result of occasional lesion bypass events during DNA replication (34). Generally, abasic sites are rapidly recognized and repaired. As a result, spontaneously generated AP sites do not play a significant role in spontaneous mutagenesis in microbes or mammals, and this is likely to be true for plants as well.

A second type of hydrolysis reaction appears to be responsible for the majority of spontaneous point mutations in human cells. Both cytosine and 5-methylcytosine are subject to hydrolytic deamination, resulting in the formaion of uracil and thymine, respectively. These two deamination products, both of which base pair with adenine and so are potentially mutagenic, do in fact differ widely in their mutagenicity. Uracil is rapidly recognized as an inappropriate base in DNA, and it is excised by uracil glycosylase in both plants and animals. Thymine, however, cannot be recognized as a DNA damage product and so is highly mutagenic, producing C:G to T:A transition mutations. A study of disease-related point mutations in the p53 tumor suppressor gene revealed that 43% were C to T transitions at 5-methyl CpG dinucleotides (101). This strongly suggests that the deamination product of 5-methylcytosine may be the most important single cause of spontaneous point mutations in the mammalian cell. The tumor DNA analyzed did not include skin cancers; the spectrum of mutations induced in tissues exposed to sunlight is different from that listed above and is discussed in the section UV-Induced Damage.

Given that plants methylate cytosine not only at CpG dinucleotides but also at CpNpG trinucleotides (37) and potentially other sites as well, and the fact that a much higher percentage of the angiosperm genome, compared with the human genome, is made up of 5-methylcytosine (approximately 10% vs less than 1%) (115), it is likely that 5-methylcytosine to thymine transitions are also a frequent spontaneous mutation event in plants. The underrepresentation of CpG vs GpC nucleotides in the plant genome supports the notion that these sequences are unstable (114).

Although the exact structure and hydration state of DNA in dried seeds is unknown (84), one would expect that seeds, when stored for long periods of time in a desiccated state in which no DNA repair occurs, would progressively accumulate AP sites and other spontaneously generated lesions. At some point the amount of damage incurred by aging seeds may exceed the repair capacity of the germinating seedling. Although desiccated seeds accumulate hydrolytic damage at a much slower rate than fully hydrated cells (21), long-term seed storage has been correlated with a delay in replicative DNA synthesis, the limited synthesis of low molecular weight, untranslated RNAs, and an increase in unscheduled (repair) DNA synthesis (83). These phenomena are consistent with a requirement for a period of genomic repair before cell division can occur in germinating seeds. At least some of the beneficial effects of "osmo-priming," a procedure involving partial hydration of seed designed to enhance early and uniform germination, may result from DNA repair activities during the priming period (4).

Alkylation Damage

Ethylmethane sulfonate (EMS), an ethylating agent, is a commonly employed artificial mutagen in plant genetics. Even in the absence of exogenously applied alkylating agents, all cells experience a biologically significant level of spontaneous DNA methylation (98b). The majority of the bonds in all four bases are susceptible to methylation, to widely varying extents, and some of the methylation products, if left unrepaired, are premutagenic and/or lethal. Most methylation damage occurs at purine bases. The most frequently generated alkylation product, 7-methyladenine, base pairs normally and is regarded as neither mutagenic nor toxic. In contrast, 3-methyladenine cannot serve as a template for DNA synthesis and therefore acts as a block to DNA replication. DNA damage products that cannot successfully base pair with any base are frequently termed "noninformational lesions" and are regarded as potentially lethal events. A third lesion, O^6-methylguanine, base pairs efficiently with thymine and therefore is a very potent premutagenic lesion. A survey of the published sequences of eight mutant alleles generated by EMS treatment of *Arabidopsis* seeds indicated that all eight were G to A transition mutants, consistent with the hypothesis that O^6-alkylguanine is the major mutagenic lesion induced by EMS in this tissue (26, 78, 82)—though see also Reference 81 for an exception to this conclusion. Because methylated bases (particularly 7-methylguanine and 3-methyladenine) are generated by endogenous methylating agents, plants, animals, and microbes have developed specialized repair pathways to reverse or excise methylation damage.

The level of genome methylation is subject to environmental perturbation. Many microbes have developed an elaborate regulatory system, termed "the adaptive response," that enables them to enhance their capacity to repair methylation damage upon exposure to a lower "challenge" dose of alkylating agents (113). The chemical nature of the significant environmental and endogenous alkylating agents is still a subject of speculation (134), although certain bacteria, fungi, and algae have been shown to produce potent inducers of the adaptive response. It is possible that plants, especially plant roots, may experience a wide range of variation in the rate of alkylation damage and therefore may also have developed an adaptive response to this class of DNA damaging agents.

Oxidative Damage

A wide variety of oxidative damage products are induced in DNA by hydroxyl radicals, superoxide, and nitric oxide (23). Some of these damaged bases, including thymine glycol and its degradation product, urea, act as blocks to DNA synthesis but are not particularly mutagenic. Oxidation products of cytosine undergo an enhanced rate of deamination (via the hydrolytic reaction discussed above) to form mutagenic uracil derivatives. Perhaps the most significant premutagenic oxidized base is 8-hydroxyguanine, which base pairs with equal facility to A and C. In addition, the nucleotide 8-hydroxydGTP can be used as a substrate for DNA synthesis by DNA polymerase. Both human and *E. coli* cells produce an enzyme that specifically degrades this deoxynucleotide triphosphate to its monophosphate form, thereby preventing its incorporation into DNA (64, 75). Because bases are easily oxidized in vitro during standard DNA purification procedures, and because some oxidation products are inherently unstable, it is difficult to determine the spontaneous rate at which certain oxidized bases arise, persist, or are repaired in the genome. It should also be noted that the bases in an intact double helix are shielded from attack by hydroxyl radicals to a large degree by their stacked, interior conformation. For this reason, a relatively large fraction of oxidation damage occurs at the sugar phosphate backbone, leading to single-stranded breaks. Such nicks are generally repaired in an efficient and error-free fashion. Because the double helix is more likely to "breathe" (become transiently single-stranded) near a nick, the bases located near a nick are substantially more accessible to hydroxyl radical attack.

The major sources of activated oxygen in the cell are almost certainly the organelles; reactions in both the chloroplast (11) and the mitochondrion (138) frequently misdirect electrons to oxygen, generating superoxide. The plastid possesses a number of enzymatic and nonenzymatic defenses against superoxide, peroxide, singlet oxygen, and hydroxyl radicals (11), designed to capture free radicals before they can interact with critical cellular components such as the photosynthetic apparatus or the genome. These defenses can be overwhelmed during periods of stress when NADP, the electron acceptor for reduced ferredoxin, becomes limiting (3). Under these "photoinhibitory" (63) conditions the production of activated oxygen species may exceed the chloroplast's extensive scavenging capacity. In addition, because hydrogen peroxide can diffuse rapidly across the lipid bilayer, no cellular compartment is completely isolated from the reactive oxidative species produced during either respiration or the light reactions of photosynthesis.

Significant extracellular sources of activated oxygen might include air pollutants such as ozone (50, 68) or perhaps radicals produced by neighboring cells during the hypersensitive response (57). Very high levels of UV-B radiation can also induce oxidative damage in DNA (40); however, it is not clear whether the amount of oxidative damage induced by the relatively low levels of UV-B radiation in solar radiation is significant in comparison with the baseline level of oxidative damage produced by normally functioning organelles. It is important to note, however, that screens for UV-sensitive *Arabidopsis* mutants (13, 22, 41, 49) have employed unnaturally intense, brief doses of UV. A screen performed in this manner may also yield mutants specifically defective in the repair of oxidative damage.

Damage Induced by Ionizing Radiation

Ionizing radiation differs from UV radiation in its complete lack of target specificity. The probability of any component of the cell directly interacting with ionizing radiation depends simply on the mass fraction it makes up of the cell. For this reason, the most frequent primary target of ionizing radiation in actively metabolizing plant cells is water, and the majority of DNA damage induced by ionizing radiation probably results from interaction of DNA with hydroxyl radicals (139). Direct absorption of radiation by the sugar phosphate backbone can also generate a nick; the sensitization of the opposing, unnicked strand may result in an increased yield of double-strand breaks. Ionizing radiation is often used to generate chromosomal breaks, inversions, duplications, and translocations in plant stocks, but it should be noted that point mutations may also be generated by this type of mutagen as a result of oxidative damage to bases. A survey of mutations induced in irradiated *Arabidopsis* seeds suggests that ionizing radiation is a fairly reliable source of chromosomal rearrangements; of nine alleles analyzed at the Southern blot level, only one (125), a fast neutron-induced mutation at *GA1*, was found to have a "point-like" mutation (116, 140, 141).

UV-Induced Damage

The cyclobutane pyrimidine dimer (CPD) and the pyrimidine (6-4) pyrimidinone dimer (the 6-4 photoproduct) make up approximately 75% and 25%, respectively, of the UV-induced DNA damage products (72). The action spectrum for the induction of pyrimidine dimers in purified DNA follows the absorbance spectrum of DNA; dimers are induced most efficiently by radiation at approximately 260 nm, i.e. by radiation in the UV-C range. Although no biologically significant UV-C (λ < 280 nm) radiation is present at the earth's surface, the small amount of UV-B (280–320 nm) and much greater flux of UV-A (320–400 nm) present in sunlight are the single most important sources of epidermal DNA damage in plants or animals. The contribution of longer (UV-A) wavelengths of the solar UV spectrum to the overall load of dimers in both human and plant tissues is further enhanced by its greater ability to penetrate through the outermost layers of cells (95). Plants are thought to produce natural sunscreens, which selectively absorb photons in the UV-B and UV-A range, and flavonoid pigments are generally regarded as UV-absorbing agents. Evidence suggests that plants defective in the synthesis of anthocyanins are shielded from the growth-inhibiting effects of UV-B (59) and from the induction of DNA damage (123). Yet, some of the UV-protective effects observed in the chalcone isomerase-defective *tt5* mutant of *Arabidopsis* may result from its defect in the synthesis of sinapic acid esters, some of which are highly UV absorbent (17, 59).

The biological effects of pyrimidine dimers have been extensively studied in microbes and mammals. Like some of the DNA damage products discussed above, pyrimidine dimers have been shown to inhibit the progress of microbial and mammalian DNA polymerases and are not directly mutagenic. Mammalian RNA polymerase II has been shown to "stall" at both CPDs and 6-4 photoproducts (74, 92). Thus, in the absence of repair, a single pyrimidine dimer is sufficient to completely eliminate expression of a transcriptional unit. In addition, evidence suggests that the stalled mammalian RNA polymerase II remains bound to the site of the obstruction (27). Thus persisting lesions may actually reduce the overall concentration of free RNA polymerase, in addition to eliminating transcription of the gene in which they are located. Every pyrimidine dimer acts as a block to transcription and replication, while only a small fraction of dimers results in a mutation. For this reason, the inhibitory effects of UV on transcription and replication in plant epidermal tissues are probably more significant (in terms of plant growth) than its mutagenic effects are.

DNA REPAIR PATHWAYS

Direct Reversal of Damage

PHOTOREACTIVATION In some organisms the biological effects of UV radiation are significantly reduced by subsequent exposure to light in the blue or UV-A range of the spectrum, a phenomenon known as photoreactivation. The photoreactivating effects of visible light usually reflect the actions of photolyase enzymes. This class of enzyme binds specifically to cyclobutane pyrimidine dimers and, upon absorption of a photon of the appropriate wavelength

(350–450 nm), directly reverses the damage in an error-free manner. Microbial photolyases carry two prosthetic groups. One chromophore (either meth enyltetrahydrofolate or 8-hydroxy-5-deazaflavin) absorbs the photoreactivating light and transfers the energy to the other chromophore, a fully reduced flavin adenine dinucleotide (FAD). The excited $FADH^-$ then transfers an electron to the dimer, inducing its reversal (107). Once photolyase has bound to a cyclobutane dimer, the efficiency of photoreactivation is extremely high; approximately one dimer is split for every blue-light photon absorbed. Microbial photolyase genes have been cloned from a variety of bacteria and fungi, and their sequences display obvious homologies (143).

Evidence for the biological effects of photoreactivation in plants is complicated by the obvious detrimental effects of growing plants in the dark. This problem can be partially alleviated by the use of appropriate controls and of filters that absorb the shorter wavelengths required for photoreactivation (450 nm and under) while transmitting photons of longer photosynthetically active wavelengths. Photoreactivation results in the reversal of several UV-induced phenomena in plants, including mutagenesis, chromosome rearrangements (47), inhibition of growth, induction of flavonoid pigments (7), and unscheduled synthesis of DNA (48). Light-enhanced repair of dimers from total cellular DNA has been documented in tobacco, *Haplopappus gracilis* (132), ginkgo (133), *Chlamydomonas* (120), *Arabidopsis* (19, 87), and wheat (130), and the action spectrum for reversal of CPDs by partially purified maize and *Arabidopsis* photolyases has been shown to be similar to that of *E. coli,* a methenyl tetrahydrofolate-type photolyase (46, 87).

The cyclobutane dimer photolyase activities of higher plants are known to be regulated by visible light. The CPD photolyase activity of the common bean is induced twofold by a brief exposure to red light; this effect is partially reversed by subsequent exposure to far red light, suggesting that the induction is phytochrome mediated (55). Similarly, the light-dependent repair of CPDs in *Arabidopsis* requires exposure to visible light prior to as well as after UV irradiation (19). Thus the repair capacity of the plant depends on the quality and timing, as well as quantity, of light in its environment. The influence of the environment on the steady-state level of pyrimidine dimers, the rate of induction of dimers, and the rate of photoreactivation of dimers has been illustrated in recent work on alfalfa (127). Researchers found that seedlings grown in an essentially UV-free environment had the same steady-state levels of cyclobutane dimers (approximately 6 dimers/megabase) as seedlings grown under unfiltered sunlight. In addition, a given dose of UV was found to induce twofold more dimers in the seedlings grown under artificial light, and these seedlings also had a lower rate of photoreactivation of CPDs than the identical strain grown under natural light. Thus both the UV transparency and the repair capacity of higher plants is altered in response to the ambient levels of UV and visible radiation. Similar effects have been observed in experiments that directly measure the effects of enhanced UV-B on yield (14, 30, 71).

A putative plant photolyase was cloned from wild mustard (6) by probing a cDNA library with a degenerate oligonucleotide specific to a conserved region of the microbial cyclobutane dimer photolyases. The clone, labeled SA-*phr1,* displays significant stretches of similarity to previously cloned microbial photolyases. Moreover, the cDNA hybridizes to an mRNA that is strongly regulated by light; seedlings grown in the dark express low levels of the mRNA, whereas light-grown seedlings express the mRNA at high levels. The protein encoded by this cDNA was expressed in *E. coli* and found to bind, like the *E. coli* photolyase, both FAD and methenyltetrahydrofolate (65). The *E. coli*–expressed mustard protein did not, however, display any photolyase activity; it neither enhanced the UV resistance of a photolyase-defective host strain nor did it split thymine dimers in vitro. For this reason, the authors concluded that the SA-*phr1* clone represents a blue light photoreceptor rather than a photolyase. This conclusion gains support because a constitutively expressed *Arabidopsis* gene (*HY4*) known to be involved in the blue light response was also found to have a region of substantial homology to the microbial photolyases (1). The *HY4* gene product, when expressed in *E. coli,* also binds both FAD and methenyltetrahydrofolate but fails to exhibit any photoreactivating activity (65).

Although the failure to find enzymatic activity in a heterologously expressed gene product is not definitive proof that a protein would lack photoreactivating activity if expressed *in planta,* it should be noted that the gene was cloned on the basis of its homology to microbial photolyases. A second class of "metazoan" photolyases, currently cloned from fish, insects, and marsupials, apparently has little sequence similarity to the more thoroughly studied microbial enzymes (146). It is conceivable that the plant photolyase is more closely related to the metazoan proteins or perhaps represents yet another class of photolyases.

In contrast with findings on microbes and mammals, experimental evidence suggests that *Arabidopsis* may have a light-dependent pathway for the repair of pyrimidine (6-4) pyrimidinone photoproducts (19). Unlike its CPD-specific photolyase activity, this repair pathway does not require induction by prior exposure to visible light. It also does not require the *UVR1* gene product (13), which is essential for dark repair of 6-4 photoproducts. Thus *Arabidopsis* has the ability to photoreactivate both of the major UV-induced DNA damage products. This ability probably extends to other plants; exposure to visible light greatly enhances the rate of removal of 6-4 photoproducts from the DNA of wheat seedlings (130). Although photoreactivation of 6-4 photoproducts has not been observed in microbial or most animals tested, a 6-4 photoproduct-specific photolyase activity has been partially characterized in extracts of

Drosophila larvae (53, 131). If 6-4 photolyase activity exists in organisms as distantly related as plants and insects, it is important to determine whether the activity is universal. The discovery of the 6-4 photolyase is particularly significant in that the biological effects of photoreactivation have previously been ascribed to the alleviation of the toxic effects of cyclobutane dimers alone.

LACK OF EVIDENCE FOR O^6-METHYLGUANINE METHYLTRANSFERASE As described above, O^6-alkylguanine base pairs directly with thymine and therefore is directly mutagenic. For this reason, most organisms produce a protein, O^6-methylguanine methyltransferase (MGMT), that removes the methyl group from the lesion, transferring it to a serine residue on the protein itself. Because no mechanism exists for the demethylation of this protein, this "enzyme" is permanently inactivated by the reaction and is sometimes termed a "suicide" DNA methyltransferase. MGMT has been identified in and cloned from bacteria (24, 91, 112), yeast (142), and mammals (43, 103, 129). No evidence for its existence in plants has been established. A careful search for the activity was performed in *Chlamydomonas,* with negative results (32). One of the two *E. coli* copies of MGMT, the *ada* gene, was recently transformed into tobacco callus, and resistance to the growth-inhibitory effects of methylating agents was enhanced in the transformed callus (2). Activity in plants grown from the callus was poor, however (135), making it difficult to determine whether the expression of the *ada* gene had an antimutator effect. It is difficult, if not impossible, to provide definitive proof that an enzymatic activity does not exist in a particular organism. The presence of MGMT in yeast was in doubt until the gene was cloned and sequenced (142). In fact, the existence of a bona fide photolyase in placental mammals is still a matter of some debate (60, 102).

Excision Repair

In contrast with photoreactivation, dark repair pathways do not directly reverse DNA damage but instead replace the damaged DNA with new, undamaged nucleotides. These "excision repair" pathways fall into two major categories: base excision repair and nucleotide excision repair.

BASE EXCISION REPAIR Base excision repair involves the removal of a single damaged base through the action of one of many lesion-specific glycosylases, which leaves the DNA sugar-phosphate backbone intact. The resulting abasic sites are then recognized by an apurinic/apyrimidinic (AP) endonuclease or AP lyase, which nicks the backbone of the DNA at the AP site (105). The nicked DNA is then restored to its original sequence through the combined actions of exonucleases, a repair polymerase, and DNA ligase. Recent evidence has suggested that the repair polymerase itself, polβ, possesses the ability to excise the 5′ deoxyribose phosphate residue that is generated by the combined actions of DNA glycosylases and class II AP endonucleases (66).

URACIL GLYCOSYLASE As described above, uracil accumulates in the genome at a rate of approximately 100 lesions per cell per day (for a genome size of 3×10^9 bp). Because this lesion is directly mutagenic, all living things probably produce a uracil glycosylase. The crystal structure of uracil glycosylase from nonplant sources has recently been solved and suggests that the protein actually binds to a uracil base that has swiveled out to the exterior of the double helix (76, 109). Although a gene corresponding to this protein has not yet been identified in plants, the activity has been purified from several plant sources (10, 128). There is some evidence that this activity is downregulated by as much as 20-fold in fully differentiated cells (38).

3-METHYLADENINE GLYCOSYLASE 3-Methyladenine is a noncoding lesion that, like uracil, occurs spontaneously at a significant rate. 3-Methyladenine glycosylases have been identified in bacteria, yeast, mammals, and *Arabidopsis* and vary in their substrate-specificity. *E. coli* expresses two 3-methyladenine glycosylases. The product of the *tag* gene is highly specific for 3-methyladenine, whereas the product of the *alkA* gene has a broad substrate specificity, cleaving the N-glycosylic bond at 7-methylguanine, 3-methylguanine, O2-methylthymine, and O2-methylcytosine, as well as 3-methyladenine (29, 51). The biological effects of an *alkA*- mutation can be suppressed by the artificial overexpression of the *tagA* gene (145), which suggests that these additional substrates do not play an important role in the lethality induced by methylating agents. The *tag* and *alkA* genes share no significant homology. All of the cloned higher eukaryote 3-methyladenine glycosylases, including one from *Arabidopsis* (108), have been isolated via complementation of the MMS-sensitivity of the *E. coli* double mutant (8, 18; summarized in 28, 80). Although the mammalian genes have a high degree of homology with one another, the overall transkingdom homology is fairly weak.

UV-ENDONUCLEASES Glycosylases and endonucleases specific for cyclobutane dimers have been observed in bacteria and bacteriophage and have been useful as diagnostic agents for the assay of UV-induced damage (31). True eukaryotic UV-endonucleases that recognize both cyclobutane dimers and 6-4 photoproducts and that generate an incision immediately 5′ to the lesion were recently identified in *Saccharomyces pombe* and *Nuerospora crassa* (12, 144). Several groups have described the partial characterization of endonucleolytic activities obtained from plant extracts that exhibit some specificity for UV-irradiated DNA (25, 77, 136). Some of these activities are particularly intriguing in that they do not appear to recognize CPDs, which suggests that the recognition site may be the 6-4 photoproduct. In only one case (the endonuclease SP purified from spinach) has a plant UV-specific endonuclease been substantially purified and characterized; this enzyme was suggested to be a single-stranded endonuclease, which apparently recognizes a single-stranded region that is induced by 6-4 photoproducts but not by CPDs (124).

NUCLEOTIDE EXCISION REPAIR (NER) NER differs from base excision repair in two ways: The spectrum of DNA damage products recognized by the repair complex is remarkably wide, and the repair complex initiates removal of the damage by generating nicks on the damaged strand. These nicks occur at a specific distance both 5′ and 3′ of the lesion, which is then excised as an oligonucleotide through the action of a helicase. The excision repair complex will, with varying efficiencies, cleave almost any abnormality in DNA structure--from very small, nondistorting lesions (such as O^6-methylguanine or abasic sites) to very bulky adducts (thymine-psoralen adducts or pyrimidine dimers). It is not likely that the cell produces a specific repair protein for every possible lesion, and nucleotide excision repair may exist, in part, to cope with the unexpected. As discussed above, placental mammals are generally thought to lack photolyase, and in mammalian cells NER is apparently the sole pathway for the repair of bulky adducts (106). It should be kept in mind, however, that most mammalian repair studies, for obvious reasons, are performed in tissue culture rather than in actual skin. It is possible, and even reasonable, that only those tissues that are normally exposed to sunlight express a specific repair pathway for UV-induced damage. Pyrimidine dimers may indeed represent an "unexpected" class of lesions to most types of cultured cells.

Light-independent ("dark") repair of CPDs, which might represent either NER or base excision repair, has been observed in several plant species. Early studies, previously reviewed by McLennan (67), involved the use of a germicidal lamp (UV-C, 254 nm) to irradiate cell suspension cultures or protoplasts (for uniformity of UV penetration) producing high concentrations of CPDs. The disappearance of dimers from the nuclear fraction was measured by hydrolyzing the nuclear DNA and assaying, via thin layer chromatography, the fraction of total thymidine bases that were present as dimers. The appearance of excised dimers in the cytosol, indicative of excision repair, was followed using similar techniques. The rate of dark repair of CPDs was found to vary widely between plant species, with high rates of repair demonstrated for carrot suspension cultures (44) and protoplasts of carrot, *Haplopappus*, petunia, and tobacco (45), whereas excision repair of CPDs was undetectable in cultured soybean cells (100). It should be stressed, however, that photoreactivation is generally a more rapid and efficient pathway for the excision of UV-induced dimers and probably provides the bulk of the protection against UV-induced DNA damage. Excision repair may, however, be essential for the repair of minor, nondimer, UV-induced photoproducts.

Recently, more sensitive techniques have been developed for the detection of UV-induced damage, including the use of lesion-specific antibodies (73), the T4 endonuclease/alkaline sucrose gradient assay (33, 87), and an exquisitely sensitive gel-electrophoresis-based method involving the extraction of intact DNA, followed by cleavage of the DNA at CPDs and the quantitative assay of various size classes of single-stranded DNA sizes to arrive at an average frequency of dimers (96). These technical advances have enabled investigators to use relatively low doses of UV to study repair in intact plants. Dark repair rates for CPDs have been assayed in 5-day-old *Arabidopsis* seedlings, where no significant repair of CPDs was detectable in 24 h, although repair of 6-4 photoproducts was efficient (13). In contrast, rapid dark repair of CPDs was observed in alfalfa (97), and an intermediate level of repair was detected in wheat seedlings (130). While these plants may actually differ in their inherent capacity for dark repair, this disparity might also result from the differing experimental conditions employed. It has recently been demonstrated that excision repair in the alfalfa seedling, while efficient and easily detectable at high levels of initial UV damage, is undetectable at lower initial damage levels (97). Extremely high doses of UV can also inhibit repair in plant tissues (44). Thus, while laboratory studies are essential for the determination of the biochemical basis of repair, caution must be used in extrapolating these results to make predictions concerning UV resistance in the field, where growth conditions, the plant tissues employed, and the levels of DNA damage induced by sunlight can radically affect both the extent of damage and the rate of repair.

Double-Strand Break Repair

Double-strand breaks (DSBs) are generated in plant DNA through a variety of mechanisms: spontaneous oxidative damage to the genome, treatment with ionizing radiation, the formation of a dicentric chromosome, cleavage with artificially introduced nucleases, and (perhaps) excision of transposable elements. To a plant molecular biologist, the most important source of DSBs is probably the recombinant DNA with which the researcher hopes to transform the plant cell. Because the DNA sequences near the ends of these breaks are rapidly degraded, DSBs generally expand into gaps that cannot simply be religated to restore the original sequence. Unlike recombination-proficient yeast cells, which will virtually always repair DSBs via homologous recombination, the cells of higher plants behave very much like those of most mammalian tissues; DSBs are simply rejoined, end to end, in what appears to be a random fashion. This end-to-end joining process is sometimes termed "illegitimate recombination." Analysis of repaired DSBs generated by ionizing radiation (116), T-DNA insertion (35, 36, 117), and transposable element excision [most recently (111)] indicates that a bias does exist toward the formation of joint molecules at regions of homology. However, this homology is extremely limited (2–5 bases) and is probably simply the result of enhanced stability of the joint for ligation, rather than the sort of extensive homology search associated with homologous recombination. In addition, DNA junctions are often characterized by multiple recombination events, such as an inversion of substantial portions of the target site (116), and novel sequences that may represent template switching by a repair polymerase (104), the transient formation of a covalently closed hairpin loop (20), or addition of nontemplate nucleotides. The natural propensity of mammalian and higher plant cells to incorporate exogenous DNAs into random rather than homologous sites [the fraction of events from homologous recombination among all integration events is approximately $1/10^{-4}$ (90, 93)] is particularly vexing to the molecular biologist, because true gene replacement is very infrequent in these systems (with the exception, for unknown reasons, of mouse embryonic stem cells).

Very little is known about the genes required for illegitimate recombination in either plants or animals. X-ray-sensitive mutant animal cell lines exist that are defective in the repair of DSBs. Some of the genes that complement these defects have been cloned [reviewed in (31)]. *scid* mice, which are severely immunodeficient as a result a defect in V(D)J recombination, are also X-ray sensitive and fail to incorporate exogenous DNAs (42). Mutants of *Arabidopsis* specifically sensitive to the growth-inhibitory effects of ionizing radiation have also been isolated (22), and some UV-sensitive *Arabidopsis* mutants also display sensitivity to ionizing radiation (49). Although these mutants have not been directly assayed for the ability to repair DSBs, recent evidence has shown that a subclass of these mutants are defective in the stable incorporation of T-DNA into their genome (122). This result is particularly significant for two reasons. First, it suggests that the mutants are indeed defective in end-to-end joining. Second, it provides the first direct evidence for the role of host enzymes in T-DNA transformation. The availability of plant mutants defective in illegitimate recombination will enable us to better understand and perhaps modify this process.

REPAIR OF THE ORGANELLAR GENOMES

Any proteins present in the organelle are either synthesized there or are specifically transported into the organelle. For this reason, the presence of a repair activity in the nucleus does not imply that the activity is present in the organelle; the presence or absence of organellar repair activities has to be established independently. For example, Chinese hamster ovary cells express only a subset of their repair activities in their mitochondria; methylated purines and interstrand crosslinks are removed efficiently, but dimer and intrastrand crosslinks are not (56). This suggests that some types of base excision repair function in mitochondria, but the more general nucleotide excision repair mechanism functions only in the nucleus. Neither the mitochondrial nor the plastid genomes encode any DNA repair proteins. Do organisms possess multiple, nuclearly encoded sets of certain repair genes, one for each genetic compartment, or are some repair proteins targeted to more than one compartment? *S. cerevisiae* has five copies of the *MSH* gene required for mismatch repair; one of the gene products is targeted to the mitochondrion (99). In contrast, *S. cerevisiae's PHR1* photolyase appears to photoreactivate both the nuclear and mitochondrial genomes, and its 5′ end can direct the transport of a *lacZ′* fusion protein to the mitochondrion (147). Similarly, the nuclearly encoded human uracil glycosylase is directed to both the mitochondrion and the nucleus (119). It is conceivable that some plant repair proteins might possess a unique targeting signal that facilitates their transport into all three compartments.

Chlamydomonas is known to photoreactivate both its nuclear and plastid genomes. The *phr1* mutant has been shown to be defective in the photoreactivation of the nuclear genome but not the plastid genome. This suggests that *Chlamydomonas* produces two distinct photolyases. Unfortunately, no studies, to my knowledge, have been published documenting repair of any kind in any higher plant organellar genome. A homologue of the *E. coli recA* gene has been cloned from *Arabidopsis* (16), and it encodes at its amino terminus a conserved recognition site for the stromal processing protease. Southern blot analysis using this cDNA as a probe suggests that there is more than one copy of this gene encoded by the *Arabidopsis* nucleus (9). This chloroplast *recA* homologue may play a role in recombinational "repair" (see below). Several other *Arabidopsis* cDNAs, cloned on the basis of their ability to partially complement the UV-sensitive and recombination-defective phenotype of *E. coli* repair-defective mutants, also appear to possess chloroplast-targeting sequences (88, 89).

Sequence analysis of the two plant organellar genomes suggests that they evolve by different mechanisms and at different paces (86). It will be interesting to determine whether some of these differences can be ascribed to differences in the mode and efficiency of their DNA repair pathways.

DNA Damage Tolerance Pathways

The excision repair pathways described above can all be divided into two steps: First the damaged base is removed, and then the undamaged strand is used as a template to fill the resulting gap. These repair pathways are essentially error free. If, however, a cell undergoes DNA replication before repair is complete, a "noninformational" DNA damage product, such as a pyrimidine dimer, will act as a block to DNA replication. DNA polymerase will normally reinitiate synthesis 3′ to the lesion, but a gap remains in the newly synthesized daughter strand at the site opposite the DNA damage product. The resulting incompletely synthesized chromosome will, as a result, no longer act as a substrate for excision repair because the sister strand is no longer available as a template. Although one would expect the persistence of such a lesion to be lethal, a variety of organisms have been shown to undergo repeated rounds of DNA synthesis and cell division in spite of the continued presence of noninformational lesions. At least two independent pathways permitting the completion of replication of damaged chromosomes exist; these are dimer bypass and recombinational "repair." These pathways are sometimes collectively termed "postreplication repair" but are better thought of as "damage tolerance pathways" because they do not involve DNA repair but instead help the cell to survive despite persisting damage.

Dimer Bypass

Although noninformational lesions normally act as blocks to DNA replication, some organisms produce a modified polymerase that is capable of performing translesion synthesis. For example, the *E. coli umuC,D* gene products are thought to bind to DNA polymerase and relax its normally stringent requirements for the stable insertion of a new base, thereby enabling it to perform translesion synthesis (98a). The altered polymerase generally installs adenine residues across from noninformational DNA damage products. As a result, UV-induced thymine dimers are not mutagenic, but cytosine-containing dimers are. Similarly, because UV radiation induces primarily pyrimidine dimers and because the *umuC,D* gene products are required for translesion synthesis, strains with defects in these genes display an enhanced sensitivity to the lethal effects of UV while completely lacking a mutagenic response to this DNA damaging agent (52). Translesion synthesis permits DNA replication (and therefore enhanced survival) at the expense of accuracy. Because of their inherent potential for generating mutations, the *umuC,D* gene products are expressed only when the cell has been exposed to a substantial dose of DNA damaging agents (5). Similarly, the *REV3* gene of *S. cerevisiae* produces a nonessential, mutagenic polymerase with a specialized ability to synthesize DNA using damaged templates (118). Humans may produce a modified polymerase with a similar tendency to install A's at pyrimidine dimers: sunlight-induced mutations in humans occur mainly at dipyrimidines and are primarily C to T or CC to TT transversions (148).

Whether mutagenesis in plants occurs as a result of lesion bypass remains to be seen. UV radiation is an excellent source of noninformational DNA damage products, and the spectrum of mutations induced by UV could provide insights into the means by which plants tolerate the persistence of DNA damage. Unfortunately, few UV-induced mutations have been generated, and to my knowledge none has been sequenced. Because the plant's germline is shielded from UV during virtually all stages of growth, studies of UV-induced mutations in higher plants have been limited to the mutagenic effects of UV irradiation of pollen. Mutagenesis of pollen has the advantage of enabling the investigator to observe the induction of mutations such as large deletions that might otherwise be nontransmissable as a result of selection during the post-meiotic mitoses and growth of the pollen tube. In fact, UV-induced mutations in maize pollen were generally found to be nontransmissable or to have reduced transmission beyond the first generation, which indicates that UV-induced lesions result in large deletions rather than point mutations (79). This finding suggests that translesion synthesis (which induces point mutations) rarely occurs during repair in pollen or during the early stages of embryonic development and that UV-induced DNA damage results in chromosome breaks and/or recombination. However, one must bear in mind that large chromosomal deletions, which result in the simultaneous loss of many genes, are simply easier to score as mutations than are single base changes, the majority of which fail to affect gene function. It is also possible that dimer bypass is preferentially employed in somatic cell lines (where mutagenesis is relatively inconsequential) but is not expressed during the critical last stage of pollen development, when mutations can no longer be eliminated through diplontic selection (54). Because of its potential role in the creation of genetic diversity (as well as in UV tolerance), more research on translesion synthesis is needed in both plants and animals.

Recombinational Repair

In contrast with lesion bypass, recombinational "repair" fills the daughter-strand gap by transferring a preexisting complementary strand from a homologous region of DNA to the site opposite the damage. As in the dimer bypass mechanism, the lesion is left unrepaired, but the cell manages to get through another round of replication, and the damaged base is now available as a substrate for excision repair. When the complementary strand is obtained from the newly replicated sister chromatid, the resulting "repair" is error free. If the information is obtained from the homologous chromosome, or perhaps from a similar DNA sequence elsewhere in the genome, there is a possibility that a change will be generated in the gene's sequence either via gene conversion or through the formation of deletions, duplications, and translocations. While UV irradiation has been shown to induce chromosomal rearrangements in plants (79), including homologous intrachromosomal recombination events (94), it remains to be seen whether the filling of daughter-strand gaps via homologous recombination is a significant UV tolerance mechanism in plants. UV radiation has been shown to induce previously quiescent transposable elements (137); it is possible that this effect is the result of chromosomal rearrangements or other repair-related activities. Conversely, some UV-induced mutations may result from the activation of transposable element activities.

Other Damage Tolerance Mechanisms

The two pathways described above permit the cell to replicate in spite of the persistence of dimers but do not reduce the deleterious effects of DNA damage on transcription. One of the most interesting recent developments in the field of DNA repair is the discovery that the template strand employed for transcription is repaired more rapidly than the untranscribed strand or untranscribed regions (39). In fact, the relationship between repair and transcription is particularly intimate--not only are some repair proteins physically coupled to RNA polymerase, but a subset of those proteins, notably the TFIIH complex, actually act independently both as transcription factors and as repair complexes (110). By selectively removing damage from actively transcribed units, targeted repair substantially reduces the toxic effect of UV. Although preferential repair of transcribed strands has been shown to exist in mammals (70), yeast (126), and *E. coli* (69), this phenomenon has not yet been investigated in plants.

Lesions opposite a daughter-strand gap are particularly problematic because the damage cannot be repaired via excision repair. If the cell is unfortunate enough to not only replicate its damaged DNA but to also undergo cell division, then the information at the site of the lesion is permanently lost because no sister chromatid is available to take part in recombinational repair. For this reason, some organisms are capable of detecting genome damage and will delay cell division until the integrity of the genome is restored. Yeasts (*S. cerevisiae, S. pombe*) damaged in G1 or S phase will cease further DNA synthesis, while G2 cells will delay mitosis (15). Cells defective in genes required for the G2 "checkpoint" will proceed with cell division in spite of the presence of gapped DNA and will therefore exhibit an increase in sensitivity to both the toxic and mutagenic effects of DNA damaging agents. Similar "checkpoint" responses to DNA damage have been observed in other fungi and in mammals (58).

Several labs are currently in the process of isolating *Arabidopsis* mutants that are hypersensitive to the growth-inhibiting effects of DNA damaging agents (13, 22, 41, 49). Unfortunately, few of these mutants have been characterized in terms of their repair capabilities. Although many of these UV-sensitive mutants will have demonstrable defects in repair, undoubtedly some fraction will display normal rates of repair. This second class of mutants is a particularly interesting one, as it may include mutants defective in damage tolerance. Thus a screen for UV-sensitivity might yield mutants defective in mutagenesis, recombination, transcription, and cell cycle control.

SUMMARY

Plants are now known to possess many of the same repair pathways as other eukaryotes; UV-induced pyrimidine dimers (both 6-4 photoproducts and cyclobutane dimers) can be removed via photoreactivation or through excision repair, and certain lesion-specific glycosylases have been shown to exist in higher plants. What may be more important is that researchers have proven that the currently available assays for repair can be applied to plants. The DNA of higher plants can be radiolabeled in vivo and can be extracted in the very intact state required for the assay of DNA damage (and its repair) induced at a very low frequency. The feasibility of employing both classical and molecular genetic approaches to DNA repair has been established in plants; at least one *Arabidopsis* mutant defective in the repair of UV-induced lesions has been isolated, and at least two radiation-sensitive mutants appear to be defective in the rejoining of double-strand breaks. Several repair-related genes have been cloned from *Arabidopsis* either via complementation of repair-defective mutants from other species or by probing for the presence of homologues to known repair genes.

Many repair-related issues remain unexplored. Although photoreactivation is undoubtedly the plant kingdom's major line of defense against UV-induced damage, the molecular nature of the two plant photolyase genes is unknown. Virtually nothing is known about organellar repair or organellar damage tolerance pathways, although the identification of an *Arabidopsis* plastid *recA* homologue should shed some light on this process. Nothing is known, in any plant species, about the mismatch repair process. We have yet to identify any DNA damage tolerance pathways in plants. Our understanding of the molecular mechanisms of both illegitimate and homologous recombination is still in its infancy. Many of these questions could be easily addressed with currently available technologies. A wide range of useful tools have been developed by researchers working on microbes and animals. These include lesion-specific antibodies, repair-defective mutants, and a multitude of cloned repair genes from a wide range of species. All of these tools can be directly applied to the study of repair and repair-related processes in plants.

The study of DNA repair and DNA damage tolerance processes in plants touches on a surprisingly wide range of subjects, including not only the effects of DNA damaging agents on plant growth and mutagenesis but also transcription, cell cycle control, and both homologous and illegitimate recombination. It also has applications beyond mutagenesis; an understanding of DNA transactions in plants is essential if we hope to progress beyond the relatively crude and haphazard level of "genetic engineering" currently available to both basic and applied plant geneticists.

ACKNOWLEDGMENTS

I thank Dr. Igor Vizir for his help in identifying published sequences of mutant alleles. I also thank several members of my lab (C.-Z. Jiang, J.-J. Chen, and N. Bence) for their very useful advice on the style and organization of this manuscript. Finally, I thank both the NSF and the USDA for their generous support of research in this area.

Literature Cited

1. Ahmad M, Cashmore AR. 1993. The *HY4* gene of *Arabidopsis thaliana* encodes a protein with characteristics of a blue-light receptor. *Nature* 11:162–66
2. Angelis K, Briza J, Satava J, Skakal I, Veleminsky J, et al. 1992. Increased resistance to the toxic effects of alkylating agents in tobacco expressing the *E. coli* DNA repair gene *ada*. *Mutat. Res.* 273: 271–80
3. Asada K, Takahasi M. 1987. Production and scavenging of active oxygen in photosynthesis. In *Photoinhibition,* ed. DJ Kyle, CB Osmond, CJ Arntzen. Amsterdam: Elsevier Sci.
4. Ashraf M, Bray CM. 1993. DNA synthesis in osmoprimed leek (*Allium porrum* L.) seeds and evidence for repair and replication. *Seed Sci. Res.* 3:15–23
5. Bagg A, Kenyon CJ, Walker GC. 1981. Inducibility of a gene product required for UV and chemical mutagenesis. *Proc. Natl. Acad. Sci. USA* 78:5749–53
6. Batschauer A. 1993. A plant gene for photolyase: an enzyme catalyzing the repair of UV-light induced DNA damage. *Plant J.* 4:705–9
7. Beggs CJ, Stolzer-Jehle A, Wellmann E. 1985. Isoflavonoid formation as an indicator of UV stress in bean (*Phaseolus vulgaris* L.) leaves. *Plant Physiol.* 79: 630–34
8. Berdal KG, Bjøras M, Bjelland S, Seeberg E. 1990. Cloning and expression in *E. coli* of a gene for an alkylbase DNA glycosylase from *S. cerevisiae:* a homolog to the bacterial *alkA* gene. *EMBO J.* 9:4563–68
9. Binet MN, Osman M, Jagendorf AT. 1993. Genomic sequence of a gene from *Arabidopsis thaliana* encoding a protein homolog of *E. coli recA*. *Plant Phys.* 103:673–74
10. Bones AM. 1993. Expression and occurrence of uracil-DNA glycosylase in higher plants. *Phys. Plant.* 88:682–88
11. Bowler C, Van Montagu M, Inzé D. 1992. Superoxide dismutase and stress tolerance. *Annu. Rev. Plant Phys. Plant Mol. Biol.* 43:83–116
12. Bowman KK, Sidik K, Smith CA, Taylor J-S, Doetsch PW, Freyer GA. 1994. A new ATP-dependent DNA endonuclease from *S. pombe* that recognizes cyclobutane pyrimidine dimers and 6-4 photoproducts. *Nucleic Acids Res.* 22:3026–32
13. Britt AB, Chen J-J, Wykoff D, Mitchell D. 1993. A UV-sensitive mutant of *Arabidopsis* defective in the repair of pyrimidine-pyrimidinone (6-4) dimers. *Science* 261: 1571–74
14. Caldwell MM, Flint SD, Searles PS. 1994. Spectral balance and UV-B sensitivity of soybean: a field experiment. *Plant Cell Environ.* 17:267–76
15. Carr AM. 1994. Radiation checkpoints in model systems. *Int. J. Radiat. Biol.* 66: S133–39
16. Cerutti H, Osman M, Grandoni P, Jagendorf AT. 1992. A homolog of *E. coli* RecA protein in plastids of higher plants. *Proc. Natl. Acad. Sci. USA* 89:8068–72
17. Chapple C, Vogt T, Ellis BE, Somerville CR. 1992. An arabidopsis mutant defective in the general phenylpropanoid pathway. *Plant Cell* 4:1413–24
18. Chen J, Derfler B, Maskati A, Samson L. 1989. Cloning a eukaryotic DNA glycosylase repair gene by the suppression of a DNA repair defect in *E. coli*. *Proc. Natl. Acad. Sci. USA* 86:7961–65
19. Chen J-J, Mitchell D, Britt AB. 1994. A light-dependent pathway for the elimination of UV-induced pyrimidine (6-4) pyrimidinone photoproducts in *Arabidopsis thaliana*. *Plant Cell* 6:1311–17
20. Coen ES, Robbins TP, Almeida J, Hudson

A, Carpenter R. 1989. Consequences and mechanisms of transposition in *Antirrhinum majus*. In *Mobile DNA,* ed. DE Berg, MM Howe, pp. 413–36. Washington, DC: Am. Soc. Microbiol.

21. Dandoy E, Schyns R, Deltour R, Verly WG. 1987. Appearance and repair of apurinic/apyrimidinic sites in DNA during early germination of *Zea mays*. *Mutat. Res.* 181: 57–60
22. Davies C, Howard D, Tam G, Wong N. 1994. Isolation of *Arabidopsis thaliana* mutants hypersensitive to gamma radiation. *Mol. Gen. Genet.* 243:660–65
23. Demple B, Harrison L. 1994. Repair of oxidative damage to DNA—enzymology and biology. *Annu. Rev. Biochem.* 63: 915–48
24. Demple B, Sedgwick B, Robins P, Totty N, Waterfield MD, Lindahl T. 1985. Active site and complete sequence of the suicidal methyltransferase that counters alkylation mutagenesis. *Proc. Natl. Acad. Sci. USA* 82:2688–92
25. Doetsch PW, McCray WH, Valenzula MRL. 1989. Partial purification and characterization of an endonuclease from spinach that cleaves ultraviolet light-damaged duplex DNA. *Biochim. Biophys. Acta* 1007:309–17
26. Dolferus R, van den Bossche D, Jacobs M. 1990. Sequence analysis of two null-mutant alleles of the single Arabidopsis *Adh* locus. *Mol. Gen. Genet.* 224:297–302
27. Donahue BA, Yin S, Taylor JS, Reines D, Hanawalt PC. 1994. Transcript cleavage by RNA polymerase II arrested by a cyclobutane pyrimidine dimer in the DNA template. *Proc. Natl. Acad. Sci. USA* 91: 8502–6
28. Engelward BP, Boosalis MS, Chen BJ, Deng Z, Siciliano MJ, Samson LD. 1993. Cloning and characterization of a mouse 3-methyladenine 7-methylguanine 3-methylguanine DNA glycosylase whose gene maps to chromosome 11. *Carcinogenesis* 14:175–81
29. Evensen G, Seeberg E. 1982. Adaptation to alkylation resistance involves the induction of a DNA glycosylase. *Nature* 296:773–75
30. Fiscus EL, Booker FL. 1995. Is increased UV-B a threat to crop photosynthesis? *Photosynth. Res.* 43:81–92
31. Friedberg EC, Walker GC, Siede W. 1995. *DNA Repair and Mutagenesis.* Washington, DC: Am. Soc. Microbiol. 698 pp.
32. Frost BF, Small G. 1987. The apparent lack of repair of O6-methylguanine in nuclear DNA of *Chlamydomonas reinhardtii*. *Mutat. Res.* 181:31–36
33. Ganesan AK, Smith CA, van Zeeland AA. 1981. Measurement of the pyrimidine dimer content of DNA in permeabilized bacterial or mammalian cells with endonuclease V of bacteriophage T4. In *DNA Repair,* ed. EC Freidberg, PC Hanawalt, 1A:89–98. New York: Dekker
34. Gentil A, Margot A, Sarasin A. 1984. Apurinic sites cause mutations in simian virus 40. *Mutat. Res.* 129:141–47
35. Gheysen G, Van Montagu M, Zambryski P. 1987. Integration of Agrobacterium tumefaciens transfer DNA (T-DNA) involves rearrangements of target plant DNA sequences. *Proc. Natl. Acad. Sci. USA* 84: 6169–73
36. Gheysen G, Villarroel R, Van Montagu M. 1991. Illegitimate recombination in plants: a model for T-DNAS integration. *Genes Dev.* 5:287–97
37. Gruenbaum Y, Naveh-Many T, Cedar H, Razin A. 1981. Sequence specificity of methylation in higher plant DNA. *Nature* 292:860–62
38. Gutierrez C. 1987. Excision repair of uracil in higher plant cells: uracil-DNA glycosylase and sister-chromatid exchange. *Mutat. Res.* 181:111–26
39. Hanawalt PC. 1994. Transcription-coupled repair and human disease. *Science* 266: 1957–58
40. Hariharan PV, Cerutti PA. 1977. Formation of products of the 5,6-dihydroxydihydrothymine type by ultraviolet light in HeLa cells. *Biochemistry* 16:2791–95
41. Harlow GR, Jenkins ME, Pittalwala TS, Mount DW. 1994. Isolation of *uvh1,* an Arabidopsis mutant hypersensitive to ultraviolet light and ionizing radiation. *Plant Cell* 6:227–35
42. Harrington J, Hsieh C-L, Gerton J, Bosma G, Lieber MR. 1992. Analysis of the defect in DNA end joining in the murine *scid* mutation. *Mol. Cell. Biol.* 12:4758–68
43. Hayakawa H, Koike G, Sekiguchi M. 1990. Expression and cloning of complementary DNA for a human enzyme that repairs O6-methylguanine in DNA. *J. Mol. Biol.* 213: 739–47
44. Howland GP. 1975. Dark-repair of ultraviolet-induced pyrimidine dimers in the DNA of wild carrot protoplasts. *Nature* 254:160–61
45. Howland GP, Hart RW. 1977. Radiation biology of cultured plant cells. In *Applied Aspects of Plant Cell Tissue and Organ Culture,* ed. J Reinert, YPS Bajaj, pp. 731–89. Berlin: Springer-Verlag
46. Ikenaga M, Kondo S, Fujii T. 1974. Action spectrum for enzymatic photoreactivation in maize. *Photochem. Photobiol.* 19: 109–13
47. Ikenaga M, Mabuchi T. 1966. Photoreactivation of endosperm mutations in maize. *Radiat. Bot.* 6:165–69
48. Jackson JF, Liskens HF. 1979. Pollen DNA repair after treatment with the mutagens 4-NQO, ultraviolet and near ultraviolet ra-

diation, and boron dependence of repair. *Mol. Gen. Genet.* 176:11–16
49. Jenkins ME, Harlow GR, Liu Z, Shotwell MA, Ma J, Mount DW. 1995. Radiation-sensitive mutants of *Arabidopsis thaliana. Genetics* 140:725–32
50. Kanofsky JR, Sima P. 1991. Singlet oxygen production from the reactions of ozone with biological molecules. *J. Biol. Chem.* 266:9039–42
51. Karran P, Hjelmgren T, Lindahl T. 1982. Induction of a DNA glycosylase for N-methylated purines is part of the adaptive response to alkylating agents. *Nature* 296: 770–73
52. Kato T, Shinoura Y. 1977. Isolation and characterization of mutants of *E. coli* deficient in induction of mutations by ultraviolet light. *Mol. Gen. Genet.* 156: 121–31
53. Kim S-T, Malhotra K, Smith CA, Taylor J-S, Sancar A. 1994. Characterization of (6-4) photoproduct DNA photolyase. *J. Biol. Chem.* 269:8535–40
54. Klekowski EJ. 1988. *Mutation, Developmental Selection, and Plant Evolution.* New York: Columbia Univ. Press. 373 pp.
55. Langer B, Wellmann E. 1990. Phytochrome induction of photoreactivating enzyme in *Phaseolus vulgaris* L. seedlings. *Photochem. Photobiol.* 52:801–3
56. LeDoux SP, Wilson GL, Beecham EJ, Stevnsner T, Wassermann K, Bohr VA. 1992. Repair of mitochondrial DNA after various types of DNA damage in Chinese hamster ovary cells. *Carcinogenesis* 13: 1967–73
57. Levine A, Tenhaken R, Dixon R, Lamb C. 1994. H2O2 from the oxidative burst orchestrates the plant hypersensitive response. *Cell* 79:583–93
58. Li JJ, Deshaies RJ. 1993. Exercising self-restraint: discouraging illicit acts of S and M in eukaryotes. *Cell* 74:223–26
59. Li JY, Ou-Lee T-M, Raba R, Amundson RG, Last RL. 1993. Arabidopsis flavonoid mutants are hypersensitive to UV-B radiation. *Plant Cell* 5:171–79
60. Li YF, Kim S-T, Sancar A. 1993. Evidence for lack of DNA photoreactivating enzyme in humans. *Proc. Natl. Acad. Sci. USA* 90: 4389–93
61. Lindahl T. 1993. Instability and decay of the primary structure of DNA. *Nature* 362: 709–15
62. Lindahl T, Nyberg B. 1972. Rate of depurination of native deoxyribonucleic acid. *Biochemistry* 11:3610–18
63. Long SP, Humphries S, Falkowski PG. 1994. Photoinhibition of photosynthesis in nature. *Annu. Rev. Plant Phys. Plant Mol. Biol.* 45:633–62
64. Maki H, Sekiguchi M. 1992. MutT protein specifically hydrolyses a potent mutagenic substrate for DNA synthesis. *Nature* 355: 273–75
65. Malhotra K, Kim S-T, Batschauer A, Dawut L, Sancar A. 1995. Putative blue-light photoreceptors from *Arabidopsis thaliana* and *Sinapus alba* with a high degree of sequence homology to DNA photolyase contain the two photolyase cofactors but lack DNA repair activity. *Biochemistry* 34:6892–99
66. Matsumoto Y, Kim K. 1995. Excision of deoxyribose phosphate residues by DNA polymerase beta during DNA repair. *Science* 269:699–102
67. McLennan AG. 1987. The repair of ultraviolet light-induced DNA damage in plant cells. *Mutat. Res.* 181:1–7
68. Mehlhorn H, Tabner B, Wellburn AR. 1990. Electron spin resonance evidence for the formation of free radicals in plants exposed to ozone. *Phys. Plant* 79:377–83
69. Mellon I, Hanawalt P. 1989. Induction of the *E. coli* lactose operon selectively increases repair of its transcribed DNA strand. *Nature* 342:95–98
70. Mellon I, Spivak G, Hanawalt PC. 1987. Selective removal of transcription-blocking DNA damage from the transcribed strand of the mammalian DHFR gene. *Cell* 51:241–49
71. Middleton EM, Teramura AH. 1994. Understanding photosynthesis, pigment and growth responses induced by UV-B and UV-A irradiances. *Photochem. Photobiol.* 60:38–45
72. Mitchell DL, Nairn RS. 1989. The biology of the (6-4) photoproduct. *Photochem. Photobiol.* 49:805–19
73. Mitchell DL, Rosenstein BS. 1987. The use of specific radioimmunoassays to determine action spectra for the photolysis of (6-4) photoproducts. *Photochem. Photobiol.* 45:781–86
74. Mitchell DL, Vaughan JE, Nairn RS. 1989. Inhibition of transient gene expression in Chinese hamster ovary cells by cyclobutane dimers and (6-4) photoproducts in transfected ultraviolet-irradiated plasmid DNA. *Plasmid* 21:21–30
75. Mo J-Y, Maki H, Sekiguchi M. 1992. Hydrolytic elimination of a mutagenic nucleotide, 8-oxodGTP, by human 18-kilodalton protein: sanitization of nucleotide pool. *Proc. Natl. Acad. Sci. USA* 89:11021–25
76. Mol CD, Arvai AS, Slupphaug G, Kavli B, Alseth I, et al. 1995. Crystal structure and mutational analysis of human uracil-DNA glycosylase: structural basis for specificity and catalysis. *Cell* 80:1–20
77. Murphy TM, Martin CP, Kami J. 1993. Endonuclease activity from tobacco nuclei specific for ultraviolet radiation-damaged DNA. *Physiol. Plant.* 87:417–25
78. Niyogi KK, Last RL, Fink GR, Keith B.

1993. Suppressors of trp1 fluorescence identify a new arabidopsis gene, TRP4, encoding the anthranilate synthase b subunit. *Plant Cell* 5:1011–27
79. Nuffer MG. 1957. Additional evidence on the effect of X-ray and ultraviolet radiation on mutation in maize. *Genetics* 42:273–82
80. O'Connor TR, Laval J. 1991. Human cDNA expressing a functional DNA glycosylase excising 3-methyladenine and 7-methylguanine. *Biochem. Biophys. Res. Commun.* 176:1170–77
81. Okagaki RJ, Neuffer MG, Wessler SR. 1991. A deletion common to two independently derived waxy mutations of maize. *Genetics* 128:425–32
82. Orozco BM, McClung CR, Werneke JM, Ogren WL. 1993. Molecular basis of the ribulose-1,5-bisphosphate carboxylase/oxygenase activase mutation in *Arabidopsis thaliana* is a guanine-to-adenine transition at the 5′-splice junction of intron 3. *Plant Physiol.* 102:227–32
83. Osborne DJ. 1983. Biochemical control systems operating in the early hours of germination. *Can. J. Bot.* 61:3568–77
84. Osborne DJ. 1994. DNA and desiccation tolerance. *Seed Sci. Res.* 4:175–85
85. Otake M. 1993. Genetic risks from exposure to the atomic bombs Hiroshima and Nagasaki. In *Genetics of Cellular Individual, Family, and Population Variability,* ed. CF Sing, CL Hanis, pp. 83–92. Oxford: Oxford Univ. Press
86. Palmer JD. 1985. Evolution of Chloroplast and mitochondrial DNA in plants and algae. In *Molecular Evolutionary Genetics,* ed. RJ MacIntyre, pp. 131–40. New York: Plenum
87. Pang Q, Hays JB. 1991. UV-B-inducible and temperature-sensitive photoreactivation of cyclobutane pyrimidine dimers in *Arabidopsis thaliana. Plant Physiol.* 95: 536–43
88. Pang Q, Hays JB, Rajagopal I. 1992. A plant cDNA that partially complements *E. coli recA* mutations predicts a polypeptide not strongly homologous to RecA proteins. *Proc. Natl. Acad. Sci. USA* 89:8073–77
89. Pang Q, Hays JB, Rajagopal I. 1993. Two cDNAs from the plant *Arabidopsis thaliana* that partially restore recombination proficiency and DNA-damage resistance to *E. coli* mutants lacking recombination-intermediate-resolution activities. *Nucleic Acids Res.* 21:1647–53
90. Paszkowski J, Baur M, Bogucki A, Potrykus I. 1988. Gene targeting in plants. *EMBO J.* 7:4021–26
90a. Plewa MJ, Wagner ED. 1993. Activation of promutagens by green plants. *Annu. Rev. Genet.* 27:93–113
91. Potter PM, Wilkinson MC, Fitton J, Carr FJ, Brennand J, et al. 1987. Characterization and nucleotide sequence of *ogt,* the O6-alkylguanine-DNA-alkyltransferase gene of *E. coli. Nucleic Acids Res.* 15:9177–93
92. Protic-Sabljic M, Kraemer KH. 1986. One pyrimidine dimer inactivates expression of a transfected gene in xeroderma pigmentosum cells. *Proc. Natl. Acad. Sci. USA* 82: 6622–26
93. Puchta H, Swoboda P, Hohn B. 1994. Homologous recombination in plants. *Experientia* 50:277–84
94. Puchta H, Swoboda P, Hohn B. 1995. Induction of intrachromosomal homologous recombination in whole plants. *Plant J.* 7: 203–10
95. Quaite FE, Sutherland BM, Sutherland JC. 1992. Action spectrum for DNA damage in alfalfa lowers predicted impact of ozone depletion. *Nature* 358:576–78
96. Quaite FE, Sutherland JC, Sutherland BM. 1994. Isolation of high molecular weight plant DNA for DNA damage quantitation: relative effects of solar 297 nm UVB and 365 nm radiation. *Plant Mol. Biol.* 24: 475–83
97. Quaite FE, Takayanagi S, Ruffini J, Sutherland JC, Sutherland BM. 1994. DNA damage levels determine cyclobutyl pyrimidine dimer repair mechanisms in alfalfa seedlings. *Plant Cell* 6:1635–41
98a. Rajagopalan M, Lu C, Woodgate R, O'Donnell M, Goodman MF, Echols H. 1992. Activity of the purified mutagenesis proteins UmuC, UmuD′, and RecA in replicative bypass of an abasic DNA lesion by DNA polymerase III. *Proc. Natl. Acad. Sci. USA* 89:10777–81
98b. Rebeck GW, Samson L. 1991. Increased spontaneous mutation and alkylation sensitivity of *E. coli* strains lacking the *ogt* O^6-methylguanine methyltransferase. *J. Bacteriol.* 173:2068–76
99. Reenan RA, Kolodner RD. 1992. Characterization of insertion mutations in the *S. cerevisiae MSH1* and *MSH2* genes: evidence for separate mitochondrial and nuclear functions. *Genetics* 132:975–85
100. Reilly JJ, Klarman WL. 1980. Thymine dimer and glyceolin accumulation in UV-irradiated soybean suspension cultures. *J. Environ. Exp. Bot.* 20:131–33
101. Rideout WM, Coetzee GA, Olumi AF, Jones PA. 1990. 5-Methylcytosine as an endogenous mutagen in the human LDL receptor and p53 genes. *Science* 249: 1288–90
102. Roza L, De Gruijl FR, Henegouwen JBAB, Guikers K, Van Weelden H, et al. 1991. Detection of photorepair of UV-induced thymine dimers in human epidermis by immunofluorescence microscopy. *J. Invest. Dermatol.* 96:903–7
103. Rydberg B, Spurr N, Karran P. 1990. cDNA

cloning and chromosomal assignment of the human O6-methylguanine-DNA methyltransferase. *J. Biol. Chem.* 265: 9563–69
104. Saedler H, Nevers P. 1985. Transposition in plants: a molecular model. *EMBO J.* 4: 585–90
105. Sakumi K, Sekiguchi M. 1990. Structures and functions of DNA glycosylases. *Mutat. Res.* 236:161–62
106. Sancar A. 1994. Mechanisms of DNA excision repair. *Science* 266:1954–56
107. Sancar A. 1994. Structure and function of DNA photolyase. *Biochemistry* 33:2–9
108. Santerre A, Britt A. 1994. Cloning of a 3-methyladenine-DNA glycosylase from *Arabidopsis thaliana. Proc. Natl. Acad. Sci. USA* 91:2240–44
109. Savva R, McAuley-Hecht K, Brown T, Pearl L. 1995. The structural basis of specific base-excision repair by uracil-DNA glycosylase. *Nature* 373:487–93
110. Schaeffer L, Roy R, Humbert S, Moncollin V, Vermeulen W, et al. 1993. DNA repair helicase: a component of BTF2 (TFIIH) basic transcription factor. *Science* 260:93
111. Scott L, LaFoe D, Weil CF. 1996. Adjacent sequences influence DNA repair accompanying transposon excision in maize. *Genetics.* In press
112. Sedgwick B. 1983. Molecular cloning of a gene which regulates the adaptive response to alkylating agents in *E. coli. Mol. Gen. Genet.* 191:466–72
113. Sedgwick B, Vaughan P. 1991. Widespread adaptive response against environmental methylating agents in microorganisms. *Mutat. Res.* 250:211–21
114. Setlow P. 1976. Nearest neighbor frequencies in deoxyribonucleic acids. In *Handbook of Biochemistry and Molecular Biology,* ed. GD Fasman, 2:313–18. Cleveland: CRC Press
115. Shapiro HS. 1976. Distribution of purines and pyrimidines in deoxyribonucleic acids. In *Handbook of Biochemistry and Molecular Biology,* ed. GD Fasman, 2:241–62. Cleveland: CRC Press
116. Shirley BW, Hanley S, Goodman H. 1992. Effects of ionizing radiation on a plant genome: analysis of two *Arabidopsis transparent* testa mutations. *Plant Cell* 4:333–47
117. Simpson RB, O'Hara PJ, Kwok W, Montoya AL, Lichtenstein C, et al. 1982. DNA from the A6S/2 crown gall tumor conteins scrambled Ti-plasmid sequences near its junctions with plant DNA. *Cell* 29: 1005–14
118. Singhal RK, Hinkle DC, Lawrence CW. 1992. The *REV3* gene of *S. cerevisiae* is transcriptionally regulated more like a repair gene than one encoding a DNA polymerase. *Mol. Gen. Genet.* 236:17–24
119. Slupphaug G, Markussen FH, Olsen LC, Aaasland R, Aarsaether N, et al. 1993. Nuclear and mitochondrial forms of human uracil-DNA glycosylase are encoded by the same gene. *Nucleic Acids Res.* 21:2579–84
120. Small GD. 1987. Repair systems for nuclear and chloroplast DNA in *Chlamydomonas reinhardtii. Mutat. Res.* 181:31–35
121. Smith KC. 1992. Spontaneous mutagenesis: experimental, genetic, and other factors. *Mutat. Res.* 277:139–62
122. Sonti RV, Chiurazzi M, Wong D, Davies CS, Harlow GR, et al. 1996. Arabidopsis mutants deficient in T-DNA integration. *Proc. Natl. Acad. Sci. USA.* In press
123. Stapleton AE, Walbot V. 1994. Flavonoids can protect maize DNA from the induction of ultraviolet radiation damage. *Plant Physiol.* 105:881–89
124. Strickland JA, Marzilli LG, Puckett JM, Doetsch PW. 1991. Purification and properties of Nuclease SP. *Biochemistry* 30: 9749–56
125. Sun T-P, Goodman HM, Ausubel FM. 1992. Cloning the Arabidopsis *GA1* locus by genomic subtraction. *Plant Cell* 4: 119–28
126. Sweder KS, Hanawalt PC. 1992. Preferential repair of cyclobutane pyrimidine dimers in the transcribed strand of a gene in yeast chromosomes and plasmids is dependent on transcription. *Proc. Natl. Acad. Sci. USA* 89:10696–700
127. Takayanagi S, Trunk JG, Sutherland JC, Sutherland BM. 1994. Alfalfa seedlings grown outdoors are more resistant to UV-induced damage than plants grown in a UV-free environmental chamber. *Photochem. Photobiol.* 60:363–67
128. Talpaert-Borlè M. 1987. Formation, detection, and repair of AP sites. *Mutat. Res.* 181:45–56
129. Tano K, Shiota S, Collier J, Foote RS, Mitra S. 1990. Isolation and structural characterization of a cDNA clone encoding the human DNA repair protein for O6-alkylguanine. *Proc. Natl. Acad. Sci. USA* 87: 686–90
130. Taylor RM, Nikaido O, Jordan BR, Rosamond J, Bray CM, Tobin AK. 1995. UV-B-induced DNA lesions and their removal in wheat (*Triticum aestivum* L.) leaves. *Plant Cell Environ.* In press
131. Todo T, Takemori H, Ryo H, Ihara M, Matsunaga T, et al. 1993. A new photoreactivating enzyme that specifically repairs ultraviolet light-induced (6-4) photoproducts. *Nature* 361:371–74
132. Trosko JE, Mansour VH. 1968. Response of tobacco and *Haplopappus* cells to ultraviolet radiation after posttreatment with photoreactivating light. *Mutat. Res.* 36: 333–43
133. Trosko JE, Mansour VH. 1969. Photoreactivation of ultraviolet light-induced

pyrimidine dimers in Ginkgo cells grown in vitro. *Mutat. Res.* 7:120–21
134. Vaughan P, Sedgwick B, Hall J, Gannon J, Lindahl T. 1991. Environmental mutagens that induce the adaptive response to alkylating agents in *E. coli. Carcinogenesis* 12: 263–68
135. Veleminsky J, Angelis K, Baburek I, Gichner T, Satava J, et al. 1993. An *E. coli ada* transgenic clone of Nicotiana tabacum var. Xanthi has increased sensitivity to the mutagenic action of alkylating agents, maleic hydrazide and gamma-rays. *Mutat. Res.* 307:193–200
136. Veleminsky J, Svachulová J, Satava J. 1980. Endonucleases for UV-irradiated and depurinated DNA in barley chloroplasts. *Nucleic Acids Res.* 8:1373–81
137. Walbot V. 1992. Reactivation of mutator transposable elements by ultraviolet light. *Mol. Gen. Genet.* 234:353–60
138. Wallace DC. 1992. Mitochondrial genetics: a paradigm for aging and degenerative diseases? *Science* 256:628–32
139. Ward JF. 1975. Molecular mechanisms of radiation-induced damage to nucleic acids. In *Advances in Radiation Biology,* Vol. 5, ed. JT Lett, H Adler. London: Academic. 277 pp.
140. Whitelam GC, Johnson E, Peng J, Carol P, Anderson ML, et al. 1993. Phytochrome A null mutants of Arabidopsis display a wild-type phenotype in white light. *Plant Cell* 5:757–68
141. Wilkinson JQ, Crawford NM. 1991. Identification of the Arabidopsis CHL3 gene as the nitrate reductase structural gene NIA2. *Plant Cell* 3:461–72
142. Xiao W, Derfler B, Chen J, Samson L. 1991. Primary sequence and biological functions of a *S. cerevisiae* O6-methylguanine/O4-methylthymine methyltransferase gene. *EMBO J.* 10:2179–86
143. Yajima H, Inoue H, Oikawa A, Yasui A. 1991. Cloning and functional characterization of a eukaryotic DNA photolyase gene from *Neurospora crassa. Nucleic Acids Res.* 19:5359–62
144. Yajima H, Takao M, Yasuhira S, Zhao JH, Ishii C, et al. 1995. A eukaryotic gene encoding an endonuclease that specifically repairs DNA damaged by ultraviolet light. *EMBO J.* 14:2393–99
145. Yamamoto Y, Sekiguchi M. 1979. Pathways for repair of DNA damaged by alkylating agent in *E. coli. Mol. Gen. Genet.* 171:251–56
146. Yasui A, Eker APM, Yasuhira S, Yajima H, Kobayashi T, et al. 1994. A new class of DNA photolyases present in various organisms including aplacental mammals. *EMBO J.* 13:6143–51
147. Yasui A, Yajima H, Kobayashi T, Ecker APM, Oikawa A. 1992. Mitochondrial repair by photolyase. *Mutat. Res.* 273:231–36
148. Ziegler A, Lefell DJ, Kunala S, Sharma HW, Gailani M, et al. 1994. Mutation hotspots due to sunlight in the p53 gene of nonmelanoma skin cancers. *Proc. Natl. Acad. Sci. USA* 90:4216–20

Annu. Rev. Plant Physiol. Plant Mol. Biol. 1996. 47:101–25

PLANT PROTEIN PHOSPHATASES

Robert D. Smith

AgBiotech Center, Rutgers University, New Brunswick, New Jersey 08903-0231

John C. Walker

Division of Biological Sciences, University of Missouri, Columbia, Missouri 65211

KEY WORDS: phosphorylation, signal transduction, protein kinase, metabolism, okadaic acid

ABSTRACT

Posttranslational modification of proteins by phosphorylation is a universal mechanism for regulating diverse biological functions. Recognition that many cellular proteins are reversibly phosphorylated in response to external stimuli or intracellular signals has generated an ongoing interest in identifying and characterizing plant protein kinases and protein phosphatases that modulate the phosphorylation status of proteins. This review discusses recent advances in our understanding of the structure, regulation, and function of plant protein phosphatases. Three major classes of enzymes have been reported in plants that are homologues of the mammalian type-1, -2A, and -2C protein serine/threonine phosphatases. Molecular genetic and biochemical studies reveal a role for some of these enzymes in signal transduction, cell cycle progression, and hormonal regulation. Studies also point to the presence of additional phosphatases in plants that are unrelated to these major classes.

CONTENTS

1040-2519/96/0601-0101$08.00

INTRODUCTION

It is well recognized that reversible phosphorylation of proteins controls many cellular processes in plants and animals. The phosphorylation status of proteins is regulated by the opposing activities of protein kinases and protein phosphatases. Phosphorylation of eukaryotic proteins occurs predominantly (97%) on serine and threonine residues and to a lesser extent on tyrosine residues (107). Phosphohistidine phosphorylation has also been reported in plants (50), fungi (86), and animals (25), but its relative contribution to the total phosphoamino acid content of eukaryotic cells is not known. In animals, protein phosphorylation plays well-known roles in diverse cellular processes such as glycogen metabolism, cell cycle control, and signal transduction (9, 23, 83, 106, 107). Clearly, there has been a similar interest in examining the role of protein phosphorylation in plant cellular regulation and in identifying the protein kinases and protein phosphatases that modulate the phosphorylation status of target substrate molecules. An increasing number of plant protein kinases have been reported in recent years. Some of these enzymes play pivotal roles in the control of plant defense mechanisms, signal transduction, and metabolism (11, 13, 50, 122). In plants, as in animals, recognition that a number of protein kinases respond directly to second messengers such as Ca^{2+} led to the view that protein kinases were primarily responsible for regulating the phosphorylation status of proteins and that protein phosphatases merely reverse the effects of protein kinases. Molecular genetic and biochemical studies have greatly advanced our knowledge of protein phosphatases and provide compelling evidence that these enzymes perform essential regulatory functions. The objective of this review is to examine recent advances in our knowledge of the structure and regulation of plant protein phosphatases as well as initial insights into their physiological roles.

Protein phosphatase activities have been reported in most plant subcellular compartments, including mitochondria, chloroplast, nuclei, and cytosol, and are associated with various membrane and particulate fractions (50, 70). Some protein phosphatases are poorly characterized and may represent novel enzymes that are unique to plants, such as the chloroplast thylakoid protein phosphatase (124). Others have biochemical properties that are very similar to well-known mammalian protein phosphatases, such as the mitochondrial pyruvate dehydrogenase phosphatase (80) and cytosolic protein serine/threonine phosphatases (69). Because some plant and animal protein phosphatases are recognized to be evolutionarily conserved, biochemical characterization and molecular cloning of distinct phosphatases in plants that correspond to the mammalian type-1 (PP1) and type-2 (PP2) protein serine/threonine phosphatases have been achieved over a short period. Understandably, the plant PP1 and PP2 represent only a subset of the total phosphatases that will subsequently be discovered in plants, but their essential roles in diverse cellular processes are already becoming evident. Biochemical and genetic studies in plants implicate PP1 and/or PP2 activity in signal transduction, hormonal regulation, mitosis, and control of carbon and nitrogen metabolism.

Classification of mammalian PP1 and PP2 is based on their unique substrate specificities and sensitivities to various inhibitors (52). PP1 dephosphorylates the β-subunit of mammalian phosphorylase kinase preferentially and is inhibited by the endogenous proteins, inhibitor-1 (I-1) and inhibitor-2 (I-2). PP2 has greater activity toward the α-subunit of phosphorylase kinase and is resistant to I-1 and I-2. PP2 is further divided into three subgroups, PP2A, PP2B, and PP2C, depending on their subunit structure, divalent cation requirements, and substrate specificities (23). PP2A is a heterotrimer of a catalytic C-subunit and two distinct regulatory A- and B-subunits and does not require divalent cations for activity. PP2B, a Ca^{2+}-activated phosphatase, exists as a heterodimer consisting of a catalytic A-subunit and a regulatory B-subunit belonging to the EF-hand family of calcium-binding proteins. PP2C is found as a monomer and activity requires Mg^{2+}. Mammalian PP1, PP2A, PP2B, and PP2C catalytic subunits are products of distinct genes, and examination of their primary structures indicates that PP1, PP2A, and PP2B are related enzymes (106). They share no structural homology with PP2C. The structure, regulation, and function of the protein serine/threonine phosphatases in animals and fungi have been the subject of many recent reviews (9, 23, 73, 83, 106, 107, 118).

Protein phosphatases are inhibited by a number of natural toxins such as okadaic acid and cyclosporin A (21, 73). These compounds have proven extremely useful in analyzing the differing actions of specific classes of phosphatases in vitro, as well as in vivo, because many of them are readily taken up by animal and plant cells (73). The marine toxin okadaic acid, for instance, is a potent inhibitor of PP2A ($IC_{50} \approx 0.1$–1.0 nM) and inhibits PP1 at 10- to 100-fold higher concentrations ($IC_{50} \approx 10$–100 nM). Okadaic acid is marginally effective against PP2B at micromolar concentrations and has no effect on PP2C. The immunosuppressant cyclosporin A complexes with endogenous immunophilins and specifically targets PP2B for downregulation (103). PP2C is insensitive to these drugs. Increasing use of these drugs to study the role of reversible protein phosphorylation in diverse cellular processes has generated new insights into the regulation and physiological function of protein serine/threonine phosphatases. Inhibitor studies also confirm the presence in plants and animals of protein phosphatases that do not belong to the PP1 and PP2 families of protein phosphatases.

PROTEIN PHOSPHATASE-1

Structure and Regulation

PP1 is a ubiquitous and highly conserved enzyme found in all eukaryotes. The native mammalian and fungal enzyme is a complex of a catalytic subunit and one or more regulatory subunits. Genetic studies have shown yeast PP1 catalytic subunit genes to be essential for cellular processes as diverse as glycogen accumulation, mitosis, and translational control (118). Regulatory subunits define specific functions of PP1 catalytic activity in vivo by controlling the subcellular location and substrate specificity of the enzyme complex. For instance, the *Saccharomyces cerevisiae* GAC1 protein, a homologue of the mammalian RG1 subunit, targets PP1 to glycogen particles and enhances dephosphorylation of glycogen-bound phosphoenzymes required for glycogen synthesis and accumulation (35). A nuclear-localized PP1 regulatory subunit, sds22^{+}, which is essential for completion of mitosis in *S. pombe,* increases phosphohistone H1 phosphatase activity of PP1 and downregulates its activity against phosphorylase a (120). Other regulatory subunits, such as mammalian I-2, have been characterized biochemically. I-2 is hypothesized to function in the cell as a chaperone that binds and activates newly synthesized PP1 catalytic subunits via a process that requires phosphorylation of I-2 by glycogen synthase kinase-3 (GSK-3) in the presence of ATP-Mg (107). The active catalytic subunit may then complex with various targeting or regulatory subunits, thereby displacing I-2. On the basis of these and additional studies in animals and fungi, a targeting subunit has been hypothesized for regulation of PP1 in vivo (24, 45). The hypothesis is that distinct regulatory subunits are present in the cell that bind transiently to the catalytic subunit and dictate its subcellular location and substrate specificity. The hypothesis is supported by the findings that in *S. pombe* PP1 is present in the cell as high molecular complexes ranging in size from 80 to 200 kD, of which only the 80-kD complex has phosphorylase a phosphatase activity (59).

Protein phosphatase activities have been reported in a number of plant species and in different tissue extracts through the use of mammalian phosphoprotein substrates. In *Brassica napus* seed extracts, over 60% of the phosphatase activity is inhibited by I-1 (IC_{50} = 0.6 nM), I-2 (IC_{50} = 2.0 nM), and okadaic acid (IC_{50} = 10 nM) and dephosphorylates the β-subunit of rabbit phosphorylase kinase preferentially, indicating that it belongs to the PP1 class of protein phosphatases (69). The PP1 activity is associated predominantly with membranes of the endoplasmic reticulum and other unidentified particulate fractions in *B. napus* seed extracts, and gel filtration analyses indicate it exists as a high molecular weight complex (70). PP1 is almost exclusively cytosolic in pea leaves and carrot cells and is associated with microsomes in wheat leaves (70). It has also been reported in isolated nuclei (104) and in plasma membranes (135). Very little is known about the structure or regulation of native PP1 in plants, nor have any physiological substrates been identified. Studies reveal a striking similarity between the biochemical properties and subcellular distribution of animal and plant PP1 and suggest the possibility that mechanisms for controlling PP1 activity and function may be equally well conserved. A key goal in attempting to understand PP1 function in plants is to identify and characterize regulatory subunits.

CATALYTIC SUBUNIT PP1 catalytic subunit cDNA and genomic clones have been reported in several plant species (Table 1). Eight distinct isoforms have been identified in *Arabidopsis* (4, 34, 84, 114, 116), and additional related genes appear to be present in its genome (114, 116). The presence of PP1 multigenes is not unique to *Arabidopsis,* and they appear to be present in most plant species. The remarkable similarity (>70%) between plant and mammalian PP1 primary sequences and the sensitivity of a bacterially expressed recombinant maize *ZmPP1* to I-2 ($IC_{50} = 0.1$ nM) and okadaic acid ($IC_{50} = 200$ nM) (113) confirms that the plant genes encode PP1 catalytic subunits. Unlike the native mammalian enzyme, however, the recombinant maize PP1 requires Mn^{2+} for activation (113). Mn^{2+}-dependent activity is also observed in recombinant rabbit PP1 or when native PP1 is converted to a Mn^{2+}-dependent form by incubation with NaF (1). The modified enzymes can be restored to their Mn^{2+}-independent forms by incubation with I-2 and GSK-3/ATP-Mg, which supports the hypothesis that I-2 acts as a chaperone to activate PP1 in vivo and raises the possibility that plant PP1 catalytic subunits may need to be activated by a plant homologue of I-2.

Multiple PP1 isogenes are present in most eukaryotes, with the exception of *S. cerevisiae* which contains a single PP1, *GLC7* (118). The physiological advantage of having multiple PP1 isogenes remains unknown. Disruption of the two *S. pombe* PP1 genes is lethal, but deletion of either gene, singly, has no effect on cell viability (85), which indicates that the yeast PP1 isoforms have overlapping functions. In contrast, loss of one of four PP1 genes (30) in *Drosophila* inhibits chromosome separation (5), which suggests a selective role for this isoform in mitosis. Unique roles for some isogenes may be governed by their spatial and temporal expression. For instance, the rat PP1γ1 is predominantly expressed in brain tissues, whereas PP1γ2 is almost exclusively expressed in testes (108, 109). Upregulation of certain PP1 isogenes has been reported in plants as well. *BoPP1* expression is enhanced at different stages of microspore development in *B. oleracea,* and mature trinucleate microspores contain a unique *BoPP1* transcript not found at other stages of the plant life cycle (100). *Arabidopsis AtPP1bg* is constitutively expressed at low levels in all tissues with upregulation in male and female tissues (4).

Table 1 Plant protein phosphatases

Species	Name	Clone	#AA	%ID[a]	Expression[b]	Database accession number	Reference(s)
Protein phosphatase-1							
Catalytic subunit							
Arabidopsis	TOPP1[c]	cDNA	318	72	R,L,S,F,C	M93408	34, 84, 114
	TOPP2[d]	cDNA	312	76	R,L,S,F,C	M93409	33, 114
	TOPP3	cDNA	322	73	R,L,S,F	M93410	114
	TOPP4	cDNA	321	72	R,L,F	M93411	114
	TOPP5[e]	cDNA	312	76	R,L,S,F,C	M93412	114
	TOPP6	cDNA	—	—	—	—	116
	AtPP1bg[f]	cDNA	322	71	F	Z46253	4, 116
	TOPP8	cDNA	—	—	—	—	116
Alfalfa	*PP1Ms*	cDNA	321	74	R,L,S,N,FB,MF	X80788	88
B. napus	PP1Bn	cDNA	—	72	—	X57438	74
B. oleracea	BoPP1	cDNA	316	71	R,L,Co,A,P	X63558	100
Maize	*ZmPP1*	cDNA	316	70	R,S,L,H,T,C	M60215	112
Acetabularia	PP1Ac1	genomic	319	71	—	Z28627	
	PP1Ac2	genomic	319	71	—	Z28632	
Protein phosphatase-2A							
Catalytic subunit							
Arabidopsis	PP2A-1	cDNA	306	82	R,L,S,F	M96732	3
	PP2A-2	cDNA	306	82	R,L,S,F	M96733	3
	PP2A-3	cDNA	308	81	R,L,S,F	M96734	3
	PP2A-4	cDNA	313	81	R,L,S,F	M96841	16
Alfalfa	*pp2aMs*	cDNA	313	79	R,L,S,FB,N,	X70399	90
B. napus	PP2ABn	cDNA	-	72	—	X57439	74
Sunflower	*PP2AHa*	cDNA	305	—	—	Z26041	
Acetabularia	PP2AAc	genomic	307	—	—	Z26654	
Regulatory A-subunit							
Arabidopsis	pDF1	cDNA	587	—	—	X82001	112
	pDF2	cDNA	—	—	—	X82002	112
	RCN1[g]	genomic	588	58	seedlings	U21557	54, 112
Pea	*PP2A-Ps*	cDNA	—	55	E,R,S,L,C	Z25888	32
Regulatory B-subunit							
Arabidopsis	AtBα	cDNA	513	46	R,L,S,Co,F,FB	U18129	102
Protein phosphatase-X							
Catalytic subunit							
Arabidopsis	PPX-At1	cDNA	305	83	R,L,S,F	Z22587	89
	PPX-At2	cDNA	307	83	R,L,S,F	Z22596	89
Protein phosphatase-2C							
Arabidopsis	ABI1	genomic	434	35	R,L,S,Si	X77116	64, 79
	KAPP	cDNA	582	19	R,L	U09505	121
	PP2C-At	cDNA	399	35	—	D38109	62

Functional differences between PP1 isoforms may also arise from structural variations in their primary sequence that control activity, substrate specificity, and/or regulatory subunit interactions. Phosphorylation of *S. pombe* PP1, dis2$^+$, at a conserved cdc2 protein kinase consensus phosphorylation site [S/T-P-X-Z: X = polar and Z = basic; (82)] downregulates its activity in vitro (141). In contrast, *S. pombe* PP1, sds21$^+$, lacks a cdc2 phosphorylation motif and is neither phosphorylated nor downregulated by cdc2 protein kinases (141). Therefore, isoforms that harbor the phosphorylation recognition site may be selectively targeted for inactivation by cdc2-like protein kinases. The phosphorylation site is found at a conserved position in about half of the known plant PP1 catalytic subunits (116), but whether they are subject to this form of regulation in vivo remains unknown.

Heterologous Expression of Plant PP1

One approach that has been taken to address the functional importance of structural variations in the primary sequences of the catalytic subunits is to express individual PP1 isoforms in heterologous systems and ask whether they can fully complement the multiple functions of their host PP1. Expression of distinct plant PP1 clones in fungal systems reveals that isoforms differ in their ability to complement various PP1 functions (Table 2). A lethal trait caused by disruption of *S. cerevisiae* PP1, *glc7*$^-$ (14), is rescued by *Arabidopsis TOPP2* but not *TOPP1* (116). Failure of *TOPP1* to complement the lethal trait cannot be attributed to improper transcription or translation because TOPP1 protein is detected in wild-type cells expressing the *TOPP1* clone. However, only *S. cerevisiae* strains expressing *TOPP2* have increased PP1 activity, which suggests that TOPP1 may be downregulated in yeast cells by an unknown post-translational mechanism. *TOPP2* is unable to suppress an *S. cerevisiae glc7-1* glycogen deficiency phenotype (14, 116, 123), which indicates that TOPP2 can dephosphorylate substrates that are essential for completion of mitosis but is unable to dephosphorylate a substrate (glycogen synthase) of GLC7 required for glycogen accumulation in yeast. Glycogen accumulation is restored

---

[a]Percent identity between plant PP1 catalytic subunits and rabbit PP1α, plant PP2A catalytic subunits and rabbit PP2α, plant PP2A regulatory A-subunits and human PR65α, plant PP2A B-regulatory subunit and human B-subunit, plant PPX catalytic subunits and rabbit PPX, plant PP2C catalytic domains and rat PP2C.

[b]Expression of plant genes in different tissues: R, root; L, leaf; S, stem; F, flower; FB, flower bud; Si, silique; N, node; MF, mature flower; Co, coleoptile; C, cell culture.

[c]*TOPP1* independently isolated and named *PP1-At* (84), and *PP1A-At2* (33).

[d]*TOPP2* independently isolated and named *PP1A-At1* (34).

[e]*TOPP5* independently isolated and named *PP1A-At3* (34).

[f]*AtPP1bg* independently isolated and named *TOPP7* (116).

[g]*RCN1* independently isolated and named *regA* (112).

Table 2 Complementation of fungal mutations by plant PP1 clones

	S. cerevisiae		S. pombe		Aspergillus nidulans
	glc7-1	glc7⁻	*dis2-11*	*wee1⁻*	*bimG11*
TOPP1	No (116)	No (116)	Yes (84)		
TOPP2	No (116)	Yes (116)	No (34)	Yes (34)	
TOPP3	No (116)	No (116)			
AtPP1bg					Yes (4)
PP1Ms			No (88)		

in *glc7-1* strains, however, by expressing a chimeric PP1 consisting of the N-terminal 1-93 amino acid residues of GLC7 fused to residues 98-312 of TOPP2 (116). This suggests that unique structural sequences located at the N-terminus may be important for controlling substrate specificity and/or regulatory subunit interactions.

Functional differences among plant PP1 isoforms have also been reported in *S. pombe* and *Aspergillus* (Table 2). A cold-sensitive *dis2-11* mutation that blocks exit from mitosis is suppressed by *TOPP1* but not *TOPP2* (34, 84). However, *TOPP2* is able to restore a temperature-sensitive *S. pombe* $cdc25^{ts}/wee1^-$ double mutation (34) that inhibits entry into mitosis (10), indicating that failure of *TOPP2* to complement the *dis2-11* mutation is not due to the absence of active protein phosphatase activity but probably results from a failure of TOPP2 to dephosphorylate substrates of $dis2^+/sds21^+$ that are required for entering mitosis. Furthermore, expression of *Arabidopsis AtPP1bg* in a temperature-sensitive *Aspergillus bimG11* PP1 mutant supports vegetative growth but not conidia development at nonpermissive temperatures (4). These results demonstrate that plant PP1 isoforms have limited ability to functionally complement fungal PP1 enzymes despite their remarkable structural and biochemical similarities and suggest that even small structural differences in their primary sequences may have profound effects on their activity and function.

PROTEIN PHOSPHATASE-2A

Structure and Regulation

Native PP2A in animals is found either as a heterodimer of a 36-kD catalytic C-subunit and a 65-kD regulatory A-subunit, or as a heterotrimer in which a variable regulatory B-subunit (50–70 kD) complexes to the core heterodimer (23, 77, 106). In *S. cerevisiae,* a PP2A A-subunit encoded by *tpd3* is essential for cytokinesis (134), and a B-subunit encoded by *cdc55* is required for cellular morphogenesis (44). Single A-subunit and B-subunit genes are present in *Drosophila* and both appear to be essential for pattern formation (132). Genetic and biochemical studies indicate that regulatory A- and B-subunits control substrate specificity of PP2A (77).

PP2A has been reported in a number of plant species and is found in most tissues (53, 66, 69–71, 91, 104). Like PP1, PP2A is found in many subcellular locations including the nucleus (104) and cytosol and is associated with various membranes and insoluble fractions (70, 135). Neither PP1 nor PP2A is found in chloroplasts (70, 125). Native PP2A from *B. napus* seed extracts dephosphorylates the α-subunit of phosphorylase kinase preferentially and is potently inhibited by okadaic acid (IC_{50} = 0.1 nM) in vitro (53, 69). A partially purified PP2A catalytic subunit from maize seedlings readily dephosphorylates phosphocasein, phosphohistone H1, and phosphorylase a; displays no appreciable activity toward *p*NPP; and fails to bind to heparin-Sepharose, a distinct binding characteristic of PP1 (53). These results indicate that plant PP2A is biochemically nearly indistinguishable from mammalian PP2A.

CATALYTIC SUBUNIT Four genes encoding PP2A catalytic subunits (C-subunit) have been isolated from *Arabidopsis,* and Southern blot analyses indicate a fifth isoform may be present in the genome (3, 16). C-subunit clones have also been identified in alfalfa (90), *B. napus* (74), sunflower, and the green alga *Acetabularia cleftonii* (Table 1). Plant and mammalian PP2A share about 80% identity. Transcripts corresponding to each of the *Arabidopsis* PP2A catalytic subunits are found in all tissues examined, although the level of expression of some isogenes appears to be developmentally regulated (16, 89). Thus, unique roles for some PP2A genes may result from enhanced expression in some tissues. Differential regulation of the C-subunits may also occur from posttranslational modifications (98). Phosphorylation of tyrosine (17, 18) or threonine (27, 41) residues downregulates in vitro activity of mammalian PP2A. Tyrosine and threonine residues are found at a similar position in all plant PP2A isoforms. Methylation of the C-terminal leucine, conserved in all PP2A, may provide an additional level of regulation (107).

REGULATORY SUBUNITS Three A-subunit clones have been isolated from *Arabidopsis* (112), and a partial cDNA clone has also been identified in pea (32) (Table 1). The plant A-subunits are approximately 58% identical to the human regulatory A-subunit. In addition, a B-subunit homologue has been reported in *Arabidopsis* that shares 46% identity with the human Bα-subunit (102). Expression of the A-subunit genes has not been thoroughly examined (112), and the B-subunit appears to be uniformly expressed throughout the plant (102). Interactions between plant catalytic and regulatory subunits have been examined, but the presence of multiple catalytic and regulatory subunits in plants suggests that plants may contain a number of different PP2A complexes with distinct physiological functions.

Insight into a potential role for PP2A in plants comes from the recent discovery that an *Arabidopsis* polar auxin transport mutant *rcn1* encodes a PP2A regulatory A-subunit homologue (54) that shares 58% identity with the human PR65α (54). *RCN1* rescues the *S. cerevisiae* temperature-sensitive *tpd3* mutant, which indicates that it is a functional A-subunit homologue capable of controlling the activity of the yeast PP2A catalytic subunit. A T-DNA insertion into the coding region of *RCN1* results in normal expression of a truncated transcript that may encode a protein lacking a significant portion of its C-terminus. This may represent a loss-of-function mutation because the C-terminal end of the mammalian regulatory A-subunit is known to be required for interaction with the catalytic domain (99). The *rcn1* recessive mutation causes an altered morphological response in seedlings to N-1-naphthylphthalamic acid (NPA), a polar auxin transport inhibitor (54). Auxin transport in stems of *rcn1* mutants also shows increased sensitivity to NPA. These results point to a unique role for *RCN1* in controlling the activity of an endogenous PP2A toward a key phosphosubstrate(s) involved in polar auxin transport.

Role in Cellular Metabolism

Control of key metabolic enzymes by reversible phosphorylation has been studied extensively in plants (13, 93). Activation of sucrose phosphate synthase (SPS), nitrate reductase (NR), and phospho*enol*pyruvate carboxylase (PEPC) in light is associated with a decrease in the phosphorylation status of SPS and NR and an increase in phosphorylation of PEPC (50). Phosphorylation of quinate dehydrogenase (QDH) and hydroxymethylglutaryl-CoA reductase kinase (HMG-CoA reductase kinase) and dephosphorylation of HMG-CoA reductase also activates these enzymes in vivo (70, 75, 95). Several lines of evidence point to PP2A activity in dephosphorylating these enzymes. Dephosphorylation-mediated changes in activity of each enzyme are blocked by PP1/PP2A inhibitors, such as okadaic acid, microcystin-LR, and calyculin A (15, 39, 47, 55, 56, 60, 78, 110). Okadaic acid also prevents in vivo activation of SPS (47, 110) and NR (46, 48) in spinach leaves. Furthermore, addition of mammalian PP2A catalytic subunit to cell extracts enhances SPS (110) and NR activities (71) and downregulates PEPC (78) and QDH (70). In contrast, addition of the mammalian PP1 catalytic subunit does not alter their activities. These studies suggest that PP2A, and not PP1, is responsible for dephosphorylating these enzymes in vivo.

SPS, PEPC, and NR are key enzymes that control nitrogen and carbon assimilation in plants. Because these pathways compete for carbon skeletons and energy sources, mechanisms must exist to regulate their relative activities in response to changing environmental and metabolic conditions. Coordination of these pathways could result from tight control of SPS, NR, and PEPC activities in the cytosol, which raises the possibility that their respective protein kinases and phosphatases respond differentially to signals in the cell. Emerging evidence indicates that the SPS PP2A may be distinct from the PP2A that dephosphorylates NR. Activation of SPS, for instance, is inhibited by Pi, sulfate, and tungstate, but NR activation is unaffected by these compounds (49). In contrast, NR activation is inhibited by Mg^{2+} and stimulated by 5′-AMP (55, 56). These results could be explained by the presence of multiple PP2A enzymes in the cytosol that respond differentially to effectors such as Pi, shown to be a potent inhibitor of the SPS phosphatase (47, 139). Alternatively, these effectors may interact directly with the protein substrates to modulate activation. Light does not appear to control the protein kinase and phosphatase activities directly because feeding mannose to excised leaves in the dark activates SPS and NR (49, 119). However, a recent study suggests that the NR PP2A may be light-activated by a process that requires de novo protein synthesis, because it can be blocked with cycloheximide (46).

PROTEIN PHOSPHATASE-2C

PP2C is the least well-characterized member of the protein serine/threonine phosphatases (106). PP2C demonstrates high activity toward enzymes of the cholesterol biosynthetic pathway in mammals (106, 107), and genetic studies in fission yeast implicate a possible role for PP2C in growth (76). Physiological substrates for PP2C remain to be identified in animals or plants. PP2C demonstrates relatively high activity toward phosphocasein, a commonly used substrate for measuring PP2C activity in vitro. Mg^{2+}-dependent and okadaic-insensitive phosphocasein phosphatase activity has been reported in several plant species (70, 75), but its distribution in plants appears more restrictive than for PP1 and PP2A. Activity is predominantly cytosolic in carrot cells, cauliflower inflorescence, and leaves from pea and wheat (70, 75). However, PP2C activity was reported to be absent in maize seedlings (53) and *B. napus* seed extracts (69). Cauliflower PP2C and PP2A readily dephosphorylate HMG-CoA reductase kinase, a key enzyme in isoprenoid biosynthesis (75). There is presently no indication whether the reductase kinase is a physiological substrate for either phosphatase.

Structure, Regulation, and Function

Three novel PP2C phosphatases have been cloned from *Arabidopsis* (Figure 1). Each contains a catalytic domain that is structurally related (20–35% identical) to mammalian PP2C. In addition, the plant genes contain N-terminal extensions of variable lengths that share no homology with one another, or to protein sequences in the data banks (Figure 1). These unique structural domains among plant PP2C are not found in any of the known fungal or animal PP2C. One of the *Arabidopsis* phosphatases, *PP2C-At,* was identified via a

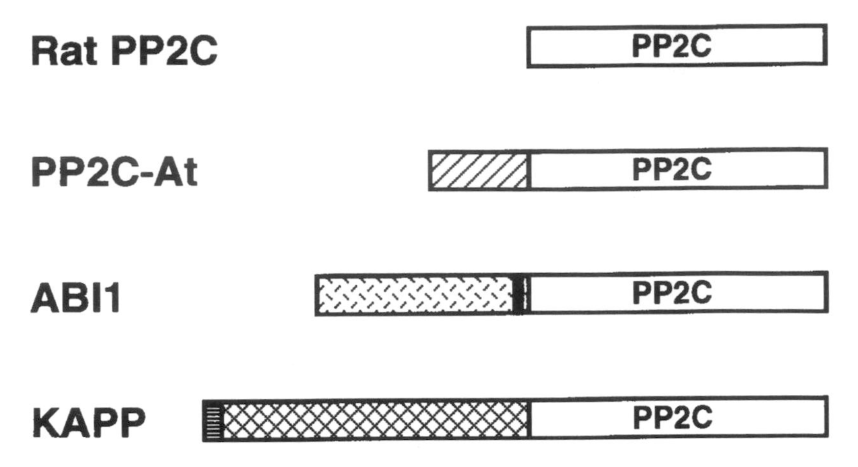

Figure 1 Structural organization of rat and *Arabidopsis thaliana* type-2C protein phosphatases. Conserved catalytic domains are indicated as PP2C. The putative EF-hand calcium binding found in ABI1 is shown as a solid box. The box containing horizontal lines in KAPP designates a signal anchor.

genetic screen for genes that complement the sterile phenotype of the *S. pombe pde1* mutant, which is defective in cAMP phosphodiesterase, a component of the cAMP-dependent protein kinase cascade (62). Among three distinct cDNA clones isolated from this screen, a PP2C that shares 35% identity with the rat PP2C was recovered. Expression of *PP2C-At* in *S. pombe pde1* mutants also restores expression of a transcription factor (*ste11*) that is required for sexual development and whose expression is inhibited by the activity of a cAMP-protein kinase (PKA). This suggests that the plant *PP2C-At* may be counteracting the activity of a PKA in *S. pombe.* Its role in *Arabidopsis,* however, remains unknown.

ABSCISIC ACID RESPONSE GENE—ABI1 Abscisic acid (ABA) regulates multiple functions in plants including embryo maturation, seed dormancy, stomatal closure, and mitosis in root meristems. Several abscisic acid–insensitive (*abi*) *Arabidopsis* mutants that are unresponsive to elevated concentrations of ABA have been extensively characterized (61). The dominant mutant, *abi1,* has been cloned and found to encode a PP2C homologue with two distinct domains: an N-terminal domain containing a putative Ca^{2+}-binding site and a C-terminal PP2C catalytic domain that is 35% identical to the rat PP2C (64, 79). This represents a novel putative Ca^{2+}-regulated protein serine/threonine phosphatase that is structurally unrelated to the PP2B class of Ca^{2+}-dependent phosphatases. Bacterially expressed ABI1 is Mg^{2+}-dependent, but Ca^{2+} regulation of ABI activity has not been reported (79). The presence of a putative Ca^{2+}-binding site suggests that ABI1 may be responsive to ABA-mediated changes in cellular Ca^{2+} levels. The involvement of Ca^{2+} in ABA-dependent stomatal closure is well established (40), and recent observations indicate that the abi1-1 mutant protein interferes with ABA-dependent regulation of K^+ channels in guard cells of transgenic tobacco (38). A Ca^{2+}-activated phosphatase activity has previously been reported in *Vicia faba* guard cells that is inhibited by cyclosporin A-cyclophilin protein complexes (CyP-CsA complex) (67). Cyclosporin A is an immunosuppressant that binds to endogenous cyclophilin proteins and inhibits Ca^{2+}/calmodulin-activated PP2B (105). It is interesting to note that the CyP-CsA complex blocks Ca^{2+}-induced inactivation of inward K^+ channel activity in *V. faba* guard cells (67), which suggests that a PP2B homologue, or possibly ABI1, may be involved in the control of K^+ channel activity. Future studies may resolve whether ABI is a target for CyP-CsA.

The lesion in the *abi1-1* mutant is a single nucleotide substitution that converts Gly-180 to Asp within the catalytic domain (64, 79). The mutation does not alter the ubiquitous expression of ABI1 (64). The dominant nature of the mutation could possibly be explained by an alteration in the response of *abi1* to ABA-mediated cellular changes in Ca^{2+} levels. Constitutive activation of the phosphatase, for instance, could inhibit ABA-stimulated signaling pathways. Alternatively, protein dephosphorylation may be required to turn on the pathways, and activation of ABI1 by Ca^{2+} may be altered in the mutant protein phosphatase.

KINASE-ASSOCIATED PROTEIN PHOSPHATASE—KAPP Additional evidence that type-2C protein phosphatases are involved in plant signaling pathways comes from the identification of a third novel PP2C in *Arabidopsis,* termed "KAPP" for kinase-associated protein phosphatase (121). KAPP was isolated via an in vitro protein interaction screen for proteins that interact with RLK5, a membrane-bound receptor-like protein kinase (RLK) (137). The predicted structure of KAPP indicates that it contains three distinct domains, an N-terminal signal anchor, a kinase interaction domain (KI), and a C-terminal PP2C domain (121). A bacterially expressed KAPP fusion protein demonstrates Mg^{2+}-dependent and okadaic-insensitive activity in vitro, consistent with the classification of this enzyme as a PP2C (121). The signal anchor predicts that KAPP is membrane localized, and translational insertion of KAPP into membrane vesicles, in vitro, supports this hypothesis (JM Stone & JC Walker, personal communication). These results suggest that PP2C is not exclusively cytosolic in the cell. Furthermore, ubiquitous expression of *KAPP* in *Arabidopsis* suggests that PP2C distribution in plants is more extensive than previously indicated from PP2C activities measured in cell extracts. A possible explanation for this difference is that native KAPP may have negligible activity toward nonphysiological substrates commonly used to measure PP2C activity.

The KI domain is both necessary and sufficient for interaction with the receptor-like protein kinase, RLK5 (121). However, interaction between

RLK5 and KAPP occurs only upon autophosphorylation of the receptor kinase domain. In vitro dephosphorylation of RLK5 with recombinant maize PP1 blocks binding by the KI domain. This indicates that sequences bearing phosphoamino acids may act as high-affinity binding sites for KAPP. RLK5 binding to KAPP is akin to the interaction between autophosphorylated protein tyrosine kinases and proteins containing src-homology-2 (SH2) domains in mammalian systems (133). However, unlike SH2 domains that bind exclusively to sites bearing phosphotyrosyl residues, the KI domain may belong to a novel class of protein-protein interacting domains that bind to sequences containing either a phosphoserine or phosphothreonine residue. The KI domain bears no sequence homology with SH2 domains.

By analogy with the regulation of SH2-phosphotyrosine phosphatase (SH2-PTP) activity in animals, it is possible that binding of KAPP to autophosphorylated RLK5 modulates KAPP phosphatase activity in vivo. Phosphorylation of the KI domain by RLK5, which has been observed in vitro, may have additional effects on its binding affinity to RLK5 or on KAPP phosphatase activity. It is not known whether RLK5 is a physiological substrate for KAPP. The structure of KAPP and its interaction with RLK5 suggests that KAPP may control early steps of the RLK5 signaling pathway. RLK5 is a member of a family of related protein kinases that participate in diverse biological functions including self-incompatibility, defense responses, and plant development (121). KAPP and related proteins, therefore, may represent a novel class of protein phosphatases in plants that are early components of RLK-mediated signaling pathways. It will be of interest to learn whether KAPP acts as a positive or a negative regulator of the pathway and whether it interacts specifically with RLK5 or additional phosphorylated receptor and nonreceptor protein kinases in the cell.

OTHER PROTEIN PHOSPHATASES

PROTEIN PHOSPHATASE-X Molecular cloning studies have revealed the presence, in animals and fungi, of a number of protein serine/threonine phosphatases that are structurally related to PP1, PP2A, and PP2B phosphatases but that cannot be placed in any of these classes because of unique structural and/or biochemical features (107). Among these is the rabbit PPX (26), a PP2A-related protein phosphatase that localizes to centrosomes and is thought to play a role in nucleation of microtubules (12). The substrate specificity of rabbit PPX and its sensitivity to various phosphatase inhibitors is similar to PP2A; however, PPX fails to bind PP2A regulatory A-subunits (12). Two cDNA clones showing 83% amino acid identity to the rabbit PPX have been isolated from *Arabidopsis* by hybridization (Table 1) (89). The *Arabidopsis* genes, *ppx1* and *ppx2*, are expressed at low levels in flowers, leaves, stems, and roots. The function of *ppx1* and *ppx2* in *Arabidopsis* is not known.

MITOCHONDRIAL PDC-PHOSPHATASE The plant mitochondrial pyruvate dehydrogenase complex (PDC) is regulated in part by reversible phosphorylation of one of its subunits, pyruvate dehydrogenase E1α (PDH) (80). Phosphorylation by an intrinsic protein kinase inactivates PDH in the light. Conversely, dephosphorylation of PDH in the dark by a loosely associated Mg^{2+}-dependent protein serine/threonine phosphatase activates the complex (81). Calmodulin, monovalent cations, and polyamines do not affect the phosphatase, and inorganic phosphate (P_i) is the only known metabolite that has a slight inhibitory effect at physiological concentrations (92). The plant PDH-phosphatase is similar in its biochemical properties to the mammalian phosphatase with the exception of its response to Ca^{2+}, which enhances Mg^{2+}-activation of the mammalian PDH-phosphatase (28) but antagonizes the plant enzyme (81). Structural characterization of the bovine PDH-phosphatase indicates that it is a novel PP2C that contains a Ca^{2+}-binding site ($K_d \approx 8\ \mu M$) (63). Recombinant bovine PDH-phosphatase expressed in *Escherichia coli* is Mg^{2+}-dependent and Ca^{2+}-stimulated. The similarity of the plant and mammalian PDH-phosphatases suggests the plant enzyme may also be a PP2C homologue, but unlike the mammalian enzyme the plant phosphatase may lack a Ca^{2+} regulatory domain.

CHLOROPLAST THYLAKOID PROTEIN PHOSPHATASE Phosphorylation of chloroplast proteins has been studied extensively over the past decade, but knowledge of the requisite protein kinases and phosphatases is limited (2, 7). Proteins are phosphorylated on serine or threonine almost exclusively, with the exception of pyruvate, P_i dikinase, which contains a phosphorylated histidine residue (97). Phosphorylation of thylakoid proteins occurs by light- and redox-dependent kinase(s), whereas redox-independent dephosphorylation in the dark is catalyzed by a membrane-bound protein phosphatase (or phosphatases), which is sensitive to sodium fluoride and molybdate ions (6, 20, 111, 124). The chloroplast thylakoid phosphatase is probably unrelated to the cytosolic serine/threonine phosphatases, because it fails to dephosphorylate phosphohistone or phosphorylase a and is not inhibited by okadaic acid or microcystin-LR (70, 125). A detailed characterization of the thylakoid phosphatase has been hampered by the lack of a suitable substrate (142). An alternate approach has been the use of synthetic peptides (126), which reveals that at least one thylakoid-bound phosphatase is able to dephosphorylate multiple thylakoid phosphoproteins in vitro (20).

PROTEIN TYROSINE PHOSPHATASE Phosphotyrosine content of proteins in eukaryotes accounts for only a fraction (<3%) of the total phosphoamino acids in the cell, yet reversible phosphorylation on tyrosine residues is essential for cellular growth and differentiation in animals (51, 127). Phosphotyrosine residues have also been reported in plants (19, 131), but little is known of the physiological role of reversible tyrosine phosphorylation in plants. Dual specificity protein kinases that phosphorylate on serine, threonine, and tyrosine residues have been reported in plants (8, 87). Some of these kinases are homologues of mammalian GSK-3, which requires phosphorylation of a specific tyrosine residue for full activation. A tyrosine residue is found at a conserved location on the plant GSK-3 homologues, suggesting that phosphorylation on the residues may also control activity in vivo. A homologue of mammalian mitogen-activated protein kinases (MAPK) has been identified in tobacco and is transiently activated and tyrosine phosphorylated in response to a fungal elicitor (128). Activation of MAPK in animals requires phosphorylation of both a threonine residue and a tyrosine residue. It is not known whether the tobacco MAPK is also phosphorylated on threonine/serine residues and whether this is important for activation. Downregulation of the tobacco MAPK is associated with tyrosine dephosphorylation and indicates that plants may contain endogenous protein tyrosine phosphatases and/or dual specificity phosphatases that can dephosphorylate both phosphoserine/threonine and phosphotyrosine residues (22). Addition of calyculin A to elicitor-treated tobacco sustains the activation and tyrosine phosphorylation of the protein kinase (128), which indicates that a PP1/PP2A may also be involved in turning off the kinase because protein tyrosine phosphatases and dual-specificity phosphatases are resistant to calyculin A. One possible explanation for these results is that a PP1/PP2A is activating an intermediate protein tyrosine phosphatase or a dual-specificity phosphatase. Inactivation of the tobacco kinase may also require dephosphorylation of phosphoserine/threonine residues directly by a PP1/PP2A. Identifying homologues of mammalian protein tyrosine phosphatases in plants has been hampered by the lack of structural information on putative tyrosine phosphatases and because phosphotyrosine phosphatase activity in plants is often attributed to acid phosphatases (36, 53). A 90-kD protein was partially purified from pea nuclei that specifically dephosphorylates phosphotyrosine-containing peptides, is inactive toward phosphoserine and phosphothreonine residues, and has little activity toward *p*NPP, a common substrate used to measure acid phosphatase activity (42). The pea nuclear phosphatase is biochemically more similar to mammalian protein tyrosine phosphatases than the acid or serine/threonine phosphatases, although its exact classification awaits structural characterization.

INHIBITOR STUDIES

The discovery of protein phosphatase inhibitors has significantly advanced our understanding of the biological processes that are controlled by reversible protein phosphorylation and has aided in the identification of physiological substrates for specific classes of protein phosphatases. Most notable among these inhibitors is okadaic acid, a marine toxin that inhibits PP2A (IC_{50} = 0.1–1.0 nM) at concentrations 10- to 100-fold lower than PP1 (IC_{50} = 10–100 nM). Okadaic acid is readily taken up by cells, which provides a useful molecular probe to study the role of PP1 and PP2A in many cellular processes. In vivo sensitivity of PP1 and PP2A to okadaic acid, however, is decreased by high protein concentrations in the cell (107). Thus, while okadaic acid sensitivity of a specific cellular event is indicative of PP1/PP2A involvement, the concentration of okadaic acid used may not distinguish between these two classes of phosphatases. Insight into the differing actions of PP1 and PP2A in particular cellular events has been furthered by the additional use of chemically distinct toxins such as calyculin A, microcystin-LR, and tautomycin (21, 73). Calyculin A and microcystin-LR inhibit PP1 and PP2A with equal potency ($IC_{50} \approx$ 0.1–0.3 nM), whereas tautomycin inhibits PP1 (IC_{50} = 0.16 nM) at concentrations fivefold lower than PP2A (IC_{50} = 8 nM). Comparison of the biological effects of these cell-permeable compounds with those of okadaic acid can provide an initial evaluation of the relative contribution of each phosphatase activity. For example, the use of calyculin A and okadaic acid to study light-dependent signaling pathways in maize protoplasts suggests that a PP1 is involved in activating multiple transcription factors involved in light-induced gene expression in plants (104). The addition of nodularin, acanthifolicin, cantharidin, and the herbicide endothal to the growing arsenal of PP1/PP2A inhibitors should provide additional confidence in our abilities to study the differing actions of PP1 and PP2A in intact cells. However, identification of novel protein phosphatases that are sensitive to these same compounds may ultimately limit identification of specific phosphatases based on inhibitor studies alone and will require additional verification by genetic and/or biochemical experiments. Nonetheless, these inhibitors provide a very powerful approach for the initial assessment of the role of protein phosphorylation in controlling numerous cellular events. More than 40 publications describing the effects of these toxins on diverse biological processes in plants have appeared within the last four years. A summary of many of these studies is outlined in Table 3 and implicates PP1/PP2A or related phosphatase activities in hormonal- (31, 94), carbohydrate- (129), light- (104), and elicitor-mediated signaling pathways; cell cycle regulation (43, 140, 143); growth and development (104, 115); ion channel control (65, 67, 130, 144); pollination (57, 101); and cellular metabolism (15, 39, 46, 47, 60, 70, 71, 75, 78, 110, 138).

A different class of phosphatase inhibitors is found in microbially derived immunosuppressants cyclosporin A and FK506, which, upon binding to endogenous immunophilin proteins, cyclophilin and FK506-binding protein, form inhibitory complexes that target Ca^{2+}/calmodulin-activated PP2B (102). Their

Table 3 Use of phosphatase inhibitors in plants

Biological function	Species	Tissue[a]	I[b]	Response	Refs
Signal transduction					
Auxin	N. plumbaginifolia	C	O	Blocks feedback inhibition of auxin-stimulated gene expression	31
Sugar	Potato	LP	O, C, M	Blocks expression of sporamin, β-amylase, ADP-glucose phosphorylase small subunit. Increases expression of sucrose synthase	129
	Rice	C	O	Increases expression of *αmy3*	68
Light	Maize	P	O, C	Blocks expression of light-inducible genes	104
Ethylene	Tobacco	L	O	Increases expression of ethylene-induced PR genes	94
Pathogens	Tomato	C	O	Blocks elicitor-induced changes in NADH oxidase and ascorbate perodixase activities	136
	Tomato	C	C	Causes growth medium alkalinization, hyperphosphorylation of cellular proteins and increased ACC-S activity	33, 117
	Tobacco	C	C	Causes sustained activation of dual-specificity kinase	128
	Potato	T	O	Increases expression of *PR-10a* and binding of PBF-1	29
	Parsley	P	O	Blocks elicitor-induced furanocoumarin accumulation	96
	Soybean	C, Ct	O, C, M	Increases PAL expression, medium alkalinization and production of isoflavonoid phytoalexins	72
	Soybean	C	O	Increases expression of PAL, CHS, and HRGP	37
Regulation of membrane channels					
K^+ channel	Vicia faba	GC	F, Cs	Blocks Ca^{2+}-induced inactivation of inward current	67, 130
	Vicia faba	GC	O, C	Blocks inward current; no effect on outward current	65
	Vicia faba	MC	O, C	Increases outward current	66
Anion channel	Tobacco		O	Alters voltage gating and kinetics	144
Control of enzyme activity					
SPS	Spinach	L	O, M	Blocks light-activation	47, 110
NR	Spinach	L	O, M	Blocks light-activation	46, 71
	Pea	R	O	Blocks light-activation	39
	B. campestris	L	C	Blocks light-activation	60
PEPCK	Cucumber	Ct	M	Blocks dephosphorylation	138

Table 3 *(continued)*

Biological function	Species	Tissue[a]	I[b]	Response	Refs
	Maize	L	O	Blocks inactivation	78
QDH	Carrot	C	O, M	Blocks inactivation	70
HMG-CoA red	Cauliflower	I	O	Blocks activation of HMG-CoA red	75
Growth and development					
Root	*Arabidopsis*	R, RH	O, C	Inhibits cell division and cortical cell elongation. Blocks root hair development	115
Chlor-oplast	Maize	L	O	Blocks light-induced chlorophyll accumulation	104
Cell cycle	Tobacco	C	O, C	Causes cell cycle arrest	43, 143
	Tradescan-tia	SH	O, M	Alters metaphase transit times and chromosome separation	140
Ca^{2+} uptake	C. roseus	C	O	Blocks Ca^{2+} uptake and callose syn	58
Pollination	*B. napus*	Pi	O, M	Blocks pollen growth	57
	B. napus; B. oleracea; B. campestris	F, FB	O, M	Inhibits pollen tube growth in cross-pollinated flowers. Alters self-incompatability in flower buds	101

[a]Tissues: A, aleurone; C, cell culture; Ct, cotyledon; F, flower; FB, flower bud; GC, guard cell; I, inflorescence; L, leaf; MC, mesophyll cell; LP, leaf petiole; P, protoplast; Pi, pistil; PM, plasma membrane; S, seed; SH, stamen hair cell; T, tuber.

[b]Phosphatase inhibitor: O, okadaic acid; C, calyculin A; M, microcystin-LR; F, FK506 binding protein-FK506 complex; Cs, cyclosporin-cyclophilin complex.

limited use in plants has been discussed above and suggests that a Ca^{2+}-activated protein phosphatase, possibly a PP2B homologue or ABI1, is involved in regulating K^+ channels in guard cells of *V. faba* (67).

CONCLUDING REMARKS

The objective of this review was to provide an overview of recent advances in the study of protein phosphatases in plants. Progress has been made in cloning a number of plant homologues belonging to three major classes of the mammalian protein serine/threonine phosphatases, and a handful of physiological substrates for PP2A have been identified. Still, very little is known about the structure and regulation of the native enzymes or their physiological roles. Molecular genetic and inhibitor studies have provided powerful avenues to begin addressing these areas, but progress on understanding the role of specific protein phosphatases in plants will depend, in large part, on the identification of physiological substrates and the purification of native enzyme complexes using traditional biochemical methods. Future studies will no doubt reveal that many novel protein phosphatases exist in plants that do not belong to the archetypal classes of serine/threonine protein phosphatases. Ultimately, identification and characterization of the protein phosphatases and protein kinases that modulate the phosphorylation status and function of key regulatory proteins in the cell will generate new insights into the role of protein phosphorylation in controlling hundreds of diverse biological events in plants.

ACKNOWLEDGMENTS

We would like to thank Drs. Doug Bush, Dieter Söll, and Sabine Rundle for communication of results prior to publication, as well as Julie Stone and Qing Lin for critical comments on the manuscript.

Literature Cited

1. Alessi DR, Street AJ, Cohen P, Cohen PTW. 1993. Inhibitor-2 functions like a chaperone to fold three expressed isoforms of mammalian protein phosphatase-1 into a conformation with the specificity and regulatory properties of the native enzyme. *Eur. J. Biochem.* 213:1055–66
2. Allen JF. 1992. Protein phosphorylation in regulation of photosynthesis. *Biochim. Biophys. Acta* 1098:275–35
3. Ariño J, Pérez-Callejón E, Cunillera N, Camps M, Posas F, Ferrer A. 1993. Protein phosphatases in higher plants: multiplicity of type 2A phosphatases in *Arabidopsis thaliana. Plant Mol. Biol.* 21:475–85
4. Arundhati A, Feiler H, Traas J, Zhang H, Lunness PA, Doonan JH. 1995. A novel *Arabidopsis* type 1 protein phosphatase is highly expressed in male and female tissues and functionally complements a conditional cell cycle mutant of *Aspergillus. Plant J.* 7:823–34
5. Axton JM, Dombradi V, Cohen PTW, Glover DM. 1990. One of the protein phosphatase 1 isoenzymes in Drosophila is essential for mitosis. *Cell* 63:33–46
6. Bennett J. 1980. Chloroplast phosphoproteins: evidence for a thylakoid-bound phosphoprotein phosphatase. *Eur. J. Biochem.* 104:85–89
7. Bennett J. 1991. Phosphorylation in green plant chloroplast. *Annu. Rev. Plant Physiol. Plant Mol. Biol.* 42:281–331
8. Bianchi MW, Guivarc'h D, Thomas M, Woodgett JR, Kreis M. 1994. *Arabidopsis* homologs of the *shaggy* and *GSK-3* protein kinases: molecular cloning and functional expression in *Escherichia coli. Mol. Gen. Genet.* 242:337–45
9. Bollen M, Stalmans W. 1992. The structure, role, and regulation of type 1 protein phosphatases. *Crit. Rev. Biochem. Mol. Biol.* 27:227–81
10. Booher R, Beach D. 1989. Involvement of a type 1 protein phoshatase encoded by bws1$^+$ in fission yeast mitotic control. *Cell* 57:1009–16
11. Bowler C, Chua N-H. 1994. Emerging themes of plant signal transduction. *Cell* 6:1529–41
12. Brewis ND, Street AJ, Prescott AR, Cohen PTW. 1993. PPX, a novel protein serine/threonine phosphatase localized to centrosomes. *EMBO J.* 12:987–96
13. Budde RJA, Randall DD. 1990. Protein kinases in higher plants. In *Inositol Metabolism in Plants,* ed. DJ Moore, WF Boss, pp. 351–67. New York: Liss
14. Cannon JF, Klemens K, Morcos P, Nair B, Pearson J, Khalil M. 1995. Type 1 protein phosphatase systems in yeast. *Adv. Protein Phosphatases* 9:215–36
15. Carter P, Nimmo H, Fewson C, Wilkins M. 1990. *Bryophyllum fedtschenkoi* protein phosphatase type 2A can dephosphorylate phospho*enol*pyruvate carboxylase. *FEBS Lett.* 263:233–36
16. Casamayor A, Pérez-Callejón E, Pujol G, Ariño J, Ferrer A. 1994. Molecular characterization of a fourth isoform of the catalytic subunit of protein phosphatase 2A from *Arabidopsis thaliana. Plant Mol. Biol.* 26:523–28
17. Chen J, Martin BL, Brautigan DL. 1992.

Regulation of protein serine/threonine phosphatase type-2A by tyrosine phosphorylation. *Science* 257:1261–64
18. Chen J, Parsons S, Brautigan DL. 1994. Tyrosine phosphorylation of protein phosphatase 2A in response to growth stimulation and v-src transformation of fibroblasts. *J. Biol. Chem.* 269:7957–62
19. Cheng HF, Tao M. 1989. Purification and characterization of a phosphotyrosyl-protein phosphatase from wheat seedlings. *Biochem. Biophys. Acta* 998:271–76
20. Cheng L, Stys D, Allen JF. 1995. Effects of synthetic peptides on thylakoid phosphoproteins. *Physiol. Plant.* 93:173–78
21. Cicirelli M. 1992. Inhibitors of protein serine/threonine phosphatases. *Focus* 14: 16–20
22. Clarke PR. 1994. Switching off MAP kinases. *Curr. Biol.* 4:647–50
23. Cohen P. 1989. Structure and regulation of protein phosphatases. *Annu. Rev. Biochem.* 58:453–508
24. Cohen PTW, Cohen P. 1989. Protein phosphatases come of age. *J. Biol. Chem.* 264: 21435–38
25. Crovello CS, Furie BC, Furie B. 1995. Histidine phosphorylation of p-selectin upon stimulation of human platelets: a novel pathway for activation-dependent signal transduction. *Cell* 82:279–86
26. da Cruz e Silva OB, da Cruz e Silva EF, Cohen PTW. 1988. Identification of a novel phosphatase catalytic subunit by cDNA cloning. *FEBS Lett.* 242:10610
27. Damuni Z, Guo H. 1993. Autophosphorylation activated protein kinase phosphorylates and inactivates protein phosphatase 2A. *Proc. Natl. Acad. Sci. USA* 90:2500–4
28. Denton RM, Randle PJ, Martin BR. 1972. Stimulation by calcium ions of the pyruvate dehydrogenase phosphatase. *Biochem. J.* 128:161–63
29. Desprès C, Subramaniam R, Matton DP, Brisson N. 1995. The activation of the potato *PR-10a* gene requires the phosphorylation of the nuclear factor PBF-1. *Plant Cell* 7:589–98
30. Dombradi V, Mann DJ, Saunders RD, Cohen PT. 1993. Cloning of the fourth functional gene for protein phosphatase 1 in *Drosophila melanogaster* from its chromosomal location. *Eur. J. Biochem.* 212: 177–83
31. Dominov JA, Stenzler L, Lee S, Schwarz JJ, Leisner S, Howell SH. 1992. Cytokinins and auxins control the expression of a gene in *Nicotiana plubaginifolia* cells by feedback regulation. *Plant Cell* 4:451–61
32. Evans IM, Fawcett T, Boulter D, Fordham-Skelton AP. 1994. A homologue of the 65-kDa regulatory subunit of protein phosphatase 2A in early pea (*Pisum sativum* L.) embryos. *Plant Mol. Biol.* 24:689–95
33. Felix G, Regenass M, Spanu P, Boller T. 1994. The protein phosphatase inhibitor calyculin A mimics elicitor action in plant cells and induces rapid hyperphosphorylation of specific proteins as revealed by pulse labeling with [^{33}P]phosphate. *Proc. Natl. Acad. Sci. USA* 91:952–56
34. Ferreira PCG, Hemerly AS, Van Montagu M, Inzé D. 1993. A protein phosphatase 1 from *Arabidopsis thaliana* restores temperature sensitivity of a *Schizosaccharomyces pombe cdc25^{ts}/wee1$^-$* double mutant. *Plant J.* 4:81–87
35. François JM, Thompson-Jaeger S, Skroch J, Zellenka U, Spevak W, Tatchell K. 1992. *GAC1* may encode a regulatory subunit for protein phosphatase type 1 in *Saccharomyces cerevisiae. EMBO J.* 11:87–96
36. Gellatly KS, Moorhead GBG, Duff SMG, Lefebvre DD, Plaxton WC. 1994. Purification and characterization of a potato tuber acid phosphatase having significant phosphotyrosine phosphatase activity. *Plant Physiol.* 106:223–32
37. Gianfagna TJ, Lawton MA. 1995. Specific activation of soybean defense genes by the phosphoprotein phosphatase inhibitor okadaic acid. *Plant Sci.* 109:165–70
38. Giraudat J. 1995. Abscisic acid signaling. *Curr. Opin. Cell Biol.* 7:232–38
39. Glaab J, Kaiser WM. 1993. Rapid modulation of nitrate reductase in pea roots. *Planta* 191:173–79
40. Guilroy S, Fricker MD, Read ND, Trewavas AJ. 1991. Role of calcium in signal transduction of *Commelina* guard cells. *Plant Cell* 3:333–44
41. Guo H, Reddy SAG, Damuni Z. 1993. Purification and characterization of a distinct autophosphorylation activated protein kinase that phosphorylates and inactivates protein phosphatase 2A. *J. Biol. Chem.* 268:11193–98
42. Guo Y-L, Roux SJ. 1995. Partial purification and characterization of an enzyme from pea nuclei with protein tyrosine phosphatase activity. *Plant Physiol.* 107:167–75
43. Hasezawa S, Nagata T. 1992. Okadaic acid as a probe to analyse the cell cycle progression in plant cells. *Bot. Acta* 105:63–69
44. Healy AM, Zolnierowicz S, Stapleton AE, Goebl M, DePaoli RA, Pringle JR. 1991. *CDC55*, a *Saccharomyces cerevisiae* gene involved in cellular morphogenesis: identification, characterization, and homology to the B subunit of mammalian type 2A protein phosphatase. *Mol. Cell. Biol.* 11: 5767–80
45. Hubbard MJ, Cohen P. 1993. On target with a new mechanism for the regulation of protein phosphorylation. *Trends Biochem. Sci.* 18:172–77
46. Huber JL, Huber SC, Campbell WH, Redinbaugh MG. 1992. Reversible light/dark

modulation of spinach leaf nitrate reductase activity involves protein phosphorylation. *Arch. Biochem. Biophys.* 296:58–65
47. Huber SC, Huber JL. 1990. Activation of sucrose-phosphate synthase from darkened spinach leaves by an endogenous protein phosphatase. *Arch. Biochem. Biophys.* 282: 421–26
48. Huber SC, Huber JL, Campbell WH, Redinbaugh MC. 1992. Comparative studies of the light modulation of nitrate reductase and sucrose-phosphate synthase activities in spinach leaves. *Plant Physiol.* 100: 706–12
49. Huber SC, Huber JL, Kaiser WM. 1994. Differential response of nitrate reductase and sucrose-phosphatase synthase-activation to inorganic and organic salts, in vitro and in situ. *Physiol. Plant.* 92:302–10
50. Huber SC, Huber JL, McMichael RW. 1994. Control of plant enzyme activity by reversible protein phosphorylation. *Int. Rev. Cytol.* 149:47–98
51. Hunter T. 1995. Protein kinases and phosphatases: the yin and yang of protein phosphorylation and signaling. *Cell* 80:225–36
52. Ingebritsen TS, Cohen P. 1983. Protein phosphatases: properties and role in cellular regulation. *Science* 221:331–38
53. Jagiello I, Donella-Deana A, Szczegielniak J, Pinna LA, Muszynska G. 1992. Identification of protein phosphatase activities in maize seedlings. *Biochim. Biophys. Acta* 1134:129–36
54. Johnson K, DeLong A, Garbers C, Simmons C, Söll D. 1995. Auxin physiology mutants rcn1 and rgr1. *Abstr. Int. Meet. Arabidopsis Res., 6th, Madison, Wis.*, p. 299
55. Kaiser WM, Huber SC. 1994. Modulation of nitrate reductase in vivo and in vitro: effects of phosphoprotein phosphatase inhibitors, free Mg^{2+} and 5′-AMP. *Planta* 193:358–64
56. Kaiser WM, Huber SC. 1994. Posttranslational regulation of nitrate reductase in higher plants. *Plant Physiol.* 106:817–21
57. Kandasamy MK, Thorsness MK, Rundle SJ, Goldberg MI, Nasrallah JB, Nasrallah ME. 1993. Ablation of papillar cell function in *Brassica* flowers results in the loss of stigma receptivity to pollination. *Plant Cell* 5:263–75
58. Kauss H, Jeblick W. 1991. Induced Ca^{2+} uptake and callose synthesis in suspension-cultured cells of *Catharanthus roseus* are decreased by the protein phosphatase inhibitor okadaic acid. *Physiol. Plant.* 81: 309–12
59. Kinoshita N, Ohkura H, Yanagida M. 1990. Distinct, essential roles of type 1 and type 2A protein phosphatases in the control of the fission yeast cell cycle. *Cell* 63: 405–15
60. Kojima M, Wu S-J, Fukui H, Sugimoto T, Nanmori T, Oji Y. 1995. Phosphorylation/dephosphorylation of Komatsuna (*Brassica campestris*) leaf nitrate reductase in vivo and in vitro in response to environmental light conditions: effects of protein kinase and protein phosphatase inhibitors. *Physiol. Plant.* 93:139–45
61. Koornneef M, Reuling G, Karssen CM. 1984. The isolation and characterization of abscisic acid–insensitive mutants of *Arabidopsis thaliana. Physiol. Plant.* 61:377–83
62. Kuromoni T, Yamamoto M. 1994. Cloning of cDNAs from *Arabidopsis thaliana* that encode putative protein phosphatase 2C and a human Dr1-like protein by transformation of a fission yeast mutant. *Nucleic Acids Res.* 22:5296–301
63. Lawson JE, Niu X-D, Browning KS, Trong HL, Yan J, Reed LJ. 1993. Molecular cloning and expression of the catalytic subunit of bovine pyruvate dehydrogenase phosphatase and sequence similarity with protein phosphatase 2C. *Biochemistry* 32: 9887–93
64. Leung J, Bouvier-Durand M, Morris P-C, Guerrier D, Chefdor F, Giraudat J. 1994. *Arabidopsis* ABA response gene *ABI1:* features of a calcium-modulated protein phosphatase. *Science* 264:1448–52
65. Li W, Luan S, Schreiber SL, Assmann SM. 1994. Evidence for protein phosphatase 1 and 2A regulation of K^+ channels in two types of leaf cells. *Plant Physiol.* 106: 963–70
66. Lin PP-C, Mori T, Key JL. 1988. Phosphoprotein phosphatase from soybean hypocotyls. *Plant Physiol.* 66:368–74
67. Luan S, Li W, Rusnak F, Assmann SM, Schreiber SL. 1993. Immunosuppressants implicate protein phosphatase regulation of K^+ channels in guard cells. *Proc. Natl. Acad. Sci. USA* 90:2202–6
68. Lue M-Y, Lee H. 1994. Protein phosphatase inhibitors enhance the expression of an α-amylase gene, α*Amy3,* in cultured rice cells. *Biochem. Biophys. Res. Commun.* 205:807–16
69. MacKintosh C, Cohen P. 1989. Identification of high levels of type 1 and type 2A protein phosphatases in higher plants. *Biochem. J.* 262:335–39
70. MacKintosh C, Coggins J, Cohen P. 1991. Plant protein phosphatase: subcellular distribution, detection of protein phosphatase 2C and identification of protein phosphatase 2A as the major quinate dehydrogenase phosphatase. *Biochem. J.* 273: 733–38
71. MacKintosh C. 1992. Regulation of spinach leaf nitrate reductase by reversible phosphorylation. *Biochim. Biophys. Acta* 1137:121–26
72. MacKintosh C, Lyon GD, MacKintosh

RW. 1994. Protein phosphatase inhibitors activate anti-fungal defence responses of soybean cotelydons and cell cultures. *Plant J.* 5:137–47
73. MacKintosh C, MacKintosh RW. 1994. Inhibitors of protein kinases and phosphatases. *Trends Biochem. Sci.* 19: 444–48
74. MacKintosh RW, Haycox G, Hardie DG, Cohen PTW. 1990. Identification by molecular cloning of two cDNA sequences from the plant *Brassica napus* which are very similar to mammalian protein phosphatases-1 and -2A. *FEBS Lett.* 276: 156–60
75. MacKintosh RW, Davies SP, Clarke PR, Weekes J, Gillespie JG, et al. 1992. Evidence for a protein kinase cascade in higher plants. *Eur. J. Biochem.* 209:923–31
76. Maeda T, Tsai AYM, Saito H. 1993. Mutations in protein tyrosine phosphatase gene (*PTP2*) and a protein serine/threonine phosphatase gene (*PTC1*) cause a synthetic growth defect in *Saccharomyces cerevisiae. Mol. Cell. Biol.* 13:5408–17
77. Mayer-Jaekel RE, Hemmings BA. 1994. Protein phosphatase 2A—a 'menage à trois.' *Trends Cell Biol.* 4:287–91
78. McNaughton GAL, MacKintosh C, Fewson CA, Wilkins MB, Nimmo HG. 1991. Illumination increases the phosphorylation state of maize leaf phospho*enol*pyruvate carboxylase by causing an increase in the activity of a protein kinase. *Biochim. Biophys. Acta* 1093:189–95
79. Meyer K, Leube MP, Grill E. 1994. A protein phosphatase 2C involved in ABA signal transduction in *Arabidopsis thaliana. Science* 264:1452–55
80. Miernyk JA, Camp PJ, Randall DD. 1985. Regulation of plant pyruvate dehydrogenase complexes. *Curr. Top. Plant Biochem. Physiol.* 4:175–90
81. Miernyk JA, Randall DD. 1987. Some properties of pea mitochondrial phospho-pyruvate dehydrogenase-phosphatase. *Plant Physiol.* 83:311–15
82. Moreno S, Nurse P. 1990. Substrates for $P34^{cdc2}$ in vivo veritas? *Cell* 61:549–51
83. Mumby M, Walter G. 1993. Protein serine/threonine phosphatases: structure, regulation and functions in cell growth. *Physiol. Rev.* 73:673–99
84. Nitschke K, Fleig U, Schell J, Palme K. 1992. Complementation of the cs *dis2*-11 cell cycle mutant of *Schizosaccharomyces pombe* by a protein phosphatase from *Arabidopsis thaliana. EMBO J.* 11:1327–33
85. Ohkura H, Kinoshita N, Miyatani S, Toda T, Yanagida M. 1989. The fission yeast $dis2^{+}$ gene required for chromosome disjoining encodes one of two putative type 1 protein phosphatases. *Cell* 57:997–1007
86. Ota IM, Varshavsky A. 1993. A yeast protein similar to bacterial two-component regulators. *Science* 262:566–69
87. Páy A, Jonak C, Bögre L, Meskiene I, Mairinger T, et al. 1993. The msK family of alfalfa protein kinase genes encodes homologues of *shaggy/glycogen synthase kinase-3* and shows differential expression patterns in plant organs and development. *Plant J.* 3:847–56
88. Páy A, Pirck M, Bögre L, Hirt H, Heberle-Bors E. 1994. Isolation and characterization of phosphoprotein phosphatase 1 from alfalfa. *Mol. Gen. Genet.* 244:176–82
89. Pérez-Callejón E, Casamayor A, Pujol G, Clua E, Ferrer A, Ariño J. 1993. Identification and molecular cloning of two homologues of protein phosphatase X from *Arabidopsis thaliana. Plant Mol. Biol.* 23: 1177–85
90. Pirck M, Páy A, Heberle-Bors E, Hirt H. 1993. Isolation and characterization of a phosphoprotein phosphatase 2A gene from alfalfa. *Mol. Gen. Genet.* 240:126–31
91. Polya GM, Haritou M. 1988. Purification and characterization of two wheat-embryo protein phosphatases. *Biochem. J.* 251: 357–63
92. Randall DD, Miernyk JA, David NR, Budde RJA, Schuller KA, et al. 1990. Phosphorylation of the leaf mitochondrial pyruvate dehydrogenase complex and inactivation of the complex in the light. *Curr. Top. Plant Biochem. Physiol.* 9:313–28
93. Ranjeva R, Boudet AM. 1987. Phosphorylation of protein in plants: regulatory effects and potential involvement in stimuli/response coupling. *Annu. Rev. Plant Physiol.* 38:73–93
94. Raz V, Fluhr R. 1993. Ethylene signal is transduced via protein phosphorylation events in plants. *Plant Cell* 5:523–30
95. Refeno R, Ranjeva R, Boudet AM. 1982. Modulation of quinate:NAD^{+} oxidoreductase activity through reversible phosphorylation in carrot cell suspensions. *Planta* 154:193–98
96. Renelt A, Colling C, Hahlbrock K, Nurnberger T, Parker JE, et al. 1993. Studies on elicitor recognition and signal transduction in plant defense. *J. Exp. Bot.* 44:257–68
97. Roeske CA, Kutny RM, Budde RJA, Chollet R. 1988. Sequence of the phosphothreonyl regulatory site peptide from inactive maize leaf pyruvate, orthophosphate dikinase. *J. Biol. Chem.* 263: 6683–87
98. Ruediger R, Hood JEV, Mumby M, Walter G. 1991. Constant expression and activity of protein phosphatase type 2A in synchronized cells. *Mol. Cell. Biol.* 11: 4282–85
99. Ruediger R, Roeckel D, Fait J, Bergqvist A, Magnusson F, Walter F. 1992. Identification of binding sites on the regulatory A

subunit of protein phosphatase 2A for the catalytic C subunit and for tumor antigens of simian virus 40 and polyomavirus. *Mol. Cell. Biol.* 12:4872–82

100. Rundle SJ, Nasrallah JB. 1992. Molecular characterization of a type 1 serine/threonine phosphatase from *Brassica oleracea. Plant Mol. Biol.* 20:367–75
101. Rundle SJ, Nasrallah ME, Nasrallah JB. 1993. Effects of inhibitors of protein serine/threonine phosphatases on pollination in *Brassica. Plant Physiol.* 103:1165–71
102. Rundle SJ, Hartung AJ, Corum JW III, O'Neill M. 1995. Characterization of the 55-kDa B regulatory subunit of *Arabidopsis* protein phosphatase 2A. *Plant Mol. Biol.* 28:257–66
103. Schreiber SL. 1992. Immunophilin-sensitive phosphatase action in cell signaling pathways. *Cell* 70:365–68
104. Sheen J. 1993. Protein phosphatase activity is required for light-inducible gene expression in maize. *EMBO J.* 12:3497–505
105. Shenolikar S. 1992. A window opens on immunosuppression. *Curr. Biol.* 2:549–51
106. Shenolikar S, Nairn AC. 1991. Protein phosphatases: recent progress. *Adv. Second Messenger Phosphoprotein Res.* 23:1–121
107. Shenolikar S. 1994. Protein serine/threonine phosphatases: new avenues for cell regulation. *Annu. Rev. Cell Biol.* 10:55–86
108. Shima H, Haneji T, Hatano Y, Kasugai I, Sugimura T, Nagao M. 1993. Protein phosphatase 1γ2 is associated with nuclei of meiotic cells in rat testis. *Biochem. Biophys. Res. Commun.* 194:930–37
109. Shima H, Hatano Y, Chun Y-S, Sugimura T, Zhang Z, et al. 1993. Identification of PP1 catalytic subunit isotypes PP1γ, PP1δ and PP1α in various rat tissues. *Biochem. Biophys. Res. Commun.* 192:1289–96
110. Siegl G, MacKintosh C, Stitt M. 1990. Sucrose-phosphatse synthase is dephosphorylated by protein phosphatase 2A in spinach leaves. *FEBS Lett.* 270:198–202
111. Silverstein T, Chang L, Allen JF. 1993. Chloroplast thylakoid protein phosphatase reactions are redox-independent and kinetically heterogeneous. *FEBS Lett.* 334:101–5
112. Slabas AR, Fordham-Skelton AP, Fletcher D, Martinez-Rivas JM, Swinhoe R, et al. 1994. Characterization of cDNA and genomic clones encoding homologues of the 65-kDa regulatory subunit of protein phosphatase 2A in *Arabidopsis thaliana. Plant Mol. Biol.* 26:1125–38
113. Smith RD, Walker JC. 1991. Isolation and expression of a maize type 1 protein phosphatase. *Plant Physiol.* 97:677–83
114. Smith RD, Walker JC. 1993. Expression of multiple type 1 protein phosphoprotein phosphatases in *Arabidopsis thaliana. Plant Mol. Biol.* 21:307–16
115. Smith RD, Wilson JE, Walker JC, Baskin TI. 1994. Protein-phosphatase inhibitors block root hair growth and alter cortical cell shape of *Arabidopsis* roots. *Planta* 194: 516–24
116. Smith RD, Lin Q, Cannon JF, Walker JC. 1995. Type-1 and Type-2C protein phosphatases of higher plants. *Adv. Protein Phosphatases* 9:105–20
117. Spanu P, Grosskopf DG, Felix G, Boller T. 1994. The apparent turnover of 1-aminocyclopropane-1-carboxylate synthase in tomato cells is regulated by protein phosphorylation and dephosphorylation. *Plant Physiol.* 106:529–35
118. Stark MJR, Black S, Sneddon AA, Andrews PD. 1994. Genetic analyses of yeast protein serine/threonine phosphatases. *FEMS Microbiol. Lett.* 117:121–30
119. Stitt M, Wilke I, Feil R, Heldt HW. 1988. Coarse control of sucrose-phosphatae synthase in leaves: alteration of the kinetic properties in response to the rate of photosynthesis and the accumulation of sucrose. *Planta* 174:217–30
120. Stone EM, Yamano H, Kinoshita N, Yanagida M. 1993. Mitotic regulation of protein phosphatases by the fission yeast sds22 protein. *Curr. Biol.* 3:13–26
121. Stone JM, Collinge MA, Smith RD, Horn MA, Walker JC. 1994. Interaction of a protein phosphatase with an *Arabidopsis* serine-threonine receptor kinase. *Science* 266: 793–95
122. Stone JM, Walker JC. 1995. Protein kinase families and signal transduction. *Plant Physiol.* 108:451–57
123. Stuart JK, Frederick DL, Varner CM, Tatchell K. 1994. The mutant type 1 protein phosphatase encoded by *glc7-1* from *Saccharomyces cerevisiae* fails to interact productively with the *GAC1*-encoded regulatory subunit. *Mol. Cell. Biol.* 14: 896–905
124. Sun G, Bailey D, Jones MW, Markwell J. 1989. Chloroplast thylakoid protein phosphatase is a membrane surface-associated activity. *Plant Physiol.* 89:238–43
125. Sun G, Markwell J. 1992. Lack of type 1 and type 2A protein serine(P)/threonine (P) phosphatase activities in chloroplasts. *Plant Physiol.* 100:620–24
126. Sun G, Sarath G, Markwell J. 1993. Phosphopeptides as substrates for thylakoid protein phosphatase activity. *Arch. Biochem. Biophys.* 304:490–95
127. Sun H, Tonks NK. 1994. The coordinated action of protein tyrosine phosphatases and kinases in cell signaling. *Trends Biochem. Sci.* 19:480–85
128. Suzuki K, Shinshi H. 1995. Transient activation and tyrosine phosphorylation of a protein kinase in tobacco cells treated with a fungal elicitor. *Plant Cell* 7:639–47
129. Takeda S, Mano S, Ohto MA, Nakamura K.

Inhibitors of protein phosphatase 1 and 2A block sugar-inducible gene expression in plants. *Plant Physiol.* 106:567–74
130. Theil G, Blatt MR. 1994. Phosphatase antagonist okadaic acid inhibits steady-state K^+ currents in guard cells of *Vicia faba. Plant J.* 5:523–30
131. Torruella M, Casano LM, Vallejos RH. 1986. Evidence of the activity of tyrosine kinase(s) and the presence of phosphotyrosine proteins in pea plantlets. *J. Biol. Chem.* 261:6651–53
132. Uemura T, Shiomi K, Togashi S, Takeichi M. 1993. Mutation of twins encoding a regulator of protein phosphatase 2A leads to pattern duplication in *Drosophila* imaginal discs. *Genes Dev.* 7:429–40
133. van der Geer P, Hunter T, Lindberg RA. 1994. Receptor protein-tyrosine kinases and their signal transduction pathways. *Annu. Rev. Cell. Biol.* 10:251–337
134. van Zyl WH, Huang W, Sneddon AA, Stark M, Camier S, et al. 1992. Inactivation of the protein phosphatase 2A regulatory subunit A results in morphological and transcriptional defects in *Saccharomyces cerevisiae. Mol. Cell. Biol.* 12:4946–59
135. Vera-Estrella R, Higgins VJ, Blumwald E. 1994. Plant defense response to fungal pathogens. *Plant Physiol.* 106:97–102
136. Viard M-P, Martin F, Pugin A, Ricci P, Blein J-P. 1994. Protein phosphorylation is induced in tobacco cells by the elicitor cryptogein. *Plant Physiol.* 104:1245–49
137. Walker JC. 1994. Structure and function of the receptor-like protein kinase of higher plants. *Plant Mol. Biol.* 26:1599–609
138. Walker RP, Leegood RC. 1995. Purification, and phosphorylation in vivo and in vitro, of phospho*enol*pyruvate carboxykinase from cucumber cotyledons. *FEBS Lett.* 362:70–74
139. Weiner H, Stitt M. 1993. Sucrose-phosphate synthase phosphatase, a type 2A protein phosphatase, changes its sensitivity towards inhibition by inorganic phosphatase in spinach leaves. *FEBS Lett.* 333:159–64
140. Wolniak SM, Larsen PM. 1992. Changes in the metaphase transit times and the pattern of sister chromatid separation in stamen hair cells of *Tradescantia* after treatment with protein phosphatase inhibitors. *J. Cell Sci.* 102:691–715
141. Yamano H, Ishii K, Yanagida M. 1994. Phosphorylation of dis2 protein phosphatase at the C-terminal cdc2 consensus and its potential role in cell cycle regulation. *EMBO J.* 13:5310–18
142. Yang C-M, Danko SJ, Markwell JP. 1987. Thylakoid acid phosphatase and protein phosphatase activities in wheat (*Triticum aestivum* L.). *Plant Sci.* 48:17–22
143. Zhang K, Tsukitani Y, John CL. 1992. Mitotic arrest in tobacco caused by the phosphoprotein phosphatase inhibitor okadaic acid. *Plant Cell Physiol.* 33:677–88
144. Zimmermann S, Thomine S, Guern J, Barbier-Brygoo H. 1994. An anion current at the plasma membrane of tobacco protoplasts shows ATP-dependent voltage regulation and is modulated by auxin. *Plant J.* 6:707–16

Annu. Rev. Plant Physiol. Plant Mol. Biol. 1996. 47:127–58

THE FUNCTIONS AND REGULATION OF GLUTATHIONE S-TRANSFERASES IN PLANTS

Kathleen A. Marrs

Department of Biological Sciences, Stanford University, Stanford California 94305-5020

KEY WORDS: glutathione S-transferase (GST), herbicides, auxins, oxidative stress

ABSTRACT

Glutathione S-transferases (GSTs) play roles in both normal cellular metabolism as well as in the detoxification of a wide variety of xenobiotic compounds, and they have been intensively studied with regard to herbicide detoxification in plants. A newly discovered plant GST subclass has been implicated in numerous stress responses, including those arising from pathogen attack, oxidative stress, and heavy-metal toxicity. In addition, plant GSTs play a role in the cellular response to auxins and during the normal metabolism of plant secondary products like anthocyanins and cinnamic acid. This review presents the current knowledge about the functions of GSTs in regard to both herbicides and endogenous substrates. The catalytic mechanism of GST activity as well as the fate of glutathione S-conjugates are reviewed. Finally, a summary of what is known about the gene structure and regulation of plant GSTs is presented.

CONTENTS

1040-2519/96/0601-0127$08.00

INTRODUCTION

Glutathione S-transferases (GSTs, E.C. 2.5.1.18) are enzymes that catalyze the conjugation of the tripeptide glutathione (GSH) to a variety of hydrophobic, electrophilic, and usually cytotoxic substrates (see 88, 110). GSTs have been found in virtually all organisms, with the first identification of corn GSTs over 25 years ago (40, 128, 130). Plant GSTs were first identified and have been intensively studied because of their ability to detoxify herbicides, and individual GSTs conferring herbicide tolerance have been characterized from most major crop species. Recently, another plant GST subclass has been implicated in numerous stress responses, including those arising from pathogen attack, oxidative stress, and heavy-metal toxicity. In addition, plant GSTs play a role in the cellular response to auxins and during the normal metabolism of plant secondary products like anthocyanins and cinnamic acid.

The subject of plant GSTs has not been reviewed in this series in the 25 years in which plant GSTs have been studied. However, this first review is quite timely: In the past three years, a vast amount of new information concerning plant GSTs has appeared. The scope of this review encompasses the current knowledge about the enzymatic activity of GSTs toward both herbicides and endogenous substrates, and the review evaluates the very recent data concerning the roles of GSTs in response to auxin, oxidative stress, lipid peroxidation, and in defense against pathogens.

FUNCTIONS OF GSTs IN CELLS: VARIATIONS ON A BASIC THEME

GSTs are most often thought of as detoxification enzymes and indeed were first discovered for their ability to metabolize a wide variety of toxic exogenous compounds (xenobiotics) via GSH conjugation (see 88). However, GSTs have since been found to function in numerous cellular processes that, despite the apparent diversity, all have in common a more basic theme—the recognition and transport of a broad spectrum of reactive electrophilic compounds, regardless of whether the compound is of an exogenous "xenobiotic" origin or of an endogenous "natural" origin.

Detoxification by GSH Conjugation

A key role played by GSTs is their ability to inactivate toxic compounds. An astonishing array of toxic defense compounds are produced by organisms in competition with one another to survive (137). For example, plants synthesize a wide variety of complex secondary products such as phytoalexins, opiates, and flavones, which function as a defense against herbivores and pathogens. Many insects secrete the bitter-tasting alkaloids they obtain from feeding on plants to deter predation from birds or other animals. The modern chemical industry has developed numerous herbicides, insecticides, and synthetic drugs to control certain so-called "pests" in our environment.

The ability to detoxify such harmful compounds is crucial to the survival of cells and organisms. Most organisms have countered exposure to toxic chemicals with the development of detoxification systems to transform, metabolize, and eliminate such compounds from tissues. Remarkably, a common pathway exists in most organisms for the detoxification of electrophilic compounds, which is mediated by three groups of enzymes (63, 119). Phase I (transformation) enzymes such as cytochrome P450 monooxygenases introduce functional groups onto substrates. Phase II (conjugation) enzymes such as UDP:glucosyltransferases and GSTs utilize the functional group as a site of further conjugation, usually resulting in a less toxic and more water-soluble conjugate. Most organisms encode multiple isoforms of both phase I and phase II enzymes, each with their own substrate specificity. Phase III (compartmentation) enzymes, ATP-dependent membrane pumps, recognize and transfer conjugates across membranes for excretion or sequestration. In both plants and animals, the glutathione pump recognizes glutathione S-conjugates for transfer across membranes; in animals, this allows for their excretion from the body (62, 63, 88, 119, 120). Plants have no excretion system, and instead, glutathione S-conjugates are either sequestered in the vacuole or transferred to the apoplast, processes termed "storage excretion" (76, 80, 93, 119, 120).

It should be stressed that these "detoxification enzymes" are remarkably versatile in the recognition of substrates. In plants, members of these same enzyme families also play normal, defined roles in secondary metabolism and thus recognize and transport both natural and xenobiotic substrates (119). For example, both herbicide metabolism as well as anthocyanin biosynthesis require the coordinate activities of various isoforms of cytochrome P450s, GSTs, and the glutathione pump (see 90, 119, and references therein).

Targeting for Transmembrane Transport

GSH conjugation "tags" numerous endogenous substrates in addition to xenobiotic substrates for recognition by the glutathione pump. GSH conjugation is an essential step in the biosynthesis of compounds like leukotrienes in animals; their extracellular biological activity requires transport by the glutathione pump (63, 118, 134). In plants, many secondary metabolites are phytotoxic, even to the cells that produce them, and thus targeting to the appropriate cellular localization, usually the vacuole, is crucial (94, 119, 120). Anthocyanin pigments require GSH conjugation for transport into the vacuole; inappropriate cytoplasmic retention of the pigment not only prevents the production of anthocyanins but is also toxic to cells (see 90 and references therein).

Protection of Tissues from Oxidative Damage

The endogenous products of oxidative damage initiated by hydroxyl radicals are highly cytotoxic. This includes membrane lipid peroxides, such as 4-hydroxyalkenals, as well as products of oxidative DNA degradation, such as base propanols. Plant and animal GSTs conjugate GSH with such endogenously produced electrophiles, which results in their detoxification (4, 6, 19, 35, 88, 110, 148). Some GSTs also function as glutathione peroxidases to detoxify such products directly (4, 148, 164).

Ligandins: Nonenzymatic Binding and Intracellular Transport

In addition to their catalytic function, GSTs serve as nonenzymatic carrier proteins (ligandins) involved in the intracellular transport of steroids, bilirubin, heme, and bile salts in animal cells (71, 84, 85). Compounds that bind GSTs as nonsubstrate ligands do so at a site other than the catalytic site of the enzyme. In plants, some GSTs apparently serve as carriers of the natural auxin indole-3-acetic acid (IAA), as at least two research groups have identified active GSTs as auxin-binding protein without detecting the formation of IAA-GSH conjugates (8, 68). This nonenzymatic binding may allow temporary storage or modulation of IAA activity or IAA uptake from membranes and trafficking to receptors (8, 68). It has been proposed that the ligandin function of GSTs prevents cytotoxic events that could result from the excessive accumulation of molecules at membranes or within cells (84).

GSTs: EVOLUTION AND CATALYTIC MECHANISM

Catalytic Mechanism of GSTs

GSTs catalyze the nucleophilic attack of the sulfur atom of glutathione (GSH, γ-glutamylcysteinylglycine) to the electrophilic center of a wide variety of substrates (see 88, 110). GSTs typically function as either heterodimers or homodimers of subunits with molecular weights in the range of 23 and 29, though microsomal GSTs in animals are trimeric or tetrameric and have subunits of ~14 kD (26, 102), and a maize GST with activity against phenylpropanoids functions as a 30-kD monomer (21). Each subunit has a GSH binding site (G-site) and an adjacent electrophilic substrate binding site (H-site) (88, 116). Substrate specificity of the G-site is high, with only GSH and structurally related molecules serving as substrates (21, 80, 88, 116). In contrast, GSTs have a broad specificity for the electrophilic substrate (Figure 1). Talalay (147) has suggested that compounds that induce GSTs or that are recognized as substrates share a common chemical signal—carbon-carbon double bonds adjacent to an electron-withdrawing group (Figure 1). This feature, termed a "Michael acceptor," is either contained naturally by GST substrates or acquired by phase I metabolism (147).

GSTs are products of gene superfamilies, each producing isozymes with broad substrate specificities. Despite the fact that they perform a similar reaction, GSTs share very little overall amino acid sequence identity, typically no more than 25–35%, though there are usually regions of much higher localized similarity within the N-terminus (34, 88). Given the array of compounds recognized as GST substrates, sequence diversity is to be expected. Well-conserved amino acids are likely to be important in common functions, i.e. GSH binding or catalysis. X-ray crystallography and site-directed mutagenesis demonstrate that a highly conserved N-terminal arginine residue is involved in GSH binding at the G-site, and an N-terminal tyrosine residue facilitates formation of the thiolate (GS-) ion (see 82, 109, 113, and references therein). Specific glutamine, proline, histidine, and aspartic acid residues bind GSH or maintain enzyme structure (see 82, 110, 113, and references therein). Site-directed mutagenesis or X-ray crystallographic structures of plant GSTs have not yet been reported, although plant GSTs also possess the well-conserved N-terminal arginine residues (34).

The standard experimental assay for GST activity utilizes 1-chloro-2,4-dinitrobenzene (CDNB, Figure 1), a model substrate for most, but not all, GSTs. Conjugation of CDNB with GSH (by chlorosubstitution) results in a change in absorbance of the compound at 340 nm, providing a simple spectrophotometric assay (89). Not all GSTs can use CDNB as a substrate, and thus the activity of certain GSTs may be underestimated or even undetected when using this assay.

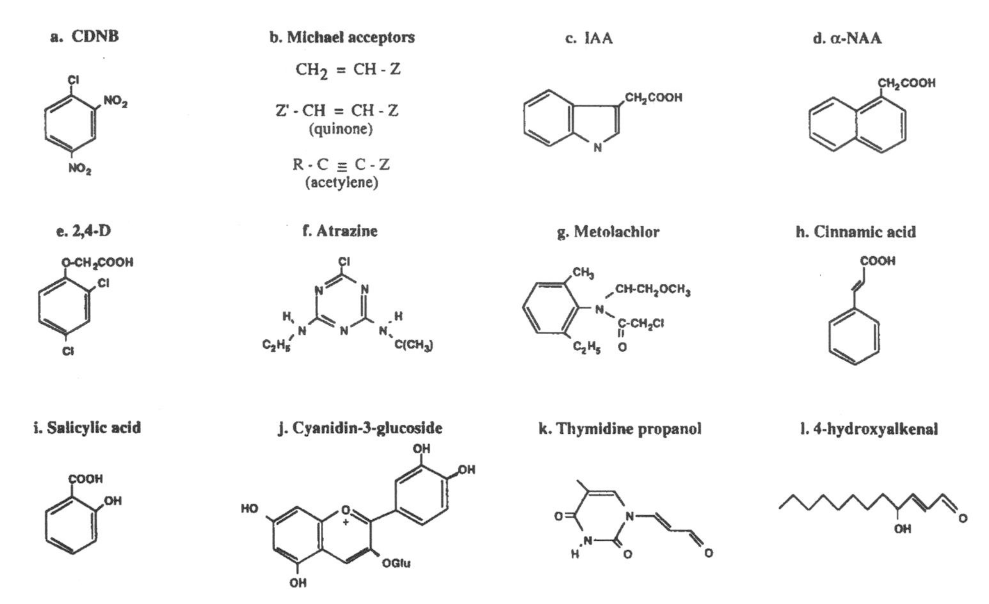

Figure 1 Compounds that are substrates of GSTs, induce GST gene expression, or bind GST noncatalytically: (*a*) CDNB: 1-chloro-2,4-dinitrobenzene, a model substrate of most GSTs used to experimentally assay GST activity; (*b*) examples of Michael acceptor structure, where Z = electron withdrawing groups such as NO_2, CHO, $COCH_3$, CN, $CONH_2$ (147); (*c*) IAA: indole-3-acetic acid, an inducer; (*d*) α-NAA: α-napthylene acetic acid, an inducer; (*e*) 2,4-D: 2,4-dichlorophenoxyacetic acid, an inducer; (*f*) atrazine, an S-triazine herbicide and GST substrate; (*g*) metolachlor, a chloroacetamide herbicide and known GST substrate; (*h*) cinnamic acid, a substrate; (*i*) SA: salicylic acid, an inducer; (*j*) cyanidin-3-glucoside, a substrate of the *bz-2* GST; (*k*) thymidine propanol, a toxic product of oxidative DNA degradation and substrate of animal GSTs; (*l*) 4-hydroxyalkenal, a toxic product of lipid peroxidation and substrate of animal GSTs.

Evolution of Glutathione and GSTs

GSH and GSTs are widely distributed in aerobic organisms and are hypothesized to have evolved in aerobic bacteria for their ability to prevent oxygen toxicity (39, 88, 109). GSH is synthesized from amino acids by the sequential action of γ-glutamylcysteine synthetase and glutathione synthetase (1). In addition to serving as the thiol substrate of GSTs, GSH is a substrate of glutathione peroxidase and glutathione reductase (1, 98, 114) and in plants is a precursor to phytochelatins, heavy-metal binding proteins (112, 135).

In mammals, GST subunits are classified into alpha, mu, pi, sigma, and theta classes on the basis of amino acid identity, immuno-crossreactivity, and substrate specificity (12, 87). The theta class is evolutionarily the most ancient group and is found in vertebrates, Drosophila, plants, and *Methylobacterium* (88, 109). This suggests that the theta class is representative of a progenitor GST that arose before the divergence of prokaryotes and eukaryotes and that these GSTs may still function in the detoxification of reactive oxidation products (109). All plant GSTs studied thus far are most similar to the theta class, although a carnation GST also shares similarities in sequence and intron position to mammalian alpha class GSTs, suggesting that the evolution of the alpha class GSTs may have begun before the plant and animal kingdoms diverged (65). The emergence of the alpha, mu, and pi GST classes may have provided animals and fungi with new detoxification enzymes to counter defense compounds produced by plants (109). Gene duplication, followed by exon shuffling, of an ancestral GSH binding protein may have been a mechanism that generated the different catalytic properties of the members of the GST superfamily (88).

CLASSES OF PLANT GSTs

Although all known plant GSTs are most similar to the theta class, they can be further classified into several subgroups, as proposed recently by Droog et al (32). This classification is attractive in that it is based on amino acid sequence identity and conservation of intron:exon placement and not on substrate recognition or antibody crossreactivity, features that are far more difficult to fully characterize. A phylogenetic tree of plant GSTs is shown in Figure 2. The only GSTs that are not classified are those for which sequence information is not yet available, and several more subgroups may be discovered as more sequences become available. For some of the genes in each subgroup, the intron:exon structure has not been determined either because a cDNA was cloned or because the gene was only partially sequenced, and thus placement in the group has been by amino acid sequence alone. As yet, there is no unified

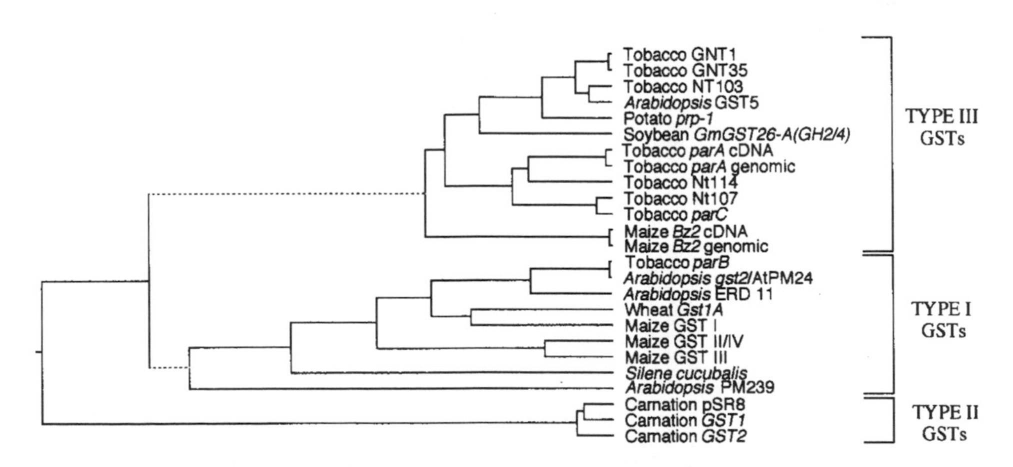

Figure 2 Phylogenetic tree of plant type I, II, and III GSTs. The tree was constructed using the DNASTAR sequence program. The sequences aligned are listed at the right hand of the figure and were obtained by retrieval of protein sequences from Genbank or PIR as listed in Table 1.

Table 1 Plant GSTs

	Protein	Amino acids per subunit	Substrates	Protein or mRNA induced by:	Database accession number	Comments	References
Type I GSTs							
Maize							
GST I	Constitutive homodimer, 29 kD	227	alachlor, atrazine, CDNB	flurazole, dichlormid	X06755	first published sequence of a plant GST	47, 100, 103, 125, 163
GST II	Heterodimer; 27-kD subunit-inducible, 29-kD subunit-constitutive	223 (27 kD), 227 aa (29 kD)	alachlor, CDNB	flurazole, dichlormid	U12679 (27 kD)	not induced by heat, drought, chilling, or pathogens	11, 56, 67, 163
GST III	Constitutive homodimer, 26 kD	222	alachlor, metolachlor, CDNB	dichlormid	X04375, X04453		47, 100, 107, 125
GST IV	Inducible homodimer, 27 kD	223	chloroacetamide and S-triazines, metolachlor	benoxacor	U12679	100% nucleotide sequence identity with GST II 27-kD subunit; CDNB not a substrate	60, 61, 67
Wheat *GstA1*(gene), cDNA WIR56	GST 29, 29 kD	229	CDNB (weak)	Transcripts induced over 20X with pathogen attack; induced by GSH	X56012	involved in onset of induced resistance to powdery mildew; not induced by xenobiotics	35, 95
Tobacco							
parB (gene)	27 kD	213	CDNB	mRNA induced by 2,4-D	D90500	Location is cytoplasmic, particularly around plastids	140, 144, 146
Arabidopsis thaliana							
PMA239X14	Constitutive 25-kD GST and glutathione peroxidase	218	CDNB (weak); high affinity for lipid peroxides	developmentally downregulated	X68304	not induced by wounding, IAA, 2,4-D, methyl jasmonate	4
AWI24	n.d.	n.d.	n.d.	mRNA induced by wounding and ethylene; developmentally regulated	n.a.	only partially sequenced	72
gst2/Atpm24	Microsomal GST and glutathione peroxidase, ABP, 24 kD	212	cumeme hydroperoxide, trans-stillbene oxide, CDNB; binds IAA	mRNA induced by ethylene; developmentally regulated	Atpm24: X75303 *gst2*: L11601	Cinnamic acid and IAA are not substrates	166

ERD11	23.5 kD	208	n.d	mRNA induced by dehydration stress	D17672	Not induced by 2,4-D, 6-BAP, ABA, GA	73
ERD13	24.2 kD	215	n.d.	mRNA induced by dehydration stress	D17673	Not induced by 2,4-D, 6-BAP, ABA, GA	73
Broccoli	26.5 kD	n.d.	CDNB, 4-nitrophenethyl bomide	n.d.	n.a.	only partially sequenced; competitively inhibited by 2,4-D	86
Sugarcane	Heterodimer, 22.5 kD and 24 kD	n.d.	dichloronitrobenzene, ethacrynic acid, and other electrophilic substrates	n.d.	n.a.	only partially sequenced; competitively inhibited by 2,4-D	131
Silene cucubalis	Constitutive	213	n.d.	n.d.	cDNA: M84968 genomic: M84969		75, 111
Hyoscyamus muticus							
Hmgst-1	25-kD auxin-binding protein	212	CDNB; also binds IAA and 2,4-D	2,4-D; 2,4,5-T; 2,3-D	X78203	binds IAA at noncatalytic site, 2,4-D at catalytic site	7, 8
Type II GSTs							
Carnation *GST1*, (pSR8), *GST2*	25 kD	220	unknown; speculate lipid peroxides	ethylene- and senescence-regulated	pSR8: M64268 GST1: L05915 GST2: L05916	Genes are 95% similar in sequence and intron placement, 10 exons, 9 introns	64, 65, 99
Type III GSTs							
Soybean *GmHsp26A* or *GH2/4*	26 kD	225	CDNB	mRNAs induced by HS, active auxins, auxin analogues, heavy metals, SA, ABA, GA, kinetin, PEG	M20363 PIR: A33654	Cadmium induces splicing failure; 14-kD protein produced	17, 48, 155, 156
Potato *prp1-1* (gene), renamed *gst1*	26-kD, ABP, and pathogenesis-related protein PRP1-1	217	CDNB; binds IAA	mRNA induced by fungal infection	J03679	mRNA not induced by abiotic stimuli (wounding or CdCl). IAA and lipid peroxides are not substrates	49, 150

Table 1 Plant GSTs (*continued*)

	Protein	Amino acids per subunit	Substrates	Protein or mRNA induced by:	Database accession number	Comments	References
Tobacco							
parA/Nt114	GST 3-1, 25.7 kD	223	CDNB (weak)	mRNA induced by 2,4-D, weakly by IAA and NAA, Cd (no effect on splicing), SA	cDNA: M29274 genomic: D90215 (*parA*)	Expressed during Go to S mRNA not induced by HS, ABA, cytokinins; localized to nucleus	10, 32, 34, 139, 140, 142, 145, 157
parC/Nt107	GST 2-1, 25 kD	221	CDNB	mRNA induced by 2,4-D and weakly by IAA and NAA; slightly heat-inducible	NT107: X56266 *parC:* X64398	not induced by Cd	10, 32, 34, 143, 157
Nt103	GST 1-1, 25.7 kD	223	CDNB	mRNA induced by 2,4-D and weakly by IAA and NAA	NT103:X56263 GNT1: X56268 GNT35: X56269		10, 32, 34, 157
Nicotiana plumbaginfolia msr1 (pLS216)	25 kD	219	n.d.	auxin- and cytokinin-regulated	n.a.		30
Maize							
Bronze-2	26 kD	241, 78-bp intron	CDNB and natural substrate cyanidin-3-glucoside (anthocyanin precursor)	Induced by Cd, ABA, arsenite	U14599 PIR: JQ0987	last gene in anthocyanin biosynthesis; not induced by 2,4-D, HS; Cd induces splicing failure	90, 91, 104, 123
Arabidopsis thaliana							
GST 5	25.9 kD	224	CDNB; IAA, 2,4-D competitively inhibit enzyme activity, SA noncompetitively	mRNA induced by wounding, HS	D44465	mRNA not induced by IAA	162

Unclassified GSTs							
Sorghum GSTs 1–6	2 GSTs constitutive, 4–5 additional GSTs safener-induced	n.d.	CDNB, metolachlor	Safeners (flurazole, oxabetranil, napthalic anhydride) induce 4-5 metolachlor GSTs	n.a.		24, 25
Chickpea							
GST I, II	Homodimers of 29 kD	n.d.	oxadiazon	oxadiazon-induced	n.a.		59
GST III, IV, IV	Heterodimers, 27kD and 29 kD	n.d.	oxadiazon	oxadiazon-induced	n.a.		59
Pea							
soluble	37 kD	n.d.	cinnamic acid	n.d.	n.a.		29, 41
soluble	47 kD	n.d.	fluorodifen	n.d.	n.a.		29, 41
microsomal	n.d.	n.d.	cinnamic acid, benzopyrene	n.d.	n.a.		29, 41
Wheat							
GST 25	Constitutive, 25 kD	n.d.	CDNB	Protein induced by Cd, atrazine, paraquat, alachlor	n.a.		95
GST 26	Inducible, 26 kD	n.d.	CDNB	Protein induced by Cd, atrazine, paraquat, alachlor	n.a.		95
Arabidopsis thaliana							
GST 1	n.d.	n.d.	n.d.	pathogen infection	n.a.		126, 168
Picea abies (Norway spruce)							
GST I	Homodimer, 26 kD	n.d.	CDNB, DCNB		n.a.		124
GST II	homodimer, 23 kD	n.d.	CDNB, DCNB, fluorodifen, alachlor		n.a.		124
Zea mays	Functions as 30-kD monomer	n.d.	*trans*-cinnamic acid, p-, o-, and m-coumaric acid, ferulic acid, coniferyl alcohol	enzyme induced by p-coumaric acid, 7-hydroxycoumarin	n.a.	does not have activity with CDNB, metolachlor, atrazine, EPTC sulfoxide or lipid peroxides; can utilize cys or GSH as thiol substrate	23
Phaseolis vulgaris (French bean)	Cytosolic	n.d.	cinnamic acid	enzyme induced by fungal elicitor	n.a.	enzyme crossreacts with corn GST antibodies	36

nomenclature for plant GSTs as there is for animal GSTs (87). A brief summary of the three types of plant GSTs follows; the reader should consult Table 1 for specific information about different properties of each GST.

Type I

These GSTs (where the gene structure is known) contain three exons and two introns (Table 1). This group includes the four distinct maize GSTs that differ with respect to subunit composition and substrate specificity toward herbicides (11, 47, 56, 60, 61, 67, 100, 103, 107, 125, 128, 130, 153, 163). Some type I GSTs appear to function as defense genes or cellular protectant genes, producing proteins in response to pathogen attack, wounding, senescence, and the resulting lipid peroxidation that accompanies these processes (4, 35, 72, 168). This is discussed in depth in the section on Oxidative Stress. Other type I GSTs are induced in response to auxins (144, 146) and may serve a ligandin function toward IAA (7, 8, 166). Type I GSTs have also been cloned from *Silene* (75, 111), sugarcane (131), broccoli (86), and *Arabidopsis* (73), though the functions of these GSTs are as yet unclear.

Type II

These GSTs contain ten exons and nine introns and have only been reported in carnation (64, 65, 99) as ethylene- and senescence-related genes expressed in floral organs (Table 1). Significant amino acid sequence homology exists with type III GSTs, but the intron:exon patterns are characteristic of mammalian alpha class GSTs.

Type III

These GSTs (where the gene structure is known) contain two exons and one intron (Table 1). This subclass was originally identified as a set of homologous genes from a variety of species that were inducible by a range of different treatments—particularly auxin, but also ethylene, pathogen infection, heavy metals, and heat shock (see Table 1). Each successive gene isolated was noted to be involved in stress responses, although the functions of the proteins were not apparent from sequence identity. The genes have alternatively been called multiple stimulus response (*msr*) genes (30), auxin-regulated genes (34, 141), or the auxin-regulated gene (ARG) subgroup (162). Only after several years were these genes recognized to encode GST enzymes.

The first gene, soybean *Gmhsp26-A* (discovered independently as *GH2/4*), was both heat shock- and auxin-inducible (17, 48) but shared only a slight resemblence to other small heat shock proteins (HSPs) and contained neither a characteristic heat shock element (HSE) or an auxin-regulated element (AuxRE) in the promoter (3, 17, 48). However, the gene was induced by a wide variety of chemicals, including 2,4-dichlorophenoxyacetic acid (2,4-D), abscisic acid (ABA), cadmium, and other heavy metals, and intron processing was inhibited by cadmium. The potato *prp-1* gene, induced during fungal infection, was noted to be highly similar to *Gmhsp26-A* (150). A number of homologous auxin-regulated tobacco genes were cloned and found to also be induced by a variety of other stress-inducing agents (10, 30, 34, 139, 140, 142, 143, 145, 157). The maize anthocyanin biosynthetic gene *Bronze-2* (*Bz-2*) was next added to this group (104, 123); as with *hsp26-A, Bz2* intron processing is inhibited by cadmium (91) and also varies with environmental conditions (104).

Droog et al (34) were the first to report that the tobacco Nt103 protein had substantial GST activity. They proposed that the entire auxin-regulated tobacco gene family, as well as the other homologous genes, may be GSTs, and they renamed the protein GST1-1. This unexpected finding prompted investigators to examine the functions of the other proteins, such that now most of the genes have been shown to encode proteins with GST activity. GST(CDNB) activity was shown for the protein enoded by the *GH2/4* gene (156), which is identical to the protein encoded by the *Gmhsp26-A* gene, now renamed *GmGST26-A* (E Czarnecka, personal communication); the potato PRP1 protein, now renamed GST1 (49); the tobacco Nt107/*parC* protein, now renamed GST2-1 (32); the Nt114/*parA* protein, now renamed GST3-1 (32, 140); the *Arabidopsis* GST5 protein (162); and the maize BZ-2 protein, which was found to also conjugate GSH to cytoplasmic anthocyanin precursors, allowing transport of anthocyanins into the vacuole (90).

At least one type III GST, the parA protein, may be localized to the nucleus (140). parA is homologous to the *E. coli* stringent starvation protein (SSP) that binds RNA polymerase during nutrient starvation and thus may function in transcriptional regulation during stress (140). Alternatively, parA may function as a GST in the nucleus to detoxify cytotoxic DNA degradation products formed during oxidative stress, which are known to be substrates for mammalian GSTs (6). Although the exact function of parA is unknown, its nuclear localization provides several interesting possibilities.

Unclassified GSTs

GST activity has been demonstrated in over 33 plant species (see 80), although in many cases individual GST proteins have not been purified. In some species, individual GSTs have been characterized, but amino acid sequence is not yet available, preventing their being grouped with any of the above classes (Table 1). In wheat, GST25 and GST26 were strongly induced by cadmium, atrazine, paraquat, and alachlor but not by pathogen attack (95). Sorghum, chickpea, and Norway spruce each contain several herbicide-active GSTs (22, 23, 46, 59, 124). In pea, there are both soluble and microsomal GSTs with activity toward herbicide safeners and the natural substrate cinnamic acid (29).

Maize and French bean each contain GSTs with activity toward cinnamic acid and other phenylpropanoid substrates (21, 36).

HERBICIDE RESISTANCE AND GSTs

The first known function for plant GSTs was in the metabolism of herbicides to nontoxic forms (40, 128, 130), and intensive research over the last 25 years has resulted in the identification of numerous GSTs conferring tolerance to various herbicides. Characteristics of individual GSTs catalyzing herbicide detoxification, selective induction of GSTs by herbicide safeners, and identification of the genes involved in herbicide metabolism are of great interest for many reasons; this research not only provides the knowledge to selectively manipulate GST activity or herbicide tolerance but can also be extended to provide a framework for understanding the metabolism of endogenous compounds in plants.

Selective Herbicides

Numerous synthetic herbicides have been developed to control weeds in important agronomic crops. Several comprehensive reviews on the mechanism of herbicide action have been published (57, 101). Selective herbicides are those that kill various weed species without causing significant damage to crop species. Herbicide selectivity can be partially attributed to differences in herbicide uptake or in the method of application. However, physiological and genetic evidence gathered over the past 25 years clearly indicates that frequently the principle determinant of herbicide selectivity in plants is the ability to metabolize and thus detoxify herbicides. There are many documented cases where the formation of the herbicide-GSH conjugate in the resistant but not in the susceptible species is responsible for herbicide selectivity (44, 74, 76, 81, 107, 127). Several specialized reviews thoroughly discuss this subject (16, 76, 79, 80, 127, 129).

It was first shown in the early 1970s that GSH conjugation is responsible for herbicide detoxification (40, 128, 130). GSTs are not equally distributed among plants, and plants with higher GST activity levels will withstand exposure to herbicides that kill susceptible species (76, 79, 127, 129). For example, maize and sorghum are tolerant to atrazine because they contain high levels of the GST that catalyzes atrazine-GSH conjugation, resulting in the conversion of atrazine to a nontoxic, water-soluble form. Susceptible species (pea, oats, wheat, barley, and many broadleaf weeds) do not have this GST activity. Similarly, GSTs have been shown to be responsible for the tolerance of maize to alachlor (107), metolachlor (22, 45, 61, 159), and EPTC sulfoxide (81). GSTs that catalyze the formation of fluorodifen-GSH conjugates are higher in resistant species (cotton, corn, peanut, pea, soybean, okra) than in susceptible species (tomato, cucumber, squash) (41, 76). Acifluoren selectivity results from a more rapid rate of GSH conjugation in soybean than in weed species (42), and high GST levels in sorghum confer tolerance to metolachlor (23, 44, 46).

Even within species, GSTs vary greatly in substrate specificity. To date, at least 6 GST isoforms have been characterized in maize. GSTI (29 kD) and GSTIII (26 kD) are constitutive homodimers with activity against CDNB and alachlor (100, 103, 107, 153). GSTII is a safener-induced heterodimer (27 and 29 kD) also active against CDNB and alachlor (56, 100, 153), and GSTIV is a safener-induced homodimer of 27 kD (identical to the 27-kD subunit of GSTII) active against alachlor and metolachlor but not CDNB (45, 61). The BZ-2 GST (26 kD) is active against CDNB and the anthocyanin precursor cyanidin-3-glucoside (90). The sixth GST is a 30-kD monomer active against phenylpropanoids but not CDNB, metolachlor, or atrazine (21). Variations of individual GSTs between different cultivars can be responsible for differential herbicide tolerance (121). For example, the atrazine-sensitive maize strain GT112 was shown to have less than 1% of the GST activity of the atrazine-resistant strain GT112 RfRf (128).

Herbicide Safeners

The usefulness of some herbicides can be extended by certain chemicals, referred to as herbicide safeners, antidotes, or protectants. Grass crops such as maize and sorghum have little intrinsic tolerance to thiocarbamate (e.g. EPTC) and chloroacetamide herbicides (e.g. metolachlor, alachlor), but pretreatment with safeners such as flurazole, dichlormid, and benoxacor greatly enhances the tolerance to these herbicides by selectively inducing GST activity and, in turn, elevating the rate of herbicide detoxification via GSH conjugation (22, 23, 44, 46, 53, 54, 60, 61, 81, 103, 159, 163). In this way, safeners decrease herbicide injury to crop species without decreasing injury to the weed species. The safeners themselves are not toxic but do exhibit structural similarities to herbicides and appear to act by inducing gene expression (53, 54, 163). In some cases, pretreatment with low doses of an herbicide enhances the capacity of tissues to metabolize the same herbicide during a following exposure, mimicking the effect of a safener (38, 66).

PHENYLPROPANOIDS AS NATURAL SUBSTRATES OF PLANT GSTs

Many plant secondary metabolites are protective compounds that can be toxic not only to herbivores and pathogens but also to the cells that produce them. The enzymes involved in biosynthesis of these compounds—cytochrome P450s, UDP glucosyltransferases, and GSTs—are also the same enzymes of xenobiotic metabolism (119). Recent evidence suggests that naturally synthesized plant metabolites are recognized, transported, and metabolized in similar ways as herbicides and other xenobiotics.

Anthocyanins and GSTs

Anthocyanin pigments are synthesized in the cytoplasm but are ultimately localized to the vacuole (58, 94, 160). In maize, mutations in the anthocyanin biosynthetic gene *Bronze-2* impart bronze pigmentation to cells that results from the inappropriate accumulation of the anthocyanin precursor cyanidin-3-glucoside in the cytoplasm, causing localized necrosis, poor vigor, or even death of plants. BZ-2 activity results in the transfer of cyanidin-3-glucoside to the vacuole, where the anthocyanin takes on its deep red or purple color. Recently, BZ-2 was shown to be a GST that catalyzed the formation of anthocyanin-GSH conjugates (90), which allows transport into vacuoles by what appears to be the glutathione pump. Thus, cyanidin-3-glucoside is a natural GST substrate. At least one other maize GST can apparently recognize cyanadin-3-glucoside as a substrate, though only weakly, as some *bz-2* mutant maize ears contain pale pink rather than bronze kernels, in contrast with the deep purple color resulting from BZ-2 action (161).

In *Petunia,* the anthocyanin biosynthetic gene *An13* has recently been cloned and shown to have significant homology with type I GSTs and, to a lesser extent, BZ-2 (E Souer & R Koes, personal communication). Complementation studies have shown that BZ-2 is able to provide the GST activity to restore red pigmentation in *an13* mutant *Petunia* flower sectors, and likewise, *An13* is able to restore purple pigmentation in *bz-2* maize tissues (MR Alfenito, personal communication). In addition, the GmGST26-A (GH2/4) GST gene complements the *bz2* mutation, producing purple pigmentation in *bz-2* maize tissues (MR Alfenito, personal communication).

Phytoalexins, Cinnamic Acid, and GSTs

One mechanism by which plants defend themselves against microbial infection or fungal attack is through the formation of phytoalexins at infection sites (20). Phytoalexins are synthesized in inclusions only after exposure to elicitors such as cell wall fragments of plants, fungi, or microorganisms (20). Because these inclusions release cytotoxic phytoalexins into the cytoplasm the cell dies, but the growth of bacteria and fungi is inhibited in the process (20). Phytoalexin synthesis in response to fungal elicitors has been shown to be strongly inhibited by cinnamic acid, a precursor to phytoalexins. Glutathione S-cinnamoyl transferases (GCSTs) have been characterized in French bean, pea, and corn (21, 22, 24, 29, 36) and can be induced in response to fungal elicitors (36). GCSTs catalyze conjugation of cinnamic acid with GSH and have been hypothesized to remove inhibitory cinnamic acid during the initial steps of phytoalexin synthesis (5, 36) and to reduce accumulation of other toxic phenolic compounds produced under stress conditions (21).

AUXINS AND GSTs

Many plant GST genes are auxin-inducible; indeed, the type III GSTs are often referred to as auxin-regulated genes (32, 34, 141). However, "auxin" is a generic term for compounds that induce shoot elongation; auxins need only resemble one another in their physiological action and not necessarily in their chemical structure (154). Since the discovery of the natural auxin IAA, numerous synthetic compounds have been developed with similar physiological activity, including α-napthoxyacetic acid (α-NAA) and the chlorophenoxyacetic acids 2,4-D and 2,4,5-T (see Figure 1). Many synthetic auxins are potent herbicides because they are very stable and not subject to destruction by the IAA-oxidase system (2). 2,4-D is also widely used in research as the auxin component of plant tissue culture media. The next section attempts to clarify the role of GSTs in binding, and perhaps detoxifying, auxins from different sources.

Some GSTs Are Auxin-Binding Proteins

In the search for cellular auxin receptors, some auxin-binding proteins (ABPs) have been identified as active GSTs (68). The Hmgst-1 protein from *Hyoscyamus muticus* bound tritiated 5-azido-IAA by photoaffinity labeling and had sequence homology with GSTs and showed GST activity in vitro (7). The *Hmgst-1* cDNA is highly similar to the tobacco type I GST *parB* (8). By examining the ability of different auxins and inactive auxin analogues to competitively inhibit GST(CDNB) activity, Bilang & Sturm (8) proposed that different auxins bind different sites on the GST, with IAA, α-NAA, and indole-3-butyric acid binding to a noncatalytic site and 2,4-D, 2,4,5,-T and inactive auxin analogues (2,3-D, 2,4,6-T) binding to a catalytic site.

Other ABPs have also been found to be GSTs. The *Arabidopsis* Atpm24 protein, isolated from plasma membrane vesicles by photoaffinity labeling, had sequence homology to type I GSTs and GST activity in vitro (166). The potato PRP1 protein, a type III GST, also bound 5-azido-IAA (49). A 65-kD ABP has recently been isolated from mung bean with N-terminal amino acid sequence homologous to Hmgst-1 and Atpm24, and also to other type I GSTs (A Jones, personal communication). Homologous genes for the 65-kD protein appear to be present also in *Arabidopsis,* maize, wheat, tomato, *Petunia,* and soybean.

Possible Functions of GSTs as Auxin-Binding Proteins

Several possible roles of GSTs as auxin-binding proteins can be envisioned. IAA may bind to GSTs as a nonsubstrate ligand and thus serve a ligandin function. In animals, binding of nonsubstrate ligands occurs at a site distinct from the catalytic site, usually in the C-terminus (113), and in a similar manner, IAA binding is proposed to occur at a noncatalytic site (8). This noncatalytic binding might allow for temporary storage, transport, or uptake of IAA. The expression of Hmgst-1 in stems, where basipetal IAA movement takes place, would suggest that this GST is involved in intracellular IAA transport (8).

IAA may also be conjugated with GSH, perhaps for modulation of activity, temporary storage as a conjugate, or cellular detoxification. In vivo, IAA is conjugated with amino acids or glucose to escape inactivation by decarboxylation (2, 15, 106). It has been proposed that IAA conjugates are "slow-release" forms of free IAA in plant tissues (51). Many GST substrates are glucosylated before GSH conjugation; however, IAA-GSH or IAA-cys conjugates have not been detected, and the Atpm24 GST and the potato PRP1-1 GST were unable to catalyze the formation of IAA-GSH in vitro (49, 166). Nevertheless, the lack of evidence does not eliminate the possibility that IAA-GSH conjugates are formed, but it does indicate that further research must be done to determine the metabolic routes of IAA in vivo.

Auxins Both Bind GST Proteins and Induce GST Gene Expression

While many plant GST genes are auxin-inducible, several recent findings have shown that, unlike other auxin-regulated genes such as the SAURs, *pGH3, Aux22, Aux28,* and *pIAA4/5* (see 141), the auxin-regulated GSTs were also induced by inactive auxin analogues and many other compounds with no hormonal activity. The soybean *GH2/4* gene is inducible not only by strong auxins (2,4-D; 2,4,5-T; α-NAA) and salicylic acid (SA) but also by numerous other electrophilic compounds, including weak auxins (β-NAA), inactive auxin analogues (2,3-D and 2,3,6-T), and inactive SA analogues (3-hydroxybenzoic acid, 4-hydroxybenzoic acid) (155, 156). The GST(CDNB) activity of the tobacco GST1-1 (Nt103) and GST2-1 (Nt107) was competitively inhibited by 2,4-D but also by compounds that are structural analogues of 2,4-D but that are not active auxins—phenoxyacetic and benzoic acid derivatives containing at least one chlorine atom—and not by the natural plant auxins IAA, indole-3-butyric acid, or by α-NAA (32). The *Hmgst1* ABP mRNA was not induced by IAA or α-NAA but was induced by 2,4-D and 2,4,5-T as well as 2,3-D (8).

Thus it is clear that different auxins induce and bind GSTs differently. These differences are not contradictory but indicate that GSTs have several functions in the response to auxins: They not only bind the natural auxin IAA, possibly for regulation of activity, transport within cells, or glutathione conjugation, but also are induced by, and possibly detoxify, numerous electrophilic substrates, some of which have auxin activity. It is very likely that auxins from various sources can be perceived by cells either as true hormones, binding GSTs as a ligand and inducing expression of some GSTs and other so-called genuine auxin-regulated genes (141), or as electrophilic xenobiotic substances to induce GST gene expression selectively.

OXIDATIVE STRESS AND GSTs

In addition to being inducible by auxins and inactive auxin analogues, many plant GSTs are induced by heavy metals, pathogen attack, wounding, ethylene, and ozone. It has been suggested that a common effect of all the processes is the generation of active oxygen species (AOS) produced during oxidative stress and that the GSTs induced respond to oxidative stress to protect cellular components from damage (83, 151, 155, 156).

Active Oxygen Species

Aerobic organisms unavoidably produce AOS, such as superoxide radicals (O_2-•), hydroxyl radicals (OH•), and hydrogen peroxide (H_2O_2), during oxygen consumption. Under normal conditions, antioxidant defenses like superoxide dismutase (SOD), catalase, peroxidase, ascorbic acid, and glutathione prevent AOS formation or scavenge those already present (1, 55, 98, 114, 122). During oxidative stress, the balance between AOS production and antioxidant scavenging is shifted in favor of AOS production (122). One particularly harmful effect of AOS production is membrane lipid peroxidation, which converts fatty acids into hydrocarbon fragments such as highly cytotoxic 4-hydroxyalkenals (Figure 1) (37, 69, 110). These reactive electrophiles have strong inhibitory effects on enzymes such as adenylate cyclase and inhibit DNA and protein synthesis (19, 37, 69). Other consequences of AOS production include the formation of base propenals, highly cytotoxic products of oxidative DNA damage (Figure 1) (6).

Sources of AOS in Plants

PATHOGEN ATTACK, SALICYLIC ACID, AND H_2O_2 Pathogen attack in plants rapidly initiates a variety of defense responses, termed the hypersensitive response (HR). During this process, AOS are produced, rather than destroyed, to kill or damage the invading pathogen; within 2–3 min of infection there is a rapid and transient generation of H_2O_2 known as the oxidative burst (13, 35, 83, 96, 115, 136, 151, 152). The H_2O_2 generated serves to reinforce cell walls, restricting fungal growth, and at higher concentrations results in hypersensitive cell death that limits pathogen spread (83, 151). In addition, H_2O_2 has been shown to act as the primary signal to rapidly and selectively stimulate the transcription of GSTs and glutathione peroxidases (83, 151). The GSTs induced in turn detoxify lipid peroxides such as 4-hydroxyalkenals or 13-hydroperoxylinoleic acid by conjugation with GSH (4, 19, 88) and function as glutathione peroxidases that act on toxic base propenals such as thymidine hydroperoxide (6). Glutathione peroxidases also catalyze the GSH-dependent reduction/inactivation of H_2O_2, forming glutathione disulfide (GSSG) and increasing GSH synthesis by feedback induction (55, 98, 132, 133). Elevated GSH levels induce transcription of other defense genes, including those for phytoalexin synthesis that kill the pathogen, pathogenesis-related (PR) proteins such as chitinases that attack fungal cell walls, and, in some cases, GSTs (31, 83, 95, 165).

Some plants initially infected by pathogens protect themselves against subsequent infection by a process termed systemic acquired resistance (SAR). Salicylic acid (SA), a phenylpropanoid derivative synthesized during pathogen attack, is involved in transducing the signal for SAR (14, 70, 117, 158) and has been proposed to prolong the production of H_2O_2 by binding and inhibiting the enzyme catalase, which normally destroys H_2O_2 (14, 70).

The GST genes known to be induced by pathogen attack include the wheat *GstA1* mRNA (35, 95), the potato *prp-1* mRNA (150), and the *Arabidopsis AWI24* mRNA (72). In addition, lipid peroxides such as 13-hydroperoxylinoleic acid and 13-hydroperoxylinolenic acid are known substrates of the *Arabidopsis* PM239 GST (4), and cumene hydroperoxide (a model substrate for glutathione peroxidases) is a substrate of Atpm24 (166). Exogenous GSH added to cells induces the wheat *GstA1* and the soybean *GH2/4* in ways almost indistinguishable from those induced by fungal elicitor (31, 83, 95, 165). SA has been shown to induce Nt103, Nt107, and Nt114 and *prp-1* and *GmGST26-A(GH2/4)* genes (10, 49, 155, 156), and H_2O_2 itself induces the soybean *GmGST26-A(GH2/4)* gene (83, 155, 156).

H_2O_2 is thus a key regulatory molecule in the response to infection (83, 151), and its ability to selectively induce a subset of defense genes (cellular protectant genes like GSTs and glutathione peroxidases) without directly inducing other defense genes indicates that other conditions that generate oxidative stress may activate GST genes through the production of H_2O_2. Figure 3 shows a model in which conditions that induce oxidative stress also induce GSTs, with the generation H_2O_2 and other AOS being a central feature of this response.

HEAVY METALS Heavy metals such as copper and cadmium induce AOS through a series of redox reactions, leading to oxidative stress and lipid peroxidation (27, 28, 50, 69). Heavy metals also induce phytochelatin (PC) synthesis, heavy metal binding peptides of the structure (γ-glu-cys)$_n$gly that are synthesized directly from glutathione (γ-glu-cys-gly) (112, 135). Consumption of GSH by PC synthesis impairs the overall AOS scavenging system, producing a rise in AOS production (25, 98). Thus, the induction of GST genes by cadmium and

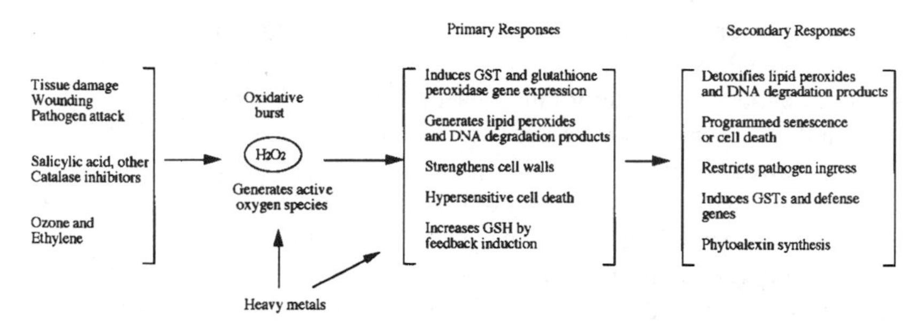

Figure 3 In response to pathogen attack, tissue damage, and other stress-inducing agents, active oxygen species are generated and lipid peroxidation and DNA damage are triggered. Active oxygen species such as H_2O_2 induce GST levels, and GSTs, in turn, metabolize the toxic products of lipid peroxidation and DNA damage. See text for further details.

other heavy metals may also be occurring through the generation of oxidative stress. Heavy metals induce the *GmGST26-A(GH2/4), parA,* NT103, NT107, NT114, and *Bz-2* genes (10, 17, 34, 91, 139, 157) and the wheat GST25 and GST26 proteins (95).

OZONE AND ETHYLENE Ozone in plant cells is rapidly dissolved in water and converted to H_2O_2 (126). Ozone exposure also results in stress-induced ethylene biosynthesis (97, 126). Ethylene plays a key role in programmed senescence, which involves membrane degradation and lipid peroxide production as well as GST gene induction (64, 65, 99, 126, 138, 152). The ethylene-regulated carnation *GST1* and *GST2* genes are induced during programmed senescence, possibly to maintain cell integrity against lipid peroxidation until death, which is necessary for proper nutrient remobilization (65, 99). The ethylene-regulated *Arabidopsis* GSTpm239 is known to be active against lipid peroxides but is downregulated at anthesis and may thus contribute to senescence by the inability to eliminate the products of lipid peroxidation (4). Ozone induces the *Arabidopsis* GST1 gene (126).

REGULATION OF GST GENE EXPRESSION

ocs Elements of GST Genes

The mechanism by which xenobiotics and other electrophiles induce animal GST genes has been intensively investigated in animals. Xenobiotic Regulatory Elements (XREs) with the core sequence GCGTG are found in multiple copies in the promoters of P450 genes and GST genes (116). The Antioxidant Response Element (ARE) or Electrophile Responsive Element (EpRE) consists of two nonoverlapping core sequences GTGACA(A/T)(A/T)GC that are binding sites for the Activator Protein-1 (AP-1) transcription factor complex (18, 43). Daniel has proposed that GST genes containing an EpRE are induced by electrophiles and conditions that generate oxidative stress (18).

Plant GST promoters have not been found to contain functional XREs or EpREs. However, several plant GST promoters were recently found to contain *ocs* (octopine synthase) elements (167). These 20-bp elements were first identified in promoters of genes from the plant pathogens cauliflower mosaic virus (CaMV) and *Agrobacterium tumefaciens* and are activated by wounding (167). There is some similarity between *ocs* elements and EpREs. Both have tandem duplication of binding sites: *ocs* elements contain a tandem core sequence of ACGT (167), and the EpRE contains tandem AP-1 sites (18, 43); and both elements are binding sites for dimeric b-ZIP transcription factors: Fos/Jun proteins bind the AP-1 site and OCSBF-1 and ASF1 bind the *ocs* element (18, 43, 167).

The only plant gene promoters containing *ocs* elements are GSTs: the soybean *GmGST26-A(GH2/4);* the wheat *GstA1;* the tobacco Nt103, Nt107, and Nt114; *parA* and *parB;* and the *Silene* GST (155) (Table 2). For at least one of these genes, the soybean *GH2/4* gene, the *ocs* elements have been

Table 2 Sequence comparisons of *ocs*-like elements in plant pathogen promoters and plant GST promoters. The *ocs*-like elements shown are the *ocs* from the octopine synthase promoter, *as-1* from the CaMV 25S promoter, *nos* from the nopaline synthase promoter, GNT25 and GNT1 from the tobacco GST1-1 (NT103) promoter, the wheat Gst1A promoter, the *Silene cucubalis* GST promoter, the tobacco *parA* promoter, and the soybean *GH2/4,hsp26-A* promoter. Core sequences are represented by arrows. Location of the element from the transcription startsite is indicated, as are the number of bases that match the 20-bp *ocs* consensus element. See text for details.

ocs	aaACGTAAGCGCTtACGTAC	-175	16/20
as-1	TGACGTAAGgGaTGACGcAc	-63	16/20
nos	TGAgcTAAGCaCatACGTcA	-111	14/20
GNT35	TtAgcTAAGtGCTtACGTAt	-323	14/20
GNT1	atAgcTAAGtGCTtACGTAt	-279	13/20
Wheat *GstA1*	atcCGTAccaaCgcACGTgt	-294	9/20
Silene GST	caACGTcAGaGtatACGTAt	-273	12/20
parA	TtACGcAAGCaaTGACaTct	-16	13/20
GH2/4, GmGST26-A	TGAtGTAAGaGaTtACGTAA	-309	16/20
Consensus	TGACGTAAGCGCTGACGTAA		

shown to confer inducibility by not only strong auxins and SA but also by weak auxins, inactive auxin analogues, and inactive SA analogues, as well as cadmium, GSH, H_2O_2, methyl jasmonate, and wounding (155, 156).

Thus, the *ocs* elements within plant GST promoters also appear to be stress-inducible elements that, like AP-1 sites, respond to a variety of electrophilic agents that includes not only biologically active hormones but also inactive hormone analogues and other agents such as heavy metals that generate conditions of oxidative stress (155, 156).

Other Promoter Elements of GST Genes

Some GST genes contain promoter elements that are more selectively induced. At least one type I GST, *parB,* does contain two domains also found in Auxin Regulatory Elements (AuxREs) (3, 146). These AuxREs flank an *ocs* site that responds only to 10–20-fold higher auxin levels, which again suggests that the *ocs* element mediates a stress response rather than a true auxin response (146). The promoter of the carnation type II GST gene contains a 126-bp ethylene-responsive element (ERE) (64). The ERE contains an 8-bp sequence found in other ethylene-regulated genes and also contains one AP-1 motif (64). The promoter of the potato *prp-1* gene (or *gst1* gene) contains a 273-bp region that is selectively induced by infection with *Phytophthora infestans* and not by other environmental cues (92). The *Bz-2* gene contains numerous promoter sequences necessary for regulation by the maize transcriptional regulators *R* and *C* during anthocyanin biosynthesis (9), and, though not yet tested functionally, *Bz-2* and *GmGST26-A(GH2/4)* (both of which are induced by heavy metals) contain promoters with putative metal regulatory elements (MREs) (17, 91).

mRNA Stability Determinants in GST Genes?

Several plant GSTs genes, including *parA, parC, GmGST26-A(GH2/4), prp-1, Bz-2,* and *msr1,* contain one or more copies of the sequence ATTTA in the 3′ untranslated region, a sequence shown to be an instability determinant that targets mRNAs for degradation by RNAses (see 144 and references therein). This instability determinant in the Drosophila *gstD21* gene confers mRNA stability in the presence of pentobarbitol, an inducer of this GST (149). Whether the putative instability determinants in the plant GSTs actually influence mRNA stability, however, has not been tested.

FATE OF GLUTATHIONE S-CONJUGATES

The Glutathione Pump

In animals, glutathione S-conjugates of xenobiotics as well as endogenous substrates such as leukotrienes are actively eliminated or secreted from the cell by an ATP-dependent transmembrane glutathione pump or GS-X pump, found in the liver, kidneys, and other organs (63). The pump has at least three essential domains: a P-domain that undergoes phosphorylation, a G-domain that recognizes GSH, and a C-domain with affinity toward the electrophilic moiety of glutathione S-conjugates (62). The GS-X pump is distinct from other ATP-dependent membrane pumps that transport toxic material such as the multidrug resistance protein (MDR), the bile salt ATPase, and the yeast heavy-metal transporter (HMT-1) (62, 63, 93, 108).

In plants, which have no excretion system, soluble glutathione S-conjugates are stored in the vacuole. The glutathione S-conjugates of *N*-ethylmaleimide and of metolachlor were taken up by vacuoles by an ATPase biochemically identical to the GS-X pump, on the basis of inhibition by vanadate, ATP-dependence, and the recognition of numerous glutathione S-conjugates (93). Anthocyanin-GSH conjugates also appear to be transported into vacuoles by this pump. Thus, anthocyanins are an endogenous substrate for the GS-X pump (90). This common mechanism allows plants to sequester structurally similar but functionally diverse molecules in the vacuole and suggests that a functionally similar mechanism for the transmembrane transport of both xenobiotics and endogenous glutathione S-conjugates operates in both plants and animals.

Metabolism of Glutathione Conjugates

In plants, glutathione S-conjugates are metabolized to other products by a complex network of processing reactions (76, 77, 79, 80). The following is a brief summary of known metabolic routes of xenobiotics in plants, as shown in Figure 4. Most herbicide-GSH conjugates in plants are rapidly metabolized by peptidases to yield to cysteine conjugates, as are many glutathione conjugates in animals (76, 77). The cysteine conjugates of herbicides in many plant species are often malonated, or they can undergo transamination to yield a thiolactic acid conjugate (78). In plants, *N*-malonylcysteine conjugates occupy the equivalent metabolic position in plants as mercuric acids (the acetylated cysteine conjugates that are excreted from the body) occupy in animals (76, 77). *N*-malonylcysteine conjugates of propachlor, metolachlor, butachlor, EPTC, PCNB, and fluorodifen have been identified from several plants, suggesting that this type of conjugate is formed from many types of glutathione conjugates (78).

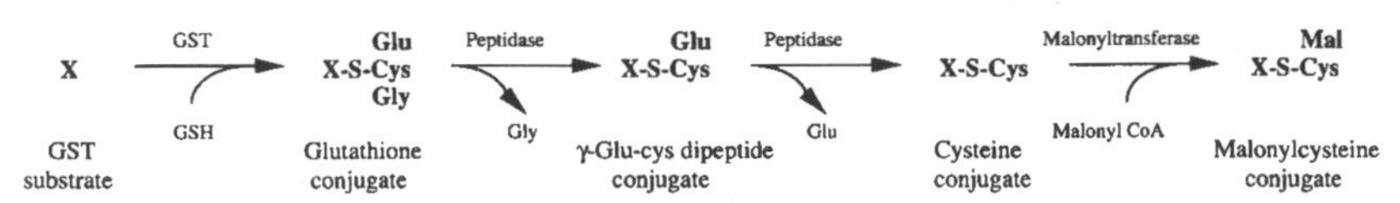

Figure 4 Metabolism of glutathione conjugates.

It is possible that the fate of anthocyanin-GSH conjugates parallels the fate of herbicide-GSH conjugates, as malonated anthocyanins are commonly found in the vacuole (52, 90). Thus, formation of malonylcysteine derivatives of glutathione S-conjugates might occur for both xenobiotic as well as endogenous compounds. In addition to the formation of soluble conjugates destined for the vacuole, glutathione S-conjugates of both herbicides and of coumarin derivatives such as phytoalexins may become associated with cell wall components such as pectin, hemicellulose, or lignins and targeted for extracellular deposition into the apoplast as an insoluble conjugate termed a "bound residue" (76, 80, 105, 120). This possibility for bidirectional transport of glutathione conjugates differs from metabolism of glutathione S-conjugates in animals, although whether a GS-X pump exists in the plant plasma membrane is unknown (120).

SUMMARY AND PERSPECTIVE

Knowledge about the functions of plant GSTs is growing at a rapid rate. We have known for over 25 years that GSTs are an integral part of the mechanism by which plants defend themselves against herbicide exposure. During the last few years GSTs have been shown to function in the response to hormones and oxidative stress, as well as toward substances such as anthocyanins and IAA, such that we may reasonably conclude that GSTs are involved in numerous processes involving the recognition and transport of electrophilic compounds in plant cells. Although identification of the exact functions of GSTs in many of these processes is still to come, the rapid progress being made in the field of plant GSTs is sure to define these roles more clearly in the near future.

Given the conservation of mechanisms allowing for the recognition and transport of glutathione S-conjugates in both animals and plants, it appears that plants utilize GSTs for as many purposes as animals do. Even within one plant species, GSTs vary greatly in amino acid sequence, regulation, and substrate specificity. It is likely that most, if not all, plants will be shown to have GSTs with interesting substrate specificities and mechanisms of gene regulation. New plant GSTs are being identified and cloned at a rapid rate, and through the use of mutant analysis, genetic engineering, or site-directed mutagenesis, the consequences of manipulating GST levels may provide another way to unravel the functions of GSTs in response to multiple environmental stimuli.

ACKNOWLEDGMENTS

I wish to thank the numerous colleagues who provided preprints and unpublished results during the preparation of this manuscript. I would also like to thank Drs. Mark Alfenito, Alan Jones, Gerald Lamoureux, Bert van der Zaal, and Virginia Walbot for critical reading of this manuscript.

Literature Cited

1. Alscher RG, Hess JL, eds. 1993. *Antioxidants in Higher Plants.* Boca Raton, FL: CRC
2. Andreae WA, Good NE. 1957. Studies on 3-indoleacetic acid metabolism. IV. Conjugation with aspartic acid and ammonia as processes in the metabolism of carboxylic acids. *Plant Physiol.* 32:566–71
3. Ballas N, Wong L-M, Theologis A. 1993. Identification of the auxin-responsive element (AuxRE) in the primary indole acetic acid–inducible gene PS-IAA4/5 of pea (*Pisum sativum*). *J. Mol. Biol.* 233: 580–96
4. Bartling D, Radzio R, Steiner U, Weiler EW. 1993. A glutathione S-transferase with glutathione peroxidase activity from *Arabidopsis thaliana:* molecular cloning and functional characterization. *Eur. J. Biochem.* 216:579–86
5. Barz W, Mackenbrock U. 1994. Constitutive and elicitation induced metabolism of isoflavones and pterocarpans in chickpea (*Cicer arietinum*) cell suspension cultures. *Plant Cell Tissue Organ Cult.* 38: 199–212
6. Berhane K, Widersten M, Engstrom A, Kozarich J, Mannervik B. 1994. Detoxication of base propenals and other a,b-unsaturated aldehyde products of radical reactions and lipid peroxidation by human gluta thione S-transferases. *Proc. Natl. Acad. Sci. USA* 91:1480–84
7. Bilang J, Macdonald H, King PJ, Sturm A. 1993. A soluble auxin-binding protein from *Hyoscyamus muticus* is a glutathione S-transferase. *Plant Physiol.* 102:29–34
8. Bilang J, Sturm A. 1995. Cloning a characterization of a glutathione S-transferase that can be photolabelled with 5-azido-indole-3-acetic acid. *Plant Physiol.* 109:253–60
9. Bodeau JP, Walbot V. 1992. Regulated transcription of the maize *Bronze-2* promoter in electroporated protoplasts requires the *C1* and *R* gene products. *Mol. Gen. Genet.* 233:379–87
10. Boot KJM, van der Zaal BJ, Velterop J, Quint A, Mennes AM, et al. 1993. Further characterization of expression of auxin-induced genes in tobacco (*Nicotiana tabacum*) cell suspension cultures. *Plant Physiol.* 102:513–20
11. Bridges I, Bright S, Greenland A, Holt D, Jepson I, Schuch W. 1993. Plant-derived enzyme and DNA sequences, and uses thereof. In *Publication No. WO 93/01294.* Geneva, Switzerland: World Intellect. Prop. Organ.
12. Buetler TM, Eaton DL. 1992. Glutathione S-transferases: amino acid sequence comparison, classification, and phylogenetic relationship. *J. Environ. Sci. Health Part C Environ. Carcinogen. Ecotoxicol. Rev.* 10: 181–203
13. Chai HB, Doke N. 1987. Superoxide anion generation: a response of potato leaves to infection with *Phytophthora infestans. Phytopathology* 77:645–49
14. Chen Z, Silva H, Klessig DF. 1993. Active oxygen species in the induction of plant systemic acquired resistance by salicylic acid. *Science* 262:1883–86
15. Cohen JD, Bandurski RS. 1982. Chemistry and physiology of the bound auxins. *Annu. Rev. Plant Physiol.* 33:403–30
16. Cole DJ, Edwards R, Owen WJ. 1987. The role of metabolism in herbicide selectivity. In *Progress in Pesticide Biochemistry and Toxicology,* ed. DH Hutson, TR Roberts, pp. 57–104. Chichester: Wiley
17. Czarnecka E, Nagao RT, Key JL. 1988. Characterization of *Gmhsp26,* a stress gene encoding a divergent heat-shock protein of soybean: heavy-metal induced inhibition of intron processing. *Mol. Cell. Biol.* 8: 1113–22
18. Daniel V. 1993. Glutathione S-transferase: gene structure and regulation of expression. *CRC Crit. Rev. Biochem.* 25:173–207
19. Danielson UH, Esterbauer H, Mannervik B. 1987. Structure-activity relationships of 4-hydroxyalkenals in the conjugation catalyzed by mammalian glutathione S-transferases. *Biochem. J.* 247:707–12
20. Darvill AG, Albersheim P. 1984. Phytoal exins and their elicitors: a defense against microbial infection in plants. *Annu. Rev. Plant Physiol.* 35:243–75
21. Dean JV, Devarenne TP, Lee I-S, Orlofsky LE. 1995. Properties of a maize glutathione S-transferase that conjugates coumaric acid and other phenylpropanoids. *Plant Physiol.* 108:985–94
22. Dean JV, Gronwald JW, Anderson MP. 1991. Glutathione S-transferase activity in nontreated and CGA-154281-treated maize shoots. *Z. Naturforsch. Teil. C.* 46:850–55

23. Dean JV, Gronwald JW, Eberlein CV. 1990. Induction of GST isozymes in sorghum by herbicide antidotes. *Plant Physiol.* 92: 467–73
24. Dean JV, Machota JH. 1993. Activation of corn glutathione S-transferase enzymes by coumaric acid and 7-hydroxycoumarin. *Phytochemistry* 34:361–65
25. Degousee N, Triantaphylides C, Montillet J-L. 1994. Involvement of oxidative processes in the signalling mechanism leading to the activation of glyceollin synthesis in soybean (*Glycine max*). *Plant Physiol.* 104: 945–52
26. DeJong JL, Morgenstern R, Jornvall H, DePierre JW, Tu C-PD. 1988. Gene expression of rat and human microsomal glutathione S-transferases. *J. Biol. Chem.* 263: 8430–36
27. De Vos CHR, ten Bookum WM, Vooijs R, Schat H, de Kok LJ. 1993. Effect of copper on fatty acid composition and peroxidation of lipids in the roots of copper tolerant and sensitive *Silene cucubalis. Plant Physiol. Biochem.* 31:151–58
28. De Vos CHR, Vonk MJ, Vooijs R, Schat H. 1992. Glutathione depletion due to copper-induced phytochelatin synthesis causes oxidative stress in *Silene cucubalis*. Plant Physiol. 98:853–58
29. Diesperger H, Sandermann H. 1979. Soluble and microsomal glutathione S-transferase activities in pea seedlings (*Pisum sativum* L.). *Planta* 146:643–48
30. Dominov JA, Stenzler L, Lee S, Schwarz JJ, Leisner S, Howell SH. 1992. Cytokinins and auxins control the expression of a gene in *Nicotiana plumbaginfolia* by feedback regulation. *Plant Cell* 4:451–61
31. Dron M, Clouse SD, Dixon RA, Lawton MA, Lamb CJ. 1988. Glutathione and fungal elicitor regulation of a plant defense gene promoter in electroporated protoplasts. *Proc. Natl. Acad. Sci. USA* 85: 6738–42
32. Droog FJN, Hooykaas PJJ, van der Zaal BJ. 1995. 2,4-Dichlorophenoxyacetic acid and related chlorinated compounds inhibit two auxin-regulated type-III tobacco glutathione *S*-transferases. *Plant Physiol.* 107: 1139–46
33. Deleted in proof
34. Droog FNJ, Hooykaas PJJ, Libbenga KR, van der Zaal EJ. 1993. Proteins encoded by an auxin-regulated gene family of tobacco share limited but functionally significant homology with glutathione S-transferases and one member indeed shows *in vitro* GST activity. *Plant Mol. Biol.* 21:965–72
35. Dudler R, Hertig C, Rebman G, Bull J, Mauch F. 1991. A pathogen-induced wheat gene encodes a protein homologous to glutathione S-transferases. *Mol. Plant Microbe Interact.* 4:4–18
36. Edwards R, Dixon RA. 1991. Glutathione S-cinnamoyl transferases in plants. *Phytochemistry* 30:79–84
37. Esterbauer H, Cheeseman KH, Dianzani MU, Poli G, Slater TF. 1982. Separation and characterization of the aldehydic products of lipid peroxidation stimulated by ADP-FE2+ in rat liver microsomes. *Biochem. J.* 208:129–40
38. Ezra G, Rusness DG, Lamoureux GL, Stephenson GR. 1985. The effect of CDAA (*N,N*-diallyl-2-chloroacetamide) pretreatment on subsequent CDAA injury to corn (*Zea mays* L.). *Pestic. Biochem. Physiol.* 23:108–15
39. Fahey RC, Sundquist AR. 1991. Biologically important thiol-disulfide reactions and the role of cysteine in proteins: an evolutionary perspective. *Adv. Enzymol. Relat. Areas Mol. Biol.* 64:1–53
40. Frear DS, Swanson HR. 1970. Biosynthesis of S-(4-ethylamino-6-isopropylamino-2-s-triazine) glutathione: partial purification and properties of glutathione S-transferase from corn. *Phytochemistry* 9: 2123–32
41. Frear DS, Swanson HR. 1973. Metabolism of substituted diphenylether herbicides in plants. I. Enzymatic cleavage of fluorodifen in peas (*Pisum sativum* L.). *Pestic. Biochem. Physiol.* 3:473–82
42. Frear DS, Swanson HR, Mansagar ER. 1983. Acifluoren metabolism in soybean: diphenylether bond cleavage and the formation of homoglutathione, cysteine, and glucose conjugates. *Pestic. Biochem. Physiol.* 20:299–316
43. Friling RS, Bergelson S, Daniel V. 1992. Two adjacent AP-1-like binding sites form the electrophile-responsive element of the murine glutathione S-transferase Ya subunit gene. *Proc. Natl. Acad. Sci. USA* 89:668–72
44. Fuerst EP, Gronwald JW. 1986. Induction of rapid metabolism of metolachlor in sorghum (*Sorghum bicolor*) shoots by CGA-92194 and other antidotes. *Weed Sci.* 34: 354–61
45. Fuerst EP, Irzyk GP, Miller KD. 1993. Partial characterization of glutathione S-transferase isozymes induced by the herbicide safener benoxacor in maize. *Plant Physiol.* 102:795–802
46. Gronwald JW, Fuerst EP, Eberlein CV, Egli MA. 1987. Effect of herbicide antidotes on glutathione content and glutathione S-transferase activity of sorghum shoots. *Pestic. Biochem. Physiol.* 29:66–76
47. Grove G, Zarlango RP, Timmerman KP, Li N-Q, Tam MF, Tu C-PD. 1988. Characterization and heterospecific expression of cDNA clones of genes in the maize GSH S-transferase multigene family. *Nucleic Acids Res.* 16:425–38

48. Hagen G, Uhrhammer N, Guilfoyle TJ. 1988. Regulation of expression of an auxin-induced soybean sequence by cadmium. *J. Biol. Chem.* 263:6442–46
49. Hahn K, Herget C, Strittmatter G. 1994. PR-gene *prp1-1* from potato: regulation of its expression and functional analysis of the gene product. In *Abstr. Int. Congr. Plant Mol. Biol., 4th,* No. 1764. Dordrecht: Kluwer
50. Halliwell B, Gutteridge JMC. 1984. Oxygen toxicity, oxygen radicals, transition metals, and disease. *Biochem. J.* 219:1–14
51. Hangarter RP, Good NE. 1981. Evidence that IAA conjugates are slow-release sources of free IAA in plant tissues. *Plant Physiol.* 68:1424–27
52. Harborne JB, Self R. 1987. Malonated cyanidin 3-glucosides in *Zea mays* and other grasses. *Phytochemistry* 26:2417–18
53. Hatzios KK. 1984. Herbicide antidotes: development, chemistry, and mode of action. *Adv. Agron.* 36:265–316
54. Hatzios KK. 1989. Mechanisms of action of herbicide safeners: an overview. In *Crop Safeners for Herbicides. Development, Uses, and Mechanisms of Action,* ed. KK Hatzios, RE Hoagland, pp. 65–101. New York/San Francisco/London: Academic
55. Hausladen A, Alscher RG. 1993. Glutathione. In *Antioxidants in Higher Plants,* ed. RG Alscher, J Hess, pp. 1–30. Boca Raton, FL: CRC
56. Holt DC, Lay VJ, Clarke ED, Dinsmore A, Jepson I, et al. 1995. Characterization of the safener-induced glutathione S-transferase isoform II from maize. *Planta* 196: 295–302
57. Holt JS. 1993. Mechanisms and agronomic aspects of herbicide resistance. *Annu. Rev. Plant Physiol. Plant Mol. Biol.* 44:203–29
58. Hrazdina G. 1992. Compartmentation in aromatic metabolism. In *Phenolic Metabolism in Plants,* ed. HA Stafford, RK Ibrahim, pp. 1–23. New York: Plenum
59. Hunaiti AA, Ali BR. 1990. Glutathione S-transferase from oxadiazon treated chickpea. *Phytochemistry* 29:2431–35
60. Irzyk GP, Potter S, Ward E, Fuerst EP. 1995. A cDNA clone encoding the 27-kilodalton subunits of glutathione S-transferase IV from *Zea mays. Plant Physiol.* 107:311–12
61. Irzyk GP, Fuerst EP. 1993. Purification and characterization of a glutathione S-transferase from benoxacor-treated maize (*Zea mays*). *Plant Physiol.* 102:803–10
62. Ishikawa T. 1990. Is the glutathione S-conjugate carrier an *mdr*1 gene product? *Trends Biol. Sci.* 15:219–20
63. Ishikawa T. 1992. The ATP-dependent glutathione S-conjugate export pump. *Trends Biol. Sci.* 17:463–68
64. Itzhaki H, Maxon JM, Woodson WR. 1994. An ethylene-responsive enhancer element is involved in the senescence-related expression of the carnation glutathione S-transferase (*GST1*) gene. *Proc. Natl. Acad. Sci. USA* 91:8925–29
65. Itzhaki H, Woodson WR. 1993. Characterization of an ethylene-responsive glutathione S-transferase gene cluster in carnation. *Plant Mol. Biol.* 22:43–58
66. Jachetta JJ, Radosevich SR. 1981. Enhanced degradation of atrazine by corn (*Zea mays*). *Weed Sci.* 29:37–44
67. Jepson I, Lay VJ, Holt DC, Bright SWJ, Greenland AJ. 1994. Cloning and characterization of maize herbicide safener-induced cDNAs encoding subunits of glutathione S-transferase isoforms I, II, and IV. *Plant Mol. Biol.* 26:1855–66
68. Jones AM. 1994. Auxin-binding proteins. *Annu. Rev. Plant Physiol. Plant Mol. Biol.* 45:393–420
69. Kappus H. 1985. Lipid peroxidation: mechanisms, analysis, enzymology, and biological relevance. In *Oxidative Stress,* ed. H Sies, pp. 273–310. London: Academic
70. Kauss H, Jeblick W. 1995. Pretreatment of parsley suspension cultures with salicylic acid enhances spontaneous and elicited production of H_2O_2. *Plant Physiol.* 108: 1171–78
71. Ketley JN, Habig WH, Jacoby WB. 1975. Binding of nonsubstrate ligands to the glutathione S-transferases. *J. Biol. Chem.* 250: 8670–73
72. Kim C-S, Kwak J-M, Nam H-G, Kim K-C, Cho B-H. 1994. Isolation and characterization of two cDNA clones that are rapidly induced during the wound response of *Arabidopsis thaliana. Plant Cell Rep.* 13: 340–43
73. Kiyosue T, Yamaguchi-Shinozaki K, Shinozaki K. 1993. Characterization of two cDNAs (ERD11 and ERD13) for dehydration-inducible genes that encode putative glutathione S-transferases in *Arabidopsis thaliana* L. *FEBS Lett.* 335:189–92
74. Komvies AV, Komvies T, Dutka F. 1985. Effects of the thiocarbamate herbicides on the activity of glutathione S-transferases in maize. *Cereal Res. Commun.* 13:253–57
75. Kutchan TM, Hochberger A. 1992. Nucleotide sequence of a cDNA encoding a constitutively expressed glutathione S-transferase from cell suspension cultures of *Silene cucubalis. Plant Physiol.* 99:789–90
76. Lamoureux GL, Rusness DG. 1989. The role of glutathione-S-transferases in pesticide metabolism, selectivity, and mode of action in plants and insects. In *Glutathione: Chemical, Biochemical, and Medical Aspects,* ed. D Dolphin, R Poulson, O Arnamovie, pp. 153–96. New York: Wiley Intersci.
77. Lamoureux GL, Rusness DG. 1981. Ca-

tabolism of glutathione conjugates of pesticides in higher plants. In *Sulfur in Pesticide Action and Metabolism,* ed. JD Rosen, PS Mager, JE Casada, pp. 133–64. Washington, DC: Am. Chem. Soc.

78. Lamoureux GL, Rusness DG. 1983. Malonylcysteine conjugates as end-products of glutathione conjugate metabolism in plants. In *Pesticide Chemistry: Human Welfare and the Environment,* ed. J Miyamoto, PC Kearney, pp. 295–300. New York: Pergamon
79. Lamoureux GL, Rusness DG. 1986. Xenobiotic conjugation in higher plants. In *Xenobiotic Conjugation Chemistry,* ed. GD Paulson, J Caldwell, DH Hutson, JJ Menn, pp. 62–105. Washington, DC: Am. Chem. Soc.
80. Lamoureux GL, Rusness DG. 1993. Glutathione in the metabolism and detoxification of xenobiotics in plants. In *Sulfur Nutrition and Assimilation in Higher Plants,* ed. LJ de Kok, I Stulen, H Rennenberg, C Brunold, WE Rauser, pp. 221–37. The Hague: SPB Academic
81. Lay MM, Casida JE. 1976. Dichloroacetamide antidotes enhance thiocarbamate sulfoxide detoxification by elevating corn root glutathione content and glutathione S-transferase activity. *Pestic. Biochem. Physiol.* 6:442–56
82. Lee H-C, Toung Y-PS, Tu Y-SL, Tu C-PD. 1995. A molecular genetic approach for the identification of essential residues in human glutathione S-transferase function in *Escherichia coli. J. Biol. Chem.* 270: 99–109
83. Levine A, Tenhaken R, Dixon R, Lamb C. 1994. H_2O_2 from the oxidative burst orchestrates the plant hypersensitive disease resistance response. *Cell* 79:583–93
84. Listowski I, Abramovitz M, Homma H, Niitsu Y. 1988. Intracellular binding and transport of hormones and xenobiotics by glutathione S-transferases. *Drug Metab. Rev.* 19:305–18
85. Litwack G, Ketterer B, Arias IM. 1971. Ligandin: a hepatic protein which binds steroids, biliruben, carcinogens, and a number of exogenous organic anions. *Nature* 234:466–67
86. Lopez MF, Patton WF, Sawlivich WB, Erdjument-Bromage H, Barry P, et al. 1994. A glutathione S-transferase (GST) isozyme from broccoli with significant sequence homology to the mammalian theta-class of GSTs. *Biochem. Biophys. Acta* 1205: 29–38
87. Mannervik B, Awasthi YC, Board PG, Hayes JD, Di Ilio C, et al. 1992. Nomenclature for human glutathione S-transferases. *Biochem. J.* 282:305–8
88. Mannervik B, Danielson UH. 1988. Glutathione transferases: structure and catalytic activity. *CRC Crit. Rev. Biochem.* 23: 283–337
89. Mannervik B, Guthenberg C. 1981. Glutathione transferase (human placenta). *Methods Enzymol.* 77:231–35
90. Marrs KA, Alfenito MR, Lloyd AM, Walbot V. 1995. A glutathione S-transferase involved in vacuolar transfer encoded by the maize gene *Bronze-2. Nature* 375: 397–400
91. Marrs KA, Walbot V. 1994. Bronze-2 gene expression and RNA splicing regulated by cadmium stress. In *Abstr. Int. Congr. Plant Mol. Biol., 4th,* No. 448. Dordrecht: Kluwer
92. Martini N, Egen M, Runtz I, Strittmatter G. 1993. Promoter sequences of a potato pathogenesis-related gene mediate transcriptional activation selectively upon fungal infection. *Mol. Gen. Genet.* 236: 179–86
93. Martinoia E, Grill E, Tommasini R, Kreuz K, Amrhein N. 1993. ATP-dependent glutathione S-conjugate 'export' pump in the vacuolar membrane of plants. *Nature* 364: 247–49
94. Matern U, Reichenbach C, Heller W. 1986. Efficient uptake of flavonoids into parsley (*Petroselinum hortense*) vacuoles requires acylated glycosides. *Planta* 167: 183–89
95. Mauch F, Dudler R. 1993. Differential induction of distinct glutathione-S-transferases of wheat by xenobiotics and by pathogen attack. *Plant Physiol.* 102: 1193–201
96. Medhy MC. 1994. Active oxygen species in plant defense against pathogens. *Plant Physiol.* 105:467–72
97. Mehlhorn H, Wellburn AR. 1987. Stress ethylene formation determines plant sensitivity to ozone. *Nature* 327:417–18
98. Meister A, Anderson ME. 1983. Glutathione. *Annu. Rev. Biochem.* 52:711–60
99. Meyer RC, Goldsbrough PB, Woodson WR. 1991. An ethylene-responsive gene from carnation encodes a protein homologous to glutathione S-transferases. *Plant Mol. Biol.* 17:277–81
100. Moore RE, Davies MS, O'Connell KM, Harding EI, Wiegand RC, Tiemeier DC. 1986. Cloning and expression of a cDNA encoding a maize glutathione S-transferase in *E. coli. Nucleic Acids Res.* 14: 7227–35
101. Moreland DE. 1980. Mechanism of action of herbicides. *Annu. Rev. Plant Physiol.* 31:597–638
102. Morgenstern R, Guthenberg C, DePierre JW. 1982. Microsomal glutathione S-transferase: purification, initial characterization, and demonstration that it is not identical to the cytosolic GSTs A, B, and C. *Eur. J. Biochem.* 128:243–48
103. Mozer TJ, Tiemeier DC, Jaworski EG. 1983. Purification and characterization of

corn glutathione S-transferase. *Biochemistry* 22:1068–72
104. Nash J, Luersen KR, Walbot V. 1990. *Bronze-2* gene of maize: reconstruction of a wild-type allele and analysis of transcription and splicing. *Plant Cell* 2:1039–49
105. Nicholson RL, Kollipara SS, Vincent JR, Lyons PC, Cadena-Gomez G. 1987. Phytoalexin synthesis by the sorghum mesocotyl in response to infection by pathogenic and nonpathogenic fungi. *Proc. Natl. Acad. Sci. USA* 1984:5520–24
106. Normanly J, Slovin JP, Cohen JD. 1995. Rethinking auxin biosynthesis and metabolism. *Plant Physiol.* 107:323–29
107. O'Connell KM, Breaux EJ, Fraley RT. 1988. Different rates of metabolism of two chloroacetanilide herbicides in Pioneer 3320 corn. *Plant Physiol.* 86:359–63
108. Ortiz DF, Ruscitti T, McCue KF, Ow DW. 1995. Transport of metal-binding peptides by HMT1, a fission yeast ABC-type vacuolar membrane protein. *J. Biol. Chem.* 270: 1–8
109. Pemble SE, Taylor JB. 1992. An evolutionary perspective on glutathione transferases inferred from class-Theta glutathione transferase cDNA sequences. *Biochem. J.* 287:957–63
110. Pickett CB, Lu AYH. 1989. Glutathione S-transferases: gene structure, regulation, and biological function. *Annu. Rev. Biochem.* 58:743–64
111. Prandl R, Kutchan TM. 1992. Nucleotide sequence of the gene for a GST from suspension cell cultures of *Silene cucubalis. Plant Physiol.* 99:1729–31
112. Rauser WE. 1990. Phytochelatins. *Annu. Rev. Biochem.* 59:61–86
113. Reinemer P, Dirr HW, Ladenstein R, Schaffer J, Gallay O, Huber R. 1991. The three-dimensional structure of class p glutathione S-transferase in complex with glutathione sulfate at 2.3 resolution. *EMBO J.* 10: 1997–2005
114. Rennenberg H. 1982. Glutathione metabolism and possible biological roles in higher plants. *Phytochemistry* 21:2771–81
115. Rogers KR, Albert F, Anderson AJ. 1988. Lipid peroxidation is a consequence of elicitor activity. *Plant Physiol.* 86: 547–53
116. Rushmore TH, Pickett CB. 1993. Glutathione S-transferases, structure, regulation, and therapeutic implications. *J. Biol. Chem.* 268:11475–78
117. Ryals J, Lawton KA, Delaney TP, Friedrich L, Kessmann H, et al. 1995. Signal transduction in systemic acquired resistance. *Proc. Natl. Acad. Sci. USA* 92: 4202–5
118. Samuelsson B. 1983. Leukotrienes: mediators of immediate hypersensitivity reactions and inflammation. *Science* 220: 568–75
119. Sandermann H. 1992. Plant metabolism of xenobiotics. *Trends Biol. Sci.* 17:82–84
120. Sandermann H. 1994. Higher plant metabolism of xenobiotics: the "green liver" concept. *Pharmacogenetics* 4:225–41
121. Sari Gorla M, Ferrario S, Rossini L, Frova C, Villa M. 1993. Developmental expression of glutathione S-transferase in maize and its possible connection with herbicide tolerance. *Euphytica* 67:221–30
122. Scandalios JG. 1990. Response of plant antioxidant defense genes to environmental stress. *Adv. Genet.* 28:1–40
123. Schmitz G, Theres K. 1992. Structural and functional analysis of the *Bz2* locus of *Zea mays:* characterization of overlapping transcripts. *Mol. Gen. Genet.* 233:269–77
124. Schroder P, Berkau C. 1993. Characterization of cytosolic glutathione S-transferase in spruce needles. *Bot. Acta* 106: 301–6
125. Shah DM, Hironaka CM, Wiegand RC, Harding EI, Krivi GG, Tiemeier DC. 1986. Structural analysis of a maize gene coding for glutathione-S-transferase involved in herbicide detoxification. *Plant Mol. Biol.* 6:203–11
126. Sharma YK, Davis KD. 1994. Ozone in duced expression of stress-related genes in *Arabidopsis thaliana. Plant Physiol.* 105: 1089–96
127. Shimabukuro RH. 1985. Detoxification of herbicides. In *Weed Physiology,* ed. SO Duke, pp. 215–40. Boca Raton, FL: CRC
128. Shimabukuro RH, Frear DS, Swanson HR, Walsh WC. 1971. Glutathione conjugation: an enzymatic basis for atrazine resistance in corn. *Plant Physiol.* 47:10–14
129. Shimabukuro RH, Lamoureux GL, Frear DS. 1978. Glutathione conjugation: a mechanism for herbicide detoxification and selectivity in plants. In *Chemistry and Action of Herbicide Antidotes,* ed. FM Pallos, JE Casida, pp. 133–49. New York: Academic
130. Shimabukuro RH, Swanson HR, Walsh WC. 1970. Glutathione conjugation: atrazine detoxification mechanism in corn. *Plant Physiol.* 46:103–7
131. Singhal SS, Tiwari NK, Ahnad H, Srivastava SK, Awasthi YC. 1991. Purification and characterization of a glutathione S-transferase from sugar cane leaves. *Phytochemistry* 30:1409–14
132. Smith IK. 1985. Stimulation of glutathione synthesis in photorespiring plants by catalase inhibitors. *Plant Physiol.* 79: 1044–47
133. Smith IK, Kendall AC, Keys AJ, Turner JC, Lea PJ. 1985. Effect of herbicides on glutathione levels in barley, tobacco, soybean, and corn. *Plant Sci.* 41:11–17
134. Soderstrom M, Mannervik B, Orning L, Hammerstrom S. 1985. Leukotriene C4

formation catalyzed by three distinct forms of human cytosolic glutathione S-transferase. *Biochem. Biophys. Res. Commun.* 128: 265–70

135. Steffens JC. 1990. The heavy-metal binding proteins of plants. *Annu. Rev. Plant Physiol. Plant Mol. Biol.* 41:553–75
136. Sutherland MW. 1991. The generation of oxygen radicals during host plant responses to infection. *Physiol. Mol. Plant Pathol.* 39:79–94
137. Swain T. 1977. Secondary compounds as protective agents. *Annu. Rev. Plant Physiol.* 28:479–501
138. Sylvestre I, Droillard M-J, Bureau J-M, Paulin A. 1989. Effects of ethylene rise on the peroxidation of membrane lipids during the senescence of cut carnations. *Plant Physiol. Biochem.* 27:407–13
139. Takahashi Y, Kusaba M, Hiroka Y, Nagata T. 1991. Characterization of the auxin-regulated *par* gene from tobacco mesophyll protoplasts. *Plant J.* 1:327–32
140. Takahashi Y, Hasezawa S, Kusaba M, Nagata T. 1995. Expression of the auxin-regulated *parA* gene in transgenic tobacco and nuclear localization of its gene products. *Planta* 196:111–17
141. Takahashi Y, Ishida S, Nagata T. 1995. Auxin-regulated genes. *Plant Cell Physiol.* 36:383–90
142. Takahashi Y, Kuroda H, Tanaka T, Machida Y, Takebe I, Nagata T. 1989. Isolation of an auxin-regulated gene cDNA expressed during the transition from Go to S phase in tobacco mesophyll protoplasts. *Proc. Natl. Acad. Sci. USA* 86:9279–83
143. Takahashi Y, Nagata T. 1992. Differential expression of an auxin-regulated gene, *parC,* and a novel related gene, *C7*, from tobacco mesophyll protoplasts in response to external stimuli in plant tissues. *Plant Cell Physiol.* 33:779–87
144. Takahashi Y, Nagata T. 1992. *parB:* an auxin-regulated gene encoding glutathione S-transferase. *Proc. Natl. Acad. Sci. USA* 89:56–59
145. Takahashi Y, Niwa Y, Machida Y, Nagata T. 1990. Location of the *cis*-acting auxin-responsive region in the promoter of the *par* gene from tobacco mesophyll protoplasts. *Proc. Natl. Acad. Sci. USA* 87: 8013–16
146. Takahashi Y, Sakai T, Ishida S, Nagata T. 1995. Identification of auxin-responsive elements of *parB* and their expression in apices of shoot and root. *Proc. Natl. Acad. Sci. USA* 92:6359–63
147. Talalay P, De Long MJ, Prochaska HJ. 1988. Identification of a common chemical signal regulating the induction of enzymes that protect against chemical carcinogenesis. *Proc. Natl. Acad. Sci. USA* 85:8261–65
148. Tan KH, Meyer DJ, Coles B, Ketterer B. 1986. Thymidine hydroperoxide: a substrate for the rat Se-dependent and Se-independent glutathione peroxidase and glutathione S-transferase isoenzymes. *FEBS Lett.* 207:231–33
149. Tang AH, Tu C-PD. 1995. Pentobarbital-induced changes in *Drosophila* glutathione S-transferase D21 mRNA stability. *J. Biol. Chem.* 270:13819–25
150. Taylor JL, Fritzemeier K-H, Hauser I, Kombrink E, Rohwer F, et al. 1990. Structural analysis and activation by fungal infection of a gene encoding a pathogenesis-related protein in potato. *Mol. Plant Microbe Interact.* 3:72–77
151. Tenhaken R, Levine A, Brisson LF, Dixon RA, Lamb C. 1995. Function of the oxidative burst in hypersensitive disease resistance. *Proc. Natl. Acad. Sci. USA* 92: 4158–63
152. Thompson JE, Legge RL, Barber RF. 1987. The role of free radicals in senescence and wounding. *New Phytol.* 105:317–44
153. Timmerman KP. 1989. Molecular characterization of corn glutathione S-transferase isozymes involved in herbicide detoxification. *Physiol. Plantarum* 77:465–71
154. Tukey HB, Went FW, Muir RM, van Overbeek J. 1954. Nomenclature of chemical plant regulators. *Plant Physiol.* 29:307–10
155. Ulmasov T, Hagen G, Guilfoyle T. 1994. The *ocs* element in the soybean *GH2/4* promoter is activated by both active and inactive auxin and salicylic acid analogues. *Plant Mol. Biol.* 26:1055–64
156. Ulmasov T, Ohmiya A, Hagen G, Guilfoyle T. 1995. The soybean *GH2/4* gene that encodes a glutathione S-transferase has a promoter that is activated by a wide range of chemical agents. *Plant Physiol.* 108: 919–27
157. van der Zaal EJ, Droog FNJ, Boot CJM, Hensgens LAM, Hoge JHC, et al. 1991. Promoters of auxin-induced genes from tobacco can lead to auxin-inducible and root tip-specific expression. *Plant Mol. Biol.* 16: 983–98
158. Vernooij B, Friedrich L, Morse A, Reist R, Kolditz-Jawhar R, et al. 1994. Salicylic acid is not the translocated signal responsible for inducing systemic acquired resistance. *Plant Cell* 6:959–65
159. Viger PR, Eberlein CV, Fuerst EP, Gronwald JW. 1991. Effects of CGA-154281 and temperature on metalochlor absorption and metabolism, glutathione content, and glutathione S-transferase activity in corn (*Zea mays*). *Weed Sci.* 39:324–28
160. Wagner GJ, Hrazdina G. 1984. Endoplasmic reticulum as the site of general phenylpropanoid metabolism in *Hippeastrum. Plant Physiol.* 74:901–6
161. Walbot V, Benito M-I, Bodeau J, Nash J. 1994. Abscisic acid induces pink pigmen-

tation in maize aleurone tissue in the absence of *Bronze-2*. *Maydica* 39:19–28

162. Watahiki MK, Mori H, Yamamoto KT. 1995. Inhibitory effects of auxins and related substances on the activity of an *Arabidopsis* glutathione S-transferase isozyme expressed in *Escherichia coli*. *Physiol. Plant.* 94:566–74
163. Wiegand RC, Shah DM, Mozer TJ, Harding EI, Diaz-Collier J, et al. 1986. Messenger RNA encoding a glutathione S-transferase responsible for herbicide tolerance in maize is induced in response to safener treatment. *Plant Mol. Biol.* 7:235–43
164. Williamson G, Beverley MC. 1987. The purification of acidic glutathione S-transferases from pea seeds and their activity with lipid peroxidation products. *Biochem. Soc. Trans.* 15:1103–4
165. Wingate VPM, Lawton MA, Lamb CJ. 1988. Glutathione causes a massive and selective induction of plant defense genes. *Plant Physiol.* 87:206–10
166. Zettl R, Schell J, Palme K. 1994. Photoaffinity labeling of *Arabidopsis thaliana* plasma membrane vesicles by 5-azido-[7–3H]indole–3-acetic acid: identification of a glutathione S-transferase. *Proc. Natl. Acad. Sci. USA* 91:689–93
167. Zhang B, Singh KB. 1994. Ocs element promoter sequences are activated by auxin and salicylic acid in *Arabidopsis*. *Proc. Natl. Acad. Sci. USA* 91:2507–11
168. Zhou J, Goldsbrough PB. 1993. An *Arabidopsis* gene with homology to glutathione S-transferases is regulated by ethylene. *Plant Mol. Biol.* 22:517–23

Annu. Rev. Plant Physiol. Plant Mol. Biol. 1996. 47:159–84

PHYSIOLOGY OF ION TRANSPORT ACROSS THE TONOPLAST OF HIGHER PLANTS

Bronwyn J. Barkla and Omar Pantoja

Departamento de Biología Molecular de Plantas, Instituto de Biotecnología, UNAM, Cuernavaca, Morelos, México, 62271

KEY WORDS: V-ATPase, V-PPase, channels, cotransporters, vacuole

ABSTRACT

The vacuole of plant cells plays an important role in the homeostasis of the cell. It is involved in the regulation of cytoplasmic pH, sequestration of toxic ions and xenobiotics, regulation of cell turgor, storage of amino acids, sugars and CO_2 in the form of malate, and possibly as a source for elevating cytoplasmic calcium. All these activities are driven by two primary active transport mechanisms present in the vacuolar membrane (tonoplast). These two mechanisms employ high-energy metabolites to pump protons into the vacuole, establishing a proton electrochemical potential that mediates the transport of a diverse range of solutes. Within the past few years, great advances at the molecular and functional levels have been made on the characterization and identification of these mechanisms. The aim of this review is to summarize these studies in the context of the physiology of the plant cell.

CONTENTS

1040-2519/96/0601-0159$08.00

INTRODUCTION

Ion transport across the tonoplast plays a central role in the control of cell homeostasis and osmoregulation, and has led to investigations of the tonoplast transport mechanisms involved in these key cellular processes. Primary transport activities requiring high-energy metabolites for their operation have been characterized at the functional and molecular levels. In contrast, secondary active transporters, as well as passive transporters, have only been defined at the functional level, with the exception of the tonoplast intrinsic protein. These studies have led to our understanding of the role of the vacuole in the physiology of plant cells. Evidence has been presented for the importance of the tonoplast transporters in sodium accumulation and salt tolerance, vacuolar Ca^{2+} release and signal transduction, and for the storage of CO_2 in the form of malate. For these processes, passive and/or secondary active transporters have been implicated, whose activities are maintained by the driving force provided by the primary active transporters.

PROTON TRANSLOCATING PUMPS

Vacuolar H^+-ATPase

It is not surprising that the first plant vacuolar transport protein(s) to be cloned belong to the vacuolar type H^+-ATPase (V-ATPase). These primary active transporters represent a ubiquitous class of proton pumps that are found on a variety of cellular organelles including lysosomes, endosomes, secretory and storage vesicles, and protein sorting organelles, as well as on the vacuolar membrane (tonoplast) of higher plants, fungi, and yeast (44, 65, 112). Because they were first characterized on the tonoplast, they were designated vacuolar type despite the misleading implication. V-ATPases have also been found as functioning enzymes on the plasma membrane of specialized vertebrate cells such as osteoblasts and neutrophils and epithelial cells of the kidney and bladder (50, 79, 123). The V-ATPases function in biological energy conversion, generating a proton-motive-force (PMF) by hydrolyzing ATP and thus providing the driving force for a wide range of secondary active and passive transport processes. These enzymes also function in cellular pH homeostasis as well as in acidification of the interior of several organelles in the case of an endomembrane distribution, or in acidification of a localized extracellular compartment in the case of the plasma membrane V-ATPase (84).

The enzyme resembles the F_0F_1-ATPases (F-ATPases) of mitochondria, chloroplasts, and eubacteria in their multimeric structure, lack of a phosphorylated intermediate, and insensitivity to vanadate (84). An evolutionary relationship between F-type and V-type ATPases is indicated by sequence homologies of some of the subunits (68, 69, 83, 122). However, V-ATPases represent a unique class of H^+-ATPases because they have a neutral rather than an alkaline pH optimum and because they show a distinct inhibitor profile. They are insensitive to azide and more sensitive than the F-ATPases to *N*-ethylmaleimide (NEM), nitrate, and bafilomycin A_1 (114).

SUBUNIT COMPOSITION The V-ATPases are large protein complexes composed of 7 to 10 different subunits that appear to copurify with the enzyme, depending on the species, with a native molecular weight between 400 and 650 kDa (113). However, in plants only the major subunits have been cloned, and much of the work on the structure and function of the V-ATPase has been carried out using yeast as a model system.

The subunits that make up the functional enzyme can be grouped into two categories based on their association with either the peripheral, hydrophilic catalytic sector (V_1) or the integral, hydrophobic membrane sector (V_0). The V_1 sector is composed of five or six soluble proteins. Among these are the two ubiquitous major subunits, A and B, present in three copies per functioning enzyme. The 70-kDa catalytic nucleotide-binding subunit A and the 60-kDa regulatory subunit B were initially cloned in carrot (122) and *Neurospora crassa* (26), respectively, with the plant sequence for the B subunit from *Arabidopsis thaliana* cloned soon after (69). Other subunits of the V_1 sector that have been cloned are subunits C and E (81). Although proteins of similar size have copurified with the plant V-ATPase (46, 89, 113), the corresponding plant sequences have yet to be obtained. Recently, two additional subunits of the V_1 sector have been cloned in yeast (subunit F) and in yeast and bovine (subunit D) (81, 82). Subunit F, a 14-kDa polypeptide necessary for the assembly and activity of the V-ATPase (81), is thought to play a role in the modulation of enzyme activity. In *Saccharomyces cerevisiae,* the subunit is released with the V_1 sector upon cold inactivation of the enzyme, a response that has only been demonstrated for the soluble subunits of the V-ATPase enzymes (81). Subunits of similar size have been shown to be associated with the V-ATPase holoenzyme from plants including mung bean and barley (46, 113). In oat, a subunit of 13 kDa was also identified; however, it was concluded that this polypeptide was associated with the V_0 sector (63). Subunit D was identified as a protein that also copurified with the cold-inactivated V_1 sector from bovine chromaffin granules. The gene encodes a protein with a calculated molecular mass of 28 kDa. Comparison of related sequences resulted in the identification of two identical partial cDNA clones from *A. thaliana* exhibiting approximately 50% identity with the internal nucleotide sequence for the bovine subunit D, which suggests that these fragments are probable candidates for the gene encoding the *A. thaliana* subunit D (82). V-ATPase-related subunits of this size have previously been identified in *Kalanchoë daigremontiana* (27), *Mesembryanthemum crystallinum* (28), and red beet (89); however, the association of these polypeptides with the V_1 sector of the V-ATPase was not demonstrated.

The V_0 sector of the V-ATPase is composed of the 16-kDa proteolipid and one to four other subunits. The proteolipid, present in six copies per functioning enzyme, is thought to function in the conduction of protons across the membrane. First cloned in mammals (68), this subunit has also been cloned in *K. daigremontiana* (DM Bartholomew, M Bettey, FM Dewey, JAC Smith, personal communication), *M. crystallinum* (MS Tsiantis, DM Bartholomew, JAC Smith, submitted for publication), and oat (64). The 16-kDa proteolipid may also act to direct the assembly of the functional enzyme by serving as a template for the assembly of the remaining V_0 subunits and thereafter for the assembly of the catalytic subunits (111). A second 16-kDa polypeptide (M16), distinct from the proteolipid, has been associated with the V_0 sector from *S. cerevisiae* (111). The gene encodes a highly charged hydrophilic protein of predicted molecular mass 13 kDa that copurifies with the V-ATPase; however, because it is not released upon cold inactivation of the enzyme, it is therefore classified as membrane associated. Another of the V_0 subunits is Ac45, cloned from bovine brain tissue and shown to encode a putative glycosylated membrane protein of 51 kDa (109). Parry et al (89) identified a V-ATPase subunit of 52 kDa in *Beta vulgaris;* however, this protein appeared to be associated with the V_1 sector rather than with the V_0 sector because it was released from the membrane following cold inactivation. Ac115 is also a glycosylated putative membrane protein associated with the V-ATPase (109) and is thought to play a role in the targeting of the enzyme to the appropriate endomembrane location (70), although it may not participate directly in the mechanism of action of the V-ATPase (111). A subunit of this size has been positively identified in the membrane sector of the tonoplast V-ATPase from *Beta* (89), and a 115-kDa polypeptide was resolved from FPLC-purified V-ATPase from barley (46). However, the purified oat V-ATPase does not appear to contain a subunit of this size (113).

ISOFORMS Isoforms of the V-ATPase subunits are suggested on the basis of several lines of evidence. The first is the variation in subunit composition among and within species. The second is the presence of multigene families for the different subunits. Plants have been shown to have isoforms for several of the different subunits of the V-ATPase. Oat has been shown to have a family of at least four genes encoding the 16-kDa proteolipid subunit (64). Isoforms of the A subunit of the V-ATPase also seem to be present in plants. In carrot, genomic fragments representing three different genes for the A subunit have been cloned and sequenced (51), and in tobacco up to four genes encoding the 70-kDa subunit have been detected (79). Isoforms have been shown for the B subunit of the V-ATPase as well. In barley, two different clones for the B subunit were identified (15). The presence of multigene families could suggest that while some genes for the V-ATPase are present as housekeeping genes, others may be under tissue or cell-specific control mechanisms or may respond to specific developmental or environmental cues, which allow each subunit to be amplified or suppressed as required.

REVERSIBILITY OF THE V-ATPASE One of the properties of the V-ATPase fundamental to our understanding of its role in the physiology of the plant cell is its coupling ratio (moles of H^+ transported per moles of ATP hydrolyzed). The coupling ratio and reversibility of the plant V-ATPase has been determined (41). Davies et al (41) demonstrated that the coupling ratio of the V-ATPase was not a fixed parameter but one that depended not only on the pH at both sides of the tonoplast but also on the absolute pH difference across the membrane. Thus, in conditions similar to those found in vivo with vacuolar pH (pH_V) = 4.8 and cytoplasmic pH (pH_C) = 7.6, a coupling ratio of 3 was calculated. Deviations from this value were observed when the pH was made more acidic at either side. The value of the coupling ratio decreased to 2 when the pH difference across the tonoplast was set at 3.68 pH units ($pH_V = 4.32$ and $pH_c = 8.0$) and decreased further to 1.75 by increasing the pH difference one additional pH unit (pH_V = 3.26). The reversibility of the V-ATPase was demonstrated under the presence of ADP and P_i, with a gradient of 2.8 pH units across the tonoplast of red beet (41). Under these conditions and employing the patch-clamp technique to whole vacuoles, an inward-directed and bafilomycin A_1–inhibitable current was recorded between ±90 mV. Possible interference from the vacuolar pyrophosphatase (V-PPase) was prevented by employing K^+-free solutions. The results presented by Davies et al (41) demonstrated the partial uncoupling of the pump through changes in pH at both sides of the tonoplast and help to explain the capacity of the V-ATPase in establishing different vacuolar pH values.

REGULATION OF THE V-ATPASE In plant cells accumulating NaCl, the driving force for secondary active Na^+ transport into the vacuole is provided by the PMF generated across the tonoplast by the two H^+ pumps, the V-ATPase and V-PPase. This suggests that the V-ATPase may play a fundamental role in energizing Na^+/H^+ antiport activity in cells accumulating significant quantities of NaCl.

Several groups have studied the regulation of V-ATPase activity and levels of expression of subunits for the enzyme during growth of the halophyte *M. crystallinum* in NaCl. Measurements of both V-ATPase H^+-transport activity and ATP hydrolytic activity were twofold higher in tonoplast vesicles isolated from the leaves of salt-treated plants (200 mM NaCl) when compared with measurements of activity in control plants of the same age (13). Bremberger et al (27) and Rockel et al (99) also reported increases in hydrolytic activity for the V-ATPase from leaves of salt-treated *M. crystallinum* plants based on comparisons with control plants. At day 23 of salt treatment, the increase was highest in plants treated with 400 mM NaCl and was seen to increase with respect to NaCl-treatment concentration (99). Evidence has been presented correlating this increase in V-ATPase activity in salt-treated *M. crystallinum* with alterations in the amount and subunit composition of the enzyme (28, 94, 99). Two polypeptides of molecular mass 31 kDa and 27 kDa were shown to be induced with salt stress in preparations of the purified V-ATPase enzyme from *M. crystallinum* (28, 94). However, though these polypeptides cross-reacted with a *K. daigremontiana* antibody against the V-ATPase holoenzyme, and were immunoprecipitated with the holoenzyme using antiserum against the V-ATPase A subunit from *M. crystallinum* (94), the possibility that these subunits may represent degradation products of higher molecular mass subunits cannot be ruled out. Changes in composition of the V-ATPase have also been implied on the basis of studies of freeze-fracture carbon replicas of tonoplast, revealing the enlargement of intramembranous particles putatively associated with the V_0 sector of the V-ATPase and leading the authors to suggest an increase in particle size with progressive salt treatment of the plants (99).

V-ATPase activity has also been studied in several species following short-term exposure of roots to NaCl (74, 78) and in salt-adapted cell lines (79, 96, 124). Matsumoto & Chung (74) reported a doubling of the H^+-transport rate of the V-ATPase following a 3-day treatment of barley roots with 200 mM NaCl. The protein synthesis inhibitors, cycloheximide and antimycin D, inhibited the salt-induced increase in V-ATPase activity, suggesting that exposure of the barley roots to NaCl induced the synthesis of V-ATPase subunits, thus accounting for the increased transport activity (74). Nakamura et al (78), who measured V-ATPase hydrolytic activity in salt-treated mung bean roots demonstrated a 1.4- and 1.3-fold increase in V-ATPase activity following 3- and 12-h exposure to 100 mM NaCl, respectively. This rapid induction led the authors to suggest posttranslational modifications of the V-ATPase rather than induction of protein synthesis, in contrast with the results presented for barley roots. Enhanced V-ATPase hydrolytic activity has also been observed in NaCl-adapted cells of *Acer pseudoplatanus* (124). Nitrate-sensitive V-ATPase specific activity was twofold higher in microsomal membranes isolated from salt-grown cells (80 mM NaCl) at the beginning of the stationary phase when compared with activities measured in the unadapted cell line at the same growth stage. In tobacco cell suspension cultures Reuveni et al (97) studied the hydrolytic and H^+-transport activities of the V-ATPase in unadapted vs NaCl-adapted cells grown in the presence of 428 mM NaCl. To overcome differences in purity of tonoplast fractions, Reuveni et al normalized results to the level of expression of the 70-kDa catalytic subunit. Using polyclonal antibodies against the red beet 70-kDa subunit, they determined that the amount of immunodetected protein in the tonoplast fraction from NaCl-adapted tobacco cells was fourfold less than in unadapted cells. Consequently, the relative H^+-transport capacity and ATP hydrolytic activity, per unit of 70-kDa subunit, from NaCl adapted cells was three to four times greater than that obtained from unadapted cells. However, to obtain 50% inhibition of transport activity in the NaCl-adapted cells a greater concentration (1.6 times) was required of 70-kDa antibody serum protein than was required for 50% inhibition of transport activity in the unadapted cells (97), suggesting that the antibody had a reduced affinity to the 70-kDa subunit in these tobacco cells. Therefore, normalization of the transport activities to the amount of 70-kDa polypeptide using this antibody would lead to possible misinterpretation of the results. Before conclusions can be drawn, it will be necessary to reexamine the normalization procedure. Regulation by NaCl of mRNA levels for the 70-kDa V-ATPase has also been studied in tobacco cells (80). The mRNA levels from unadapted cells and adapted cells grown continually in the presence of NaCl (428 mM) showed no significant difference. However, the amount of immunodetected 70-kDa polypeptide was fourfold less in the adapted cells (97). When adapted cells grown in the absence of NaCl were exposed to NaCl for 24 h, the levels of the 70-kDa polypeptide mRNA increased 2.3 times over the control levels, but only when treated cells were in the midlinear stage of growth (80). This was attributed to the enhanced ability of the adapted cells, upon reexposure to salt, to regulate mechanisms necessary for growth in saline medium.

Vacuolar Pyrophosphatase

The V-PPase belongs to a fourth category of primary ion translocases distinct from the F-, P-, and V-type H^+-ATPases (101). Like the V-ATPase, the V-PPase catalyzes electrogenic tonoplast H^+ translocation. However, unlike the V-ATPase, the V-PPase derives energy from the hydrolysis of PPi and appears to be present only in plants and phototrophic bacteria (see section on Relationship to Other PPases). Since the isolation and sequencing of the first cDNA clone for the 64-kDa to 67-kDa substrate-binding subunit from *A. thaliana* (101), work on the structure and function of the V-PPase has advanced rapidly. cDNA sequences for the substrate-binding subunit of the enzyme have now been cloned in *Hordeum vulgare* (115) and *B. vulgaris* (61). In addition, a partial N-terminal amino acid sequence has been obtained from *Vigna radiata* (67). In *B. vulgaris,* cDNA sequence analysis and Southern and Northern blot analysis indicated multiple genes encoding this subunit, and two different cDNA clones were obtained (61). Genomic Southern analysis of *A. thaliana* indicated the presence of only a single copy of the gene (101). From the plant sequences cloned to date, it appears that the V-PPase catalytic subunit is highly conserved. The barley sequence showed 86% homology at the amino acid level to the deduced sequence from *A. thaliana* (115), and the two *B. vulgaris* sequences exhibited 89% identity with each other and 89% identity with the corresponding polypeptide from *A. thaliana* (61).

Evidence for a catalytic function of the 64–67 kDa V-PPase subunit and its participation in substrate binding has been implied by the kinetics of inhibition and labeling by the sulfhydryl reagent [^{14}C]-NEM (95). Inhibition studies demonstrated that NEM needed only to modify, via alkylation, a single cysteine residue, presumed to be the Mg^{2+} + PPi-binding site, to irreversibly inactivate the V-PPase (95). To identify the implicated cysteine, the 64–67 kDa subunit was specifically labeled with [^{14}C]-NEM, purified, and subsequently digested with V8 protease. From the generated protein fragments only a single labeled band of 14 kDa was obtained (121). The microsequence of this band aligned to the carboxy-terminal segment of the deduced amino acid sequence for the 64–67 kDa subunit. Within this region only a single cysteine (Cys^{634}) was conserved among the known plant sequences, which suggested that alkylation of this specific residue was responsible for enzyme inactivation. Moreover, the location of Cys^{634}, on hydrophilic loop X, indicated that this loop was orientated to the cytoplasmic face of the tonoplast (121). These results indirectly implicated the 64–67 kDa subunit as the catalytic subunit of the enzyme; however, the heterologous expression of the *A. thaliana* V-PPase subunit in *S. cerevisiae* clearly established its function in both PPi hydrolysis and H^+ transport (59).

HETEROLOGOUS EXPRESSION AND SITE-DIRECTED MUTAGENESIS OF THE V-PPASE The heterologously expressed 64–67 kDa subunit showed identical characteristics to the native plant V-PPase with respect to PPi hydrolysis, H^+ translocation, K^+ and Ca^{2+} regulation, and inhibitor sensitivity (59), presenting unequivocal evidence that the catalytic functions of the native V-PPase can be directly explained by the functioning of the 64–67 kDa subunit. These results invalidated the proposal that putative subunits of 21 and 20 kDa may be necessary for the H^+ translocation properties of the enzyme as observed for the reconstituted 64–67 kDa subunit (29).

The demonstrated facility of heterologous expression of the V-PPase in yeast, together with the evidence that the expressed enzyme functioned as the native plant enzyme, opened the way for site-directed mutagenesis studies aimed at understanding the structure and function of the V-PPase. Kim et al (60) individually mutated the 9 Cys residues conserved among the known plant V-PPase sequences to determine the role of these amino acids in enzyme inactivation by NEM and substrate binding. Mutations were Cys to Ser, allowing for charge conservation, or Cys to Ala, maintaining the size of the side group (60). With the exception of Cys^{634}, the previously identified NEM reactive residue (121), the other eight mutations had no effect on enzyme activity. Mutations in Cys^{634} to either Ser or Ala resulted in an NEM-insensitive enzyme (60). The Cys^{634} NEM-insensitive mutant was still capable of both PPi hydrolysis and H^+ translocation. This indicated that the site for NEM inhibition is not the substrate binding site as previously postulated (95) and suggested that the inhibitory action of NEM and related compounds on the V-PPase is due to a conformational change in the substrate binding site (60). Indications of a possible substrate binding site have come from site-directed mutagenesis studies and analysis of alignments for all the sequenced soluble PPases (39) with the *A. thaliana* and *B. vulgaris* sequences (61). Although demonstrating little overall sequence homology, a single motif of acidic and basic residues, D or $E(X)_7KXE$, appears to be conserved between the V-PPases and among the soluble PPases (61) and may participate in substrate binding.

RELATIONSHIP TO OTHER PPASES Although exhibiting no apparent relationship to the soluble PPases, the V-PPase does appear to share common features with the PPi synthase from the purple, nonsulfur bacterium *Rhodospirillum* (95). It has been proposed that these energy conserving/transducing PPases belong to the same category of ion-translocase (120). Both showed the same inhibitor sensitivity profile to the pyrophosphate analogs 1,1-diphosphonates, with aminomethylenediphosphonate in particular being a highly specific and potent competitive inhibitor of both enzymes (14, 120). In addition, both enzymes appear to be made up of a single immunologically cross-reactive subunit of similar molecular weight, and Southern analysis of *Rhodospirillum* genomic DNA also demonstrated sequence similarities to the *Arabidopsis* and *Beta* sequences (95).

ROLE OF THE V-PPASE Several roles for the V-PPase in plant cells have been suggested (95). The enzyme may function as an energy conservation system through the establishment of a pH gradient across the tonoplast that is utilized to energize secondary active transport, and it may also function as a mechanism for the regulation of cytosolic pH. The role of the V-PPase as a K^+ pumping mechanism into the vacuole has also been defined (42), and the implications for this enzyme in turgor regulation have been postulated (95).

REGULATION OF THE V-PPASE Evidence for different isoforms of the V-PPase may be an indication of differential regulation of this enzyme by developmental or tissue-specific cues or of enzyme regulation by environmental factors. Although providing necessary insight into the physiological importance of the PPase in the tonoplast of plant cells, few studies have concentrated on this field of research. In *B. vulgaris,* Northern analysis of a wide range of tissues from different developmental stages using sequence-specific probes for the two V-PPase isoforms indicated that steady state levels of expression of the isoforms were the same (61). However, as the authors suggested, more comprehensive studies employing techniques with higher sensitivity and resolution are required to determine differential expression of these two *Beta* isoforms.

Several studies have focused on the regulation of the V-PPase activity by growth in NaCl. In contrast with the general sodium-induced increase in V-ATPase activity, there appears to be a decrease in V-PPase activity with exposure to NaCl (28, 74, 78). The V-PPase hydrolytic activity in tonoplast-enriched microsomes from barley roots treated with 200 mM NaCl for 3 days was only 50% of the activity seen in the control roots (74). In *M. crystallinum,* the activity of the V-PPase in the tonoplast from salt-treated plants (400 mM) was lower than in the nontreated plants at all stages of treatment, and in both control and salt-treated plants the V-PPase activity decreased with plant age (28). Mung bean V-PPase activity was severely inhibited when mung roots were exposed to NaCl (100 mM) (78). Moreover, this inhibition of enzyme activity was also observed on isolated tonoplast vesicles upon the addition of NaCl to the reaction medium. From these results, it was concluded that the inhibitory effects of Na^+ were directly on the V-PPase, which accounts for the decrease in activity following exposure to NaCl and accumulation of this ion by the plant (78).

Evidence for exceptions to this Na^+-induced decrease in V-PPase activity has been presented (38, 124). In NaCl-adapted cells of *A. pseudoplatanus,* both the V-ATPase and V-PPase activities were higher than in the unadapted cell line, with the V-PPase activity increasing by 100% in the cells adapted to 80 mM NaCl (124). Unadapted cells of *Daucus carota* grown in the presence of 50 mM NaCl for 10 days also showed a doubling of V-PPase activity when compared with the activity of the control cells (38). The discrepancy in results between studies may be due to differences in the Na^+ concentrations employed or may simply be due to species differences.

Regulation of the V-PPase by environmental factors other than NaCl has also been investigated. Conditions of anoxia have been demonstrated to result in an increased V-PPase specific activity of up to 75-fold in tonoplast vesicles isolated from rice seedlings grown in the presence of nitrogen (31). V-PPase levels decreased to control values following the return of the seedlings to air. This anoxic-induced increase in V-PPase specific activity was mirrored by increases in transcript and protein levels (31). During anoxia, levels of cellular ATP were severely reduced while levels of PPi showed no variation (95). Therefore, upregulation of the V-PPase during anoxia would function to conserve levels of ATP and maintain tonoplast energization and cytoplasmic pH homeostasis, essential requirements for survival of the plants under oxygen stress (31, 95).

Regulation of V-PPase activity would also be essential under conditions of chilling stress where, concomitant with reduced levels of cellular ATP, there is cold-inactivation of the V-ATPase (95). Rice seedlings exposed to temperatures of 10°C for several days showed increases in both V-PPase-specific activity (20-fold increase) and amount of immunoreactive enzyme (31).

ION COTRANSPORTERS

Na^+/H^+ Antiport

The presence of a tonoplast Na^+/H^+ antiporter, or exchanger, involved in vacuolar Na^+ sequestration has been well documented in several plant species (25, 48, 52, 108, and reviewed in 10). Secondary active transport of Na^+ into the vacuole via the tonoplast Na^+/H^+ antiporter would be energized by the PMF generated by the activity of the V-ATPase and/or the V-PPase. As a fundamental mechanism in salt tolerance, an active antiport would function to sequester Na^+ into the vacuole, which results in avoidance of cytoplasmic Na^+ toxicity and maintenance of a high cytoplasmic K^+/Na^+ ratio. In parallel, vacuolar Na^+ would serve as an osmoticum necessary for cellular H_2O homeostasis. Evidence for a role of the tonoplast Na^+/H^+ antiport in salt tolcrancc has been provided from several studies demonstrating induction of antiport activity upon exposure to NaCl. In sugar beet cell suspensions, increases in constitutive antiport activity were observed upon growth of cells in NaCl (25) and in barley roots; although Na^+/H^+ antiport activity was not detected in plants grown in the absence of NaCl, antiport activity was rapidly induced in plants treated with NaCl (48). Further evidence was provided by comparison of two *Plantago* species showing marked differences in their sensitivity to NaCl. Studies revealed that tonoplast Na^+/H^+ antiport activity could be detected only in tonoplast vesicles of the salt-tolerant *P. maritima* but not in the salt-sensitive *P. media* (108).

More recently, evidence for a Na^+/H^+ antiport has been presented in the halophyte *M. crystallinum* (13). When plants of *M. crystallinum* were exposed to salt for several weeks, measurements of bulk tissue concentrations of NaCl were as high as 1.0 M, and in the large epidermal bladder cells that cover the shoot surface, Na^+ concentrations were shown to exceed this value (1). The demonstrated efficiency of Na^+ sequestration in this plant presents it as a model system with which to study the mechanism employed for vacuolar Na^+ accumulation. Electroneutral Na^+/H^+ antiport activity was detected in tonoplast vesicles isolated from leaves of both control and salt-treated (200 mM) *M. crystallinum* plants (13) and indicates constitutive levels of expression and activity of this protein in agreement with results presented for cell suspension cultures of *B. vulgaris* (25). Initial rates of Na^+/H^+ exchange were 2.1 times higher from vesicles of salt-treated plants compared with vesicles from control plants and indicated a specific induction of antiport activity following salt treatment. As demonstrated for the Na^+/H^+ antiport from *B. vulgaris* (23), the *M. crystallinum* antiport was inhibited by amiloride (13). The NaCl-induced increase in Na^+/H^+ antiport activity in *M. crystallinum* was closely correlated with an increase in V-ATPase activity in the salt-treated plants and implicated the V-ATPase in energizing vacuolar Na^+ accumulation via the Na^+/H^+ antiport.

Although biochemical studies aimed at the identification of the tonoplast Na^+/H^+ antiport have been carried out (10, 12) and a 170-kDa tonoplast polypeptide has been associated with the antiporter (11), no recent advances have been made in this field.

Ca^{2+}/H^+ Antiport

The accumulation of vacuolar Ca^{2+} in plants is now well recognized to be explained by the activity of a tonoplast Ca^{2+}/H^+ antiport (24, 104, 105), and preliminary results on the reconstitution of this exchanger have been reported (107). The stoichiometry of the Ca^{2+}/H^+ antiport has been suggested to be equal to 3 and to be stimulated by vacuolar positive potentials, leading to the thermodynamically possible accumulation of Ca^{2+} inside the vacuoles (19). Pharmacological studies have demonstrated that the Ca^{2+}/H^+ antiport is inhibited by several agents known to affect other Ca^{2+} transporters including ruthenium red, verapamil, La^{3+}, Cd^{2+}, and other divalent cations (33, 34). In view of these reports it is interesting to bring into context the results concerning the presence of a Cd^{2+}/H^+ antiport in the tonoplast of oat roots (100). Salt & Wagner (100) demonstrated the uptake of Cd^{2+} by tonoplast-enriched vesicles from oat roots energized either by ATP or by K^+/nigericin. Initial rates of Cd^{2+} accumulation showed saturation kinetics with a $K_m = 5.5$ μM and a $V_{max} = 14$ nmol mg protein^{-1} min^{-1}. These results led the authors to suggest a detoxificating role for the Cd^{2+}/H^+ antiport by which Cd^{2+} is removed from the cytoplasm to prevent its potential toxic reaction with -SH-containing compounds (116). However, it is possible that the activity of the Cd^{2+}/H^+ antiport may be the same as that for the Ca^{2+}/H^+ antiport, as indicated by the results of Chanson, who demonstrated the inhibition of the Ca^{2+}/H^+ antiport by Cd^{2+} (34). This view was also suggested by Salt & Wagner, who recorded the activity of a Ca^{2+}/H^+ antiport in the oat tonoplast vesicles (100).

ION CHANNELS

The third category of ion transporters present in the tonoplast corresponds to the ion channels that mediate the movement of ions down their electrochemical potential gradient.

Slow-Activating Channels

The slow-activating (SV) channels were the first channels described in the tonoplast of plant cells (53) and are present in all species studied. The distinguishing characteristics of the SV channels are a slow activation time constant, a marked outward rectification at positive tonoplast potentials[1], activation by cytoplasmic Ca^{2+} ($Ca^{2+}{}_c$) at levels higher than 10^{-6} M, and a low selectivity for monovalent cations. However, because these channels are mainly open at nonphysiological positive tonoplast potentials, their role in the physiology of plant cells is not clear, although attempts have been made to assign a function to these channels. Lowering pH_v from 7.2 to 5.5 had no effect on the SV channel activity from onion guard cells (7). However, acidic pH_v reduced the SV activity from *Vicia faba* guard cells without affecting the single channel conductance (103). Decreasing pH_c also resulted in a diminished activity of *V. faba* vacuolar SV channels with a pK_d of 6.8, with no change in the single-channel conductance (103).

In order to gain more information on the role of tonoplast SV channels, their regulation by other cytoplasmic factors has been addressed. The regulation of SV channels by calmodulin (CaM) has been demonstrated (118) and appears to be dependent upon $Ca^{2+}{}_c$. Although addition of CaM in the presence of 0.1 mM $Ca^{2+}{}_c$ did not increase the vacuolar currents of *V. faba* guard cells (103), CaM stimulated the SV currents from storage protein vacuoles from barley aleurone (18), but only at $Ca^{2+}{}_c$ below 10^{-5} M. This indicates that SV channels may be regulated by the association Ca^{2+}-CaM at $Ca^{2+}{}_c$ below 10^{-5} M, a mechanism that may be overridden by $Ca^{2+}{}_c$ above 10^{-5} M. The CaM antagonists [*N*-(6-aminohexyl)-5-chloro-1-naphtalenesulfonamide] (W7), trifluoperazine, and calmidazolium directly inhibited the SV channels by affecting their gating mechanism without changing the channel conductance (18, 103). In vacuoles from barley aleurone a partial recovery of the SV currents was observed upon the addition of 3.5 μM CaM in the presence of W7 (18), leading the authors to conclude that the stimulation of vacuolar SV channels is through tonoplast-associated CaM activated by $Ca^{2+}{}_c$.

[1] The direction of the currents and the tonoplast voltage are according to Bertl et al (17).

Changes in the levels of cytoplasmic chloride (Cl^-_c) have also been reported to regulate SV channel activity. In vacuoles from sugar beet cell cultures, decreasing levels of Cl^-_c reduced the activity of the SV channels, with a linear relationship between whole vacuole current and Cl^-_c from 10 to 100 mM (86). Studies at the single-channel level demonstrated that the effect of decreasing Cl^-_c was a reduction in the mean open time and the possible induction of an additional closed state, with no effect on the single-channel conductance (86). Similar results were reported for the SV channels of *V. faba* guard cells (103), although these authors concluded that the reduction in the magnitude of the currents was due to blocking of the SV channels by gluconate. If this conclusion were correct, the reduction in the SV channel currents seen with other anions employed as substitutes for Cl^- (86) would also have to be due to the blockage of the SV channels.

Regulation of the red beet taproot vacuolar SV channels by Mg^{2+}_c has also been reported (43). The induction of SV-like currents was observed in vacuoles isolated in a Mg^{2+}-free medium by including 0.4 mM Mg^{2+} on the cytoplasmic side. When the vacuoles were isolated in a medium containing 0.4 mM Mg^{2+}, raising the concentration of Mg^{2+} to 2.4 mM resulted in a further increase in magnitude of the SV currents, an effect that was greatest after 20 min (43). However, these results differ from early reports where only the activity of FV channels was recorded at low Ca^{2+}_c in the presence of 1–2 mM Mg^{2+}_c (54, 66, 87). Why the SV channels were stimulated by Mg^{2+}_c under the conditions reported by Davies & Sanders (43) is not clear, but it is possible that isolation of the vacuoles in a Mg^{2+}-free or Mg^{2+}-low medium may be of relevance.

Fast-Activating Channels

The fast-activating (FV) channels in plant vacuoles were first identified by Hedrich & Neher (54). These channels are active at physiological levels of Ca^{2+}_c, are voltage independent, and have a selectivity of 6:1 between K^+ and Cl^-. Similar characteristics have been reported for the vacuoles of sugar beet cell cultures (87), *V. faba* guard cells (117) and red beet storage tissue (43). The FV channels from beet were as selective for K^+ as for Na^+, with a permeability ratio P_{c+}/P_{Cl-} of 7. A higher selectivity was reported for the FV channels from guard cells with a permeability sequence $K^+ > Rb^+ > NH_4^+ >> Cs^+ \approx Na^+ \approx Li^+$ (117). The FV channels from guard cell vacuoles were also stimulated by small increases in Ca^{2+}_c. Recordings of FV channel activity at the whole vacuole and single channel level suggest that these channels may function in the release and uptake of K^+ during cellular osmoregulation, and in the particular case of guard cells during the opening and closing of stomata (117). Insight into the role of the FV channels has been obtained by studying the regulation of these channels by cytoplasmic factors. Davies & Sanders (43) have reported the stimulation of the vacuolar conductance in red beet through FV channels by cytoplasmic ATP or ATPγS, in the absence of either $Ca^{2+}{}_c$ or $Mg^{2+}{}_c$, indicating this effect was independent of protein phosphorylation. ATP was also reported to stimulate the movement of arginine through these channels (43). Regulation of the FV channels by pH, independent of ATP, was suggested by the stimulation of the FV channels by increasing pH_c from 7.3 to 8.0 (43). However, inhibition of FV channels by increasing pH_c was reported for vacuoles of *V. faba* guard cells (117), suggesting that regulation of FV channels by pH_C may be tissue or species specific. Further stimulation of red beet FV channels was observed with $Mg^{2+}{}_c$ between 0.4 and 2.4 mM in conditions of low $Ca^{2+}{}_c$ (43). Regulation of red beet FV channels by ATP, however, must be taken with care in view of the results that were obtained under more physiological conditions. Davies & Sanders showed that with pH_V set between 5.5 and 4.3, stimulation of the inward FV channels by ATP was prevented (43). Thus, regulation of the FV channels by physiological vacuolar pH may suppress the effects of cytoplasmic ATP. In spite of this, the roles ascribed to the FV channels are convincing (43). It is proposed that these channels serve as the shunt conductance for the V-ATPase, allowing the release of positive charges from the vacuolar interior facilitating the establishment of a ΔpH. The FV channels could also function as a mechanism for the release of $K^+{}_v$ into the cytoplasm under conditions of low K^+ availability and could thus play an important role in $K^+{}_c$ homeostasis (43). An additional role for the FV channels was proposed by Ward & Schroeder (117) who suggested the involvement of these channels in tonoplast depolarization (see next section).

Ca^{2+}-Selective Channels

It is now well established that $Ca^{2+}{}_c$ plays an important role in stimulus-response coupling in plants. Some of the stimulus-response coupling processes in which changes in $Ca^{2+}{}_c$ have been implicated are those induced by plant growth regulators (76), touch and cold (62), as well as in the regulation of protein kinases (20) and ion channels (54, 102). Because of the size of the vacuole and the concentration of Ca^{2+} within, it is argued that this organelle serves as the source for the increases in $Ca^{2+}{}_c$ (58). Two classes of tonoplast Ca^{2+}-selective channels have been described to date: voltage-dependent channels and agonist-activated channels.

VOLTAGE-DEPENDENT Ca^{2+} CHANNELS The first evidence for the existence of Ca^{2+}-selective channels in plants was reported in the tonoplast of sugar beet employing Ba^{2+} as the charge carrier (88). Under bi-ionic conditions with $Ba^{2+}{}_c$ and $K^+{}_v$ as the charge carriers, Pantoja et al (88) reported the activation of putative outward Ca^{2+} channels with a permeability ratio $P_{Ba^{2+}}/P_{K^+}$ of 5 to 7. These currents, observed at the whole vacuole and isolated patch level, showed a conductance of 40 pS with 100 mM $Ba^{2+}{}_c$, and were inhibited by La^{3+} and verapamil, two Ca^{2+} channel blockers identified in animal cells. A further characteristic of these currents was the saturation observed with increasing $Ba^{2+}{}_c$ with a dissociation constant (K_d) = 16 mM (88). Similar channel activity was obtained with vacuoles of red beet and tobacco. Under comparable experimental conditions, Ping et al (90) recorded an outward directed channel with a conductance of 30 pS with 50 mM $Ba^{2+}{}_v$, which also carried Sr^{2+}. Employing experimental conditions amenable for the detection of inward Ca^{2+} channels, Ping et al failed to record any channel activity (90). Similar outward Ca^{2+}-selective channels have been reported for the tonoplast of guard cells from *V. faba* (103, 117). Employing symmetrical $CaCl_2$, SV-like currents were recorded with whole vacuoles and isolated channels. These currents reversed near the equilibrium potential for Ca^{2+} with a calculated permeability ratio $P_{Ca^{2+}}/P_{K^+}$ of 5; single-channel recordings showed a conductance of 16 pS for these channels with 50 mM $Ca^{2+}{}_v$ (117).

Voltage-dependent inward Ca^{2+} channels have been described for the tonoplast of red beet taproot (57, 58) and *V. faba* guard cells (6). These channels are proposed to function as a release mechanism for Ca^{2+}. Employing isolated inside-out tonoplast patches with 50 mM $K^+{}_c$ and 5 mM $Ca^{2+}{}_v$ as charge carriers, researchers found that negative potentials stimulated the activity of inward channels with a conductance of 12 pS in red beet (57, 58) and 27 pS in *V. faba* (6), which showed saturation at 5 mM $Ca^{2+}{}_c$. Positive potentials induced the opening of outward channels with a conductance of between 120 and 200 pS (6, 57). Analysis of these current-voltage relationships shows a resemblance to the outward rectification of the SV channels. The inward channels showed a selectivity towards Ca^{2+} with a permeability ratio $P_{Ca^{2+}}/P_{K^+}$ in the range of 6 to 20, with the upper value corresponding to the red beet channels. Increasing $Ca^{2+}{}_v$ between 5 and 20 mM shifted the activation potential of the currents toward less-negative potentials and caused a concomitant increase in the open-channel probability in both species (6, 57). Regulation of the Ca^{2+}-release channels was effected by pH_v. Under physiological levels of pH_V, a marked decrease in the open probability of the channels was observed with respect to that recorded at pH 7.3 (6, 57). A characteristic of the red beet tonoplast Ca^{2+} channels was a "noticeable rundown" in the channel activity at physiological pH_V (58). Both Ca^{2+} channels were inhibited by Gd^{3+} with a half-maximal inhibition ($K_{1/2}$) of 10–20 μM (6, 57). In addition, red beet channels were insensitive to $Ca^{2+}{}_c$, the alkaloid ryanodine, inositol 1,4,5-triphosphate (IP_3), or heparin but were inhibited by Zn^{3+} (57), whereas the *V. faba* channels were reversibly blocked by the dihydropyridine, nifedipine, with a $K_{1/2}$ of 77 μM (6).

Allen & Sanders (6) also reported the presence of a second type of voltage-dependent Ca^{2+} release channel in the tonoplast of guard cells. This channel had a single channel conductance of 14 pS, a permeability ratio $P_{Ca^{2+}}/P_{K^+}$ of 4, and was found only in 8% of the patches analyzed. The voltage dependence of the open probability was similar to the 27 pS channel. The presence of two Ca^{2+}-release channels in the tonoplast of guard cells was suggested to confer the capacity to respond to a variety of stimuli known to control the physiology of the guard cells (6).

Permeability of the red beet inward vacuolar Ca^{2+} channel toward K^+ was demonstrated by Johannes et al (58). Using K^+ as the only charge carrier, a linear current-voltage relationship was obtained. Upon elevating $Ca^{2+}{}_v$ a decrease in the single-channel current at negative potentials with half-maximal inhibition at 0.3 mM $Ca^{2+}{}_v$ was observed (58). This result suggests that $Ca^{2+}{}_v$ may be blocking the channels at negative tonoplast potentials and thus causing an outward rectification similar to that observed for the SV channels (see section on Slow-Activating Channels). This inhibition by $Ca^{2+}{}_v$ resembles that observed for the inward rectification caused by $Mg^{2+}{}_c$ in cardiac K^+ channels (72, 73). Johannes et al (58) also reported that $Ca^{2+}{}_v$ was required for the gating of the inward Ca^{2+} channels. At nonphysiological levels of $Ca^{2+}{}_v$ (30 μM), channel openings were slow and recorded only at very negative tonoplast potentials, with increased $Ca^{2+}{}_v$ causing a shift in the activation potential toward positive potentials as well as a faster activation (58). However, observations on the effects of $Ca^{2+}{}_v$ on the single-channel current and gating mechanism of the Ca^{2+} channels—with K^+ as the charge carrier—were only obtained at the single-channel level, making direct comparison with the vacuolar SV channels difficult.

The results reported for vacuolar Ca^{2+} channels lead us to suggest that most if not all of these activities could correspond to the functioning of SV channels in the tonoplast. Evidence for this is best demonstrated by the results of Ward & Schroeder (117) and Schulz-Lessdorf & Hedrich (103). Ward & Schroeder have proposed a model by which Ca^{2+}-induced Ca^{2+} release from the vacuole may occur through the synchronized activity of FV and SV channels (117). In this model, increases in $Ca^{2+}{}_c$ above 10^{-6} M would activate FV channels and stimulate the release of K^+ from the vacuole, following its electrochemical potential, and cause a depolarization of the tonoplast membrane potential. This depolarization, together with the increase in $Ca^{2+}{}_c$, would in turn activate the SV channels and thus cause the release of Ca^{2+} into the cytoplasm. The validity of this proposal rests on the postulated capacity of SV channels to release Ca^{2+}. For these channels to be involved in stimulus-response coupling, a tight regulatory mechanism must exist to prevent uncontrolled release of Ca^{2+} to the cytoplasm. One such mechanism could be vacuolar pH. Acidic vacuolar pH has been shown to down-regulate the channel-mediated Ca^{2+} release (6, 58). This together with the high concentration ratio K^+/Ca^{2+} in the $P_{Ca^{2+}}/P_{K^+}$ vacuole (around 20:1) and the low permeability ratio of the Ca^{2+}

channels would reduce the potentially constant release of Ca^{2+} into the cytoplasm.

Although the activity of the Ca^{2+} channels in the inward direction was not detected by several groups (88, 90), it is possible that the channels were inactivated or inhibited by the experimental conditions employed. Pantoja et al (88) did not observe outward or inward currents employing symmetrical Ba^{2+} solutions, which leads to the conclusion of a possible inhibition of outward currents by $Ba^{2+}{}_v$ and inhibition of inward currents by $Ba^{2+}{}_c$. On the other hand, under similar experimental conditions to those employed by Johannes et al (57, 58) and Allen & Sanders (7), Ping et al (90) also failed to record inward Ca^{2+} currents, employing Ca^{2+}-gluconate in the cytoplasm, a condition that has been demonstrated to inactivate SV channels (85, 103).

LIGAND-GATED Ca^{2+} CHANNELS *IP_3-gated channels* Although stimulus-response coupling can be mediated directly by changes in $Ca^{2+}{}_c$, this coupling may require the participation of intermediate signaling molecules including CaM, IP_3, or cyclic ADP ribose (cADPR). The presence of endogenous CaM (8) and IP_3 (55) has been reported in plant cells. However, the presence of cADPR in plants has not been demonstrated. CaM has been shown to regulate the activity of Ca^{2+}-ATPases from the plasma membrane (45, 92, 98) and endoplasmic reticulum (9, 49a), as well as the ion channels in the tonoplast (18, 103, 118). IP_3 on the other hand, has been shown to cause stomatal closing, a response thought to be mediated by increasing levels of $Ca^{2+}{}_c$ (22, 49). The source for this increase in $Ca^{2+}{}_c$ has been suggested to be intracellular pools including the endoplasmic reticulum and/or the vacuole (49). Evidence for a possible role of IP_3 in the release of vacuolar Ca^{2+} has been presented (30, 91, 106). Employing purified tonoplast vesicles from red beet (30), oat roots (106), and isolated vacuoles from *Acer* (91), researchers showed release of $^{45}Ca^{2+}$ upon the application of IP_3. Applying the patch-clamp technique to isolated vacuoles of red beet, researchers have also shown IP_3 to directly activate Ca^{2+}-release channels (2, 3). Attempts by several investigators to repeat these results have failed, which throws into question the existence of such a $Ca^{2+}{}_c$-release mechanism (35). However, Allen & Sanders (5) recently showed that addition of 1 μM IP_3 to isolated vacuoles from red beet roots exposed to 1 mM $Ca^{2+}{}_c$ and 200 μM Zn^{2+}—conditions that should inhibit SV and FV channels—stimulated the magnitude of the Ca^{2+} inward currents. The reversal potential of the IP_3-induced currents indicated a permeability ratioP_C $P_{Ca^{2+}}/P_{K^+}$ of 200 for these channels. Interestingly, preplasmolysis of the vacuoles at high osmotic pressures (851 mOsmol) was required to observe the stimulation of Ca^{2+} currents by IP_3. Further stimulation of these currents was observed when vacuoles—under the whole-vacuole configuration—were exposed to hypoosmotic solutions (5). Thus, an almost twofold increase in the magnitude of the IP_3-induced currents was recorded when a difference of 250 mOsmol kg^{-1} was established across the tonoplast (5). The IP_3-induced currents were independent of the levels of $Ca^{2+}{}_c$ between 0.1 μM and 1 mM, and the single-channel conductance was variable between 11 and 182 pS. Although it was concluded that hyperosmotic stress was a prerequisite to record IP_3-induced currents (5), Ping et al (90) failed to observe the same response employing similar hyperosmotic conditions for vacuole isolation. From the many reports that have failed to detect the activity of IP_3-induced Ca^{2+} currents (35) and the particular conditions necessary to record these channels (3, 5), it is clear that more exhaustive work is required.

Cyclic ADP ribose-gated Ca^{2+} channels The presence of a second mechanism, independent of IP_3, for the release of Ca^{2+} from intracellular sources has been reported in animal cells and is suggested to be associated with the ryanodine receptor (16, 47). Reports by Galione et al (47) and Mészáros et al (77) suggested that the ryanodine receptors may be under the control of cADPR. However, it appears that cADPR does not have any effect on some animal cells, and direct demonstration that cADPR levels change in response to certain stimuli is still lacking (16). Stimulation of Ca^{2+} release from vacuoles by cADPR has recently been demonstrated in plants (4). Addition of cADPR to a red beet microsome tonoplast-rich preparation, previously loaded with $^{45}Ca^{2+}$, induced the release of 15% of the total $^{45}Ca^{2+}$ uptake. This activity was inhibited by ruthenium red, an inhibitor of the ryanodine receptor. Ryanodine also stimulated the release of $^{45}Ca^{2+}$ and prevented the effect of cADPR, if previously added, indicating that these two ligands bind to a common receptor on the tonoplast. Employing isolated vacuoles and the patch-clamp technique, researchers found that addition of cADPR induced an increase in the magnitude of inward instantaneous currents with a K_m of 20–25 nM and a permeability ratio $P_{Ca^{2+}}/P_{K^+}$ between 9 to 27. Specificity for cADPR was demonstrated by the insensitivity to the noncyclic analog, adenosine 5′-diphosphoribose. Similar to findings from reports on animal cells (47), addition of IP_3 increased the magnitude of the control inward currents, and a further rise was induced with the addition of cADPR (4), indicating that the effects of cADPR were independent of the IP_3-induced Ca^{2+} release. Although this report suggests the presence of a second signaling pathway involved in vacuolar Ca^{2+} release, reservations must be considered before cADPR is accepted as a plant second messenger. First, it is necessary to demonstrate that cADPR and the enzyme involved in its synthesis are present in plants, and second, it is important to demonstrate that cytoplasmic levels of cADPR are affected by a specific external signal.

Malate-Selective Channels

The organic anion malate is accumulated in the vacuole of most plant cells. The role of malate in the physiology of plant cells is manifold: as a storage form of CO_2, as a charge balance, and as an osmolyte involved in the maintenance of cell turgor. The transport of malate into the vacuole is important for the regulation of cytoplasmic pH and the control of cellular metabolism, particularly in plants showing Crassulacean acid metabolism (CAM), where large fluxes of malate occur during the day/night cycle. Previously, it had been considered that the mechanism mediating the accumulation of malate into the vacuole was a carrier (85, 119), and its reconstitution from barley and *K. daigremontiana* vacuoles has been reported by Martinoia et al (71) and Ratajczack et al (93), respectively. In both cases, two polypeptides of molecular mass 20–30 kDa were associated with the reconstituted activity.

More direct evidence on the nature of the malate transporter has been obtained by employing the patch-clamp technique. Using Ca^{2+}_c below 10^{-7} M to eliminate SV channel activity, and employing potassium malate as the main electrolyte, the activity of voltage-activated and inward rectifying channels selective to malate in the tonoplast of the CAM plant *Graptopetalum paraguayense* (55) and sugar beet cell cultures (86) has been demonstrated. In both cases pH_c was 7.5, indicating that the divalent form of the anion is the species moving through the channels. Studies at the single-channel level with vacuoles from *G. paraguayense* demonstrated that the channel open probability was slightly voltage dependent with a rundown observed within 10–15 min (56). Characterization at the whole vacuole level in red beet cell suspensions demonstrated the presence of slow-activating inward malate currents (87). The selectivity of the slow-activating channels was 6 to 10 times higher for malate over K^+ (87). Malate-selective channels have also been found in the vacuoles of the CAM plant *K. daigremontiana* (O Pantoja & JAC Smith, manuscript in preparation) and in vacuoles of *A. thaliana* (32). The whole vacuole currents from *K. daigremontiana* and *A. thaliana* showed similar kinetics to those from sugar beet with a clear inward rectification. Varying the levels of Ca^{2+}_c had no effect on the malate currents from the two species; however, inhibition by cytoplasmic acidification within the physiological range of pH 7.0–6.5 was observed in *K. daigremontiana.* This independent electrophysiological evidence demonstrated that the plausible mechanism for vacuolar malate accumulation is an anion channel that has the particular property of being selective to an organic ion, rather than to inorganic ions. From these reports on malate channels, the diversity of species that have been employed, and the importance of this organic anion in plant cell physiology, it is tempting to speculate that plant vacuolar malate channels may be ubiquitous.

Tonoplast Intrinsic Protein

The tonoplast of plant cells also contains a channel selectively permeable to water, the tonoplast intrinsic protein (TIP). Thorough reviews on this and related membrane intrinsic proteins (MIP) have appeared recently (36, 37, 96).

The role of γTIP as a water channel has been demonstrated through heterologous expression of the protein in *Xenopus laevis* oocytes. Oocytes injected with γTIP mRNA showed rapid swelling and burst within 6 min following exposure to hypoosmotic solutions. Uninjected or water-injected oocytes swelled very slowly upon the same treatment (75). From these results, the authors calculated that γTIP caused a six- to eightfold increase in oocyte water permeability, similar to results obtained with CHIP28, a well-characterized water channel in human erythrocytes (90a). γTIP, on the other hand, did not affect the oocyte's glycerol uptake rate or result in the appearance of additional ionic currents when assayed in Bart's medium. Here we suggest the possibility that under the experimental conditions employed (75), the potential permeability of γTIP to particular ions may have been overlooked. For this, two conditions need to be considered. The first is that during heterologous expression studies the protein will insert in the *Xenopus laevis* oocyte membrane in the same orientation in which it is found in the tonoplast, i.e. with its vacuolar side facing the extracellular side of the oocyte, and the second is that γTIP is an inward rectifying anion channel, for example, the malate channel. If these two conditions are met, the activity of γTIP as a malate channel would only be recorded if this anion were present in the interior of the oocyte, a condition that has not been tested (75). Therefore, it would be interesting to assay under these conditions the activity of the heterologous expressed γTIP. To prevent potential effects of malate on the oocyte, the employment of macropatches could be a better approach to assist in resolving this concern.

CONCLUDING REMARKS

In the next few years it will be important to address the molecular characterization of ion cotransporters and channels at the tonoplast in order to advance the understanding of the role of the vacuole in the physiology of plant cells. Further studies aimed at unraveling the regulation of the transporters by biotic and abiotic factors should also be emphasized. While research over the past few years has enlightened our understanding of tonoplast ion transport, questions that remain unanswered include: Are the SV channels responsible for the release of vacuolar Ca^{2+}? Is cADPR present in plants? Is it regulated by external stimuli? Is there a Ca^{2+}-ATPase in the tonoplast? Are channels the only mechanism by which malate is accumulated? How is malate released from the vacuole? What is the mechanism involved in the transport of Cl^- into the vacuole? Does the Ca^{2+}/H^+ antiport participate in the sequestration of heavy metals?

ACKNOWLEDGMENTS

We thank all the colleagues who helped us by making available reprints and preprints of their work and by responding so quickly. We also thank Dr. Federico Sánchez for his encouragement to undertake this task. We wish to dedicate this to our parents.

Literature Cited

1. Adams P, Thomas JC, Vernon DM, Bohnert HJ, Jensen RG. 1992. Distinct cellular and organismic responses to salt. *Plant Cell Physiol.* 33:1215–23
2. Alexandre J, Lasalles JP. 1990. Effect of D-myo-inositol 1,4,5-trisphosphate on the electrical properties of the red beet vacuole membrane. *Plant Physiol.* 93:837–40
3. Alexandre J, Lasalles JP, Kado RT. 1990. Opening of Ca^{2+} channels in isolated red beet root vacuole membrane by inositol 1,4,5-trisphosphate. *Nature* 343: 567–70
4. Allen GJ, Muir SR, Sanders D. 1995. Release of Ca^{2+} from individual plant vacuoles by both $InsP_3$ and cyclic ADP-ribose. *Science* 268:735–37
5. Allen GJ, Sanders D. 1994. Osmotic stress enhances the competence of *Beta vulgaris* vacuoles to respond to inositol 1,4,5-trisphosphate. *Plant. J.* 6(5):687–95
6. Allen GJ, Sanders D. 1994. Two voltage-gated, calcium release channels coreside in the vacuolar membrane of broad bean guard cells. *Plant Cell* 6:685–94
7. Amodeo G, Escobar A, Zeiger E. 1994. A cationic channel in the guard cell tonoplast of *Allium cepa. Plant Physiol.* 105: 999–1006
8. Anderson JM, Carbonneau H, Jones HP, McCann RO, Cormier MJ. 1980. Characterization of plant nicotinamide adenine dinucleotide kinase activator protein and its identification as calmodulin. *Biochemistry* 19(13):3113–20
9. Askerlund P, Evans DE. 1992. Reconstitution and characterization of a calmodulin-stimulated Ca^{2+}-pumping ATPase purified from *Brassica oleracea* L. *Plant Physiol.* 100:1670–81
10. Barkla BJ, Apse MP, Manolson MF, Blumwald E. 1994. The plant vacuolar Na^+/H^+ antiport. See Ref. 21, pp. 141–53
11. Barkla BJ, Blumwald E. 1991. Identification of a 170-kDa protein associated with the vacuolar Na^+/H^+ antiport of *Beta vulgaris. Proc. Natl. Acad. Sci. USA* 88: 11177–81
12. Barkla BJ, Charuk JHM, Cragoe EJ Jr, Blumwald E. 1990. Photolabelling of tonoplast from sugar beet cell suspensions by [^{3}H]5-(*N*-methyl-*N*-isobutyl)-amiloride, an inhibitor of the vacuolar Na^+/H^+ antiport. *Plant Physiol.* 93:924–30
13. Barkla BJ, Zingarelli L, Blumwald E, Smith JAC. 1995. Tonoplast Na^+/H^+ antiport activity and its energization by the vacuolar H^+-ATPase in the halophytic plant *Mesembryanthemum crystallinum* L. *Plant Physiol.* 109:549–56
14. Baykov AA, Dubnova EB, Bakuleva NP, Evtushenko OA, Zhen R-G, Rea PA. 1993. Differential sensitivity of membrane-associated pyrophosphatases to inhibition by diphosphonates and fluoride delineates two classes of enzyme. *FEBS Lett.* 327(2): 199–202
15. Berkelman T, Houtchens KA, DuPont FM. 1994. Two cDNA clones encoding isoforms of the B subunit of the vacuolar ATPase from barley roots. *Plant Physiol.* 104:287–88
16. Berridge MJ. 1993. A tale of two messengers. *Nature* 365:388–89
17. Bertl A, Blumwald E, Coronado R, Eisenberg R, Findlay G, et al. 1992. Electrical measurements on endomembranes. *Science* 258:873–74
18. Bethke PC, Jones R. 1994. Ca^{2+}-calmod ulin modulates ion channel activity in storage protein vacuoles of barley aleurone cells. *Plant Cell* 6:277–85
19. Blackford S, Rea PA, Sanders D. 1990. Voltage sensitivity of H^+/Ca^{2+} antiport in higher plant tonoplast suggests a role in vacuolar calcium accumulation. *J. Biol. Chem.* 265(17):9617–20
20. Blackshear PJ, Nairn AC, Kuo JF. 1988.

Protein kinases 1988: a current perspective. *FASEB J.* 2:2957–69
21. Blatt MR, Leigh RA, Sanders D, eds. 1994. *Membrane Transport in Plants and Fungi: Molecular Mechanisms and Control.* Cambridge: Company of Biologists. 248 pp.
22. Blatt MR, Thiel G, Trentham DR. 1990. Reversible inactivation of K^+ channels of *Vicia* stomatal guard cells following the photolysis of caged inositol 1,4,5-trisphosphate. *Nature* 346:766–69
23. Blumwald E, Poole RJ. 1985. Na^+/H^+ antiport in isolated tonoplast vesicles from storage tissue of *Beta vulgaris. Plant Physiol.* 78:163–67
24. Blumwald E, Poole RJ. 1986. Kinetics of Ca^{2+}/H^+ antiport in isolated tonoplast vesicles from storage tissue of *Beta vulgaris. Plant Physiol.* 80:727–31
25. Blumwald E, Poole RJ. 1987. Salt tolerance in suspension cultures of sugar beet: induction of Na^+/H^+ antiport activity at the tonoplast by growth in salt. *Plant Physiol.* 83: 884–87
26. Bowman BJ, Allen R, Wechser MA, Bowman EJ. 1988. Isolation of genes encoding the *Neurospora* vacuolar ATPase: analysis of *vma-2* encoding the 57-kDa polypeptide and comparison to *vma-1. J. Biol. Chem.* 263(28):14002–7
27. Bremberger C, Haschke H-P, Lüttge U. 1988. Separation and purification of the tonoplast ATPase and pyrophosphatase from plants with constitutive and inducible Crassulacean acid metabolism. *Planta* 175: 465–70
28. Bremberger C, Lüttge U. 1992. Dynamics of tonoplast proton pumps and other tonoplast proteins of *Mesembryanthemum crystallinum* L. during the induction of crassulacean acid metabolism. *Planta* 188: 575–80
29. Britten CJ, Zhen R-G, Kim EJ, Rea PA. 1992. Reconstitution of transport function of vacuolar H^+-translocating inorganic pyrophosphatase. *J. Biol. Chem.* 267(30): 21850–55
30. Brosnan JM, Sanders D. 1990. Inositol tris phosphate-mediated Ca^{2+} release in beet microsomes is inhibited by heparin. *FEBS Lett.* 260(1):70–72
31. Carystinos GD, MacDonald HR, Monroy AF, Dhindsa RS, Poole RJ. 1995. Vacuolar H^+-translocating pyrophosphatase is induced by anoxia or chilling in seedlings of rice. *Plant Physiol.* 108:641–49
32. Cerana R, Giromini L, Colombo R. 1995. Malate-regulated channels permeable to anions in vacuoles of *Arabidopsis thaliana. Aust. J. Plant Physiol.* 22:115–21
33. Chanson A. 1991. A Ca^{2+}/H^+ antiport system driven by the tonoplast pyrophosphate-dependent proton pump from maize roots. *J. Plant Physiol.* 137:471–76
34. Chanson A. 1994. Characterization of the Ca^{2+}/H^+ antiport system from maize roots. *Plant Physiol. Biochem.* 32:341–46
35. Chasan R, Schroeder JI. 1992. Excitation in plant membrane biology. *Plant Cell* 4: 1180–88
36. Chrispeels MJ, Agre P. 1994. Aquaporins: water channel proteins of plant and animal cells. *Trends Biochem. Sci.* 19:421–25
37. Chrispeels MJ, Maurel C. 1994. Aquaporins: the molecular basis of facilitated water movement through living plant cells? *Plant Physiol.* 105:9–13
38. Colombo R, Cerana R. 1993. Enhanced activity of tonoplast pyrophosphatase in NaCl-grown cells of *Daucus carota. J. Plant Physiol.* 142:226–29
39. Cooperman BS, Baykov AA, Lahti R. 1992. Evolutionary conservation of the active site of soluble inorganic pyrophosphatase. *Trends Biochem Sci.* 17:262–66
40. Dainty J, de Michelis MI, Marrè E, Rasi-Caldogno F, eds. 1989. *Plant Membrane Transport: The Current Position.* Amsterdam: Elsevier. 712 pp.
41. Davies JM, Hunt I, Sanders D. 1994. Vacuolar H^+-pumping ATPase variable transport coupling ratio controlled by pH. *Proc. Natl. Acad. Sci. USA* 91:8547–51
42. Davies JM, Poole RJ, Rea PA, Sanders D. 1992. Potassium transport into vacuoles energized directly by a proton-pumping inorganic pyrophosphatase. *Proc. Natl. Acad. Sci. USA* 89:11701–5
43. Davies JM, Sanders D. 1995. ATP, pH and Mg^{2+} modulate a cation current in *Beta vulgaris* vacuoles: a possible shunt conductance for the vacuolar H^+-ATPase. *J. Membr. Biol.* 145:75–86
44. Depta H, Holstein SEH, Robinson DG, Lützelschwab M, Michalke W. 1991. Membrane markers in highly purified clathrin-coated vesicles from *Cucurbita* hypocotyls. *Planta* 183:434–42
45. Dieter P, Marmé D. 1981. A calmodulin-dependent, microsomal ATPase from corn (*Zea mays* L.) *FEBS Lett.* 125:245–48
46. DuPont FM, Morrissey PJ. 1992. Subunit composition and Ca^{2+}-ATPase activity of the vacuolar ATPase from barley roots. *Arch. Biochem. Biophys.* 294(2):341–46
47. Galione A, Lee HC, Busa WB. 1991. Ca^{2+}-induced Ca^{2+} release in sea urchin egg homogenates: modulation by cyclic ADP-ribose. *Science* 253:1143–46
48. Garbarino J, DuPont FM. 1988. NaCl induces a Na^+/H^+ antiport in tonoplast vesicles from barley roots. *Plant Physiol.* 86: 231–36
49. Gilroy S, Read ND, Trewavas AJ. 1990. Elevation of cytoplasmic calcium by caged calcium or caged inositol trisphosphate initiates stomatal closure. *Nature* 346: 769–71

49a. Gilroy S, Jones RL. 1993. Calmodulin stimulation of unidirectional calcium uptake in the endoplasmic reticulum of barley aleurone. *Planta* 190:289–96
50. Gluck S. 1992. V-ATPases of the plasma membrane. *J. Exp. Biol.* 172:29–37
51. Gogarten JP, Fichmann J, Braun Y, Morgan L, Styles P, et al. 1992. The use of antisense mRNA to inhibit the tonoplast H^+ ATPase in carrot. *Plant Cell* 4:851–64
52. Guern J, Mathieu Y, Kurkdjian A, Manigault P, Gillet B, et al. 1989. Regulation of vacuolar pH of plant cells. II. A ^{31}P NMR study of the modifications of vacuolar pH in isolated vacuoles induced by proton pumping and cation/H^+ exchangers. *Plant Physiol.* 89:27–36
53. Hedrich R, Flügge UI, Fernandez JM. 1986. Patch-clamp studies of ion transport in isolated plant vacuoles. *FEBS Lett.* 204(2):228–32
54. Hedrich R, Neher E. 1987. Cytoplasmic calcium regulates voltage-dependent ion channels in plant vacuoles. *Nature* 329: 833–36
55. Heim S, Wagner KG. 1989. Inositol phosphates in the growth cycle of suspended cultured plant cells. *Plant Sci.* 63:159–65
56. Iwasaki I, Arata H, Kijima H, Nishimura M. 1994. Two types of channels involved in the malate ion transport across the tonoplast of a crassulacean acid metabolism plant. *Plant Physiol.* 98:1494–97
57. Johannes E, Brosnan JM, Sanders D. 1992. Calcium channels in the vacuolar membrane of plants: multiple pathways for intracellular calcium mobilization. *Philos. Trans. R. Soc. London Ser. B* 338:105–12
58. Johannes E, Brosnan JM, Sanders D. 1992. Parallel pathways for intracellular Ca^{2+} release from the vacuole of higher plants. *Plant J.* 2:97–102
59. Kim EJ, Zhen R-G, Rea PA. 1994. Heterologous expression of plant vacuolar pyrophosphatase in yeast demonstrates sufficiency of the substrate-binding subunit for proton transport. *Proc. Natl. Acad. Sci. USA* 91:6128–32
60. Kim EJ, Zhen R-G, Rea PA. 1995. Site directed mutagenesis of vacuolar H^+-pyrophosphatase. Necessity of Cys^{634} for inhibition by maleimides but not catalysis. *J. Biol. Chem.* 270(6):2630–35
61. Kim Y, Kim EJ, Rea PA. 1994. Isolation and characterization of cDNAs encoding the vacuolar H^+-pyrophosphatase of *Beta vulgaris*. *Plant Physiol.* 106:375–82
62. Knight MR, Campbell AK, Smith SM, Trewavas AJ. 1991. Transgenic plant aequorin reports the effects of touch and cold shock and elicitors on cytoplasmic calcium. *Nature* 353:524–26
63. Lai S, Randall SK, Sze H. 1988. Peripheral and integral subunits of the tonoplast H^+-ATPase from oat roots. *J. Biol. Chem.* 263(32):16731–37
64. Lai S, Watson JC, Hansen JN, Sze H. 1991. Molecular cloning and sequencing of cDNAs encoding the proteolipid subunit of the vacuolar H^+-ATPase from a higher plant. *J. Biol. Chem.* 266(24):16078–84
65. Läuger P. 1991. *Electrogenic Ion Pumps.* Sunderland, MA: Sinauer Assoc. 313 pp.
66. Maathuis FJM, Prins HBA. 1991. Inhibition of inward rectifying tonoplast channels by a vacuolar factor; physiological and kinetic implications. *J. Membr. Biol.* 122: 251–58
67. Maeshima M, Yoshida S. 1989. Purification and properties of vacuolar membrane proton-translocating inorganic pyrophosphatase from mung bean. *J. Biol. Chem.* 264(33):20068–73
68. Mandel M, Moriyama Y, Hulmes JD, Pan Y-CD, Nelson H, Nelson N. 1988. cDNA sequence encoding the 16-kDa proteolipid of chromaffin granules implies gene duplication in the evolution of H^+-ATPases. *Proc. Natl. Acad. Sci. USA* 5:5521–24
69. Manolson MF, Ouellete BFF, Filion M, Poole RJ. 1988. cDNA sequence and homologies of the "57-kDa" nucleotide-binding subunit of the vacuolar ATPase from *Arabidopsis*. *J. Biol. Chem.* 263(34): 17987–94
70. Manolson MF, Wu B, Proteau D, Taillon BE, Roberts BT, et al. 1994. *STV1* gene encodes functional homologue of 95-kDa yeast vacuolar H^+-ATPase subunit Vph1p. *J. Biol. Chem.* 269(19):14064–74
71. Martinoia E, Vogt E, Rentsch D, Amrhein N. 1991. Functional reconstitution of the malate carrier of barley mesophyll vacuoles in liposomes. *Biochim. Biophys. Acta* 1062:271–78
72. Matsuda H. 1988. Open-state substructure of inwardly rectifying potassium channels revealed by magnesium block in guinea-pig heart cells. *J. Physiol.* 397:237–58
73. Matsuda H, Saigusa A, Irisawa H. 1987. Ohmic conductance through the inwardly rectifying K channel and blocking by internal Mg^{2+}. *Nature* 325:156–59
74. Matsumoto H, Chung GC. 1988. Increase in proton-transport activity of tonoplast vesicles as an adaptive response of barley roots to NaCl stress. *Plant Cell Physiol.* 29(7):1133–40
75. Maurel C, Reizer J, Schroeder JI, Chrispeels MJ. 1993. The vacuolar membrane protein γ-TIP creates water specific channels in *Xenopus* oocytes. *EMBO J.* 12(6): 2241–47
76. McAinsh MR, Brownlee C, Hetherington AM. 1990. Abscisic acid–induced elevation of guard cell cytosolic Ca^{2+} precedes stomatal closure. *Nature* 343:186–88
77. Mészáros LG, Bak J, Chu A. 1993. Cyclic

ADP-ribose as an endogenous regulator of the non-skeletal type ryanodine receptor Ca^{2+} channel. *Nature* 364:76–79

78. Nakamura Y, Kasamo K, Shimosato N, Sakata M, Ohta E. 1992. Stimulation of the extrusion of protons and H^+-ATPase activities with the decline in pyrophosphatase activity of the tonoplast in intact mung bean roots under high-NaCl stress and its relation to external levels of Ca^{2+} ions. *Plant Cell Physiol.* 33(2):139–49
79. Nanda A, Gukovskaya A, Tseng J, Grinstein S. 1992. Activation of vacuolar-type proton pumps by protein kinase C. Role in neutrophil pH regulation. *J. Biol. Chem.* 267(32):22740–46
80. Narasimhan ML, Binzel ML, Perez-Prat E, Chen Z, Nelson DE, et al. 1991. NaCl regulation of tonoplast ATPase 70-kilodalton subunit mRNA in tobacco cells. *Plant Physiol.* 97:562–68
81. Nelson H, Mandiyan S, Nelson N. 1994. The *Saccharomyces cerevisiae* VMA7 gene encodes a 14-kDa subunit of the vacuolar H^+-ATPase catalytic sector. *J. Biol. Chem.* 269(39):24150–55
82. Nelson H, Mandiyan S, Nelson N. 1995. A bovine cDNA and a yeast gene (VMA8) encoding the subunit D of the vacuolar H^+-ATPase. *Proc. Natl. Acad. Sci. USA* 92:497–501
83. Nelson H, Nelson N. 1989. The progenitor of ATP synthases was closely related to the current vacuolar H^+-ATPase. *FEBS Lett.* 247(1):147–53
84. Nelson N. 1992. The vacuolar H^+-ATPase: one of the most fundamental pumps in nature. *J. Exp. Biol.* 172:19–27
85. Nishida K, Tominaga O. 1987. Energy dependent uptake of malate into vacuoles isolated from CAM plant, *Kalanchoë daigremontiana*. *J. Plant Physiol.* 127:385–93
86. Pantoja O, Dainty J, Blumwald E. 1992. Cytoplasmic chloride regulates cation channels in the vacuolar membrane of plant cells. *J. Membr. Biol.* 125:219–29
87. Pantoja O, Gelli A, Blumwald E. 1992. Characterization of vacuolar malate and K^+ channels under physiological conditions. *Plant Physiol.* 100:1137–41
88. Pantoja O, Gelli A, Blumwald E. 1992. Voltage-dependent calcium channels in plant vacuoles. *Science* 255:1567–70
89. Parry RV, Turner JC, Rea PA. 1989. High purity preparations of higher plant vacuolar H^+-ATPase reveal additional subunits. Revised subunit composition. *J. Biol. Chem.* 264(33):20025–32
90. Ping Z, Yabe I, Muto S. 1992. Identification of K^+, Cl^-, and Ca^{2+} channels in the vacuolar membrane of tobacco cell suspension cultures. *Protoplasma* 171:7–18

90a. Preston GM, Carroll TP, Guggino WB, Agre P. 1992. Appearance of water channels in *Xenopus* oocytes expressing red cell CHIP28 protein. *Science* 256:385–87

91. Ranjeva R, Carrasco A, Boudet AM. 1988. Inositol-trisphosphate stimulates the release of calcium from intact vacuoles from *Acer* cells. *FEBS Lett.* 230:137–41
92. Rasi-Caldogno F, Carnelli A, De Michelis MI. 1992. Plasma membrane Ca-ATPase of radish seedlings. II. Regulation by calmodulin. *Plant Physiol.* 98:1202–6
93. Ratajczak R, Kemna I, Lüttge U. 1994. Characteristics, partial purification and reconstitution of the vacuolar malate transporter of the CAM plant *Kalanchoë daigremontiana* Hamet et Perrier de la Bâthie. *Planta* 195:226–36
94. Ratajczak R, Richter J, Lüttge U. 1994. Adaptation of the tonoplast V-type H^+-ATPase of *Mesembryanthemum crystallinum* to salt stress, C_3-CAM transition and plant age. *Plant Cell Environ.* 17:1101–12
95. Rea PA, Poole RJ. 1993. Vacuolar H^+-translocating pyrophosphatase. *Annu. Rev. Plant Physiol. Plant Mol. Biol.* 44:157–80
96. Reizer J, Reizer A, Saier MH Jr. 1993. The MIP family of integral membrane channel proteins: sequence comparisons, evolutionary relationships, reconstructed pathway of evolution, and proposed functional differentiation of the two repeated halves of the proteins. *Crit. Rev. Biochem. Mol. Biol.* 28:235–57
97. Reuveni M, Bennett AB, Bressan RA, Hasegawa PM. 1990. Enhanced H^+ transport capacity and ATP hydrolysis activity of the tonoplast H^+-ATPase after NaCl adaptation. *Plant Physiol.* 94:524–30
98. Robinson C, Larsson C, Buckhout TJ. 1988. Identification of a calmodulin stimulated (Ca^{2+}+Mg^{2+})-ATPase in a plasma membrane fraction isolated from maize (*Zea mays*) leaves. *Physiol. Plant.* 72: 177–84
99. Rockel B, Ratajczak R, Becker A, Lüttge U. 1994. Changed particle and diameters of intra-membrane tonoplast particles of *Mesembryanthemum crystallinum* in correlation with NaCl-induced salt stress. *J. Plant Physiol.* 143:318–24
100. Salt DE, Wagner GJ. 1993. Cadmium transport across tonoplast of vesicles from oat roots. *J. Biol. Chem.* 268(17):12297–302
101. Sarafian V, Kim Y, Poole RJ, Rea PA. 1992. Molecular cloning and sequence of cDNA encoding the pyrophosphatase-energized vacuolar membrane proton pump of *Arabidopsis thaliana*. *Proc. Natl. Acad. Sci. USA* 89:1775–79
102. Schroeder JI, Hagiwara S. 1989. Cytosolic calcium regulates ion channels in the plasma membrane of *Vicia faba* guard cells. *Nature* 338:427–30
103. Schulz-Lessdorf B, Hedrich R. 1995. Protons and calcium modulate SV-type chan-

nels in the vacuolar-lysosomal compartment: channel interaction with calmodulin inhibitors. *Planta* 178:In press
104. Schumaker KS, Sze H. 1985. A Ca^{2+}/H^+ antiport system driven by the proton electrochemical gradient of a tonoplast H^+-ATPase from oat roots. *Plant Physiol.* 79: 1111–17
105. Schumaker KS, Sze H. 1986. Calcium transport into the vacuole of oat roots. Characterization of a H^+/Ca^{2+} exchange activity. *J. Biol. Chem.* 261:12172–87
106. Schumaker KS, Sze H. 1987. Inositol-1,4,5-trisphosphate releases Ca^{2+} from vacuolar membrane vesicles of oat roots. *J. Biol. Chem.* 262(9):3944–46
107. Schumaker KS, Sze H. 1990. Solubilization and reconstitution of the oat root vacuolar H^+/Ca^{2+} exchanger. *Plant Physiol.* 92: 340–45
108. Staal M, Maathuis FJM, Elzenga JTM, Overbeek JHM, Prins HBA. 1991. Na^+/H^+ antiport activity in tonoplast vesicles from roots of the salt-tolerant *Plantago maritima* and the salt-sensitive *Plantago media. Physiol. Plant.* 82:179–84
109. Supek F, Supeková L, Mandiyan S, Pan Y-CE, Nelson H, Nelson N. 1994. A novel accessory subunit for vacuolar H^+-ATPase from chromaffin granules. *J. Biol. Chem.* 269(39):24102–6
110. Deleted in proof
111. Supeková L, Supek F, Nelson N. 1995. The *Saccharomyces cerevisiae* VMA10 is an intron-containing gene encoding a novel 13-kDa subunit of vacuolar H^+-ATPase. *J. Biol. Chem.* 270(23):13726–32
112. Sze H, Ward JM, Lai S. 1992. Vacuolar H^+-translocating ATPases from plants: structure, function and isoforms. *J. Bioenerg. Biomembr.* 24(4):371–81
113. Sze H, Ward JM, Lai S, Perera I. 1992. Vacuolar-type H^+-translocating ATPases in plant endomembranes: subunit organization and multigene families. *J. Exp. Biol.* 172:123–35
114. Taiz L, Gogarten JP, Kibak H, Struve I, Bernasconi P, et al. 1989. Studies on the structure and evolution of the vacuolar H^+-ATPase. See Ref. 40, pp. 131–37
115. Tanaka Y, Chiba K, Maeda M, Maeshima M. 1993. Molecular cloning of cDNA for vacuolar membrane proton-translocating inorganic pyrophosphatase in *Hordeum vulgare. Biochem. Biophys. Res. Commun.* 190(3):1110–14
116. Wagner GJ. 1993. Accumulation of cadmium in crop plants and its consequence to human health. *Adv. Agron.* 51:173–212
117. Ward JM, Schroeder JI. 1994. Calcium-activated K^+ channels and calcium-induced calcium release by slow vacuolar ion channels in guard cell vacuoles implicated in the control of stomatal closure. *Plant Cell* 6: 669–83
118. Weiser T, Blum W, Bentrup FW. 1991. Calmodulin regulates the Ca^{2+}-dependent slow-vacuolar ion channels in the tonoplast of *Chenopodium rubrum* suspension cells. *Planta* 185:440–42
119. White PJ, Smith JAC. 1989. Proton and anion transport at the tonoplast in crassulacean acid metabolism plants: specificity of the malate influx system of *Kalanchoë daigremontiana. Planta* 179:265–74
120. Zhen R-G, Baykov AA, Bakuleva NP, Rea PA. 1994. Aminomethylenediphosphonate: a potent type-specific inhibitor of both plant and phototrophic bacterial H^+ pyrophosphatases. *Plant Physiol.* 104:153–59
121. Zhen R-G, Kim EJ, Rea PA. 1994. Localization of cytosolically oriented maleimide-reactive domain of vacuolar H^+-pyrophosphatase. *J. Biol. Chem.* 269(37): 23342–50
122. Zimniak L, Dittrich P, Gogarten JP, Kibak H, Taiz L. 1988. The cDNA sequence of the 69-kDa subunit of the carrot vacuolar H^+-ATPase: homology to the ß-chain of F_0F_1-ATPases. *J. Biol. Chem.* 263(19):9102–12
123. Zimolo Z, Montrose MH, Murer H. 1992. H^+ extrusion by an apical vacuolar-type H^+-ATPase in the rat renal proximal tubules. *J. Membr. Biol.* 126:19–26
124. Zingarelli L, Anzani P, Lado P. 1994. Enhanced K^+-stimulated pyrophosphatase activity in NaCl-adapted cells of *Acer pseudoplatanus. Physiol. Plant.* 91:510–16

Annu. Rev. Plant Physiol. Plant Mol. Biol. 1996. 47:185–214

THE ORGANIZATION AND REGULATION OF PLANT GLYCOLYSIS

William C. Plaxton

Departments of Biology and Biochemistry, Queen's University, Kingston, Ontario K7L 3N6, Canada

KEY WORDS: compartmentation, metabolic regulation, isozymes, pyrophosphate, gluconeogenesis, respiration

ABSTRACT

This review discusses the organization and regulation of the glycolytic pathway in plants and compares and contrasts plant and nonplant glycolysis. Plant glycolysis exists both in the cytosol and plastid, and the parallel reactions are catalyzed by distinct nuclear-encoded isozymes. Cytosolic glycolysis is a complex network containing alternative enzymatic reactions. Two alternate cytosolic reactions enhance the pathway's ATP yield through the use of pyrophosphate in place of ATP. The cytosolic glycolytic network may provide an essential metabolic flexibility that facilitates plant development and acclimation to environmental stress. The regulation of plant glycolytic flux is assessed, with a focus on the fine control of enzymes involved in the metabolism of fructose-6-phosphate and phosphoenolpyruvate. Plant and nonplant glycolysis are regulated from the "bottom up" and "top down," respectively. Research on tissue- and developmental-specific isozymes of plant glycolytic enzymes is summarized. Potential pitfalls associated with studies of glycolytic enzymes are considered. Some glycolytic enzymes may be multifunctional proteins involved in processes other than carbohydrate metabolism.

CONTENTS

1040-2519/96/0601-0185$08.00

INTRODUCTION

In about 1940, glycolysis became the first major metabolic pathway to be fully elucidated, an achievement that was instrumental in the development of many experimental and conceptual aspects of modern biochemistry. Subsequent studies of glycolysis have shown that it is the "central" metabolic pathway that is present, at least in part, in all organisms. Glycolysis is an excellent example of how (*a*) a ubiquitous metabolic pathway can show significant differences in terms of its roles, structure, regulation, and compartmentation in different phyla, and even within different cells of the same species; (*b*) the function of the ATP/ADP system in biological energy transduction can be replaced or augmented by the pyrophosphate/orthophosphate (PPi/Pi) system; and (*c*) the various mechanisms of metabolic regulation apply to a specific pathway in vivo. Glycolysis is also directly involved in many biochemical adaptations of plant and nonplant species to environmental stresses such as nutrient limitation, osmotic stress, drought, cold/freezing, and anoxia.

Several reviews concerning plant glycolysis have appeared in the past decade (3, 4, 12, 14, 24, 32, 37, 39, 40, 68, 92, 108, 111, 135, 136, 140). This review considers the organization and regulation of the glycolytic pathway in plants and compares and contrasts plant glycolysis with its counterpart in nonplant systems.

The Functions of Glycolysis

Glycolysis evolved as a catabolic anaerobic pathway to fulfil two fundamental roles. It oxidizes hexoses to generate ATP, reductant, and pyruvate, and it produces building blocks for anabolism. Glycolysis is also an amphibolic pathway because it can function in reverse to generate hexoses from various low-molecular-weight compounds in energy-dependent gluconeogenesis. Much attention has been devoted to determining the mechanisms by which the opposing processes of glycolysis and gluconeogenesis are reciprocally regulated in vivo.

In contrast with animal mitochondria, which frequently respire fatty acids and glycolytically derived pyruvate, plant mitochondria rarely respire fatty acids (5). Glycolysis is thus of crucial importance in plants because it is the predominant pathway that "fuels" plant respiration. Moreover, a significant proportion of the carbon that enters the plant glycolytic and tricarboxylic acid (TCA) cycle pathways is not oxidized to CO_2 but is utilized in the biosynthesis of numerous compounds such as secondary metabolites, isoprenoids, amino acids, nucleic acids, and fatty acids. The biosynthetic role of glycolysis and respiration is particularly important in actively growing autotrophic tissues (5).

THE ORGANIZATION OF PLANT GLYCOLYSIS

So-called classical glycolysis, which occurs in most nonplant organisms (Figure 1*a*), is the cytosolic linear sequence of 10 enzymatic reactions that catalyze the net reaction:

$$\text{glucose} + 2\ \text{ADP} + 2\ \text{Pi} + 2\ \text{NAD}^+ \rightarrow 2\ \text{pyruvate} + 2\ \text{ATP} + 2\ \text{NADH}.$$

Although higher plants use sucrose and starch as the principal substrates for glycolysis, they are known to metabolize the immediate products of sucrose and starch breakdown via the classic intermediates of glycolysis (Figure 1*b*) (3, 4, 32, 136). Nevertheless, as outlined in Figure 1, there are several profound differences in structural and associated bioenergetic features of the glycolytic pathway in plant vs nonplant organisms.

Compartmentation of Glycolysis

The sequential conversion of hexoses to pyruvate in plants can occur independently in either of two subcellular compartments, the cytosol and plastid (Figure 1*b*). Compartmentation concentrates enzymes of a pathway and their associated metabolites and prevents the simultaneous occurrence of potentially incompatible metabolic processes (38). The integration of cellular metabolism necessitates controlled interactions between pathways sequestered in various subcellular compartments. Thus, plastidic and cytosolic glycolysis can interact through the action of highly selective transporters present in the inner plastid envelope (Figure 1*b*) (46). The prime functions of glycolysis in chloroplasts in the dark and in nonphotosynthetic plastids are to participate in the breakdown of starch as well as to generate carbon skeletons, reductant, and ATP for anabolic pathways such as fatty acid synthesis (14, 37, 40, 46). Although plastids from several nonphotosynthetic tissues, including developing wheat and castor seeds, have been found to possess all the enzymes of glycolysis from glucose to pyruvate, some chloroplasts may lack one or several of the enzymes of the lower half of glycolysis (e.g. enolase and phosphoglyceromutase) (40, 46, 92). In contrast, the cytosol of many unicellular green algae appears to lack a complete glycolytic pathway (and is generally deficient in the enzymes of the upper half of glycolysis), whereas their chloroplasts seem to contain the entire suite of glycolytic enzymes (68, 128). This is because starch is the dominant respiratory fuel for algae, and sucrose is relatively unimportant in algal carbon metabolism (68).

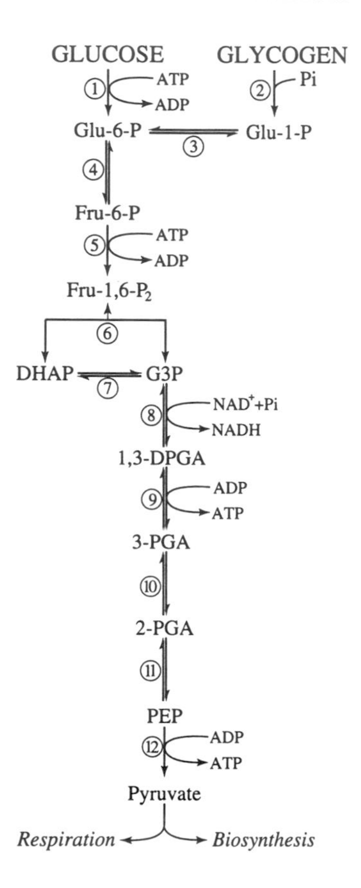

Figure 1 A comparison of the organization of nonplant (A) vs plant (B) glycolysis. The enzymes that catalyze the numbered reactions are as follows: 1, hexokinase; 2, phosphorylase; 3, phosphoglucomutase; 4, phosphoglucose isomerase; 5, PFK; 6, ALD; 7, triose phosphate isomerase; 8, NAD-dependent GAPDH (phosphorylating); 9, 3-PGA kinase; 10, phosphoglyceromutase; 11, enolase; 12, PK; 13, invertase; 14, sucrose synthase; 15, UDP-glucose pyrophosphorylase; 16, nucleoside diphosphate kinase; 17, α- and β-amylases; 18, PFP; 19, NADP-dependent GAPDH (nonphosphorylating); 20, PEPase; 21, PEPC; 22, MDH; 23, ME. Abbreviations are as in the text or as follows: Glu-1-P, glucose-1-phosphate; DHAP, dihydroxyacetone phosphate; G3P, glyceraldehyde-3-phosphate; 1,3-DPGA, 1,3-diphosphoglycerate; 2-PGA, 2-phosphoglycerate; OAA, oxaloacetate. $\rightarrow$, indicates physiologically irreversible reactions; $\rightleftharpoons$ or $\leftrightarrow$ indicate physiologically reversible reactions. Note that the number of substrate and product molecules in all reactions from G3P to pyruvate should be doubled because two molecules of G3P are formed from one molecule of hexose.

B. PLANT GLYCOLYSIS

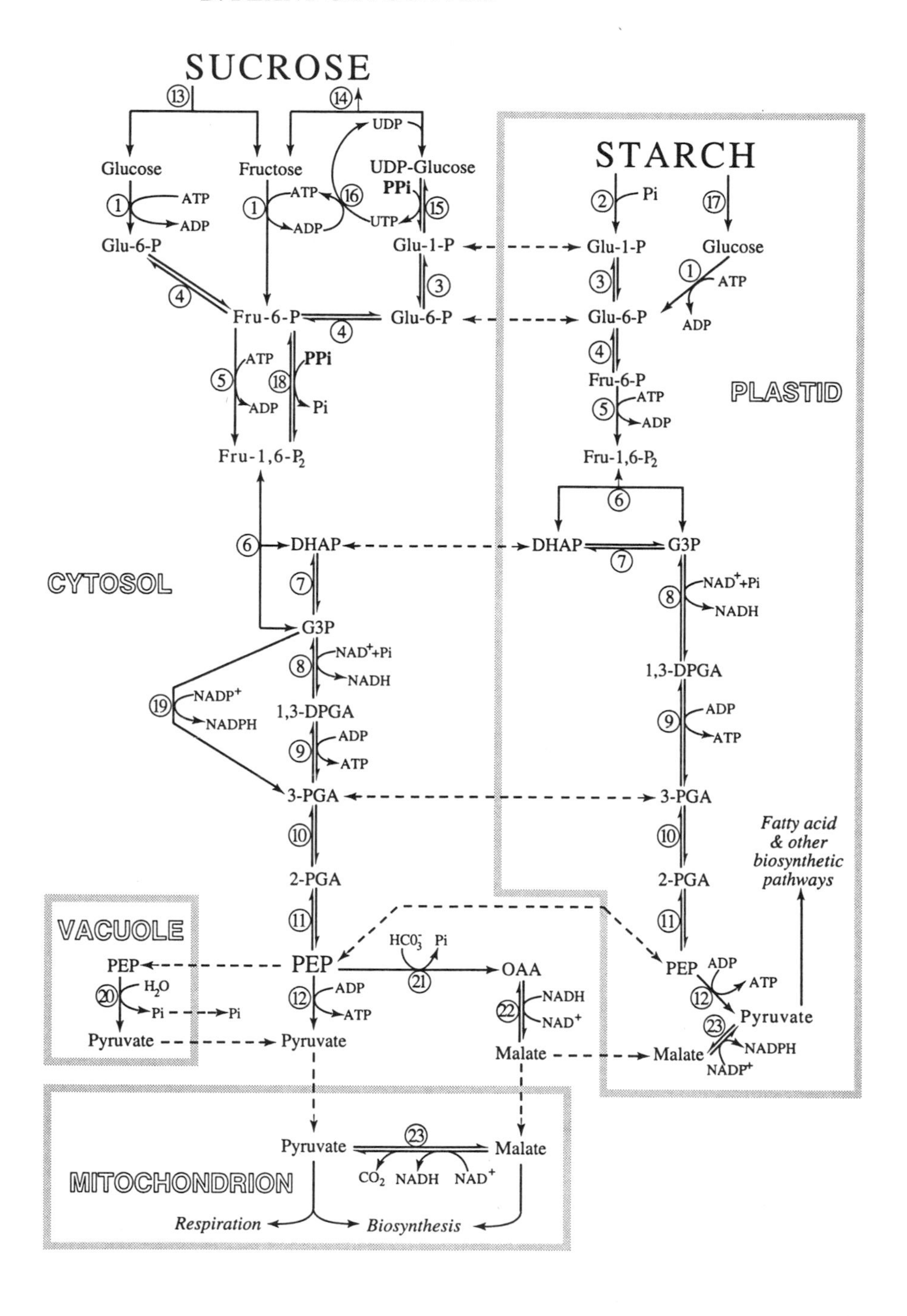

The parallel plastidic and cytosolic glycolytic reactions are believed to be catalyzed by isozymes encoded by distinct nuclear genes (14, 32, 57, 58, 61, 91, 111). The relative proportions of the isozymes can vary, not only according to the type of tissue and its developmental stage but also as a result of alterations in the plant's environmental or nutritional status (24, 37, 38, 40, 52, 93, 126). The cytosolic and plastidic isozymes can be nearly identical, except for differences in charge, or, in the case of key regulatory enzymes such as ATP-dependent phosphofructokinase (PFK) (22, 29, 39, 78) and pyruvate kinase (PK) (52, 85, 86, 106, 110a, 111, 126), the differences in physical, immunological, and kinetic properties can be profound. Plastidic glycolytic enzymes are thought to be synthesized as inactive precursors on ribosomes in the cytosol, followed by their import into the organelle with concomitant cleavage of an *N*-terminal transit peptide (15, 48).

EVOLUTION OF CYTOSOLIC AND PLASTIDIC GLYCOLYTIC ISOZYMES Gene pairs encoding plastidic and cytosolic isozymes could have evolved from duplications of common ancestral nuclear genes, or the genes encoding plastidic isozymes may have been transferred to the nucleus from the genome of the prokaryotic symbiont thought to have given rise to plastids (46). The former possibility appears to be the case for plastidic and cytosolic fructose bisphosphate aldolases (ALDs), which both appear to belong to the eukaryotic family of class I ALDs (128). Likewise, DNA sequence analyses of triose phosphate isomerase and NAD^+-dependent glyceraldehyde-3-phosphate dehydrogenase (NAD-GAPDH) demonstrated that the plastidic isozymes arose via duplication of the preexisting nuclear counterparts for the corresponding cytosolic isozyme (61, 91). In contrast, the gene and/or derived amino acid sequences encoding cytosolic and plastidic isozymes of PK (PK_c and PK_p, respectively) and hexose-phosphate isomerase show closer respective homologies to their eukaryotic and prokaryotic counterparts (16, 17, 57, 58). Similarly, immunoblot and peptide mapping analyses demonstrated that PK_p from the green alga *Selenastrum minutum* is more closely related to a bacterial PK than it is to *S. minutum* PK_c or rabbit muscle PK (76).

The Glycolytic Network of the Plant Cytosol

The cytosolic glycolytic pathway is a complex network containing parallel enzymatic reactions at the level of sucrose, fructose-6-phosphate (Fru-6-P), glyceraldehyde-3-phosphate, and phosphoenolpyruvate (PEP) metabolism (136, 140) (Figure 1*b*). An ongoing and challenging problem has been to elucidate the respective role(s), regulation, and relative importance of the various alternative reactions of plant cytosolic glycolysis. The cytosolic glycolytic network is proposed to furnish plants with the requisite metabolic options needed to facilitate their development and to acclimate to unavoidable environmental stresses such as anoxia and Pi starvation (90, 136, 140). A fundamental characteristic of cytosolic glycolysis is the existence of two alternative reactions that can utilize PPi rather than nucleoside triphosphates (NTPs) as a phosphoryl donor (Figure 1*b*).

PPi: AN AUTONOMOUS ENERGY DONOR OF THE PLANT CYTOSOL PPi is a by-product of a host of reactions involved in macromolecule biosynthesis. One dogma of cellular bioenergetics is that the anhydride bond of PPi is never utilized and that PPi produced in anabolism is always removed by the hydrolytic action of an inorganic alkaline pyrophosphatase (PPiase), thereby providing a thermodynamic "pull" for biosynthetic processes. Macromolecule biosynthesis, however, remains thermodynamically favorable no matter how the low concentration of PPi is maintained, whether by hydrolysis or by some other means, including the utilization of the high energy of the PPi bond (148).

The importance of PPi in the glycolytic metabolism of some organisms was discovered in research on so-called energy-poor anaerobic microorganisms such as the bacteria *Priopionibacterium shermanii* and parasitic amoeba *Entamoeba histolytica* (148). These species have no PFK but instead convert Fru-6-P to fructose-1,6-bisphosphate (Fru-1,6-P_2) via a PPi-dependent phosphofructokinase (PFP). They also lack PK but employ PPi to convert PEP and AMP into pyruvate, ATP and Pi via pyruvate, Pi dikinase (PPDK), thereby converting the bond energy of PPi into a high-energy phosphate of ATP. Owing to their use of PFP and PPDK, these organisms are theoretically able to yield 5 ATP instead of 2 per glucose degraded to pyruvate. This obviously represents a considerable energetic advantage for obligate anaerobes such as *P. shermanii* and *E. histolytica.*

The discovery in 1979 of the strictly cytosolic PFP in plants (27) and the subsequent observation of its potent activation by μM levels of the regulatory metabolite fructose-2,6-bisphosphate (Fru-2,6-P_2) (125) led to a surge of research on the role of PPi in plant sugar-phosphate metabolism. Because PFP catalyzes a reaction that is readily reversible (K_{eq} = 3.3) and is close to equilibrium in vivo (24, 82, 135), it could theoretically catalyze a net flux in the direction of glycolysis or gluconeogenesis. Possible functions of PFP as a glycolytic or gluconeogenic enzyme have been reviewed (12, 14, 24, 39, 135, 136, 140). PFP activity and molecular composition depend on a variety of environmental, developmental, species- and tissue-specific cues. Recent reports describe the production of transgenic potato plants exhibiting significantly lower expression of PFP (56) or overexpression of mammalian 6-phosphofructo-2-kinase (81). Metabolite studies of the transgenic plants led both groups to conclude that PFP catalyzes a net glycolytic flux in potato tubers.

The plant cytosol lacks soluble inorganic alkaline PPiase and consequently contains PPi concentrations of up to 0.3 mM (3, 4, 135, 145). Moreover, PPi levels of plant cells are remarkably insensitive to environmental perturbations that elicit significant decreases in NTP pools (35, 44, 94, 140). The significance of PPi in plant metabolism was recently demonstrated by the introduction of the *Escherichia coli* inorganic PPiase gene into tobacco and potato plants under control of a constitutive promoter (70). Expression of the *E. coli* PPiase in the cytosol reduced PPi levels by up to threefold and led to a dramatic inhibition of plant growth (70). To assess the relative importance of PPi vs ATP as an energy donor in the plant cytosol, Davies et al (36) computed the standard free energy changes for PPi and ATP hydrolysis under a variety of cytosolic conditions. The results indicate that PPi would be particularly favored as a phosphoryl donor, relative to ATP, under cytosolic conditions known to accompany stresses such as anoxia or nutritional Pi deprivation. These results underscore the importance of PPi as an autonomous energy donor of the plant cytosol. In support, tolerance of acclimated maize root tips to anoxia is not critically dependent upon high energy charge (150).

Apart from PFP, PPi could be employed as an energy donor for two other cytosolic reactions: (*a*) the conversion of UDP-glucose to UTP and glucose-1-phosphate catalyzed by UDP-glucose pyrophosphorylase (Figure 1*b*), and (*b*) the PPi-dependent H^+-pump of the tonoplast. The H^+-PPiase is one of two primary H^+-pumps residing at the tonoplast; the other is an H^+-ATPase (36). That PPi-powered processes may be a crucial facet of the metabolic adaptations of plants to environmental stresses that cause depressed NTP (but not PPi) pools is indicated by the significant induction of: (*a*) sucrose synthase (55, 122), PFP (90), and the tonoplast H^+-PPiase (28) by anoxia in rice seedlings; and (*b*) PFP of *Brassica nigra, B. napus,* and related crucifers by Pi starvation (44, 137, 139, 140; VL Murley, CC Carswell & WC Plaxton, unpublished observations). Recent evidence (55) indicates that sucrose metabolism of anoxic rice seedlings occurs mainly through sucrose synthase with nucleoside diphosphate kinase facilitating the cycling of uridilates needed for operation of this pathway (Figure 1*b*). Assuming that PPi is a by-product of anabolism, no ATP is needed for the conversion of sucrose to hexose-Ps via the sucrose synthase pathway, whereas 2 ATPs are needed for the invertase pathway. Mertens (90) argued that PFP functions in glycolysis in anoxic rice seedlings because both PFP activity and the level of Fru-2,6-P_2 increase, while PFK activity declines. Thus, the net yield of ATP obtained during glycolytic fermentation of sucrose is increased from 4 to 8 if sucrose is metabolized via the sucrose synthase and PFP bypasses, relative to the invertase and PFK pathways (Figure 1*b*).

THE UNIQUE FLEXIBILITY OF PLANT PHOSPHOENOLPYRUVATE METABOLISM

PEP is able to generate a large amount of energy for anabolism (it occupies the highest position on the thermodynamic scale of known phosphorylated metabolites) and participates in a wide range of reactions by enzymatic cleavage on either side of its enol oxygen atom (111). It is also both a key allosteric effector of a number of plant enzymes as well as a major branchpoint leading into a variety of primary and secondary metabolic pathways.

Plant cells have been proposed to employ two alternative metabolic routes to indirectly or directly circumvent the reaction catalyzed by PK_c (Figure 1*b*). PEP carboxylase (PEPC), a ubiquitous plant cytosolic enzyme, plays the important role in C_3 plants and nonphotosynthetic tissues of C_4 and Crassulacean acid metabolism plants of replenishing TCA cycle intermediates consumed in biosynthesis (30, 67, 68). However, together with cytosolic malate dehydrogenase (MDH) and the mitochondrial NAD-dependent malic enzyme (ME), PEPC can also function as a glycolytic enzyme by indirectly bypassing the reaction catalyzed by PK_c (Figure 1*b*). On the basis of pulse-chase radiolabeling experiments using $NaH^{14}CO_3$, it was concluded that this bypass operates in roots of *Pisum sativum* (26). The PEPC-MDH-ME bypass of PK_c is thought to be important during nutritional Pi deprivation when PK_c activity may become ADP limited (102, 138, 140). Compared to nutrient-sufficient controls, PEPC activity was five- and threefold greater in extracts of Pi-deficient *B. nigra* and *Catharanthus roseus* suspension cells, respectively (44, 102). Metabolite determinations and kinetic studies of PK_c and PEPC in *C. roseus* suggested that the contribution of PEPC to the metabolism of PEP increased in Pi-starved cells in vivo (102). Further evidence for the operation of this PK_c bypass in Pi starved *C. roseus* was provided by the rapid release of $^{14}CO_2$ from organic compounds derived from fixed $NaH^{14}CO_3$ and from [4-^{14}C]malate (102). In addition, dark CO_2 fixation rates of *S. minutum* and proteoid roots of *Lupinus albus* were found to increase with Pi limitation, while respiration declined (71, 138). This suggested that the Pi-starvation-dependent in vitro and in vivo elevations of *C. roseus, S. minutum,* and *L. albus* PEPC activities were in response to increased demands for pyruvate and/or Pi recycling.

A variation on this scheme occurs in developing oil seeds where PEPC, in concert with cytosolic MDH and plastidic NADP-ME, may provide an alternative to PK_p to supply pyruvate for fatty acid biosynthesis (111, 127, 133) (Figure 1*b*). Malate is an excellent exogenous substrate for fatty acid biosynthesis by isolated leucoplasts of developing *Ricinus communis* seeds (133). Relative to most nonphotosynthetic tissues, the PEPC activity in developing *R. communis* and *B. napus* seeds is substantial (127). Furthermore, the most significant increase in the activity and concentration of PEPC in developing *R. communis* endosperm coincides with the tissue's most active phase of storage oil accumulation (127). All of the reductant required for carbon incorporation into fatty acids is produced as a consequence of the metabolism of malate to acetyl-CoA through the leucoplast-localized NADP-ME and pyruvate dehydrogenase complex (133). However, fatty acid synthesis also requires a source of ATP (for acetyl-CoA carboxylase), and serving as this source may be an indispensable function for the 3-phosphoglycerate kinase and PK_p of leucoplastic glycolysis (Figure 1*b*). The respective contributions of the leucoplastic glycolysis vs PEPC-MDH-ME pathways for supporting oilseed fatty acid synthesis require clarification. It has been suggested that the relative importance of each route varies diurnally according to the rate of photosynthate import from source tissues (111).

PEP phosphatase: a glycolytic enzyme? It has long been recognized that a PEP phosphatase (PEPase) activity often interferes with the determination of plant PK activity (3, 108, 111). PEPase activity was generally attributed to a nonspecific acid phosphatase (APase). However, two dissimilar plant phosphatases exhibiting exceptional selectivity for PEP have been isolated (42, 88). Although a PEP-specific alkaline phosphatase was purified and characterized from germinating mung beans (88), a subsequent study provided immunological and kinetic evidence suggesting that this activity arose from PK_c during its purification (114). In contrast, a PEP-specific APase that is physically and immunologically distinct from PK_c was purified to homogeneity from heterotrophic *B. nigra* suspension cells (42). Although its substrate specificity was nonabsolute, this APase was designated as a PEPase because of its extremely low K_m for PEP of about 50 μM (a value equivalent to those reported for various plant PK_cs) (86, 115, 116, 119, 149) and because its specificity constant (V_{max}/K_m) for PEP was at least sixfold higher than that obtained for any of the other of 14 nonsynthetic substrates identified (42).

B. nigra PEPase demonstrated potent inhibition by Pi (42), and its specific activity was increased more than ten-fold following Pi-deprivation (44). Similar results have been reported for the PEPase of Pi-limited *S. minutum* (138). PEPase may function to directly bypass the ADP-limited PK_c and to recycle intracellular Pi during Pi starvation of *B. nigra* and *S. minutum* (44, 138, 140). Immunoquantification studies have revealed that the significant induction of *B. nigra* PEPase activity during Pi stress arises from de novo synthesis of PEPase protein (45). According to the "glycolytic bypass" theory for PEPase, as *B. nigra* PEPase is localized in the cell vacuole (43) Pi-depleted *B. nigra* suspension cells transport PEP into—and pyruvate out of—it (Figure 1*b*). Future studies must evaluate whether the tonoplast of Pi-starved plant cells is permeable to PEP and pyruvate. However, movement of PEP into the vacuole is electrogenically feasible, because at the alkaline pH of the cytosol it would carry three net negative charges (E Blumwald, personal communication).

Transgenic tobacco plants lacking cytosolic pyruvate kinase in their leaves Transformation of tobacco (*Nicotiana tabacum*) with a vector that was designed to examine the impact of overexpressing potato PK_c unexpectedly led to the production of plants that specifically lacked PK_c in their leaves as a result of cosuppression (53). The primary tobacco transformants showed no alteration in shoot growth, morphology, photosynthesis, or respiration (53). Analysis of the homozygous offspring of these transformants, which either lacked PK_c in their leaves (PK_c–) or were genetically equivalent to wild type (PK_c+), led to these major (77) results and conclusions:

1. The absence of leaf PK_c had a particularly deleterious effect on root development, which may be the first example where a specific "knockout" of an enzyme in one tissue exerts a maximal detrimental effect in a different tissue where the same enzyme is expressed normally.
2. Partially purified tobacco leaf PK_c displayed regulatory properties consistent with PK_c's involvement in controlling respiratory C-flow to support N-assimilation by glutamine synthetase/glutamate-oxoglutarate aminotransferase. That the regulation of N-metabolism was impaired in eight-week-old PK_c– plants was indicated by the elevated free amino acid content of their leaves. The PK_c– plants also had significantly reduced starch, glucose, and hexose-P levels in leaves and roots accompanied by enhanced rates of photosynthesis and dark respiration in leaves. Although the metabolic basis for these observations remains to be determined, the analysis proves that plants contain metabolic bypasses to PK_c. In contrast, PK bypasses are absent in yeast, *E. coli,* and animals (cited in 53).
3. Leaves of the PK_c– plants contained twofold more PEPase activity than did the PK_c+ controls, whereas activities of PEPC, PFK, and PFP were unaffected. This indicates that PEPase might be acting to circumvent the PK_c reaction in the PK_c– leaves.

Overall, these results illustrate the remarkable flexibility of plant PEP metabolism. This flexibility is probably an evolutionary adaptation to the stresses that plants are exposed to in a fluctuating environment.

THE REGULATION OF PLANT GLYCOLYSIS

The magnitude of metabolite flux through any metabolic pathway depends upon the activities of the individual enzymes involved. "Coarse" and/or "fine" metabolic controls can vary the reaction velocity of a particular enzyme in vivo (32, 107). Coarse control is achieved through varying the total population of enzyme molecules via alterations in the rates of enzyme biosynthesis or proteolysis; it most frequently comes into play during tissue differentiation or long-term environmental (adaptive) changes (107). By modulating the activity of preexisting enzymes, fine controls function as "metabolic transducers" that sense the momentary metabolic requirements of the cell and adjust the rate of metabolite flux through the various pathways accordingly.

Coarse Metabolic Control of Plant Glycolysis

Some of the best evidence for coarse control of plant glycolytic enzymes comes from studies of developing and germinating seeds (9, 13, 18–21, 23, 24, 40, 52, 66, 93, 106, 109, 118, 127), cell and tissue cultures (8a, 44, 101, 126, 134, 137), and anaerobically induced proteins such as cytosolic ALD (ALD_c) and enolase (49, 73, 92). Although an understanding of the transcriptional and translational regulation of genes encoding enzymes such as PFP, phosphoglyceromutase, enolase, PK_c, PK_p, and PEPC is beginning to emerge (13, 15, 30, 47, 49, 52, 66, 83), very little is known about how the proteolytic turnover of plant glycolytic enzymes is regulated. Analysis of PK_c and PK_p at the mRNA and protein levels in developing tobacco seeds demonstrated that the expression of these isozymes may be controlled by independent transcriptional and posttranscriptional mechanisms, and that PK_p has a much greater rate of turnover than PK_c (52). An asparginyl endopeptidase has been suggested to be involved in the turnover of PK_p in developing *R. communis* seeds (34, 109), whereas *Vicia fabia* PEPC appears to be degraded via ubiquitin-dependent proteolysis (131).

Fine Metabolic Control of Plant Glycolysis

Diverse fine control mechanisms have evolved to regulate the interplay of glycolysis with its associated pathways and to coordinate glycolytic flux with cellular needs for energy and anabolic precursors. Quantitative and qualitative approaches have been used to identify these control mechanisms (6, 32, 48).

CONTROL ANALYSIS One important approach to metabolic regulation is the control analysis theory and the analytical tools used to implement it (6). The goal of control analysis is to elucidate key points of regulation in vivo without relying on preconceived notions as to which pathway enzymes are "rate-determining-steps." Control analysis of nonplant glycolysis clearly shows that the contribution of the various enzymes to the overall flux control is highly dependent upon the organism, tissue, and physiological condition (48). Changes in physiological conditions may cause a redistribution of the control that each enzyme exerts on glycolytic flux, and this undoubtedly applies to plant glycolysis. However, because plant glycolysis is a complex network rather than a strictly cytosolic linear pathway, the practical use of control analysis for analyzing glycolytic pathway regulation in plants may prove difficult. Readers are referred to a recent review (6) for further insights concerning the application of control analysis to plant metabolism.

KEY REGULATORY ENZYMES OF GLYCOLYSIS The traditional (or qualitative) approach to metabolic regulation involves identifying the so-called key regulatory enzymes of a pathway and the mechanisms whereby their activities are controlled in vivo. These reactions are often characterized by their strategic position in a pathway—by being greatly displaced from equilibrium in vivo—and they were classically recognized by showing that their substrate(s) to product(s) concentration ratio changed in the opposite direction to the flux when the latter was varied (6, 32). Elucidation of the fine control of plant glycolysis is complicated by the alternative glycolytic reactions that exist in the plant cytosol (Figure 1*b*). Flexibility in the structure of cytosolic glycolysis necessarily implies flexibility in glycolytic regulation, and the regulation will vary according to the specific tissue, its developmental stage, and the external environment. Nevertheless, various approaches have demonstrated that, as in nonplant systems, fine control of plant glycolysis is primarily exerted by those enzymes that catalyze reactions involved in the conversion of hexose to hexose-P, Fru-6-P to Fru-1,6-P_2, and PEP to pyruvate (32, 82, 92). It has been argued that the first committed step of plant glycolysis is the conversion of Fru-6-P to Fru-1,6-P_2 (39). I concentrate on the fine control of the glycolytic sequence from Fru-6-P to pyruvate. The study of plant glycolytic regulation has been facilitated by advances in protein purification technologies that have made it possible to fully purify—and thus more thoroughly characterize—PFP (87, 98, 104), PEPC (30, 83, 84), cytosolic PFK (PFK_c) (29, 60, 143, 146), plastidic PFK (PFK_p) (25, 78), PK_c (65, 105), and PK_p (76, 103, 110a) from various plant sources.

SPECIFIC MECHANISMS OF FINE CONTROL AS APPLIED TO PLANT GLYCOLYTIC REGULATION At least six mechanisms of fine control can modulate the activities of preexisting enzymes: variation in substrate(s) and cofactor concentrations, variation in pH, metabolite effectors, subunit association-disassociation, reversible covalent modification, and reversible associations of metabolically sequential enzymes (107). These fine controls often interact with or may actually be dependent upon one another. Examples of how each mechanism may apply to the regulation of plant glycolysis are briefly considered below.

Fine control #1: variation in substrate(s) or cofactor concentration Changes in substrate concentrations that normally occur in vivo are generally not significant in metabolic regulation (107). Nevertheless, regulation of several plant glycolytic enzymes may be achieved, in part, through changes in a substrate. For example, the initial activation of PK_c that accompanies N- or P-resupply to nutrient-limited *S. minutum* has been proposed to arise from a release of ADP-limitation of the enzyme (50, 68). Similarly, the low affinity of soybean nodule cytosolic NAD-GAPDH for its cosubstrate Pi (K_m = 9 mM) has been postulated to make it responsive to changes in Pi levels that could occur in vivo (33).

During long-term Pi deprivation, cellular pools of adenylates and Pi become severely depressed (8b, 35, 43, 138, 140). This process has been proposed to restrict the activities of the adenylate-dependent (e.g. PFK, 3-phosphoglycerate kinase, and PK) and Pi-dependent (e.g. NAD-GAPDH) glycolytic enzymes as well as the phosphorylating (cytochrome) pathway of mitochondrial respiration. As a consequence, the flux of hexose-P through inducible adenylate- and Pi-independent cytosolic glycolytic bypasses (i.e. PFP, irreversible nonphosphorylating NADP-GAPDH, PEPase, and PEPC) (Figure 1*b*) may be promoted, concomitant with an increased participation of the nonphosphorylating cyanide- and rotenone-insensitive alternative pathways of mitochondrial electron transport (44, 102, 137–140).

Fine control #2: variation in pH The pH dependence of enzyme activity may be an important aspect of the fine control of plant glycolysis because the pH of the cytosolic and plastidic compartments can change in response to a variety of environmental factors. Inhibition of leaf PFK_p activity may arise from alkalinization of the stroma that follows illumination (32). The shift from lactate to ethanol production that occurs during fermentation in many plant cells is hypothesized to be mediated by anoxia-induced reductions in cytosolic pH, which lead to stimulation of pyruvate decarboxylase (73, 92). PEPC activity and response to metabolite effectors are well known to be sensitive to cytosolic pH changes that may occur in vivo. PEPCs from various sources including *S. minutum* (J Rivoal, R Dunford, WC Plaxton & DH Turpin, unpublished data), soybean nodules (130), maize leaves (83, 112), banana fruit (84), and cotyledons of germinating *R. communis* seeds (119) demonstrate a greater activity and weaker response to various metabolite inhibitors as assay pH is increased from about pH 7 to 8. This has given rise to the proposed "pH stat" function for PEPC because its activity should increase in response to cytosolic alkalinization owing to direct pH effects on its activity as well as the enzyme's desensitization to inhibitory metabolites. It was recently reported that light-dependent alkalinization and dark-acidification of cytosol-enriched cell sap are respectively correlated with the light-activation and dark-inactivation of leaf PEPC activity, particularly in C_4 plants (120). The regulation of the PK_c isoforms from endosperm and cotyledons of germinating *R. communis* seeds may also arise from pH-dependent alterations in the enzyme's response to a variety of metabolite inhibitors (115, 119). It was proposed that an enhancement in PK_c activity of germinating *R. communis* endosperms occurs during anaerobiosis through concerted decreases in cytosolic pH and concentrations of several key inhibitors (115), and that an enhancement in PK_c activity of germinating *R. communis* cotyledons will arise from reduced cytosolic pH and ATP levels caused by operation of a plasmalemma H^+-symport that powers the uptake of endosperm-derived sucrose and amino acids from the apoplast (119).

Fine control #3: metabolite effectors Tissue-specific expression of kinetically distinct isoforms of key enzymes such as PEPC and PK_c (see below) gives rise to several important tissue-specific differences with respect to what metabolites are key effectors of plant glycolytic enzymes. Two fundamental facets of this regulation emerge. First, adenine nucleosides (and hence energy charge) do not always play as prominent a role in regulating plant glycolysis as they do in many nonplant systems. Recall that PPi can function as an autonomous energy donor in the plant cytosol, and glycolysis in many plant tissues is key to supplying biosynthetic precursors, rather than ATP production per se. Thus, ATP and AMP are usually not considered to be critical effectors of PFK_c or PFK_p (39, 78, 143), PFP (87, 104, 135), PEPC (68, 83, 119, 129, 130), and PK_c or PK_p (65, 69a, 86, 119, 149). One exception to this view may be in anoxia-tolerant tissues such as germinating seeds. For example, MgATP is a key inhibitor of the PFK_c from germinating *Cucumis sativus* seeds (22). Similarly, the switch from gluconeogenesis to glycolysis that follows anoxia stress in the endosperm of germinating *R. communis* seeds is thought to arise in part from the release of inhibition of the endosperm-specific PK_c isoform by MgATP (115, 117).

A second notable attribute of plant glycolytic regulation by metabolite effectors is the ubiquitous role played by PEP, Pi, Fru-2,6-P_2, and TCA/glyoxylate cycle intermediates. Virtually all PFK_cs and PFK_ps examined to date show potent inhibition by PEP, and this inhibition is relieved by the activator Pi. It is thus the concentration ratio of Pi:PEP that is believed to be critical in regulating PFK activity in vivo (4, 32, 39, 60, 143). PEP may also allosterically inhibit ALD_c, because it is a mixed-type inhibitor of the enzyme from carrot storage roots (97). In contrast with PFK, Pi is a potent inhibitor of PFP in the forward direction (87, 135) and PEPase (42). The large (up to 50-fold) reductions in intracellular Pi levels that follow long-term Pi starvation have, therefore, been proposed to promote the activity of PFP and PEPase while curtailing the activity of PFK_c (42, 140).

The cytosolic regulatory metabolite Fru-2,6-P_2 reciprocally regulates liver glycolysis and gluconeogenesis owing to its potent activation and inhibition of PFK and FBPase, respectively (Figure 2*a*) (48). In plants, Fru-2,6-P_2 activates and inhibits PFP and cytosolic FBPase, respectively, but has no effect on PFK (Figure 2*b*). The metabolism and possible functions of Fru-2,6-P_2 in plants have been reviewed (135). Although the roles of Fru-2,6-P_2 have not been fully resolved, in at least some instances it probably operates to regulate the opposing processes of gluconeogenesis and glycolysis in the plant cytosol. A rise in the cytosolic concentration ratio of Pi:3-phosphoglycerate (3-PGA) favors Fru-2,6-P_2 synthesis owing to reciprocal effects of Pi (activator) and 3-PGA (inhibitor) on 6-phosphofructo-2-kinase (135).

The TCA cycle intermediates citrate, 2-oxoglutarate, succinate, and/or malate are effective feedback inhibitors of many plant PK_cs (10, 65, 86, 115, 119, 149) and PEPCs (68, 83, 84, 112, 119, 124, 129, 130). Glycolytic flux rises whenever TCA cycle intermediates are consumed via anabolism or respiration. All PEPCs examined to date display varying degrees of inhibition by malate, and this is usually relieved by the activator glucose-6-phosphate (30, 84, 112, 119, 129, 130) or through protein kinase-mediated phosphorylation (30, 67, 83).

The amino acids aspartate (Asp) and/or glutamate (Glu) are important tissue-specific effectors of several PEPCs and PK_cs. Potent inhibition by Asp and Glu has been reported for PEPC from the green alga *S. minutum* (129), soybean root nodules (130), cotyledons of germinated *R. communis* seeds (119), and ripened banana fruit (84). Likewise, Glu is a potent allosteric inhibitor of PK_c from the green algae *S. minutum* (86) and *Chlamydomonas reinhardtii* (149), cotyledons of germinated *R. communis* seeds (119), and spinach and *R. communis* leaves (10, 65). In contrast, the activity of PK_c from germinating endosperm of *R. communis* shows no response to any amino acid (115). The regulatory differences in germinating *R. communis* seed PK_c isoforms are consistent with the role of the endosperm as a substrate exporter and the cotyledons as a biosynthetic tissue active in sucrose and amino acid import. The inhibition of some PEPCs and PK_cs by Asp and/or Glu provides a tight feedback control that could closely balance their overall activity with the production of carbon skeletons (e.g. oxaloacetate and 2-oxoglutarate) required for NH_4^+ assimilation and transamination reactions in tissues active in amino acid and protein synthesis (68).

Fine control #4: subunit association-dissociation Regulatory enzymes invariably exist as oligomers. Many can reversibly dissociate, usually in response to effector binding. Because dissociation is often accompanied by a change in enzyme activity, it provides a mechanism for regulation (107, 142). A recent review (142) summarized evidence indicating that this form of fine control may be important in vivo for at least 13 enzymes of nonplant carbohydrate metabolism, including mammalian PFK and PK. Likewise, subunit association-disassociation has been suggested to regulate plant PFP (12, 39, 104, 135), PFK_c (25, 39, 69b, 79, 146), PEPC (83, 112), and PK_c (79, 124). If an enzyme can be shown to reversibly dissociate in vitro, no assumption should be made that the same process will occur in vivo until appropriate experiments indicate that the concentrations of enzyme and effectors are physiological. Although such rigorous proof is generally lacking for the aforementioned plant glycolytic enzymes, it appears likely that subunit association-disassociation will prove to be an important facet of the fine control of plant glycolytic flux in vivo.

Fine control #5: reversible covalent modification Enzyme regulation by reversible covalent modification is the major mechanism whereby extracellular stimuli such as hormones or light coordinate the regulation of intermediary metabolism. Disulfide-dithiol and phosphorylation-dephosphorylation interconversions are the most important types of reversible covalent modification used in higher eukaryote enzyme regulation (107).

Covalent modification by disulfide-dithiol exchange links photosynthetic electron transport flow to the light regulation of several key enzymes of the chloroplast stroma via the thioredoxin system (46, 107). This process is critical for the light-dependent activation and inhibition of the Calvin cycle and oxidative pentose phosphate pathways, respectively (46, 68). That disulfide-dithiol interconversion may directly participate in the fine control of plant plastidic glycolysis was suggested by the observation that pea leaf PFK_p is light inactivated, an effect that could be mimicked by the addition of dithiothreitol to darkened chloroplasts (62). The discovery of a cytosolic-specific thioredoxin *h* suggests that this form of covalent modification could potentially regulate cytosolic glycolytic enzymes (107). Of note are reports that reduced thiol groups (*a*) cause a sixfold activation of cytosolic NAD-GAPDH from roots of *Mesembyanthemum crystallinum* (2b), (*b*) elicit maximal activation of tomato fruit and wheat endosperm PFP by Fru-2,6-P_2 (75), and (*c*) protect potato tuber PFP from dilution-dependent declines in intrinsic fluorescence and activity through stabilization of the enzyme's native 460-kDa heterooctameric form (113).

Protein phosphorylation-dephosphorylation plays a central role in regulation of cellular metabolism in all cells (67, 107). Of relevance here is the hormonal regulation of animal glycolytic enzymes such as PK by phosphorylation (48). Moreover, at least ten plant enzymes, including the cytosolic-localized sucrose phosphate synthase and PEP carboxykinase have been shown, or strongly suggested, to be controlled by reversible phosphorylation in vivo (67, 144). As summarized in this volume (30) and elsewhere (67, 83), this mechanism of regulation appears to apply to all higher plant PEPCs examined to date. Phosphorylation has also been invoked as a possible explanation for the disparity between enolase protein levels and activities found in developing *R. communis* seeds and *M. crystallinum* roots (47, 93). Although the anaerobically induced enolase from *Echinochloa phyllopogon* has been phosphorylated in vitro by an endogenous protein kinase, it remains to be established whether this process also occurs in vivo (100). ^{32}P-labeling and kinetic studies have determined that PK_c is not phosphorylated in vivo in aerobic or anoxic germinating *R. communis* endosperms (115). This is consistent with the absence in potato PK_c of a phosphorylation consensus sequence that is found in yeast PK (16). Although the phosphorylation site of the yeast PK does align with a similar motif on the deduced amino acid sequence for a *R. communis* developing seed PK_p (15), attempts to phosphorylate *R. communis* PK_p by incubating isolated intact leucoplasts or leucoplast lysates with [γ-^{32}P]-ATP have proved unsuccessful (FB Negm & WC Plaxton, unpublished observations). Nevertheless, these results do not eliminate the possibility of plant PK_c or PK_p being phosphorylated in other tissues or under other physiological conditions. Relatively little research has been done with regard to phosphorylation of other plant glycolytic enzymes, particularly PFK and PFP.

Fine control #6: reversible associations of metabolically sequential enzymes Association of glycolytic enzymes into multienzyme complexes (or "meta bolons") has also been proposed as a mechanism to control glycolytic flux (48, 92, 107). In particular, glycolytic enzymes in muscle cells are thought to form transient complexes on contractile proteins during contraction-induced stimulation of glycolysis (48, 107). Thus, restricted diffusion or even direct transfer (or "channeling") of intermediates can occur between active sites of sequential enzymes. In addition to the possible kinetic advantages, channeling may reduce the concentration of the channeled intermediates in the bulk solution, thus sparing the limited solvent capacity of the cell. Channeling may also alter enzyme kinetic properties because of conformation changes occurring during binding (64, 107). Although the existence of a complete glycolytic metabolon is doubtful, it seems evident that physical interactions between groups of sequential glycolytic enzymes such as PFK, ALD, and/or NAD-GAPDH can occur in vivo in animal cells (48). The recent development of a metabolic control theory for muscle concluded that flux control coefficients for enzymes of channeled pathways are usually larger than those in the corresponding nonchanneled pathway (74).

This mechanism of regulation also appears to be involved in the control of several plant metabolic pathways, including the Calvin cycle and aromatic biosynthesis (2a, 64, 107). Yet, few researchers have considered how it might apply to plant glycolysis. Kinetic and physical studies of homogeneous enzymes indicated that NADP-GAPDH, triose phosphate isomerase, and ALD interact during glycolysis in the chloroplast of *Pisum sativum* (2a) and that NAD-GAPDH and 3-PGA kinase form a specific complex in the cytosol of germinating mung beans (89). Furthermore, the well-characterized stimulation of respiration that accompanies aging of carrot and sugar beet storage root slices was suggested to arise, in part, from an interaction between cytosolic glycolytic enzymes (96). A subsequent study applied the techniques of immunoaffinity chromatography and immunoblotting to demonstrate that ALD_c may specifically interact with the metabolically sequential PFK_c and PFP in carrot storage roots (99). Evidence also indicates that ALD_c specifically associates with cytosolic FBPase in the gluconeogenic endosperm of germinating *R. communis* seeds (95). Whether interactions between plant glycolytic and gluconeogenic enzymes occur in vivo remains to be established.

OVERVIEW OF THE FINE METABOLIC CONTROL OF NONPLANT VS PLANT GLYCOLYSIS A notable discrepancy in the fine control of nonplant vs plant glycolysis is that the overall regulation exerted by respective enzymes involved in Fru-6-P and PEP metabolism appears to be reversed (Figure 2). In nonplant systems such as mammalian liver (Figure 2*a*), primary control of glycolytic flux to pyruvate is believed to be mediated by PFK, with secondary control at PK

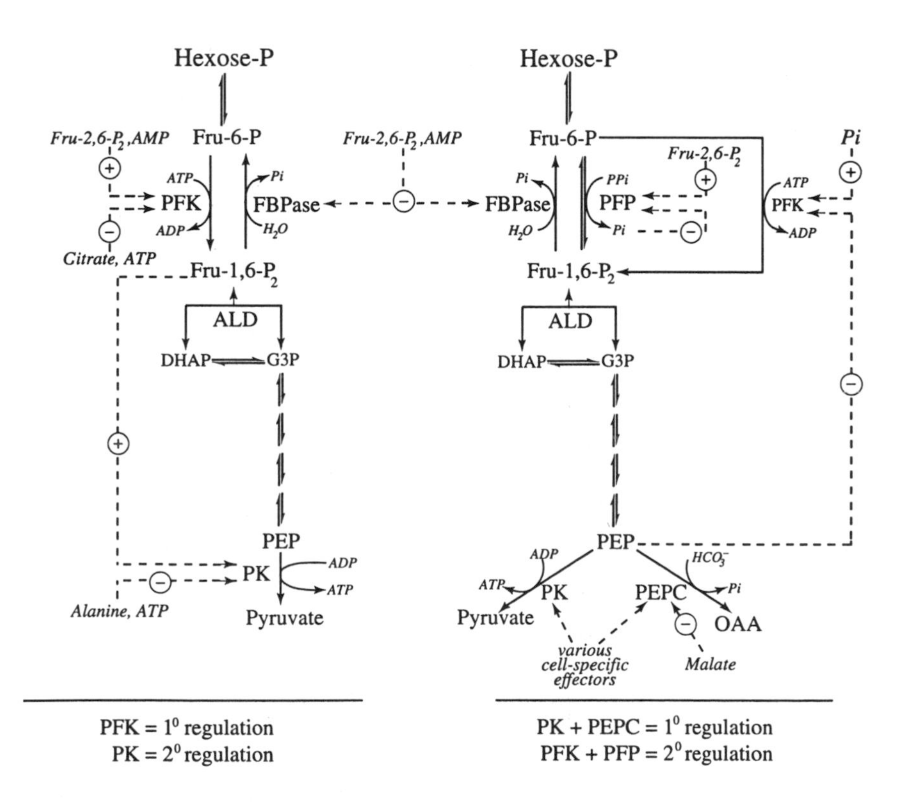

Figure 2 A comparison of the metabolite regulation of glycolytic flux from hexose-monophosphates to pyruvate in mammalian liver (A) vs the plant cytosol (B). ⊕ and ⊖ denote activation and inhibition, respectively. The abbreviations are as in the text.

(48, 111). Activation of PFK enhances the level of its product, Fru-1,6-P_2, which is a potent feed-forward allosteric activator of the majority of nonplant PKs examined to date (Figure 2*a*). In contrast, quantification of changes in levels of glycolytic intermediates that occur following stimulation of glycolytic flux in green algae (50, 68), ripening fruit (11), aged storage root slices (1), *R. communis* cotyledons (51), and *Chenopodium rubrum* suspension cell cultures (59) consistently demonstrate that plant glycolysis is controlled from the "bottom up" with primary and secondary regulation exerted at the levels of PEP and Fru-6-P utilization, respectively (Figure 2*b*). These findings are compatible with the potent allosteric inhibition by PEP of many isolated plant PFK_ps and PFK_cs (32, 39, 60, 143). Enhancement in the activity of PK or PEPC (or other enzymes that metabolize PEP) relieves the PEP inhibition of PFK and thereby allows the glycolysis of hexose-P to proceed (Figure 2*b*). Reduced cytosolic PEP levels also cause elevated Fru-2,6-P_2 levels (and thus possible PFP activation) because a drop in PEP results in a fall in 3-PGA (these metabolites are at equilibrium in vivo) (51, 59). 3-PGA is a potent inhibitor of 6-phosphofructo-2-kinase (135). This regulatory scenario is consistent with the fact that, unlike many nonplant PKs, no plant PK has ever been found to be activated by Fru 1,6-P_2 (10, 69a, 86, 111, 115, 119, 149). In fact, homogeneous PK_c from endosperm of germinating *R. communis* seeds exhibits a kinetic reaction mechanism similar to that described for the Fru-1,6-P_2-activated yeast PK (116). One conceivable benefit of bottom up regulation of glycolysis is that it permits plants to regulate net glycolytic flux to pyruvate independent of related metabolic processes such as the Calvin cycle and sucrose-triose phosphate-starch interconversion.

TISSUE- AND DEVELOPMENTAL-SPECIFIC ISOZYMES OF KEY GLYCOLYTIC ENZYMES In animals, glycolytic isozymes are often expressed in a tissue- and/or developmental-specific manner. Such isozymes are frequently characterized by unique kinetic and regulatory properties, closely matching the cell's metabolic requirements in each situation. For example, different isozymes of mammalian hexokinase, PFK, and PK are separately expressed in tissues such as brain, kidney, liver, and muscle (48). These isozymes can be encoded by independent genes or by one gene. In the latter case the different isozymes arise via posttranscriptional or posttranslational processes (48).

Relatively little is known about the existence, functions, or genetic basis for tissue- or developmental-specific isozymes of plant glycolytic enzymes. Several lines of evidence suggest that these isozymes could be important in contributing to cell-specific metabolism in higher plants. For example, organ- and developmental-specific changes in the proportion of several hexokinase isozymes have been proposed to contribute to the regulation of hexose metabolism in the potato plant (121). Similarly, PEPC is encoded by a small multigene family; the expression of each member is controlled by exogenous and/or endogenous stimuli in a tissue-specific fashion (30, 83). In addition, endosperm-, cotyledon-, and leaf-specific isozymes of *R. communis* PK_c that exhibit marked differences in their respective physical and kinetic/regulatory characteristics have been purified and characterized (65, 115, 116, 119). These results are consistent with Southern blot studies indicating the existence of at least six PK_c genes in potato (31). Likewise, kinetic and/or immunological studies suggested that distinct isozymes of PFK_p and PK_p are expressed in leaves and developing seeds of *R. communis* (76, 78, 110a). Different isoforms of PFP showing unique physical and kinetic characteristics are expressed during seed germination (19, 21, 24, 118), fruit ripening (147), or following exposure to environmental stresses such as anoxia (18), or Pi starvation (137, 139, 140). Evidently, elucidation of plant glycolytic regulation requires the purification and characterization of the various control enzymes on a tissue-by-tissue basis.

PRACTICAL ASPECTS OF PLANT GLYCOLYTIC ENZYMOLOGY

Formulation of models for glycolytic control has traditionally been based on combining quantification of in vivo concentrations of enzymes, substrates, products, and effectors with in vitro studies of the kinetic and regulatory properties of the purified key enzymes. The need to ensure that measurements of metabolites and enzymes are reliable and authenticated, particularly if these values are to be used in any application of metabolic control analysis, was recently stressed (6). Nevertheless, determining what glycolytic enzymes characterized in vitro actually do in situ remains a problem.

Errors and Artifacts in Assays of Glycolytic Enzymes

The in vitro activities of most glycolytic enzymes are usually assessed with coupled spectrophotometric assays based upon the differential absorbance of NAD(P) and NAD(P)H. These assays are fraught with potential artifacts that may arise from contamination of substrates, cofactors, and/or coupling enzymes. As a result, erroneous biochemical and physiological conclusions concerning plant enzymes such as PFK_c, PFK_p, and PFP (80) have been published. Kruger (80) recently summarized important criteria for establishing optimal assay conditions, ensuring that the coupled assay accurately reflects enzyme activity, and identifying artifacts resulting from contaminants that exist in coupled assay components.

Protease and Dilution Problems

Large multimeric regulatory proteins such as PFK, PFP, PK, and PEPC are susceptible to artifactual posttranslation modifications such as proteolysis as well as dilution-dependent alterations in their oligomeric structure. Very minor proteolysis can have profound effects on the allosteric properties of a plant regulatory enzyme (110b), indicating that kinetic data obtained with partially degraded enzymes are of dubious significance. Moreover, many different protease inhibitors may need to be screened to identify one that effectively suppresses proteolysis of a specific enzyme in vitro (109, 110b). Plant glycolytic enzymes that have been shown to be vulnerable to partial degradation by endogenous proteases during their extraction, purification at 4°C, and/or storage at –20°C include developing *R. communis* seed PK_p (103, 109) and potato tuber and *B. nigra* suspension cell PFP (113; ME Theodorou & WC Plaxton, unpublished data). Similarly, the *N*-terminal phosphorylation domain of various PEPCs is very prone to proteolysis during the enzyme's purification in the absence of specific protease inhibitors (30, 83). Advice concerning the diagnosis and prevention of artifactual proteolysis of plant enzymes and the preparation and use of protease inhibitors was provided by Gray (54). Monitoring enzyme subunit size is facilitated by the availability of monospecific antibodies against the enzyme of interest. Such antibodies allow verification of the enzyme's subunit composition via immunoblotting of extracts prepared under completely denaturing conditions (84, 95, 109, 110b).

The effect of protein concentration must also be considered because enzymes are present in vivo at far higher concentrations than they are during in vitro assays. Concentration-dependence is thought to be particularly significant for enzymes important in metabolic regulation, because their structure, and hence their kinetic properties, may be affected by protein-protein interactions (7, 63, 113, 117, 142). The interactions between enzyme subunits that normally exist at the high protein concentrations prevailing in vivo can be specifically promoted in vitro by the addition of compatible solutes such as glycerol or polyethylene glycol (PEG) to the reaction mixture. The mechanism involves exclusion of the protein from the binary solvent, thus increasing the local enzyme concentration and favoring protein:protein interactions (7). The in vitro activities of rat liver PFK and PK, for instance, are enhanced by the presence of PEG (7). Few studies have examined the influence of enzyme concentration on the oligomeric and catalytic properties of plant glycolytic or gluconeogenic enzymes. Protein concentration does markedly influence the activity and/or aggregation state of various PEPCs (83) and *Chlorella pyrensoidosa* PFK (72). The presence of 10% (v/v) glycerol, a compound that stabilizes the native homotetrameric structure of maize leaf PEPC (112), activated homogeneous banana fruit PEPC by decreasing the K_m(PEP) by threefold and the K_a values for hexose-Ps by up to 7.5-fold, while greatly amplifying the ability of glucose-6-phosphate to relieve malate inhibition (84). Similarly, the addition of 5% (w/v) PEG stabilized the native heterotetrameric structure of homogeneous *R. communis* endosperm PK_c in dilute solutions, caused a 2.6-fold increase in V_{max} and 12.5- and twofold reductions in K_m values for PEP and ADP, respectively, and enhanced the enzyme's inhibition by MgATP (117). It was concluded that *R. communis* PK_c activity and regulation are modified by extreme dilution in the assay medium lacking PEG as a result of partial dissociation of the native tetrameric enzyme (117). A rapid decline in the intrinsic fluorescence of homogeneous potato tuber PFP occurred in response to dilution. This was paralleled by a loss in activity and a concomitant disassociation of the native $\alpha_4\beta_4$ heterooctamer into the inactive free subunits; dissociation was followed by random aggregation of the subunits into an inactive, high-molecular-weight conglomerate (113). These dilution-dependent processes were prevented by the presence of 5% (w/v) PEG (113). The addition of PEG has also been shown to increase the substrate affinity of homogeneous cytosolic FBPase from germinating *R. communis* endosperm, through stabilization of the enzyme's native tetrameric structure (63). Thus, compatible solutes appear to aid examination of catalytic properties of regulatory oligomers in an in vitro environment that may be closer to the conditions prevailing in vivo.

OTHER FUNCTIONS FOR GLYCOLYTIC ENZYMES

Several animal proteins with functions not related to glycolysis have turned out to be encoded by genes identical to those for glycolytic enzymes. For example, (*a*) yeast hexokinase exhibits protein kinase activity; (*b*) the sequence of the growth factor neuroleukin is identical to that of mouse hexose-phosphate isomerase; (*c*) enolase is a structural protein of the eye lens and in yeast is a heat shock protein that may confer thermotolerance; (*d*) the monomeric form of human NAD-GAPDH is a uracil DNA glycosylase, an enzyme involved in DNA repair; (*e*) yeast PK is involved in cell cycle control; and (*f*) the monomeric form of mammalian muscle-type PK is a thyroid hormone-binding protein, and the conversion of inactive PK monomer to active PK tetramer is promoted by its allosteric activator, Fru-1,6-P_2 (48). Thus some plant glycolytic enzymes may also have nonglycolytic functions in vivo. Enolase from *E. phyllopogon* is induced by anoxia as well as by cold and heat shock and may act as a general stress protein that protects cellular components at the structural level (49, 132). Furthermore, immunolocalization studies indicated that in addition to their cytosolic and chloroplastic compartments, 3-PGA kinase, NAD-GAPDH, and ALD are located in the nucleus of leaf mesophyll cells of *P. sativum* (2c). It was postulated that these enzymes are nuclear proteins with secondary roles not directly related to their enzymatic functions in carbohydrate metabolism.

CONCLUDING REMARKS

Despite the remarkable progress in analysis of the organization and regulation of plant glycolysis, many complex issues remain. Developments in molecular genetics have led to exciting new strategies. Five years ago, little was known about the primary structure of the key regulatory enzymes. Since then, the primary structures of the subunit(s) of PFP, PK_c, PK_p, and PEPC have been deduced from the respective cDNA sequences. This information in turn has provided new insights into molecular mechanisms of catalysis and regulation of these proteins. Isolation of cDNAs for enzymes has also permitted the quantification of specific mRNAs, analyses of their gene expression, and elucidation of requirements for import into plastids. Nuclear genes encoding several of these enzymes have been sequenced, and their promoter regions are being characterized. Transformation of plants with sense or antisense cDNA constructs provides new information on the functions and control of these enzymes in vivo. In addition, there is a practical interest in genetically modifying crops to over- or underexpress glycolytic enzymes in order to redirect the flux of photosynthate into economically important endproducts such as starch, triglycerides, and protein.

Although significant advances in our understanding of plant glycolysis are facilitated by molecular genetics, it is imperative that future workers continue to examine physiological aspects of glycolysis. We also require further purification and characterization of the cell- and/or developmental-specific isozymes that regulate cytosolic and plastidic glycolytic flux. Not only does enzyme purification lead to the production of crucial tools for the molecular biologist (e.g. antibodies), but together with the appropriate kinetic and physiological studies it also serves to establish the control mechanisms that regulate glycolysis in vivo. Only when these basic mechanisms are understood will the ongoing efforts of plant molecular biologists be fully realized.

ACKNOWLEDGMENTS

Research in my laboratory is supported by grants from The Natural Sciences and Research Council of Canada. I also thank Drs. H. Ashihara, D. T. Dennis, N. J. Kruger, T. C. Fox, B. L. Miki, and M. E. Rumpho for sending me their unpublished manuscripts and preprints. I am indebted to the current members of my laboratory as well as to Drs. S. D. Blakeley, M. Kuzma, and J. Rivoal for critical reading of the manuscript. The many contributions and collaborations provided by my fellow plant biology colleagues at Queen's University are gratefully acknowledged.

Literature Cited

1. Adams PB, Rowan KS. 1970. Glycolytic control of respiration during aging of carrot root tissue. *Plant Physiol.* 45:490–95

2a. Anderson LE, Goldhaber-Gordon IM, Li D, Tang X, Xiang M, Prakash N. 1995. Enzyme-enzyme interaction in the chloroplast: glyceraldehyde-3-phosphate dehydrogenase, triose phosphate isomerase and aldolase. *Planta* 196:245–55

2b. Anderson LE, Li D, Prakash N, Stevens FJ. 1995. Identification of potential redox-sensitive cysteines in cytosolic forms of fructosebisphosphatase and glyceraldehyde-3-phosphate dehydrogenase. *Planta* 196:118–24

2c. Anderson LE, Wang X, Gibbons JT. 1995. Three enzymes of carbon metabolism or their antigenic analogs in pea leaf nuclei. *Plant Physiol.* 108:659–67

3. ap Rees T. 1985. The organization of glycolysis and the oxidative pentose phosphate pathway in plants. In *Enclylopedia of Plant Physiology,* ed. R Douce, DA Day, 18:391–414. Berlin: Springer-Verlag

4. ap Rees T. 1988. Hexose phosphate metabolism by nonphotosynthetic tissues of higher plants. In *The Biochemistry of Plants,* ed. J Preiss, 14:1–14. New York: Academic

5. ap Rees T. 1990. See Ref. 41, pp. 106–33

6. ap Rees T, Hill SA. 1994. Metabolic control analysis of plant metabolism. *Plant Cell Environ.* 17:587–99

7. Aragón JJ, Sols A. 1991. Regulation of enzyme activity in the cell: effect of enzyme concentration. *FASEB J.* 5:2945–50

8a. Ashihara H, Horikosi T, Li X-N, Sagishima K, Yamashita Y. 1988. Profiles of enzymes involved in glycolysis in *Catharanthus roseus* cells in batch suspension culture. *J. Plant Physiol.* 133:38–45

8b. Ashihara H, Li X-N, Ukaji T. 1988. Effect of inorganic phosphate on the biosynthesis of purine and pyrimidine nucleotides in suspension-cultured cells of *Catharanthus roseus. Ann. Bot.* 61:225–32

9. Ashihara H, Sato F. 1993. Pyrophosphate: fructose-6-phosphate 1-phosphotransferase and biosynthetic capacity during differentiation of hypocotyls of *Vignas* seedlings. *Biochim. Biophys. Acta* 1156: 123–27

10. Baysdorfer C, Bassham JA. 1984. Spinach pyruvate kinase isoforms: partial purification and regulatory properties. *Plant Physiol.* 74:374–79

11. Beaudry RM, Severson RF, Black CC, Kays SJ. 1989. Banana ripening: implications of changes in glycolytic intermediate concentrations, glycolytic and gluconeogenic carbon flux, and fructose 2,6-bisphosphate concentration. *Plant Physiol.* 91:1436–44

12. Black CC, Mustardy L, Sung SS, Kormanik PP, Xu D-P, Paz N. 1987. Regulation and roles for alternative pathways of hexose metabolism in plants. *Physiol. Plant.* 69: 387–94

13. Blakeley SD, Crews L, Todd JF, Dennis DT. 1992. Expression of the genes for the α- and β-subunits of pyrophosphate-dependent phosphofructokinase in germinating and developing seeds from *Ricinus communis. Plant Physiol.* 69:387–94

14. Blakeley SD, Dennis DT. 1993. Molecular approaches to the manipulation of carbon allocation in plants. *Can. J. Bot.* 71:765–78

15. Blakeley S, Gottlob-McHugh S, Wan J, Crews L, Miki B, et al. 1995. Molecular characterization of plastid pyruvate kinase from castor and tobacco. *Plant Mol. Biol.* 27:79–89

16. Blakeley SD, Plaxton WC, Dennis DT. 1990. The isolation, sequencing and characterization of cDNA clones for the cytosolic isozyme of plant pyruvate kinase. *Plant Mol. Biol.* 15:665–69

17. Blakeley SD, Plaxton WC, Dennis DT. 1991. Relationship between the subunits of leucoplast pyruvate kinase from *Ricinus communis* and a comparison with the enzyme from other sources. *Plant Physiol.* 96:1283–88

18. Botha A-M, Botha FC. 1991. Effect of anoxia on the expression and molecular form of the pyrophosphate dependent phosphofructokinase. *Plant Cell Physiol.* 32: 1299–302

19. Botha A-M, Botha FC. 1991. Pyrophosphate dependent phosphofructokinase of *Citrullus lanatus:* molecular forms and expression of subunits. *Plant Physiol.* 96: 1185–92

20. Botha A-M, Botha FC. 1993. Induction of pyrophosphate dependent phosphofructokinase in watermelon (*Citrullus lanatus*)

cotyledons coincides with insufficient cytosolic D-fructose-1,6-bisphosphate 1-phosphohydrolase to sustain gluconeogenesis. *Plant Physiol.* 101:1385–90
21. Botha A-M, Botha FC. 1993. Effect of the radicle and hormones on the subunit composition and molecular form of pyrophosphate-dependent phosphofructokinase in the cotyledons of *Citrullus lanatus. Aust. J. Plant Physiol.* 20:265–73
22. Botha FC, Cawood MC, Small JGC. 1988. Kinetic properties of the ATP-dependent phosphofructokinase isoenzymes from cucumber seeds. *Plant Cell Physiol.* 29: 415–21
23. Botha FC, de Vries C, Small JGC. 1989. Changes in the activity and concentration of the pyrophosphate-dependent phosphofructokinase during germination of *Citrullus lanatus* seeds. *Plant Physiol. Biochem.* 27:75–80
24. Botha FC, Potgieter GP, Botha A-M. 1992. Respiratory metabolism and gene expression during seed germination. *Plant Growth Regul.* 11:211–24
25. Botha FC, Turpin DH. 1990. Molecular, kinetic, and immunological properties of the 6-phosphofructokinase from the green alga *Selenastrum minutum. Plant Physiol.* 93:871–79
26. Bryce JH, ap Rees T. 1985. Rapid decarboxylation of the products of dark fixation of CO_2 in roots of *Pisum* and *Plantago. Phytochemistry* 24:1635–38
27. Carnal NW, Black CC. 1979. Pyrophosphate-dependent 6-phosphofructokinase, a new glycolytic enzyme in pineapple leaves. *Biochem. Biophys. Res. Commun.* 86: 20–26
28. Carystinos GD, MacDonald HR, Monroy AF, Dhindsa RS, Poole RJ. 1995. Vacuolar H^+-translocating pyrophosphatase is induced by anoxia or chilling in seedlings of rice. *Plant Physiol.* 108:641–49
29. Cawood ME, Botha FC, Small JGC. 1988. Molecular properties of the ATP:D-fructose-6-phosphate 1-phosphotransferase isoenzymes from *Cucumis sativus. Plant Cell Physiol.* 29:195–99
30. Chollet R, Vidal J, O'Leary MH. 1996. Phosphoenolpyruvate carboxylase: a ubiquitous, highly regulated enzyme in plants. *Annu. Rev. Plant Physiol. Plant Mol. Biol.* 47:273–97
31. Cole KP, Blakeley SD, Dennis DT. 1992. Structure of the gene encoding potato cytosolic pyruvate kinase. *Gene* 122:255–61
32. Copeland L, Turner JF. 1987. The regulation of glycolysis and the pentose phosphate pathway. In *The Biochemistry of Plants,* ed. PK Stumpf, EE Conn, 11: 107–29. San Diego: Academic
33. Copeland L, Zammit A. 1994. Kinetic properties of NAD-dependent glyceraldehyde-3-phosphate dehydrogenase from the host fraction of soybean root nodules. *Arch. Biochem. Biophys.* 312:107–13
34. Cornel FA, Plaxton WC. 1994. Characterization of asparaginyl endopeptidase activity in endosperm of developing and germinating castor oil seeds. *Physiol. Plant.* 91:599–604
35. Dancer J, Veith R, Feil R, Komor E, Stitt M. 1990. Independent changes of inorganic pyrophosphate and the ATP/ADP or UTP/UDP ratios in plant cell suspension cultures. *Plant Sci.* 66:59–63
36. Davies JM, Poole RJ, Sanders D. 1993. The computed free energy change of hydrolysis of inorganic pyrophosphate and ATP: apparent significance of inorganic-pyrophosphate-driven reactions of intermediary metabolism. *Biochim. Biophys. Acta* 1141: 29–36
37. Dennis DT, Blakeley S, Carlisle S. 1991. Isozymes and compartmentation in leucoplasts. In *Compartmentation of Plant Metabolism in Non Photosynthetic Tissues,* ed. MJ Emes, pp. 77–94. Cambridge: Cambridge Univ. Press
38. Dennis DT, Emes MJ. 1990. See Ref. 41, pp. 45–56
39. Dennis DT, Greyson MF. 1987. Fructose 6-phosphate metabolism in plants. *Physiol. Plant.* 69:395–404
40. Dennis DT, Miernyk JA. 1982. Compartmentation of nonphotosynthetic carbohydrate metabolism. *Annu. Rev. Plant Physiol.* 33:27–50
41. Dennis DT, Turpin DH, eds. 1990. *Plant Physiology, Biochemistry and Molecular Biology.* Singapore: Longman. 529 pp.
42. Duff SMG, Lefebvre DD, Plaxton WC. 1989. Purification and characterization of a phosphoenolpyruvate phosphatase from *Brassica nigra* suspension cell cultures. *Plant Physiol.* 90:734–41
43. Duff SMG, Lefebvre DD, Plaxton WC. 1991. Purification, characterization, and subcellular localization of an acid phosphatase from black mustard cell-suspension cultures: comparison with phosphoenolpyruvate phosphatase. *Arch. Biochem. Biophys.* 286:226–32
44. Duff SMG, Moorhead GBG, Lefebvre DD, Plaxton WC. 1989. Phosphate starvation inducible 'bypasses' of adenylate and phosphate-dependent glycolytic enzymes in *Brassica nigra* suspension cells. *Plant Physiol.* 90:1275–78
45. Duff SMG, Plaxton WC, Lefebvre DD. 1991. Phosphate starvation response in plant cells: de novo synthesis and degradation of acid phosphatases. *Proc. Natl. Acad. Sci. USA* 88:9538–42
46. Emes MJ, Tobin AK. 1993. Control of metabolism and development in higher plant plastids. *Int. Rev. Cytol.* 145:149–216

47. Forsthoefel NR, Cushman MF, Cushman JC. 1995. Posttranscriptional and post-translational control of enolase expression in the facultative crassulacean acid metabolism plant *Mesembryanthemum crystallinum* L. *Plant Physiol.* 108:1185–95
48. Fothergill-Gilmore LA, Michels PAM. 1993. Evolution of glycolysis. *Prog. Biophys. Mol. Biol.* 59:105–235
49. Fox TC, Mujer CV, Andrews DL, Williams AS, Cobb BG, et al. 1995. Identification and gene expression of anaerobically induced enolase in *Echinochloa phyllopogons* and *E. crus-pavonis*. *Plant Physiol.* 109:433–43
50. Gauthier DA, Turpin DH. 1994. Interactions between H^+/Pi cotransport, the plasmalemma H^+-ATPase, and dark respiratory carbon flow. *Plant Physiol.* 104: 629–37
51. Geigenberger P, Stitt M. 1991. Regulation of carbon partitioning between sucrose and nitrogen assimilation in cotyledons of germinating *Ricinus communis* L. seedlings. *Planta* 185:563–68
52. Gottlob-McHugh S, Knowles VL, Blakeley SD, Sangwan RS, Dennis DT, et al. 1995. Differential expression of cytosolic and plastidic pyruvate kinases in tobacco. *Physiol. Plant.* 95:507–14
53. Gottlob-McHugh S, Sangwan RS, Blakeley SD, Vanlerberghe GC, Turpin DH, et al. 1992. Normal growth of tobacco plants in the absence of cytosolic pyruvate kinase. *Plant Physiol.* 100:820–25
54. Gray JC. 1982. Use of proteolytic inhibitors during the isolation of plastid proteins. In *Methods in Chloroplast Molecular Biology*, ed. M Edelman, R Hallick, N Chu, pp. 1903–102. Amsterdam: Elsevier
55. Guglielminetti L, Perata P, Alpi A. 1995. Effect of anoxia on carbohydrate metabolism in rice seedlings. *Plant Physiol.* 108: 735–41
56. Hajirezaei M, Sonnewald U, Viola R, Carlisle S, Dennis D, Stitt M. 1994. Transgenic potato plants with strongly decreased expression of pyrophosphate: Fructose-6-phosphate phosphotransferase show no visible phenotype and only minor changes in metabolic fluxes in their tubers. *Planta* 192:16–30
57. Hattori J, Baum BR, McHugh SG, Blakeley SD, Dennis DT, Miki BL. 1995. Pyruvate kinase isozymes: ancient diversity retained in modern plant cells. *Biochem. Syst. Ecol.* In press
58. Hattori J, Baum BR, Miki BL. 1995. Ancient diversity of the glucose-6-phosphate isomerase genes. *Biochem. Syst. Ecol.* 23: 33–38
59. Hatzfeld WD, Stitt M. 1991. Regulation of glycolysis in heterotrophic cell suspension cultures of *Chenopodium rubrum* in response to proton fluxes at the plasmalemma. *Physiol. Plant.* 81:103–10
60. Häusler RE, Holtum JAM, Latzko E. 1989. Cytosolic phosphofructokinase from spinach leaves. *Plant Physiol.* 90:1498–505
61. Henze K, Schnarrenberger C, Kellermann J, Martin W. 1994. Chloroplast and cytosolic triosephosphate isomerases from spinach: purification, microsequencing and cDNA cloning of the chloroplast enzyme. *Plant Mol. Biol.* 26:1961–73
62. Heuer B, Hansen MJ, Anderson LE. 1982. Light modulation of phosphofructokinase in pea leaf chloroplasts. *Plant Physiol.* 69: 1404–6
63. Hodgson RJ, Plaxton WC. 1995. Effect of polyethylene glycol on the kinetic, fluorometric and oligomeric properties of castor seed cytosolic fructose-1,6-bisphosphatase. *FEBS Lett.* 368:559–62
64. Hrazdina G, Jensen RA. 1992. Spatial organization of enzymes in plant metabolic pathways. *Annu. Rev. Plant Physiol. Plant Mol. Biol.* 43:241–67
65. Hu Z-H, Podestá FE, Plaxton WC. 1995. Tissue-specific isoforms of cytosolic pyruvate kinase in the castor oil plant. *Plant Physiol.* 108:S66 (Abstr.)
66. Huang U, Blakeley SD, McAleese SM, Fothergill-Gilmore LA, Dennis DT. 1993. Higher-plant cofactor-independent phosphoglyceromutase: purification, molecular characterization and expression. *Plant Mol. Biol.* 23:1039–53
67. Huber SC, Huber JL, McMichael RW. 1994. Control of plant enzyme activity by reversible protein phosphorylation. *Int. Rev. Cytol.* 149:47–98
68. Huppe HC, Turpin DH. 1994. Integration of carbon and nitrogen metabolism in plant and algal cells. *Annu. Rev. Plant Physiol. Plant Mol. Biol.* 45:577–607
69a. Ireland RJ, DeLuca V, Dennis DT. 1980. Characterization and kinetics of isozymes of pyruvate kinase from developing castor bean endosperm. *Plant Physiol.* 65: 1188–93
69b. Iyer MG, Kaimal KS, Nair PM. 1989. Correlation between increase in 6-phosphofructokinase activity and appearance of three multiple forms in ripening banana. *Plant Physiol. Biochem.* 27:99–106
70. Jellito T, Sonnewald U, Willmitzer L, Hajirezaei MR, Stitt M. 1992. Inorganic pyrophosphate content and metabolites in leaves and tubers of potato and tobacco plants expressing *E. coli* pyrophosphatase in their cytosol: biochemical evidence that sucrose metabolism has been manipulated. *Planta* 188:238–44
71. Johnson JF, Allan DL, Vance CP. 1994. Phosphorus stress-induced proteoid roots show altered metabolism in *Lupinus albus*. *Plant Physiol.* 104:657–65

72. Kelly GJ, Latzko E. 1981. Contrasting sensitivities of *Chlorella* and higher plant phosphofructokinases to dilution. *Plant Cell Physiol.* 22:207–12
73. Kennedy RA, Rumpho ME, Fox TC. 1992. Anaerobic metabolism in plants. *Plant Physiol.* 100:1–6
74. Kholodenko BN, Cascante M, Westerhoff HV. 1994. VI-1 control theory of metabolic channelling. *Mol. Cell. Biochem.* 133/134: 313–31
75. Kiss F, Wu M-X, Wong JH, Balogh A, Buchanan BB. 1991. Redox active sulfhydryls are required for fructose 2,6-bisphosphate activation of plant pyrophosphate fructose-6-phosphate 1-phosphotransferase. *Arch. Biochem. Biophys.* 287:337–40
76. Knowles VL, Dennis DT, Plaxton WC. 1989. Purification and characterization of a novel type of pyruvate kinase from the green alga, *Selenastrum minutum. FEBS Lett.* 259:130–32
77. Knowles VL, Gottlob-McHugh S, Hu Z-H, Gauthier D, Falk S, et al. 1995. Altered growth under limiting light of transgenic tobacco plants lacking leaf cytosolic pyruvate kinase. *Plant Physiol.* 108:S151 (Abstr.)
78. Knowles VL, Greyson MF, Dennis DT. 1990. Characterization of ATP-dependent fructose 6-phosphate 1-phosophotransferase isozymes from leaf and endosperm tissues of *Ricinus communis. Plant Physiol.* 92:155–59
79. Kowallik W, Grotjohann N, Ruyters G. 1990. Oligomeric forms of glycolytic enzymes in *Chlorella* grown in different light qualities. *Bot. Acta* 103:197–202
80. Kruger NJ. 1995. Errors and artifacts in coupled spectrophotometric assays of enzyme activity. *Phytochemistry* 38:1065–71
81. Kruger NJ, Scott P. 1994. Manipulation of fructose-2,6-bisphosphate levels in transgenic plants. *Biochem. Soc. Trans.* 22: 904–9
82. Kubota K, Ashihara H. 1990. Identification of non-equilibrium glycolytic reactions in suspension-cultured plant cells. *Biochim. Biophys. Acta* 1036:138–42
83. Lepiniec L, Vidal J, Chollet R, Gadal P, Crétin C. 1994. Phosphoenolpyruvate carboxylase: structure, regulation and evolution. *Plant Sci.* 99:111–24
84. Law RD, Plaxton WC. 1995. Purification and characterization of a novel phosphoenolpyruvate carboxylase from banana fruit. *Biochem. J.* 307:807–16
85. Lin M, Turpin DH, Plaxton WC. 1989. Pyruvate kinase isozymes from a green alga, *Selenastrum minutum.* I. Purification, physical, and immunological characterization. *Arch. Biochem. Biophys.* 269: 219–27
86. Lin M, Turpin DH, Plaxton WC. 1989. Pyruvate kinase isozymes from a green alga *Selenastrum minutum.* II. Kinetic and regulatory properties. *Arch. Biochem. Biophys.* 269:228–38
87. Mahajan R, Singh R. 1989. Properties of pyrophosphate: fructose-6-phosphate phosphotransferase from endosperm of developing wheat (*Triticum aestivum* L.) grains. *Plant Physiol.* 91:421–26
88. Malhotra OP, Kayastha AM. 1990. Isolation and characterization of phosphoenolpyruvate phosphatase from germinating mung beans (*Vigna radiata*). *Plant Physiol.* 93:194–200
89. Malhotra OP, Kumar A, Tikoo K. 1987. Isolation and quaternary structure of a complex of glyceraldehyde 3-phosphate dehydrogenase and phosphoglycerate kinase. *Ind. J. Biochem. Biophys.* 24:16–20
90. Mertens E. 1991. Pyrophosphate-dependent phosphofructokinase: an anaerobic glycolytic enzyme? *FEBS Lett.* 285: 1–5
91. Meyer-Gauen G, Schnarrenberger C, Cerff R, Martin W. 1994. Molecular characterization of a novel, nuclear encoded, NAD^+-dependent glyceraldehyde-3-phosphate dehydrogenase in plastids of the gymnosperm *Pinus sylvestris* L. *Plant Mol. Biol.* 26:1155–66
92. Miernyk JA. 1990. See Ref. 41, pp. 77–100
93. Miernyk JA, Dennis DT. 1992. A developmental analysis of the enolase isozymes from *Ricinus communis. Plant Physiol.* 99: 748–50
94. Mohanty B, Wilson PM, ap Rees T. 1993. Effects of anoxia on growth and carbohydrate metabolism in suspension cultures of soybean and rice. *Phytochemistry* 34: 75–82
95. Moorhead GBG, Hodgson RJ, Plaxton WC. 1994. Co-purification of cytosolic fructose-1,6-bisphosphatase and cytosolic aldolase from endosperm of germinating castor oil seeds. *Arch. Biochem. Biophys.* 312:326–35
96. Moorhead GBG, Plaxton WC. 1988. Binding of glycolytic enzymes to a particulate fraction in carrot and sugar beet storage roots: dependence on metabolic state. *Plant Physiol.* 86:348–51
97. Moorhead GBG, Plaxton WC. 1990. Purification and characterization of cytosolic aldolase from carrot storage roots. *Biochem. J.* 269:133–39
98. Moorhead GBG, Plaxton WC. 1991. High yield purification of potato tuber pyrophosphate: fructose-6-phosphate 1-phosphotransferase. *Protein Express. Purif.* 2: 29–33.
99. Moorhead GBG, Plaxton WC. 1992. Evidence that cytosolic aldolase specifically

interacts with the ATP- and PPi-dependent phosphofructokinases in carrot storage roots. *FEBS Lett.* 313:277–80
100. Mujer CV, Fox TC, Williams AS, Andrews DL, Kennedy RA, Rumpho ME. 1995. Purification, properties and phosphorylation of anaerobically induced enolase in *Echinochloa phyllopogon* and *E. crus-pavonis. Plant Cell. Physiol.* In press
101. Nagano M, Ashihara H. 1993. Long-term phosphate starvation and respiratory metabolism in suspension-cultured *Catharanthus roseus* cells. *Plant Cell Physiol.* 34: 1219–28
102. Nagano M, Hachiya A, Ashihara H. 1994. Phosphate starvation and a glycolytic bypass catalyzed by phosphoenolpyruvate carboxylase in suspension cultured *Catharanthus roseus* cells. *Z. Naturforsch.* 49c:742–50
103. Negm FB, Cornel FA, Plaxton WC. 1995. Suborganellar localization and molecular characterization of non-proteolytic degraded leucoplast pyruvate kinase from developing castor oil seeds. *Plant Physiol.* 109:1461–69
104. Nielsen TH. 1994. Pyrophosphate: fructose-6-phosphate 1-phosphotransferase from barley seedlings: isolation, subunit composition and kinetic characterization. *Physiol. Plant.* 92:311–321
105. Plaxton WC. 1988. Purification of pyruvate kinase from germinating castor bean endosperm. *Plant Physiol.* 86:1064–69
106. Plaxton WC. 1989. Molecular and immunological characterization of plastid and cytosolic pyruvate kinase isozymes from castor-oil-plant leaf and endosperm. *Eur. J. Biochem.* 181:443–51
107. Plaxton WC. 1990. See Ref. 41, pp. 28–44
108. Plaxton WC. 1990. Glycolysis. In *Methods in Plant Biochemistry,* ed. P Lea, 3:145–73. New York: Academic
109. Plaxton WC. 1991. Leucoplast pyruvate kinase from developing castor oil seeds: characterization of the enzyme's degradation by a cysteine endopeptidase. *Plant Physiol.* 97:1334–38
110a. Plaxton WC, Dennis DT, Knowles VL. 1990. Purification of leucoplast pyruvate kinase from developing castor bean endosperm. *Plant Physiol.* 94:1528–34
110b. Plaxton WC, Preiss J. 1987. Purification and properties of nonproteolytic degraded ADPglucose pyrophosphorylase from maize endosperm. *Plant Physiol.* 83: 105–12
111. Plaxton WC, Sangwan RS, Singh N, Gauthier DA, Turpin DH. 1993. Phosphoenolpyruvate metabolism of developing oil seeds. In *Seed Oils for the Future,* ed. SL MacKenzie, DC Taylor, pp. 35–43. Champaign: Am. Oil Chem. Soc.
112. Podestá FE, Andreo CS. 1989. Maize leaf phosphoenolpyruvate carboxylase: oligomeric state and activity in the presence of glycerol. *Plant Physiol.* 90:427–33
113. Podestá FE, Moorhead GBG, Plaxton WC. 1994. Potato tuber PPi-dependent phosphofructokinase: effects of thiols and polyalcohols on the enzyme's intrinsic fluorescence, activity and quaternary structure in dilute solutions. *Arch. Biochem. Biophys.* 313:50–57
114. Podestá FE, Plaxton WC. 1991. Association of phosphoenolpyruvate phosphatase activity with the cytosolic pyruvate kinase of germinating mung beans. *Plant Physiol.* 97:1329–33
115. Podestá FE, Plaxton WC. 1991. Kinetic and regulatory properties of cytosolic pyruvate kinase from germinating castor oil seeds. *Biochem J.* 279:495–501
116. Podestá FE, Plaxton WC. 1992. Plant cytosolic pyruvate kinase: a kinetic study. *Biochim. Biophys. Acta* 1160:213–20
117. Podestá FE, Plaxton WC. 1993. Activation of cytosolic pyruvate kinase by polyethylene glycol. *Plant Physiol.* 103: 285–88
118. Podestá FE, Plaxton WC. 1994. Regulation of carbon metabolism in germinating *Ricinus communis* cotyledons. I. Developmental profiles for the activity, concentration, and molecular structure of the pyrophosphate- and ATP-dependent phosphofructokinases, phosphoenolpyruvate carboxylase, and pyruvate kinase. *Planta* 194:374–80
119. Podestá FE, Plaxton WC. 1994. Regulation of carbon metabolism in germinating *Ricinus communis* cotyledons. II. Properties of phosphoenolpyruvate carboxylase and cytosolic pyruvate kinase associated with the regulation of glycolysis and nitrogen assimilation. *Planta* 194:381–87
120. Rajagopalan AV, Tirumala Devi M, Raghavendra AS. 1993. Patterns of phosphoenolpyruvate carboxylase activity and cytosolic pH during light activation and dark deactivation in C_3 and C_4 plants. *Photosynth. Res.* 38:51–60
121. Renz A, Merlo L, Stitt M. 1993. Partial purification from potato tubers of three fructokinases and three hexokinases which show differing organ and developmental specificity. *Planta* 190:156–65
122. Ricard B, Rivoal J, Spiteri A, Pradet A. 1991. Anaerobic stress induced the transcription and translation of sucrose synthase in rice. *Plant Physiol.* 95:669–74
123. Deleted in proof
124. Ruyters G, Grotjohann N, Kowallik W. 1991. Oligomeric forms of pyruvate kinase from *Chlorella* with different kinetic properties. *Z. Naturforsch* 46c:416–22
125. Sabularse DC, Anderson RL. 1981. D-fructose-2,6-bisphosphate: a naturally occur-

ring activator for inorganic pyrophosphate: D-fructose-6-phosphate 1-phosphotransferase. *Biochem. Biophys. Res. Commun.* 103:848–55

126. Sangwan RS, Gauthier DA, Turpin DH, Pomeroy MK, Plaxton WC. 1992. Pyruvate kinase isoenzymes from zygotic and microspore derived embryos of *Brassica napus:* developmental profiles and subunit composition. *Planta* 187:198–202
127. Sangwan RS, Singh N, Plaxton WC. 1992. Phosphoenolpyruvate carboxylase activity and concentration in the endosperm of developing and germinating castor oil seeds. *Plant Physiol.* 99:445–49
128. Schnarrenberger C, Pelzer Reith B, Yatsuki H, Freund S, Jacobshagen S, Hori K. 1994. Expression and sequence of the only detectable aldolase in *Chlamydomonas reinhardtii. Arch. Biochem. Biophys.* 313: 173–78
129. Schuller KA, Plaxton WC, Turpin DH. 1990. Regulation of phosphoenolpyruvate carboxylase from the green alga *Selenastrum minutum. Plant Physiol.* 93:1303–11
130. Schuller KA, Turpin DH, Plaxton WC. 1990. Metabolite regulation of partially purified soybean nodule phosphoenolpyruvate carboxylase. *Plant Physiol.* 94: 1429–35
131. Schulz M, Klockenbring T, Hunte C, Schnabl H. 1993. Involvement of ubiquitin in phosphoenolypyruvate carboxylase degradation. *Bot. Acta* 106:143–45
132. Schweisguth A, Maier CV, Fox TC, Rumpho ME. 1995. Is the anaerobic stress protein enolase a general stress protein? *Plant Physiol.* 106:S104 (Abstr.)
133. Smith RG, Gauthier DA, Dennis DT, Turpin DH. 1992. Malate- and pyruvate-dependent fatty acid synthesis in leukoplasts from developing castor endosperm. *Plant Physiol.* 99:1233–38
134. Spilatro SR, Anderson JM. 1988. Carbohydrate metabolism and activity of pyrophosphate: fructose-6-phosphate phosphotransferase in photosynthetic soybean (*Glycine max,* Merr.) suspension cells. *Plant Physiol.* 88:862–68
135. Stitt M. 1990. Fructose 2,6-bisphosphate as a regulatory molecule in plants. *Annu. Rev. Plant Physiol. Plant Mol. Biol.* 41:153–85
136. Sung S-JS, Xu D-P, Galloway CM, Black CC. 1988. A reassessment of glycolysis and gluconeogenesis in higher plants. *Physiol. Plant.* 72:650–54
137. Theodorou ME, Cornel FA, Duff SMG, Plaxton WC. 1992. Phosphate starvation inducible synthesis of the α-subunit of pyrophophosphate: fructose-6-phosphate phosphotransferase in black mustard cell-suspension cultures. *J. Biol. Chem.* 267:21901–5
138. Theodorou ME, Elrifi IR, Turpin DH, Plaxton WC. 1991. Effects of phosphorus limitation on respiratory metabolism in the green alga *Selenastrum minutum. Plant Physiol.* 95:1089–95
139. Theodorou ME, Plaxton WC. 1994. Induction of PPi-dependent phosphofructokinase by phosphate starvation in seedlings of *Brassica nigra. Plant Cell. Environ.* 17: 287–94
140. Theodorou ME, Plaxton WC. 1995. Adaptations of plant respiratory metabolism to nutritional phosphate deprivation. In *Environment and Metabolism: Flexibility and Acclimation,* ed. N Smirnoff, pp. 79–109. London: BIOS Sci.
141. Deleted in proof
142. Traut TW. 1994. Dissociation of enzyme oligomers: a mechanism for allosteric regulation. *Crit. Rev. Biochem. Mol. Biol.* 29: 125–63
143. Vella J, Copeland L. 1993. Phosphofructokinase from the host fraction of soybean nodules. *J. Plant Physiol.* 141:398–404
144. Walker RP, Leegood RC. 1995. Purification, and phosphorylation in vivo and in vitro, of phosphoenolpyruvate carboxykinase from cucumber cotyledons. *FEBS Lett.* 362:70–74
145. Weiner H, Stitt M, Heldt HW. 1987. Subcellular compartmentation of pyrophosphate and alkaline pyrophosphatase in leaves. *Biochem. Biophys. Acta* 893:13–21
146. Wong JH, Yee BC, Buchanan BB. 1987. A novel type of phosphofructokinase from plants. *J. Biol. Chem.* 262:3185–92
147. Wong JH, Kiss F, Wu MX, Buchanan BB. 1990. Pyrophosphate fructose-6-phosphate 1-phosphotransferase from tomato fruit: evidence for change during ripening. *Plant Physiol.* 94:499–506
148. Wood HC. 1985. Inorganic pyrophosphate and polyphosphates as sources of energy. *Curr. Top. Cell. Regul.* 26:355–69
149. Wu HB, Turpin DH. 1992. Purification and characterization of pyruvate kinase from the green alga *Chlamydomonas reinhardtii. J. Phycol.* 28:472–81
150. Xia J-H, Saglio P, Roberts JKM. 1995. Nucleotide levels do not critically determine survival of maize root tips acclimated to a low oxygen environment. *Plant Physiol.* 108:589–95

Annu. Rev. Plant Physiol. Plant Mol. Biol. 1996. 47:215–43

LIGHT CONTROL OF SEEDLING DEVELOPMENT

Albrecht von Arnim and Xing-Wang Deng

Department of Biology, Yale University, New Haven, Connecticut 06520-8104

KEY WORDS: seedling, development, light, *Arabidopsis*, morphogenesis

ABSTRACT

Light control of plant development is most dramatically illustrated by seedling development. Seedling development patterns under light (photomorphogenesis) are distinct from those in darkness (skotomorphogenesis or etiolation) with respect to gene expression, cellular and subcellular differentiation, and organ morphology. A complex network of molecular interactions couples the regulatory photoreceptors to developmental decisions. Rapid progress in defining the roles of individual photoreceptors and the downstream regulators mediating light control of seedling development has been achieved in recent years, predominantly because of molecular genetic studies in *Arabidopsis thaliana* and other species. This review summarizes those important recent advances and highlights the working models underlying the light control of cellular development. We focus mainly on seedling morphogenesis in *Arabidopsis* but include complementary findings from other species.

1040-2519/96/0601-0215$08.00

INTRODUCTION[1]

Seed germination and seedling development interpret the body plan with its preformed embryonic organs, the cotyledons, the hypocotyl, and the root. Whereas embryo and seed development take place in the shelter of the parental ovule, rather independently of the environmental conditions, seed germination and seedling development are highly sensitive and dramatically responsive to environmental conditions. Higher plants have evolved a remarkable plasticity in their developmental pathways with respect to many environmental parameters. First, the embryonic axes of the root and the shoot orient themselves in opposite directions in the field of gravity. Second, seedling development is highly responsive to light fluence rates over approximately six orders of magnitude. The direction of incident light entices shoots and roots to respond phototropically. Light intensity and wavelength composition are important factors in determining the speed of cell growth, of pigment accumulation, and of plastid differentiation.

Angiosperms, in particular, choose between two distinct developmental pathways, according to whether germination occurred in darkness or in light. The light developmental pathway, known as photomorphogenesis, leads to a seedling morphology that is optimally designed to carry out photosynthesis. The dark developmental pathway, known as etiolation, maximizes cell elongation in the shoot with little leaf or cotyledon development as the plant attempts to reach light conditions sufficient for photoautotrophic growth. Although the exact patterns of seedling morphogenesis vary widely among different taxa, the light-regulated developmental decision between the etiolated and the photomorphogenic pathway transcends phylogenetic classification. *Arabidopsis thaliana,* for example, is programmed to switch between the photomorphogenic and the etiolation pathway in the hypocotyl and the cotyledons. In the legumes, such as *Pisum sativum,* which use the cotyledons as storage organs, the corresponding developmental switch is programmed in the epicotyl and the primary leaves. Comparable phenomena of developmental plasticity are observed in the mesocotyl, the coleoptile, and the first leaf of monocotyledonous seedlings such as oat and rice (50, 99, 136). It is reasonable to speculate that the fundamental mechanism is similar in different plant species, though de-

[1] Abbreviations: UV-A, 320–400 nm light; UV-B, 280–320 nm light; HIR, high-irradiance response; LF, low fluence response; VLF, very low fluence response; Pr, phytochrome form with absorbance maximum in the red; Pfr, phytochrome form with absorbance maximum in far-red. We follow the phytochrome nomenclature proposed in Reference 141.

fined light signals are frequently coupled to individual responses in a species-dependent manner.

The contrasting developmental patterns are thought to be mediated primarily by changes in the expression level of light-regulated genes (reviewed in 5, 90, 168). The cellular events leading to the activation of these genes have been studied intensively, and extensive progress in elucidating some of the downstream steps in phytochrome and blue light receptor signaling has been reviewed during the past two years (15, 115, 138, 140, 152). Use of *Arabidopsis thaliana* as a model organism has led to profound advances in understanding the light control of seedling development. We attempt to put these recent genetic advances in perspective with traditional physiological data from other species and focus on the overlap and the mutual fertilization between these two lines of research.

Many valuable perspectives on photomorphogenesis have been described in a series of recent reviews on, for example, responses to blue light (75, 97), responses to UV light and high light stress (7, 40), ecological considerations (154, 155, 157), progress based on transgenic plants (188), phytochrome (138–140, 157, 176, 177), and mutational analysis (28, 45, 78, 86, 97, 134, 143, 188).

COMPLEXITY OF LIGHT RESPONSES AND PHOTORECEPTORS

In a given species, the specific effects of light can differ drastically from one organ or cell type to the other and even between neighboring cells. In *Arabidopsis,* light treatment of a dark-grown seedling will reduce the cell elongation rate in the hypocotyl while inducing cell expansion and cell division in the cotyledon and the shoot apex. Meanwhile, stomata differentiation is promoted by light in both the growth-arrested hypocotyl and the expanding cotyledon. In light-exposed shoots, plastids differentiate from proplastids or etioplasts into chloroplasts, whereas in the adjacent cells of the root, plastids differentiate instead into amyloplasts under both light and dark conditions.

Light responses are also known to exhibit fundamental changes in sensitivity as development progresses, e.g. from responsiveness to red light to blue light. This effect is conceptually most easily demonstrated at the level of gene expression (18, 61, 64, 73).

Finally, the kinetics of light-mediated effects allow for fast responses (in minutes), such as the inhibition of hypocotyl elongation (37, 162) or the induction of early light-inducible transcripts (114), but also accommodate slow responses (from hours to days), e.g. the entrainment of circadian rhythms by light (3a, 12, 103, 106, 121, 192).

Perhaps the most important component for encoding the complexity of responses is the multiple families of photoreceptors. Seedling responses to light are mediated by at least three classes of regulatory photoreceptors: (*a*) phytochromes, which respond mainly to red and far-red light but which also absorb blue and UV-light; (*b*) photoreceptors that are specific for blue and UV-A light; and (*c*) UV-B photoreceptors. Photosynthetic pigments, for example chlorophylls and carotenoids, have important roles as screening agents for the regulatory photoreceptors. Surprisingly little is known about their direct effects on developmental responses (169).

Molecular cloning of the *Arabidopsis CRY1/HY4* gene (3, 91a, 91b, 103a) as well as characterization of photoreceptor mutants (80, 81, 83, 92) may help to reveal the well-kept secrets of the blue light photoreceptors, which have long been referred to as "cryptochromes." The basis for the wide absorption band of *CRY1* in the blue and UV-A region of the spectrum has been resolved by molecular analysis, which showed that *CRY1* can bind two types of chromophores simultaneously, a flavin and a pterin (91b, 103a). It is conceivable that both chromophore binding sites can be mutated independently, thus separating sensitivity to UV-A and blue light genetically, as has been demonstrated in Reference 191. In *Arabidopsis* (3, 87), and also in transgenic tobacco (103a), *CRY1* is responsible for sensitivity of hypocotyl elongation to green light, in addition to blue and UV-A. A possible explanation is provided by the tendency of *CRY1* to stabilize a reduced, green-light absorbing form of the flavin chromophore (91b).

The phytochromes (phys), a family of dimeric, approximately 240-kDa chromoproteins, are by far the best studied of all photoreceptors (62, 138–141, 176, 177). In all known phytochromes, absorption of light leads to a conformational shift (photoconversion) of a red light absorbing form (Pr, absorption maximum around 660 nm) into a form with increased sensitivity to far-red light (Pfr, 730 nm). Absorption of far-red light converts the molecule back to the Pr form. Overlap in the absorption spectra of Pr and Pfr ensures that no light condition can convert all phytochrome into exclusively one form. Only phytochrome that is newly synthesized in complete darkness is present exclusively as Pr. It is generally approximated that the concentration of Pfr, rather than Pr, is responsible for all of the photomorphogenic effects of phytochrome, although this view has recently been challenged (95, 142, 151, 154).

Phytochrome responses allow an operational distinction among different levels of sensitivity: Low fluence (LF) responses are saturated by pulses with a red fluence component of 1 to 1000 μmol photons/m^2. They are at least partially reversible by a subsequent far-red light pulse. The effectiveness of the far-red pulse is a function of the intervening period of darkness. This function is known as the escape kinetic, a powerful tool to probe the chain of downstream signaling events. Very low fluence (VLF) responses (<1 μmol/m^2 red) are inducible by pulses of far-red light alone. High-irradiance responses (HIR) (typically >1 μmol/m^2/s) require continuous irradiation, which precludes analysis of far-red reversibility.

Phytochrome apoproteins are encoded by a small gene family with five members in *Arabidopsis* (*PHYA, PHYB, PHYC, PHYD,* and *PHYE*) (34, 139). Typically, phytochrome A (phyA) holoprotein is abundant in etiolated tissue and is greatly reduced in green tissue because it is degraded with a half life of approximately 1 h when in the Pfr form (177). Expression of the *PHYA* gene in *Arabidopsis* and some other species is downregulated by light (38, 66, 149, 159). *PHYB, C, D,* and *E* are moderately expressed in both etiolated and green tissues (34, 141, 158). Mutational analysis has revealed that phyA and phyB have distinct, partially complementary and partially overlapping functions in *Arabidopsis* (31, 71, 140, 142, 154, 155, 188). The absorption spectra of at least phyA and phyB appear to be almost identical and can therefore not be responsible for the different physiological functions (179, 180). In addition, the phytochromes that have been examined at the seedling stage seem to be expressed ubiquitously, albeit at different levels, and may contribute only quantitatively to the tissue specificity of many light responses (1, 34, 158).

Functionally distinct phytochromes have been similarly observed in many other species (139, 188). Apart from *Arabidopsis* (Table 1) (86, 87), mutants deficient in specific phytochromes have been described in cucumber (*lh*) (2, 102), *Brassica* (*ein*) (49), pea (*lv*) (186), tomato (85) (*tri*) (172, 173), and sorghum (27).

RESPONSES

Seed Germination

Seminal experiments on the light control of seed germination and its action spectrum revealed the photoreversibility of the phytochrome photoreceptor (13, 153). The recent availability of mutants affecting specific phytochromes in *Arabidopsis* has allowed a critical evaluation of the roles of individual phytochromes on seed germination (35, 142, 151). Rapid germination of *Arabidopsis* seeds imbibed in darkness is controlled primarily by phyB, stored in the Pfr form (PfrB) in the seed, whereas phyA contributes little to the decision to germinate in darkness. The seed germination rate in darkness is particularly high in seeds overexpressing phyB (108). Moreover, short pulses of red light increase the germination efficiency of the wild type further, which suggests that a significant level of phyB is also present in the Pr form in at least a fraction of seeds (151). A prominent role for a light-stable phytochrome such as phyB is consistent with the slow escape of the red response from reversibility by a pulse of far-red light (14).

Table 1 *Arabidopsis* mutants defective in aspects of photomorphogenesis[a]

Name	Other names	Gene cloned	Phenotype	References
Mutants in photoreceptor apoprotein genes				
phyA	hy8, fre1	Yes	long hy in FR	41, 120, 133, 142, 189
phyB	hy3	Yes	long hy in R and W	87, 118, 142, 160
cry1	hy4	Yes	long hy in B and W	3, 87, 91a,b, 98, 103a
Mutants in genes for pleiotropic regulators				
Negative regulators			*cop1* to *fus 12* in D:	
cop1	fus1, emb168	Yes	–short hy	46–48, 112, 116
cop8	fus8, emb134	No	–open cotyledons	116, 185
cop9	fus7, emb143	Yes	–green root	116, 183, 184
cop10	fus9, emb144	No	–thylakoid membranes	116, 185
cop11	fus6, emb78	Yes	–cell differentiation	24, 116, 185
det1	fus2	Yes	–derepression of light-inducible genes	30, 32, 33, 116, 135
fus4, -5, -11, -12		No		88a, 116
Positive regulators				
fhy1, fhy3		No	long hy in FR	189
hy5		No	long hy under R and B	87
Mutants in genes regulating specific responses				
Negative regulators				
amp1		No	hook open, derepressed gene expression in D	25
cop2		No	hook open	68
cop3	hls1	No	hook open	51, 68
cop4		No	de-etiolated, nuclear gene expression agravitropic	68
det2		No	de-etiolated, nuclear gene expression	29
det3		No	short hy, open cotyledons	19a
doc1–doc3		No	elevated *CAB* expression in D	89
gun1–gun3		No	nuclear gene expression uncoupled from plastid	165
icx1		No	*CHS* expression sensitized to light	68a
Positive regulators				
cue1		No	low CAB expression	89a

Table 1 (*continued*)

Name	Other names	Gene cloned	Phenotype	References
Phototropism mutants				
nph1	*JK224*	No	nonphototropic	80, 81, 83, 92
nph2, nph4		No	"-"	92
nph3	JK218	No	"-"	80, 81, 92
rpt1–rpt3		No	nonphototropic root	129–131

Abbreviations: hy, hypocotyl; FR, far-red light; R, red light; B, blue light; W, white light; D, darkness; CAB, chlorophyll a/b–binding protein gene; CHS, chalcone synthase gene.

Other phytochromes, such as phyA, gradually accumulate in the imbibed seed, promoting germination particularly under continuous far-red light. Under far-red light, *phyB* mutant seeds germinate more efficiently than wild type. Thus, the wild-type *PHYB* allele appears to negate the action of phyA, which provides an example for antagonistic control of the same response by the two types of phytochrome (142, 151). Because the inhibitory effect of *PHYB* is clearly retained in the *phyA* mutant (142), phyB probably does not act directly on phyA but rather on a downstream effector of phyA. The level of PfrB is extremely low in far-red light and suggests that the negative effect of phyB on phyA-mediated seed germination may be conferred by its Pr form (142, 151).

Sketch of Seedling Photomorphogenesis

Seedling development in complete darkness differs markedly from that in continuous white light or under a light-dark cycle. Under light conditions, the *Arabidopsis* seedling emerges from the seed coat, consisting of a hypocotyl with an apical hook, two small folded cotyledons, and a short main root. After an initial increase in volume, which is attributable primarily to hydration, the apical hook starts to straighten, and, at the same time, the cotyledons open and expand by cell division and cell expansion, accompanied by chlorophyll accumulation and greening. Meanwhile, the apical meristem gives rise to the first pair of true leaves, which carry trichomes. The hypocotyl translates positive phototropic and negative gravitropic cues into growth by differential cell elongation, in order to position the cotyledons for optimal exposure to the light source. In contrast, the root interprets the same signals in exactly the opposite manner, showing positive gravitropism and negative phototropism.

In darkness, the embryonic organs develop according to a completely different program. After an initial phase of moderate cell expansion due to hydration, the hypocotyl cells elongate rapidly while the apical hook persists for over a week and the cotyledons remain folded. The apical meristem remains arrested in the dark. Only gravitropic cues position the hypocotyl upright in anticipation of a light source.

At the subcellular level, plastids undergo drastic changes in morphology and protein composition in both darkness and light. In darkness, proplastids in the cotyledons develop into etioplasts, characterized by a paracrystalline prolamellar body and a poorly developed endomembrane system. Upon exposure to light the prolamellar body quickly disappears and is replaced by an extensive network of unstacked and grana-stacked thylakoid membranes, similar to those of the light-grown seedlings.

In contrast with the aerial portion of the seedling, roots develop in a similar pattern in both light and darkness. During the first week of seedling development, chloroplast development remains repressed under both light and dark conditions. However, if plants continue to expose their roots to light beyond the seedling stage and into adulthood, chloroplast development can be observed in the older part of the root system.

Formation of New Cell Types

Under laboratory conditions, light-dependent decisions that affect cell shape and differentiation during *Arabidopsis* seedling development only become apparent two days after germination and then lead quickly into diverging pathways (185). The first two days in light or darkness are accompanied by overall cell enlargement in hypocotyl, root, and cotyledon, and root hairs develop. During the following day in darkness, hypocotyl cells elongate enormously, and the hypocotyl epidermis forms a smooth surface. In the light-grown sibling hypocotyl, cell enlongation is inhibited and cell-type differentiation proceeds. The most obvious examples are the initiation and maturation of stomata in the epidermal layer of the hypocotyl, which are absent from the hypocotyl in the dark (185). In *Arabidopsis,* epidermal hairs are confined to true leaves, which are themselves dependent on light. The related crucifer *Sinapis alba* (white mustard), however, shows that differentiation of epidermal hairs on the hypocotyl can be under direct light control (181).

The cotyledon development in darkness is essentially arrested after two days, whereas in the light, cotyledon cells differentiate into distinct cell types, such as vascular, mesophyll, and epidermal cells. The most conspicuous change in the expanding mesophyll cells is the differentiation of numerous proplastids into green chloroplasts, although little differentiation between palisade and spongy mesophyll cells occurs (32, 46). Meanwhile, the epidermal pavement cells expand into their characteristic lobed, jigsaw puzzle shape. A set of guard-cell initials is present in the embryonic cotyledons and will differentiate into guard-cell precursors regardless of light conditions during the first two days (185). While development of guard-cell precursors then arrests in darkness, it continues to completion under light conditions. In addition, only in the light-grown cotyledons do new guard-cell initials form, following polarized cell divisions of epidermal cells (185). Finally, the vasculature in dark-grown seedlings is rather undeveloped but is quite extensive in light-grown seedlings. This has been clearly documented in *Zinnia* seedlings by using three early markers for developing xylem and phloem cells (44).

Cell Autonomy and Cell-Cell Interactions

What is the role of cell-cell communication and the extent of cell autonomy during the light control of seedling development? After microinjection of phytochrome into tomato hypocotyl cells, it was shown that the anthocyanin accumulation and the phytochrome-mediated gene expression pathways are cell autonomous (123). Similar conclusions were reached for the anthocyanin accumulation pathway by irradiating mustard cotyledons with a microbeam of red or far-red light (124). Microirradiation at one site, specifically on a vascular bundle or on a site in the lamina, could suppress anthocyanin accumulation at a site distant from the irradiated site, such as the leaf margin. This demonstrates that long-range inhibitory signals are transmitted through the leaf upon irradiation (124). Even under uniform irradiation, individual cells in the mustard cotyledons responded in an all-or-none fashion, which resulted in patchy patterns for both anthocyanin synthesis and for the mRNA of the key enzyme chalcone synthase (CHS). However, only a subset of those cells that accumulated *CHS* mRNA accumulated a significant level of anthocyanin. This indicates that, in addition to *CHS* mRNA, additional rate-limiting factors must be distributed unevenly over the leaf as well. If those factors are themselves light-dependent, then the mustard seedling's competence for the stochastic patterning may be interpreted as a means to preserve responsiveness over a wide range of fluence rates, which is typical for light responses in plants (124).

Unhooking and Separation of the Cotyledons

The separation of the cotyledons and the opening of the apical hook are stimulated by red light (23, 94); in some species, such as *Arabidopsis,* hook opening is also promoted by far-red or blue light, which indicates that multiple photoreceptors mediate this response (94, 96). Cotyledons that are folded back on the hypocotyl are formed initially during the late stage of embryogenesis, and the apical hook is maintained at the top of the hypocotyl during etiolation. A developmental problem is that the hypocotyl, but not the hook, responds to directional light with differential cell elongation and phototropic curvature, while the hook manages to perform differential cell elongation in the complete absence of any directional light signal. Mutations in a number of *Arabidopsis* loci, such as *hookless1* (*hls1*)/*constitutively photomorphogenic* 3 (*cop3*), *amp1,* and *cop2* (25, 51, 68), result in a loss of the apical hook in darkness. On the other hand, mutation to ethylene overproduction (*eto*) or to a constitutive ethylene response (*ctr1*), or external application of ethylene, all result in an exaggerated hook. Therefore it is likely that ethylene, perhaps in conjunction with auxins, is required to maintain the apical hook (51). The effect of ethylene on hook formation may well be counterbalanced by cytokinins, as suggested by the hookless phenotype of the cytokinin overproducing mutant *amp1* (25), although external application of cytokinins can also facilitate hook formation via ethylene (21). Light, however, overrides the effect of the ethylene signal (82).

In *Arabidopsis,* the response of the hook to both low fluence red and blue light is phytochrome mediated and far-red reversible, whereas intense far-red and blue light act through high-irradiance responses, which lack fluence reciprocity and far-red reversibility (94, 96). The latter involve phyA and a blue light–specific pathway that is also involved in hypocotyl elongation (97). During the transition of dark-grown seedlings to light, unhooking and arrest of hypocotyl elongation follow similar fluence rate-response curves in *Arabidopsis* (94). Both also require a functional *HY5* gene (96) and may therefore depend on similar signal transduction pathways, but in other species the two processes are more easily separated (94).

Cotyledon Development and Expansion

The same light signals that inhibit cell elongation in the hypocotyl also promote cell expansion in the cotyledons and the leaf. How these differential responses to the same light environment are regulated remains poorly understood (39, 50, 174, 175). Blue and red light–mediated leaf expansion begins after distinct short and long lag times, respectively, in leaves of *Phaseolus,* which is reminiscent of the short and long lag times that pass between blue and red irradiation, respectively, and the inhibition of hypocotyl cell elongation in cucumber (10, 37). Perhaps promotion and inhibition of cell expansion are induced by similar pathways in the two organs. Under white light, phytochromes A and B appear almost dispensable individually for cotyledon expansion in *Arabidopsis* (142). Full cotyledon expansion in bright red and blue light, however, depends on signals perceived by phytochrome B, as indicated by reduced cotyledon expansion in the *phyB* mutant (142). This phytochrome-B-mediated cotyledon expansion is organ autonomous (122). The *cry1/hy4* mutants show a decrease in cotyledon expansion, which confirms the involvement of a blue light receptor (11, 87, 93). Detaching the cotyledons from the hypocotyl rescues the cotyledon defect of *cry1/hy4* in blue light, suggesting that the *cry1/hy4* mutant hypocotyl may exert an inhibitory effect on cotyledon expansion (11). The importance of light perception in the hypocotyl or the hook for events in the cotyledons (126), and vice versa (9), has long been known. Overexpression of the photomorphogenic repressor COP1 leads to an inhibition of cotyledon cell expansion under blue light only, similar to the effect in *cry1/hy4* (113). This result indicates that COP1 contributes to the inhibition of cotyledon cell expansion that is alleviated by CRY1/HY4 activity.

Inhibition of Hypocotyl Elongation

The inhibition of hypocotyl elongation in response to illumination is one of the most easily quantifiable developmental processes and has greatly facilitated our understanding of the roles of regulatory components (16, 65, 85, 87, 93, 113, 182). Inhibition of hypocotyl elongation shows a complex fluence dependence that combines inductive, i.e. phytochrome-photoreversible (50, 127), and high-irradiance responses (HIR) (6, 89, 191). The considerable quantitative variation in the spectral sensitivity among different species is not easily explained (104).

Multiple photoreceptors can control hypocotyl elongation, although with distinct kinetics. In cucumber hypocotyls, for instance, cell wall extensibility and subsequent cell elongation are inhibited extremely rapidly in response to blue light, starting after a lag of less than 30 s (37, 161, 162). Such a fast response is unlikely to require gene expression. Following a lag of over 10 min, the phytochrome-mediated response is considerably delayed (157).

In *Arabidopsis,* phytochromes A and B, the blue light receptor *CRY1/HY4* and a genetically separable UV-A receptor have been shown to contribute to the inhibition of hypocotyl cell elongation (65, 87, 188, 191). Different from, for example, *Sinapis alba,* which is most sensitive to red and far-red light (6, 67), *Arabidopsis* hypocotyl elongation is most easily inhibited under blue light, where phytochromes and blue light receptors appear to act additively, specializing in low and high fluence responses, respectively (97, 91a). Whether the species-specific differences in sensitivity reflect subtle differences in the spectral properties or concentrations of the photoreceptors, the light environment inside the plant tissue, the mutual coupling of different phototransduction chains, or merely differences in the experimental protocol remains to be addressed.

Arabidopsis phyB mutants show a moderately elongated hypocotyl under white light (87, 118, 144, 160), where *phyA* mutants exhibit no such defect (41, 120, 133, 189). A drastic exaggeration of the long hypocotyl phenotype is seen in *phyA/phyB* double mutants (142). Therefore, under white light, inhibition of hypocotyl elongation can be mediated not only by blue and UV-A light but also by both phyA and phyB in an arrangement suggesting multiple, but only partially redundant, control (22, 140, 142).

Under continuous red light, phyB is the active phytochrome species inhibiting hypocotyl elongation, whereas under continuous far-red light, phyA is primarily responsible. However, supplementing a seedling growing in constant red or white light with far-red light causes the hypocotyl to elongate more rapidly than in constant red or white or in far-red light alone (110, 190). This effect is apparent in many species and is part of the "shade avoidance" response in adult plants, because a high ratio of far-red to red light is typical for the light environment under a plant canopy (154, 155).

One simple model attempts to explain the rate of hypocotyl elongation solely on the basis of the distinct steady-state levels of PfrB and PfrA, as estimated from their differential stability and expression levels: As discussed in the section on Complexity of Light Responses and Photoreceptors, phyB is relatively stable as Pfr and is expressed continuously at a moderate level, fairly independent of light. In contrast, phyA expression is high in darkness and low in the light, and PfrA is rapidly degraded (160, 177). Continuous red-light irradiation will therefore rapidly deplete the plant of phyA but not of phyB. Little phyA-mediated signaling is expected under these conditions, and continuous red light effects are attributed mainly to phyB. Continuous far-red light, however, can lead to a relatively high steady-state level of PfrA and hence can cause a pronounced developmental effect, because the low ratio of active Pfr to inactive Pr may be compensated for by the high overall phyA level. The low level of phyB in combination with the low Pfr-to-Pr ratio makes signaling through phyB negligible under the continuous far-red light. This model approximates the different roles of phyA and phyB on the basis of their expression levels and degradation rates alone and does not take into account some possible differences in the coupling of phyA and phyB to regulators or downstream effectors.

Consistent with the overlapping roles of phyA and phyB as inferred from mutant analysis, overexpression of cereal phyA in tobacco (26, 72, 76, 77, 107, 110, 111, 119), tomato (16), and *Arabidopsis* (17) or of phyB in *Arabidopsis* (108, 180) increases the seedlings' sensitivity to light, as judged by an extremely short hypocotyl under white light. Overexpression of heterologous phyA increases the sensitivity to far-red, white, and red light, which suggests that the elevated level of phyA persists even under prolonged irradiation with red light, a condition that leads to rapid degradation of the endogenous phyA (16, 25, 110, 111, 119). On the other hand, is sensitivity to far-red light also increased in seedlings overexpressing the photostable phyB? Such an increase would be expected according to the model put forward, on the basis of the essentially identical absorption spectra of phyA and phyB. However, this does not seem to be the case. Overexpression of phyB does not sensitize hypocotyl inhibition to far-red light (108, 109, 187). Therefore, phyA and phyB may interact with distinct sets of regulator or effector molecules. Similarly, phyB overexpression does not disable the shade avoidance response in adult plants, in contrast with the effect of phyA (110).

Phototropism

Seedlings of higher plants orient their growth with regard to the direction of light in an attempt to optimize exposure of the photosynthetic organs to light (55). The phototropic curvature of the shoot is achieved by differential growth, with cells on the shaded side elongating more strongly than cells on the surface exposed to light. Phototropism in *Arabidopsis* is superimposed over the light inhibition of hypocotyl elongation and over the negative gravitropic regulation of cell elongation in the hypocotyl (19). Therefore, a single cell elongation event may integrate signals from three different transduction chains simultaneously. The role of light on tropic responses is complicated further by the interaction of a phytochrome-mediated pathway with the gravitropic pathway in both the hypocotyl (63) and the root (54, 95).

Typically, phototropism is mediated by blue and UV-A light, but green light is effective in *Arabidopsis* as well (163). Red light absorbed by phytochrome stimulates phototropism directly in pea (132) and modulates the phototropic response to blue light in *Arabidopsis* (69, 70). Transduction of phototropic stimuli involves an early phosphorylation of a 120-kDa membrane-associated protein (152). Mutational analysis has identified four complementation groups of nonphototropic hypocotyl (*nph*) mutants (80, 92) and three complementation groups for root phototropism (*rpt*) mutants (129–131).

Mutants defective in blue light–mediated hypocotyl cell elongation (*cry1/hy4*) retain normal phototropic responses, and phototropic mutants (JK218/*nph3* and *nph1-1*) respond normally to light by inhibition of hypocotyl cell elongation. Those results not only confirm that the two processes are genetically separable but also suggest that they are almost certainly mediated by distinct photoreceptors (92, 98), because *CRY1/HY4* encodes a blue light receptor (3) and *nph1* mutants are defective in an early signaling step of blue and green light–mediated phototropism, most likely light perception (92). A conceptual similarity among the distinct roles of individual phytochrome family members and the members of the family of blue light receptors becomes apparent.

Although *nph1* mutants are defective in both shoot and root phototropism, other phototropism mutants, such as *rpt1* and *rpt2*, are specifically defective in root—but not shoot—phototropism only (129, 130). This suggests that *nph1*-mediated signals can be fed into at least two different pathways, depending on cell type, and that the signaling mechanisms in shoot and root are distinct. In support of this interpretation, shoot but not root phototropism is responsive to green light in *Arabidopsis* (131).

Plant hormones, especially auxins, have been implicated in differential cell elongation processes, such as during phototropism and gravitropism in hypocotyls, coleoptiles, and roots. Whether phototropism requires redistribution of auxins or modulation of auxin sensitivity is not well established (55, 125, 130), but investigation of phototropism in auxin-deficient, resistant, or overproducing strains could differentiate between different models. In fact, because the phototropic response is unaffected by the auxin resistance mutation *aux1* (130), root phototropism may not necessarily be compromised by defects in auxin signaling.

Roots and hypocotyls of *Arabidopsis* also exhibit differential cell elongation responses after gravitropic stimulation (129). Phototropism mutants now enable us to dissect the relationship between phototropism and gravitropism in *Arabidopsis* roots. Roots exhibit positive gravitropism and negative phototropism, and both responses rely on differential cell elongation. Three mutants have been described that affect both photo- and gravitropism (80, 81), but many mutations that affect one of the responses specifically are available. Both *rpt1* and *rpt2* have apparently normal shoot gravitropism, which also suggests that a specific signaling pathway for light exists in the root. However, the agravitropic mutants *aux1* and *agr1* show normal root phototropism (130). In the shoot, photo- and gravitropism are clearly separable genetically, because four different *nph* mutants, including *nph1,* are gravitropically normal (92).

Previous photophysiological analysis demonstrated separate photoreceptor systems for blue/UV-A light (PI) and green light (PII) (75, 83, 84), but an allelic series of *nph1* mutants suggests a new interpretation. The weak allele *nph1-2* (JK224) is merely defective in first (low light) positive phototropism in blue light but is wild-type-like in green light, whereas more severe alleles, for example *nph1-1,* are additionally defective in second positive curvature in both blue light and green light. The allelic series of *NPH1* indicates that both types of signals are processed by a single gene product for a photoreceptor apoprotein, possibly carrying two independently disruptable chromophores or one chromophore in alternative chemical states (92). In this respect it is interesting to note that the CRY1/HY4 gene product may also bind two distinct chromophores, namely a pterin and a flavin (3, 91b, 103a). A detailed comparison of the relationship between the two photoreceptor systems awaits molecular analysis of the *NPH1* gene.

Plastid Development

Chlorophyll accumulation and the concomitant transition of the proplastid or etioplast to the chloroplast are arguably the most striking responses to light and certainly require the intricate cooperation between multiple biochemical assembly and disassembly processes. In several species, pretreatment with red light, perceived at least partially by phyA and phyB, potentiates the accumulation of chlorophyll (74, 91). A feedback mechanism that conditions nuclear gene expression for a subset of plastid proteins operates via a biochemically undefined plastid signal, which is depleted by photooxidative damage to the plastid (36, 166, 167). At least part of this regulatory circuit involves an inhibitory element, because nuclear gene expression has escaped from the control by the plastid in recessive mutant alleles of at least three *Arabidopsis* complementation groups (*GUN1–GUN3*) (165). These mutants have few morphological defects, but have the tendency to green inefficiently when they are transferred from darkness to light in the etiolated state (165).

Mutant analysis has further demonstrated a panel of negative regulators, which keep chloroplast development suppressed in the cotyledons of dark-grown seedlings. Mutations at each of 10 loci known as *Constitutively Photomorphogenic* (*COP1, COP8–COP11*), *De-etiolated* (*DET1),* and *Fusca* (*FUS4, 5, 11,* and *12*) result in the absence of etioplasts and in partial chloroplast development in complete darkness, accompanied by cotyledon expansion, arrest of hypocotyl elongation, and light-specific cell type differentiation (24, 30, 32, 46, 48, 88a, 112, 116, 184, 185). The same loci also suppress chloroplast development in the roots of light-grown seedlings, because the mutants' roots develop chloroplasts and turn green.

Mutations at several other loci affect overall organ development and expression of light-inducible genes during seedling morphogenesis without resulting in plastid differentiation. For examples, *cop4* and *det2,* both of which display significant derepression of light-inducible genes in darkness, have no impact on chloroplast development (29, 68). A group of mutants at three loci, known as *DOC1, DOC2,* and *DOC3* (dark overexpressors of CAB) accumulate mRNA from the chlorophyll a/b binding protein gene (*CAB*) promoter in darkness, without a conspicuous morphological phenotype (89). The complementary mutant phenotype, namely opening of the cotyledons on a short hypocotyl accompanied by continued repression of light-inducible genes in darkness, has been described for the *det3* mutant (19a). In this mutant chloroplast, development remains tightly controlled by light.

The particular combinations of phenotypic defects in the various mutants described to date present a dilemma for the geneticist, because no simple linear sequence of gene action is consistent with all the data. While it is tempting and straightforward to invoke feedback circuits in order to solve the problem, detailed evidence for them is scarce. Elucidating the logic of the light signal transduction events may therefore require the mechanistic dissection of the function of the various gene products.

REGULATORS OF LIGHT RESPONSES IN SEEDLINGS

Positive Regulators

It is a recurring theme that predominantly negative regulators of photomorphogenic seedling development are uncovered in various genetic screens, whereas genetic identification of the positive regulators is sparse (see Figure 1) (28, 45).

The prevalent coaction between pigment systems (52, 70, 117, 128, 171) and, in particular, the partial redundancy inherent in multiple photoreceptor systems may be two reasons for the difficulty of establishing mutants in positive regulators. Remarkable exceptions are the genes *FHY1* and *FHY3*, which result in phenotypes similar to *phyA* mutants. These gene products may act downstream of *PHYA* in transmitting a signal specific for phytochrome A (71, 189). A fruitful approach to identifying specific regulators genetically may include screening for modifiers, suppressors or enhancers, of established photomorphogenic mutations. As an example, a suppressor of the *Arabidopsis* phytochrome-deficient *hy2* mutation (*shy1*) with some specificity for suppressing the phenotype under red rather than far-red light has been uncovered (BC Kim, MS Soh, BJ Kang, M Furuya & HG Nam, personal communication).

A second promising line of research is to screen for mutants defective in the light-dependent expression of specific genes. Following this rationale, the *CUE1* gene has been identified as a positive regulator of light-dependent nuclear and plastid gene expression (89a). Similarly, expression of the gene for chalcone synthase (*CHS*) and genes for related enzymes in the anthocyanin synthesis pathway is sensitized to light in the *Arabidopsis* mutant *icx1* (68a).

The HY5 gene product integrates signals from multiple photoreceptors to mediate inhibition of hypocotyl cell elongation in response to blue, red, and far-red light, but not UV-B light (4, 28, 87). These light conditions have also been shown to cooperate in abolishing the activity of COP1, a negative regulator of light responses (113). Mutations of *COP1* and the positive light regulatory component *HY5* exhibit an allele-specific interaction. Although a strong allele of *cop1* is epistatic over a mutant *hy5* allele, *hy5* can suppress the phenotype of the weak allele *cop1-6*. The suppression of the *cop1-6* phenotype by *hy5* could be attributed to compensating alterations in the surfaces of two

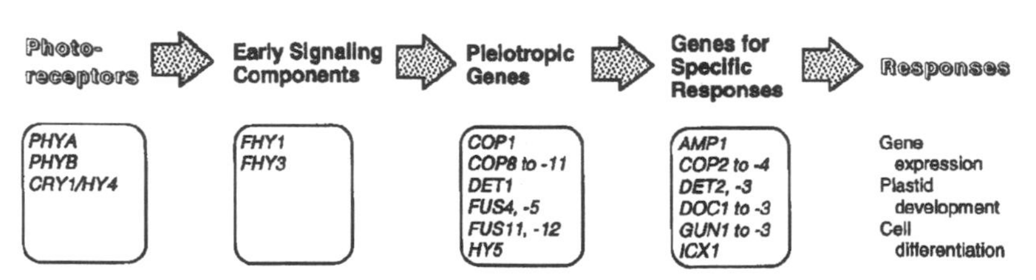

Figure 1 Scheme change of the sequence of events leading from light perception to developmental response. Examples for relevant gene products that have been identified by mutation are indicated. Please note that developmental and hormonal signals can modulate the light response by affecting any step of the sequence.

interacting proteins (4). Molecular cloning of the *HY5* gene should allow a direct test of this hypothesis.

Possible Roles of Plant Hormones

Many of the light-regulated seedling developmental responses, such as the inhibition of hypocotyl cell elongation (21, 101, 150), stem elongation (8), the cell division in the shoot apex (33), opening of the apical hook (51), and the induction or repression of nuclear gene expression, also respond to treatment with one or more of a variety of plant hormones (38, 42, 56). We have briefly alluded to possible contributions of ethylene and auxins to the control of tropic responses (51, 55, 125, 130). The question arises how hormonal and light effects are integrated with each other: Are hormones second messengers in light responses? Or do they transmit another signal that shares a common target with light? Or do they serve as integrators of distinct signaling pathways by "cross talk?"

CYTOKININS The relationship between light on the one hand and cytokinins and giberellins on the other has received the most thorough attention. For example, control of a wheat kinase gene homolog (*WPK4*) by light may require the presence of cytokinin, because a putative cytokinin antagonist (2-chloro-4-cyclobutylamino-6-ethylamino-s-triazine) specifically prevented the induction of *WPK4* mRNA (148). Even more intriguing is that de-etiolation of dark-grown *Arabidopsis* seedlings can be mimicked partially by supplementing seedlings with cytokinins under defined laboratory conditions (33). Moreover, the typically light-inducible genes for a chlorophyll a/b binding protein (*CAB*) (64, 73) and for chalcone synthase (*CHS*) (53, 88) were moderately activated by cytokinin treatment, although the effective levels of applied cytokinins were about tenfold higher than the endogenous level in these experiments. Tentative links between a high cytokinin level and a suppressed etiolation response are suggested because of the correlation of a hook opening phenotype with a significantly elevated cytokinin level in the *amp1* mutant, epistasis of *amp1* over *hy2* (25), and the aberrant response of *det1* and *det2* to cytokinin (33). However, endogenous cytokinin levels in dark- and light-grown seedlings and in *det1* and *det2* mutants were not consistent with the notion that light signals are transmitted simply by a change in cytokinin concentration (33). Furthermore, an additive and probably independent co-action of cytokinin and light has been demonstrated for the control of hypocotyl elongation (163a).

Cytokinin may act posttranscriptionally on *CAB* mRNA levels in *Lemna gibba,* while light clearly has a transcriptional component, indicating independent and probably additive action of cytokinin and light (56, 57). Stimulatory effects by moderate concentrations of applied cytokinins on the expression of light-inducible, circadian clock–regulated genes have also been observed under light conditions, which again suggests additive effects (42, 43).

In conclusion, although light effects may not be mediated primarily by cytokinin levels, any drastic alteration of the cellular cytokinin level by any stimulus other than light would be expected to modulate the seedling's responsiveness to light.

GIBERELLINS Although both light and giberellins control hypocotyl elongation, e.g. in cucumber (101), and mesocotyl elongation, for example, in rice (170), few genetic studies support the simple hypothesis that light acts through altering the levels of active giberellins (145–147). However, alterations in phytochrome levels have been shown to affect giberellin levels in tobacco (72), *Sorghum* (58, 59), and *Brassica* (49). One phenotype of the long hypocotyl mutant *lh* of cucumber, which has a deficiency in a B-type phytochrome and a constitutive shade avoidance response (100–102, 156), is an increase in the responsiveness to giberellin. Conversely, in the wild type, depletion of Pfr increases the responsiveness to giberellin (101). Phytochrome may similarly inhibit the sensitivity of the rice mesocotyl to giberellin (170).

The careful dissection of the signaling chains for light on the one hand and hormones on the other by use of defined mutations and by molecular and physiological analysis will allow researchers to pry apart the intertwined pathways and will finally reveal the role of cytokinin, giberellins, and other hormones in the light control of seedling development.

Repressors of Light-Mediated Development

PLEIOTROPIC COP, DET, AND FUS GENES Combined efforts in photomorphogenic mutant screens (*cop or det* mutants) and purple seed color screens (*fusca* mutants) have yielded a set of at least ten loci that are required for the full establishment of the dark developmental pathway and suppression of photomorphogenic development in darkness. The ten loci have been designated as *COP1/FUS1, DET1/FUS2, COP8/FUS8, COP9/FUS7, COP10/FUS9, COP11/FUS6, FUS4, FUS5, FUS11, and FUS12,* and mutations in each result in dark-grown seedlings with a pleiotropic de-etiolated or constitutively photomorphogenic phenotype (24, 32, 46, 88a, 116, 184, 185). The mutants display chloroplast-like plastids containing thylakoid membrane systems and lacking a prolamellar body in the dark-grown cotyledons and show a pattern of light-regulated gene expression in the dark resembling their light-grown siblings. The wild-type *COP/DET/FUS* genes also contribute to the repression of photomorphogenic responses in seedling roots under light conditions, because greening and chloroplast development have been consistently observed in the light-grown mutants. The elevated expression of nuclear- and plastid-encoded light-inducible genes in the dark-grown mutants is a direct or indirect loss in the ability to repress nuclear gene promoter activity as indicated by analysis of transgenic promoter-reporter gene fusions in the mutant backgrounds (29, 46, 184, 185). Consistent with a widespread evolution of COP/DET/FUS-like functions, a pea mutant with light-independent photomorphogenesis (*lip*) has been described (60). On the basis of the recessive nature of those mutations and their pleiotropic phenotype, it has been proposed that the photomorphogenic pathway constitutes the default route of development, whereas the etiolation pathway is an evolutionarily recent adaptation pioneered by the angiosperms (185).

It is important to point out that for certain light-regulated processes, the pleiotropic *COP/DET/FUS* loci are clearly not required. For instance, severe loss-of-function alleles of all ten loci seem to retain normal phytochrome control of seed germination (46, 88a, 185).

The immediate downstream target of any of the COP, DET, and FUS repressor molecules remains unknown. It is possible that they directly inhibit the transcription of photosynthetic genes or that they modulate expression of an intermediate set of regulatory genes, which in turn regulate the final target genes. Some indirect evidence is consistent with the latter possibility. For example, COP1 is a regulator of the *PHYA* gene itself (46). Further, it was recently reported that the normally light-inducible homeobox gene *ATH1* was derepressed in dark-grown *cop1* and *det1* mutants (137). Together, *ATH1* and two other homeobox genes, *ATHB-2* and *-4,* exhibit light-regulated expression patterns (20, 137) and may act as downstream effectors of the COP/DET/FUS proteins. In addition, the DNA binding factor CA-1, which binds to a region of the *CAB* promoter important for light regulation, was absent in the *det1-1* mutant allele, consistent with its function as a downstream target of inactivation by DET1 (79, 164).

OVEREXPRESSION STUDIES The molecular cloning of four pleiotropic *COP/DET/FUS* loci (24, 47, 135, 183) permitted a direct test using overexpression studies of the genetic model that their gene products act as repressors of the photomorphogenic pathway. To date, overexpression of COP1 has produced the clearest evidence for a photomorphogenic suppressor (113). Moderate two- to fourfold overexpression resulted in two responses typical for the etiolation pathway, namely, an elongation of the hypocotyl under blue and far-red light and under a dark/light cycle and a reduction in cotyledon expansion under blue light. These effects of COP1 overexpression under far-red or blue light resemble those of the *phyA* or *cry1/hy4* mutants, which supports the notion that COP1 acts downstream of multiple photoreceptors (11, 113). Overexpression of COP9 resulted in a similar but less dramatic elongation of the hypocotyl, again most pronounced under blue and far-red light, although no inhibitory effect on cotyledon cell expansion could be discerned (N Wei & X-W Deng, unpublished data). Therefore, the overexpression results in general confirm the conclusion that the pleiotropic COP/DET/FUS gene products are repressors of photomorphogenic development.

REGULATION OF COP/DET/FUS ACTIVITY BY LIGHT Limited sequence identity of COP9, COP11/FUS6, and DET1 to other eukaryotic protein sequences submitted in databases suggests that they participate in a highly conserved, but as yet undefined, cellular process (24a). For COP1, sequence similarity to the Ring-finger class of zinc binding proteins and the β-subunit of trimeric G proteins is consistent with a role as a nuclear regulator of gene expression (47). This alone, however, does not clarify the mechanism of light regulation. COP1 and COP9 protein are detected in dark-grown seedlings, when the proteins are active, and also in the light-grown seedlings, when they are photomorphogenically inactive (112, 183). Similarly, *DET1* mRNA is expressed in both dark- and light-grown seedlings (135). Therefore, the light inactivation of DET1, COP1, and COP9 is probably not mediated by the expression level but by some other means of posttranslational regulation. Recent studies with COP1 and COP9 point out some interesting insights. Our laboratory has shown that COP9 is a subunit of a large nuclear-localized protein complex whose apparent molecular weight can be shifted upon exposure of dark-grown seedlings to light (183). This may reflect light regulation of COP9 activity through modulation of protein-protein interactions in the COP9 complex. While DET1 (135) and the COP9 complex (N Wei & X-W Deng, unpublished data) appear to have the potential for nuclear localization, examination of the subcellular localization of fusion proteins between COP1 and β-glucuronidase (GUS) reporter enzyme suggested that the nucleocytoplasmic partitioning of COP1 is regulated by light in a cell type–dependent manner (178). In root cells, where COP1 is constitutively active and helps to repress chloroplast development (48), GUS-COP1 is found in the nucleus under all light conditions. In hypocotyl cells, however, GUS-COP1 is nuclear in darkness but excluded from the nucleus under constant white light. Upon a switch in illumination conditions from darkness to light, the protein slowly repartitions in the hypocotyl from a nuclear to a cytoplasmic location, and vice versa (178). Complementation of the *cop1*-mutant phenotype by the *GUS-COP1* transgene suggests that it is a functional substitute of *COP1* (our unpublished data). GUS-COP1 nucleocytoplasmic repartitioning coincides with light-regulated developmental redifferentiation processes, but their initiation clearly precedes any noticeable shift of GUS-COP1 (178). Further, the kinetics of repartitioning seem independent of the presence or absence of wild-type COP1 in the genetic background (our unpublished data). It is therefore likely that COP1 relocalization serves to maintain a developmental commitment that has previously been communicated to the cell by means other than COP1 partitioning.

Evidence for a role of the cytoskeleton in modulating COP1 partitioning has been obtained following the isolation of a COP1-interacting protein, CIP1 (105). The subcellular localization of CIP1 is cell type–dependent. While CIP1 shows a fibrous, cytoskeleton-like distribution in protoplasts derived from hypocotyl and cotyledons, it is localized to discrete foci in protoplasts derived from roots. The cell type–dependent localization pattern of CIP1 may provide a mechanism to regulate access of COP1 to the nucleus by specific protein-protein interactions with a cytoplasmic anchor protein (105).

CONCLUSION

In the past few years a combination of molecular genetic, biochemical, and cell biological studies has brought us remarkable progress in defining the critical players and their specific roles in mediating the light control of seedling development. The basic network of signaling events seems remarkably conserved throughout the angiosperm species examined. Therefore the focus on a model organism will prove to be continuously fruitful.

How light signals perceived by distinct photoreceptors are integrated to control cellular development and differentiation decisions may become a major focus in the coming years. A comprehensive understanding will not only reveal the players involved but also how those players interact with other stimuli, such as hormones, to reach a precise response according to the exact light environmental cues. In particular, many decisions in the natural environment are not simply all-or-none but quantitative. Therefore, physiological studies will probably play increasingly greater roles in the future work. Moreover, individual light responses vary significantly among different species. Explaining the evolutionary flexibility on the basis of conserved modules of protein function will be a major challenge in the future. In addition, most of the key players identified to date seem to be present in most if not all cell types, though each cell type produces a distinct response to a particular light stimulus. Answers to the questions presented here will bring further insights into the light control of seedling development.

ACKNOWLEDGMENTS

We thank Jeffrey Staub for a critical review of the manuscript and Peter Quail, Ning Wei, Steve Kay, and Hong Gil Nam for communicating unpublished results. Research in this laboratory was supported by grants from the National Science Foundation and by the National Institutes of Health.

Literature Cited

1. Adam E, Szell M, Szekeres M, Schäfer E, Nagy F. 1994. The developmental and tissue-specific expression of tobacco phytochrome A genes. *Plant J.* 6:283–93
2. Adamse P, Jaspers PAPM, Kendrick RE, Koorneef M. 1987. Photomorphogenic responses of a long hypocotyl mutant of *Cucumis sativus. J. Plant Physiol.* 127:481–91
3. Ahmad M, Cashmore AR. 1993. The *HY4* gene of *Arabidopsis thaliana* encodes a protein with characteristics of a blue light photoreceptor. *Nature* 366:162–66

3a. Anderson SL, Kay SA. 1995. Phototransduction and circadian clock pathways regulating gene transcription in higher plants. *Adv. Genet.* In press

4. Ang L-H, Deng X-W. 1994. Regulatory hierarchy of photomorphogenic loci: allele-specific and light-dependent interaction between *HY5* and *COP1* loci. *Plant Cell* 6:613–28
5. Batschauer A, Gilmartin PM, Nagy F, Schäfer E. 1994. The molecular biology of photoregulated genes. See Ref. 78a, pp. 559–600
6. Beggs CJ, Holmes MG, Jabben M, Schäfer E. 1980. Action spectra for the inhibition of hypocotyl growth by continuous irradiation in light and dark-grown *Sinapis alba* L. seedlings. *Plant Physiol.* 66:615–18
7. Beggs CJ, Wellmann E. 1994. Photocontrol of flavonoid biosynthesis. See Ref. 78a, pp. 733–52
8. Behringer FJ, Davies PJ, Reid JB. 1992. Phytochrome regulation of stem growth and indole-3-acetic acid levels in the *lv* and *Lv* genotypes of Pisum. *Photochem. Photobiol.* 56:677–84
9. Black M, Shuttleworth JE. 1974. The role of the cotyledons in the photocontrol of hypocotyl extension in *Cucumis sativus* L. *Planta* 117:57–66
10. Blum DE, Elzenga JTM, Linnemeyer PA, van Volkenburgh E. 1992. Stimulation of growth and ion uptake in bean leaves by red and blue light. *Plant Physiol.* 100:1968–75
11. Blum DE, Neff MM, van Volkenburgh E. 1994. Light-stimulated cotyledon expansion in the *blu3* and *hy4* mutants of *Arabidopsis thaliana. Plant Physiol.* 105: 1433–36
12. Boldt R, Scandalios JG. 1995. Circadian regulation of the *Cat3* catalase gene in maize (*Zea mays* L.): entrainment of the circadian rhythm of *Cat3* by different light treatments. *Plant J.* 7:989–99
13. Borthwick HA, Hendricks SB, Parker MW, Toole EH, Toole VK. 1952. A reversible photoreaction controlling seed germination. *Proc. Natl. Acad. Sci. USA* 38:662–66
14. Borthwick HA, Hendricks SB, Toole EH, Toole VK. 1954. Action of light on lettuce seed germination. *Bot. Gaz.* 115:205–25
15. Bowler C, Chua N. 1994. Emerging themes of plant signal transduction. *Plant Cell* 6: 1529–41
16. Boylan MT, Quail PH. 1989. Oat phytochrome is biologically active in transgenic tomatoes. *Plant Cell* 1:765–73
17. Boylan MT, Quail PH. 1991. Phytochrome A overexpression inhibits hypocotyl elongation in transgenic *Arabidopsis. Proc. Natl. Acad. Sci. USA* 88:10806–10
18. Brusslan JA, Tobin EM. 1992. Light-independent developmental regulation of gene expression in *Arabidopsis thaliana* seedlings. *Proc. Natl. Acad. Sci. USA* 89: 7791–95
19. Bullen BL, Best TR, Gregg MM, Barsel SE, Poff KL. 1990. A direct screening procedure for gravitropism mutants in *Arabidopsis thaliana* (L.) Heynh. *Plant Physiol.* 93:525–31

19a. Cabrera y Poch HL, Peto CA, Chory J. 1993. A mutation in the *Arabidopsis* DET3 gene uncouples photoregulated leaf development from gene expression and chloroplasts biogenesis. *Plant J.* 4:671–82

20. Carabelli M, Sessa G, Baima S, Morelli G, Ruberti I. 1993. The *Arabidopsis Athb-2* and *-4* genes are strongly induced by far-red-rich light. *Plant J.* 4:469–79
21. Cary AJ, Liu W, Howell SH. 1995. Cytokinin action is coupled to ethylene in its effects on the inhibition of root and hypocotyl elongation in *Arabidopsis thaliana* seedlings. *Plant Physiol.* 107:1075–82
22. Casal JJ. 1995. Coupling of phytochrome B to the control of hypocotyl growth in *Arabidopsis. Planta* 196:23–29
23. Casal JJ, Sanchez RA, Vierstra RD. 1994. *Avena* phytochrome A overexpressed in transgenic tobacco seedlings differentially affects red/far-red reversible and very-low-fluence responses (cotyledon unfolding) during de-etiolation. *Planta* 192:306–9
24. Castle L, Meinke D. 1994. A *FUSCA* gene of *Arabidopsis* encodes a novel protein essential for plant development. *Plant Cell* 6:25–41

24a. Chamovitz D, Deng X-W. 1995. The novel components of the Arabidopsis light signaling pathway may define a group of general developmental regulators shared by both animal and plant kingdoms. *Cell* 82: 353–54

25. Chaudhury AM, Letham S, Craig S, Dennis ES. 1993. *amp1,* a mutant with high cytokinin levels and altered embryonic pattern, faster vegetative growth, constitutive photomorphogenesis, and precocious flowering. *Plant J.* 4:907–16

26. Cherry JR, Hershey HP, Vierstra RD. 1991. Characterization of tobacco expressing functional oat phytochrome. *Plant Physiol.* 96:775–85
27. Childs KL, Cordonnier-Pratt MM, Pratt LH, Morgan PW. 1992. Genetic regulation of development in *Sorghum bicolor.* VII. The *ma3R* mutant lacks a phytochrome that predominates in green tissue. *Plant Physiol.* 99:765–70
28. Chory J. 1993. Out of darkness: mutants reveal pathways controlling light-regulated development in plants. *Trends Genet.* 9: 167–72
29. Chory J, Nagpal P, Peto CA. 1991. Phenotypic and genetic analysis of *det2,* a new mutant that affects light-regulated seedling development in *Arabidopsis. Plant Cell* 3: 445–59
30. Chory J, Peto CA. 1990. Mutations in the *DET1* gene affect cell-type-specific expression of light-regulated genes and chloroplast development in *Arabidopsis. Proc. Natl. Acad. Sci. USA* 87:8776–80
31. Chory J, Peto C, Ashbaugh M, Saganich R, Pratt LH, Ausubel F. 1989a. Different roles for phytochrome in etiolated and green plants deduced from characterization of *Arabidopsis thaliana* mutants. *Plant Cell* 1:867–80
32. Chory J, Peto C, Feinbaum R, Pratt L, Ausubel F. 1989. *Arabidopsis thaliana* mutant that develops as a light grown plant in the absence of light. *Cell* 58:991–99
33. Chory J, Reinecke D, Sim S, Washburn T, Brenner M. 1994. A role for cytokinins in de-etiolation in *Arabidopsis: det* mutants have an altered response to cytokinins. *Plant Physiol.* 104:339–47
34. Clack T, Mathews S, Sharrock RA. 1994. The phytochrome apoprotein family in *Arabidopsis* is encoded by five genes: the sequences and expression of *PHYD* and *PHYE. Plant Mol. Biol.* 25:413–27
35. Cone JW, Kendrick RE. 1985. Fluence-response curves and action spectra for promotion and inhibition of seed germination in wild type and long-hypocotyl mutants of *Arabidopsis thaliana* L. *Planta* 163:45–54
36. Conley TR, Shih M-C. 1995. Effects of light and chloroplast functional state on expression of nuclear genes encoding chloroplast glyceraldehyde-3-phosphate dehydrogenase in long hypocotyl (*hy*) mutants and wild type *Arabidopsis thaliana. Plant Physiol.* 108:1013–22
37. Cosgrove D. 1994. Photomodulation of growth. See Ref. 78a, pp. 631–58
38. Cotton J, Ross C, Byrne D, Colbert J. 1990. Down-regulation of phytochrome mRNA abundance by red light and benzyladenine in etiolated cucumber cotyledons. *Plant Mol. Biol.* 14:707–14
39. Dale JE. 1988. The control of leaf expansion. *Annu. Rev. Plant Physiol. Plant Mol. Biol.* 39:267–95
40. Demmig-Adams B, Adams WW. 1992. Photoprotection and other responses of plants to high light stress. *Annu. Rev. Plant Physiol. Plant Mol. Biol.* 43:599–26
41. Dehesh K, Franci C, Parks BM, Seeley KA, Short TW, et al. 1993. *Arabidopsis HY8* locus encodes phytochrome A. *Plant Cell* 5:1081–88
42. Deikman J, Hammer PE. 1995. Induction of anthocyanin accumulation by cytokinins in *Arabidopsis thaliana. Plant Physiol.* 108:47–57
43. Deikman J, Ulrich M. 1995. A novel cytokinin resistant mutant of *Arabidopsis* with abbreviated shoot development. *Planta* 195:440–49
44. Demura T, Fukuda H. 1994. Novel vascular cell-specific genes whose expression is regulated temporally and spatially during vascular system development. *Plant Cell* 6:967–81
45. Deng X-W. 1994. Fresh view of light signal transduction in plants. *Cell* 76:423–26
46. Deng X-W, Caspar T, Quail PH. 1991. *cop1:* a regulatory locus involved in light-controlled development and gene expression in *Arabidopsis. Genes Dev.* 5: 1172–82
47. Deng X-W, Matsui M, Wei N, Wagner D, Chu AM, et al. 1992. *COP1,* an *Arabidopsis* regulatory gene, encodes a novel protein with both a Zn-binding motif and a Gβ homologous domain. *Cell* 71:791–801
48. Deng X-W, Quail PH. 1992. Genetic and phenotypic characterization of *cop1* mutants of *Arabidopsis thaliana. Plant J.* 2: 83–95
49. Devlin DF, Rood SB, Somers DE, Quail PH, Whitelam GC. 1992. Photophysiology of the elongated internode (*ein*) mutant of *Brassica rapa:* the *ein*-mutant lacks a detectable phytochrome B-like protein. *Plant Physiol.* 100:1442–47
50. Downs RJ. 1955. Photoreversibility of leaf and hypocotyl elongation of dark grown red kidney bean seedlings. *Plant Physiol.* 30:468–73
51. Ecker JR. 1995. The ethylene signal transduction pathway in plants. *Science* 268:667–75
52. Elmlinger MW, Bolle C, Batschauer A, Oelmüller R, Mohr H. 1994. Coaction of blue light and light absorbed by phytochrome in control of glutamine synthetase gene expression in Scots pine (*Pinus sylvestris* L.) seedlings. *Planta* 192: 189–94
53. Feinbaum RL, Storz G, Ausubel F. 1991. High intensity, blue light regulated expression of chimeric chalcone synthase genes in transgenic *Arabidopsis thaliana* plants. *Mol. Gen. Genet.* 226:449–56

54. Feldman LJ, Briggs WR. 1987. Light-regulated geotropism in seedling roots of maize. *Plant Physiol.* 83:241–43
55. Firn RD. 1994. Phototropism. See Ref. 78a, pp. 659–81
56. Flores S, Tobin E. 1986. Benzyladenine modulation of the expression of two genes for nuclear-encoded chloroplasts proteins in *Lemna gibba:* apparent post-transcriptional regulation. *Planta* 168:340–49
57. Flores S, Tobin E. 1988. Cytokinin modulation of LHCP mRNA levels: the involvement of post-transcriptional regulation. *Plant Mol. Biol.* 11:409–15
58. Foster KR, Miller FR, Childs KL, Morgan PW. 1994. Genetic regulation of development in *Sorghum bicolor.* VIII. Shoot growth, tillering, flowering, giberellin biosynthesis, and phytochrome levels are differentially affected by dosage of the ma_3^R allele. *Plant Physiol.* 105:941–48
59. Foster KR, Morgan PW. 1995. Genetic regulation of development in *Sorghum bicolor.* IX. The ma_3^R allele disrupts diurnal control of giberellin biosynthesis. *Plant Physiol.* 108:337–43
60. Frances S, White MJ, Edgerton MD, Jones AM, Elliott RC, Thompson WF. 1992. Initial characterization of a pea mutant with light-independent photomorphogenesis. *Plant Cell* 4:1519–30
61. Frohnmeyer H, Ehmann B, Kretsch T, Rocholl M, Harter K, et al. 1992. Differential usage of photoreceptors for chalcone synthase gene expression during plant development. *Plant J.* 2:899–906
62. Furuya M. 1993. Phytochromes: their molecular species, gene families, and functions. *Annu. Rev. Plant Physiol. Plant Mol. Biol.* 44:617–46
63. Gaiser JC, Lomax TL. 1993. The altered gravitropic response of the *lazy-2* mutant of tomato is phytochrome regulated. *Plant Physiol.* 102:339–44
64. Gao J, Kaufman LS. 1994. Blue-light regulation of the *Arabidopsis thaliana Cab1* gene. *Plant Physiol.* 104:1251–57
65. Goto N, Yamamoto KT, Watanabe M. 1993. Action spectra for inhibition of hypocotyl growth of wild-type plants and of the *hy2* long-hypocotyl mutant of *Arabidopsis thaliana* L. *Photochem. Photobiol.* 57: 867–71
66. Higgs DC, Colbert JT. 1994. Oat phytochrome A mRNA degradation appears to occur via two distinct pathways. *Plant Cell* 6:1007–19
67. Holmes MG, Schäfer E. 1981. Action spectra for changes in the 'high irradiance reaction' in hypocotyls of *Sinapis alba* L. *Planta* 153:267–72
68. Hou Y, von Arnim AG, Deng X-W. 1993. A new class of *Arabidopsis* constitutive photomorphogenic genes involved in regulating cotyledon development. *Plant Cell* 5: 329–39
68a. Jackson JA, Fuglevand G, Brown BA, Shaw MJ, Jenkins GI. 1995. Isolation of *Arabidopsis* mutants altered in the light-regulation of chalcone synthase gene expression using a transgenic screening approach. *Plant J.* 8:369–80
69. Janoudi AK, Poff KL. 1991. Characterization of adaptation in phototropism of *Arabidopsis thaliana. Plant Physiol.* 95: 517–22
70. Janoudi AK, Poff KL. 1992. Action spectrum for enhancement of phototropism by *Arabidopsis thaliana* seedlings. *Photochem. Photobiol.* 56:655–59
71. Johnson E, Bradley M, Harberd NP, Whitelam GC. 1994. Photoresponses of light-grown phyA mutants of *Arabidopsis:* phytochrome A is required for the perception of daylength extensions. *Plant Physiol.* 105:141–49
72. Jordan ET, Hatfield PM, Hondred D, Talon M, Zeevaart JAD, Vierstra RD. 1995. Phytochrome A overexpression in transgenic tobacco: correlation of dwarf phenotype with high concentrations of phytochrome in vascular tissue and attenuated giberellin levels. *Plant Physiol.* 107:797–805
73. Karlin-Neumann GA, Sun L, Tobin E. 1988. Expression of light harvesting chlorophyll a/b binding protein genes is phytochrome-regulated in etiolated *Arabidopsis thaliana* seedlings. *Plant Physiol.* 88: 1323–31
74. Kasemir H, Oberdorfer U, Mohr H. 1973. A two-fold action of phytochrome in controlling chlorophyll a accumulation. *Photochem. Photobiol.* 18:481–86
75. Kaufman LS. 1993. Transduction of blue-light signals. *Plant Physiol.* 102:333–37
76. Kay SA, Nagatani A, Keith B, Deak M, Furuya M, Chua N-H. 1989. Rice phytochrome is biologically active in transgenic tobacco. *Plant Cell* 1:775–82
77. Keller JM, Shanklin J, Vierstra RD, Hershey HP. 1989. Expression of a functional monocotyledonous phytochrome in tobacco. *EMBO J.* 8:1005–12
78. Kendrick RE, Kerkhoffs LHJ, Pundsnes AS, van Tuinen A, Koorneef M, et al. 1994. Photomorphogenic mutants of tomato. *Euphytica* 79:227–34
78a. Kendrick RE, Kronenberg GHM, eds. 1994. *Photomorphogenesis in Plants.* Dordrecht: Kluwer
79. Kenigsbuch D, Tobin EM. 1995. A region of the *Arabidopsis* Lhcb1*3 promoter that binds to CA-1 activity is essential for high expression and phytochrome regulation. *Plant Physiol.* 108:1023–27
80. Khurana JP, Poff KL. 1989. Mutants of *Arabidopsis thaliana* with altered phototropism. *Planta* 178:400–506

81. Khurana JP, Ren Z, Steinitz B, Parks B, Best TR, Poff KL. 1989. Mutants of *Arabidopsis thaliana* with decreased amplitude in their phototropic response. *Plant Physiol.* 91:685–89
82. Kieber JJ, Rothenberg M, Roman G, Feldman KA, Ecker JR. 1993. CTR1, a negative regulator of the ethylene response pathway in *Arabidopsis,* encodes a member of the Raf family of protein kinases. *Cell* 72: 427–41
83. Konjevic R, Khurana JP, Poff KL. 1992. Analysis of multiple photoreceptor pigments for phototropism in a mutant of *Arabidopsis thaliana. Photchem. Photobiol.* 55:789–92
84. Konjevic R, Steinitz B, Poff KL. 1989. Dependence of the phototropic response of *Arabidopsis thaliana* on fluence rate and wavelength. *Proc. Natl. Acad. Sci. USA* 86:9876–80
85. Koornneef M, Cone JW, Dekens RG, O'Herne-Robers EG, Spruit CJP, Kendrick RE. 1985. Photomorphogenic responses of long hypocotyl mutants of tomato. *J. Plant Physiol.* 120:153–65
86. Koornneef M, Kendrick RE. 1994. Photomorphogenic mutants of higher plants. See Ref. 78a, pp. 601–28
87. Koornneef M, Rolff E, Spruit CJP. 1980. Genetic control of light inhibited hypocotyl elongation in *Arabidopsis thaliana* (L.) Heynh. *Z. Pflanzenphysiol.* 100: 147–60
88. Kubasek WL, Shirley BW, McKillop A, Goodman HM, Briggs W, Ausubel FM. 1992. Regulation of flavonoid biosynthetic genes in germinating *Arabidopsis* seedlings. *Plant Cell* 4:1229–36
88a. Kwok SF, Piekos B, Misera M, Deng X-W. 1996. A complement of ten essential and pleiotropic *Arabidopsis* COP/DET/FUS genes are necessary for repression of photomorphogenesis in darkness. *Plant Physiol.* In press
89. Li H, Altschmied L, Chory J. 1994. *Arabidopsis* mutants define downstream branches in the phototransduction pathway. *Genes Dev.* 8:339–49
89a. Li H, Culligan K, Dixon RA, Chory J. 1995. *CUE1:* a mesophyll cell–specific regulator of light-controlled gene expression in *Arabidopsis. Plant Cell* 7: 1599–620
90. Li H, Washburn T, Chory J. 1993. Regulation of gene expression by light. *Opin. Cell. Biol.* 5:455–60
91. Lifschitz S, Gepstein S, Horwitz BA. 1990. Photoregulation of greening in wild type and long hypocotyl mutants of *Arabidopsis thaliana. Planta* 181:234–38
91a. Lin C, Ahmad M, Gordon D, Cashmore A. 1995. Expression of an *Arabidopsis* cryptochrome gene in transgenic tobacco results in hypersensitivity to blue, UV-A, and green light. *Proc. Natl. Acad. Sci. USA* 92:8423–27
91b. Lin C, Robertson DE, Ahmad M, Raibekas AA, Schuman-Jorns M, et al. 1995. Association of flavin adenine dinucleotide with the *Arabidopsis* blue light receptor CRY1. *Science* 269:968–70
92. Liscum E, Briggs WR. 1995. Mutations in the NPH1 locus of *Arabidopsis* disrupt the perception of phototropic stimuli. *Plant Cell* 7:473–85
93. Liscum E, Hangarter RP. 1991. *Arabidopsis* mutants lacking blue light dependent inhibition of hypocotyl elongation. *Plant Cell* 3:685–94
94. Liscum E, Hangarter RP. 1993. Light-stimulated apical hook opening in wild-type *Arabidopsis* thaliana seedlings. *Plant Physiol.* 101:567–72
95. Liscum E, Hangarter RP. 1993. Genetic evidence that the Pr form of phytochrome B modulates gravitropism in *Arabidopsis thaliana. Plant Physiol.* 103:15–19
96. Liscum E, Hangarter RP. 1993. Photomorphogenic mutants of *Arabidopsis thaliana* reveal activities of multiple photosensory systems during light-stimulated apical hook opening. *Planta* 191:214–21
97. Liscum E, Hangarter RP. 1994. Mutational analysis of blue-light sensing in *Arabidopsis. Plant Cell Environ.* 17:639–48
98. Liscum E, Young JC, Poff KL, Hangarter RP. 1992. Genetic separation of phototropism and blue light inhibition of stem elongation. *Plant Physiol.* 100:267–71
99. Loercher L. 1966. Phytochrome changes correlated to mesocotyl inhibition in etiolated Avena seedlings. *Plant Physiol.* 41: 932–36
100. Lopez-Juez E, Buurmeijer WF, Heeringa GH, Kendrick RE, Wesselius JC. 1990. Response of light grown wild-type and long hypocotyl mutant cucumber plants to end-of-day far-red light. *Photochem. Photobiol.* 52:143–49
101. Lopez-Juez E, Kobayashi M, Sakurai A, Kamiya Y, Kendrick RE. 1995. Phytochrome, giberellins, and hypocotyl growth. *Plant Physiol.* 107:131–40
102. Lopez-Juez E, Nagatani A, Tomizawa K-I, Deak M, Kern R, et al. 1992. The cucumber long hypocotyl mutant lacks a light stable phyB-like phytochrome. *Plant Cell* 4: 241–51
103. Lumsden PJ. 1991. Circadian rhythms and phytochrome. *Annu. Rev. Plant Physiol. Plant Mol. Biol.* 42:351–71
103a. Malhotra K, Kim ST, Batschauer A, Dawut L, Sancar A. 1995. Putative blue-light photoreceptors from *Arabidopsis thaliana* and *Sinapis alba* with a high degree of homology to DNA photolyase contain the two photolyase cofactors but lack

DNA repair activity. *Biochemistry* 34: 6892–99

104. Mancinelli AI. 1994. The physiology of phytochrome action. See Ref. 78a, pp. 211–70

105. Matsui M, Stoop CD, von Arnim AG, Wei N, Deng X-W. 1995. *Arabidopsis* COP1 protein specifically interacts in vitro with a cytoskeleton-associated protein, CIP1. *Proc. Natl. Acad. Sci. USA* 92: 4239–43

106. McClung CR, Kay SA. 1994. Circadian rhythms in *Arabidopsis thaliana*. In *Arabidopsis*, ed. CR Somerville, EM Meyerowitz, pp. 615–38. Cold Spring Harbor, NY: Cold Spring Harbor Lab. Press

107. McCormac AC, Cherry JR, Hershey HP, Vierstra RD, Smith H. 1991. Photoresponses of transgenic tobacco plants expressing an oat phytochrome gene. *Planta* 185: 162–70

108. McCormac AC, Smith H, Whitelam GC. 1993. Photoregulation of germination in seed of transgenic lines of tobacco and *Arabidopsis* which express an introduced cDNA encoding phytochrome A or phytochrome B. *Planta* 191:386–93

109. McCormac AC, Wagner D, Boylan MT, Quail PH, Smith H, Whitelam GC. 1993. Photoresponses of transgenic *Arabidopsis* seedlings expressing introduced phytochrome B-encoding cDNAs: evidence that phytochrome A and phytochrome B have distinct photoregulatory roles. *Plant J.* 4: 19–27

110. McCormac AC, Whitelam GC, Boylan MT, Quail PH, Smith H. 1992. Contrasting responses of etiolated and light-adapted seedlings to red:far-red ratio: a comparison of wild type, mutant, and transgenic plants has revealed differential functions of members of the phytochrome family. *J. Plant Physiol.* 140:707–14

111. McCormac AC, Whitelam GC, Smith H. 1992. Light grown plants of transgenic tobacco expressing an introduced phytochrome A gene under the control of a constitutive viral promoter exhibit persistent growth inhibition by far-red light. *Planta* 188:173–81

112. McNellis T, von Arnim AG, Araki T, Komeda Y, Misera S, Deng X-W. 1994. Genetic and molecular analysis of an allelic series of *cop1* mutants suggests functional roles for the multiple protein domains. *Plant Cell* 6:487–500

113. McNellis T, von Arnim AG, Deng X-W. 1994. Overexpression of *Arabidopsis* COP1 results in partial suppression of light-mediated development: evidence for a light-inactivable repressor of photomorphogenesis. *Plant Cell* 6:1391–400

114. Meyer G, Kloppstech K. 1984. A rapidly light-induced chloroplast protein with a high turnover coded for by pea nuclear DNA. *Eur. J. Biochem.* 138:201–7

115. Millar AJ, McGrath RB, Chua N-H. 1994. Phytochrome phototransduction pathways. *Annu. Rev. Genet.* 28:325–49

116. Miséra S, Müller AJ, Weiland-Heidecker U, Jürgens G. 1994. The *FUSCA* genes of *Arabidopsis:* negative regulators of light responses. *Mol. Gen. Genet.* 244: 242–52

117. Mohr H. 1994. Coaction between pigment systems. See Ref. 78a, pp. 353–73

118. Nagatani A, Chory J, Furuya M. 1991. Phytochrome B is not detectable in the *hy3* mutant of *Arabidopsis* which is deficient in responding to end-of-day far-red light treatment. *Plant Cell Physiol.* 32:1119–22

119. Nagatani A, Kay SA, Deak M, Chua N, Furuya M. 1991. Rice type I phytochrome regulates hypocotyl elongation in transgenic tobacco seedlings. *Proc. Natl. Acad. Sci. USA* 88:5207–11

120. Nagatani A, Reed J, Chory J. 1993. Isolation and initial characterization of *Arabidopsis* mutants deficient in phytochrome A. *Plant Physiol.* 102:269–77

121. Nagy F, Fejes E, Wehmeyer B, Dallmann G, Schäfer E. 1993. The circadian oscillator is regulated by a very low fluence response of phytochrome in wheat. *Proc. Natl. Acad. Sci. USA* 90:6290–94

122. Neff MM, van Volkenburgh E. 1994. Light-stimulated cotyledon expansion in *Arabidopsis* seedlings: the role of phytochrome B. *Plant Physiol.* 104:1027–32

123. Neuhaus G, Bowler C, Kern R, Chua N-H. 1993. Calcium/calmodulin-dependent and independent phytochrome signal transduction pathways. *Cell* 73:937–52

124. Nick P, Ehmann B, Furuya M, Schäfer E. 1993. Cell communication, stochastic cell responses, and anthocyanin pattern in mustard cotyledons. *Plant Cell* 5:541–52

125. Nick P, Schäfer E. 1995. Polarity induction versus phototropism in maize: auxin cannot replace blue light. *Planta* 195:63–69

126. Oelze-Karow H, Mohr H. 1988. Rapid transmission of a phytochrome signal from hypocotyl hook to cotyledons in mustard (*Sinapis alba* L.). *Photochem. Photobiol.* 47:447–50

127. Oelze-Karow H, Schopfer P. 1971. Demonstration of a threshold regulation by phytochrome in the photomodulation of longitudinal growth of the hypocotyl of mustard seedlings (*Sinapis alba* L.). *Planta* 100: 167–75

128. Ohl S, Hahlbrock K, Schäfer E. 1989. A stable, blue light–derived signal modulates UV-light-induced activation of the chalcone synthase gene in cultured parsley cells. *Planta* 177:228–36

129. Okada K, Shimura Y. 1992. Aspects of recent developments in mutational studies

of plant signaling pathways. *Cell* 70: 369–72

130. Okada K, Shimura Y. 1992. Mutational analysis of root gravitropism and phototropism of *Arabidopsis thaliana* seedlings. *Aust. J. Plant Physiol.* 19:439–48
131. Okada K, Shimura Y. 1994. Modulation of root growth by physical stimuli. In *Arabidopsis,* ed. CR Somerville, EM Meyerowitz, pp. 665–84. Cold Spring Harbor, NY: Cold Spring Harbor Lab. Press
132. Parker K, Baskin TI, Briggs WR. 1989. Evidence for a phytochrome-mediated phototropism in etiolated pea seedlings. *Plant Physiol.* 89:493–97
133. Parks BM, Quail PH. 1993. *hy8,* a new class of *Arabidopsis* long hypocotyl mutants deficient in functional phytochrome A. *Plant Cell* 5:39–48
134. Pepper A, Delaney TP, Chory J. 1993. Genetic interactions in plant photomorphogenesis. *Semin. Dev. Biol.* 4:15–22
135. Pepper A, Delaney T, Washburn T, Poole D, Chory J. 1994. *DET1,* a negative regulator of light-mediated development and gene expression in *Arabidopsis,* encodes a novel nuclear-localized protein. *Cell* 78:109–16
136. Pjon C-J, Furuya M. 1967. Phytochrome action in *Oryza sativa* L. I. Growth responses of etiolated coleoptiles to red, far-red and blue light. *Plant Cell. Physiol.* 8: 709–18
137. Quaedvlieg N, Dockx J, Rook F, Weisbeek P, Smeekens S. 1995. The homeobox gene *ATH1* of *Arabidopsis* is derepressed in the photomorphogenic mutants *cop1* and *det1. Plant Cell* 7:117–29
138. Quail PH. 1994. Photosensory perception and signal transduction in plants. *Curr. Opin. Genet. Dev.* 6:613–28
139. Quail PH. 1994. Phytochrome genes and their expression. See Ref. 78a, pp. 71–104
140. Quail PH, Boylan MT, Parks BM, Short TW, Xu Y, Wagner D. 1995. Phytochromes: photosensory perception and signal transduction. *Science* 268:675–80
141. Quail PH, Briggs WR, Chory J, Hangarter RP, Harberd NP, et al. 1994. Letter to the editor: spotlight on phytochrome nomenclature. *Plant Cell* 6:468–72
142. Reed JW, Nagatani A, Elich T, Fagan M, Chory J. 1994. Phytochrome A and phytochrome B have overlapping but distinct functions in *Arabidopsis* development. *Plant Physiol.* 104:1139–49
143. Reed JW, Nagpal P, Chory J. 1992. Searching for phytochrome mutants. *Photochem. Photobiol.* 56:833–38
144. Reed JW, Nagpal P, Poole DS, Furuya M, Chory J. 1993. Mutations in the gene for the red/far-red light receptor phytochrome B alter cell elongation and physiological responses throughout *Arabidopsis* development. *Plant Cell* 5:147–57
145. Reid JB, Hasan O, Ross JJ. 1990. Internode length in *Pisum:* giberellins and the response to far-red rich light. *J. Plant Physiol.* 137:46–52
146. Reid JB, Ross JJ, Swain SM. 1992. Internode length in *Pisum:* a new, slender mutant with elevated levels of C(19) giberellins. *Planta* 188:462–67
147. Ross JJ, Willis CL, Gaskin P, Reid JB. 1992. Shoot elongation in *Lathyrus elongatus* L.: giberellin levels in light and dark-grown tall and dwarf seedlings. *Planta* 187: 10–13
148. Sano H, Youssefian S. 1994. Light and nutritional regulation of transcripts encoding a wheat protein kinase homolog is mediated by cytokinins. *Proc. Natl. Acad. Sci. USA* 91:2582–86
149. Sharrock RA, Quail PH. 1989. Novel phytochrome sequences in *Arabidopsis thaliana:* structure, evolution, and differential expression of a plant regulatory photoreceptor family. *Genes Dev.* 3:1745–57
150. Shibaoka H. 1994. Plant hormone induced changes in the orientation of cortical microtubules: alterations in the crosslinking between microtubules and the plasma membrane. *Annu. Rev. Plant Physiol. Plant Mol. Biol.* 45:527–44
151. Shinomura T, Nagatani A, Chory J, Furuya M. 1994. The induction of seed germination in *Arabidopsis thaliana* is regulated principally by phytochrome B and secondarily by phytochrome A. *Plant Physiol.* 104:363–71
152. Short TW, Briggs WR. 1994. The transduction of blue light signals in higher plants. *Annu. Rev. Plant Physiol. Plant Mol. Biol.* 45:143–72
153. Shropshire WJ, Klein WH, Elstad VB. 1961. Action spectra for photomorphogenic induction and photoinactivation of germination in *Arabidopsis thaliana. Plant Cell Physiol.* 2:63–69
154. Smith H. 1994. Sensing the light environment: the functions of the phytochrome family. See Ref. 78a, pp. 377–416
155. Smith H. 1995. Physiological and ecological function within the phytochrome family. *Annu. Rev. Plant Phys. Plant Mol. Biol.* 46:289–315
156. Smith H, Turnbull M, Kendrick RE. 1992. Light grown plants of the cucumber long hypocotyl mutant exhibit both long-term and rapid elongation growth responses to irradiation with supplementary far-red light. *Photochem. Photobiol.* 56: 607–10
157. Smith H, Whitelam G. 1990. Phytochrome, a family of photoreceptors with multiple physiological roles. *Plant Cell Environ.* 13: 695–708
158. Somers DE, Quail PH. 1995. Temporal and spatial expression patterns of phyA and

phyB genes in *Arabidopsis*. *Plant J.* 7: 413–28

159. Somers DE, Quail PH. 1995. Phytochrome-mediated light regulation of phyA- and phyB-GUS transgenes in *Arabidopsis thaliana* seedlings. *Plant Physiol.* 107: 523–34
160. Somers DE, Sharrock RA, Tepperman JM, Quail PH. 1991. The *hy3* long hypocotyl mutant of *Arabidopsis* is deficient in phytochrome B. *Plant Cell* 3:1263–74
161. Spalding E, Cosgrove DJ. 1989. Large plasma-membrane depolarization precedes rapid blue-light-induced growth inhibition in cucumber. *Planta* 178:407–10
162. Spalding E, Cosgrove DJ. 1992. Mechanism of blue-light induced plasma membrane depolarization in etiolated cucumber hypocotyls. *Planta* 188:199–205
163. Steinitz B, Ren Z, Poff KL. 1985. Blue-and green-light induced phototropism in *Arabidopsis thaliana* and *Lactuca sativa* L. seedlings. *Plant Physiol.* 77:248–51

163a. Su W, Howell SH. 1995. The effects of cytokinin and light on hypocotyl elongation in *Arabidopsis* seedlings are independent and additive. *Plant Physiol.* 108:1420–30

164. Sun L, Doxsee RA, Harel E, Tobin EM. 1993. CA-1, a novel phosphoprotein, interacts with the promoter of the *cab140* gene in *Arabidopsis* and is undetectable in *det1* mutant seedlings. *Plant Cell* 5:109–21
165. Susek RE, Ausubel FM, Chory J. 1993. Signal transduction mutants of *Arabidopsis* uncouple nuclear CAB and RBCS gene expression from chloroplast development. *Cell* 74:787–99
166. Susek RE, Chory J. 1992. A tale of two genomes: role of a chloroplast signal in coordinating nuclear and plastid gene expression. *Aust. J. Plant Physiol.* 19:387–99
167. Taylor WC. 1989. Regulatory interactions between nuclear and plastid genomes. *Annu. Rev. Plant Physiol. Plant Mol. Biol.* 40:211–33
168. Terzaghi WB, Cashmore AR. 1995. Light-regulated transcription. *Annu. Rev. Plant Phys. Plant Mol. Biol.* 46:445–74
169. Thompson WF, White MJ. 1991. Physiological and molecular studies of light-regulated nuclear genes in higher plants. *Annu. Rev. Plant Physiol. Plant Mol. Biol.* 42:423–66
170. Toyomasu T, Yamane H, Murofushi N, Nick P. 1994. Phytochrome inhibits the effectiveness of giberellins to induce cell elongation in rice. *Planta* 194:256–63
171. Tripathy BC, Brown CS. 1995. Root-shoot interaction in the greening of wheat seedlings grown under red light. *Plant Physiol.* 107:407–11
172. van Tuinen A, Kerckhoffs LHJ, Nagatani A, Kendrick RE, Koorneef M. 1995. Far-red light-insensitive, phytochrome A-deficient mutants of tomato. *Mol. Gen. Genet.* 246:133–41
173. van Tuinen A, Kerckhoffs LHJ, Nagatani A, Kendrick RE, Koorneef M. 1995. A temporarily red light–insensitive mutant of tomato lacks a light-stable, B-like phytochrome. *Plant Physiol.* 108:939–47
174. van Volkenburgh E, Cleland RE. 1990. Light-stimulated cell expansion in bean (*Phaseolus vulgaris* L.) leaves. I. Growth can occur without photosynthesis. *Planta* 182:72–76
175. van Volkenburgh E, Cleland RE, Watanabe M. 1990. Light-stimulated cell expansion in bean (*Phaseolus vulgaris* L.) leaves. II. Quantity and quality of light required. *Planta* 182:77–80
176. Vierstra RD. 1993. Illuminating phytochrome functions: there is light at the end of the tunnel. *Plant Physiol.* 103: 679–84
177. Vierstra RD. 1994. Phytochrome degradation. See Ref. 78a, pp. 141–62
178. von Arnim AG, Deng X-W. 1994. Light inactivation of *Arabidopsis* photomorphogenic repressor COP1 involves a cell type specific modulation of its nucleocytoplasmic partitioning. *Cell* 79:1035–45
179. Wagner D, Quail PH. 1995. Mutational analysis of phytochrome B identifies a small COOH-terminal-domain region critical for regulatory activity. *Proc. Natl. Acad. Sci. USA* 92:8596–600
180. Wagner D, Tepperman JM, Quail PH. 1991. Overexpression of phytochrome B induces a short hypocotyl phenotype in transgenic *Arabidopsis. Plant Cell* 3:1275–88
181. Wagner E, Mohr H. 1966. Primary and secondary differentiation in connection with photomorphogenesis in seedlings of Sinapis alba L. *Planta* 71:204–21
182. Wall JK, Johnson CB. 1983. An analysis of phytochrome action in the 'high irradiance response.' *Planta* 159:387–97
183. Wei N, Chamovitz D, Deng X-W. 1994. *Arabidopsis* COP9 is a component of a novel signaling complex mediating light control of plant development. *Cell* 78: 117–24
184. Wei N, Deng X-W. 1992. COP9: a new genetic locus involved in light-regulated development and gene expression in *Arabidopsis. Plant Cell* 4:1507–18
185. Wei N, Kwok SF, von Arnim AG, Lee A, MacNellis T, et al. 1994. *Arabidopsis COP*8, *COP*10 and *COP*11 genes are involved in repression of photomorphogenic developmental pathway in darkness. *Plant Cell* 6:629–43
186. Weller JL, Nagatani A, Kendrick RE, Murfet IC, Reid JB. 1995. New lv mutants of pea are deficient in phytochrome B. *Plant Physiol.* 108:525–32
187. Wester L, Somers DE, Clack T, Sharrock

RA. 1994. Transgenic complementation of the hy3 phytochrome B mutation and response to phyB gene copy number in *Ara bidopsis. Plant J.* 5:261–72

188. Whitelam GC, Harberd NP. 1994. Action and function of phytochrome family members revealed through the study of mutant and transgenic plants. *Plant Cell Environ.* 17:615–25
189. Whitelam GC, Johnson E, Peng J, Carol P, Anderson ML, et al. 1993. Phytochrome A null mutants of *Arabidopsis* display a wild type phenotype in white light. *Plant Cell* 5:757–68
190. Whitelam GC, Smith H. 1991. Retention of phytochrome-mediated shade avoidance responses in phytochrome-deficient mutants of *Arabidopsis,* cucumber and tomato. *J. Plant Physiol.* 139:119–25
191. Young JC, Liscum E, Hangarter RP. 1992. Spectral dependence of light-inhibited hypocotyl elongation in photomorphogenic mutants of *Arabidopsis:* evidence for a UV-A photosensor. *Planta* 188:106–14
192. Zhong HH, Young AC, Pease EA, Hangarter RP, McClung CR. 1994. Interactions between light and the circadian clock in the regulation of CAT2 expression in *Arabidopsis. Plant Physiol.* 104:889–98

Annu. Rev. Plant Physiol. Plant Mol. Biol. 1996. 47:245–71

DIOXYGENASES: Molecular Structure and Role in Plant Metabolism

Andy G. Prescott

Department of Applied Genetics, John Innes Centre, Norwich Research Park, Colney Lane, Norwich NR4 7UH, United Kingdom

Philip John

Department of Agricultural Botany, Plant Science Laboratories, The University of Reading, Reading RG6 2AS, United Kingdom

KEY WORDS: flavonoid biosynthesis, lipoxygenase, gibberellin biosynthesis, alkaloid biosynthesis, ethylene biosynthesis

ABSTRACT

Dioxygenases are nonheme iron-containing enzymes important in the biosynthesis of plant signaling compounds such as abscisic acid, gibberellins, and ethylene and also of secondary metabolites, notably flavonoids and alkaloids. Plant dioxygenases fall into two classes: lipoxygenases and 2-oxoacid-dependent dioxygenases. The latter catalyze hydroxylation, epoxidation, and desaturation reactions; some enzymes catalyze more than one type of reaction in successive steps in a biosynthetic pathway. This review highlights recent discoveries on both enzyme groups, particularly in relation to gibberellin biosynthesis, in vivo activity of 1-aminocyclopropane-1-carboxylate oxidase, and molecular structure/function relationships. Similarities between the roles of monooxygenases and dioxygenases are also discussed.

CONTENTS

1040-2519/96/0601-0245$08.00

INTRODUCTION

Plant dioxygenases fall into two categories: lipoxygenases (LOXs) and 2-oxoacid-dependent dioxygenases (2-ODDs); the former has been reviewed extensively (42, 109, 125, 137), and the latter has been reviewed by Prescott (121). The purpose of this review is to compare features of plant dioxygenases, set these enzymes within the wider context of oxygenases as a whole, summarize recently gained knowledge in this field, and highlight the role of dioxygenases in plant metabolism.

Oxygenases are enzymes that catalyze the incorporation of O from O_2 into an organic substrate. They commonly use iron (Fe) as a cofactor either as part of a heme group or in a nonheme form. Oxygenases are involved in secondary metabolism and therefore acquire particular importance in plants. There are two classes of oxygenase: dioxygenases in which both O atoms of O_2 are incorporated into the substrate(s), and monooxygenases in which one O atom is incorporated into the substrate while the other is reduced to water. These are precise definitions, applicable to many of the enzymes we discuss, but some oxygenases do not conform to these rules. It has already been noted (121) that the family of dioxygenases that we describe is a more intuitive grouping. A blurring of boundaries between groups of enzymes is not restricted to dioxygenases. Among the cytochrome P-450 monooxygenase superfamily, which is defined by sequence characteristics (107), are two plant enzymes that do not use O_2 and thus are not biochemically "monooxygenases": allene oxide synthase, which uses a fatty acid hydroperoxide as a source of the O atom (145), and berbamunine synthase, which catalyzes the formation of a C-O phenol couple without the concomitant incorporation of an O atom (77).

LIPOXYGENASE (LOX, EC 1.13.11.12)

LOXs are nonheme Fe-containing dioxygenases that catalyze the introduction of O_2 into unsaturated fatty acids that contain a *cis,cis*-1,4-pentadiene moiety (reviewed in 137). LOXs are described by the carbon atom of arachidonic or linoleic acid that is predominantly oxygenated during the reaction. Most plant LOXs are 15-LOXs; however, there are some exceptions (135). The fatty acid hydroperoxides produced by the action of LOX in plants are the precursors of traumatic acid and jasmonic acid (137), though LOX also appears to catalyze co-oxidation of carotenoids to form abscisic acid (reviewed in 125). Products of LOX action are also partially responsible for imparting flavors and aromas and causing the loss of essential nutrients by co-oxidation (125).

Plant LOXs are monomeric enzymes with a molecular mass of 95–100 kDa and are predominantly located in the cytoplasm. LOXs are also associated with vacuoles, mitochondria, chloroplasts, microsomal membranes, and the plasmalemma (89 and references therein). The best characterized plant LOX is that from soybean, which contains a single atom of Fe^{2+} per enzyme molecule. The products of soybean LOX action on linoleic acid are 9-hydroperoxy-10,12-octadecadienoic acid and 13-hydroperoxy-9,11-octadecadienoic acid (13-HPOD). The ratio between these two products varies among the isoenzymes that have been identified. LOX also catalyzes double dioxygenations of fatty acids to form dihydroperoxides and anaerobic reactions generating free radicals (reviewed in 42).

2-OXOACID-DEPENDENT DIOXYGENASES (2-ODDs)

Enzymes in this group have also been called 2-oxoglutarate-dependent dioxygenases (1, 121). They catalyze a range of substrate conversions including hydroxylation, desaturation, and epoxidation, with most enzymes showing an absolute requirement for 2-oxoglutarate as a cosubstrate. In some cases other compounds take over the function of 2-oxoglutarate (29). Thus the term 2-oxoacid-dependent dioxygenase (2-ODD) has been adopted to describe this group (121). All 2-ODDs are soluble enzymes that require Fe^{2+} and ascorbate for optimal substrate conversion in vitro. Some enzymes have an absolute requirement for ascorbate, but many do not. Often the role of ascorbate is undefined and is likely to be indirect and unrelated to the reaction mechanism (34). With most 2-ODDs, addition of catalase to the reaction medium at the relatively high concentrations of around 1 mg/ml optimizes substrate turnover, especially during the later stages of purification. Presumably catalase protects the enzyme from H_2O_2 generated in the reaction medium (8). Table 1 lists the 2-ODDs identified in plants.

The biochemical unity of the 2-ODDs is reflected in the presence of sequence homology. Their sequence is characterized by conserved residues principally clustered in the carboxy-terminal half of the protein (Figure 1; see also 16, 121, 124).

2-ODDs in Flavonoid Biosynthesis

Flavonoid biosynthesis involves oxygenation and desaturation by both soluble 2-ODDs and insoluble P-450-dependent enzymes. Figure 2 illustrates the roles of the 2-ODDs.

Table 1 Plant 2-oxoacid-dependent dioxygenases. 2-OG, 2-oxoglutarate; asc, ascorbate; F3H, flavanone 3β-hydroxylase; FSI, flavone synthase; FLS, flavonol synthase; GA, gibberellin; H6H, hyoscyamine 6β-hydroxylase; D4H, desacetoxyvindoline 4-hydroxylase; P4H, prolyl 4-hydroxylase; ACCO, aminocyclopropane-l-carboxylate oxidase; 4HPPD, 4-hydroxyphenylpyruvate dioxygenase

Enzyme (reaction)	Cofactors	Molecular mass (kDa)		Source for purification	Recombinant expression	References
		Gel	SDS/PAGE			
F3H (hydroxylation)	2-OG, Fe, asc	74	37–39	*P. hybrida*	*E. coli*	15, 17, 18
FLS (desaturation)	2-OG, Fe, asc	—	—	*P. hybrida*	yeast, tobacco[a] *P. hybrida*[a]	60
FSI (desaturation)	2-OG, Fe, asc	48	48	parsley	—	14
GA 7-oxidase (oxidation)	2-OG, Fe, asc?	35	40–50	*C. maxima*	—	82
GA 13-hydroxylase (hydroxylation)	2-OG, Fe, asc	28	—	spinach	—	43
GA 20-oxidase (3 consecutive oxidations)	2-OG, Fe, asc	45	44	*C. maxima*	E. coli	81, 82, 117, 162
GA 3β-hydroxylase (hydroxylation, dehydrogenation)	2-OG?, Fe, asc	58	40–50	*C. maxima*	—	68, 82, 143
GA 2β-hydroxylase (hydroxylation)	2-OG, Fe, asc	44 26, 42	45 —	*P. sativum* *P. vulgaris*	—	144 47

H6H (hydroxylation, epoxidation)	2-OG, Fe, asc	41	38	*H. niger*	E. coli *A. belladonna*	54, 55, 166
D4H (hydroxylation)	2-OG, Fe, asc	45	45	*Catharanthus roseus*	—	25
ACCO (? N-hydroxylation)	Fe, asc, CO_2	39, 40	35–40	various fruit	yeast, *E. coli* *Xenopus,* grape cells, tomato[a]	3, 29, 31, 50, 51 101, 146, 167
Deguelin cyclase? (oxidative cyclization ?)	Fe?	78?	—	*Tephrosia vogellii*	—	23
P4H (hydroxylation)	2-OG, Fe, asc	>300 40, 250	65 60, 65	*P. vulgaris* algae	—	10, 70 and references therein
4HPPD (hydroxylation, decarboxylation)	Fe, asc	—	—	—		132, 133

[a] Designates antisense expression

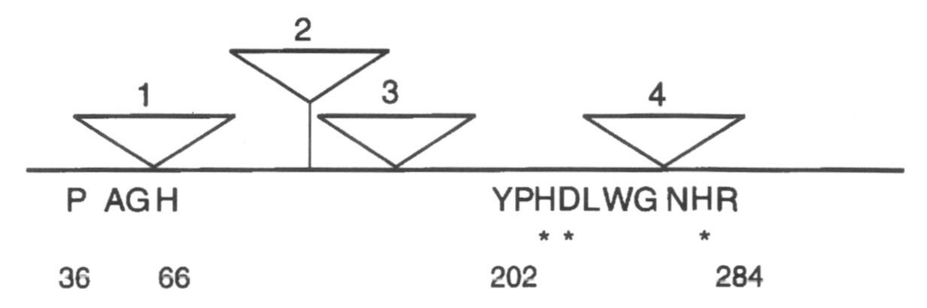

Figure 1 The positions of the four introns (1–4) found in different plant 2-ODDs (see section on Evolution) are marked in relation to the two clusters of conserved amino acid residues. The numbering of the amino acids refers to their position in hyoscyamine 6β-hydroxylase. The three putative Fe ligands are marked (*).

FLAVANONE 3β-HYDROXYLASE (F3H) Initially, purification of F3H was characterized by low recoveries (15, 17), which was attributable to degradation of the enzyme during purification (15) that was probably due to oxidative damage and associated limited proteolysis (148). Genes encoding F3Hs have been cloned from a number of species (16). The coding sequence of one of these clones has been expressed in *E. coli* (18) to yield extracts with specific activities two orders of magnitude higher than those found in plant extracts, making recombinant expression a useful source for purification of large quantities of F3H.

FLAVONE SYNTHASE (FSI) FSI catalyzes the conversion of flavanones to flavones by the introduction of a double bond between the C-2 and C-3 carbons of the C ring (14). This 2-ODD is designated as FSI to distinguish it from FSII. The latter is found in microsomal fractions (150), requires NADPH, and is a P-450-dependent monooxygenase. FSI has been found to date only in parsley (14), whereas FSII has been found in all flavone-containing flowers investigated (150).

FLAVONOL SYNTHASE (FLS) FLS introduces a double bond between the C-2 and C-3 in the C-ring of dihydroflavonols (60). A cDNA encoding FLS has been cloned from petunia. The identity of the cDNA was confirmed by expression in yeast to give FLS activity (60). Reducing FLS by the introduction of genes expressing antisense FLS RNA in petunia and tobacco enhanced the redness of the flowers, owing to a lower flavonol content (60).

ANTHOCYANIDIN SYNTHASE (ANS) The precise reaction catalyzed by this enzyme has not been determined. Mutations in this function have been identified in two species; both mutants are characterized by a loss of anthocyanins (90, 122). The block in anthocyanin biosynthesis appears to occur after dihydroflavonol reductase (90, 122), thus indicating that ANS is involved in the transfor-

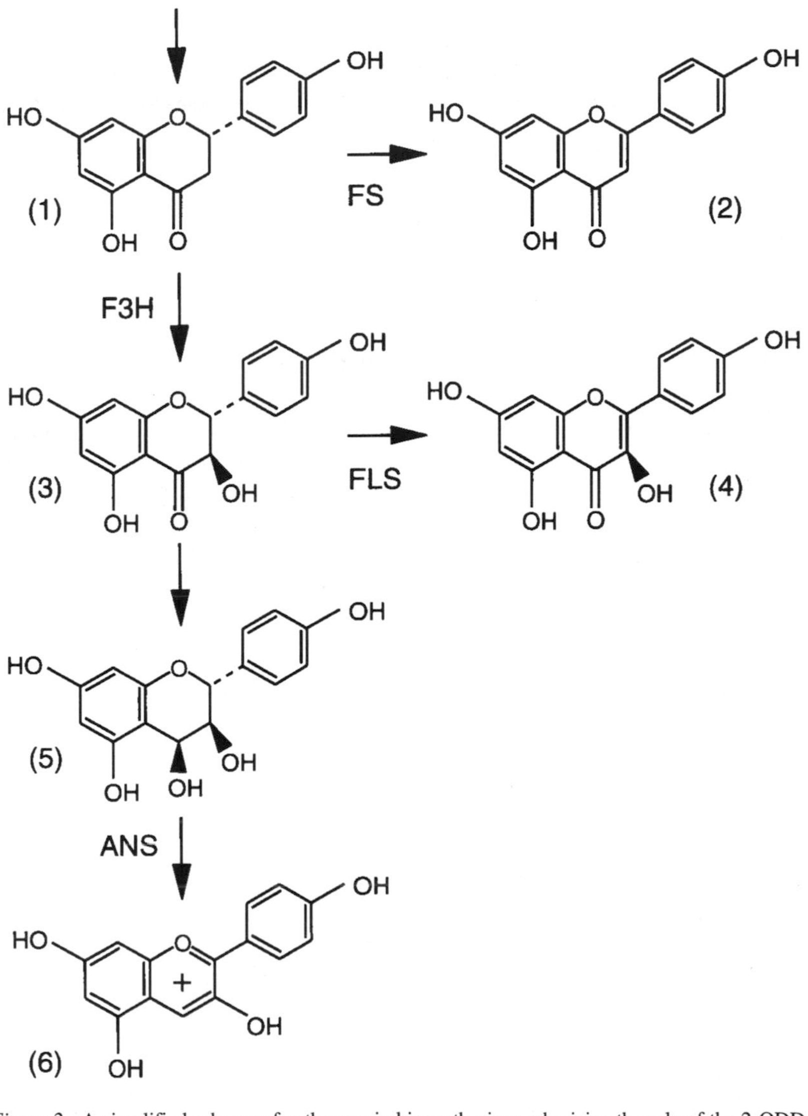

Figure 2 A simplified scheme of anthocyanin biosynthesis emphasizing the role of the 2-ODDs. F3H = flavanone 3β-hydroxylase, FS = flavone synthase I, FLS = flavonol synthase, ANS = anthocyanidin synthase. 1 = flavanone, 2 = flavone, 3 = dihydroflavonol, 4 = flavonol, 5 = leucoanthocyanidin, 6 = anthocyanidin pseudobase.

mation of *cis*-leucoanthocyanidins to anthocyanidins. Genes encoding ANS have been cloned from maize (97), *Antirrhinum* (90), petunia (159), grape (147), and apple (24).

2-ODDs in Gibberellin Biosynthesis

Gibberellins (GAs) control many aspects of plant growth and development (reviewed in 45). Several of the later steps in GA biosynthesis are catalyzed by 2-ODDs (Figure 3).

GIBBERELLIN 7-OXIDASES GA 7-oxidases catalyze the oxidation of the aldehyde group at C-7 of the B ring (e.g. GA_{12}-aldehyde) to a carboxylic acid (e.g. GA_{12}). This reaction is unusual because both a microsomal P-450-dependent monooxygenase (57) and a 2-ODD (56, 82) have been shown to catalyze the conversion in *Cucurbita maxima* endosperm. It is not known whether this enzyme duality occurs in other species. It has been suggested (82) that the soluble 7-oxidase and the microsomal 7-oxidase are part of separate, parallel pathways of GA biosynthesis leading to different GA products. GA_{12} is a major branchpoint within GA biosynthesis because it is the substrate for a number of 2-ODDs. The relative partitioning of GA_{12} between the disparate metabolic fates varies among different species (and possibly among different developmental stages within a given tissue). One of these 2-ODDs catalyzes hydroxylation at C-13; this is another example of a reaction that is catalyzed by a 2-ODD (43, 44) or a microsomal P-450-dependent monooxygenase (57, 67).

GIBBERELLIN 20-OXIDASES GA 20-oxidases catalyze successive oxidations of the C-20 culminating in its removal to form the C-19 GAs. When expressed as recombinant protein in *E. coli*, the 20-oxidase from *C. maxima* catalyzed the conversions $GA_{12} \rightarrow GA_{15} \rightarrow GA_{24} \rightarrow GA_{25}$, GA_9 and, to a lesser extent, the conversions $GA_{53} \rightarrow GA_{44} \rightarrow GA_{19} \rightarrow GA_{17}$, GA_{20} and $GA_{23} \rightarrow GA_{28}$, GA_1 (81). This demonstrated that a single 20-oxidase accepts a range of substrates with different specificities. Recombinant 20-oxidases from *A. thaliana* differed in that two oxidized GA_{12} predominantly to GA_9, whereas a third also produced substantial amounts of GA_{15} and GA_{24}, and all three enzymes oxidized GA_{53} to GA_{44} to a lesser degree (117, 162). *A. thaliana* shoots accumulate high levels of 13-deoxy GAs (152), which may be explained by the relative activities of specific 2-ODDs (117). The multifunctionality of 2-ODDs required for catalyzing sequential steps is also seen with thymine 7-hydroxylase (154) and hyoscyamine 6β-hydroxylase (54), although in the latter case only two reactions, a hydroxylation and epoxidations, are catalyzed.

A multigene family encoding GA 20-oxidases has been identified in *A. thaliana* (117). The three *A. thaliana* 20-oxidases are 64–76% identical to one another and 53% identical to the *C. maxima* cDNA at the amino acid level (117). One of the cDNAs isolated from *A. thaliana* corresponds to the *GA5* locus (152). A point mutation in the gene from this mutant has been identified (162). Three distinct 20-oxidases (GA_{53} oxidase, GA_{44} oxidase, and GA_{19} oxidase) have been purified from young spinach leaves. These enzymes dif-

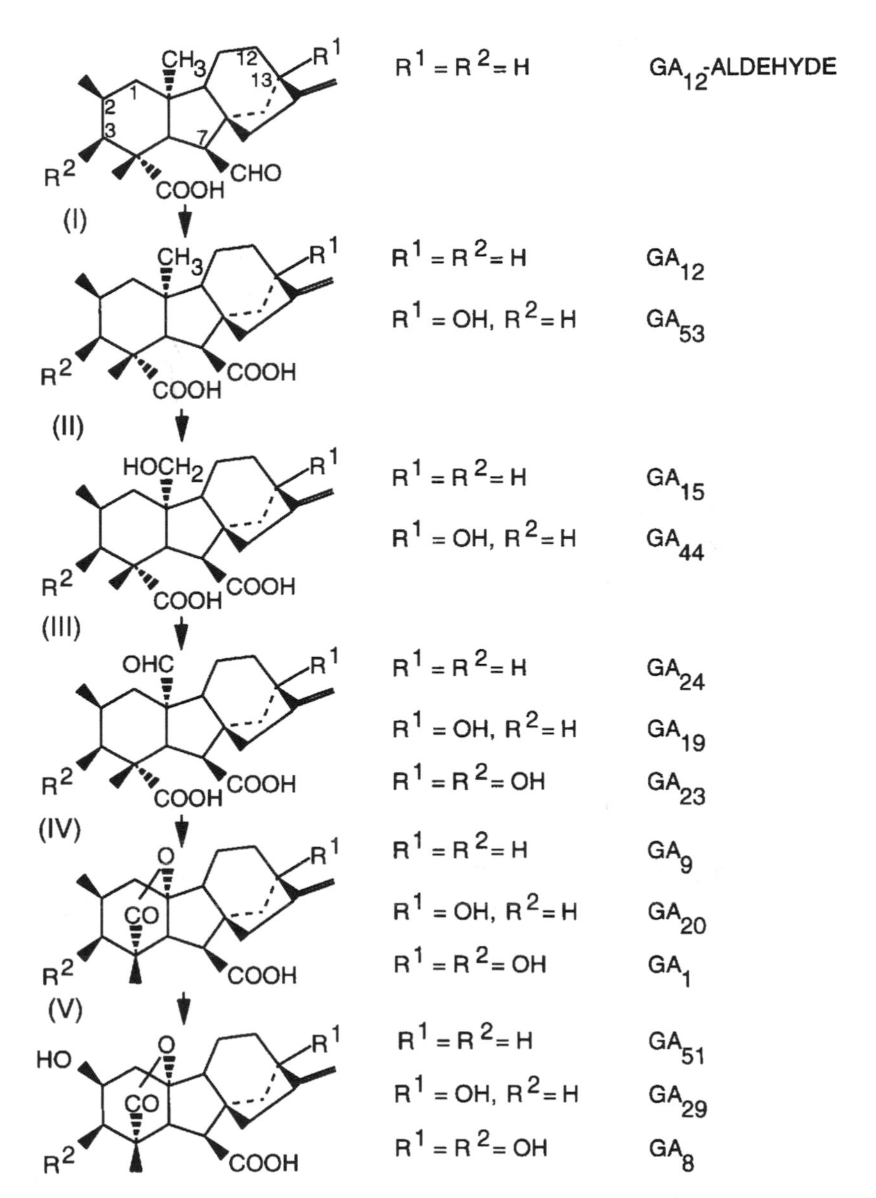

Figure 3 A generalized scheme of gibberellin biosynthesis in higher plants. Reaction I is catalyzed by GA 7-oxidase; reactions II, III, and IV by GA 20-oxidase; and reaction V by GA 2β-hydroxylase. GA 13-hydroxylase acts on R^1, and GA 3β-hydroxylase acts on R^2. Other GAs mentioned in the text are GA_{25}, GA_{17}, and GA_{28}, the tricarboxylic acid derivatives of GA_{24}, GA_{19}, and GA_{23}, respectively, resulting from oxidation of C-20 to a carboxylic acid, and GA_5, GA_6, and GA_3, which are derivatives of GA_{20}.

fered in their photoregulation, apparent molecular mass, and pH optima (43, 44). Thus the regulation of the oxidation at C-20 in GA biosynthesis is extremely complex, involving a number of enzymes that are differentially regulated and that show a range of substrate specificities.

GIBBERELLIN 3β-HYDROXYLASES GA 3β-hydroxylases catalyze the addition of a hydroxyl group to the C-3 position of the A ring. The 3-hydroxylated C-19 GAs are considered the most physiologically active. A partially purified 3β-hydroxylase preparation from *Phaseolus vulgaris* also catalyzed the epoxidation of GA_5 to GA_6 (68). It was impossible to determine whether the epoxidase was a separate 2-ODD which copurified with the 3β-hydroxylase activity or whether a single enzyme catalyzed both reactions. A partially purified GA_{20} 3β-hydroxylase from *P. vulgaris* also catalyzed the dehydrogenation of GA_{20} to GA_5 and the 2β-hydroxylation of GA_{20} to GA_{29} (143). These reports (68, 143) suggest that GA 3β-hydroxylases may be capable of catalyzing several reactions in an analogous manner to GA 20-oxidases, but this remains to be confirmed. A second controversial aspect of the biochemistry of GA 3β-hydroxylases revolves around the use of 2-oxoglutarate as a cosubstrate (79, 82, 143).

The cloning of a cDNA encoding a putative GA 3β-hydroxylase from *A. thaliana* (21) provides an opportunity to investigate the substrate specificities of GA 3β-hydroxylases. A mutation at the *GA4* locus of *A. thaliana* impairs the 3β-hydroxylation of a number of GA substrates (152). The gene corresponding to the *GA4* locus encodes a protein similar to known plant 2-ODDs, which supports the suggestion that the mutation, which results in the alteration of a conserved cysteine residue to a tyrosine, impairs 3β-hydroxylase activity (21). The cDNA has not yet been shown to encode a GA 3β-hydroxylase, but it is anticipated that rapid progress will be made in the understanding of the part 3β-hydroxylases play in the regulation of GA biosynthesis.

GIBBERELLIN 2β-HYDROXYLASES Hydroxylation at the 2β position of the A ring results in the production of physiologically inactive GAs. Two 2β-hydroxylases have been reported in seeds of both *Pisum sativum* (144) and *P. vulgaris* (47), which suggests that GA 2β-hydroxylases may be encoded by multigene families in legume species. Several 2-ODDs involved in the biosynthesis of less well known GAs have been identified, including 12α-hydroxylases from *C. maxima* (80) and enzymes catalyzing the synthesis of GA_3 from GA_5 in *Marah macrocarpus* (142). These may represent "side" reactions of other 2-ODDs involved in GA biosynthesis.

2-ODDs in Alkaloid Biosynthesis

HYOSCYAMINE 6β-HYDROXYLASE (H6H) H6H is responsible for the conversion of hyoscyamine to 6β-hydroxyhyoscyamine and the subsequent epoxidation to scopolamine in tropane alkaloid biosynthesis. Interest in H6H arises in part from the commercial demands for scopolamine, an anticholinergic agent (166). H6H activity is limited to a small number of the Solanaceae; in *Hyoscyamus niger* roots it is localized in the pericycle (53). The enzyme hydroxylates a range of hyoscyamine analogs (55). Confirmation that H6H catalyzes both hydroxylation and epoxidation reactions has come from enzyme inhibition studies with a monoclonal antibody and from expression and analysis of a recombinant protein in *E. coli* (54). Following the cloning of the cDNA for H6H (92) from *H. niger*, it was expressed in *Atropa belladonna* (166), which enabled the leaves to convert hyoscyamine to the commercially more valuable scopolamine. This demonstrates how biosynthetic pathways to medicinal compounds can be usefully manipulated in plants.

DESACETOXYVINDOLINE 4-HYDROXYLASE (D4H) D4H is involved in the synthesis of two drugs of commercial importance, vinblastine and vincristine. Their precursor, vindoline, is synthesized from tabersonine in a series of steps including three hydroxylations; the last of these is catalyzed by D4H (25).

1-Aminocyclopropane-1-Carboxylate Oxidase (ACCO)

ACCO catalyzes the following reaction: ACC + ascorbate + $O_2 \rightarrow C_2H_4$ + CO_2 + HCN + $2H_2O$ + dehydroascorbate, which constitutes the last step in ethylene synthesis by plants. Our discussion concentrates on aspects of ACCO that have emerged since the review by Kende (72). ACCO is an unusual 2-ODD in that it does not use 2-oxoglutarate as a reductant; this function is apparently carried out by ascorbate (29, 140). The evidence for the substrate role of ascorbate comes from a demonstration of 1:1 stoichiometry between ethylene synthesis and dehydroascorbate production (29). Since the initial extraction from melon (157), in vitro ACCO activity has been studied with the enzyme from apple (29, 31, 35, 118, 120), avocado (93), banana (100, 101), pear (158), winter squash (63), and sunflower (36).

CARBON DIOXIDE ACTIVATION CO_2 is an essential cofactor for ACCO activity (29, 120, 138, 139). CO_2 itself and not bicarbonate appears to be the active species (35, 120, 138). Raising the CO_2 concentration from ambient (0.03%) to levels in the range of 5 to 20% not only increases the V_{max} but also increases the apparent K_m toward ACC (35, 36, 99, 120, 138), oxygen (36, 99, 138), and ascorbate (36, 99). In addition, raised CO_2 levels lowered the pH optimum of ACCO about 0.5 pH units to pH 6.5–6.7 (36, 99). The finding that maximal activity of ACCO depends on the concentration of CO_2 has a number of implications. First it implies that early work (31, 93, 157) on ACCO activity employed suboptimal assay conditions. Nevertheless the catalytic turnover number of the ACCO described by Dupille et al (31) at about 1 mol substrate/mol enzyme/min is among the highest reported, even though no CO_2 was added.

This indicates that the potential activity of this ACCO under saturating CO_2 levels is probably an order of magnitude higher than that of other comparable purified ACCOs (29, 101, 117, 167), all of which were assayed in the presence of near-saturating levels of CO_2. The potentially high ACCO activity described by Dupille et al (31) may in part be attributable to the use of 1,10-phenanthroline to chelate Fe during purification of the enzyme, thus avoiding Fe-catalyzed oxidative damage.

Manipulation of CO_2 concentration is used commercially to help in the conservation of fresh produce. An examination of the effect of CO_2 levels on ACCO activity assayed in tissue discs of kiwifruit showed that the apparent K_m for ACC was raised from 15 μM to 72.5 μM and that the V_{max} was almost doubled, when the CO_2 level was increased from 0.2% to 28.5% (127). It was apparent that where the ACC concentration was greater than 55 μM, high levels of CO_2 stimulated ethylene synthesis, whereas at lower concentrations of ACC, high levels of CO_2 inhibited ethylene synthesis (127). Thus the effect of CO_2 is crucially dependent on the internal ACC level. The mechanism by which CO_2 stimulates ACCO activity is unknown. It has been proposed that CO_2 activation of ACCO occurs via carbamylation of an amino group of the enzyme (35). However, interaction with the substrates cannot be ruled out (for example, see 94).

CATALYTIC INACTIVATION Catalytic inactivation is caused by preincubation of the enzyme in the presence of ACC, Fe, and ascorbate (118, 138, 139, 141). Thus inactivation is probably due to catalytic turnover, because omission of any of the reactants prevented its occurrence. A second type of inactivation is observed on incubating ACCO with ascorbate in the absence of ACC and added Fe^{2+} (141). This inactivation was prevented by catalase and was attributed (141) to H_2O_2 generated from the autoxidation of ascorbate by O_2.

ACCO ACTIVITY IN VIVO There are two potential sites of ACCO activity found in vivo, the apoplastic space and the cytosol (12, 126). The relative proportion of ACCO activity attributable to either site varies in different plant tissues (3, 12, 83, 123, 126). It is notable that none of the ACCO cDNAs appears to predict a signal sequence encoding a peptide that would facilitate transport to the apoplast (4, 11, 126).

ACC levels have been estimated in many plant tissues, but the subcellular distribution of the ACC has not been determined; thus, the estimated ACC levels may reflect those of the vacuole, because it is the largest subcellular compartment, though one in which ACCO is unlikely to be located (114, 126). The ascorbate concentration in the apoplast of spinach leaves is about 0.4 mM, whereas that of the intracellular ascorbate is nearer to 4.0 mM (86). Dehydroascorbate is also found in the apoplast but not in significant quantities within plant cells. Electron flow across the plasma membrane from inside to outside can occur via high-potential cytochrome *b* located in the membrane (22, 61, 128), providing a route for an externally orientated flow of electrons driven by the plasma membrane potential, the polarity of which is external positive. Thus an apoplastic ACCO could be supplied with reduced ascorbate via an outward flow of electrons from the cytosol, where ascorbate could be regenerated from its oxidized forms by intracellular glutathione. Regeneration of ascorbate would occur more readily in the cytoplasm than in the apoplast, which is devoid of the reductases of the cytosolic ascorbate-glutathione cycle (37). An attractive feature of this proposal is that it explains the apparent membrane requirement for ACCO activity in vivo (72).

John (66) noted previously that although ACCO is inactive in vitro in the absence of ascorbate, recombinant ACCO expressed in *Saccharomyces cerevisiae* shows significant activity without the addition of ascorbate (50, 160). The concentration of ascorbate found in yeast cells is normally very low (66, 84). Thus, yeast may utilize a different reductant in place of ascorbate. Other 2-ODDs are also active in vivo in the absence of ascorbate, but in these cases ascorbate is not a substrate but has an indirect role. For example, P4H is active in a variety of cultured animal cells, which are devoid of ascorbate (1, 34). In this case the reductant substituting for ascorbate was a component of microsomal membranes (91). 4-hydroxyphenylpyruvate dioxygenase (4HPPD) can be induced in *Pseudomonas* sp. (34), where it is presumably active but where it would not find any ascorbate. In both cases, in vitro activity requires ascorbate, but in vivo activity apparently occurs in its absence.

The H^+ activity in the cytosol is equivalent to a pH of 7.2, whereas the pH of the apoplast varies with the physiological state of the tissue—and from plant to plant—though it generally lies between 5.0 and 6.5 (48). The CO_2-induced acid shift of the pH optimum for ACCO activity (36, 99) indicates that a pH of 6.5 is consistent with maximal rates of ethylene evolution.

ETHYLENE-FORMING ENZYME (EFE) Higher plant ACCOs are related by structure and function to EFE from the plant pathogen, *Pseudomonas syringae* (reviewed in 39). EFE has a molecular mass of 36 kDa, uses O_2 to break down 2-oxoglutarate with the release of ethylene, and requires Fe^{2+}, and its activity is stimulated by catalase (105). EFE differs from ACCO in that it does not require ascorbate. The *Pseudomonas* EFE can also catalyze the oxidative decarboxylation of 2-oxoglutarate to succinate and the formation of guanidine and δ^1-pyrroline 5-carboxylate from arginine. The last reaction probably involves hydroxylation of arginine. Thus EFE shows a remarkable parallel with 2-oxoglutarate-dependent dioxygenases where hydroxylation of the substrate is accompanied by a concomitant oxidative decarboxylation of 2-oxoglutarate to succinate. This parallel has been supported by sequence comparisons of EFE and a variety of 2-ODDs. Heterologous expression of EFE in *E. coli* (38) and in a cyanobacterium (40) opens the way for the development of large-scale processes for the generation of ethylene as an industrial feedstock from organic wastes and from atmospheric CO_2 via photosynthesis, respectively.

The relationship between EFE and ACCOs raises questions concerning the evolutionary origin of ACCO in higher plants. In general, all plants produce and respond to ethylene. However, bryophytes and most pteridophytes do not convert exogenously supplied ACC to ethylene, whereas gymnosperms can do so (112). The acquisition of an ACCO appears to have occurred at an evolutionary stage represented today by the Gnetales (112).

Other 2-ODDs

MUGINEIC ACIDS Cereals respond to a deficiency of available Fe in the soil by secreting Fe-chelating compounds called mugineic acids. These are formed from methionine, which is converted to 2′-deoxymugineic acid; subsequently, in barley two sequential hydroxylation steps give rise to 3-epihydroxymugineic acid and mugineic acid (71, 87). The hydroxylases have not been studied biochemically, but there is indirect evidence that they are 2-ODDs (106, 111). Differential screening of cDNA libraries from Fe-sufficient and Fe-deficient barley roots identified two 2-ODDs induced by Fe-deficiency: *Ids2* (111) and *Ids3* (106). Both clones show root-specific and Fe-deficiency-specific expression, and thus their involvement in the response to Fe-deficiency seems to be clear (106, 111); however, there is no direct evidence yet for their involvement in mugineic acid biosynthesis.

PROLYL 4-HYDROXYLASE (P4H) Vertebrate P4H (EC 1.14.11.2) is the most intensively studied 2-ODD and, as the first 2-ODD to be identified, has long been considered the type member of the group (75). P4H catalyzes the introduction of an hydroxyl group into the prolyl residues of certain peptides that are incorporated into collagens in animals and into cell wall proteins in higher plants and algae. Little work has been reported on plant P4H since the subject was last reviewed (75, 121). P4H isolated from vertebrates is a tetramer composed of 2α and 2β subunits. The Fe-binding site is located mainly on the α subunits (75). The β subunit is identical to the enzyme protein disulphide isomerase (PDI). PDI has been cloned from *Medicago sativa* (136), but it is not known whether it forms an active complex with plant or vertebrate P4H α subunits. Current data indicate that algal P4Hs do not have the 2α2β subunit structure of the vertebrate enzyme (see 121), yet they are functionally alike with similar inhibition by the histidine reagent, diethyl pyrocarbonate (102). Sequences with homology to vertebrate P4H α subunits are present in the databases of plant expressed sequence tags (EST) (A Prescott, personal observation).

E8 The gene encoding the E8 protein (28)—also called pTOM99 (46)—is another that shows high homology to 2-ODDs at the amino acid sequence level but for which biochemical characterization is lacking. E8 is a member of a small multigene family specified by two genetic loci in tomato (74). Expression of E8 shows temporal and spatial control; as a result, the mRNA is abundant in ripening fruit but not detectable in leaves or unripe fruit (85). During fruit ripening, E8 expression coincides with the increase in ethylene synthesis and can be induced by the application of ethylene (28, 85). Antisense experiments have been interpreted to indicate a role for E8 in the regulation of ethylene biosynthesis in the ripening process (115). However, there is increasing evidence that E8 is not directly involved in ethylene synthesis or reception during fruit ripening. For example, the tomato mutant *ripening inhibitor,* which does not synthesize ethylene, has E8 mRNA levels that are 60% of those of the wild-type (reviewed in 46). It now appears more likely that the role of E8 impinges on ethylene-linked processes indirectly and that E8, or its equivalents, function in tissues other than ripening fruits.

4-HYDROXYPHENYLPYRUVATE DIOXYGENASE (4HPPD) 4HPPD has been characterized from vertebrate liver (33) and *Pseudomonas* (129), but little is known of the plant enzyme. However, an activity has been observed in extracts of etiolated corn seedlings (132) and barnyard grass (133). 4HPPD catalyzes the introduction of a second ring hydroxyl group into its substrate and the carboxylation of the side chain to form 2,5-dihydroxyphenylacetate (homogentisate). Unlike in reactions catalyzed by 2-oxoglutarate-dependent dioxygenases, the substrate itself becomes decarboxylated. 4HPPD also requires ascorbate and Fe^{2+} as cofactors. In plants, 4HPPD has been identified as the site of action of a bleaching herbicide (132, 133). ESTs that may encode the subsequent enzyme in the phenylalanine catabolic pathway, homogentisate dioxygenase, have been identified in *Arabidopsis* and *Ricinus communis* (34a).

DEGUELIN CYCLASE Isoflavonoids of the rotenoid group have applications as insecticides. During rotenoid biosynthesis in seeds of *Tephrosia vogellii*, a prenyl group undergoes an oxidative cyclization catalyzed by deguelin cyclase to form deguelin (23). This enzyme has been partially purified (23) but not well characterized biochemically. However, it is possible that it is related to the 2-ODDs because it is soluble, inhibited by 1,10-phenanthroline, does not require NADPH, and is relatively insensitive to inhibitors of P-450-dependent monooxygenases.

There are a number of 2-ODDs that have been described from microorganisms or animals, which would, from our knowledge of plant metabolism, be expected to occur in plants. These include γ-butyrobetaine hydroxylase (BBH, EC 1.14.11.1), ε-*N*-trimethyl-L-lysine hydroxylase (EC 1.14.11.8), thymine

7-hydroxylase (EC 1.14.11.6), thymidine 2′-hydroxylase (EC 1.14.11.3), aspartyl β-hydroxylase (AspH, see references in 121), and proline hydroxylase (5). The sequences of mammalian BBH and AspH are known, but searches of the plant EST databases have not revealed any cDNAs with significant homology to these proteins (A Prescott, unpublished observation). There are sequences of three 2-ODDs in Genbank (release 89), for which the isolation and characterization have not yet been published. Two of these cDNAs (accession numbers X77368 and U21800) isolated from seedling hypocotyls of *Solanum melongena* and shoots of tomato, respectively, are 87% identical at the amino acid level, which suggests that they may encode the same function. Both genes also show a higher than average similarity to H6H (Figure 4).

FEATURES OF DIOXYGENASES

Gene Expression

Plant LOXs are encoded by multigene families. Molecular genetic studies of the expression of LOX in a number of species have shown that individual genes are expressed in a tissue-specific manner and may also be controlled in terms of development and stress responsiveness (2, 6, 32, 95). The complex control of LOX expression probably results from the multiplicity of functions of this enzyme and its fundamental role in lipid metabolism and the synthesis of signaling molecules. Three soybean seed LOX isoenzymes have been shown to have different substrate specificities (reviewed in 137), and recent evidence suggests that individual isoforms may have different functions (88, 116).

Some plant 2-ODDs are encoded by multigene families, e.g. GA 20-oxidase (117), E8 (28), and ACCO (11, 59, 119, 153); others are encoded by single genes, e.g. F3H (26, 27) and H6H (69). The expression of 3 *Arabidopsis* GA 20-oxidases was found to be spatially-regulated (117) and rapidly down-regulated by the application of GA_3, in common with that of a putative GA 3β-hydroxylase (21, 117, 162). The spatial and temporal expression of ACCO transcripts in plant development has been studied in a variety of floral structures (30, 104, 119, 160). ACCO is also expressed in response to wounding (4, 59, 63, 73), senescence (65, 104, 119, 161), abscission, and the application of IAA (113). As yet, unlike LOXs, there is only limited evidence (119) to indicate that specific ACCOs are expressed in response to different stimuli.

Reaction Mechanism

The hydroxylation of organic compounds is favored thermodynamically and is irreversible under normal conditions. It is known that Fenton chemistry involving ascorbate and Fe^{2+}, in the absence of enzymes, can give rise to the

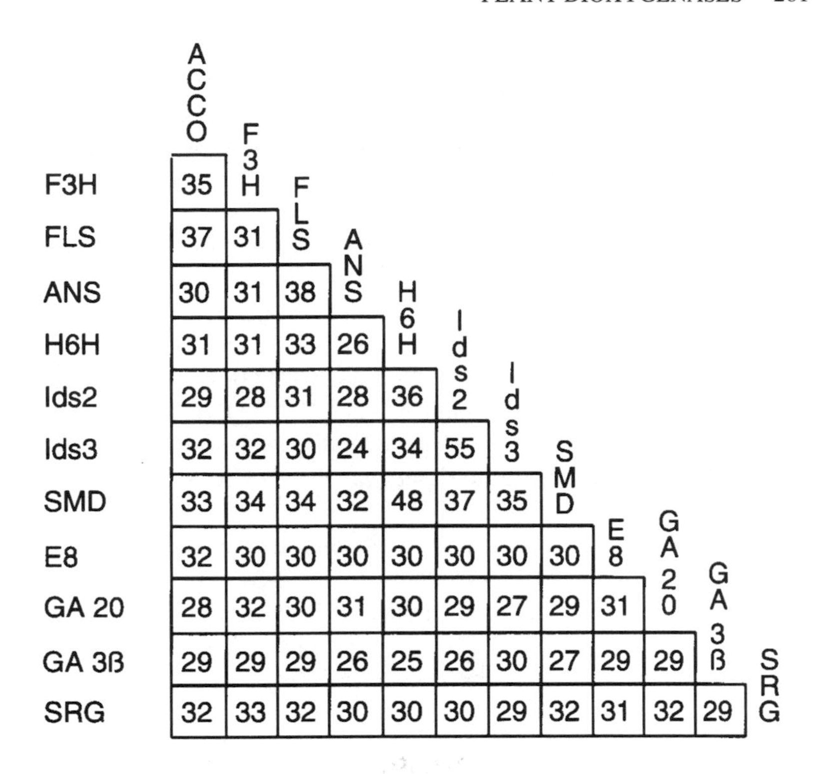

	ACCO	F3H	FLS	ANS	H6H	Ids2	Ids3	SMD	E8	GA20	GA3β	SRG
F3H	35											
FLS	37	31										
ANS	30	31	38									
H6H	31	31	33	26								
Ids2	29	28	31	28	36							
Ids3	32	32	30	24	34	55						
SMD	33	34	34	32	48	37	35					
E8	32	30	30	30	30	30	30	30				
GA 20	28	32	30	31	30	29	27	29	31			
GA 3ß	29	29	29	26	25	26	30	27	29	29		
SRG	32	33	32	30	30	30	29	32	31	32	29	

	ACCO	F3H	FLS	ANS	H6H	Ids2	Ids3	SMD	E8	GA20	GA3β	SRG
WITHIN SPECIES	93		NA			NA	NA	NA	NA	64	NA	67
BETWEEN SPECIES	70	74	NA	47	NA	NA	NA	87	NA	53	NA	NA

Figure 4 A presentation of the degree of sequence identity between plant 2-ODDs, for which either sequences from the same species were compared or, when this was not possible, a sequence from the most closely related species was utilized. Note the variation in conservation of sequence of an enzyme both within a species and between species. F3H = flavanone 3β-hydroxylase, FLS = flavonol synthase, ANS = anthocyanidin synthase, H6H = hyoscyamine 6β-hydroxylase, Ids2 and Ids3 are involved in the response to Fe deficiency in barley, SMD = enzyme of unknown function from *S. melongena*, E8 = enzyme of unknown function from tomato, GA 20 = GA 20-oxidase, GA 3β = GA 3β-hydroxylase, SRG = enzyme of unknown function in *Arabidopsis*, NA = data unavailable. The third cDNA (SRG, accession number X79052) was isolated from a suspension culture of *A. thaliana* and is a member of a small multigene family (A Prescott, unpublished observation). A survey of the *A. thaliana* EST database for 2-ODDs has revealed the presence of at least 10 other 2-ODDs that appear to encode novel enzymes (A Prescott, unpublished observation), which indicates that the family is more diverse than previously anticipated.

hydroxylation of aromatic compounds (155). Moreover, the products of Fenton reactions can be identical to those formed enzymatically in vivo (19). Thus the role of oxygenases is one of harnessing the catalytic capability of Fe, maintained in the reduced state by ascorbate, so that oxygenation takes place efficiently and damaging side reactions caused by the production of H_2O_2 and free radicals are minimized.

The precise mechanism of action of LOX is unknown. Two models have been put forward to explain the initial lag in the reaction rate of Fe^{2+} forms of LOX (Fe^{2+}-LOX, reviewed in 109). The first model proposes that the product of LOX action, 13-HPOD, is required to activate the enzyme by oxidizing the native Fe^{2+} to Fe^{3+} and that only Fe^{3+}-containing LOX (Fe^{3+}-LOX) is active. A second model proposes that both Fe^{2+}- and Fe^{3+}-LOXs are active to differing extents and are inhibited by substrate binding to noncatalytic sites. Recent evidence supports the suggestion that only Fe^{3+}-LOX is active (108, 131, 156). Four reaction mechanisms (109, 130) have been put forward for LOX. None of the proposals involves higher oxidation states of Fe than Fe^{3+}.

In P-450-dependent monooxygenases (9) and 2-ODDs (7) an ($Fe{=}O^{V}$) and an ($Fe{=}O^{IV}$) group respectively are thought to be the oxidizing species capable of inserting an O atom into a carbon-hydrogen bond of the organic substrate. An interesting problem with regard to 2-ODDs is how different types of reaction, e.g. hydroxylation and desaturation, originate from the same reaction mechanism. Hydroxylation reactions are thought to proceed by abstraction of a hydrogen atom from the substrate to yield a Fe^{3+}-hydroxyl complex and a carbon-centered radical that undergo recombination (7). Desaturations have been proposed to involve either a hydroxyl intermediate (58) or a free radical-type mechanism (14, 124). Another feature of 2-ODD reactions is the ability to decarboxylate 2-oxoglutarate in the absence of substrate. This uncoupled activity appears to involve stoichiometric consumption of ascorbate (103).

Protein Structure

Three-dimensional structures have been published for soybean LOX-1 (13, 98) and for the microbial 2-ODD, isopenicillin N synthase (IPNS, 124). Although IPNS shows relatively low sequence identity to plant 2-ODDs, it shares a number of conserved residues/motifs and structural elements and has been proposed as a model for the structure of plant 2-ODDs (124). LOX and IPNS show no overall sequence homology. However, both enzymes have a jellyroll motif, a comparatively rare structure in enzymes but one that is more commonly found in viral capsid proteins (151). In the case of IPNS, the jellyroll forms the core of the enzyme (124), whereas in LOX it is found as a domain separate from that containing the active site (13). The active site Fe is buried within the core of both enzymes. This is probably a requirement for the isolation of highly reactive Fe-O species from the external environment during catalysis (62, 124).

The structure for IPNS (124) was determined using complexed Mn in place of the active-site Fe; the structure around the Mn showed a distorted octahedral geometry with the ligands provided by two water molecules, and by the side chains of 2 histidines, 1 aspartate, and the glutamine next to the carboxy-terminal. The coordination geometry around the Fe center of LOX is presently unresolved. Two structures are proposed, both of which indicate a potential distorted octahedral geometry. Boyington et al (13) identified 3 histidines and the carboxy-terminal isoleucine as ligands, leaving two unoccupied ligand positions. Minor et al (98) identified the side-chain oxygen of an asparagine as a ligand, in addition to those found by Boyington and coworkers. The modification of histidines in the plant 2-ODDs, ACCO, and F3H by diethyl pyrocarbonate, causing abolition of activity, supports their role as Fe ligands (16, 167). Mutagenesis studies on the putative Fe ligands of mammalian 5-LOX have confirmed the identity of the histidine and isoleucine residues (52, 64) but suggest that the asparagine residue is essential for efficient catalytic activity and not for Fe binding (76).

The 3-D structure of IPNS (124) has also allowed examination of a putative leucine zipper motif identified in a number of 2-ODD sequences (11, 16, 72). The putative leucine zipper forms part of a conserved α-helix that lies on the surface of the enzyme. The leucines face the center of the molecule and are therefore not available to interact with other proteins. Roach et al (124) have proposed that the function of this α-helix is to stabilize the distorted jellyroll core of 2-ODDs. This hypothesis does not preclude the possibility that 2-ODDs may be involved in hydrophobic interactions with other proteins; however, such an interaction would not proceed via a leucine zipper.

Evolution

Affiliations between 2-ODDs can be assessed using two independent lines of evidence: namely, degree of sequence similarity and the number and position of introns. This information must be considered in conjunction with biochemical considerations as to when these enzyme activities are likely to have evolved. A guiding principle is that enzyme activities confined to relatively few higher plant species, e.g. H6H, have evolved much later than those widespread throughout the plant kingdom.

ACCO activity is found in present-day gymnosperms (112), while GAs have been identified in ferns (164). Flavones and flavonols have been identified in bryophytes and ferns, while anthocyanins are found in gymnosperms and angiosperms (149). On this basis, we suggest that FSI and F3H were among the earliest 2-ODDs to arise, followed by FLS, and subsequently by enzymes of the GA biosynthetic pathway, and ACCO and ANS. Figure 4 shows the identity between amino acid sequences of plant 2-ODDs. It is important to note that the degree of identity found among isoenzymes within a species varies from enzyme to enzyme. For example, the 3 ACCOs of petunia or tomato are 90% identical at the amino acid level, whereas the 3 GA 20-oxidases of *Arabidopsis* are only 64% identical.

In general, it appears that the amino acid sequences of enzymes with different functions show approximately 27–32% identity. Thus figures higher than this value may indicate a closer evolutionary relationship. For example, there is an obvious relationship between the sequences of H6H and the dioxygenase of unknown function from *Solanum melongena* and a weaker relationship with the *Ids2* and *Ids3* enzymes from barley. Similarly, the FLS and ANS genes of petunia are more similar to each other than either is to the F3H from the same species. Given the order of enzymes in anthocyanin biosynthesis (Figure 2), it is probable that F3H evolved before ANS or FLS; however, on the basis of sequence homology, F3H is no more appropriate as representing a direct progenitor of these two enzymes than other 2-ODDs, despite belonging to the same biosynthetic pathway.

The high degree of conservation of intron position among 2-ODDs has been noted previously (69). Up to 3 introns at 4 varying positions (Figure 1) have been found in different dioxygenase genes. Genes encoding the ACCO (153), H6H (69), Ids2 (111), Ids3 (S Mori, personal communication), and SRG (A Prescott, unpublished data) enzymes have three introns at positions 1, 3, and 4; those encoding GA 20-oxidase (162) and E8 (28) contain 2 introns at positions 3 and 4, and those encoding F3H contain 2 introns at positions 2 and 4 (26). Genomic clones encoding ANS (159) and GA 3β-hydroxylase (21) have not been completely characterized, but both have been reported to contain an intron at the 3 position. Thus intron position, like sequence homology, does not indicate that F3H is a likely progenitor for ANS. These data also suggest that the ancestral 2-ODD had three introns at positions 1, 3, and 4.

LOX activity has been found in higher plants, animals, fungi, algae, and a cyanobacterium (reviewed in 137). Animal and plant LOXs are 26–36% identical at the amino acid level and differ in that mammalian enzymes lack the N-terminal domain (146 residues) found in plants and also contain several internal deletions (13). Plant LOXs generally contain 8 introns (96, 134, 165), whereas mammalian LOXs contain 13 introns (reviewed in 163). The intron positions are not conserved between plant and animal LOXs (41).

CONCLUDING REMARKS

This review has highlighted the importance of LOX and 2-ODDs to secondary metabolism and the synthesis of plant signaling molecules. Recent advances in the study of plant dioxygenases have come about by application of recombinant techniques such as PCR, the use of antisense transgenic plants, high-level heterologous expression in *E. coli*, and initial steps toward the metabolic engineering of medicinal plants. In particular, high-level expression in *E. coli* has made available relatively large amounts of enzyme for both biochemical and structural studies.

The overlapping functions of 2-ODDs, LOX, and P-450-monooxygenases are also noteworthy. The reactions catalyzed by the 2-ODDs, FSI (14), GA 7-oxidase (57), and GA 13-oxidase (43, 57) are also catalyzed by cytochrome P-450 enzymes. There is also evidence that a low-level P-450-dependent activity can generate ethylene from ACC (49, 78). Study of thymine hydroxylase, a 2-ODD from yeast, revealed the reaction to be "strikingly similar" (154) to P-450-dependent systems. Similarly there are P-450 enzymes that hydroxylate arachidonic acid; both LOX and the P-450 enzymes produce mixtures of *R*- and *S*- hydroxyacids (20, 110). Thus, together with their extensive functional complementarity, dioxygenases and monooxygenases also overlap in terms of their catalytic capabilities.

ACKNOWLEDGMENTS

We thank all those authors who provided preprints and reprints, and we thank R. Briddon, R. Casey, R. B. Flavell, C. J. Schofield, P. Thomas, and N. J. Walton for comments on the manuscript. Grant support from Zeneca/BBSRC and the EU is also gratefully acknowledged.

Literature Cited

1. Abbott MT, Udenfriend S. 1974. α-ketoglutarate-coupled dioxygenases. In *Molecular Mechanisms of Oxygen Activation*, ed. O Hayaishi, pp. 167–214. New York: Academic
2. Altschuler M, Grayburn WS, Collins GB, Hildebrand DF. 1989. Developmental expression of lipoxygenases in soybeans. *Plant Sci.* 63:151–58
3. Ayub RA, Rombaldi C, Petitprez M, Latché A, Pech JC, Lelièvre JM. 1993. Biochemical and immunocytological characterization of ACC oxidase in transgenic grape cells. See Ref. 112a, pp. 98–99
4. Balagué C, Watson CF, Turner AJ, Rouge P, Picton S, et al. 1993. Isolation of a ripening and wound-induced cDNA from *Cucumis melo* L. encoding a protein with homology to the ethylene-forming enzyme. *Eur. J. Biochem.* 212:27–34
5. Baldwin JE, Field RA, Lawrence CC, Merritt KD, Schofield CJ. 1993. Proline 4-hydroxylase: stereochemical course of the reaction. *Tetrahedron Lett.* 34:7489–92
6. Bell E, Mullet JE. 1993. Characterization of an *Arabidopsis* lipoxygenase gene responsive to methyl jasmonate and wounding. *Plant Physiol.* 103:1133–37
7. Blanchard JS, Englard S. 1983. γ-butyrobetaine hydroxylase: primary and secondary tritium kinetic isotope effects. *Biochemistry* 22:5922–29
8. Blanchard JS, Englard S, Kondo A. 1982. γ-butyrobetaine hydroxylase: a unique protective effect of catalase. *Arch. Biochem. Biophys.* 219:327–34
9. Bolwell GP, Bozak K, Zimmerlin A. 1994. Plant cytochrome P450. *Phytochemistry* 37:1491–506
10. Bolwell GP, Robbins MP, Dixon RA. 1985.

Elicitor-induced prolyl hydroxylase from French bean (*Phaseolus vulgaris*). *Biochem. J.* 229:693–99
11. Bouzayen M, Cooper W, Barry C, Zegouti H, Hamilton AJ, Grierson D. 1993. EFE multigene family in tomato plants: expression and characterization. See Ref. 112a, pp. 76–81
12. Bouzayen M, Latché A, Pech JC. 1990. Subcellular localization of the sites of conversion of 1-aminocyclopropane-1-carboxylic acid into ethylene in plant cells. *Planta* 180:175–80
13. Boyington JC, Gaffney BJ, Amzel LM. 1993. The three-dimensional structure of an arachidonic acid 15-lipoxygenase. *Science* 260:1482–86
14. Britsch L. 1990. Purification and characterization of flavone synthase I, a 2-oxoglutarate-dependent desaturase. *Arch. Biochem. Biophys.* 282:152–60
15. Britsch L. 1990. Purification of flavanone 3β-hydroxylase from *Petunia hybrida:* antibody preparation and characterization of a chemogenetically defined mutant. *Arch. Biochem. Biophys.* 276:348–54
16. Britsch L, Dedio J, Saedler H, Forkmann G. 1993. Molecular characterization of flavanone 3β-hydroxylases. *Eur. J. Biochem.* 217:745–54
17. Britsch L, Grisebach H. 1986. Purification and characterization of (2*S*)-flavanone 3-hydroxylase from *Petunia hybrida. Eur. J. Biochem.* 156:569–77
18. Britsch L, Ruhnau-Brich B, Forkmann G. 1992. Molecular cloning, sequence analysis, and *in vitro* expression of flavanone 3β-hydroxylase from *Petunia hybrida. J. Biol. Chem.* 267:5380–87
19. Brodie BB, Axelrod J, Shore PA, Udenfriend S. 1954. Ascorbic acid in aromatic hydroxylation II. Products formed by reaction of substrates with ascorbic acid, ferrous ion, and oxygen. *J. Biol. Chem.* 208: 741–50
20. Capdevila J, Yadagiri P, Manna S, Falck JR. 1986. Absolute configuration of the hydroxyeicosatetraenoic acids (HTETEs) formed during catalytic oxygenation of ara chidonic acid by microsomal cytochrome P-450. *Biochem. Biophys. Res. Commun.* 141:1007–11
21. Chiang H-H, Hwang I, Goodman HM. 1995. Isolation of the Arabidopsis *GA4* locus. *Plant Cell* 7:195–201
22. Córdoba F, González-Reyes JA. 1994. Ascorbate and plant cell growth. *J. Bioenerg. Biomembrane.* 26:385–91
23. Crombie L, Rossiter JT, Van Bruggen N, Whiting DA. 1992. Deguelin cyclase, a prenyl to chromen transforming enzyme from *Tephrosia vogellii. Phytochemistry* 31:451–61
24. Davies KM. 1993. A *Malus* cDNA with homology to the *Antirrhinum Candica* and *Zea A2* genes. *Plant Physiol.* 103:1015
25. De Carolis E, De Luca V. 1993. Purification, characterization, and kinetic analysis of a 2-oxoglutarate-dependent dioxygenase involved in vindoline biosynthesis from *Catharanthus roseus. J. Biol. Chem.* 268: 5504–11
26. Deboo GB, Albertsen MC, Taylor LP. 1995. Flavanone 3-hydroxylase transcripts and flavonol accumulation are temporally coordinate in maize anthers. *Plant J.* 7: 703–13
27. Dedio J, Saedler H, Forkmann G. 1995. Molecular cloning of the flavanone 3β-hydroxylase gene (FHT) from carnation (*Dianthus caryophyllus*) and analysis of stable and unstable FHT mutants. *Theor. Appl. Genet.* 90:611–17
28. Deikman J, Fischer RL. 1988. Interaction of a DNA binding factor with the 5′-flanking region of an ethylene-responsive fruit ripening gene from tomato. *EMBO J.* 7: 3315–20
29. Dong JG, Fernández-Maculet JC, Yang SF. 1992. Purification and characterization of 1-aminocyclopropane-1-carboxylate oxidase from apple fruit. *Proc. Natl. Acad. Sci. USA* 89:9789–93
30. Drory A, Mayak S, Woodson WR. 1993. Expression of ethylene biosynthetic pathway mRNAs is spatially regulated within carnation flower petals. *J. Plant Physiol.* 141:663–67
31. Dupille E, Rombaldi C, Lelièvre J-M, Cleyet-Marel J-C, Pech JC, Latché A. 1993. Purification, properties and partial amino-acid sequence of 1-aminocyclopropane-1-carboxylic acid oxidase from apple fruits. *Planta* 190:65–70
32. Eiben HG, Slusarenko AJ. 1994. Complex spatial and temporal expression of lipoxygenase genes during *Phaseolus vulgaris* (L.) development. *Plant J.* 5:123–35
33. Endo F, Awata H, Tanoue A, Ishiguro M, Eda Y, et al. 1992. Primary structure deduced from complementary DNA sequence and expression in cultured cells of mammalian 4-hydroxyphenylpyruvic acid dioxygenase. *J. Biol. Chem.* 267:24235–40
34. Englard S, Seifter S. 1986. The biochemical functions of ascorbic acid. *Annu. Rev. Nutr.* 6:365–406
34a. Fernández-Cañón JM, Peñalva MA. 1995. Molecular characterisation of a gene encoding homogentisate dioxygenase from *Aspergillus nidulans* and identification of its human and plant homologues. *J. Biol. Chem.* 270:21199–205
35. Fernández-Maculet JC, Dong JG, Yang SF. 1993. Activation of 1-aminocyclopropane-1-carboxylate oxidase by carbon dioxide. *Biochem. Biophys. Res. Commun.* 193: 1168–73

36. Finlayson SA, Reid DM. 1994. Influence of CO_2 on ACC oxidase activity from roots of sunflower (*Helianthus annuus*) seedlings. *Phytochemistry* 35:847–51
37. Foyer CH. 1993. Ascorbic acid. In *Antioxidants in Higher Plants,* ed. RG Alscher, JL Hess, pp. 31–58. Boca Raton: CRC
38. Fukuda H, Ogawa T, Ishihara K, Fujii T, Nagahama K, et al. 1992. Molecular cloning in *Escherichia coli,* expression, and nucleotide sequence of the gene for the ethylene-forming enzyme of *Pseudomonas syringae* pv. *phaseolicola* PK2. *Biochem. Biophys. Res. Commun.* 188: 826–32
39. Fukuda H, Ogawa T, Tanase S. 1993. Ethylene production by micro-organisms. *Adv. Microb. Physiol.* 35:275–306
40. Fukuda H, Sakai M, Nagahama K, Fujii T, Matsuoka M, et al. 1994. Heterologous expression of the gene for the ethylene-forming enzyme from *Pseudomonas syringae* in the cyanobacterium *Synechococcus. Biotech. Lett.* 16:1–6
41. Funk CD, Hoshiko S, Matsumoto T, Rådmark O, Samuelsson B. 1989. Characterization of the human 5-lipoxygenase gene. *Proc. Natl. Acad. Sci. USA* 86: 2587–91
42. Gardner HW. 1991. Recent investigations into the lipoxygenase pathway of plants. *Biochim. Biophys. Acta* 1084:221–39
43. Gilmour SJ, Bleecker AB, Zeevaart JAD. 1987. Partial purification of gibberellin oxidases from spinach leaves. *Plant Physiol.* 85:87–90
44. Gilmour SJ, Zeevaart JAD, Schwenen L, Graebe JE. 1986. Gibberellin metabolism in cell-free extracts from spinach leaves in relation to photoperiod. *Plant Physiol.* 82: 190–95
45. Graebe JE. 1987. Gibberellin biosynthesis and control. *Annu. Rev. Plant Physiol.* 38: 419–65
46. Gray J, Picton S, Shabbeer J, Schuch W, Grierson D. 1992. Molecular biology of fruit ripening and its manipulation with antisense genes. *Plant Mol. Biol.* 19:69–87
47. Griggs DL, Hedden P, Lazarus CM. 1991. Partial purification of two gibberellin 2β-hydroxylases from cotyledons of *Phaseolus vulgaris. Phytochemistry* 30: 2507–12
48. Grignon C, Sentenac H. 1991. pH and ionic conditions in the apoplast. *Annu. Rev. Plant Physiol. Plant Mol. Biol.* 42:103–28
49. Grossmann K, Siefert F, Kwiatkowski J, Schraudner M, Langebartels C, Sandermann H Jr. 1993. Inhibition of ethylene production in sunflower cell suspensions by the plant growth retardant BAS 111..W: possible relations to changes in polyamine and cytokinin contents. *J. Plant Growth Regul.* 12:5–11
50. Hamilton AJ, Bouzayen M, Grierson D. 1991. Identification of a tomato gene for the ethylene-forming enzyme by expression in yeast. *Proc. Natl. Acad. Sci. USA* 88:7434–37
51. Hamilton AJ, Lycett GW, Grierson D. 1990. Antisense gene that inhibits synthesis of the hormone ethylene in transgenic plants. *Nature* 346:284–87
52. Hammarberg T, Zhang Y-Y, Lind B, Rådmark O, Samuelsson B. 1995. Mutations at the C-terminal isoleucine and other potential iron ligands of 5-lipoxygenase. *Eur. J. Biochem.* 230:401–7
53. Hashimoto T, Hayashi A, Amano Y, Kohno J, Iwanari H, et al. 1991. Hyoscyamine 6β-hydroxylase, an enzyme involved in tropane alkaloid biosynthesis, is located at the pericycle of the root. *J. Biol. Chem.* 266:4648–53
54. Hashimoto T, Matsuda J, Yamada Y. 1993. Two-step epoxidation of hyoscyamine to scopolamine is catalyzed by bifunctional hyoscyamine 6β-hydroxylase. *FEBS Lett.* 329:35–39
55. Hashimoto T, Yamada Y. 1987. Purification and characterization of hyoscyamine 6β-hydroxylase from root cultures of *Hyoscyamus niger* L. *Eur. J. Biochem.* 164: 277–85
56. Hedden P, Graebe JE. 1982. Cofactor requirements for the soluble oxidases in the metabolism of the C_{20}-gibberellins. *J. Plant Growth Regul.* 1:105–16
57. Hedden P, Graebe JE, Beale MH, Gaskin P, MacMillan J. 1984. The biosynthesis of 12α-hydroxylated gibberellins in a cell-free system from *Cucurbita maxima* endosperm. *Phytochemistry* 23:569–74
58. Heller W, Forkmann G. 1988. Biosynthesis. In *The Flavonoids,* ed. JB Harborne, pp. 399–425. London: Chapman & Hall
59. Holdsworth MJ, Schuch W, Grierson D. 1988. Organisation and expression of a wound/ripening-related small multigene family from tomato. *Plant Mol. Biol.* 11: 81–88
60. Holton TA, Brugliera F, Tanaka Y. 1993. Cloning and expression of flavonol synthase from *Petunia hybrida. Plant J.* 4: 1003–10
61. Horemans N, Asard H, Caubergs RJ. 1994. The role of ascorbate free radical as an electron acceptor to cytochrome *b*-mediated trans-plasma membrane electron transport in higher plants. *Plant Physiol.* 104:1455–58
62. Howard JB, Rees DC. 1991. Perspectives on non-heme iron protein chemistry. *Adv. Protein Chem.* 42:199–280
63. Hyodo H, Hashimoto C, Morozumi S, Hu W, Tanaka K. 1993. Characterization and induction of the activity of 1-aminocyclopropane-1-carboxylate oxidase in the

wounded mesocarp tissue of *Cucurbita maxima*. *Plant Cell Physiol.* 34:667–71
64. Ishii S, Noguchi M, Miyano M, Matsumoto T, Noma M. 1992. Mutagenesis studies on the amino acid residues involved in the iron-binding and the activity of human 5-lipoxygenase. *Biochem. Biophys. Res. Commun.* 182:1482–90
65. John I, Drake R, Farrell A, Cooper W, Lee P, et al. 1995. Delayed leaf senescence in ethylene-deficient ACC-oxidase antisense tomato plants: molecular and physiological analysis. *Plant J.* 7:483–90
66. John P. 1995. Oxidation of 1-aminocyclopropane-1-carboxylic acid (ACC) in the generation of ethylene by plants. In *Amino Acids and Their Derivatives in Higher Plants,* ed. RM Wallsgrove, pp. 51–58. Cambridge: Cambridge Univ. Press
67. Kamiya Y, Graebe JE. 1983. The biosynthesis of all major pea gibberellins in a cell-free system from *Pisum sativum. Phytochemistry* 22:681–89
68. Kamiya Y, Kwak S-S. 1990. Partial characterization of the gibberellin 3β-hydroxylase from immature seeds of *Phaseolus vulgaris.* In *Gibberellins,* ed. N Takahashi, BO Phinney, J MacMillan, pp. 72–81. New York: Springer-Verlag
69. Kanegae T, Kajiya H, Amano Y, Hashimoto T, Yamada Y. 1994. Species-dependent expression of the hyoscyamine 6β-hydroxylase gene in the pericycle. *Plant Physiol.* 105:483–90
70. Kaska DD, Myllylä R, Günzler V, Gibor A, Kivirikko KI. 1988. Prolyl 4-hydroxylase from *Volvox carteri. Biochem. J.* 256: 257–63
71. Kawai S, Itoh K, Takagi S. 1993. Incorporation of ^{15}N and ^{14}C of methionine into the mugineic acid family of phytosiderophores in iron-deficient barley roots. *Physiol. Plant.* 88:668–74
72. Kende H. 1993. Ethylene biosynthesis. *Annu. Rev. Plant Physiol. Plant Mol. Biol* 44:283–307
73. Kim WT, Yang SF. 1994. Structure and expression of cDNAs encoding 1-aminocyclopropane-1-carboxylate oxidase homologs isolated from excised mung bean hypocotyls. *Planta* 194:223–29
74. Kinzer SM, Schwager SJ, Mutschler MA. 1990. Mapping of ripening-related or -specific cDNA clones of tomato (*Lycopersicon esculentum*). *Theor. Appl. Genet.* 79: 489–96
75. Kivirikko KI, Myllylä R, Pihlajaniemi T. 1992. Hydroxylation of proline and lysine residues in collagens and other animal and plant proteins. In *Post-translational Modifications of Proteins*, ed. JJ Harding, MJC Crabbe, pp. 1–51. Boca Raton: CRC
76. Kramer JA, Johnson KR, Dunham WR, Sands RH, Funk MO Jr. 1994. Position 713 is critical for catalysis but not iron binding in soybean lipoxygenase 3. *Biochemistry* 33:15017–22
77. Kraus PFX, Kutchan TM. 1995. Molecular cloning and heterologous expression of a cDNA encoding berbamunine synthase, a C-O phenol-coupling cytochrome P450 from the higher plant *Berberis stolonifera. Proc. Natl. Acad. Sci. USA* 92: 2071–75
78. Kraus TE, Murr DP, Hofstra G, Fletcher RA. 1992. Modulation of ethylene synthesis in acotyledonous soybean and wheat seedlings. *J. Plant Growth Regul.* 11: 47–53
79. Kwak S-S, Kamiya Y, Sakurai A, Takahashi N, Graebe JE. 1988. Partial purification and characterization of gibberellin 3β-hydroxylase from immature seeds of *Phaseolus vulgaris* L. *Plant Cell Physiol.* 29:935–43
80. Lange T, Hedden P, Graebe JE. 1993. Biosynthesis of 12α- and 13-hydroxylated gibberellins in a cell-free system from *Cucurbita maxima* endosperm and the identification of new endogenous gibberellins. *Planta* 189:340–49
81. Lange T, Hedden P, Graebe JE. 1994. Expression cloning of a gibberellin 20-oxidase, a multifunctional enzyme involved in gibberellin biosynthesis. *Proc. Natl. Acad. Sci. USA* 91:8552–56
82. Lange T, Schweimer A, Ward DA, Hedden P, Graebe JE. 1994. Separation and characterisation of three 2-oxoglutarate-dependent dioxygenases from *Cucurbita maxima* L. endosperm involved in gibberellin biosynthesis. *Planta* 195:98–107
83. Latché A, Dupille E, Rombaldi C, Cleyet-Marel JC, Lelièvre JM, Pech JC. 1993. Purification, characterization and subcellular localization of ACC oxidase from fruits. See Ref. 112a, pp. 39–45
84. Leung CT, Loewus FA. 1985. Concerning the presence and formation of ascorbic acid in yeasts. *Plant Sci.* 38:65–69
85. Lincoln JE, Fischer RL. 1988. Diverse mechanisms for the regulation of ethylene-inducible gene expression. *Mol. Gen. Genet.* 212:71–75
86. Luwe MWF, Takahama U, Heber U. 1993. Role of ascorbate in detoxifying ozone in the apoplast of spinach (*Spinacia oleracea* L.) leaves. *Plant Physiol.* 101: 969–76
87. Ma JF, Nomoto K. 1993. Two related biosynthetic pathways of mugineic acids in Gramineous plants. *Plant Physiol.* 102: 373–78
88. Maccarrone M, van Aarle PGM, Veldink GA, Vliegenthart JFG. 1994. *In vitro* oxygenation of soybean biomembranes by lipoxygenase-2. *Biochim. Biophys. Acta* 1190:164–69

89. Macrì F, Braidot E, Petrussa E, Vianello A. 1994. Lipoxygenase activity associated to isolated soybean plasma membranes. *Biochim. Biophys. Acta* 1215:109–14
90. Martin C, Prescott A, Mackay S, Bartlett J, Vrijlandt E. 1991. Control of anthocyanin biosynthesis in flowers of *Antirrhinum majus. Plant J.* 1:37–49
91. Mata JM, Assad R, Peterkofsky B. 1981. An intramembranous reductant which participates in the proline hydroxylation reaction with intracisternal prolyl hydroxylase and unhydroxylated procollagen in isolated microsomes from L-929 cells. *Arch. Biochem. Biophys.* 206:93–104
92. Matsuda J, Okabe S, Hashimoto T, Yamada Y. 1991. Molecular cloning of hyoscyamine 6β-hydroxylase, a 2-oxoglutarate-dependent dioxygenase, from cultured roots of *Hyoscyamus niger. J. Biol. Chem.* 266:9460–64
93. McGarvey DJ, Christoffersen RE. 1992. Characterization and kinetic parameters of ethylene-forming enzyme from avocado fruit. *J. Biol. Chem.* 267:5964–67
94. McRae DG, Coker JA, Legge RL, Thompson JE. 1983. Bicarbonate/CO_2-facilitated conversion of 1-aminocyclopropane-1-carboxylic acid to ethylene in model systems and intact tissues. *Plant Physiol.* 73:784–90
95. Melan MA, Dong X, Endara ME, Davis KR, Ausubel FM, Peterman TK. 1993. An *Arabidopsis thaliana* lipoxygenase gene can be induced by pathogens, abscisic acid, and methyl jasmonate. *Plant Physiol.* 101: 441–50
96. Melan MA, Nemhauser JL, Peterman TK. 1994. Structure and sequence of the *Arabidopsis thaliana* lipoxygenase 1 gene. *Biochim. Biophys. Acta* 1210:377–80
97. Menssen A, Höhmann S, Martin W, Schnable PS, Peterson PA, et al. 1990. The En/Spm transposable element of *Zea mays* contains splice sites at the termini generating a novel intron from a dSpm element in the *A2* gene. *EMBO J.* 9:3051–57
98. Minor W, Steczko J, Bolin JT, Otwinowski Z, Axelrod B. 1993. Crystallographic determination of the active site iron and its ligands in soybean lipoxygenase L-1. *Biochemistry* 32:6320–23
99. Mizutani F, Dong JG, Yang SF. 1995. Effect of pH on CO_2-activated 1-aminocyclopropane-1-carboxylate oxidase activity from apple fruit. *Phytochemistry* 39: 751–55
100. Moya-León MA, John P. 1994. Activity of 1-aminocyclopropane-1-carboxylate (ACC) oxidase (ethylene-forming enzyme) in the pulp and peel of ripening bananas. *J. Hort. Sci.* 69:243–50
101. Moya-León MA, John P. 1995. Purification and biochemical characterization of 1-aminocyclopropane-1-carboxylate oxidase from banana fruit. *Phytochemistry* 39: 15–20
102. Myllylä R, Günzler V, Kivirikko KI, Kaska DD. 1992. Modification of vertebrate and algal prolyl 4-hydroxylases and vertebrate lysyl hydroxylase by diethyl pyrocarbonate. *Biochem. J.* 286:923–27
103. Myllylä R, Majamaa K, Günzler V, Hanauske-Abel HM, Kivirikko KI. 1984. Ascorbate is consumed stoichiometrically in the uncoupled reactions catalyzed by prolyl 4-hydroxylase and lysyl hydroxylase. *J. Biol. Chem.* 259:5403–5
104. Nadeau JA, Zhang XS, Nair H, O'Neill SD. 1993. Temporal and spatial regulation of 1-aminocyclopropane-1-carboxylate oxidase in the pollination-induced senescence of orchid flowers. *Plant Physiol.* 103: 31–39
105. Nagahama K, Ogawa T, Fujii T, Tazaki M, Tanase S, et al. 1991. Purification and properties of an ethylene-forming enzyme from *Pseudomonas syringae* pv. *phaseolicola* PK2. *J. Gen. Microb.* 137:2281–86
106. Nakanishi H, Okumura N, Umehara Y, Nishizawa N-K, Chino M, Mori S. 1993. Expression of a gene specific for iron deficiency (*Ids3*) in the roots of *Hordeum vulgare. Plant Cell Physiol.* 34:401–10
107. Nebert DW, Adesnik M, Coon MJ, Estabrook RW, Gonzales FJ, et al. 1987. The P450 gene superfamily: recommended nomenclature. *DNA* 6:1–11
108. Nelson MJ, Chase DB, Seitz SP. 1995. Photolysis of "purple" lipoxygenase: implications for the structure of the chromophore. *Biochemistry* 34:6159–63
109. Nelson MJ, Seitz SP. 1994. The structure and function of lipoxygenase. *Curr. Opin. Struct. Biol.* 4:878–84
110. Nikolaev V, Reddanna P, Whelan J, Hildenbrandt G, Reddy CC. 1990. Stereochemical nature of the products of linoleic acid oxidation catalyzed by lipoxygenases from potato and soybean. *Biochem. Biophys. Res. Commun.* 170:491–96
111. Okumura N, Nishizawa N-K, Umehara Y, Ohata T, Nakanishi H, et al. 1994. A dioxygenase gene (*Ids2*) expressed under iron deficiency conditions in the roots of *Hordeum vulgare. Plant Mol. Biol.* 25:705–19
112. Osborne DJ. 1989. The control of ethylene in plant growth and development. In *Biochemical and Physiological Aspects of Ethylene Production in Lower and Higher Plants,* ed. H Clijsters, M De Proft, R Marcelle, M van Poucke, pp. 1–11. Dordrecht: Kluwer
112a. Pech JC, Latché A, Balagué C, eds. 1993. *Cellular and Molecular Aspects of the Plant Hormone Ethylene*. Dordrecht: Kluwer
113. Peck SC, Kende H. 1995. Sequential induc-

tion of the ethylene biosynthetic enzymes by indole-3-acetic acid in etiolated peas. *Plant Mol. Biol.* 28:293–301
114. Peck SC, Reinhardt D, Olson DC, Boller T, Kende H. 1992. Localization of the ethylene-forming enzyme from tomatoes, 1-aminocyclopropane-1-carboxylate oxidase, in transgenic yeast. *J. Plant Physiol.* 140:681–86
115. Peñarrubia L, Aguilar M, Margossian L, Fischer RL. 1992. An antisense gene stimulates ethylene hormone production during tomato fruit ripening. *Plant Cell* 4:681–87
116. Peng Y-L, Shirano Y, Ohta H, Hibino T, Tanaka K, Shibata D. 1994. A novel lipoxygenase from rice. *J. Biol. Chem.* 269: 3755–61
117. Phillips AL, Ward DA, Uknes S, Appleford NEJ, Lange T, et al. 1995. Isolation and expression of three gibberellin 20-oxidase cDNA clones from *Arabidopsis*. *Plant Physiol.* 108:1049–57
118. Pirrung MC, Kaiser LM, Chen J. 1993. Purification and properties of the apple fruit ethylene-forming enzyme. *Biochemistry* 32:7445–50
119. Pogson BJ, Downs CG, Davies KM. 1995. Differential expression of two 1-aminocyclopropane-1-carboxylic acid oxidase genes in broccoli after harvest. *Plant Physiol.* 108:651–57
120. Poneleit LS, Dilley DR. 1993. Carbon dioxide activation of 1- aminocyclopropane-1-carboxylate (ACC) oxidase in ethylene synthesis. *Postharvest Biol. Technol.* 3: 191–99
121. Prescott AG. 1993. A dilemma of dioxygenases (or where biochemistry and molecular biology fail to meet). *J. Exp. Bot.* 44:849–61
122. Reddy GM, Coe EH Jr. 1962. Inter-tissue complementation: a simple technique for direct analysis of gene-action sequence. *Science* 138:149–50
123. Reinhardt D, Kende H, Boller T. 1994. Subcellular localization of 1-aminocyclopropane-1-carboxylate oxidase in tomato cells. *Planta* 195:142–46
124. Roach PL, Clifton IJ, Fülöp V, Harlos K, Barton GJ, et al. 1995. Crystal structure of isopenicillin *N* synthase is the first from a new structural family of enzymes. *Nature* 375:700–4
125. Robinson DS, Wu Z, Domoney C, Casey R. 1995. Lipoxygenases and the quality of foods. *Food Chem.* 54:33–43
126. Rombaldi C, Lelièvre J-M, Latché A, Petitprez M, Bouzayen M, Pech J-C. 1994. Immunocytolocalization of 1-aminocyclopropane-1-carboxylic acid oxidase in tomato and apple fruit. *Planta* 192:453–60
127. Rothan C, Nichols J. 1994. High CO_2 levels reduce ethylene production in kiwifruit. *Physiol. Plant.* 92:1–8
128. Rubinstein B. 1994. The action of ascorbate in vesicular systems. *J. Bioenerg. Biomembr.* 26:385–91
129. Rüetschi U, Odelhög B, Lindstedt S, Barros-Söderling J, Persson B, Jörnvall H. 1992. Characterisation of 4-hydroxyphenylpyruvate dioxygenase: primary structure of the *Pseudomonas* enzyme. *Eur. J. Biochem.* 205:459–66
130. Scarrow RC, Trimitsis MG, Buck CP, Grove GN, Cowling RA, Nelson MJ. 1994. X-ray spectroscopy of the iron site in soybean lipoxygenase-1: changes in coordination upon oxidation or addition of methanol. *Biochemistry* 33:15023–35
131. Schilstra MJ, Veldink GA, Vliegenthart JFG. 1994. The dioxygenation rate in lipoxygenase catalysis is determined by the amount of iron (III) lipoxygenase in solution. *Biochemistry* 33:3974–79
132. Schulz A, Ort O, Beyer P, Kleinig H. 1993. SC-0051, a 2-benzoyl-cyclohexane-1,3-dione bleaching herbicide is a potent inhibitor of the enzyme *p*-hydroxyphenylpyruvate dioxygenase. *FEBS Lett.* 318:162–66
133. Secor J. 1994. Inhibition of barnyardgrass 4-hydroxyphenylpyruvate dioxygenase by sulcotrione. *Plant Physiol.* 106:1429–33
134. Shibata D, Kato T, Tanaka K. 1991. Nucleotide sequences of a soybean lipoxygenase gene and the short intergenic region between an upstream lipoxygenase gene. *Plant Mol. Biol.* 16:353–59
135. Shimizu T, Rådmark O, Samuelsson B. 1984. Enzyme with dual lipoxygenase activities catalyzes leukotriene A_4 synthesis from arachidonic acid. *Proc. Natl. Acad. Sci. USA* 81:689–93
136. Shorrosh BS, Dixon RA. 1991. Molecular cloning of a putative plant endomembrane protein resembling vertebrate protein disulphide-isomerase and a phosphatidylinositol-specific phospholipase C. *Proc. Natl. Acad. Sci. USA* 88:10941–45
137. Siedow JN. 1991. Plant lipoxygenase: structure and function. *Annu. Rev. Plant Physiol. Plant Mol. Biol* 42:145–88
138. Smith JJ, John P. 1993. Activation of 1-aminocyclopropane-1-carboxylate oxidase by bicarbonate/carbon dioxide. *Phytochemistry* 32:1381–86
139. Smith JJ, John P. 1993. Maximising the activity of the ethylene-forming enzyme. See Ref. 112a, pp. 33–38
140. Smith JJ, Ververidis P, John P. 1992. Characterization of the ethylene-forming enzyme partially purified from melon. *Phytochemistry* 31:1485–94
141. Smith JJ, Zhang ZH, Schofield CJ, John P, Baldwin JE. 1994. Inactivation of 1-aminocyclopropane-1-carboxylate (ACC) oxidase. *J. Exp. Bot.* 45:521–27
142. Smith VA, Albone KS, MacMillan J. 1990. Enzymatic 3β-hydroxylation of gibberel-

lins A_{20} and A_5. In *Gibberellins,* ed. N Takahashi, BO Phinney, J MacMillan, pp. 62–71. New York: Springer-Verlag

143. Smith VA, Gaskin P, MacMillan J. 1990. Partial purification and characterization of the gibberellin A_{20} 3β-hydroxylase from seeds of *Phaseolus vulgaris. Plant Physiol.* 94:1390–401
144. Smith VA, MacMillan J. 1986. The partial purification and characterisation of gibberellin 2β-hydroxylases from seeds of *Pisum sativum. Planta* 167:9–18
145. Song W-C, Funk CD, Brash AR. 1993. Molecular cloning of an allene oxide synthase: a cytochrome P450 specialized for the metabolism of fatty acid hydroperoxides. *Proc. Natl. Acad. Sci. USA* 90: 8519–23
146. Spanu P, Reinhardt D, Boller T. 1991. Analysis and cloning of the ethylene-forming enzyme from tomato by functional expression of its mRNA in *Xenopus laevis* oocytes. *EMBO J.* 10:2007–13
147. Sparvoli F, Martin C, Scienza A, Gavazzi G, Tonelli C. 1994. Cloning and molecular analysis of structural genes involved in flavonoid and stilbene biosynthesis in grape (*Vitis vinifera* L.). *Plant Mol. Biol.* 24: 743–55
148. Stadtman ER, Oliver CN. 1991. Metal-catalyzed oxidation of proteins. *J. Biol. Chem.* 266:2005–8
149. Stafford HA. 1991. Flavonoid evolution: an enzymic approach. *Plant Physiol.* 96: 680–85
150. Stotz G, Forkmann G. 1981. Oxidation of flavanones to flavones with flower extracts of *Antirrhinum majus* (snapdragon). *Z. Naturforsch.* 36c:737–41
151. Stuart D. 1993. Viruses. *Curr. Opin. Struct. Biol.* 3:167–74
152. Talon M, Koornneef M, Zeevaart JAD. 1990. Endogenous gibberellins in *Arabidopsis thaliana* and possible steps blocked in the biosynthetic pathways of the semidwarf *ga4* and *ga5* mutants. *Proc. Natl. Acad. Sci. USA* 87:7983–87
153. Tang X, Wang H, Brandt AS, Woodson WR. 1993. Organization and structure of the 1-aminocyclopropane-1-carboxylate oxidase gene family from *Petunia hybrida. Plant Mol. Biol.* 23:1151–64
154. Thornburg LD, Lai M-T, Wishnok JS, Stubbe J. 1993. A non-heme iron protein with heme tendencies: an investigation of the substrate specificity of thymine hydroxylase. *Biochemistry* 32:14023–33
155. Udenfriend S, Clark CT, Axelrod J, Brodie BB. 1954. Ascorbic acid in aromatic hydroxylation I: a model system for aromatic hydroxylation. *J. Biol. Chem.* 208:731–39
156. van der Heijdt LM, Schilstra MJ, Feiters MC, Nolting H-F, Hermes C, et al. 1995. Changes in the iron coordination sphere of Fe(II) lipoxygenase-1 from soybeans upon binding of linoleate or oleate. *Eur. J. Biochem.* 231:186–91
157. Ververidis P, John P. 1991. Complete recovery *in vitro* of ethylene-forming enzyme activity. *Phytochemistry* 30:725–27
158. Vioque B, Castellano JM. 1994. Extraction and biochemical characterization of 1-aminocyclopropane-1-carboxylic acid oxidase from pear. *Physiol. Plant.* 90:334–38
159. Weiss D, van der Luit AH, Kroon JTM, Mol JNM, Kooter JM. 1993. The petunia homologue of the *Antirrhinum majus candi* and *Zea mays A2* flavonoid genes; homology to flavanone 3-hydroxylase and ethylene-forming enzyme. *Plant Mol. Biol.* 22: 893–97
160. Wilson ID, Zhu Y, Burmeister DM, Dilley DR. 1993. Apple ripening-related cDNA clone pAP4 confers ethylene-forming ability in transformed *Saccharomyces cerevisiae. Plant Physiol.* 102:783–88
161. Woodson WR, Park KY, Drory A, Larsen PB, Wang H. 1992. Expression of ethylene biosynthetic pathway transcripts in senescing carnation flowers. *Plant Physiol.* 99: 526–32
162. Xu Y-L, Li L, Wu K, Peeters AJM, Gage DA, Zeevaart JAD. 1995. The *GA5* locus of *Arabidopsis thaliana* encodes a multifunctional gibberellin 20-oxidase: molecular cloning and functional expression. *Proc. Natl. Acad. Sci. USA* 92:6640–44
163. Yamamoto S. 1992. Mammalian lipoxygenases: molecular structures and functions. *Biochim. Biophys. Acta* 1128:117–31
164. Yamane H, Fujioka S, Spray CR, Phinney BO, MacMillan J, et al. 1988. Endogenous gibberellins from sporophytes of two tree ferns, *Cibotium glaucum* and *Dicksonia antarctica. Plant Physiol.* 86:857–62
165. Yenofsky RL, Fine M, Liu C. 1988. Isolation and characterization of a soybean (*Glycine max*) lipoxygenase-3 gene. *Mol. Gen. Genet.* 211:215–22
166. Yun D-J, Hashimoto T, Yamada Y. 1992. Metabolic engineering of medicinal plants: transgenic *Atropa belladonna* with an improved alkaloid composition. *Proc. Natl. Acad. Sci. USA* 89:11799–803
167. Zhang Z, Schofield CJ, Baldwin JE, Thomas P, John P. 1995. Expression, purification and characterization of 1-aminocyclopropane-1-carboxylate oxidase from tomato in *Escherichia coli. Biochem. J.* 307: 77–85

Annu. Rev. Plant Physiol. Plant Mol. Biol. 1996. 47:273–98

PHOSPHO*ENOL*PYRUVATE CARBOXYLASE: A Ubiquitous, Highly Regulated Enzyme in Plants

Raymond Chollet

Department of Biochemistry, University of Nebraska-Lincoln, Lincoln, Nebraska 68588-0664

Jean Vidal

Institut de Biotechnologie des Plantes, UA CNRS D-1128, Université de Paris-Sud, 91405 Orsay Cedex, France

Marion H. O'Leary

Department of Biochemistry, University of Nebraska-Lincoln, Lincoln, Nebraska 68588-0664

KEY WORDS: PEP carboxylase (PEPC); catalytic reaction mechanism; regulatory protein phosphorylation; gene structure, expression, and evolution

ABSTRACT

Since plant phospho*enol*pyruvate carboxylase (PEPC) was last reviewed in the *Annual Review of Plant Physiology* over a decade ago (O'Leary 1982), significant advances have been made in our knowledge of this oligomeric, cytosolic enzyme. This review highlights this exciting progress in plant PEPC research by focusing on the three major areas of recent investigation: the enzymology of the protein; its posttranslational regulation by reversible protein phosphorylation and opposing metabolite effectors; and the structure, expression, and molecular evolution of the nuclear PEPC genes. It is hoped that the next ten years will be equally enlightening, especially with respect to the three-dimensional structure of the plant enzyme, the molecular analysis of its highly regulated protein-Ser/Thr kinase, and the elucidation of its associated signal-transduction pathways in various plant cell types.

1040-2519/96/0601-0273$08.00

CONTENTS

INTRODUCTION

Phospho*enol*pyruvate carboxylase (PEPC; EC 4.1.1.31) is a ubiquitous cytosolic enzyme in higher plants and is also widely distributed in bacteria, cyanobacteria, and green algae (68, 114). It catalyzes the irreversible β-carboxylation of phospho*enol*pyruvate (PEP) in the presence of HCO_3^- and Me^{2+} to yield oxaloacetate (OAA) and Pi and thus is involved intimately in C_4-dicarboxylic acid metabolism in plants. Besides its cardinal roles in the initial fixation of atmospheric CO_2 during C_4 photosynthesis and Crassulacean acid metabolism (CAM), PEPC functions anaplerotically in a variety of nonphotosynthetic systems such as C/N partitioning in C_3 leaves, seed formation and germination, and fruit ripening (66, 68). Nonphotosynthetic isoforms of PEPC also play specialized roles in guard-cell C metabolism during stomatal opening (90) and plant host–cell C_4-acid formation in N_2-fixing legume root nodules (19, 115).

Since 1982, when PEPC was reviewed last in the *Annual Review of Plant Physiology* (87), many new and significant findings about this oligomeric enzyme have been made. In addition to the further elucidation of its catalytic reaction mechanism and the initiation of structure-function analyses by site-directed mutagenesis, there has been an explosion in research related to the posttranslational regulation of the enzyme's activity and allosteric properties by reversible protein phosphorylation and to PEPC gene (*Ppc*) structure, expression, and molecular evolution. These exciting new advances in plant PEPC research are the primary focus of this review, while only limited reference will be made to the microbial enzyme. The interested reader should consult earlier reviews on PEPC for additional breadth and detail (3, 19, 35, 54, 68, 84, 87, 90, 96, 104, 114, 117).

ENZYMOLOGY OF PEP CARBOXYLASE

Comments on Isolation of PEPC

It is now amply documented that native leaf and recombinant forms of PEPC are highly susceptible to limited proteolysis near the N-terminus during extraction and subsequent purification (6, 9, 23, 77, 82, 121). While such modification has no major influence on the enzyme's electrophoretic mobility, V_{max}, and carbon-isotope effects, removal of this plant-invariant N-terminal domain markedly decreases the in vitro phosphorylatability and sensitivity of PEPC to its negative allosteric effector L-malate. Thus, it is our view that many earlier kinetic analyses of purified or commercial plant PEPC have probably been compromised by this N-terminal truncation (see comments in 23, 54, 68). More recent studies have preserved the enzyme's integrity during isolation by the inclusion of glycerol, L-malate, and proteinase inhibitors (especially chymostatin) and by the use of rapid purification protocols that exploit fast-protein liquid chromatography, HPLC, or immunochromatography (4, 7, 9, 23, 58, 77, 82, 119, 121, 136). With such strategies, preparations of intact, N-blocked leaf (C_4, CAM, C_3), nodule, and recombinant PEPC are readily obtained.

Carboxylation and Hydrolysis of PEP Analogs

A variety of PEP analogs have been examined as substrates for C_4 PEPC (see 35 and Table 1). Although a number of compounds are processed by the enzyme, most are not carboxylated but instead are hydrolyzed to pyruvate

Table 1 Activity of PEP analogs with PEPC[a]

Compound	V_{max} (rel)[b]	% Carboxylation	% Hydrolysis	References
PEP	100	97	3	6
(*E*)-3-fluoro-PEP	5	86	14	32, 50
(*Z*)-3-fluoro-PEP	5	3	97	32, 50
(*Z*)-3-chloro-PEP	25	25	75	71
Alleno-PEP	90	0	100	126
Thio-PEP	9	0	100	103
(*Z*)-3-methyl-PEP	4	0	100	31, 33, 34, 86
(*Z*)-3-bromo-PEP	25	0	100	21
3,3-dimethyl-PEP	2	0	100	31

[a]Values given are for the maize leaf enzyme in the presence of Mg^{2+}.

[b]Carboxylation plus hydrolysis.

derivatives (Equation 1) by a mechanism that shares several steps with catalysis (see section on Catalytic Mechanism of PEPC). This phosphatase activity is probably not related to the much slower bicarbonate-independent hydrolysis of phosphoglycolate and phospholactate that is also catalyzed by the enzyme (48, 50).

$$R-CH=\overset{\overset{\displaystyle O-PO_3^{2-}}{|}}{C}-CO_2^- + HCO_3^- \rightarrow {}^-O_2C-CHR-\overset{\overset{\displaystyle O}{\|}}{C}-CO_2^- + R-CH_2-\overset{\overset{\displaystyle O}{\|}}{C}-CO_2^- + Pi \qquad 1.$$

PEP itself also undergoes a few percent of an HCO_3^--dependent pyruvate formation. This hydrolysis is a minor component of the overall reaction flux with Mg^{2+} under in vivo conditions (<5%), but it increases with other metal ions and constitutes over 50% of the total reaction flux when Ni^{2+} is used (6). Interestingly, the PEP analog in which the phosphate has been replaced by a sulfate is not a substrate for the enzyme and, in fact, this compound does not bind to the active site [but it is a substrate for pyruvate kinase (93)].

Functional analogs for CO_2 and HCO_3^- are rare in enzymatic reactions. In the case of PEPC, HCO_2^- can replace HCO_3^-, forming formyl-P and pyruvate at a rate that is about 1% of that for PEP carboxylation (48).

Kinetic and Isotopic Studies

Early thinking about the catalytic mechanism was dominated by the seminal observation of Maruyama et al (72) [recently confirmed by O'Leary & Hermes (88)] that ^{18}O-labeled HCO_3^- gives products containing one equivalent of ^{18}O in Pi and two in the γ-carboxyl of OAA. This isotope transfer persists with a number of other substrates, including those that undergo hydrolysis rather than carboxylation. (*Z*)-3-methyl-PEP gives more than one equivalent of ^{18}O in Pi and also gives ^{18}O incorporation into reisolated starting material after partial reaction (29, 86). A similar phenomenon is observed with 3-fluoro-PEP; exchange is eight times faster than substrate consumption (50). These observations indicate that the initial steps in the carboxylation mechanism are reversible (see section on Catalytic Mechanism of PEPC).

PEPC has been subjected to a variety of kinetic studies over the years, but these have generally been qualitative in nature because investigators failed to rigorously control HCO_3^- concentrations and to account for the presence of PEP-metal complexes. Recent studies of initial velocity patterns varying the two substrates and Me^{2+} indicate that there is a high level of synergism in the binding of substrates (49). Mg^{2+} binds first, and this binding is at equilibrium; PEP binds second; HCO_3^- binds third; and all three have to be present before the reaction begins.

The small carbon-isotope effect (k^{12}/k^{13} = 1.003) that accompanies the carboxylation of PEP by PEPC has been of interest in connection with studies of isotope fractionation in plants (28). The carbon-isotope fractionation by PEPC is independent of the phosphorylation state of the enzyme and the presence or absence of the N-terminal phosphorylation domain, and nearly independent of pH (50, 89, 124; P Paneth & S Madhavan, unpublished data). This fractionation is small compared to what would be expected if C-C bond formation were simply rate determining. Instead, some step prior to C-C bond formation must be rate limiting.

The oxygen-isotope effect for the bridging oxygen of PEP is large (k^{16}/k^{18} = 1.0056) when the HCO_3^- concentration is low, but the value decreases to 0.994 at high [HCO_3^-], consistent with the ordered stepwise mechanism given below (30). Deuterium-isotope effects for PEP-3,3-d_2 are 0.94 on V and 0.95 on V/K, also consistent with the stepwise mechanism (D Arnelle & MH O'Leary, unpublished data).

Carbon-isotope effects on the (*E*) and (*Z*) isomers of 3-fluoro-PEP provide an interesting contrast (50). The (*E*) isomer has a small carbon-isotope effect (1.009), consistent with rate-determination phosphate transfer. However, the (*Z*) isomer [which mostly gives hydrolysis rather than carboxylation (Table 1)] shows a large isotope effect (1.049), which is apparently associated with the loss of CO_2 from the complex during catalysis.

Several stereochemical probes have been used to define PEPC catalysis. Early work by Rose et al (97) demonstrated that carboxylation of PEP occurs on the si face of the substrate, and carboxylation of the two isomers of 3-fluoro-PEP occurs on the same face (50). When (*Z*)-3-methyl-PEP is hydrolyzed by PEPC in D_2O, the 3-D-α-ketobutyrate that is produced is racemic, which indicates that protonation of the enolate occurs in solution rather than on the surface of the enzyme (33). The stereochemistry of substitution at phosphorus can be determined by using PEP containing S, ^{16}O, and ^{17}O in nonbridging positions of the phosphate ester. Carboxylation in $H_2{}^{18}O$ produces a chiral thiophosphate with inversion of configuration at phosphorus (39). Thus, substitution at phosphorus occurs by an in-line mechanism.

Active-Site Structure

Mn-EPR studies of PEPC with PEP and various substrate analogs suggest that PEP itself is bidentate coordinated to the metal. Metal coordination in the enolate intermediate is to the enolate oxygen, the carboxyl oxygen, and a phosphate oxygen (5).

Results of chemical modification studies on various plant PEPCs with group-selective reagents have suggested that Cys, His, Arg, and Lys are essential for activity (3, 96, 104). To date, only one such residue has been identified in the plant primary structure—Lys-606 in maize PEPC (57). Furthermore, the complete absence of Cys in PEPC from *Thermus* sp., a thermophilic bacterium, excludes the direct involvement of these residues in catalysis (79a). Site-directed mutagenesis studies of the active-site domain of PEPC have thus far been performed only with the enzyme from *Escherichia coli*. His-138 (*E. coli* numbering) is required for carboxylation, but the mutant H138N is able to catalyze PEP hydrolysis to pyruvate in the presence of HCO_3^- (109, 112). His-579 is not obligatory for catalysis, in spite of the fact that it is species-invariant (111). Replacement of conserved Arg-587 by Ser also gives an enzyme that catalyzes hydrolysis, but not carboxylation (112, 134). Figure 1 indicates these targeted, species-invariant Lys, His, and Arg residues in the deduced primary structure of *Sorghum* C_4 PEPC.

Along with site-directed mutagenesis, X-ray crystallography has become the sine qua non of enzymology. Alas, PEPC does not yet seem to have yielded to the efforts of crystallographers. The *E. coli* enzyme has been reported to give crystals that diffract X-rays (46). We are also aware of attempts in other laboratories to obtain diffraction-quality crystals of recombinant PEPC from various plant sources, but no substantial progress in this area has been reported.

Catalytic Mechanism of PEPC

The information cited above permits presentation of a relatively convincing mechanism for action of PEPC (Figure 2). Substrates and Me^{2+} bind in the preferred order metal, PEP, HCO_3^-. The first chemical step is phosphate transfer to form carboxyphosphate and the enolate of pyruvate, as perhaps first suggested by Walsh (118). Stereochemical studies require that the transition state for this step is linear at phosphorus; thus, the carbonyl carbon in the intermediate carboxyphosphate following transfer is quite far from carbon-3 of the enolate, and a conformational change is required to place the two carbons near each other. The most parsimonious way to accomplish this is to have an enzyme base deprotonate the carboxyl group of carboxyphosphate, after which carboxyphosphate decomposes to form enzyme-bound CO_2 and Pi. Earlier mechanisms (87) did not recognize this aspect. This step brings CO_2 above the plane of the enolate and within bonding distance of its carbon-3. CO_2 in this intermediate is sequestered so that under optimum catalytic conditions it seldom escapes [3% (6)], but under other circumstances CO_2 is lost easily, as when the metal ion is changed in such a way as to lower the reactivity of the enolate. In the case of a variety of PEP analogs, loss of CO_2 competes effectively with carboxylation (cf Table 1). In some cases, the formation of enzyme-bound CO_2 must be reversible. Isotope exchange studies on (*Z*)-3-methyl-PEP (29, 86) and 3-fluoro-PEP (50) require that CO_2 is formed reversibly and can scramble isotopes and return to starting material. It is not clear

1 *
MASERHHSIDAQLRALAPGKVSEELIQYDALLVDRFLDILQDLHGPSLREFVQECYEVSADYEGKKDTSKLGELG

76
AKLTGLAPADAILVASSILHMLNLANLAEEVELAHRRRNSKLKHGDFSDEGSATTESDIEETLKRLVSLGKTPAE

151 •
VFEALKNQSVDLVFTAHPTQSARRSLLQKNARIRNCLTQLSAKDVTVEDKKELDEALHREIQAAFRTDEIRRAQP

226 1
TPQDEMRYGMSYIHETVWNGVPKFLRRVDTALKNIGINERLPYDVPLIKFCSWMGGDRDGNPRVTPEVTRDVCLL

301 2
SRMMAANLYINQVEDLMFELSMWRCNDELRARAEEVQSTPASKKVTKYYIEFWKQIPPNEPYRVILGAVRDKLYN

376 3 4 5
TRERARHLLATGFSEISEDAVFTKIEEFLEPLELCYKSLCECGDKAIADGSLLDLLRQVFTFGLSLVKLDIRQES

451
ERQTDVIDAITTHLGIGSYRSWPEDKRMEWLVSELKGKRPLLPPDLPMTEEIADVIGAMRVLAELPIDSFGPYII

526 •
SMCTAPSDVLAVELLQREMWHSPAVPVVPLFERLADLQAAPASVEKLFSTDWYINHINGKQQVMVGYSDSGKDAG

601 • •
RLSAAWQLYVAQEEMAKVAKKYGVKLTLFHGRGGTVGRGGGPTHLAILSQPPDTINGSIRVTVQGEVIEFMFGEE

676
NLCFQSLQRFTAATLEHGMHPPVSPKPEWRKLMEEMAVVATEEYRSVVVKEPRFVEYFRSATPETEYGKMNIGSR

751
PAKRRPGGGITTLRAIPWIFSWTQTRFHLPVWLGVGAAFKWAIDKDIKNFQKLKEMYNEWPFFRVTLDLLEMVFA

826
KGDPGIAGLYDELLVAEELKPFGKQLRDKYVETQQLLLQIAGHKDILEGDPYLKQGLRLRNPYITTLNVFQAYTL

901 960
KRIRDPSFKVTPQPPLSKEFADENKPAGLVKLNGERVPPGLEDTLILTMKGIAAGMQNTG

Figure 1 Deduced amino-acid sequence of the C_4-PEPC isoform from *Sorghum* (67). The plant-invariant phosphorylation domain, with its target Ser (*), is underlined twice, whereas the species-invariant functional regions identified to date are underlined singly. The specific His, Lys, and Arg residues targeted by site-directed mutagenesis or chemical modification (see text) are indicated within these three domains (•). C (1–5), plant-invariant Cys residues.

whether CO_2 formation is reversible in the case of PEP. Isotope-effect results suggest that it is not.

In the final chemical step of the overall reaction, CO_2 combines with the metal-stabilized enolate. It is interesting to note that there is no evidence that

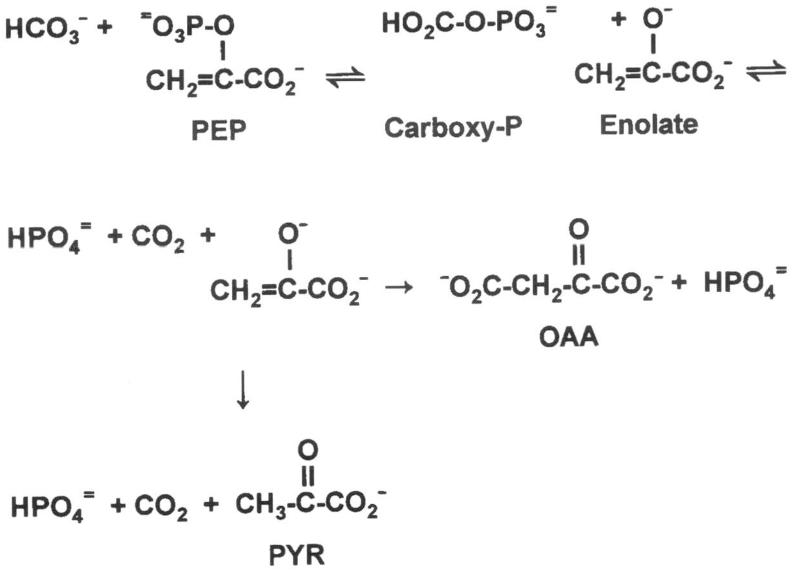

$$HCO_3^- + {}^{=}O_3P\text{-}O\text{-}C(=CH_2)\text{-}CO_2^- \;(\text{PEP}) \rightleftharpoons HO_2C\text{-}O\text{-}PO_3^{=} \;(\text{Carboxy-P}) + {}^-O\text{-}C(=CH_2)\text{-}CO_2^- \;(\text{Enolate}) \rightleftharpoons$$

$$HPO_4^{=} + CO_2 + {}^-O\text{-}C(=CH_2)\text{-}CO_2^- \rightarrow {}^-O_2C\text{-}CH_2\text{-}C(=O)\text{-}CO_2^- \;(\text{OAA}) + HPO_4^{=}$$

$$\downarrow$$

$$HPO_4^{=} + CO_2 + CH_3\text{-}C(=O)\text{-}CO_2^- \;(\text{PYR})$$

Figure 2 Mechanism of carboxylation and hydrolysis of PEP by PEPC. PYR, pyruvate.

this step is reversible, even though the reverse reaction (decarboxylation of metal-chelated OAA) is well known in other systems. Finally, Pi and OAA are released.

POSTTRANSLATIONAL REGULATION OF PEP-CARBOXYLASE ACTIVITY

It is well documented that the activity of the various isoforms of plant PEPC are subject to allosteric control by a variety of positive [e.g. glucose 6-P (G6P), triose-P] and negative (e.g. L-malate, Asp) metabolite effectors, especially when assayed at suboptimal pH values that approximate that of the cytosol (e.g. 3, 23, 65, 66, 68, 90, 100, 101, 104). For example, the K_i(L-malate) of the intact recombinant C_4 enzyme from *Sorghum* is decreased about 25-fold at pH 7.3 compared with pH 8.0 (23). Although changes in the cytosolic levels of these opposing allosteric effectors and H^+ likely contribute to the overall regulation of PEPC activity in vivo (22, 26, 54, 66, 90), research over the past decade has focused primarily on the reversible phosphorylation of the enzyme. In fact, cytosolic PEPC and sucrose-P synthase presently represent the two best-defined examples of in vivo regulatory enzyme phosphorylation in plants (40a, 41).

Regulatory Phosphorylation of Photosynthetic PEPC

The regulatory phosphorylation of photosynthetic PEPC has been intensively studied and recently reviewed (41, 54, 68, 84, 96, 117) since the initial observations were published about ten years ago on the CAM and C_4 isoforms (10, 38, 51, 80–82). As an important prelude to these protein-phosphorylation studies, several reports had appeared that indicated that both photosynthetic PEPC isoforms were subject to a striking diel regulation in vivo that altered the enzyme's activity and/or sensitivity to L-malate under near-physiological assay conditions, without accompanying changes in V_{max} or PEPC amount (e.g. 42, 60, 81, 125). It thus became evident that the CAM enzyme was upregulated at night and downregulated during the day, thereby paralleling the classical changes in CAM physiology (e.g. leaf atmospheric CO_2 fixation and titratable acidity) (66). Related investigations of several CAM plants under *continuous* night or day conditions indicated that CAM physiology, as well as the L-malate sensitivity of PEPC, was controlled by an endogenous circadian rhythm rather than by light or dark signals per se (83, 125). In marked contrast, C_4 PEPC was shown to be reversibly light activated in vivo by a mechanism that was dependent, either directly or indirectly, on photosynthesis and modulated by the incident photosynthetic photon flux density above a minimum threshold of about 300 μmol m^{-2} s^{-1} (7, 36, 55, 60, 78, 98).

It is now established unequivocally by a wealth of in vivo and in vitro data that this striking diel regulation is caused by changes in the phosphorylation state of a single serine residue near the ~110-kDa subunit's N-terminus (e.g. Ser-8 and Ser-15 in the *Sorghum* and maize C_4 enzymes, respectively, and Ser-11 in PEPC from the facultative CAM plant *Mesembryanthemum crystallinum*) (9, 23, 53, 58, 110, 121). Upregulation/phosphorylation of the target enzyme is catalyzed by a highly regulated (see below) protein kinase and downregulation/dephosphorylation by a typical mammalian-type protein phosphatase 2A (7, 11, 12, 12a, 27, 52, 53, 55, 56, 78). It is notable that this target Ser resides in a plant-invariant motif [E/DR/KxxSIDAQL/MR (see Figure 1)] that is absent in the bacterial and cyanobacterial primary structures deduced to date (67, 68, 79a, 96a, 114). Moreover, in vitro studies with the intact, recombinant *Sorghum* C_4 enzyme have established that phosphorylation of this N-terminal domain not only renders PEPC considerably less sensitive to inhibition by L-malate under near-physiological assay conditions (~sevenfold increase in K_i) but, conversely, both more active and more sensitive to activation by G6P (~fivefold decrease in K_a) (23, 26). Thus, this reversible means of fine tuning the activity and allosteric properties of PEPC is unique to the plant enzyme.

The molecular mechanism by which protein phosphorylation regulates C_4 PEPC has recently been addressed by site-directed mutagenesis and chemical modification. The introduction of a monoanionic residue at position 8 in the recombinant *Sorghum* enzyme by directed mutagenesis (S8D) or sequential mutagenesis (S8C) and *S*-carboxymethylation functionally mimics the specific effects of regulatory phosphorylation on the target enzyme. In contrast, various neutral substitutions (S8T, S8Y, S8C, *S*-carboxamidomethylated S8C) are without major influence (23, 25, 121; GB Maralihalli, V Pacquit, B Li, JA Jiao, G Sarath, et al, unpublished data). Consequently, addition of negative charge to this N-terminal domain by reversible phosphorylation appears crucial to this regulatory mechanism, but the exact details must await the high-resolution crystal structures of the dephospho and phospho (or S8D) enzyme-forms.

Recent research on the phosphorylation of C_4 and CAM PEPC has focused on the physiologically relevant protein kinase and its requisite signal-transduction chain. This work took on special significance with the near-simultaneous discoveries that the C_4 and CAM PEPC kinases were both activated reversibly in vivo by some mechanism involving cytosolic protein turnover, thereby resulting in the upregulation of the kinase and, thus, its target enzyme in the light (C_4) or at night (CAM) (7, 12, 27, 55, 56, 78). Not only is the CAM kinase activated at night under the control of a circadian rhythm, but it is also coinduced with its protein-substrate during C_3 to CAM switching in the facultative CAM species *M. crystallinum* (12, 12a, 70). In contrast, the activity state of the type 2A PEPC-phosphatase catalytic subunit appears to be relatively constant during light-dark (C_4) or day-night (CAM) transitions (12, 56, 78), further underscoring the critical role of the kinase in the PEPC-phosphorylation cycle.

Following the initial report by Jiao & Chollet (52), the extremely low-abundance PEPC kinase has been partially purified about 4000-fold. It is likely to be a monomer of ~37/30-kDa (C_4) or ~39/32-kDa (CAM) polypeptides (69, 70, 120). As isolated, this protein kinase catalyzes neither autophosphorylation nor the phosphorylation of heterologous substrates (e.g. casein, histone III-S, BSA, leaf sucrose-P synthase). Similarly, position-8 *Sorghum* C_4-PEPC mutants (e.g. S8Y, S8D, S8C) are not phosphorylated except for the Thr substitution (70, 120, 121; GB Maralihalli, V Pacquit, B Li, JA Jiao, G Sarath, et al, unpublished data). In contrast, all plant PEPC isoforms examined to date (C_4, CAM, C_3-leaf, root nodule) serve as substrates in vitro (70, 119, 120), but with a distinct preference for the corresponding PEPC kinase (B Li, XQ Zhang & R Chollet, unpublished data). Considerable effort has been expended to (re)investigate the Ca^{2+}-dependency of this protein kinase. It is our view that although a variety of other protein-Ser/Thr kinases, including C_4-leaf calmodulin-like domain protein kinase (CDPK) and mammalian protein kinase A, specifically phosphorylate the single target Ser in plant PEPC in vitro (7, 53, 69, 85, 110), *only* the Ca^{2+}-independent, 30- to 39-kDa PEPC kinase has been shown to be light-dark (C_4) or day-night (CAM) regulated in vivo (69, 70). Notably, these differential activity states of the kinase are maintained throughout chromatography on various matrices and even following SDS-PAGE and subsequent renaturation (69; B Li & R Chollet, unpublished data). Thus, PEPC kinase is likely up/downregulated in vivo by some mechanism that modulates its amount (7, 12, 56, 69) or else by covalent modification rather than by some noncovalent means (e.g. regulatory subunit, tight-binding effector). Repeated attempts to demonstrate an effect of in vitro dephosphorylation by alkaline phosphatase on the activity states of the light (active) and dark (inactive) C_4 kinase and its component ~37/30-kDa polypeptides have proven unsuccessful (B Li & R Chollet, unpublished data).

The signal-transduction chains that impinge upon the highly regulated PEPC kinases are also a focus of current research. Initial studies using a chemical inhibitor-based approach with detached leaves (7, 12, 55, 56, 69, 78) have been supplanted by in situ analyses with isolated C_4 mesophyll cells and protoplasts and cell biology techniques (24a, 30a, 94, 117). It is now established that the light-induced C_4 transduction cascade is initiated in the illuminated chloroplast by photosynthesis and likely involves some "signal" from the light-activated Calvin cycle in the neighboring bundle sheath, possibly 3-P-glycerate (Figure 3). In addition, there is mounting in situ evidence for the involvement of increases in mesophyll-cytosol pH and [Ca^{2+}], the latter perhaps modulating an upstream protein kinase (24a, 30a, 94, 117), together with the inhibitor-based data that implicate a key role for a cytosolic protein-synthesis event (7, 8, 30a, 56, 69, 94). In contrast, not much is known about the CAM PEPC kinase signal-transduction pathway other than its light independency and the involvement of a circadian rhythm and cytosolic protcin turnover (Figure 3) (12, 12a, 83, 84). Clearly, this area would benefit from detailed in situ analyses of intact mesophyll protoplasts isolated from night and day leaves performing CAM.

Finally, the results from leaf CO_2-exchange studies have underscored the impact of the PEPC regulatory-phosphorylation cycle on C_4 photosynthesis and dark CO_2 fixation during CAM (8, 12, 12a). For example, when the activity states of PEPC kinase and, thus, its target enzyme were downregulated in vivo by short-term pretreatment with cytosolic protein-synthesis inhibitors in the light (C_4) or prior to the night period (CAM), net leaf CO_2 uptake was diminished markedly. In contrast, no effects were observed on the activation states of other nuclear-encoded, photosynthesis-related enzymes, stomatal conductance, or CO_2 uptake by a C_3 leaf (7, 8, 56). Thus, the phosphorylation of photosynthetic PEPC is a cardinal regulatory event that influences atmospheric CO_2 fixation; this mechanism enables this primary carboxylase to function in the leaf cytosol even in the presence of the millimolar levels of C_4 acids (e.g. L-malate) required for C_4 photosynthesis and CAM.

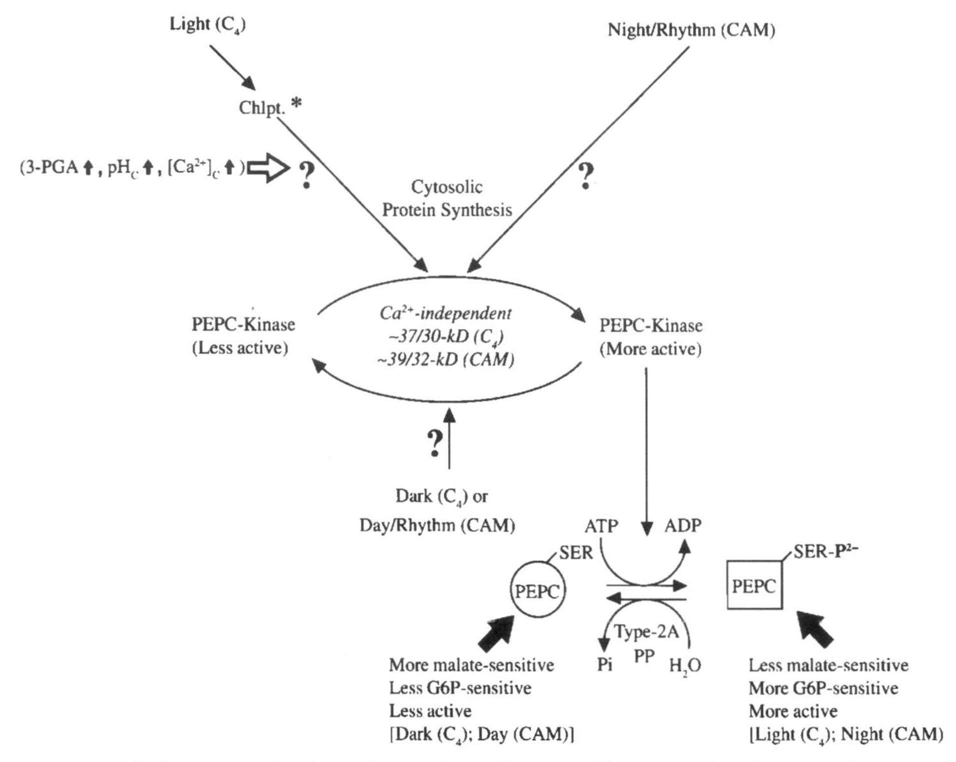

Figure 3 Proposed molecular mechanism for the light-dark (C_4) or night-day (CAM) regulation of the effector sensitivity [L-malate (negative), G6P (positive)] and activity of photosynthetic PEPC in the leaf mesophyll cell by reversible phosphorylation of a single target serine near the subunit's N-terminus [e.g. Ser-8 in *Sorghum* (see Figure 1)]. Chlpt.*, illuminated chloroplast; pH_c and $[Ca^{2+}]_c$, mesophyll cytosolic pH and $[Ca^{2+}]$, respectively; 3-PGA, 3-P-glycerate; PP, protein phosphatase. [Updated from Jiao & Chollet (54).]

Regulatory Phosphorylation of Nonphotosynthetic PEPC Isoforms

There is now convincing evidence that the reversible phosphorylation of the N-terminal domain of plant PEPC is widespread, if not ubiquitous. In vivo studies with ^{32}Pi have demonstrated the reversible phosphorylation of nonphotosynthetic PEPC in soybean root nodules (136) and in wheat leaves excised from N-deficient seedlings (24, 116). Complementary measurements of in vivo changes in PEPC activity and/or malate sensitivity under near-physiological assay conditions (i.e. low pH, low [PEP] relative to K_m) have underscored the regulatory nature of this covalent modification in nodules (136), illuminated C_3 leaves (24, 116; B Li, XQ Zhang & R Chollet, unpublished data), and *Vicia faba* guard cells microdissected from opening stomata (135).

Furthermore, related in vitro studies have established that PEPC kinase activity is present in soybean and alfalfa root nodules (102, 115), wheat and tobacco leaves (24, 119), and *Sorghum* roots (91), and have demonstrated this kinase's similarity to the C_4 and CAM enzymes with respect to its Ca^{2+} independency, chromatographic properties, and catalytic subunit(s) (24, 91, 119). The activity state of this PEPC kinase is modulated reversibly in vivo by a complex interaction between photosynthesis and N (C_3 leaves) or photosynthate supply to N_2-fixing root nodules (24; B Li, XQ Zhang & R Chollet, unpublished data). Thus, the phosphorylation of cytosolic PEPC by a highly regulated protein-Ser/Thr kinase is likely the major posttranslational mechanism for altering the allosteric properties and activity of this "multifaceted" plant enzyme in vivo.

Other Proposed Regulatory Mechanisms

Two other mechanisms have been proposed for the diel regulation of C_4 and CAM PEPC activity and/or sensitivity to L-malate based wholly on in vitro observations.

DIMER-TETRAMER INTERCONVERSION The Wedding laboratory found that CAM PEPC purified from day- and night-adapted *Crassula argentea* leaves exists as kinetically distinct but interconvertible oligomers (128). The day enzyme was mainly a malate-sensitive homodimer (α_2) and the night form a malate-"insensitive" homotetramer (α_4), with about a twofold higher K_i. PEP, G6P, Mg^{2+}, or a higher [PEPC] favors conversion of α_2 to α_4, whereas L-malate or a lower [PEPC] shifts the equilibrium toward the dimeric form (79, 128, 129). Similar in vitro association/dissociation properties have been reported for the active C_4 homotetramer from maize (123, 127). There is no evidence, however, to support the involvement of these aggregation-state changes in the diel regulation of the CAM and C_4 isoforms in vivo. On the contrary, several reports document that the phospho and dephospho C_4 and CAM enzyme forms are isolated in the same aggregation state while retaining the characteristic differential sensitivity to L-malate (4, 63, 77, 82, 122). Thus, it is our opinion that there is not a significant regulation of photosynthetic PEPC in vivo by changes in its aggregation state.

REDOX REGULATION Even more speculative in our view is the proposal that the regulation of cytosolic C_4 PEPC may be primarily under the control of the redox state of certain critical cysteines (13, 45). While there are, indeed, five plant-invariant Cys residues in the various PEPC isoforms that are absent in the microbial enzymes (Figure 1) (67, 68, 79a, 114), none of them have been shown specifically to be involved in regulation of activity or L-malate sensitivity. On the contrary, related observations with the dephospho maize enzyme indicate no effect of reduced cytosolic thioredoxin *h* on the properties of C_4 PEPC in vitro (52).

PEPC GENE STRUCTURE, EXPRESSION, AND MOLECULAR EVOLUTION

Multigene Families

PEPC isoforms have been characterized in both photosynthetic and nonphotosynthetic tissues of various plants (reviewed in 68, 114). Consistent with the enzyme's functional diversity, small multigene families have been found. For example, three PEPC nuclear genes—*SvC3, SvC3RI*, and *SvC4*—have been characterized in *Sorghum* that encode the C_3-like housekeeping and root forms and the C_4-photosynthetic isoform, respectively (67). The maize family possesses at least five genes (37) that can be classified into three distinct groups (99). The C_4-PEPC gene is unique and is located near the centromere of chromosome 9. Three other genes have been mapped to different loci on chromosomes 4L, 5, and 7 (37, 47, 61). Both C_3 and C_4 species in the dicot genus *Flaveria* contain very similar families of distinct *Ppc* subgroups (40, 95). The C_4 isoform in *Flaveria trinervia* is encoded by the *PpcA* subgroup of the family. Homologous *PpcA* genes are found in the C_3 species *Flaveria pringlei;* however, they are weakly expressed, and their transcripts do not show the strict leaf-specific accumulation pattern found in the related C_4 species (40). In the facultative CAM plant *M. crystallinum*, two isogenes (*Ppc1*, *Ppc2*) have been described, and another distinct member might exist (17, 18); the transcriptional activity of *Ppc1* is strongly and selectively enhanced during C_3 to CAM switching induced by salt stress (18). The *Brassica napus* genome contains more than four highly similar PEPC genes, but some of them lack specific introns (133). PEPC gene families have also been found or suggested to exist in sugarcane, *Amaranthus,* tobacco, alfalfa, rice, wheat (reviewed in 68), and *Arabidopsis* (79b).

Ppc and PEPC Sequence Comparisons

The plant PEPC genes contain nine introns (with the exception reported in 133) of variable length but identical location with respect to the coding regions. Consensus intron/exon splice sites (aG*GT*aag—tgc*AG*g) are conserved. Generally, a classical gene organization is observed, although in some C_4- and C_3-type *Ppc* genes there is no typical TATA box, and multiple polyadenylation sites are found in the 3′-untranslated region (15, 43, 74, 132).

In alignments of all the deduced PEPC amino acid sequences reported, several highly conserved residues and motifs are found, and these likely contribute to the domains involved in the active site and/or regulation of the enzyme (see sections on Active-Site Structure and Regulatory Phosphorylation) (67, 68, 79a, 96a, 114). Figure 1 exemplifies these structural features in the deduced sequence of the C_4-isoform from *Sorghum* (SvC4). The phosphorylation motif near the N-terminus (E/DR/Kxx**S**IDAQL/MR), including the target Ser, and five cysteine residues, some of which have been proposed to be involved in redox regulation and/or stabilization of the tetrameric structure of the holoenzyme (13, 45), are specific to plant PEPC (68, 79a, 114). In addition, there are several species-invariant motifs in all PEPCs examined to date (TA**H**PT, VMxGYSDSx**K**DxG, F**H**GRGxxxx**R**GxxP) that contain specific His, Lys, and Arg residues implicated in the active-site domain (see section on Active-Site Structure; 3, 57, 79a, 96, 96a, 112). In general, the C-terminal half of the ~110-kDa PEPC polypeptide contains most of these presumed active-site determinants, whereas the N-terminal half appears to include the motifs that are regulatory in nature (53, 57, 110, 114). Further insight into the structure/function relationships of PEPC must await continued mutagenesis of these and other (114) highly conserved domains and, most importantly, high-resolution crystallographic analysis of the plant and microbial proteins.

Ppc Promoter Analysis and Transcription

The C_4-PEPC gene is expressed in photosynthetic tissues during greening via a phytochrome-mediated response (113). Expression of this gene is not necessarily coupled to the development of Kranz leaf anatomy because, in maize, it also occurs in such tissues as the inner leaf sheaths and tassels (43). In addition to light, cytokinins upregulate the transcriptional activity of the C_4-PEPC gene in maize leaves recovering from N deficiency (106), whereas in *Sorghum* abscisic acid (ABA) stimulates specific *Ppc* mRNA accumulation (2). In *M. crystallinum*, CAM-PEPC gene expression is induced by salt stress and/or ABA during C_3 to CAM switching, and these effects are moderated by light (76). In the CAM plant *Kalanchoë blossfeldiana*, changes in photoperiod and ABA are also involved in the induction of the photosynthetic PEPC gene (108). Lastly, C_3-type PEPC mRNAs are accumulated during the development of alfalfa root nodules (92, 115) and in recovering roots of N-deficient *Sorghum* (P Gadal, L Lepiniec & S Santi, unpublished data).

Light-responsive elements corresponding to those in the nuclear genes encoding the small subunit of Rubisco are lacking in the C_4-*Ppc* promoters of maize and *Sorghum*. Other conserved, direct repeated sequences (TTACCACTAGCTA), or the light-responsive element (CCTTATCCT) characterized previously in the promoter of light-inducible phytochrome genes, could play such a role, at least in part (15, 68, 74). The maize nuclear factor (MNF) (see below; 131) and *SV40* Sp1 (15, 68, 74) binding sites—(AAGG) and (CCGCCC), respectively—are also found in C_4-*Ppc* promoters. In addition, the presence of CpG islands (68) is consistent with the possible regulation of specific sites in the promoters of both C_4 and C_3 PEPC genes by changes in DNA methylation status (64). In the *Sorghum* SvC3 and SvC3RI *Ppc* promoters, sequences homologous to the light-responsive element AT-1 (AATATTTTTATA) and nod- (TCTACGTAGA) and G-boxes (CCACGTGG) are found (68). Both C_4 and C_3 species of *Flaveria* have orthologous C_4 genes (*PpcA* subgroup), the 5′-flanking regions of which are essentially homologous and share CCAAT, AT-1, and GT-1 III/IIIα boxes and an octameric motif known to confer cell-type specificity (40). It has been suggested that certain specific features of the C_4-PEPC gene promoter in *F. trinervia* could account for the much higher expression level in this C_4 species, including a light-responsive box II element, the microheterogeneity of the sequence around the TATA box, and the presence of a putative scaffold attachment region near the promoter that is often associated with highly expressed genes (40). Recent experiments using transgenic tobacco plants have shown that the sequences responsible for the enhanced, leaf-specific expression of C_4 *Ppc* in *F. trinervia* are located between positions −2118 and −500 relative to the transcription start site in the *PpcA* promoter (105); whether these sequences involve the above-mentioned proximal elements is not known.

Three leaf-specific DNA-binding proteins (MNF1, MNF2a, MNF2b) have been shown to interact specifically with the promoter of the maize C_4-PEPC gene (130, 131). Among these nuclear factors, MNF2a is presumed to act as a negative transcriptional effector (130). Two cDNA clones (MNB1a, MNB1b) encoding proteins that bind to an AAGG motif at the MNF1 site have been identified (131). Two other clones (designated 281, 282) may encode PEP1, a light-dependent factor interacting with the promoter of the maize C_4-PEPC gene (59). In *M. crystallinum*, salt stress causes three protein factors (PCAT-1, -2, and -3) to differentially recognize two AT-rich regions in the *Ppc1* promoter (16). Recently, several salt-responsive enhancer regions and one silencer region have been identified in this promoter (98a).

From the above, it is clear that data on nuclear *trans*-acting factors and the corresponding regulatory *cis*-acting DNA sequences of the *Ppc* promoters are still relatively scarce. Thus, no clear picture has emerged concerning the regulatory mechanisms that control the transcription rate of the different classes of PEPC genes in plants.

Transgenic Plants

In transgenic tobacco transformed with maize C_4-*Ppc1* genes containing the upstream regulatory region (about 2 kb), a low level of PEPC transcripts was produced; although their size was aberrantly large, accumulation still required light (44). These transformants possessed a twofold increase in PEPC activity that was correlated with the appearance of a high-K_m(PEP) C_4 form of the enzyme and an elevated level of leaf malate. However, these biochemical changes did not result in any detectable physiological effects with respect to the rate of leaf net photosynthesis in air and to the CO_2 compensation concentration. In a related study, the maize C_4-PEPC gene was placed under the control of a CaMV 35S promoter (62). Although the transgenic tobacco plants contained *Ppc* transcripts of the correct size and about twice as much PEPC protein, their growth rate was retarded relative to that of the nontransformed plants. Transgenic tobacco plants transformed with either the C_4-PEPC gene from *Sorghum* or chimeric constructs containing the promoter of the C_4 gene from maize fused to the *gus*A reporter gene showed a high expression of transcripts as well as leaf mesophyll-cell specificity (75, 107). Similar results have been reported recently in transgenic rice using the same experimental strategy (73). Transgenic tobacco plants also expressed constructs containing various parts of the 5′-flanking region of the *PpcA1* (C_4-type) genes from both C_4 and C_3 species of *Flaveria* (105). In this heterologous system, only the C_4-*Ppc* promoter from the C_4 species conferred a high level of reporter gene expression, thus showing that it contains regulatory *cis*-elements responsible for abundant expression. In addition, a leaf palisade mesophyll-cell specificity was partially maintained in these transgenic tobacco plants. Hence, it appears that most of the regulatory elements that control the light-inducible expression of *Ppc* in C_4 leaves are also present in C_3 plants. On the other hand, although the CAM-specific *Ppc* promoter from the *M. crystallinum* gene is highly active in transgenic tobacco, it directs transcript synthesis in most cell types and lacks the salt inducibility found in its natural cellular environment (17). Finally, in homologous transient-expression systems using leaf-, stem-, and root-derived protoplasts from maize, a cell-specific expression pattern is largely dependent on the specific *Ppc* promoter used (99). In this system, transcript accumulation is not immediate but rather is related to light-dependent developmental changes, in contrast with other photosynthetic genes. This latter observation has led to the suggestion that distinct transduction pathways operate for the coordination of light-dependent genes encoding photosynthetic enzymes (99).

Molecular Evolution

Phylogenetic trees have been constructed using unambiguously aligned sites from the available PEPC amino acid sequences as well as on the basis of parsimony or distance analyses (1, 47, 61, 67, 68, 114). The cyanobacterial and bacterial PEPCs consistently group with prokaryotic phylogenetic relationships (68). As for the plant enzymes, phylogenetic relationships have been studied with particular emphasis on the molecular mechanisms that have shaped the expression characteristics and kinetic properties of PEPC during the evolutionary transition from C_3 into CAM and C_4 plants. The acquisition of these new photosynthetic strategies by a wide variety of plant species indicates that they have originated independently and on many separate occasions during the evolution of flowering plants, with CAM being the antecedent of C_4 (47, 61, 67). Thus, an obvious question is how to account for the polyphyletic evolution of C_4 plants. From the various independently derived trees it can be inferred that all plant PEPC sequences diverged from a single common ancestral gene. On the other hand, the presence of different genes could have preceded angiosperm diversification and perhaps also that of higher plants. C_4-PEPC genes could have arisen from a duplication event long before the monocot-dicot divergence and thus prior to the appearance of C_4 plants. In this manner, the PEPC gene for C_4 photosynthesis could have

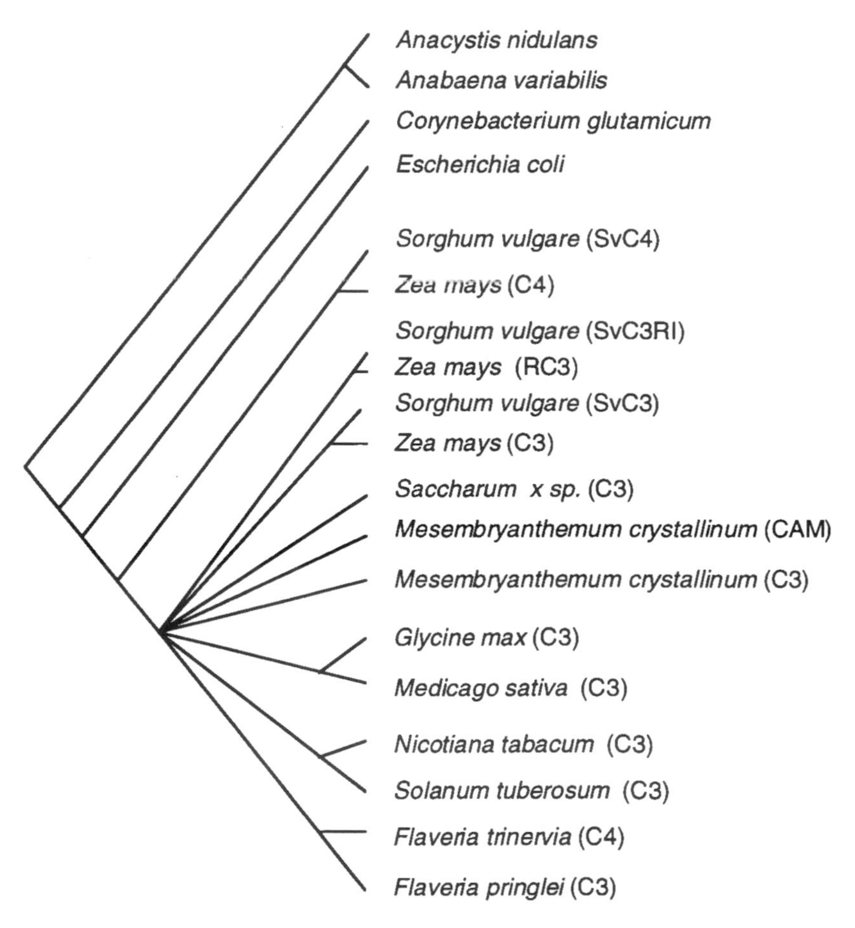

Figure 4 Consensus phylogenetic tree of 19 microbial and plant PEPCs. Branch lengths have no significance (see 67 and 68 for details). SvC4, SvC3RI, and SvC3 are the photosynthetic, root, and housekeeping isoforms of *Sorghum vulgare* PEPC, respectively; *Zea mays* C4, RC3, and C3 are the corresponding isoforms in maize. [Redrawn from Lepiniec et al (68), with permission from Elsevier Science Ireland Ltd.]

evolved in a limited number of species while disappearing in others (47, 61, 67). In the consensus tree depicted in Figure 4, C_4-PEPCs from the monocots *Sorghum* and maize are clearly distinguishable from the various C_3 and CAM isoforms and also from their indigenous C_3 counterparts (e.g. SvC3RI, SvC3). In contrast, the photosynthetic enzyme in the C_4-dicot *F. trinervia* is more closely related to the various isoforms in C_3 and CAM dicots (40, 95) than to the two monocot C_4-PEPCs (68). Furthermore, because the promoters of the C_4-PEPC gene in *F. trinervia* and the orthologous gene in *F. pringlei* (C_3) are very similar, it has been suggested that a C_3 promoter could have been "tuned" to meet the special demands of C_4 photosynthesis (40). The possibility that an alternative evolution has led to the formation of C_4 enzymes in the various genera containing C_4 species could account for the observed divergence between monocots and dicots (40). Finally, it is not clear why a homologous form of C_4-PEPC is not found in dicots because, as mentioned above, a primordial PEPC form could have arisen before the divergence of monocots and dicots (68). Further investigations involving PEPC sequences from different genera are required to refine the phylogenetic relationships of the microbial and plant enzymes, including sequence analysis of green algal PEPCs and additional gymnosperm species (*Picea abies*) (96a).

CONCLUSIONS AND FUTURE PROSPECTS

While the past decade has seen a number of truly impressive revelations concerning PEPC, future research awaits the results of three-dimensional structure studies that will provide another important chapter in PEPC mechanism, regulation, allosteric effects, and other areas. In addition, the emerging pictures of the highly regulated PEPC kinase, together with its requisite signal-transduction cascades, must be completed. Related work on the heteromeric intracellular form of the type 2A protein phosphatase that dephosphorylates plant PEPC in the cytosol will also be important (cf 122a). With the recent generation of the first C_4 PEPC-deficient mutant in the dicot *Amaranthus edulis* (20) and the development of an efficient, *Agrobacterium*-mediated transformation system for C_4 dicots (14), the stage is finally set for the genetic manipulation of C_4 photosynthesis in vivo by engineering the regulatory properties and amount of PEPC in the leaf cytosol. We anticipate that these and other fertile avenues for future research on PEPC will continue to deepen our understanding of this "multifaceted" enzyme in plants. Finally, we hope that this survey has reminded the reader that there is, indeed, another CO_2-fixing enzyme in plants besides Rubisco that is worthy of detailed study.

ACKNOWLEDGMENTS

R. C. gratefully acknowledges the continued support of his laboratory's research on plant PEPC by NSF, USDA/NRI, and the Nebraska Agricultural Research Division (in which this article is published as Journal Series No. 11, 284).

Literature Cited

1. Albert HA, Martin T, Sun SSM. 1992. Structure and expression of a sugarcane gene encoding a housekeeping phosphoenolpyruvate carboxylase. *Plant Mol. Biol.* 20:663–71
2. Amzallag GN, Lerner HR, Poljakoff-Mayber A. 1990. Exogenous ABA as a modulator of the response of *Sorghum* to high salinity. *J. Exp. Bot.* 41:1529–34
3. Andrco CS, González DH, Iglesias AA. 1987. Higher plant phospho*enol*pyruvate carboxylase: structure and regulation. *FEBS Lett.* 213:1–8
4. Arrio-Dupont M, Bakrim N, Echevarria C, Gadal P, Le Maréchal P, Vidal J. 1992. Compared properties of phosphoenolpyruvate carboxylase from dark- and light-adapted *Sorghum* leaves: use of a rapid purification technique by immunochromatography. *Plant Sci.* 81:37–46
5. Ausenhus SL. 1993. *Phosphoenolpyruvate carboxylase from maize: function of the divalent metal ion in binding and catalysis.* PhD thesis. Univ. Wis., Madison. 160 pp.
6. Ausenhus SL, O'Leary MH. 1992. Hydrolysis of phosphoenolpyruvate catalyzed by phosphoenolpyruvate carboxylase from *Zea mays*. *Biochemistry* 31:6427–31
7. Bakrim N, Echevarria C, Crétin C, Arrio-Dupont M, Pierre JN, et al. 1992. Regulatory phosphorylation of *Sorghum* leaf phospho*enol*pyruvate carboxylase: identification of the protein-serine kinase and some elements of the signal-transduction cascade. *Eur. J. Biochem.* 204:821–30
8. Bakrim N, Prioul JL, Deleens E, Rocher JP, Arrio-Dupont M, et al. 1993. Regulatory phosphorylation of C_4 phospho*enol*pyruvate carboxylase: a cardinal event influencing the photosynthesis rate in *Sorghum* and maize. *Plant Physiol.* 101:891–97
9. Baur B, Dietz KJ, Winter K. 1992. Regulatory protein phosphorylation of phospho*enol*pyruvate carboxylase in the facultative crassulacean-acid-metabolism plant *Mesembryanthemum crystallinum* L. *Eur. J. Biochem.* 209:95–101
10. Budde RJA, Chollet R. 1986. *In vitro* phosphorylation of maize leaf phosphoenolpyruvate carboxylase. *Plant Physiol.* 82:1107–14
11. Carter PJ, Nimmo HG, Fewson CA, Wilkins MB. 1990. *Bryophyllum fedtschenkoi* protein phosphatase type 2A can dephosphorylate phosphoenolpyruvate carboxylase. *FEBS Lett.* 263:233–36
12. Carter PJ, Nimmo HG, Fewson CA, Wilkins MB. 1991. Circadian rhythms in the activity of a plant protein kinase. *EMBO J.* 10:2063–68
12a. Carter PJ, Wilkins MB, Nimmo HG, Fewson CA. 1995. Effects of temperature on the activity of phosphoenolpyruvate carboxylase and on the control of CO_2 fixation in *Bryophyllum fedtschenkoi*. *Planta* 196:375–80
13. Chardot TP, Wedding RT. 1992. Role of cysteine in activation and allosteric regulation of maize phospho*enol*pyruvate carboxylase. *Plant Physiol.* 98:780–83
14. Chitty JA, Furbank RT, Marshall JS, Chen Z, Taylor WC. 1994. Genetic transformation of the C_4 plant, *Flaveria bidentis*. *Plant J.* 6:949–56
15. Crétin C, Santi S, Keryer E, Lepiniec L, Tagu D, et al. 1991. The phosphoenolpyruvate carboxylase gene family of *Sorghum*: promoter structures, amino acid sequences and expression of genes. *Gene* 99:87–94
16. Cushman JC, Bohnert HJ. 1992. Salt stress alters A/T-rich DNA-binding factor interactions within the phosphoenolpyruvate carboxylase promoter from *Mesembryanthemum crystallinum*. *Plant Mol. Biol.* 20:411–24
17. Cushman JC, Meiners MS, Bohnert HJ. 1993. Expression of a phosphoenolpyruvate carboxylase promoter from *Mesembryanthemum crystallinum* is not salt-inducible in mature transgenic tobacco. *Plant Mol. Biol.* 21:561–66

18. Cushman JC, Meyer G, Michalowski CB, Schmitt JM, Bohnert HJ. 1989. Salt stress leads to differential expression of two isogenes of phosphoenolpyruvate carboxylase during Crassulacean acid metabolism induction in the common ice plant. *Plant Cell* 1:715–25
19. Deroche ME, Carrayol E. 1988. Nodule phosphoenolpyruvate carboxylase: a review. *Physiol. Plant.* 74:775–82
20. Dever LV, Blackwell RD, Fullwood NJ, Lacuesta M, Leegood RC, et al. 1995. The isolation and characterization of mutants of the C_4 photosynthetic pathway. *J. Exp. Bot.* 46:1363–76
21. Díaz E, O'Laughlin JT, O'Leary MH. 1988. Reaction of phosphoenolpyruvate carboxylase with (*Z*)-3-bromophosphoenolpyruvate and (*Z*)-3-fluorophosphoenolpyruvate. *Biochemistry* 27:1336–41
22. Doncaster HD, Leegood RC. 1987. Regulation of phosphoenolpyruvate carboxylase activity in maize leaves. *Plant Physiol.* 84: 82–87
23. Duff SMG, Andreo CS, Pacquit V, Lepiniec L, Sarath G, et al. 1995. Kinetic analysis of the non-phosphorylated, *in vitro* phosphorylated, and phosphorylation-site-mutant (Asp8) forms of intact recombinant C_4 phospho*enol*pyruvate carboxylase from sorghum. *Eur. J. Biochem.* 228:92–95
24. Duff SMG, Chollet R. 1995. In vivo regulation of wheat-leaf phospho*enol*pyruvate carboxylase by reversible phosphorylation. *Plant Physiol.* 107:775–82
24a. Duff SMG, Giglioli-Guivarc'h N, Pierre J-N, Vidal J, Condon SA, Chollet R. 1996. *In situ* evidence for the involvement of calcium and bundle sheath–derived photosynthetic metabolites in the C_4 phospho*enol*pyruvate-carboxylase kinase signal-transduction chain. *Planta*. In press
25. Duff SMG, Lepiniec L, Crétin C, Andreo CS, Condon SA, et al. 1993. An engineered change in the L-malate sensitivity of a site-directed mutant of sorghum phospho*enol*pyruvate carboxylase: the effect of sequential mutagenesis and S-carboxymethylation at position 8. *Arch. Biochem. Biophys.* 306:272–76
26. Echevarria C, Pacquit V, Bakrim N, Osuna L, Delgado B, et al. 1994. The effect of pH on the covalent and metabolic control of C_4 phospho*enol*pyruvate carboxylase from *Sorghum* leaf. *Arch. Biochem. Biophys.* 315:425–30
27. Echevarría C, Vidal J, Jiao JA, Chollet R. 1990. Reversible light activation of the phospho*enol*pyruvate carboxylase protein-serine kinase in maize leaves. *FEBS Lett.* 275:25–28
28. Farquhar GD, Ehleringer JR, Hubick KT. 1989. Carbon isotope discrimination and photosynthesis. *Annu. Rev. Plant Physiol. Plant Mol. Biol.* 40:503–37
29. Fujita N, Izui K, Nishino T, Katsuki H. 1984. Reaction mechanism of phosphoenolpyruvate carboxylase: bicarbonate-dependent dephosphorylation of phospho enol-α-ketobutyrate. *Biochemistry* 23: 1774–79
30. Gawlita E, Caldwell WS, O'Leary MH, Paneth P, Anderson VE. 1995. Kinetic isotope effects on substrate association: reactions of phosphoenolpyruvate with phosphoenolpyruvate carboxylase and pyruvate kinase. *Biochemistry* 32:2577–83
30a. Giglioli-Guivarc'h N, Pierre J-N, Brown S, Chollet R, Vidal J, Gadal P. 1996. The light-dependent transduction pathway controlling the regulatory phosphorylation of C_4 phospho*enol*pyruvate carboxylase in protoplasts from *Digitaria sanguinalis*. *Plant Cell* 8:In press
31. González DH, Andreo CS. 1986. Phospho enolpyruvate carboxylase from maize leaves: studies using β-methylated phosphoenolpyruvate analogues as inhibitors and substrates. *Z. Naturforsch.* 41C: 1004–10
32. González DH, Andreo CS. 1988. Carboxylation and dephosphorylation of phosphoenol-3-fluoropyruvate by maize leaf phosphoenolpyruvate carboxylase. *Biochem. J.* 253:217–22
33. González DH, Andreo CS. 1988. Identification of 2-enolbutyrate as the product of the reaction of maize leaf phosphoenolpyruvate carboxylase with (*Z*)- and (*E*)-2-phosphoenolbutyrate: evidence from NMR and kinetic measurements. *Biochemistry* 27:177–83
34. Gonzalez DH, Andreo CS. 1988. Stereoselectivity of the interaction of *E*- and *Z*-2-phospho*enol*butyrate with maize leaf phospho*enol*pyruvate carboxylase. *Eur. J. Biochem.* 173:339–43
35. González DH, Andreo CS. 1989. The use of substrate analogues to study the active-site structure and mechanism of PEP carboxylase. *Trends Biochem. Sci.* 14: 24–27
36. Grammatikopoulos G, Manetas Y. 1990. Diurnal changes in phosphoenolpyruvate carboxylase and pyruvate,orthophosphate dikinase properties in the natural environment: interplay of light and temperature in a C_4 thermophile. *Physiol. Plant.* 80: 593–97
37. Grula JW, Hudspeth RL. 1987. The phosphoenolpyruvate carboxylase gene family of maize. In *Plant Gene Systems and Their Biology*, ed. JL Key, L McIntosh, pp. 207–16. New York: Liss
38. Guidici-Orticoni MT, Vidal J, Le Maréchal P, Thomas M, Gadal P, Rémy R. 1988. *In vivo* phosphorylation of sorghum leaf phos-

phoe*nol*pyruvate carboxylase. *Biochimie* 70:769–72

39. Hansen DE, Knowles JR. 1982. The stereochemical course at phosphorus of the reaction catalyzed by phosphoenolpyruvate carboxylase. *J. Biol. Chem.* 257:14795–98
40. Hermans J, Westhoff P. 1992. Homologous genes for the C_4 isoform of phosphoenolpyruvate carboxylase in a C_3 and a C_4 *Flaveria* species. *Mol. Gen. Genet.* 234: 275–84

40a. Huber SC, Huber JL. 1996. Role and regulation of sucrose-phosphate synthase in higher plants *Annu. Rev. Plant Physiol. Plant Mol. Biol.* 47:431–44

41. Huber SC, Huber JL, McMichael RW Jr. 1994. Control of plant enzyme activity by reversible protein phosphorylation. *Int. Rev. Cytol.* 149:47–98
42. Huber SC, Sugiyama T. 1986. Changes in sensitivity to effectors of maize leaf phosphoenolpyruvate carboxylase during light/dark transitions. *Plant Physiol.* 81:674–77
43. Hudspeth RL, Grula JW. 1989. Structure and expression of the maize gene encoding the phosphoenolpyruvate carboxylase isozyme involved in C_4 photosynthesis. *Plant Mol. Biol.* 12:579–89
44. Hudspeth RL, Grula JW, Dai Z, Edwards GE, Ku MSB. 1992. Expression of maize phospho*enol*pyruvate carboxylase in transgenic tobacco: effects on biochemistry and physiology. *Plant Physiol.* 98:458–64
45. Iglesias AA, Andreo CS. 1984. On the molecular mechanism of maize phosphoenolpyruvate carboxylase activation by thiol compounds. *Plant Physiol.* 75:983–87
46. Inoue M, Hayashi M, Sugimoto M, Harada S, Kai Y, et al. 1989. First crystallization of a phosphoenolpyruvate carboxylase from *Escherichia coli*. *J. Mol. Biol.* 208:509–10
47. Izui K, Kawamura T, Okumura S, Toh H. 1992. Molecular evolution of phospho*enol*pyruvate carboxylase for C_4 photosynthesis in maize. In *Research in Photosynthesis,* ed. N Murata, 3:827–30. Dordrecht: Kluwer
48. Janc JW, Cleland WW, O'Leary MH. 1992. Mechanistic studies of phosphoenolpyruvate carboxylase from *Zea mays* utilizing formate as an alternate substrate for bicarbonate. *Biochemistry* 31:6441–46
49. Janc JW, O'Leary MH, Cleland WW. 1992. A kinetic investigation of phosphoenolpyruvate carboxylase from *Zea mays*. *Biochemistry* 31:6421–26
50. Janc JW, Urbauer JL, O'Leary MH, Cleland WW. 1992. Mechanistic studies of phosphoenolpyruvate carboxylase from *Zea mays* with (*Z*)- and (*E*)-3-fluorophosphoenolpyruvate as substrates. *Biochemistry* 31:6432–40
51. Jiao JA, Chollet R. 1988. Light/dark regulation of maize leaf phospho*enol*pyruvate carboxylase by *in vivo* phosphorylation. *Arch. Biochem. Biophys.* 261:409–17
52. Jiao JA, Chollet R. 1989. Regulatory serylphosphorylation of C_4 phospho*enol*pyruvate carboxylase by a soluble protein kinase from maize leaves. *Arch. Biochem. Biophys.* 269:526–35
53. Jiao JA, Chollet R. 1990. Regulatory phosphorylation of serine-15 in maize phospho*enol*pyruvate carboxylase by a C_4-leaf protein-serine kinase. *Arch. Biochem. Biophys.* 283:300–5
54. Jiao JA, Chollet R. 1991. Posttranslational regulation of phospho*enol*pyruvate carboxylase in C_4 and Crassulacean acid metabolism plants. *Plant Physiol.* 95:981–85
55. Jiao JA, Chollet R. 1992. Light activation of maize phospho*enol*pyruvate carboxylase protein-serine kinase activity is inhibited by mesophyll and bundle sheath–directed photosynthesis inhibitors. *Plant Physiol.* 98:152–56
56. Jiao JA, Echevarría C, Vidal J, Chollet R. 1991. Protein turnover as a component in the light/dark regulation of phospho*enol*pyruvate carboxylase protein-serine kinase activity in C_4 plants. *Proc. Natl. Acad. Sci USA* 88:2712–15
57. Jiao JA, Podestá FE, Chollet R, O'Leary MH, Andreo CS. 1990. Isolation and sequence of an active-site peptide from maize leaf phospho*enol*pyruvate carboxylase inactivated by pyridoxal 5′-phosphate. *Biochim. Biophys. Acta* 1041:291–95
58. Jiao JA, Vidal J, Echevarría C, Chollet R. 1991. *In vivo* regulatory phosphorylation site in C_4-leaf phospho*enol*pyruvate carboxylase from maize and sorghum. *Plant Physiol.* 96:297–301
59. Kano-Murakami Y, Matsuoka M. 1992. Gene expression of PEP carboxylase gene. See Ref. 47, 3:843–46
60. Karabourniotis G, Manetas Y, Gavalas NA. 1983. Photoregulation of phosphoenolpyruvate carboxylase in *Salsola soda* L. and other C_4 plants. *Plant Physiol.* 73: 735–39
61. Kawamura T, Shigesada K, Toh H, Okumura S, Yanagisawa S, Izui K. 1992. Molecular evolution of phosphoenolpyruvate carboxylase for C_4 photosynthesis in maize: comparison of its cDNA sequence with a newly isolated cDNA encoding an isozyme involved in the anaplerotic function. *J. Biochem.* 112:147–54
62. Kogami H, Shono M, Koike T, Yanagisawa S, Izui K, et al. 1994. Molecular and physiological evaluation of transgenic tobacco plants expressing a maize phospho*enol*pyruvate carboxylase gene under the control of the cauliflower mosaic virus 35S promoter. *Transgenic Res.* 3:287–96
63. Krüger I, Kluge M. 1987. Diurnal changes in the regulatory properties of phos-

phoenolpyruvate carboxylase in plants: Are alterations in the quaternary structure involved? *Bot. Acta* 101:24–27

64. Langdale JA, Taylor WC, Nelson T. 1991. Cell-specific accumulation of maize phosphoenolpyruvate carboxylase is correlated with demethylation at a specific site >3kb upstream of the gene. *Mol. Gen. Genet.* 225:49–55
65. Law RD, Plaxton WC. 1995. Purification and characterization of a novel phosphoenolpyruvate carboxylase from banana fruit. *Biochem. J.* 307:807–16
66. Leegood RC, Osmond CB. 1990. The flux of metabolites in C_4 and CAM plants. In *Plant Physiology, Biochemistry and Molecular Biology*, ed. DT Dennis, DH Turpin, pp. 274–98. Essex: Longman Sci. Tech.
67. Lepiniec L, Keryer E, Philippe H, Gadal P, Crétin C. 1993. *Sorghum* phosphoenolpyruvate carboxylase gene family: structure, function and molecular evolution. *Plant Mol. Biol.* 21:487–502
68. Lepiniec L, Vidal J, Chollet R, Gadal P, Crétin C. 1994. Phosphoenolpyruvate carboxylase: structure, regulation and evolution. *Plant Sci.* 99:111–24
69. Li B, Chollet R. 1993. Resolution and identification of C_4 phospho*enol*pyruvate-carboxylase protein-kinase polypeptides and their reversible light activation in maize leaves. *Arch. Biochem. Biophys.* 307: 416–19
70. Li B, Chollet R. 1994. Salt induction and the partial purification/characterization of phospho*enol*pyruvate carboxylase protein-serine kinase from an inducible Crassulacean-acid-metabolism (CAM) plant, *Mesembryanthemum crytallinum* L. *Arch. Biochem. Biophys.* 314:247–54
71. Liu J, Peliska JA, O'Leary MH. 1990. Synthesis and study of (*Z*)-3-chlorophosphoenolpyruvate. *Arch. Biochem. Biophys.* 277:143–48
72. Maruyama H, Easterday RL, Chang HC, Lane MD. 1966. The enzymatic carboxylation of phosphoenolpyruvate. I. Purification and properties of phosphoenolpyruvate carboxylase. *J. Biol. Chem.* 241: 2405–12
73. Matsuoka M, Kyozuka J, Shimamoto K, Kano-Murakami Y. 1994. The promoters of two carboxylases in a C_4 plant (maize) direct cell-specific, light-regulated expression in a C_3 plant (rice). *Plant J.* 6:311–19
74. Matsuoka M, Minami E. 1989. Complete structure of the gene for phospho*enol*pyruvate carboxylase from maize. *Eur. J. Biochem.* 181:593–98
75. Matsuoka M, Sanada Y. 1991. Expression of photosynthetic genes from the C_4 plant, maize, in tobacco. *Mol. Gen. Genet.* 225: 411–19
76. McElwain EF, Bohnert HJ, Thomas JC. 1992. Light moderates the induction of phospho*enol*pyruvate carboxylase by NaCl and abscisic acid in *Mesembryanthemum crystallinum. Plant Physiol.* 99:1261–64
77. McNaughton GAL, Fewson CA, Wilkins MB, Nimmo HG. 1989. Purification, oligomerization state and malate sensitivity of maize leaf phosphoenolpyruvate carboxylase. *Biochem. J.* 261:349–55
78. McNaughton GAL, MacKintosh C, Fewson CA, Wilkins MB, Nimmo HG. 1991. Illumination increases the phosphorylation state of maize leaf phospho*enol*pyruvate carboxylase by causing an increase in the activity of a protein kinase. *Biochim. Biophys. Acta* 1093:189–95
79. Meyer CR, Willeford KO, Wedding RT. 1991. Regulation of phospho*enol*pyruvate carboxylase from *Crassula argentea:* effect of incubation with ligands and dilution on oligomeric state, activity, and allosteric properties. *Arch. Biochem. Biophys.* 288: 343–49

79a. Nakamura T, Yoshioka I, Takahashi M, Toh H, Izui K. 1995. Cloning and sequence analysis of the gene for phosphoenolpyruvate carboxylase from an extreme thermophile, *Thermus* sp. *J. Biochem.* 118: 319–24

79b. Newman T, de Bruijn FJ, Green P, Keegstra K, Kende H, et al. 1994. Genes galore: a summary of methods for accessing results from a large-scale partial sequencing of anonymous *Arabidopsis* cDNA clones. *Plant Physiol.* 106:1241–55

80. Nimmo GA, McNaughton GAL, Fewson CA, Wilkins MB, Nimmo HG. 1987. Changes in the kinetic properties and phosphorylation state of phospho*enol*pyruvate carboxylase in *Zea mays* leaves in response to light and dark. *FEBS Lett.* 213:18–22
81. Nimmo GA, Nimmo HG, Fewson CA, Wilkins MB. 1984. Diurnal changes in the properties of phosphoenolpyruvate carboxylase in *Bryophyllum* leaves: a possible covalent modification. *FEBS Lett.* 178: 199–203
82. Nimmo GA, Nimmo HG, Hamilton ID, Fewson CA, Wilkins MB. 1986. Purification of the phosphorylated night form and dephosphorylated day form of phosphoenolpyruvate carboxylase from *Bryophyllum fedtschenkoi. Biochem. J.* 239:213–20
83. Nimmo GA, Wilkins MB, Fewson CA, Nimmo HG. 1987. Persistent circadian rhythms in the phosphorylation state of phosphoenolpyruvate carboxylase from *Bryophyllum fedtschenkoi* leaves and in its sensitivity to inhibition by malate. *Planta* 170:408–15
84. Nimmo HG. 1993. The regulation of phosphoenolpyruvate carboxylase by reversible phosphorylation. In *Society for Experimen-*

tal Biology Seminar Series 53: Post-translational Modifications in Plants, ed. NH Battey, HG Dickinson, AM Hetherington, pp. 161–70. London: Cambridge Univ. Press

85. Ogawa N, Okumura S, Izui K. 1992. A Ca^{2+}-dependent protein kinase phosphorylates phosphoenolpyruvate carboxylase in maize. *FEBS Lett.* 302:86–88
86. O'Laughlin JT. 1988. *Mechanistic probes of the catalytic activity of the enzyme phosphoenolpyruvate carboxylase.* PhD thesis. Univ. Wis., Madison. 135 pp.
87. O'Leary MH. 1982. Phosphoenolpyruvate carboxylase: an enzymologist's view. *Annu. Rev. Plant Physiol.* 33:297–315
88. O'Leary MH, Hermes JD. 1987. Determination of substrate specificity of carboxylases by nuclear magnetic resonance. *Anal. Biochem.* 162:358–62
89. O'Leary MH, Rife JE, Slater JD. 1981. Kinetic and isotope effect studies of maize phosphoenolpyruvate carboxylase. *Biochemistry* 20:7308–14
90. Outlaw WH Jr. 1990. Kinetic properties of guard-cell phospho*enol*pyruvate carboxylase. *Biochem. Physiol. Pflanzen* 186: 317–25
91. Pacquit V, Santi S, Crétin C, Bui VL, Vidal J, Gadal P. 1993. Production and properties of recombinant C_3-type phospho*enol*pyruvate carboxylase from *Sorghum vulgare*: *in vitro* phosphorylation by leaf and root PyrPC protein serine kinases. *Biochem. Biophys. Res. Commun.* 197:1415–23
92. Pathirana SM, Vance CP, Miller SS, Gantt JS. 1992. Alfalfa root nodule phosphoenolpyruvate carboxylase: characterization of the cDNA and expression in effective and plant-controlled ineffective nodules. *Plant Mol. Biol.* 20:437–50
93. Peliska JA, O'Leary MH. 1989. Sulfuryl transfer catalyzed by pyruvate kinase. *Biochemistry* 28:1604–11
94. Pierre JN, Pacquit V, Vidal J, Gadal P. 1992. Regulatory phosphorylation of phospho*enol*pyruvate carboxylase in protoplasts from *Sorghum* mesophyll cells and the role of pH and Ca^{2+} as possible components of the light-transduction pathway. *Eur. J. Biochem.* 210:531–37
95. Poetsch W, Hermans J, Westhoff P. 1991. Multiple cDNAs of phosphoenolpyruvate carboxylase in the C_4 dicot *Flaveria trinervia*. *FEBS Lett.* 292:133–36
96. Rajagopalan AV, Devi MT, Raghavendra AS. 1994. Molecular biology of C_4 phospho*enol*pyruvate carboxylase: structure, regulation and genetic engineering. *Photosynth. Res.* 39:115–35

96a. Relle M, Wild A. 1994. *EMBL/GenBank/ DDBJ databases.* Accession number X79090

97. Rose IA, O'Connell EL, Noce P, Utter MF, Wood HG, et al. 1969. Stereochemistry of the enzymatic carboxylation of phosphoenolpyruvate. *J. Biol. Chem.* 244:6130–33
98. Samaras Y, Manetas Y, Gavalas NA. 1988. Effects of temperature and photosynthetic inhibitors on light activation of C_4-phosphoenolpyruvate carboxylase. *Photosynth. Res.* 16:233–42

98a. Schaeffer HJ, Forsthoefel NR, Cushman JC. 1995. Identification of enhancer and silencer regions involved in salt-responsive expression of Crassulacean acid metabolism (CAM) genes in the facultative halophyte *Mesembryanthemum crystallinum*. *Plant Mol. Biol.* 28:205–18

99. Schäffner AR, Sheen J. 1992. Maize C_4 photosynthesis involves differential regulation of phosphoenolpyruvate carboxylase genes. *Plant J.* 2:221–32
100. Schuller KA, Plaxton WC, Turpin DH. 1990. Regulation of phospho*enol*pyruvate carboxylase from the green alga *Selenastrum minutum*: properties associated with replenishment of tricarboxylic acid cycle intermediates during ammonium assimilation. *Plant Physiol.* 93:1303–11
101. Schuller KA, Turpin DH, Plaxton WC. 1990. Metabolite regulation of partially purified soybean nodule phospho*enol*pyruvate carboxylase. *Plant Physiol.* 94: 1429–35
102. Schuller KA, Werner D. 1993. Phosphorylation of soybean (*Glycine max* L.) nodule phospho*enol*pyruvate carboxylase in vitro decreases sensitivity to inhibition by L-malate. *Plant Physiol.* 101:1267–73
103. Sikkema KD, O'Leary MH. 1988. Synthesis and study of phosphoenolthiopyruvate. *Biochemistry* 27:1342–47
104. Stiborová M. 1988. Phosphoenolpyruvate carboxylase: the key enzyme of C_4-photosynthesis. *Photosynthetica* 22:240–63
105. Stockhaus J, Poetsch W, Steinmüller K, Westhoff P. 1994. Evolution of the C_4 phosphoenolpyruvate carboxylase promoter of the C_4 dicot *Flaveria trinervia:* an expression analysis in the C_3 plant tobacco. *Mol. Gen. Genet.* 245:286–93
106. Suzuki I, Crétin C, Omata T, Sugiyama T. 1994. Transcriptional and posttranscriptional regulation of nitrogen-responding expression of phospho*enol*pyruvate carboxylase gene in maize. *Plant Physiol.* 105: 1223–29
107. Tagu D, Crétin C, Bergounioux C, Lepiniec L, Gadal P. 1991. Transcription of a *Sorghum* phosphoenolpyruvate carboxylase gene in transgenic tobacco leaves: maturation of monocot pre-mRNA by dicot cells. *Plant Cell Rep.* 9:688–90
108. Taybi T, Sotta B, Gehrig H, Guclu S, Kluge M, Brulfert J. 1995. Differential effects of abscisic acid on phosphoenolpyruvate car-

boxylase and CAM operation in *Kalanchoë blossfeldiana*. *Bot. Acta* 108:240–46

109. Terada K, Izui K. 1991. Site-directed mutagenesis of the conserved histidine residue of phospho*enol*pyruvate carboxylase. His138 is essential for the second partial reaction. *Eur. J. Biochem.* 202:797–803
110. Terada K, Kai T, Okuno S, Fujisawa H, Izui K. 1990. Maize leaf phosphoenolpyruvate carboxylase: phosphorylation of Ser^{15} with a mammalian cyclic AMP-dependent protein kinase diminishes sensitivity to inhibition by malate. *FEBS Lett.* 259:241–44
111. Terada K, Murata T, Izui K. 1991. Site-directed mutagenesis of phosphoenolpyruvate carboxylase from *E. coli:* the role of His^{579} in the catalytic and regulatory functions. *J. Biochem.* 109:49–54
112. Terada K, Yano M, Izui K. 1992. Functional analysis of PEP carboxylase by site-directed mutagenesis. See Ref. 47, pp. 3: 823–26
113. Thomas M, Crétin C, Vidal J, Keryer E, Gadal P, Monsinger E. 1990. Light-regulation of phosphoenolpyruvate carboxylase mRNA in leaves of C_4 plants: evidence for phytochrome control on transcription during greening and for rhythmicity. *Plant Sci.* 69:65–78
114. Toh H, Kawamura T, Izui K. 1994. Molecular evolution of phospho*enol*pyruvate carboxylase. *Plant Cell Environ.* 17: 31–43
115. Vance CP, Gregerson RG, Robinson DL, Miller SS, Gantt JS. 1994. Primary assimilation of nitrogen in alfalfa nodules: molecular features of the enzymes involved. *Plant Sci.* 101:51–64
116. Van Quy L, Foyer C, Champigny ML. 1991. Effect of light and NO_3^- on wheat leaf phospho*enol*pyruvate carboxylase activity: evidence for covalent modulation of the C_3 enzyme. *Plant Physiol.* 97:1476–82
117. Vidal J, Pierre J-N, Echevarria C. 1996. The regulatory phosphorylation of C_4 phospho*enol*pyruvate carboxylase: a cardinal event in C_4 photosynthesis. In *Plant Gene Research*, ed. ES Dennis, B Hohn, PJ King, J Schell, DPS Verma. New York: Springer-Verlag. In press
118. Walsh C. 1979. *Enzymatic Reaction Mechanisms*, pp. 705–7. San Francisco: Freeman
119. Wang YH, Chollet R. 1993. In vitro phosphorylation of purified tobacco-leaf phospho*enol*pyruvate carboxylase. *FEBS Lett.* 328:215–18
120. Wang YH, Chollet R. 1993. Partial purification and characterization of phospho*enol*pyruvate carboxylase protein-serine kinase from illuminated maize leaves. *Arch. Biochem. Biophys.* 304:496–502
121. Wang YH, Duff SMG, Lepiniec L, Crétin C, Sarath G, et al. 1992. Site-directed mutagenesis of the phosphorylatable serine (Ser^8) in C_4 phospho*enol*pyruvate carboxylase from sorghum: the effect of negative charge at position 8. *J. Biol. Chem.* 267: 16759–62
122. Weigend M, Hincha DK. 1992. Quaternary structure of phosphoenolpyruvate carboxylase from CAM-C_4-and C_3-plants: no evidence for diurnal changes in oligomeric state. *J. Plant Physiol.* 140:653–60

122a. Wera S, Hemmings BA. 1995. Serine/threonine protein phosphatases. *Biochem. J.* 311:17–29

123. Willeford KO, Wedding RT. 1992. Oligomerization and regulation of higher plant phospho*enol*pyruvate carboxylase. *Plant Physiol.* 99:755–58
124. Winkler FJ, Schmidt H-L, Wirth E, Latzko E, Lenhart B, Ziegler H. 1983. Temperature, pH and enzyme-source dependence of the HCO_3^- carbon isotope effect on the phosphoenolpyruvate carboxylase reaction. *Physiol. Vég.* 21:889–95
125. Winter K. 1982. Properties of phospho enolpyruvate carboxylase in rapidly prepared, desalted leaf extracts of the Crassulacean acid metabolism plant *Mesembryanthemum crystallinum* L. *Planta* 154: 298–308
126. Wirsching P, O'Leary MH. 1988. 1-Carboxyallenyl phosphate, an allenic analogue of phosphoenolpyruvate. *Biochemistry* 27: 1355–60
127. Wu MX, Meyer CR, Willeford KO, Wedding RT. 1990. Regulation of the aggregation state of maize phospho*enol*pyruvate carboxylase: evidence from dynamic light-scattering measurements. *Arch. Biochem. Biophys.* 281:324–29
128. Wu MX, Wedding RT. 1985. Regulation of phosphoenolpyruvate carboxylase from *Crassula* by interconversion of oligomeric forms. *Arch. Biochem. Biophys.* 240: 655–62
129. Wu MX, Wedding RT. 1987. Regulation of phosphoenolpyruvate carboxylase from *Crassula argentea: further evidence on the dimer-tetramer interconversion. Plant Physiol.* 84:1080–83
130. Yanagisawa S, Izui K. 1992. Maize nuclear factors interacting with the C_4 photosynthetic phosphoenolpyruvate carboxylase gene promoter. See Ref. 47, 3:839–42
131. Yanagisawa S, Izui K. 1993. Molecular cloning of two DNA-binding proteins of maize that are structurally different but interact with the same sequence motif. *J. Biol. Chem.* 268:16028–36
132. Yanagisawa S, Izui K, Yamaguchi Y, Shigesada K, Katsuki H. 1988. Further analysis of cDNA clones for maize phospho*enol*pyruvate carboxylase involved in C_4 photosynthesis: nucleotide sequence of entire open reading frame and evidence for

polyadenylation of mRNA at multiple sites in vivo. *FEBS Lett.* 229:107–10

133. Yanai Y, Okumura S, Shimada H. 1994. Structure of *Brassica napus* phospho*enol*pyruvate carboxylase genes: missing introns causing polymorphisms among gene family members. *Biosci. Biotech. Biochem.* 58:950–53
134. Yano M, Terada K, Umiji K, Izui K. 1995. Catalytic role of an arginine residue in the highly conserved and unique sequence of phosphoenolpyruvate carboxylase. *J. Biochem.* 117:1196–1200
135. Zhang SQ, Outlaw WH Jr, Chollet R. 1994. Lessened malate inhibition of guard-cell phospho*enol*pyruvate carboxylase velocity during stomatal opening. *FEBS Lett.* 352: 45–48
136. Zhang XQ, Li B, Chollet R. 1995. In vivo regulatory phosphorylation of soybean nodule phospho*enol*pyruvate carboxylase. *Plant Physiol.* 108:1561–68

Annu. Rev. Plant Physiol. Plant Mol. Biol. 1996. 47:299–325

XYLOGENESIS: INITIATION, PROGRESSION, AND CELL DEATH

Hiroo Fukuda

Botanical Gardens, Faculty of Science, University of Tokyo, Hakusan, Tokyo 112, Japan

KEY WORDS: gene expression, programmed cell death, secondary cell walls, tracheary element, xylem

ABSTRACT

Xylem cells develop from procambial or cambial initials in situ, and they can also be induced from parenchyma cells by wound stress and/or a combination of phytohormones in vitro. Recent molecular and biochemical studies have identified some of the genes and proteins involved in xylem differentiation, which have led to an understanding of xylem differentiation based on comparisons of events in situ and in vitro. As a result, differentiation into tracheary elements (TEs) has been divided into two processes. The "early" process involves the origination and development of procambial initials in situ. In vitro, the early process of transdifferentiation involves the dedifferentiation of cells and subsequent differentiation of dedifferentiated cells into TE precursor cells. The "late" process, observed both in situ and in vitro, involves a variety of events specific to TE formation, most of which have been observed in association with secondary wall thickenings and programmed cell death. In this review, I summarize these events, including coordinated expression of genes that are involved in secondary wall formation.

CONTENTS

1040-2519/96/0601-0299$08.00

INTRODUCTION

The formation of xylem, the water-conducing tissue, has been a focus of many studies of differentiation in higher plants, not only because the function of xylem is essential to the existence of vascular plants but also because xylem formation appears to be a good model system for the analysis of differentiation in higher plants (2, 45, 51, 109, 124). Xylem is composed of tracheary elements (TE), parenchyma cells, and fibers. TEs, which are the distinctive cells of the xylem, are characterized by the formation of a secondary cell wall with annular, spiral, reticulate, or pitted wall thickenings. At maturity, TEs lose their nuclei and cell contents and leave a hollow tube that is part of a vessel or tracheid. The ease with which these differentiated cells can be identified by their morphological features, as well as the relatively straightforward induction of TE differentiation in vitro and the presence of many biochemical and molecular markers, is very advantageous for studies of cell differentiation (45). At the final stage of TE differentiation, cell death occurs, and therefore this differentiation also focuses our attention on a typical example of programmed cell death in higher plants. Furthermore, xylogenesis is important from an applied and biotechnological perspective, because biomaterials, such as cellulose and lignin in xylem, represent the prominent part of the terrestrial biomass and therefore play an important role in the carbon cycle (17).

Recent progress in studies of xylem differentiation has come from experiments both in vivo and in vitro and has allowed us to dissect the process of differentiation. Therefore, in this review, I discuss recent findings related to xylem differentiation—in particular, those derived from molecular approaches. The focus is mainly on the formation of the primary xylem, because of space limitation. For further information, I refer the reader to recent reviews on vascular differentiation, including secondary xylem formation and phloem formation (3, 4, 14, 23, 45, 124).

INITIATION

Plant Hormones

IN VIVO Considerable evidence indicates that plant hormones are involved in the initiation of vascular differentiation (2, 4, 45, 99, 123, 129). The continuity of vascular tissues along the plant axis may be a result of the steady polar flow of auxin from leaves to roots (2, 129). Jacobs (75) showed that indoleacetic acid (IAA), produced by expanding leaves and transported basipetally, was the limiting and controlling factor in the regeneration of xylem strands around a wound. Sachs (129) proposed the "canalization" hypothesis, which suggests that auxin flow that starts initially by diffusion induces the formation of a polar auxin transport cell system, which in turn promotes auxin transport and leads to canalization of auxin flow along a narrow file of cells. This continuous polar transport of auxin through cells finally results in the differentiation of a xylem strand. The differentiation of circular vessels in *Agrobacterium*-inducing crown galls also reflects circular movement of auxin that is produced additionally in the galls (5). The circular vessels also occur normally in healthy tissues such as suppressed buds (7) and branch junctions (94). Aloni & Zimmermann (8) proposed the "six-point" hypothesis, which suggests that high auxin levels near the young leaves induce many TEs that remain small because of their rapid differentiation, while low auxin concentrations further down result in slow differentiation and therefore in fewer and larger TEs. The overproduction of the product of an *Agrobacterium* auxin biosynthetic gene in transgenic *Petunia* plants caused an increase in the number of TEs and a decrease in their size (84). The inactivation of endogenous auxin in tobacco plants, transformed with *Pseudomonas* IAA-lysine synthase gene, decreased the number of TEs and increased their diameter (126). These results supported the hypothesis that the endogenous level of auxin plays a key role in controlling the initiation of TE differentiation and the size of TEs. However, the genes mentioned were expressed universally in various cells of transgenic plants under the control of the cauliflower mosaic virus (CaMV) 35S promoter. More directed modification of IAA levels at specific stages in given cells by use of stage-specific and cell-specific promoters will provide clearer details of the function of IAA in xylem differentiation.

Cytokinin from roots may also be a controlling factor in vascular differentiation (4, 45), although the relatively high levels of endogenous cytokinin in tissues often mask the requirement for cytokinin of xylem differentiation (45). Cytokinin promotes the vessel regeneration in the acropetal direction in the presence of IAA, which suggests that cytokinin increases the sensitivity of cambial initials and derivatives to auxin, which stimulates them to differentiate xylem cells (4, 13). The effect of the overproduction of cytokinin on xylem formation in transgenic plants is still controversial: In some cases, overproduction inhibits xylem formation (101); in other cases, it promotes the formation of a thicker vascular cylinder (97).

IN VITRO Induction of xylem cells has been achieved in callus, in suspension-cultured cells, and in excised tissues and cells (45, 51, 124). In vitro, xylem cells often differentiate as single or clustered TEs or as nested, discrete bands of TEs, but not as strands of vessels (51). In vitro systems are very useful for studies of TE differentiation at the cellular level because of easy induction, easy external control of differentiation, and in some cases, a high degree of synchrony. Auxin is a prerequisite for the induction of TE differentiation in vitro (4, 45, 123). Auxin is necessary both for the induction of differentiation and for the progression of such differentiation (27, 51).

Cytokinin promotes TE differentiation in a variety of cultured tissues only when present with auxin, but the requirement for cytokinin varies in different cultures, ranging from nonexistent to essential (45, 123). Mizuno & Komamine (105), using cultured phloem slices of carrot, found that the difference among requirements for cytokinin of TE differentiation in different cultivars reflected endogenous levels of cytokinin, which suggests that cytokinin may be essential for induction of differentiation. Cytokinin is necessary for the progression of TE differentiation as well as for its induction, but the requirement for cytokinin may be limited at a much earlier stage of differentiation than that of auxin (27, 51). Molecular approaches to elucidation of the mechanism of induction of differentiation by auxin and cytokinin have just recently been initiated.

Although an exogenous supply of other hormones does not induce xylem differentiation in vitro, endogenous gibberellin and ethylene seem to contribute to xylem differentiation as a hidden promoter or inducer (45). In addition, a role of endogenous brassinosteroid seems likely during the process of TE differentiation in *Zinnia* cells (74). The identification of brassinosteroids from the cambial region of pine trees (83) is consistent with the requirement for brassinosteroids of TE differentiation.

Wounding

Mechanical wounding often induces transdifferentiation of parenchyma cells into TEs, which is typically demonstrated by the formation of wound vessel members around wounding sites (75). Wounding causes the interruption of vascular bundles, which disturbs hormonal transport along the bundles. This disturbed hormonal transport may result in the formation of new vascular tissues around the wound (4). Church & Galston (28) have shown that TE formation from mesophyll cells is substantially promoted in *Zinnia* leaf disks whose upper or lower epidermis is peeled off. This result is consistent with the observation that mechanically isolated mesophyll cells of *Zinnia* differentiate at high frequency (48). In fact, during the first 12 h after their isolation, isolated *Zinnia* mesophyll cells do not seem to react to plant hormones (51), and most of the genes that are expressed preferentially during the first 12 h of culture appear to be induced by the isolation or culture of *Zinnia* cells and not by plant hormones (106). Mechanically damaged cells are known to stimulate rapid induction of wound responses in suspension-cultured cells of *Lycopersicon peruvianum,* such as a transient alkalinization of the culture medium and ethylene synthesis (41). Plants react to localized mechanical wounding both locally and systemically. Systemin, which is involved in systemic signaling that leads to the induction of protease inhibitors, was able to replace the mechanically damaged cells in the *L. peruvianum* system (41). Similarly, systemic wound signals might be involved in the initiation of transdifferentiation of parenchyma and epidermal cells into TEs.

EARLY PROCESS

The differentiation of TEs is divided into four ontogenetic processes: cell origination, cell elongation, secondary wall deposition and lignification, and wall lysis and cell autolysis (151). In this review, the process of TE differentiation is considered first to be divided into an early process, which includes all the events that occur before TE-specific events such as secondary wall formation and autolysis, and the late process, which includes TE-specific events. These two processes are further dissected below. I first consider the early process of differentiation both in situ and in vitro.

Origin and Development of Procambial Initials

Procambial cells originate in the early embryo. Precursor cells of the procambium of *Nicotiana* and *Trifolium* appear in embryos at the 20-cell stage and the 120-cell stage, respectively (54). In *Arabidopsis,* the procambial primordium, which consists of eight narrow cells, is formed by the late globular stage of the embryo, and it forms the primary xylem tissues in the torpedo-shaped embryo (19). The strong interconnections between cells of the embryo through developed plasmodesmata suggest the presence of diffusible morphogens that are responsible for the organized differentiation of a variety of cells, including vascular cells (76). In plants, the apical meristem serves as a continuous source of the procambial initials (151).

Langdale et al (91) analyzed the cell lineage of vascular tissues in maize leaves using six spontaneous striping mutants. They found that the formation of lateral and intermediate veins was initiated most often by divisions that contributed daughter cells to both the procambium and the ground meristem. They also showed that intermediate veins are multiclonal in both the transverse and the lengthwise directions. Clonal analysis of many plants has shown that cells are not assigned a particular fate early in development (54). The data of Langdale et al (91, 92) also suggest that cells can differentiate according to their position in the leaf, irrespective of their clonal history. Procambial initials elongate and differentiate into different types of vascular cell such as TEs, xylem parenchyma cells, sieve elements, and phloem parenchyma cells. There is little evidence as to when the determination of differentiation into such different types of cell occurs. It is unknown, for example, whether the determination occurs in cambial initials before elongation or in mature procambial cells, and it is not known what signals are involved in such determination.

Early ultrastructural observations indicated that the development of intracellular organelles, i.e. the enlargement of the vacuole and nuclei, and the increases in numbers of ribosomes and the extent of the rough endoplasmic reticulum occur during differentiation along the file of cell lineage from procambial initials to xylem cells (51, 90). To date, however, a cytological description is the only information that is related to the origin of procambial precursor cells and even the procambial cells themselves (54). We are anxiously awaiting the identification of molecular and biochemical markers that will allow us to specify particular stages in the development of procambial cells. Some genes that have been isolated from differentiating *Zinnia* cells may provide such markers (31, 32, 158, 159).

Mutants are promising tools for elucidation of details of the initiation and progression of xylem differentiation. However, because a severe defect in vascular tissues is likely to cause the death of the plant, we need a novel strategy for generation of useful mutants. Embryo-lethal mutants of *Arabidopsis,* maize, and rice (71, 102) may help us to analyze the early process of xylem development in the embryo. Jürgens and colleagues (76), who isolated mutants that affect body organization in the *Arabidopsis* embryo, found that in the *gnom* mutant from which the root was absent and in which the cotyledon was strongly reduced in size or eliminated, TEs were present but were not interconnected to form strands. Instead, they were arranged in clusters or as single isolated cells. This mutant may be useful for studies of the involvement of intercellular interactions in vessel formation.

Early Events in Transdifferentiation

The *Zinnia* experimental system in which single mesophyll cells transdifferentiate synchronously into TEs without cell division has proved to be extremely useful for studies of the sequence of events in TE differentiation (26, 45, 46, 124, 144). Detailed analyses of the transdifferentiation have provided a number of cytological, biochemical, and molecular markers that should be useful for the dissection of the process of transdifferentiation (45, 46). The various markers indicate that the process of transdifferentiation can be divided into Stages I, II, and III (Figure 1).

Stage III corresponds to the late process of TE differentiation in situ. At Stage III, various enzymes and structural proteins associated with the secondary wall thickenings and autolysis, as well as the corresponding genes, are expressed.

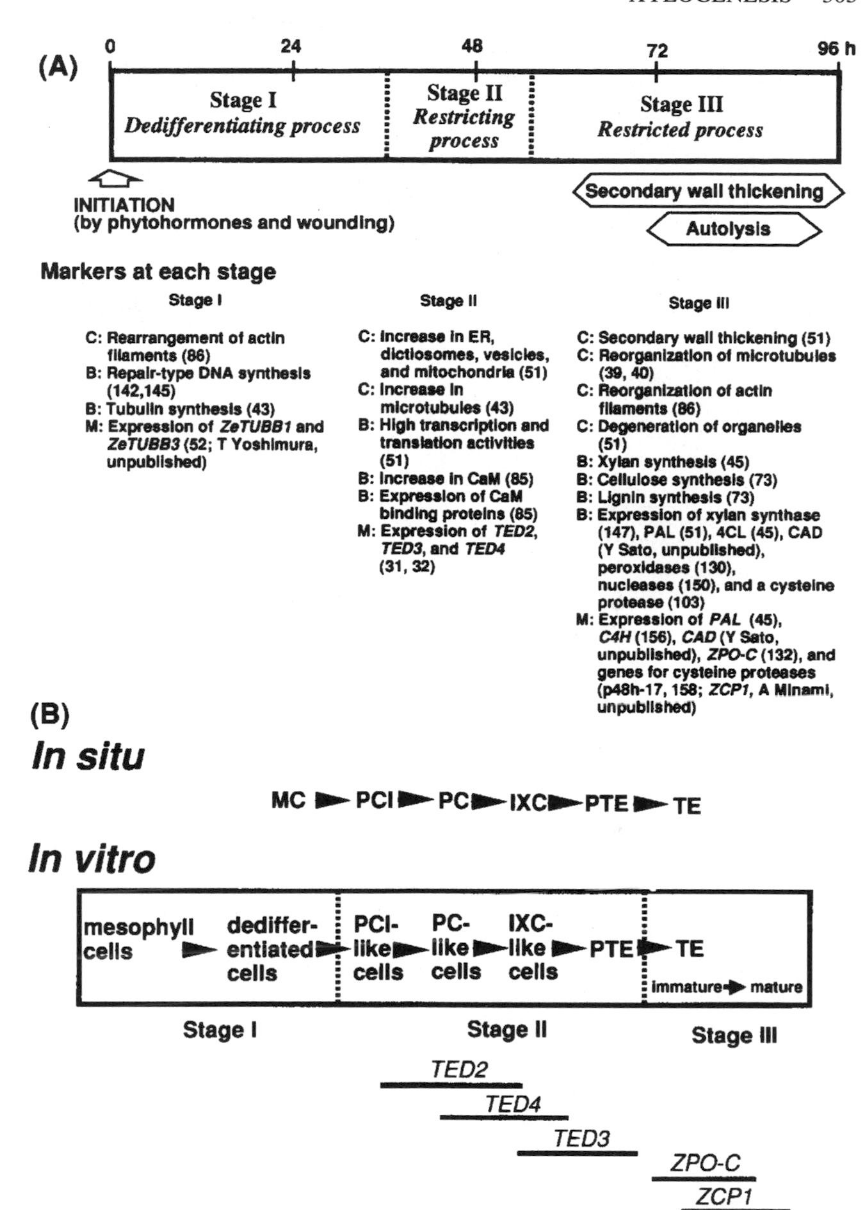

Figure 1 Sequence of events during transdifferentiation of single mesophyll cells of *Zinnia elegans*. (A) Sequence of events during transdifferentiation. C, B, and M represent cytological, biochemical, and molecular markers, respectively. (B) A comparison between the sequences of events in TE differentiation in situ and in vitro. *TED2, TED4, TED3, ZPO-C* (a gene for a peroxidase), and *ZCP1* (a gene for a cysteine protease) are expressed in this order. Bars represent the periods of their expression. MC, meristematic cells; PCI, procambial initials; PC, procambial cells; IXC, immature xylem cells; PTE, precursors of TEs; TE, tracheary elements.

Stage I, which immediately follows the induction of differentiation, corresponds to the dedifferentiation process during which isolated mesophyll cells lose their potential to function as photosynthetic cells and acquire the ability to grow and differentiate in a new environment. For example, at this stage, the reticulate arrays of actin filaments around chloroplasts, which may anchor chloroplasts to the plasma membrane, turn into a three-dimensional network over the entire length of the cell (86), which causes chloroplasts to leave the vicinity of the plasma membrane and the mesophyll cells to lose their photosynthetic capacity. It should be emphasized that this process of dedifferentiation is not accompanied by cell division. This stage seems not to be specific to TE differentiation but to be necessary for transdifferentiation into TEs (46).

The characterization of molecular markers, isolated as genes (*TED2, TED3, TED4*) that are expressed prior to the secondary wall thickenings, led to the definition of a new stage of transdifferentiation—Stage II (Figure 1) (31, 32)—between the very early stage (Stage I) and the late stage (Stage III). These genes are expressed preferentially in cells in cultures in which differentiation is induced 12–24 h before the secondary wall thickenings (31, 32). Ye & Varner (158) also isolated a cDNA clone identical to *TED4* from *Zinnia* cells. In situ hybridization demonstrated that the expression of these genes was restricted to cells that were involved in vascular differentiation even during the development of intact plants. *TED3* transcripts were expressed specifically in differentiating TEs or in the cells that were expected soon to differentiate into TEs. The expression of *TED4* transcripts was restricted to vascular cells or future vascular cells and, in particular, to immature xylem cells that did not show any morphological changes. *TED2* transcripts were restricted to procambial regions of root tips as well as to immature phloem and immature xylem cells at the boundary between the hypocotyl and root. Both in situ and in vitro *TED* transcripts were expressed in the same order as follows: *TED2,* then *TED4,* and finally, *TED3.* This experiment suggested that Stage II of transdifferentiation corresponds to the process of differentiation from procambial initials to precursors of TEs in situ (Figure 1) (32, 46). Fukuda (46) proposed the hypothesis that multistep determination occurs during differentiation into TEs from cells that have not been committed, such as meristematic cells and dedifferentiated cells. Their potency to differentiate is restricted from pluripotency to single potency, such that they only form TEs (Figure 1). According to this hypothesis, the final determination occurs just before secondary walls are formed. Therefore, Stages I, II, and III in *Zinnia* can be defined as the dedifferentiating, restricting, and restricted stages, respectively. Gahan & Rana (53, 119), also using a carboxylesterase as a marker of stele tissues, found evidence that multistep programming occurs during the transdifferentiation from parenchyma cells to TEs in wounded roots of pea, namely, the determination of the differentiation of parenchyma cells into stele cells and the subsequent determination of the differentiation of stele cells into TEs. Further information on the sequence of events during transdifferentiation of *Zinnia* mesophyll cells can be found in earlier reviews (45, 46, 52)

Factors Affecting the Early Process

CELL DIVISION AND DNA SYNTHESIS Cell division is obviously important for the continuous supply of xylem cells. The regeneration of xylem tissues around wound sites involves ordered cell division for generation of well-organized vessels. Therefore, the fine control of the organization of cell division must also be important for the formation of xylem tissues.

The role of cell division as an initiator of transdifferentiation into TEs has been discussed (34, 51). A requirement of transdifferentiation for cell division or DNA replication during the S phase of the cell cycle has been proposed mainly on the basis of results with inhibitors of cell division or DNA synthesis (137). However, transdifferentiation can occur without the progression of cell division or the S phase in *Zinnia* and Jerusalem artichoke, which indicates that cell cycling is not essential for the initiation of transdifferentiation (49, 116). On the basis of a detailed analysis of the time course of cell division and differentiation in cultured *Zinnia* cells, Fukuda & Komamine (49) suggested that the early process of transdifferentiation (Stages I and II) may be compatible with the progression of the cell cycle. Nevertheless, it was also evident early on in their study that a variety of inhibitors of DNA synthesis prevented transdifferentiation (50), even in the *Zinnia* system in which replication of DNA in the S phase is not required (49). Thus, there might be a requirement for some minor DNA synthesis of transdifferentiation (50). Sugiyama and colleagues (141, 143, 145) suggested that this minor DNA synthesis may be repair-type DNA synthesis, a suggestion that is consistent with the fact that inhibitors of poly(ADP-ribose) polymerase, which plays an important role in DNA excision repair, prevent TE differentiation in *Zinnia* (142), pea, and Jerusalem artichoke (117). This repair-type DNA synthesis may be involved in the early process of transdifferentiation from parenchyma cells to TEs.

CALCIUM IONS AND CALMODULIN The essential role of calcium ions and calmodulin (CaM) in certain types of signaling has been established in a variety of organisms including plants (20, 122). It has also been suggested that TE differentiation may be regulated by the calcium/CaM system, because differentiation can be inhibited specifically by a decrease in the intracellular concentration of calcium ions, a calcium channel blocker, and CaM antagonists (85, 121). CaM antagonists were only effective early in Stage I and late in Stage II of the transdifferentiation of *Zinnia* mesophyll cells into TEs (85, 121). At Stage II, CaM levels increase transiently and, subsequently, a few CaM-binding proteins start to be expressed in a differentiation-specific manner (85), which suggests that the calcium/CaM system is involved in the progression of Stage II, i.e. in the differentiation from procambial initials to TE precursor cells. The possibility that calcium/CaM is involved in the progression of Stage II is also supported by higher levels of membrane-associated calcium ions in TE precursor cells than in control *Zinnia* cells (120). The formation of the secondary walls might itself also be under the control of calcium/CaM (47, 121).

LATE PROCESS

The most striking features of TE formation are the development of patterned secondary walls and the autolysis of cell contents and walls, which occur in the late process of differentiation. The secondary cell wall is composed of cellulose microfibrils arranged parallel to one another and to the bands of the secondary wall and of cementing substances that contain lignin, hemicellulose, pectin, and protein, which add strength and rigidity to the wall. These substances are synthesized and deposited cooperatively during secondary wall formation. The autolysis is a typical example of programmed cell death and involves activation of the expression of degradative enzymes such as nucleases and proteases.

Pattern Formation of Secondary Walls

Microtubules in differentiating TEs are located as bands over the thickenings of secondary walls in various plants (59). Disturbance of microtubule organization by colchicine and taxol caused the formation of unusual secondary wall thickenings (40). Such cytological studies led to the suggestion that microtubules determine the wall pattern by defining the position and orientation of secondary walls, probably by guiding the movement of the cellulose-synthesizing complex in the plasma membrane (59). Microtubule arrays change dramatically from a longitudinal to a transverse orientation prior to secondary wall formation in *Zinnia* (39, 40). This rearrangement of microtubule arrays is under the control of actin filaments (47).

The number of cortical microtubules increases prior to the reorganization of microtubules during TE differentiation in cultured *Zinnia* cells (43) and *Azolla* (63). The increase is caused by an increase of tubulin levels, which results from tubulin synthesis de novo. This synthesis starts as early as 4 to 8 h after the induction of differentiation (43). The degradation of tubulin is also associated with the reorganization of microtubules (44). The small genome of *Arabidopsis* contains at least six genes for α-tubulin and at least nine genes for β-tubulin (55). Recent studies have provided evidence of the differential expression of various isotypes of plant tubulin genes in specific organs and tissues during development, in addition to the regulation of their expression by environmental factors such as temperature and phytohormones (55). Each tubulin gene seems to have a unique spatial and temporal pattern of expression. Three different cDNAs for β-tubulin—*ZeTUBB1, ZeTUBB2, ZeTUBB3*—were isolated from differentiating cells of *Zinnia* (52). RNA gel blot analysis indicated that the levels of expression of *ZeTUBB1* and *ZeTUBB3* transcripts increased rapidly late in Stage I. The very recent results of in situ hybridization suggest the preferential expression of *ZeTUBB1* transcripts in differentiating xylem cells, as well as in actively dividing cells (T Yoshimura, T Demura & H Fukuda, unpublished data).

Polysaccharide Synthesis

During the formation of secondary walls, levels of cellulose and hemicellulose increase, and the deposition of pectin ceases (109). Putative complexes of cellulose-synthesizing enzymes, visible as rosettes, are localized in the plasma membrane over regions of secondary wall thickenings in TEs (61). The rosettes are also present in the Golgi apparatus throughout the enlargement of secondary wall thickening, which suggests that new rosettes are continuously inserted into the plasma membrane via Golgi vesicles during secondary wall formation (61). Northcote et al (110), using an antiserum against β1-4 oligoxylosides, demonstrated the presence of xylose not only in the secondary thickening but also in the Golgi vesicles of TEs. Hogetsu (69), in experiments with fluorescein-conjugated wheat germ agglutinin (WGA), also found that a hemicellulose(s), probably xylan, was localized in the secondary walls of TEs in many angiosperm plants.

Polysaccharide synthesis is regulated in the level of the availability of nucleoside phosphate sugars as precursors by their location, transport, and feedback controls during secondary wall formation (108). In addition, control of the activity of the membrane-bound enzymes involved in polysaccharide synthesis is critical to secondary wall formation during xylem differentiation (109). Increased activity of xylosyltransferase (16, 147) and decreased activities of polygalacturonic acid synthase (15) and arabinosyltransferase (16) have been noted in conjunction with secondary wall formation during TE differentiation. Rodgers & Bolwell (125) succeeded in the partial purification of Golgi-bound arabinosyltransferase and xylosyltransferase from French bean. The xylosyltransferase with a molecular mass of 40 kDa, with increased activity during secondary wall formation, was suggested to be a xylan synthase. Particulate membrane preparations isolated from differentiating xylem cells of *Pinus sylvestris* included an enzyme complex that was active in the synthesis of hemicellulose (30). This enzyme complex, resembling that for cellulose synthesis, is constructed in part on the endoplasmic reticulum and is then transferred to the Golgi apparatus, where it is modified and sorted.

Synthesis of Cell Wall Proteins

Recent progress in molecular approaches has permitted us to identify structural proteins, many of which are insoluble and therefore difficult to study directly, on the cell walls of vascular tissues (22, 77, 109). They belong to four classes: hydroxyproline-rich glycoproteins (HRGPs), proline-rich proteins (PRPs), glycine-rich proteins (GRPs), and arabinogalactan proteins (AGPs) (22).

One family of HRGPs contains a characteristic Ser-$(Hyp)_4$ pentapeptide sequence. Extensins are the best-characterized proteins among the HRGPs. Extensins contribute to the mechanical strength of cell walls (22) and also to the control of intracellular organization, e.g. the control of microtubule organization (1). Extensins have been located in the cell wall of specific types of cell, including vascular cells (21). The expression of mRNAs for HRGPs is regulated developmentally, as well as by environmental cues such as wounding and infection. Some *HRGP* genes are expressed strongly in vascular cells, including cambial and phloem cells, in soybean tissues (157), and in regions where vascular cells and sclerenchyma are initiated in embryos, leaves, and roots of maize (139). However, such expression is not specific to vascular tissues. Recently, Bao et al (9) isolated an extensin-like protein from the xylem of loblolly pine, and they indicated that the protein was present in secondary cell walls of xylem cells during lignification and remained as a structural component of the cell walls in the wood.

GRPs are a class of proteins with a very repetitive primary structure. About 60% of their residues are glycine residues that are arranged predominantly in $(Gly\text{-}X)_n$ repeats (77). Some GRPs such as the *Petunia* GRP1 and French bean GRP1.8 are present on cell walls of vascular cells (77). GRP1.8 is localized on the cell walls of the primary xylem elements and primary phloem of many plant species (80, 157). Detailed immunoelectron microscopic studies suggested that GRP1.8 is produced by xylem parenchyma cells that export the protein to the walls of protoxylem vessels (127).

AGPs are characterized by high levels of arabinose and galactose in the carbohydrate moiety. The protein part, which contributes only 2–10% of the molecular mass, is commonly characterized by high levels of hydroxylproline, serine, and alanine (22). Two antibodies against AGPs—JIM13 and JIM14—bind to developing xylem cells and sclerenchyma cells in maize coleoptiles (134). In carrot, JIM13e binds to the future xylem as well as the epidermis of the root apex, whereas JIM14e binds to all cells of the root.

WGA, which has a strong affinity for a sequence of three β-(1-4)-linked *N*-acetyl-D-glucosamine residues, binds specifically to the secondary walls of TEs (69, 155). Wojtaszek & Bolwell (155) isolated three novel glycoproteins that bind to WGA. An antibody against one of the glycoproteins, SWGP90, revealed that this glycoprotein was localized in the secondary walls of TEs, xylem fibers, and phloem fibers. Another novel type of wall protein, Tyr- and Lys-rich protein from tomato, has also been shown to be localized in the secondary walls (35). Thus, evidence is accumulating that some specific proteins are localized in secondary walls and function in the formation of the networks of the secondary walls.

Lignin Synthesis

Lignin is one of the most characteristic components of secondary walls and has been more intensively studied than any of the other macromolecules whose synthesis is closely associated with secondary wall formation. However, the synthesis of lignin is also induced by wounding or infection. I focus here on the lignification that occurs in association with xylem differentiation. The details of lignification can be found in several excellent reviews (17, 68, 95).

Lignins are cross-linked polymers composed of *p*-hydroxyphenyl (H), guaiacyl (G), and syringyl (S) units (68). Gymnosperm lignins consist essentially of G units. Angiosperm lignins are mainly mixtures of G and S units. Grass lignins are composed of H, G, and S units. The distribution of the G and S units is modified in the xylem cells and phloem fibers of *Coleus* by auxin and gibberellic acid (6). Terashima et al (149) indicated that H and S units are deposited, respectively, at the initial stage of lignification within middle lamella and, later, in the secondary walls of xylem cells. Thus, the composition of lignins also involves developmental and spatial events. An inhibitor of phenylalanine ammonia-lyase (PAL) activity, α-aminooxy-β-phenylpropionic acid not only blocked lignin synthesis but also caused the release of polysaccharides containing xylose into the culture medium of *Zinnia* cells (73). Inhibitors of cellulose synthesis such as 2,6-dichlorobenzonitrile dispersed deposits of lignin and reduced the level of xylan in the thickenings of differentiating TEs of *Zinnia* (146, 148). These results suggest that there is a mechanism whereby some components of secondary walls mediate patterning of deposition of other components. Indeed, lignin can make cross-links with many cell wall polymers such as polysaccharides and proteins through ester, ether, and glycosidic linkages (72). The cross-links, together with the hydrophobicity and chemical stability of lignins, reinforce and waterproof the walls, providing mechanical support, a pathway for solute conductance, and protection against pathogens.

THE GENERAL PHENYLPROPANOID PATHWAY The biosynthesis of lignin involves three pathways, known as the shikimate, the general phenylpropanoid, and the specific lignin pathways (95). The general phenylpropanoid pathway involves PAL, cinnamate hydroxylase (C4H), *O*-methyltransferase (OMT), and

4-coumarate:CoA ligase (4CL), all of which seem to be expressed in association with xylogenesis and have been suggested as markers of lignification during xylem differentiation (45). Ye et al (156) indicated that the activity and transcripts of caffeoyl-CoA-3-*O*-methyltransferase (CCoAOMT) were expressed specifically in differentiating TEs both in the in vitro culture system and in intact plants of *Zinnia,* whereas the expression of caffeic acid–OMT transcripts was not correlated with lignification during TE differentiation in vitro or in the plant (160). These results strongly suggest that CCoAOMT is involved in an alternative methylation pathway to lignin that is predominant in differentiating *Zinnia* cells, although the question remains as to whether this pathway is predominant in differentiating TEs in many plant species. The end products of this general pathway, the hydroxycinnamoyl CoAs, however, are precursors not only of lignins but also of other phenolic compounds such as flavonoids and stilbenes (60), a fact that complicates the analysis of lignin synthesis.

A gene for *S*-adenosylmethionine synthase (sam-1), which serves as a donor of methyl groups in numerous transmethylation reactions, was reported to be expressed preferentially in vascular tissues of *Arabidopsis* (115). This finding suggests that during the differentiation of xylem cells, genes involved in the synthesis of compounds that are not specific to but are necessary for differentiation are also expressed preferentially.

THE SPECIFIC LIGNIN PATHWAY Hydroxycinnamoyl-CoA esters are converted to their corresponding alcohols by cinnamoyl CoA reductase (CCR) and cinnamyl alcohol dehydrogenase (CAD), which are specific to the lignin pathway. CCR, isolated from differentiating xylem of *Eucalyptus gunnii,* had approximately equal affinity for *p*-coumaroyl CoA, feruloyl CoA, and sinapoyl CoA, which suggests that CCR may not contribute to the regulation of monomer composition of lignins in plants (56).

CADs and their corresponding cDNA clones have been isolated from various angiosperms and gymnosperms (17). Two different isoforms with different molecular masses and different substrate-specificities exist in angiosperms such as tobacco (57), while gymnosperms have a single CAD isoform (112). This difference between angiosperms and gymnosperms has been discussed in terms of the progressive evolution of lignins from G lignin to S lignin (17). The expression of β-glucuronidase (GUS), driven by the *Eucalyptus* CAD promoter, indicated that CAD genes are expressed preferentially in the primary xylem vessels in young regions of the stem and in the xylem rays in more mature regions of the stem (42).

In differentiating TEs, the cinnamyl alcohols are delivered to the cell walls by vesicles that are derived from the Golgi apparatus or the endoplasmic reticulum. At the cell walls, they are polymerized into lignin by wall-bound peroxidase isoenzymes via free-radical formation in the presence of H_2O_2 (95). In differentiating *Zinnia* cells, a cationic isozyme of peroxidase (P5), bound ionically to the cell walls, was shown to be involved in lignin synthesis (130, 131). It was suggested that the enzyme was fixed tightly into the secondary walls with the progression of lignin deposition (130). A histochemical test for H_2O_2 (114) and observations by electron microscopy (29) indicated that H_2O_2 is localized preferentially in cells that are undergoing lignification such as TEs, although Schopfer (135) also showed the presence of H_2O_2 in the phloem and sclerenchyma, but not in the xylem, by tissue blots of soybean. Many genes for peroxidases have been isolated from a variety of plant species (89). However, there is no direct evidence for the isolation of a peroxidase gene that is specifically responsible for lignin synthesis during the differentiation of TEs (45). A cDNA clone that may correspond to P5 was recently isolated from a cDNA library of *Zinnia* differentiating cells (132). The characterization of this clone may provide interesting insights.

Laccase, which is a ubiquitous polyphenol oxidase and oxidizes polyphenols in the absence of H_2O_2, is another candidate for the enzyme that catalyzes the oxidation of cinnamyl alcohols. Laccases or laccase-like oxidase are colocalized with lignin in the secondary walls of the xylem in various plant species, and they can catalyze the polymerization of lignin precursors in vitro (10, 37). However, there is also a report that xylem cells are only capable of oxidizing diaminobenzidine on the secondary walls in the presence of H_2O_2 (11). In addition, a serious problem is related to the difficulties encountered in attempts to separate laccases from other oxidases (17). The laccases and laccase-like oxidases that have been isolated from different plant species have different molecular masses, e.g. 38 kDa for the protein from mung bean (24), 90 kDa for that from *Pinus* (10), and 56 or 110 kDa for that from sycamore (37). O'Malley et al (113) speculated that the heterogeneity of laccases might result in the formation of heterologous lignins. The isolation of laccases specific for lignin synthesis and their genes are now of considerable importance.

Lignification proceeds even after TEs have lost their cell contents (138). The expression of GUS under the control of promoters of *CAD, PAL,* and *4CL* suggested that genes for these lignin-related enzymes are expressed in cells adjacent to differentiating TEs, as well as in differentiating TEs (42, 66). Immunolocalization studies have also confirmed that PAL and 4CL are accumulated in developing metaxylem tissues and in cells adjacent to the metaxylem (138). It appears, therefore, that lignin precursors can be supplied by parenchyma cells adjacent to differentiating TEs. During lignin deposition in conifers, the monolignol precursors may be transported to the cell walls of TEs as glucosides such as coniferin (133). A β-glucosidase specific for coniferin has been isolated from lodgepole pine, and its activity seems to be localized to the area of differentiating xylem (33). Thus, in conifers, lignin precursors may be supplied as β-glucosides from the adjacent parenchyma cells to the secondary walls of differentiating TEs. However, lignification in mature

TEs occurs even in isolated single *Zinnia* cells in the absence of direct cell-to-cell interactions (H Fukuda, unpublished observation). It is very important now to identify the cells in which lignin precursors are synthesized, how these precursors are transported, and whether the site of their synthesis changes with developmental stage or from tissue to tissue.

MUTANTS IN LIGNIN BIOSYNTHESIS Mutants that have a defect in the pathway to lignin are useful not only for the analysis of the complex regulation of lignin synthesis but also for the improvement of biomass via modifications of lignin content and composition (17). The *brown midrib* mutants of maize have reddish-brown pigmentation in the leaf midrib, which is associated with lower levels and a change in quality of lignins (88). Vignols et al (152) demonstrated that the *brown midrib3* mutation of maize was located in the gene for OMT. Pillonel et al (118) also found that the *brown midrib6* mutant of *Sorghum* resulted from a genetic lesion at a single locus, which resulted in decreased activities of both CAD and OMT. This result suggests that the mutated gene may be a regulatory gene that is responsible for control of the expression of CAD and OMT genes. The *Arabidopsis* mutant *fah1,* which produces abnormal lignins that lack syringyl units, sccms to havc a genetic lesion that results in a malfunctioning feruloyl-5-hydroxylase (25).

Genetic manipulation of lignin metabolism by antisense and sense suppression of gene expression has been attempted for several enzymes including PAL (12), OMT (38), and CAD (62). The analysis of transgenic plants in which PAL expression was suppressed indicated that the reaction catalyzed by PAL became the dominant rate-limiting step in lignin biosynthesis when PAL levels were below 20–25% of wild-type levels (12). Antisense expression of a homologous CAD gene caused strong suppression of CAD activity in tobacco plants, with the resultant production of modified lignin in which less cinnamyl alcohol monomers and more cinnamyl aldehyde monomers were incorporated into lignin than in control plants, with little quantitative change in lignin (62). The "antisense plants" had red-brown xylem tissues, resembling the *brown midrib* mutants.

Coordinated Regulation of Gene Expression

Lignin-related enzymes are coordinately expressed (87). Recent progress in promoter analyses of their genes is helping to elucidate the mechanism of such coordination. PAL is encoded by a single-copy gene in gymnosperms (140) but by a small family of genes in angiosperms (100). The isotypes of *PAL* genes in angiosperms are expressed differently in different organs and in response to different environment stimuli, such as wounding, infection, and light (98). 4CL is encoded in parsley by two highly homologous genes, which show no evidence of differential regulation (36), and it is encoded in *Arabi-*

dopsis by a single gene (93). Histochemical analysis of transgenic tobacco plants with a bean *PAL2* promoter-GUS chimeric gene and a parsley *4CL-1* promoter-GUS chimeric gene revealed that the promoters were active in differentiating xylem cells but were also active in other types of cell. Both promoters appear to transduce a complex set of developmental and environmental stimuli into an integrated spatial and temporal program of gene expression (65, 96).

Analysis of the regulatory properties of promoters of the bean *PAL2* and parsley *4CL-1* in transgenic tobacco plants showed that functionally redundant *cis* elements, located in regions proximal to the transcription start site, are essential for expression in xylem (64, 65, 96). A sequence in the *PAL* promoter, (T)CCACCAACCCC (C) (ACII), contained a negative element that suppresses the activity of a cryptic *cis* element required for expression in phloem (64, 96). This sequence corresponds to a conserved motif in a number of *PAL* promoters from bean, parsley, and *Arabidopsis,* which have a consensus sequence of CCA(A/C)C(A/T)AAC(C/T)CC (100). A similar element was also found in the negative regulatory sequence of 4CL-1 (65). Thus, the two enzymes in the general phenylpropanoid pathway seem to be regulated coordinately by a negative *cis* element. However, this motif has been found in a number of other stress-inducible promoters (60, 64), and moreover, the motif is also involved in gene expression specific to petals (64). Thus, other factors may be necessary for gene expression specific to xylem cells. The ACI motif in *PAL2* promoter has also been suggested to suppress gene expression in phloem (64). These negative elements also promote the gene expression in xylem tissues. The maize Myb protein P binds to overlapping hexameric repeats of CC(T/A)ACC and CCACC (58). The ACI and ACII elements are both composed of the overlapping hexamers, suggesting that tissue-specific *trans*-acting elements that bind to AC elements may be Myb proteins (64, 128).

The expression of the bean *GRP1.8* promoter has also been analyzed in detail (77). Transgenic tobacco plants containing a chimeric gene of a fragment starting at nucleotide –427 of the *GRP1.8* promoter fused to the GUS reporter gene expressed GUS activity in the vascular tissue of roots, stems, leaves, and flowers (81). The promoter contained at least five different regulatory elements that are involved in tissue-specific expression: SE1, SE2, RSE, VSE, and NRE (78, 79). RSE was essential for the promoter activity in roots and induced vascular tissue–specific expression. Keller & Heierli (79) also indicated that five tandem aligned RSE elements could confer vascular tissue–specific expression on the CaMV 35S minimal promoter. However, it is unknown which cells in vascular tissues express this gene at present. VSE was also essential for vascular tissue–specific expression. NRE seems to be a unique negative regulatory sequence that represses promoter activity in nonvascular cells of leaves and stems. Thus, negative control of tissue-specific expression of genes could be a general mechanism in vascular differentiation. The *GRP1.8* promoter is regulated by a combinatorial mechanism that can integrate the actions of different regulatory elements to yield a vascular tissue–specific pattern of expression (79). Keller's group isolated a cDNA clone for a bZIP-type transcription factor that binds to the NRE element (77). It has also been reported that a nuclear factor that binds to the promoter of an extensin gene, EGBF-1, is present in the phloem and xylem of carrot roots (70).

Cell Death

Genetically programmed cell death plays an important role in the development of multicellular organisms. Differentiation into TEs is a typical example of programmed cell death in higher plants, and mature TEs are completed by the loss of cell contents including the nucleus, plastids, mitochondria, Golgi apparatus, and the endoplasmic reticulum, and by the partial digestion of the primary walls (51, 111, 124). In animals, programmed cell death is often apoptotic (18), and the death of TEs has been discussed in terms of apoptosis (104, 134). However, many observations by electron microscopy have failed to reveal any substantial similarity between apoptosis and the autolysis that occurs during TE differentiation (90, 111). There is no direct evidence of the occurrence of apoptosis-specific events in the autolytic process, such as DNA laddering, nuclear shrinkage, cellular shrinkage, and the formation of apoptotic bodies. [For details of apoptosis-specific events, see Kerr & Harmon (82).] Recently, Mittler & Lam (104) reported the presence of fragmented nuclear DNA in differentiating TEs. However, they did not provide direct evidence of the DNA laddering. Because of increases in levels of several different nucleases, as shown by Thelen & Northcote (150), which may be a common feature of differentiating TEs, the results of Mittler & Lam might also be explainable in terms of DNA fragmentation by such nucleases, rather than by apoptosis-specific endonucleases. The autolytic process seems to be a type of programmed cell death that is unique to higher plants. Figure 2 summarizes the process of degradation for each type of organelle during the autolysis of differentiating *Zinnia* TEs (153). This figure indicates that no signs of autolysis are apparent before the disruption of the tonoplast and that the degeneration of all organelles starts after its disruption. The disruption of the tonoplast occurs several hours after secondary wall thickenings become visible in *Zinnia* cells. After the disruption of the tonoplast, organelles with a single membrane such as Golgi bodies and the endoplasmic reticulum become swollen and then disrupted. Subsequently, organelles with double membranes are degraded (Figure 2). Therefore, the disruption of the tonoplast may be an irreversible step toward cell death.

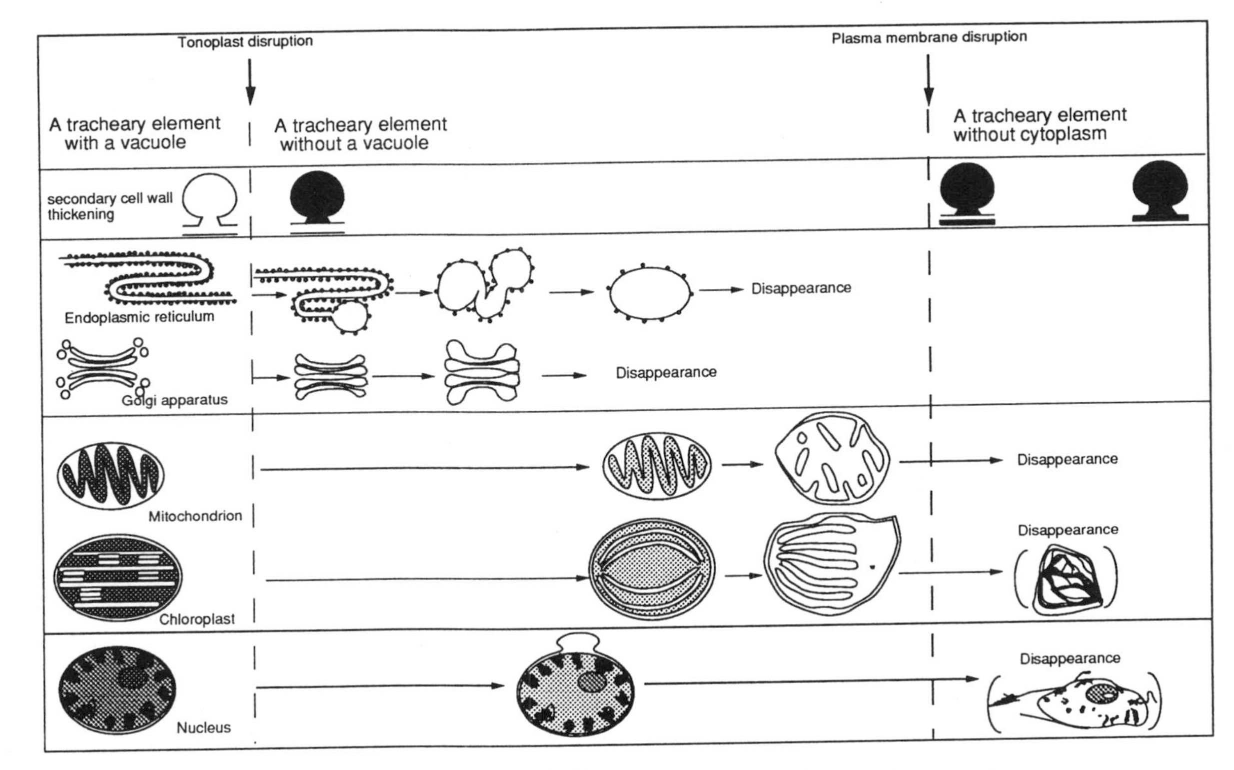

Figure 2 Structural changes in organelles during autolysis of differentiating *Zinnia* TEs. [Courtesy of Y Watanabe.]

Genetic analysis of the nematode *Caenorhabditis elegans* has provided the most direct evidence that the initiation of cell death during development is controlled by specific genes such as *ced-3, ced-4,* and *ced-9* (67, 161). The nematode *ced-3* and *ced-9* genes have mammalian counterparts, the gene for interleukin-1β-converting enzyme/CPRE-32 and bcl2, respectively, which are also responsible for the initiation of mammalian apoptosis (154). The molecular and biochemical analysis of the autolytic events in higher plants is just beginning, and no genes or proteins responsible for the initiation of the autolysis during TE differentiation have been identified. Is the initiation of autolysis coupled with the initiation or the progression of secondary wall formation? This is an important question because nobody has yet succeeded in separating the two processes.

The degeneration of organelles during autolysis may involve an increase in the activity of various hydrolytic enzymes that degrade large intracellular molecules. Thelen & Northcote (150) found that DNase and RNase activities were closely associated with TE differentiation in *Zinnia* cells. Nucleases—43, 22, and 25 kDa—appeared 12 h prior to the visible formation of TEs, and their levels increased conspicuously during the maturation phase of differentiation. An RNase of 17 kDa was detected transiently during the initial induction of other nucleolytic enzymes. One of the nucleases appeared before the onset of secondary wall thickenings. Minami & Fukuda (103) also found that the activity of a cysteine protease increased transiently at the start of autolysis and was specific to differentiating TEs in *Zinnia.* The mRNA corresponding to the cysteine protease was shown to be expressed specifically in differentiating TEs in *Zinnia* hypocotyls by in situ hybridization (A Minami, T Demura & H Fukuda, unpublished observation). Ye & Varner (158) also reported that the deduced amino acid sequence of a cDNA whose corresponding mRNA was expressed specifically in differentiating cells was homologous to papain, a typical cysteine protease. The inhibition of the activity of intracellular cysteine proteases by a specific inhibitor suppressed the loss of the nucleus, which suggests that a cysteine protease(s) plays a key role in nuclear degeneration (153). The addition of the inhibitor just before the start of secondary wall thickenings prevented TE differentiation itself, and therefore it seems possible that some cysteine protease(s) might be involved in the final determination of TE differentiation. CED3 protein, which is responsible for the initiation of cell death in animals, is a cysteine protease responsible for the cleavage of poly(ADP-ribose) polymerase (107). Our speculations are not contradicted by this result.

Sheldrake & Northcote (136) suggested that auxin and cytokinin, produced as a consequence of cellular autolysis, may induce the formation of new TEs. Other substances released from autolyzing TEs might also promote new cytodifferentiation. Thus, autolysis may be not only the final step in differentiation, but it might also be the first step toward differentiation of an adjacent cell.

CONCLUDING REMARKS

TE differentiation is induced by a combination of auxin and cytokinin and, in some cases, by wounding. The differentiation can be divided into early and late processes. Each process has been dissected by use of genes and proteins that have been identified as markers associated with TE differentiation. The early process is complex and seems to involve several different stages. The early stage of transdifferentiation into TEs in vitro appears to involve functional dedifferentiation and the conversion of dedifferentiated cells into TE precursors via procambium-like cells. The latter process may be common to the process of meristematic cells, which may correspond to dedifferentiated cells, into TE precursor cells during differentiation of the primary xylem in situ.

The late process is common to systems in vitro and in situ and involves a variety of TE-specific events that are associated with secondary wall formation and cell death. Molecular analysis of genes that are expressed in the late process suggested the presence of putative *cis*-elements that direct the xylem-specific expression of a certain set of genes. Cell death begins with the disruption of the tonoplast, and the process differs from apoptotic cell death in animals.

Many problems remain to be solved. In particular, the following questions require our urgent attention. What are the different stages into which the early process can be subdivided by use of new cytological, biochemical, and molecular markers, and what is the signal(s) responsible for the transition from one stage to the next? At what point are the thickening of cell walls and autolysis separated in the putative signal cascade? What is the mechanism of coordinated gene expression? How is the coordinated differentiation of different types of xylem cell regulated?

To answer these questions, we need to develop: (*a*) a method by which many differentiation stage-specific marker genes can easily be isolated, (*b*) a cultured cell system in which gene transfer is easy and in which the effects of an introduced gene on TE differentiation can easily be analyzed, and (*c*) a mutation system in which mutants that have a defect in xylem differentiation can easily be isolated.

ACKNOWLEDGMENTS

The author thanks Drs. R. Aloni and M. Sugiyama for critical reading of the manuscript, as well as colleagues who sent preprints of manuscripts and reprints, namely, G. P. Bolwell, A. M. Catesson, R. A. Dixon, C. J. Douglas, B.

Keller, C. Lamb, D. W. Meinke, L. W. Roberts, R. A. Savidge, R. Sederoff, and J. E. Varner. Financial support from the Ministry of Education, Science, and Culture of Japan, from the Science and Technology Agency of the Japanese Government, and from the Toray Science Foundation is gratefully acknowledged.

Literature Cited

1. Akashi T, Kawasaki S, Shibaoka H. 1990. Stabilization of cortical microtubules by the cell wall in cultured tobacco cells: effects of extensin on the cold-stability of cortical microtubules. *Planta* 182:363–69
2. Aloni R. 1987. Differentiation of vascular tissues. *Annu. Rev. Plant Physiol.* 38: 179–204
3. Aloni R. 1991. Wood formation in deciduous hardwood trees. In *Physiology of Trees,* ed. AS Raghavendra, pp. 175–97. New York: Wiley
4. Aloni R. 1995. The induction of vascular tissues by auxin and cytokinin. In *Plant Hormones, Physiology, Biochemistry and Molecular Biology,* ed. PJ Davies, pp. 531–46. Dordrecht/Boston/London: Kluwer
5. Aloni R, Pradel K, Ullrich CI. 1995. The three-dimensional structure of vascular tissues in *Agrobacterium tumefaciens*-induced crown galls and in the host stems of *Ricinus communis* L. *Planta* 196:597–605
6. Aloni R, Tollier MT, Monties B. 1990. The role of auxin and gibberellin in controlling lignin formation in primary phloem fibers and in xylem of *Coleus blumei* stems. *Plant Physiol.* 94:1743–47
7. Aloni R, Wolf A. 1984. Suppressed buds embedded in the bark across the bole and the occurrence of their circular vessels in *Ficus religiosa. Am. J. Bot.* 71:1060–66
8. Aloni R, Zimmermann MH. 1983. The control of vessel size and density along the plant axis: a new hypothesis. *Differentiation* 24:203–8
9. Bao W, O'Malley DM, Sederoff RR. 1992. Wood contains a cell-wall structural protein. *Proc. Natl. Acad. Sci. USA* 89:6604–8
10. Bao W, O'Malley DM, Whetten R, Sederoff RR. 1993. A laccase associated with lignification in loblolly pine xylem. *Science* 260:672–74
11. Barceló AR. 1995. Peroxidase and not laccase is the enzyme responsible for cell wall lignification in the secondary thickening of xylem vessels in *Lupinus. Protoplasma* 186:41–44
12. Bate NJ, Orr J, Ni W, Heromi A, Nadler-Hassar T, et al. 1994. Quantitative relationship between phenylalanine ammonia-lyase levels and phenylpropanoid accumulation in transgenic tobacco identifies a rate-determining step in natural product synthesis. *Proc. Natl. Acad. Sci. USA* 91: 7608–12
13. Baum SF, Aloni R, Peterson CA. 1991. The role of cytokinin in vessel regeneration in wounded Coleus internodes. *Ann Bot.* 67: 543–48
14. Beheke HD, Sjolund RD, eds. 1990. *Sieve Elements.* Berlin: Springer-Verlag. 305 pp.
15. Bolwell GP, Dalessandro G, Northcote DH. 1985. Decrease of polygalacturonic acid synthase during xylem differentiation in sycamore. *Phytochemistry* 24:699–702
16. Bolwell GP, Northcote DH. 1981. Control of hemicellulose and pectin synthesis during differentiation of vascular tissue in bean (*Phaseolus vulgaris*) callus and in bean hypocotyl. *Planta* 152:225–33
17. Boudet AM, Lapierre C, Grima-Pettenati J. 1995. Tansley review No. 80: biochemistry and molecular biology of lignification. *New Phytol.* 129:203–36
18. Bowen ID. 1993. Apoptosis or programmed cell death? *Cell Biol. Int.* 17: 365–80
19. Bowman J. 1993. *Arabidopsis.* Berlin: Springer-Verlag. 450 pp.
20. Bush DS. 1995. Calcium regulation in plant cells and its role in signaling. *Annu. Rev. Plant Physiol. Plant Mol. Biol.* 46:95–122
21. Cassab GI, Varner JE. 1987. Immunocytolocalization of extensins in developing soybean seedcoat by immunogold-silver staining and by tissue printing on nitrocellulose paper. *J. Cell Biol.* 105:2581–88

22. Cassab GI, Varner JE. 1988. Cell wall proteins. *Annu. Rev. Plant Physiol.* 39: 321–53
23. Catesson AM. 1994. Cambial ultrastructure and biochemistry: changes in relation to vascular tissue differentiation and the seasonal cycle. *Int. J. Plant Sci.* 155:251–61
24. Chabanet A, Goldberg R, Catesson A-M, Quinet-Szély M, Delaunay A-M, Faye L. 1994. Characterization and localization of a phenoloxidase in mung bean hypocotyl cell walls. *Plant Physiol.* 106:1095–102
25. Chapple CCS, Vogt T, Ellis BE, Somerville CR. 1992. An *Arabidopsis* mutant defective in the general phenylpropanoid pathway. *Plant Cell* 4:1413–24
26. Chasan R. 1994. Tracing tracheary element development. *Plant Cell* 6:917–19
27. Church DL, Galston AW. 1988. Kinetics of determination in the differentiation of isolated mesophyll cells of *Zinnia elegans* to tracheary elements. *Plant Physiol.* 88: 92–96
28. Church DL, Galston AW. 1989. Hormonal induction of vascular differentiation in cultured *Zinnia* leaf disks. *Plant Cell Physiol.* 30:73–78
29. Czaninski Y, Sachot RM, Catesson AM. 1993. Cytochemical localization of hydrogen peroxide in lignifying cell walls. *Ann. Bot.* 72:547–50
30. Dalessandro G, Piro G, Northcote DH. 1988. A membrane-bound enzyme complex synthesizing glucan and glucomannan in pine tissues. *Planta* 175:60–70
31. Demura T, Fukuda H. 1993. Molecular cloning and characterization of cDNAs associated with tracheary element differentiation in cultured *Zinnia* cells. *Plant Physiol.* 103:815–21
32. Demura T, Fukuda H. 1994. Novel vascular cell-specific genes whose expression is regulated temporally and spatially during vascular system development. *Plant Cell* 6:967–81
33. Dharmawardhana DP, Ellis BE, Carlson JE. 1995. A β-glucosidase from lodgepole pine xylem specific for the lignin precursor coniferin. *Plant Physiol.* 107:331–39
34. Dodds JH. 1981. Relationship of the cell cycle to xylem cell differentiation: a new model. *Plant Cell. Environ.* 4:145–46
35. Domingo C, Gomez MD, Canas L, Hernandez-Yago J, Conejero V, Vera P. 1994. A novel extracellular matrix protein from tomato associated with lignified secondary cell walls. *Plant Cell* 6:1035–47
36. Douglas CJ, Hoffmann H, Schulz W, Hahlbrock K. 1987. Structure and elicitor or u.v.-light-stimulated expression of two 4-coumarate:CoA ligase genes in parsley. *EMBO J.* 6:1189–95
37. Driouich A, Lainé A-C, Vian B, Faye L. 1992. Characterization and localization of laccase forms in stem and cell cultures of sycamore. *Plant J.* 2:13–24
38. Dwivedi UN, Campbell WH, Yu J, Datla RSS, Bugos RC, et al. 1994. Modification of lignin biosynthesis in transgenic tobacco through expression an antisense *O*-methyltransferase gene from *Populus. Plant Mol. Biol.* 26:61–71
39. Falconer MM, Seagull RW. 1985. Immuno fluorescent and Calcofluor white staining of developing tracheary elements in *Zinnia elegans* L. suspension cultures. *Protoplasma* 125:190–98
40. Falconer MM, Seagull RW. 1985. Xylogenesis in tissue culture: taxol effects on microtubule reorientation and lateral association in differentiating cells. *Protoplasma* 128:157–66
41. Felix G, Boller T. 1995. Systemin induces rapid ion fluxes and ethylene biosynthesis in *Lycopersicon peruvianum* cells. *Plant J.* 7:381–89
42. Feuillet C, Lauvergeat V, Deswarte C, Pilate G, Boudet A, Grima-Pettenati J. 1995. Tissue- and cell-specific expression of a cinnamyl alcohol dehydrogenase promoter in transgenic poplar plants. *Plant Mol. Biol.* 27:651–67
43. Fukuda H. 1987. A change in tubulin synthesis in the process of tracheary element differentiation and cell division of isolated *Zinnia* mesophyll cells. *Plant Cell Physiol.* 28:517–28
44. Fukuda H. 1989. Regulation of tubulin degradation in isolated *Zinnia* mesophyll cells in culture. *Plant Cell Physiol.* 30:243–52
45. Fukuda H. 1992. Tracheary element formation as a model system of cell differentiation. *Int. Rev. Cytol.* 136:289–332
46. Fukuda H. 1994. Redifferentiation of single mesophyll cells into tracheary elements. *Int. J. Plant Sci.* 155:262–71
47. Fukuda H, Kobayashi H. 1989. Dynamic organization of the cytoskeleton during tracheary-element differentiation. *Dev. Growth Differ.* 31:9–16
48. Fukuda H, Komamine A. 1980. Establishment of an experimental system for the tracheary element differentiation from single cells isolated from the mesophyll of *Zinnia elegans. Plant Physiol.* 65:57–60
49. Fukuda H, Komamine A. 1981. Relationship between tracheary element differentiation and the cell cycle in single cells isolated from the mesophyll of *Zinnia elegans. Physiol. Plant.* 52:423–30
50. Fukuda H, Komamine A. 1981. Relationship between tracheary element differentiation and DNA synthesis in single cells isolated from the mesophyll of *Zinnia elegans:* analysis by inhibitors of DNA synthesis. *Plant Cell Physiol.* 22:41–49
51. Fukuda H, Komamine A. 1985. Cytodifferentiation. In *Cell Culture and Somatic Cell*

Genetics of Plants, ed. IK Vasil, 2: 149–212. New York: Academic
52. Fukuda H, Yoshimura T, Sato Y, Demura T. 1993. Molecular mechanism of xylem differentiation. *J. Plant Res.* 3:97–107
53. Gahan PB. 1981. An early cytochemical marker of commitment to stelar differentiation in meristems from dicotyledonous plants. *Ann. Bot.* 48:69–775
54. Gahan PB. 1988. Xylem and phloem differentiation in perspective. See Ref. 124, pp. 1–21
55. Goddard RH, Wick SM, Silflow CD, Snustad DP. 1994. Microtubule components of the plant cell cytoskeleton. *Plant Physiol.* 104:1–6
56. Goffner D, Campbell MM, Campargue C, Clastre M, Borderies G, et al. 1994. Purification and characterization of cinnamoyl-coenzyme A:NADP oxidoreductase in *Eucalyptus gunnii. Plant Physiol.* 106: 625–32
57. Goffner D, Joffroy I, Grima-Pettenati J, Halpin C, Knight ME, et al. 1992. Purification and characterization of isoforms of cinnamyl alcohol dehydrogenase (CAD) from *Eucalyptus* xylem. *Planta* 188:48–53
58. Grotewold E, Drummond B, Bowen B, Peterson T. 1994. The *Myb*-homologous *P* gene controls phlobaphene pigmentation in maize floral organs by directly activating a subset of flavonoid biosynthetic genes. *Cell* 76:543–53
59. Gunning BES, Hardham AR. 1982. Microtubules. *Annu. Rev. Plant Physiol.* 33: 651–98
60. Hahlbrock K, Scheel D. 1989. Physiology and molecular biology of phenylpropanoid metabolism. *Annu. Rev. Plant Physiol. Plant Mol. Biol.* 40:347–69
61. Haigler CH, Brown RM Jr. 1986. Transport of rosettes from the Golgi apparatus to the plasma membrane in isolated mesophyll cells of *Zinnia elegans* during differentiation to tracheary elements in suspension culture. *Protoplasma* 134:111–20
62. Halpin C, Knight ME, Foxon GA, Campbell MM, Boudet AM, et al. 1994. Manipulation of lignin quality by down regulation of cinnamyl alcohol dehydrogenase. *Plant J.* 6:339–50
63. Hardham AR, Gunning BES. 1979. Interpolation of microtubules into cortical arrays during cell elongation and differentiation in roots of *Azolla pinnata. J. Cell Sci.* 37:411–42
64. Hatton D, Sablowski R, Yung M-H, Smith C, Schuch W, Bevan M. 1995. Two classes of *cis* sequences contribute to tissue-specific expression of a *PAL2* promoter in transgenic tobacco. *Plant J.* 7:859–76
65. Hauffe KD, Lee SP, Subramaniam R, Douglas CJ. 1993. Combinatorial interactions between positive and negative *cis*-acting elements control spatial patterns of *4CL-1* expression in transgenic tobacco. *Plant J.* 4:235–53
66. Hauffe KD, Paszkowski U, Schulze-Lefert P, Hahlbrock K, Dangl JL, Douglas CJ. 1991. A parsley 4CL-1 promoter fragment specifies complex expression patterns in transgenic tobacco. *Plant Cell* 3:435–43
67. Hengartner MO, Ellis RE, Horovitz HR. 1992. *Caenorhabdilis elegans* gene *ced-9* protects cells from programmed cell death. *Nature* 356:494–99
68. Higuchi T. 1985. Biosynthesis of lignin. In *Biosynthesis and Biodegradation of Wood Components,* ed. T Higuchi, pp. 141–60. New York: Academic
69. Hogetsu T. 1990. Detection of hemicelluloses specific to the cell wall of tracheary elements and phloem cells by fluorescein-conjugated lectins. *Protoplasma* 156: 67–73
70. Holdsworth MJ, Laties GG. 1989. Site-specific binding of a nuclear factor to the carrot extensin gene is influenced by both ethylene and wounding. *Planta* 179:17–23
71. Hong SK, Aoki T, Kitano H, Satoh H, Nagato Y. 1995. Phenotypic diversity of 188 rice embryo mutants. *Dev. Genet.* 16: 298–310
72. Iiyama K, Lam TBT, Meikle PJ, Ng K, Rhodes D, Stone BA. 1994. Covalent cross-links in the cell wall. *Plant Physiol.* 104:315–20
73. Ingold E, Sugiyama M, Komamine A. 1990. L-α-aminooxy-β-phenylpropionic acid inhibits lignification but not the differentiation to tracheary elements of isolated mesophyll cells of *Zinnia elegans. Physiol. Plant.* 78:67–74
74. Iwasaki T, Shibaoka H. 1991. Brassinosteroids act as regulators of tracheary-element differentiation in isolated *Zinnia* mesophyll cells. *Plant Cell Physiol.* 32: 1007–14
75. Jacobs WP. 1952. The role of auxin in differentiation of xylem around a wound. *Am. J. Bot.* 39:301–9
76. Jürgens G, Mayer U, Ruiz RAT, Berleth T. 1991. Genetic analysis of pattern formation in the *Arabidopsis* embryo. *Dev. Suppl.* 1: 27–38
77. Keller B. 1994. Gene expression of plant extracellular proteins. In *Genetic Engineering,* ed. JK Setlow, 16:255–70. New York: Plenum
78. Keller B, Baumgartner C. 1991. Vascular-specific expression of the bean GRP 1.8 gene is negatively regulated. *Plant Cell* 3: 1051–61
79. Keller B, Heierli D. 1994. Vascular expression of the *grp1.8* promoter is controlled by three specific regulatory elements and one unspecific activating sequence. *Plant Mol. Biol.* 26:747–56
80. Keller B, Sauer N, Lamb CJ. 1988. Gly-

cine-rich cell wall proteins in bean: gene structure and association of the protein with the vascular system. *EMBO J.* 7:3625–33
81. Keller B, Schmid J, Lamb CJ. 1989. Vascular expression of a bean cell wall glycine-rich protein-β-glucuronidase gene fusion in transgenic tobacco. *EMBO J.* 8:1309–14
82. Kerr JFR, Harmon BV. 1991. Definition and incidence of apoptosis: an historical perspective. In *Apoptosis: The Molecular Basis of Cell Death,* ed. LD Tomei, FO Cope, pp. 5–29. New York: Cold Spring Harbor
83. Kim S-K, Abe H, Little CHA, Pharis RP. 1990. Identification of two brassinosteroids from the cambial region of Scots pine (*Pinus silverstris*) by gas chromatography-mass spectrometry, after detection using a dwarf rice lamina inclination bioassay. *Plant Physiol.* 94:1709–13
84. Klee HJ, Horsch RB, Hinchee MA, Hein MB, Hoffmann NL. 1987. The effects of overproduction of two *Agrobacterium tumefaciens* T-DNA auxin biosynthetic gene products in transgenic petunia plants. *Genes Dev.* 1:86–96
85. Kobayashi H, Fukuda H. 1994. Involvement of calmodulin and calmodulin-binding proteins in the differentiation of tracheary elements in *Zinnia* cells. *Planta* 194:388–94
86. Kobayashi H, Fukuda H, Shibaoka H. 1987. Reorganization of actin filaments associated with the differentiation of tracheary elements in *Zinnia* mesophyll cells. *Protoplasma* 138:69–71
87. Kuboi T, Yamada Y. 1978. Regulation of the enzyme activities related to lignin synthesis in cell aggregates of tobacco cell culture. *Biochim. Biophys. Acta* 542:181–90
88. Kuc J, Nelson OE. 1964. The abnormal lignins produced by the Brown-Midrib mutants of maize. *Arch. Biochem. Biophys.* 105:103–13
89. Lagrimini LM, Burkhart W, Moyer M, Rothstein S. 1987. Molecular cloning of complementary DNA encoding the lignin-forming peroxidase from tobacco: molecular analysis and tissue-specific expression. *Proc. Natl. Acad. Sci. USA* 84:7542–46
90. Lai V, Srivastava LM. 1976. Nuclear changes during differentiation of xylem vessel elements. *Cytobiologie* 12:220–43
91. Langdale JA, Lane B, Freeling M, Nelson T. 1989. Cell lineage analysis of maize bundle sheath and mesophyll cells. *Dev. Biol.* 133:128–39
92. Langdale JA, Metzler MC, Nelson T. 1987. The *Argentia* mutation delays normal development of photosynthetic cell-types in *Zea mays. Dev. Biol.* 122:243–55
93. Lee D, Ellard M, Wanner LA, Davis KR, Douglas CJ. 1995. The *Arabidopsis thaliana* 4-coumarate: CoA ligase (4CL) gene: stress and developmentally regulated expression and nucleotide sequence of its cDNA. *Plant Mol. Biol.* 28:871–84
94. Lev-Yadun S, Aloni R. 1990. Vascular differentiation in branch junctions of trees: circular patterns and functional significance. *Trees* 4:49–54
95. Lewis NG, Yamamoto E. 1990. Lignin: occurrence, biogenesis and biodegradation. *Annu. Rev. Plant Physiol. Plant Mol. Biol.* 41:455–96
96. Leyva A, Liang X, Pintor-Toro JA, Dixon RA, Lamb CJ. 1992. *cis*-Element combinations determine phenylalanine ammonia-lyase gene tissue-specific expression patterns. *Plant Cell* 4:263–71
97. Li Y, Hagen G, Guilfoyle TJ. 1992. Altered morphology in transgenic tobacco plants that overproduce cytokinins in specific tissues and organs. *Dev. Biol.* 153:386–95
98. Liang X, Dron M, Schmid J, Dixon RA, Lamb CJ. 1989. Differential regulation of phenylalanine ammonia-lyase genes during plant development and by environmental cues. *J. Biol. Chem.* 264:14486–92
99. Little CHA, Savidge RA. 1987. The role of plant growth regulators in forest tree cambial growth. *Plant Growth Regul.* 6: 137–69
100. Lois R, Dietrich A, Hahlbrock K, Schulz W. 1989. A phenylalanine ammonia-lyase gene from parsley: structure, regulation, and identification of elicitor and light-responsive *cis*-acting elements. *EMBO J.* 8: 1641–48
101. Medford JI, Horgan R, El-Sawi Z, Klee HJ. 1989. Alterations of endogenous cytokinins in transgenic plants using a chimeric isopentenyl transferase gene. *Plant Cell* 1: 403–13
102. Meinke DW. 1995. Molecular genetics of plant embryogenesis. *Annu. Rev. Plant Physiol. Plant Mol. Biol.* 46:369–94
103. Minami A, Fukuda H. 1995. Transient and specific expression of a cysteine endopeptidase during autolysis in differentiating tracheary elements from *Zinnia* mesophyll cells. *Plant Cell Physiol.* 36:1599–606
104. Mittler R, Lam E. 1995. In situ detection of nDNA fragmentation during the differentiation of tracheary elements in higher plants. *Plant Physiol.* 108:489–93
105. Mizuno K, Komamine A. 1978. Isolation and identification of substances inducing formation of tracheary elements in cultured carrot-root slices. *Planta* 138:59–62
106. Nagata M, Demura T, Fukuda H. 1995. Analysis of genes expressed in the early process of transdifferentiation of *Zinnia* mesophyll cells into tracheary elements. *Plant Cell Physiol.* 36(Suppl.):26
107. Nicholson DW, Ali A, Thornberry NA, Vaillancourt JP, Ding CK, et al. 1995. Identification and inhibition of the ICE/CED-3

protease necessary for mammalian apoptosis. *Nature* 376:37–43

108. Northcote DH. 1989. Control of plant cell wall biogenesis. In *Cell Wall Polymers: Biogenesis and Biodegradation,* ed. NG Lewis, MG Paice, pp. 1–15. Washington, DC: Am. Chem. Soc.
109. Northcote DH. 1995. Aspects of vascular tissue differentiation in plants: parameters that may be used to monitor the process. *Int. J. Plant Sci.* 156:145–56
110. Northcote DH, Davey R, Lay J. 1989. Use of antisera to localize callose, xylan and arabinogalactan in the cell-plate, primary and secondary walls of plant cells. *Planta* 178:353–66
111. O'Brien TP. 1981. The primary xylem. In *Xylem Cell Development,* ed. JR Barnett, pp. 14–46. Tunbridge Wells: Castle House
112. O'Malley DM, Porter S, Sederoff RR. 1992. Purification, characterization and cloning of cinnamyl alcohol dehydrogenase in loblolly pine (*Pinus taeda* L.). *Plant Physiol.* 98:1364–71
113. O'Malley DM, Whetten R, Bao W, Chen CL, Sederoff RR. 1993. The role of laccase in lignification. *Plant J.* 4:751–57
114. Olson PD, Varner JE. 1993. Hydrogen peroxide and lignification. *Plant J.* 4:887–92
115. Peleman J, Boerjan W, Engler G, Seurinck J, Botterman J, et al. 1989. Strong cellular preference in the expression of a housekeeping gene of *Arabidopsis thaliana* encoding *S*-adenosylmethionine synthase. *Plant Cell* 1:81–93
116. Phillips R. 1981. Direct differentiation of tracheary elements in cultured explants of gamma-irradiated tubers of *Helianthus tuberosus. Planta* 153:262–66
117. Phillips R, Hawkins SW. 1985. Characteristics of the inhibition of induced tracheary element differentiation by 3-aminobenzamide and related compounds. *J. Exp. Bot.* 36:119–28
118. Pillonel C, Mudler MM, Boon JJ, Forster B, Binder A. 1991. Involvement of cinnamyl alcohol dehydrogenase in the control of lignin formation in *Sorghum bicolor* L. Moench. *Planta* 185:538–44
119. Rana MA, Gahan PB. 1983. A quantitative cytochemical study of determination for xylem-element formation in response to wounding in roots of *Pisum sativum* L. *Planta* 157:307–16
120. Roberts AW, Haigler CH. 1989. Rise in chlortetracycline accompanies tracheary element differentiation in suspension cultures of *Zinnia. Protoplasma* 152:37–45
121. Roberts AW, Haigler CH. 1990. Tracheary-element differentiation in suspension-cultured cells of *Zinnia* requires uptake of extracellular Ca^{2+}. *Planta* 180:502–9
122. Roberts DM, Harmon AC. 1992. Calcium-modulated proteins: targets of intracellular calcium signals in higher plants. *Annu. Rev. Plant Physiol. Plant Mol. Biol.* 43:375–14
123. Roberts LW. 1988. Hormonal aspects of vascular differentiation. See Ref. 124, pp. 22–38
124. Roberts LW, Gahan PB, Aloni R. 1988. *Vascular Differentiation and Plant Growth Regulators.* Berlin: Springer-Verlag. 154 pp.
125. Rodgers MW, Bolwell GP. 1992. Partial purification of golgi-bound arabinosyltransferase and two isoforms of xylosyltransferase from French bean (*Phaseolus vulgaris* L.). *Biochem. J.* 288:817–22
126. Romano CP, Hein MB, Klee HJ. 1991. Inactivation of auxin in tobacco transformed with the indoleacetic acid-lysine synthetase gene of *Pseudomonas savastanoi. Genes Dev.* 5:438–46
127. Ryser U, Keller B. 1992. Ultrastructual localization of a bean glycine-rich protein in unlignified primary walls of protoxylem cells. *Plant Cell* 4:773–83
128. Sablowski RWM, Moyano E, Culianez-Macia FA, Schuch W, Martin C, Bevan M. 1994. A flower-specific Myb protein activates transcription of phenylpropanoid biosynthetic genes. *EMBO J.* 13:128–37
129. Sachs T. 1981. The control of the patterned differentiation of vascular tissues. *Adv. Bot. Res.* 9:151–262
130. Sato Y, Sugiyama M, Gorecki RJ, Fukuda H, Komamine A. 1993. Interrelationship between lignin deposition and the activities of peroxidase isozymes in differentiating tracheary elements of *Zinnia:* Analysis with L-α-aminooxy-β-phenylpropionic acid and 2-aminoindan-2-phosphonic acid. *Planta* 189:584–89
131. Sato Y, Sugiyama M, Komamine A, Fukuda H. 1995. Separation and characterization of the isozymes of wall-bound peroxidase from cultured *Zinnia* cells during tracheary element differentiation. *Planta* 196:141–47
132. Sato Y, Takagi T, Sugiyama M, Fukuda H. 1994. Molecular cloning of a cDNA encoding peroxidase expressed specifically during differentiation into tracheary elements from *Zinnia* mesophyll cells. *Plant Cell Physiol.* 35(Suppl.):28
133. Savidge RA. 1989. Coniferin, a biochemical indicator of commitment to tracheid differentiation in conifers. *Can. J. Bot.* 67: 2663–68
134. Schindler T, Bergfeld R, Schopfer P. 1995. Arabinogalactan proteins in maize coleoptiles: developmental relationship to cell death during xylem differentiation but not to extension growth. *Plant J.* 7:25–36
135. Schopfer P. 1994. Histological demonstration and localization of H_2O_2 in organs of higher plants by tissue printing on nitrocellulose paper. *Plant Physiol.* 104:1269–75

136. Sheldrake AR, Northcote DH. 1968. The production of auxin by tobacco internode tissues. *New Phytol.* 67:1–13
137. Shininger TL. 1975. Is DNA synthesis required for the induction of differentiation in quiescent root cortical parenchyma? *Dev. Biol.* 45:137–50
138. Smith CG, Rodgers MW, Zimmerlin A, Ferdinando D, Bolwell GP. 1994. Tissue and subcellular immunolocalization of enzymes of lignin synthesis in differentiating and wounded hypocotyl tissue of French bean (*Phaseolus vulgaris* L.). *Planta* 192: 155–64
139. Stiefel V, Avila LR, Raz R, Vallès MP, Gómez J, et al. 1990. Expression of a maize cell wall hydroxyproline-rich glycoprotein gene in early leaf and root vascular differentiation. *Plant Cell* 2:785–93
140. Subramaniam R, Reinold S, Molitor EK, Douglas CJ. 1993. Structure, inheritance, and expression of hybrid poplar (*Populus trichocarpa* x *Populus deltoides*) phenylalanine ammonia-lyase genes. *Plant Physiol.* 102:71–83
141. Sugiyama M, Fukuda H, Komamine A. 1994. Characteristics of the inhibitory effects of aphidicolin on transdifferentiation into tracheary elements of isolated mesophyll cells of *Zinnia elegans. Plant Cell Physiol.* 35:519–22
142. Sugiyama M, Komamine A. 1987. Effect of inhibitors of ADP-ribosyltransferase on the differentiation of tracheary elements from isolated mesophyll cells of *Zinnia elegans. Plant Cell Physiol.* 28:541–44
143. Sugiyama M, Komamine A. 1987. Relationship between DNA synthesis and cytodifferentiation to tracheary elements. *Oxford Surv. Plant Mol. Cell Biol.* 4:343–46
144. Sugiyama M, Komamine A. 1990. Trans differentiation of quiescent parenchymatous cells into tracheary elements. *Cell Differ. Dev.* 31:77–87
145. Sugiyama M, Yeung EC, Shoji Y, Komamine A. 1995. Possible involvement of DNA-repair events in the transdifferentiation of mesophyll cells of *Zinnia elegans* into tracheary elements. *J. Plant Res.* 108: 351–61
146. Suzuki K, Ingold E, Sugiyama M, Fukuda H, Komamine A. 1992. Effects of 2,6-dichlorobenzonitrile on differentiation to tracheary elements of isolated mesophyll cells of *Zinnia elegans* and formation of secondary cell walls. *Physiol. Plant.* 86:43–48
147. Suzuki K, Ingold E, Sugiyama M, Komamine A. 1991. Xylan synthase activity in isolated mesophyll cells of *Zinnia elegans* during differentiation to tracheary elements. *Plant Cell Physiol.* 32:303–6
148. Taylor JG, Owen TP Jr, Koonce LT, Haigler CH. 1992. Dispersed lignin in tracheary elements treated with cellulose synthesis inhibitors provides evidence that molecules of the secondary cell wall mediate wall patterning. *Plant J.* 2:959–70.
149. Terashima N, Fukushima K, He LF, Takabe K. 1993. Comprehensive model of the lignified plant cell wall. In *Forage Cell Wall Structure and Digestibility.* ed HG Jung, DR Buxton, RD Hatfield, J Ralph, pp. 247–69. Madison, WI
150. Thelen MP, Northcote DH. 1989. Identification and purification of a nuclease from *Zinnia elegans* L.: a potential molecular marker for xylogenesis. *Planta* 179:181–95
151. Torrey JG, Fosket DE, Hepler PK. 1971. Xylem formation: a paradigm of cytodifferentiation in higher plants. *Am. Sci.* 59: 338–52
152. Vignols F, Rigau J, Torres MA, Capellades M, Puigdomenech P. 1995. The *brown midrib$_3$* (*bm$_3$*) mutation in maize occurs in the gene encoding caffeic acid *O*-methyltransferase. *Plant Cell* 7:407–16
153. Watanabe Y, Fukuda H. 1995. Autolysis during tracheary element differentiation: analysis with inhibitors. *Plant Cell Physiol.* 36(Suppl.):87
154. Whyte M, Evan G. 1995. The last cut is the deepest. *Nature* 376:17–18
155. Wojtaszek P, Bolwell GP. 1995. Secondary cell-wall-specific glycoprotein(s) from French bean (*Phaseolus vulgaris* L.) hypocotyls. *Plant Physiol.* 108:1001–12
156. Ye Z-H, Kneusel RE, Matern U, Varner JE. 1994. An alternative methylation pathway in lignin biosynthesis in *Zinnia. Plant Cell* 6:1427–39
157. Ye Z-H, Song Y-R, Marcus A, Varner JE. 1991. Comparative localization of three classes of cell wall proteins. *Plant J.* 1: 175–83
158. Ye Z-H, Varner JE. 1993. Gene expression patterns associated with in vitro tracheary element formation in isolated single mesophyll cells of *Zinnia elegans. Plant Physiol.* 103:805–13
159. Ye Z-H, Varner JE. 1994. Expression of an auxin- and cytokinin-regulated gene in cambial region in *Zinnia. Proc. Natl. Acad. Sci. USA* 91:6539–43
160. Ye Z-H, Varner JE. 1995. Differential expression of *O*-methyltransferases in lignin biosynthesis in *Zinnia elegans. Plant Physiol.* 108:459–67
161. Yuan J, Horvitz HR. 1990. The *Caenorhabditis elegans* genes *ced-3* and *ced-4* act autonomously to cause programmed cell death. *Dev. Biol.* 138:33–41

Annu. Rev. Plant Physiol. Plant Mol. Biol. 1996. 47:327–50

COMPARTMENTATION OF PROTEINS IN THE ENDOMEMBRANE SYSTEM OF PLANT CELLS

Thomas W. Okita

Institute of Biological Chemistry, Washington State University, Pullman, Washington 99164-6340

John C. Rogers

Department of Biochemistry, University of Missouri, Columbia, Missouri 65211

KEY WORDS: endomembrane system, endoplasmic reticulum, ER subdomains, protein body formation vacuoles, vacuole biogenesis

ABSTRACT

This review focuses on four interrelated processes in the plant endomembrane system: compartmentation of proteins in subdomains of the endoplasmic reticulum, mechanisms that determine whether storage proteins are retained within the ER lumen or transported out, the origin and function of biochemically distinct vacuoles or prevacuolar organelles, and the cellular processes by which proteins are sorted to vacuolar compartments. We postulate that ER-localized protein bodies are formed by a series of orderly events of protein synthesis, protein concentration, and protein assembly in subdomains of the ER. Protein concentration, which facilitates protein-to-protein interactions and subsequent protein assembly, may be achieved by the interactions with chaperones and by the localization of storage protein mRNAs. We also describe recent developments on the coexistence of two biochemically distinguishable vacuolar compartments, the possible direct role of the ER in vacuole biogenesis, and proposed mechanisms for transport of proteins from the ER or Golgi apparatus to the vacuole.

1040-2519/96/0601-0327$08.00

CONTENTS

INTRODUCTION

Our review focuses on processes that affect concentration, retention, and sorting of proteins within the endomembrane system of plant cells. We direct the reader to previous reviews of other aspects of secretory pathway function (9, 19, 114, 124). The secretory pathway begins with synthesis of proteins on the endoplasmic reticulum (ER), at which time they are cotranslationally translocated into the lumen (Figure 1). These proteins are then transported through an ordered set of intracellular compartments that include the ER-to-Golgi intermediate compartment (90), the *cis*-Golgi network (CGN), the Golgi stacks, and the *trans*-Golgi network (TGN) (114). Proteins are concentrated up to 200-fold between the ER and TGN compartments (102), and the process of concentration may begin as the proteins are packaged to exit the ER (5, 77). Within the Golgi or TGN, proteins are sorted away from those that will be secreted from the cell and are instead directed to a pathway leading to the vacuole.

In plant cells, sorting away from the default pathway that leads to secretion (25, 91) requires the presence of specific peptide sequence information within the protein, which may be composed of a contiguous set of amino acid residues (e.g. signal peptides, propeptides of vacuolar proteins) or a noncontiguous arrangement of residues that comprises a three-dimensional structure. In

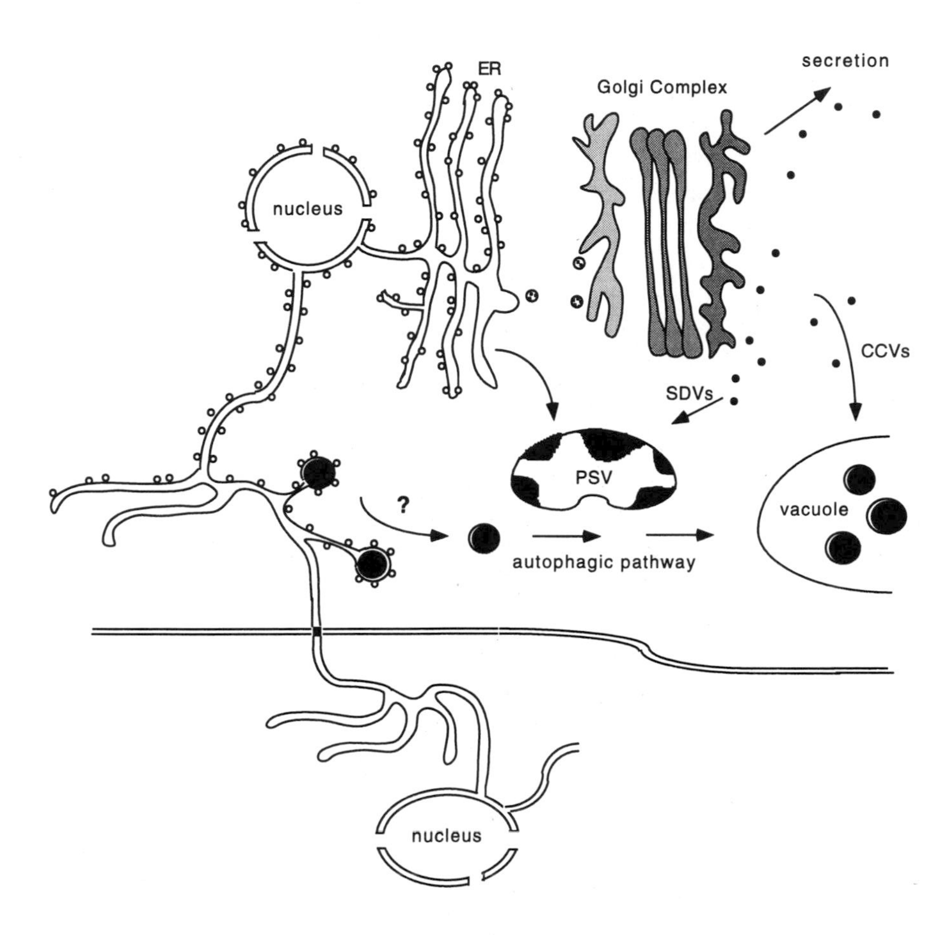

Figure 1 Diagram of the endomembrane system in plants and various models of protein transport to the vacuole. Vacuolar and secretory proteins are synthesized on the ER, where they first enter the secretory pathway by their translocation to the ER lumen. These proteins are transported from the ER to the Golgi apparatus, where they are concentrated and sorted to the cell surface (secretion) or vacuole. As discussed in more detail in the text, specific vacuolar proteins are sorted in the Golgi by two different biochemically defined pathways (73). In this hypothetical model, one pathway would use smooth "dense" vesicles (SDVs) to transport proteins to the protein storage vacuole (PSV) or its equivalent, while the other would use clathrin-coated vesicles (CCVs) to transport proteins to an acidified lytic compartment that classically defines the vacuole. Also depicted is the proposed route of protein body formation in pea and wheat. Recent evidence (49) shows that pea storage proteins are packaged in PSV that is biochemically distinct from a coexisting vegetative vacuole. The PSV is formed de novo, and its membrane may arise directly from the ER during the early stages of protein body formation (98). A novel mechanism has been proposed (67) that wheat storage proteins are initially assembled in the ER lumen, released into the cytoplasm, and transported to the vacuole by an autophagy-like process.

some cases, a peptide determinant may be recognized by a specific receptor that targets, retains, or retrieves proteins to specific compartments, though the role of receptor-mediated sorting of proteins to the vacuole in plants remains unclear (see section on Biochemical Definition of Two Pathways). Sorting determinants localize specific proteins within each of the compartments comprising the endomembrane system.

The endomembrane system of plant cells has unique features not seen in yeast or animal cells. These features include the capacity to localize and assemble large protein inclusion bodies in either the ER lumen or vacuole (110, 124, 125) and the potential to communicate between cells at the molecular level via ER-containing plasmodesmata (70).

THE ER IS COMPOSED OF SUBDOMAINS

The ER is classically divided into three subcompartments—the rough ER (RER), smooth ER (SER), and nuclear envelope (NE)—that can be distinguished morphologically. The RER is studded with ribosomes actively engaged in the synthesis of proteins that are to be sorted by the secretory pathway or that are residents of this endomembrane system. The SER is the site of lipid synthesis and serves as a reservoir of Ca^{2+} (112). The NE is also a site of protein synthesis but can be easily discerned by the presence of lamin receptors and components of the nuclear pore complex. All three compartments contain soluble ER resident proteins called reticuloplasmins (61), including the lumenal chaperones binding protein (BiP), protein disulfide isomerase (PDI), GRP94, and calnexin. The chaperones facilitate proper folding and assembly of newly synthesized polypeptides in the lumen (35). BiP also acts as a quality control monitor in binding and retaining unassembled or malfolded proteins in the ER (39). Although these chaperones are not always evenly distributed along the continuous lumen of the three subcompartments, they nevertheless are ubiquitous and useful biochemical markers of the ER compartment.

The tripartite structure of the ER is now considered an oversimplified view; this membrane compartment is much more complex than envisioned. The ER is composed of a number of different subdomains that are distinguished by the presence of specific biochemical properties and/or morphological features (99, 112). For example, the RER complex contains a transitional area of smooth membranes that mediate vesicular transport between the ER and Golgi (85). An obvious example of the existence of ER subdomains in plants is their capacity to store reserve protein as ER-localized protein bodies (see section on Plant Protein Bodies).

The Formation of Intracisternal Granules in the ER

The ER can assemble proteins as inclusion bodies within its lumen. Formation of intracisternal granules in animal cells occurs only when protein export rates from the ER are unable to keep pace with the rate of protein synthesis (120), whereupon the protein granules, devoid of the chaperones BiP and PDI, are formed by the aggregation of misfolded or unfolded proteins. A high concentration of unfolded or malfolded proteins appears to favor the occurrence of abnormal protein-protein interactions and ultimately results in the formation of a protein aggregate. Such a situation may also occur when the wheat prolamines are overexpressed in yeasts (92) and insect cells (107, 124). In contrast, plant cells routinely use the ER to store proteins as part of a programmed developmental process that appears to involve the lumenal chaperones. As a consequence, intracisternal granule (protein body) formation in plants is carried out by a series of orderly events of protein synthesis and assembly of properly folded proteins (8, 66).

Plant Protein Bodies

Storage proteins of seeds are packaged in organelles called protein bodies (PBs) (110, 125). Seed storage proteins are divided into two main groups, the salt-soluble globulins and alcohol-soluble prolamines. Globulins are stored in a vacuolar compartment (17), whereas prolamines are stored in multiple subcellular compartments (Figure 1). The prolamines of maize, rice, and *Sorghum* are packaged in ER-delimited PBs (110), whereas the bulk of the prolamines of wheat, barley, and oat are stored in vacuolar compartments (58, 65, 67, 94). The mechanisms by which prolamines are transported to the vacuole have been debated for several decades and continue to be an issue. Readers are directed to reviews of the biochemistry of prolamines and protein body assembly (32, 33, 78, 83, 105, 106, 108, 110).

THE MECHANISM OF ER-LOCALIZED PLANT PROTEIN BODIES

Retention by Protein-Protein Interactions

None of the maize or rice prolamines isolated to date contains the well-defined ER retrieval signal, KDEL, or its functional derivatives (28, 80, 124) at their C-termini to explain ER localization. Retention of prolamines in the ER is not a plant cell–specific phenomenon, however, because when a maize poly(A^+)-RNA fraction enriched in zein mRNAs or synthetic α-zein transcripts was injected into *Xenopus* oocytes, the synthesized zeins cosedimented with a density similar, though not identical, to that exhibited by maize PBs (56, 127).

These results could be due to an ER retention signal in zeins or to other protein-protein interactions.

Studies to identify such retention signals have been conducted on wheat gliadins and maize γ-zeins in heterologous systems (1, 2, 100, 111). When transcripts encoding the intact γ-gliadin polypeptide were microinjected into *Xenopus* oocytes, much of the protein was slowly secreted (2), but significant amounts remained intracellularly and sedimented in metrizamide gradients at a density close to that exhibited by wheat PBs. When the N-terminal region was deleted, the resulting C-terminal half of the protein was secreted into the medium at a much faster rate than the intact polypeptide. In contrast, deletion of the C-terminal region resulted in an N-terminal polypeptide that was completely retained. These observations are consistent with the view that sorting of γ-gliadin in *Xenopus* oocytes is determined by the action of two opposing signals: an N-terminal region composed of a tandem repeat of PQQPFPQ that functions in ER localization and subsequent PB formation, and a C-terminal region that renders the polypeptide competent for protein export through the Golgi apparatus (2). A similar approach was taken to identify possible retention signals of γ-zein in both *Xenopus* oocytes (121) and transgenic *Arabidopsis* (34). Intact γ-zein and mutated versions that contained an N-terminal region with a tandem repeat of PPPVHL were retained, whereas deletion of this N-terminal region resulted in the secretion of the truncated polypeptide.

These studies indicate that an N-terminal region, containing a tandem repeat of a conserved peptide, is responsible for localization of these prolamines within the ER. Other than being rich in proline residues and overall hydrophobicity, however, the conserved tandem repetitive peptides are dissimilar. Moreover, rice prolamines and β-zeins lack tandem repeats. In view of the differences in structures of these repetitive peptides, and their absence in several major prolamine classes of maize and rice, it is questionable whether a specific peptide retention signal and corresponding receptor actually exist. A more likely mechanism for ER retention is that these proteins interact with ER resident proteins, especially the chaperones (33, 121), even after they have been properly folded. Such interactions are likely to occur with prolamines as a whole because BiP preferentially interacts with peptides rich in hydrophobic and aromatic amino acids (10), residues enriched in the prolamines. Interaction with ER resident proteins may be one of several factors (see section on The Role of Chaperones in Protein Body Formation) that may explain why some prolamines are retained in the ER lumen while others are exported to the Golgi apparatus (32, 33, 83, 121). The conserved repeats of different prolamines display different consensus sequences, and it is possible that interactions of these conserved peptide regions with chaperones may vary considerably. Prolamines with weaker interactions may be dissociated from the putative chaperone complex before initiation of PB formation and hence exported to the Golgi (33). Alternatively, prolamines that are less dependent on chaperones for proper polypeptide folding may attain transport competence much sooner, such that a critical concentration necessary for PB formation is never achieved in the ER lumen. The latter explanation is more consistent with the observation that formation of ER protein bodies is not limited only to hydrophobic proteins such as prolamines, but also can include hydrophilic globular proteins as well if conditions of synthesis or transport lead to lumenal concentrations high enough for protein assembly. For example, pea vicilin, normally stored in the vacuole, is also capable of forming intracisternal ER protein granules when the KDEL ER retrieval signal is attached to its C-terminus (128).

The Role of Chaperones in Protein Body Formation

Plants have all of the lumenal chaperones observed in other organisms (3, 12, 26–29, 31, 36, 42, 69, 109, 134), and evidence indicates that these chaperones participate in folding and assembly of plant proteins. Plant BiP can complement a mutation in the yeast gene (29) and is induced when plant cells are treated with tunicamycin, an inhibitor of glycosylation, or are subjected to stress (23, 26, 29). In keeping with its proposed function, BiP has been shown to be bound to malfolded (26), unglycosylated (23), or assembly-defective (89) proteins in an ATP-dependent manner. The plant homologs of calreticulin (26) and PDI (13) may also participate in polypeptide folding.

BiP is also involved in prolamine PB formation in rice and maize. Li et al (69) showed that the ER of developing rice endosperm cells is enriched in BiP but that the distribution was not uniform. About 90% of the BiP was associated with a subcellular fraction enriched in PBs, and the remainder was associated with a microsomal fraction enriched in RER vesicles. The nonrandom localization of BiP was also supported by immunocytochemical analysis (X Li, T Okita & R Boston, unpublished data). BiP was observed mainly at the periphery of the PBs and not in the cisternal membranes that connect these organelles. A similar localization of BiP has also been made in maize endosperm cells (134). In addition, in developing rice endosperm cells, BiP is observed associated with prolamine-containing polysomes and specifically with the nascent polypeptide chain, with mature prolamine polypeptides, and with the prolamine inclusion body itself (69). These observations are consistent with genetic studies of yeasts (103, 126) and in vitro transport studies of animals (82) demonstrating that BiP facilitates translocation of the nascent chain across the ER membrane. In the lumen, BiP helps fold prolamine polypeptide chains into a competent state for subsequent assembly onto the growing surface of the PB. Direct involvement of BiP in PB formation is also supported by the elevated levels of BiP in the *floury-2* mutant of maize (12, 31, 71, 134), where the *floury-2* locus codes for α-zein polypeptide with an uncleavable signal peptide (20) that disrupts the ordered assembly process of α-zeins within the central core of the PB.

Storage of Prolamines—ER vs Vacuole

Why are maize and rice prolamines stored in the ER while those from wheat (7, 58, 67, 87), barley (15, 94), and oat (65) are packaged in the vacuole? It is not entirely because of the overall hydrophobic character of the proteins (although this is a contributing factor), because expression of β-zein in transgenic tobacco resulted in localization to different intracellular compartments, depending on the study. This storage protein was observed to be either retained in the ER lumen in leaves (4), sorted to vacuolar PBs in seeds (48), or localized to both compartments in seeds (4). Although it is not clear why such varied results were obtained, the observations indicate that even proteins that are normally localized in the ER have the capacity to exit this compartment. On the other hand, barley hordeins, which are normally packaged in the vacuole, are also capable of forming ER-localized PBs. In the barley line Nevsky, which is deficient in γ-hordeins, ER-delimited PBs of the same size as those found within the vacuoles of other barley lines are readily observed (94). The presence and frequency of these ER-delimited PBs is developmentally regulated. In older cells deep within the endosperm that exhibit high rates of hordein synthesis, almost all of the hordeins are deposited within the RER, whereas in younger subaleurone cells, where rates of hordein synthesis are low, all of the hordeins are localized to vacuoles. In cells between these two extreme types, PB formation occurs in both compartments. Overall, these observations indicate that the intracellular site of PB formation may be dictated by the relative rates of hordein synthesis and export from the ER (D Simpson, personal communication).

Thus, the concentration of prolamines in the ER lumen may determine whether PBs are formed there. A critical concentration, defined by relative rates of protein synthesis and ER export, would increase the likelihood of protein-protein interactions and the assembly of these proteins into a PB. If ER residence times for rice and maize prolamines were much longer than those for the wheat and barley proteins, differences in behavior between the two groups of prolamines would be partially explained. Increased residence time could be accomplished, as discussed above, by interactions with chaperones (33, 69, 83, 121). A second mechanism that would effectively increase protein concentration would be for PB formation to occur at junctions where multiple ER cisternae converged (43, 64); proteins synthesized over the large ER complex would then effectively be concentrated at these junction points during their transport by bulk flow. Another possible mechanism to increase the effective concentration of these proteins in the ER would be to localize their mRNAs to subdomains of the ER, a process evident in rice (68).

mRNA Localization in Rice

Rice is unusual in that it not only accumulates both prolamines and glutelins, proteins homologous to the 11S globulins of legumes, in the endosperm, but also packages these proteins into separate PBs (62, 116, 133). Prolamines and glutelins are synthesized on the ER but then are routed by different cellular processes into separate PBs as the latter proteins are packaged in a vacuolar compartment. Therefore, rice cells must efficiently segregate prolamines from glutelins within the ER so that export of glutelins is not impeded by prolamine PBs (78, 83). As viewed by electron microscopy, developing rice endosperm cells possess two morphologically distinct RER membranes and, hence, two possible subdomains: the cisternal ER (C-ER), consisting of single lamellar membranes distributed throughout the cell, and the protein body–containing ER (PB-ER), which delimits prolamine PBs (68). Biochemical analyses of subcellular fractions enriched for these membrane types are consistent with this model. Polyadenylated RNAs isolated from PB-ER directed the in vitro synthesis of predominantly prolamines, which suggests that the PB-ER is enriched for prolamine mRNAs (132). In contrast, glutelin mRNAs were present at more than twofold excess over prolamine transcripts in membrane-bound polysome fractions throughout seed development (59), which suggests that C-ER is enriched for glutelin transcripts. Unequivocal evidence for the segregation of the rice storage protein mRNAs on distinct ER types was obtained by Li et al (68). Through the use of blot hybridization analysis of RNAs obtained from subcellular fractions enriched for these ER membranes, in vitro hybridization analysis of PBs, and single and double in situ hybridization analysis of endosperm thin sections, it was shown that glutelins are present at a twofold molar excess relative to prolamine transcripts on the C-ER, whereas prolamine transcripts are present at a seven- to tenfold molar excess over the glutelin transcripts on the PB-ER. Thus, the initial targeting process of the glutelins and prolamines into their distinct PBs is facilitated by segregation of their transcripts on the C-ER and PB-ER, respectively. Localization of prolamine transcripts to PB-ER would be an effective mechanism of concentrating newly synthesized polypeptides in the lumen and may help to specify where PB formation occurs within the ER complex. These findings raise the question as to whether such a phenomenon occurs in maize. Although localization of zein mRNAs has not been directly assessed, available evidence (24, 84) supports that A-type mRNAs for γ-zein, an essential component for zein PB formation, may be targeted and/or specifically anchored to PBs (BA Larkins, personal communication).

Nonrandom distribution of mRNAs on the ER has been well described in other systems (6, 88, 115, 118, 122). For a discussion on the mechanisms of mRNA localization, see References 83, 113, 129.

THE VACUOLE AND PREVACUOLAR ORGANELLES

For an overview of vacuolar function, see References 11, 44, 131. This section considers the coexistence of functionally distinct vacuoles, their origins, and the mechanisms by which proteins reach these subcellular compartments. Little consideration has been given to the possibility that different types of vacuoles may be present in the same cell. This is important because the coexistence of two functionally distinct types of vacuoles in one cell means that they must be distinguished by mechanisms directing proteins to their tonoplast membranes and to their interiors.

The classic micrograph in textbooks invariably shows a cell with most of the cytoplasm occupied by a large, clear central vacuole, but vacuole morphology can vary greatly. Variability may reflect the state of maturity or differentiation of the cell and, in turn, the developmental state of the vacuole. A dramatic example is provided by observations of the autofluorescent vacuoles in differentiating, living stomatal cells of *Allium* (86). In progenitor cells the vacuole consists of globular cisternae. In cells intermediate in the differentiation pathway, these cisternae are replaced by a network of interlinked tubules and small chambers that demonstrates great plasticity and undergoes complex movements and rearrangements (86). In mature stomatal cells, vacuoles reassume a large globular appearance. Similarly, the vacuolar compartment in undifferentiated root tip cells also consists of a network of connecting tubules and small chambers, while mature cells have a clearly identifiable, large central vacuole (46). For these reasons, the term "vacuole" in our discussion is assigned to a membrane-bound structure where there is reason to believe that it represents an existing acidic compartment with lytic activity (11, 131), or that is likely to be an immediate precursor to such a compartment. Such a definition allows inclusion of structures with morphologies different from the classic large central vacuole.

NOT ALL VACUOLES ARE EQUAL

Evidence for the presence of distinct types of vacuoles in one cell derives from morphologic studies where immunological or biochemical markers were used to distinguish vacuole types (38, 49, 51, 130).

Peroxidase-Containing Vacuoles in Soybean Suspension Culture Cells and Protoplasts

Peroxidases in vacuoles and the distribution of the enzyme activity can be studied using histochemistry with electron microscopy. Griffing & Fowke (38) observed that frequently the tonoplast of the central vacuole stained positive for peroxidase activity, whereas smaller vacuoles in the same cell did not show activity. They did not measure the presence of peroxidase protein directly, and thus their observations could be explained by inactivation of the enzyme in some vacuoles. An alternative explanation—that some vacuoles contained peroxidase protein and some did not—suggests that vacuoles at different developmental stages, or functionally different types of vacuoles, could exist in the same cell (38).

Aleurain- and Storage Protein–Containing Vacuoles in Barley Aleurone Cells

Aleurain is a barley vacuolar cysteine protease that is synthesized as a proenzyme and transported through the Golgi to the vacuole, where it is processed to mature form (51, 52, 54). This processing requires an environment with acidic pH, and the enzymatic activity of the mature enzyme has a pH optimum of about 5 (51, 53).

Immunogold EM studies using an affinity-purified polyclonal antibody raised to recombinant mature aleurain showed aleurain to be present in numerous small vacuoles in barley aleurone cells from grains imbibed but not germinated (51). Surprisingly, these "aleurain-containing vacuoles" were morphologically distinct from vacuoles containing storage protein bodies and phytic acid crystals. The latter were not labeled with the antialeurain antibody (51). Similar results with the same antibody were obtained by Hillmer and coworkers (S Hillmer, P Bethke & RL Jones, personal communication). These investigators also found that in aleurone cells treated with gibberellic acid to activate synthesis and secretion of hydrolytic enzymes, protein markers for the two vacuole types merged and were detected together as larger, clear vacuoles were formed. In an extension of this work, P Bethke and RL Jones (personal communication) isolated protein body–containing vacuoles from aleurone cell protoplasts not treated with gibberellin. Consistent with the findings from electron microscopy, these protein body–containing vacuoles contained little or no immunologically detectable aleurain but did contain substantial amounts of the barley vacuolar aspartic proteinase (101).

These results indicate that two separate vacuole types exist in aleurone cells prior to their activation by gibberellin. The two types might be considered intermediate compartments in vacuole ontogeny because, during germination when proteolysis of storage proteins is required to provide amino acids for high-level synthesis of secreted hydrolytic enzymes, larger vacuoles with characteristics of central vacuoles are formed. These larger vacuoles apparently represent a merger of the two previously separate vacuole types. Existence of two separate vacuole types would be reasonable if storage products (protein bodies and phytic acid crystals) are to be protected from an environment that would promote their degradation (aleurain-containing vacuoles) at some stage during cell development. Although the pH of a central vacuole is known to be acidic (11, 131), storage protein vacuoles isolated from barley aleurone cells not treated with gibberellin have an internal pH of about 7, and acidification only occurs after the cells are treated with gibberellin (SJ Swanson & RL Jones, personal communication). The pH of aleurain-containing vacuoles is assumed to be acidic because proteolytic processing of the proenzyme, a pH-dependent event, is observed when proaleurain reaches that compartment (51). Identification on a biochemical level of other enzymes present in aleurain-containing vacuoles is necessary before a clear picture of that compartment can be formed.

Two Distinct Vacuole Types During Differentiation of Cotton Seed Trichomes

Cotton seed trichomes differentiate from the outer epidermis of the ovule. The epidermal cells in immature ovules contain spherical, clear-appearing vacuoles that are typical in their morphology of those present in other plant cells. As trichome cells begin differentiation, however, a morphologically distinct "type 2" vacuole becomes prominent (130). These vacuoles are highly pleomorphic and consist of a complex anastomosed network filled with electron-dense material that strongly stains for the presence of polysaccharides and reacts with a monoclonal antibody to arabinogalactans. The two vacuole types fuse as cell expansion begins, and the mature cell is left with a single, homogeneous central vacuole (130).

Identification of arabinogalactans and polysaccharides in the type 2 vacuoles provides biochemical markers for these structures and distinguishes them from central vacuoles. Their function is unclear, but it is interesting that arabinogalactans, prominent components of cell walls, have been identified in intracellular multivesicular bodies and in partially degraded multivesicular bodies within the central vacuole of tobacco leaf parenchymal cells (45). Multivesicular bodies are thought to be participants in the pathway of endocytosis (45, 95, 117). Arabinogalactan-containing multivesicular bodies may represent intermediates in a pathway for internalizing cell wall materials for disposal in the vacuole (45). The volume of type 2 vacuoles in cotton seed trichomes appears to be vastly larger than the volume of multivesicular bodies in tobacco leaf cells, however, and the origin of their contents remains to be determined.

Protein Storage Vacuoles and Vegetative Vacuoles in Developing Pea Cotyledons

Development of protein storage vacuoles in legume cotyledon cells has been described (21, 22, 44) as a process where storage proteins are deposited within the central vacuole of an undifferentiated cell. (It is not clear how these proteins are protected from the active proteases present in a central vacuole.)

As this process continues, the central vacuole is thought to subdivide into protein storage vacuoles, which results in its replacement by thousands of small, densely filled protein storage vacuoles (21, 22, 44). However, electron microscopic analysis of young developing pea cotyledons, coupled with immunolocalization of specific proteins as biochemical markers and subcellular fractionation experiments (49), have offered a compelling alternative model.

Hoh et al (49) demonstrated that storage protein deposits are first visualized at the periphery of clumps of vegetative vacuoles (equivalent to the central vacuole) but are actually contained within a tubular membrane system that is separate from the vegetative vacuole tonoplast. As protein deposition continues, the tube-like membrane system dilates and compresses adjacent vegetative vacuoles. Immunogold localization using antivicilin antibodies confirmed that the morphologically identified protein bodies within the tube-like membrane system contained vicilin. Immunogold studies using antibodies to an integral membrane protein specific for storage protein vacuoles in bean cotyledons, α-TIP, localized that protein specifically to the tubular membranes enclosing the protein bodies; no α-TIP was detected in the tonoplast of vegetative vacuoles. Conversely, antibodies to an integral tonoplast protein of the γ-TIP family localized that protein to the tonoplast of vegetative vacuoles but not to the membranes surrounding protein bodies. When antibodies against pyrophosphatase and vacuolar ATPase were used, they labeled both protein body vacuole and vegetative vacuole membranes. Thus, the two TIP proteins served as specific markers for membranes that otherwise contained general markers (pyrophosphatase and ATPase) characteristic for tonoplast. When subcellular fractions were prepared and separated on sucrose gradients, mature legumin, another storage protein, localized to the portion of the gradient containing α-TIP antigen, which was separate from the portion containing γ-TIP antigen (49).

With these results, it may be strongly argued that protein bodies in developing pea cotyledon cells are deposited within a membrane system that is morphologically and biochemically distinct from the central (vegetative) vacuole. Presumably the existence of a second storage-type vacuole separates the storage protein bodies from a lytic environment and protects them from degradation during development of the cotyledon. Root tip cells from barley and pea also have two different vacuolar compartments identified separately by α-TIP and γ-TIP antibodies (N Paris & JC Rogers, unpublished data). Therefore, it is likely that many different types of plant cells contain these two compartments.

VACUOLE BIOGENESIS

In *Saccharomyces cerevisiae,* soluble proteins are sorted from the TGN to a prevacuolar endosome-like compartment (93, 123). Movement from the TGN to the prevacuolar compartment occurs rapidly, whereas transport—by as yet uncharacterized mechanisms—from the prevacuolar compartment to the vacuole occurs much more slowly (123). It is believed that this organism has only one type of vacuole, and there is no evidence to suggest alternate pathways from the ER or Golgi to the vacuole. In plant cells, however, different types of vacuoles may coexist in the same cell. Different types of vacuoles may arise, or be maintained, if separate and distinct pathways to the different vacuole types exist. This concept may explain why events leading to initial biogenesis of the tonoplast remain poorly understood. In one model, the initial vacuole compartment arises from the TGN (72). A second model proposes that the vacuole originates from dilation of a branched portion of the smooth endoplasmic reticulum (14, 47). Recent studies provide evidence to support both models (46, 55, 98).

Role of the ER

Herman et al (46) used a monoclonal antibody to the ~60-kDa regulatory subunit (subunit B) of oat vacuolar ATPase (V-ATPase) to study the distribution of this enzyme in differentiating root tip cells that initially lack a large central vacuole. This subunit of V-ATPase is synthesized in the cytoplasm and, together with the catalytic subunit A, comprises part of the peripheral complex; the peripheral complex then associates with the membrane integral sector to form the active V-ATPase enzyme (46). Detection of subunits A or B in association with an organelle is therefore evidence for the presence of an intact enzyme in the membrane of that organelle. These major peripheral complex subunits were predominantly found to be associated with ER-containing subcellular fractions on gradients. Parallel results were obtained when the monoclonal antibody to subunit B was used in immunogold electron microscopy experiments. In these experiments, gold particle labeling was observed on the membranes of small vesicles, provacuoles, and ER, whereas the Golgi apparatus itself remained unlabeled. The authors observed that ER labeling appeared to be localized to certain regions, which suggests that there might be regional specialization of that compartment, a specific subdomain, that was directly responsible for vacuole formation (46). The presence of the presumably intact enzyme on ER membranes might allow it to be incorporated directly into the tonoplast during formation of vacuoles from the ER. Alternatively, it would also be possible for the V-ATPase to be sorted via the Golgi to its final destination. Although localization of B-subunit antigen to the Golgi was not observed in this study, antibodies to the A subunit of the peripheral V-ATPase complex labeled the Golgi complex in maize root tip cells (57). Biogenesis, ontogeny, and maturation of the vacuole may be dependent upon pathways direct from the ER as well as from the Golgi, and the relative contributions from each pathway may influence the function of the resulting organelles.

Studies by Robinson et al (98) indicate that protein storage vacuoles might be formed de novo during pea cotyledon development from sections of endoplasmic reticulum lacking ribosomes, contiguous with rough ER, where osmiophilic deposits of protein were present. Later during development they observed greatly dilated cisternae of smooth ER almost completely filled with osmiophilic material, indistinguishable in appearance from material present in protein bodies in protein storage vacuoles. The authors also observed heavy labeling of protein bodies in protein storage vacuoles with antibodies to the ER chaperone, BiP. They hypothesized that a major portion of protein storage vacuoles in developing pea cotyledons derives directly from specific domains of the ER, where storage protein body formation initially occurs (98).

The concept of direct traffic between ER and a protein storage vacuole compartment is further supported by experiments with brefeldin A and monensin, compounds that prevent sorting of soluble proteins from the Golgi to the vacuole (37, 51, 53). Neither prevented newly synthesized α-TIP from reaching the tonoplast when it was expressed in transgenic tobacco plants (37). Formation of vacuole membrane directly from the ER would provide a brefeldin A–insensitive pathway to the tonoplast for its integral membrane proteins.

These results indicate that one pathway for vacuole formation may occur directly from the ER. Functional tonoplast transmembrane proteins in specific regions of the ER might lead to local pH change (H^+-ATPase function) and dilution of lumenal contents [TIP function (74)], and these changes themselves could be a trigger for vacuolar differentiation.

Role of the Golgi

The complexity of processes that might be involved in vacuole formation is illustrated by other morphologic data from Herman et al (46), who found that electron translucent vesicles of a size consistent with provacuoles appeared to be budding from the *trans*-Golgi or from tubular elements radiating from the Golgi apparatus. Their results indicate that provacuoles also may be derived from Golgi membranes (44, 46).

MULTIPLE PATHWAYS TO THE VACUOLE FROM THE GOLGI

Two distinct types of vesicles are associated with Golgi-to-vacuole traffic.

Smooth Dense Vesicles

Vesicles lacking a visible cytoplasmic coat, with an osmiophilic or dense appearance, have been commonly reported in association with or budding off from the Golgi apparatus. Immunogold labeling has confirmed that they contain storage proteins (46, 58, 98, 135), and the dense appearance is characteristic of that for protein bodies. The vesicles are hypothesized to carry concentrated packages of storage protein to the vacuole. Their biochemical purification and characterization would help us to understand their function. In a preliminary approach to this goal, Hara-Nishimura et al (40) isolated a subcellular fraction with large amounts of the precursor unprocessed forms of the pumpkin storage proteins that were thought to be enriched in smooth dense vesicles (SDVs). This preparation would have also contained both ER and Golgi vesicles, however, but separation of those organelles from vesicles containing concentrated forms of the storage protein precursors was not attempted.

Clathrin-Coated Vesicles

Clathrin-coated vesicles (CCVs) are smaller than SDVs and have a distinct cytoplasmic coat (97). The precursor unprocessed forms of different pea storage proteins have been identified in highly enriched CCV preparations (41, 96), and results from reconstruction experiments indicated that these proteins did not result from contamination of the vesicle preparations by ER or Golgi membranes (50). It is likely that SDVs and CCVs represent vehicles for two different pathways of transport from the Golgi.

Biochemical Definition of Two Pathways

Morphological evidence for two vesicle types involved in Golgi-to-vacuole traffic is supported by biochemical evidence for two pathways. Soluble proteins destined for the vacuole must carry sorting determinants that allow them to be identified and sorted from the secretory pathway. In general, two types of sorting determinants have been identified (18, 79). One, carried within N-terminal propeptides on barley proaleurain and sweet potato prosporamin, has a conserved amino acid motif and may be recognized in a specific manner by a sorting receptor (60). The other, localized to C-terminal propeptides, has little apparent requirement for sequence conservation but may be sensitive to the distribution of hydrophobic and charged residues (30, 81). Matsuoka et al (73) established a system where they could study the function of either N-terminal or C-terminal sorting determinants independent of the type of protein to which they were attached. Either the prosporamin N-terminal propeptide (NTPP) or the barley lectin C-terminal propeptide (CTPP), when attached to either sporamin or barley lectin, directed each protein to the vacuole of tobacco suspension culture cells. Wortmannin, an inhibitor of phosphatidylinositol kinases, caused almost complete inhibition of CTPP-mediated transport to the vacuole at concentrations that had little or no effect on NTPP-mediated transport to the vacuole. This result strongly supports the concept that different pathways must be used for the two types of vacuolar targeting (73). If so, what are the destinations of the two pathways?

When barley prolectin and prosporamin were expressed together in the same transgenic tobacco plant, the mature forms of both proteins were localized together in large protein aggregates in central vacuoles in leaf and root cells (104), which suggests that two different pathways lead directly to the same vacuole. However, the two pathways may also lead to two different intermediate prevacuolar compartments, and subsequent fusion with the large central vacuole may have resulted in their colocalization. The experiments (104) did not focus on less-differentiated cells where a separation of the two hypothetical compartments might be expected to be seen. In addition, the two antigens were detected only in intravacuolar protein aggregates where they would be present in highly concentrated form, and they were not detected elsewhere along the secretory pathway in compartments such as the ER or Golgi, through which the proteins are known to pass. Thus, it is possible that the assay for their detection was not sensitive enough to allow identification of intermediate compartments that might merge in later stages of cell differentiation to form a central vacuole.

A protein marker has been identified that may be specific for the CCV pathway. A protein with binding specificity for the N-terminal propeptide determinants of proaleurain and prosporamin has been purified from pea CCV membranes (60). This protein, BP-80, has characteristics consistent with a function of a plant vacuolar sorting receptor (60). BP-80 is also present in much greater abundance in a lower density fraction from gradients used to isolate CCVs (60), and this lower density fraction can now be clearly separated from IDPase activity, which indicates that it is not derived from Golgi (T Kirsch & L Beevers, unpublished data). It is of interest to determine whether this fraction is part of a prevacuolar or endosomal compartment analogous to that identified in yeast. At present, direct functional evidence that BP-80 acts as a receptor is lacking, but the function of BP-80 and its homologues may be of importance in intracellular trafficking of proteins in the plant endomembrane system.

It is also possible that receptor-independent sorting of proteins to the vacuole may occur. Physical aggregation in the Golgi could be a receptor-independent signal for sorting into either SDVs or CCVs to direct the aggregate to a vacuole (125). Presumably, large aggregates would be directed to the larger SDVs. Aggregation would be more likely with certain types of proteins, in particular storage proteins where their physical characteristics predispose them to interact with other similar molecules during protein body formation. CTPP vacuolar targeting determinants, where hydrophobic residues appear to be important for function (30, 81), might also participate in a process of aggregation to direct proteins into this hypothetical pathway. The available data, however, cannot exclude the possible role of some sort of receptor protein in vacuolar sorting in the Golgi of storage proteins and of those with C-terminal propeptides.

Formation of Wheat Protein Bodies

It is well accepted that the Golgi apparatus plays a role in wheat PB formation, although it remains uncertain whether processing through this organelle is the major pathway (7, 58, 67, 87). Recently, Levanony et al (67) suggested that the Golgi-mediated pathway plays only a minor role in PB formation. Instead, the major pathway involves formation of a protein inclusion body in the ER which then becomes disconnected from the cisternal ER. The protein inclusion is then surrounded by electron-lucent vesicles that eventually fuse together to form a small vacuole containing a PB inclusion. The small vacuole containing a PB is then engulfed by the large vacuoles by an autophagy-like process (67).

This autophagy pathway is consistent with many of the observations made in previous studies with wheat (16, 63, 75, 76, 87). Similar observations, however, have not been reported in studies of protein body and protein storage vacuole formation in legumes and in other cereals (see sections on Storage of Prolamines—ER vs Vacuole and Protein Storage Vacuoles), and it is puzzling that the process appears to be limited only to wheat. The localization of the lumenal chaperones BiP and PDI in mature PBs is also consistent with the view that the ER is a site of PB formation [67, 109; see Takemoto et al (115a) for a possible alternative explanation]. Moreover, an autophagy pathway would account for the extensive membrane-like remnants observed within and adjacent to vacuoles containing PBs.

It is unclear how ER-localized PBs would be released from the ER network in this pathway. In animals, ER-derived intracisternal granules are degraded by autophagic vacuoles by a multistep process (119). The portion of RER containing a protein inclusion body is enclosed within a multilayered smooth membrane of unknown origin to yield the phagophore, whose function is to sequester a specific region of the ER from the rest of the membrane network. The phagophore containing the intracisternal granule fuses with endosomes as the degradation process continues (119). The granule is always enclosed within a membrane-bound compartment, and there are no precedents in this system for release of lumenal contents from an organelle in the secretory pathway directly into the cytoplasm. Structures analogous to phagophores have not been reported in studies of wheat PB formation. If the functional equivalent of phagophores does not exist in plants, then there must be an alternative process.

In summary, evidence that supports an autophagy pathway is incomplete, but, at the same time, direct evidence in support of a Golgi-mediated pathway is almost equally lacking. The problem is technically very difficult to resolve because it has not yet been possible to follow the progress of a protein body from synthesis to deposition in a vacuole in real time using biochemical methods, or in a single living cell using microscopy.

CONCLUSION

Synthesis and transport of storage proteins and their incorporation into protein bodies provides a focus for understanding unique features of protein compartmentation in the endomembrane system in plant cells. These processes are profoundly affected by interactions of the storage proteins with ER chaperones, by concentrations of newly synthesized storage proteins within subdomains of the lumen of the ER, and by transport mechanisms that may cause soluble storage proteins to be greatly concentrated at specific points along the secretory pathway. At each step, the ultimate fate of the proteins (and the location of protein body formation) depends upon whether the proteins remain soluble or whether they form physical aggregates. Although protein body formation is ordered and not due to random precipitation of component proteins, physical association of those proteins into an insoluble form is at the heart of the process. Transport processes from different endomembrane compartments in plant cells may have evolved an ability to detect and sort physical aggregates of protein, so that both biophysical recognition as well as receptor-mediated recognition of specific peptide sequences may be determinants of a protein's traffic and destination.

The problem of assembling and protecting protein bodies from degradation must affect the development and organization of the vacuolar compartment. Plant cells, at least early in differentiation, may have two separate, different types of vacuoles. Recent data support previous hypotheses that vacuoles may derive directly from the ER and that certain storage proteins may be sorted directly from ER to vacuole. In addition, sorting of proteins from Golgi to vacuole clearly involves two separate pathways, and each pathway may lead to a different type of vacuolar compartment. In fully differentiated cells, evidence suggests that the two different vacuolar compartments merge into a single, central vacuole. The next few years of plant cell biology should be exciting, as mechanisms for originating and maintaining the vacuolar types, mechanisms for targeting vesicles to each type, and mechanisms for ultimately merging the compartments are better defined.

ACKNOWLEDGMENTS

We thank Brian Larkins, Rebecca Boston, David Simpson, Leonard Beevers, and Russell Jones for discussions and for providing unpublished results. This work was supported in part by grants from DOE #DE-FG95ER20165 (JCR), NIH #GM52427 (JCR), USDA-NRICG 94-37304-1174 (TWO), and Rockefeller Foundation Program in Rice Biotechnology (TWO).

Literature Cited

1. Altschuler Y, Galili G. 1994. Role of conserved cysteines of a wheat gliadin in its transport and assembly into protein bodies in *Xenopus* oocytes. *J. Biol. Chem.* 269: 6677–82
2. Altschuler Y, Rosenberg N, Harel R, Galili G. 1993. The N- and C-terminal regions regulate the transport of wheat γ-gliadin through the endoplasmic reticulum in Xenopus oocytes. *Plant Cell* 5:443–50
3. Anderson JV, Haskell DW, Guy CL. 1994. Differential influence of ATP on native spinach 70-kilodalton heat-shock cognates. *Plant Physiol.* 104:1371–80
4. Bagga S, Adams H, Kemp JD, Sengupta-Gopalan C. 1995. Accumulation of 15-kilodalton zein in novel protein bodies in transgenic tobacco. *Plant Physiol.* 107: 13–23
5. Balch WE, McCaffery JM, Plutner H, Farquhar MG. 1994. Vesicular stomatitis virus glycoprotein is sorted and concentrated during export from the endoplasmic reticulum. *Cell* 76:841–52
6. Banerji U, Renfranz PJ, Pollock JA, Benzer S. 1987. Molecular characterization and expression of *sevenless,* a gene involved in neuronal pattern formation in the *Drosophila* eye. *Cell* 49:281–91
7. Bechtel DB, Gaines RL, Pomeranz Y. 1982. Early stages in wheat endosperm formation and protein body initiation. *Ann. Bot.* 50: 507–18
8. Bechtel DB, Juliano BO. 1980. Formation of protein bodies in the starchy endosperm of rice (*Oryza sativa* L.) a reinvestigation. *Ann. Bot.* 45:503–9
9. Bednarek SY, Raikhel NV. 1992. Intracellular trafficking of secretory proteins. *Plant Mol. Biol.* 20:133–50
10. Blond-Elguindi S, Cwirla SE, Dower WJ, Lipshutz RJ, Sprang SR, et al. 1993. Affinity panning of a library of peptides displayed on bacteriophages reveals the binding specificity of BiP. *Cell* 75:717–28
11. Boller T, Wiemken A. 1986. Dynamics of vacuolar compartmentation. *Annu. Rev. Plant Physiol.* 37:137–64
12. Boston RS, Fontes EBP, Shank BB, Wrobel RL. 1991. Increased expression of the maize immunoglobulin binding protein homolog b-70 in three zein regulatory mutants. *Plant Cell* 3:497–505
13. Bullied NJ, Freedman RB. 1988. Defective co-translational formation of disulphide bonds in protein disulphide isomerase deficient microsomes. *Nature* 335:649–51
14. Burgess J, Lawrence W. 1985. Studies of the recovery of tobacco mesophyll protoplasts from an evacuolation treatment. *Protoplasma* 126:140–46
15. Cameron-Mills V, von Wettstein D. 1980. Protein body formation in the developing barley endosperm. *Carlsberg Res. Commun.* 45:577–94
16. Campbell WP, Lee JW, O'Brien TP, Smart MG. 1981. Endosperm morphology and protein body formation in developing wheat grains. *Aust. J. Plant Physiol.* 8: 5–19
17. Chrispeels MJ. 1991. Sorting of proteins in the secretory system. *Annu. Rev. Plant Physiol. Plant Mol. Biol.* 42:21–53
18. Chrispeels MJ, Raikhel NV. 1992. Short peptide domains target proteins to plant vacuoles. *Cell* 68:613–16
19. Chrispeels MJ, Von Schaewen A. 1992. Sorting of proteins in the secretory system of plant cells. *Antonie van Leeuwenhoek Int. J. Gen. Mol. Microbiol.* 61:161–65
20. Coleman CE, Lopes MA, Gillikin JW, Boston RS, Larkins BA. 1995. A defective signal peptide in the maize high-lysine mutant *floury 2. Proc. Natl. Acad. Sci. USA* 92: 6828–31
21. Craig S, Goodchild DJ, Hardham AR. 1979. Structural aspects of protein accumulation in developing pea cotyledons. I. Qualitative and quantitative changes in parenchyma cell vacuoles. *Aust. J. Plant Physiol.* 6:81–98
22. Craig S, Goodchild DJ, Miller C. 1980. Structural aspects of protein accumulation in developing pea cotyledons. II. Three-dimensional reconstructions of vacuoles and protein bodies from serial sections. *Aust. J. Plant Physiol.* 7:329–37
23. D'Amico L, Valsasina B, Daminati MG, Fabbrini MS, Nitti G, et al. 1992. Bean homologs of the mammalian glucose-regulated proteins: induction by tunicamycin and interaction with newly synthesized seed storage proteins in the endoplasmic reticulum. *Plant J.* 2:443–55
24. Dannenhoffer JM, Bostwick DE, Or E, Larkins BA. 1995. Opaque-15, a maize mutation with properties of a defective opaque-2 modifier. *Proc. Natl. Acad. Sci. USA* 92:1931–35
25. Denecke J, Botterman J, Deblaere R. 1990.

Plant secretion in plant cells can occur via a default pathway. *Plant Cell* 2:51–59
26. Denecke J, Carlsson LE, Bidal S, Hogland A-S, Elk B, et al. 1995. The tobacco homolog of mammalian calreticulum is present in protein complexes *in vivo*. *Plant Cell* 7:391–406
27. Denecke J, DeRycke R, Botterman J. 1992. Plant and mammalian sorting signals for protein retention in the endoplasmic reticulum contain a conserved epitope. *EMBO J.* 11:2345–55
28. Denecke J, Ek B, Caspers M, Sinjorgo KMC, Palva ET. 1993. Analysis of sorting signals responsible for the accumulation of soluble reticuloplasmins in the plant endoplasmic reticulum. *J. Exp. Bot.* 44 (Suppl.):213–21
29. Denecke J, Goldman MHS, Demolder J, Seurinck J, Botterman J. 1991. The tobacco luminal binding protein is encoded by a multigene family. *Plant Cell* 3:1025–35
30. Dombrowski JE, Schroeder MR, Bednarek SY, Raikhel NV. 1993. Determination of the functional elements within the vacuolar targeting signal of barley lectin. *Plant Cell* 5:587–96
31. Fontes EBP, Shank BB, Wrobel RL, Moose SP, O'Brian GR, et al. 1991. Characterization of an immunoglobulin binding protein homolog in the maize *floury-2* endosperm mutant. *Plant Cell* 3:483–96
32. Galili G, Altschuler Y, Levanony H. 1993. Assembly and transport of seed storage proteins. *Trends Cell Biol.* 3:437–42
33. Galili G, Altschuler Y, Levanony H, Giorini-Silfen S, Shimoni Y, et al. 1995. Assembly and transport of wheat storage proteins. *J. Plant Physiol.* 145:626–31
34. Geli MI, Torrent M, Ludevid D. 1994. Two structural domains mediate two sequential events in γ-zein targeting: protein endoplasmic reticulum retention and protein body formation. *Plant Cell* 6:1911–22
35. Gething MJ, Sambrook J. 1992. Protein folding in the cell. *Nature* 335:33–45
36. Giorini S, Galili G. 1991. Characterization of HSP-70 cognate proteins from wheat. *Theor. Appl. Genet.* 82:615–20
37. Gomez L, Chrispeels MJ. 1993. Tonoplast and soluble vacuolar proteins are targeted by different mechanisms. *Plant Cell* 5: 1113–24
38. Griffing LR, Fowke LC. 1985. Cytochemical localization of peroxidase in soybean suspension culture cells and protoplasts: intracellular vacuole differentiation and presence of peroxidase in coated vesicles and multivesicular bodies. *Protoplasma* 128:22–30
39. Hammond C, Helenius A. 1995. Quality control in the secretory system. *Curr. Opin. Cell Biol.* 7:523–29
40. Hara-Nishimura I, Inoue K, Nishimura M. 1991. A unique vacuolar processing enzyme responsible for conversion of several precursors into mature forms. *FEBS Lett.* 294:89–93
41. Harley SM, Beevers L. 1989. Coated vesicles are involved in the transport of storage proteins during seed development in *Pisum sativum* L. *Plant Physiol.* 91:674–78
42. Helm KW, LaFayette PR, Nagao RT, Key JL, Vierling E. 1993. Localization of small heat shock proteins to the higher plant endomembrane system. *Mol. Cell. Biol.* 13: 238–47
43. Hepler PK, Palevitz BA, Lancelle SA, McCauley MM, Lichtscheibl I. 1990. Cortical endoplasmic reticulum in plants. *J. Cell Sci.* 96:355–73
44. Herman EM. 1994. Multiple origins of intravacuolar protein accumulation of plant cells. In *Advances in Structural Biology,* ed. S Malhotra, pp. 243–83. Greenwich, Conn: JAI
45. Herman EM, Lamb CJ. 1992. Arabino galactan-rich glycoproteins are localized on the cell surface and in intravacuolar multivesicular bodies. *Plant Physiol.* 98: 264–72
46. Herman EM, Li XH, Su RT, Larsen P, Hsu HT, Sze H. 1994. Vacuolar-type H^+-ATPases are associated with the endoplasmic reticulum and provacuoles of root tip cells. *Plant Physiol.* 106:1313–24
47. Hilling B, Amelunxen F. 1985. On the development of the vacuole. II. Further evidence for endoplasmic reticulum origin. *Eur. J. Cell Biol.* 38:195–200
48. Hoffman LM, Donaldson DD, Bookland R, Rashka K, Herman EM. 1987. Synthesis and protein deposition of maize 15-kD zein in transgenic tobacco seeds. *EMBO J.* 6: 3213–21
49. Hoh B, Hinz G, Jeong B-K, Robinson DG. 1995. Protein storage vacuoles form de novo during pea cotyledon development. *J. Cell Sci.* 108:299–310
50. Hoh B, Schauermann G, Robinson DG. 1991. Storage protein polypeptides in clathrin-coated vesicle fractions from developing pea cotyledons are not due to endomembrane contamination. *J. Plant Physiol.* 138:309–16
51. Holwerda BC, Galvin NJ, Baranski TJ, Rogers JC. 1990. In vitro processing of aleurain, a barley vacuolar thiol protease. *Plant Cell* 2:1091–106
52. Holwerda BC, Padgett HS, Rogers JC. 1992. Proaleurain vacuolar targeting is mediated by short contiguous peptide interactions. *Plant Cell* 4:307–18
53. Holwerda BC, Rogers JC. 1992. Purification and characterization of aleurain: a plant thiol protease functionally homologous to mammalian cathepsin H. *Plant Physiol.* 99:848–55

54. Holwerda BC, Rogers JC. 1993. Structure, functional properties and vacuolar targeting of the barley thiol protease, aleurain. *J. Exp. Bot.* 44(Suppl.):321–39
55. Hörtensteiner S, Martinoia E, Amrhein N. 1992. Reappearance of hydrolytic activities and tonoplast proteins in the regenerated vacuole of evacuolated protoplasts. *Planta* 187:113–21
56. Hurkman WJ, Smith LD, Richter J, Larkins BA. 1981. Subcellular compartmentation of maize storage proteins in *Xenopus* oocytes injected with zein messenger RNAs. *J. Cell Biol.* 89:292–99
57. Hurley D, Taiz L. 1989. Immunocytochemical localization of the vacuolar H^+-ATPase in maize root tip cells. *Plant Physiol.* 89:391–95
58. Kim WT, Franceschi VR, Krishnan HB, Okita TW. 1988. Formation of wheat protein bodies: involvement of the Golgi apparatus in gliadin transport. *Planta* 173: 173–82
59. Kim WT, Li X, Okita TW. 1993. Expression of storage protein multigene families in developing rice endosperm. *Plant Cell Physiol.* 34:595–603
60. Kirsch T, Paris N, Butler JM, Beevers L, Rogers JC. 1994. Purification and initial characterization of a potential plant vacuolar targeting receptor. *Proc. Natl. Acad. Sci. USA* 91:3403–7
61. Koch GLE. 1987. Reticuloplasmins: a novel group of proteins in the endoplasmic reticulum. *J. Cell Sci.* 92:491–92
62. Krishnan HB, Franceschi VR, Okita TW. 1986. Immunochemical studies on the role of the Golgi complex in protein body formation in rice seeds. *Planta* 169: 471–80
63. Krishnan HB, White JA, Pueppke SG. 1991. Immunocytochemical localization of wheat prolamins in the lumen of the rough endoplasmic reticulum. *Can. J. Bot.* 69: 2574–77
64. Lee C, Chen LB. 1988. Dynamic behavior of endoplasmic reticulum in living cells. *Cell* 54:37–46
65. Lending CR, Chestnut RS, Shaw KL, Larkins BA. 1989. Immunolocalization of avenin and globulin storage proteins in developing endosperm of *Avena sativa* L. *Planta* 178:315–24
66. Lending CR, Larkins BA. 1989. Changes in the zein composition of protein bodies during maize endosperm development. *Plant Cell* 1:1011–23
67. Levanony H, Rubin R, Altschuler Y, Galili G. 1992. Evidence for a novel route of wheat storage proteins to vacuoles. *J. Cell Biol.* 119:1117–28
68. Li X, Franceschi VR, Okita TW. 1993. Segregation of storage protein mRNAs on the rough endoplasmic reticulum membranes of rice endosperm cells. *Cell* 72: 869–79
69. Li X, Wu Y, Zhang D-Z, Gillikin JW, Boston R, et al. 1993. Rice prolamine protein body biosynthesis: a BiP-mediated process. *Science* 262:1054–56
70. Lucas WJ, Wolf S. 1993. Plasmodesmata: the intercellular organelles of green plants. *Trends Cell Biol.* 3:308–15
71. Marocco A, Santucci A, Cerioli S, Motto M, Di Fonzo N, et al. 1991. Three high-lysine mutations control the level of ATP-binding HSP70-like proteins in the maize endosperm. *Plant Cell* 3:507–15
72. Marty F, Branton D, Leigh RA. 1980. Plant vacuoles. In *The Biochemistry of Plants: The Plant Cell,* ed. NE Tolbert, 1:625–58. New York: Academic
73. Matsuoka K, Bassham DC, Raikhel N, Nakamura K. 1995. Different sensitivity to wortmannin of two vacuolar sorting signals indicates the presence of distinct sorting machineries in tobacco cells. *J. Cell. Biol.* 130:1307–18
74. Maurel C, Reizer J, Schroeder JI, Chrispeels MJ. 1993. The vacuolar membrane protein γ-TIP creates water specific channels in *Xenopus* oocytes. *EMBO J.* 12: 2241–48
75. Miflin BJ, Burgess SR, Shewry PR. 1981. The development of protein bodies in the storage tissues of seeds: subcellular separation of homogenates of barley, maize and wheat endosperm and pea cotyledon. *J. Exp. Bot.* 32:199–19
76. Miflin BJ, Field JM, Shewry PR. 1983. Cereal storage proteins and their effect on technological properties. In *Seed Proteins,* ed. J Daussant, J Mosse, J Vaughan, pp. 255–319. London: Academic
77. Mizuno M, Singer SJ. 1993. A soluble secretory protein is first concentrated in the endoplasmic reticulum before transfer to the Golgi apparatus. *Proc. Natl. Acad. Sci. USA* 90:5732–36
78. Muench DG, Okita TW. 1996. The storage proteins of rice and oat. In *Cellular and Molecular Biology of Plant Development,* ed. BA Larkins, IK Vasil. Netherlands: Kluwer. In press
79. Nakamura K, Matsuoka K. 1993. Protein targeting to the vacuole in plant cells. *Plant Physiol.* 101:1–6
80. Napier RM, Fowke LC, Hawes C, Lewis M, Pelham HRB. 1992. Immunological evidence that plants use both HDEL and KDEL for targeting proteins to the endo plasmic reticulum. *J. Cell Sci.* 102: 261–71
81. Neuhaus J-M, Pietrzak M, Boller T. 1994. Mutation analysis of the C-terminal vacuolar targeting peptide of tobacco chitinase: low specificity of the sorting system, and gradual transition between intracellular re-

tention and secretion into the extracellular space. *Plant J.* 5:45–54
82. Nicchitta CV, Blobel G. 1993. Lumenal proteins of the mammalian endoplasmic reticulum are required to complete protein translocation. *Cell* 73:989–98
83. Okita TW, Li X, Roberts MW. 1994. Targeting of mRNAs to domains of the endoplasmic reticulum. *Trends Cell Biol.* 4: 91–96
84. Or E, Boyer SK, Larkins BA. 1993. *opaque2* Modifiers act post-transcriptionally and in a polar manner on γ-zein gene expression in maize endosperm. *Plant Cell* 5:1599–609
85. Palade GE. 1975. Intracellular aspects of the processing of protein synthesis. *Science* 189:347–58
86. Palevitz BA, O'Kane DJ, Kobres RE, Raikhel NV. 1981. The vacuole system in stomatal cells of *Allium:* vacuole movements and changes in morphology in differentiating cells as revealed by epifluorescence, video and electron microscopy. *Protoplasma* 109:23–55
87. Parker ML. 1982. Protein accumulation in developing endosperm of a high-protein line of *Triticum dicoccoides. Plant Cell Environ.* 5:237–43
88. Pathak RK, Luskey KL, Anderson RGW. 1986. Biogenesis of the crystalloid endoplasmic reticulum in UT-1 cells: evidence that newly formed endoplasmic reticulum emerges from the nuclear envelope. *J. Cell Biol.* 102:2158–68
89. Pedrazzini E, Giovinazzo G, Bollini R, Ceriotti A, Vitale A. 1994. Binding of BiP to an assembly-defective protein in plant cells. *Plant J.* 5:103–10
90. Pelham HRB. 1995. Sorting and retrieval between the endoplasmic reticulum and Golgi apparatus. *Curr. Opin. Cell Biol.* 7: 530–35
91. Pfeffer SR, Rothman JE. 1987. Biosynthetic protein transport and sorting by the endoplasmic reticulum and Golgi. *Annu. Rev. Biochem.* 56:828–52
92. Pratt KA, Madgwick PJ, Shewry PR. 1991. Expression of a wheat gliadin protein in yeasts (*Saccharomyces cerevisiae*). *J. Cereal Sci.* 14:223–29
93. Raymond CK, Howald-Stevenson I, Vater CA, Stevens TH. 1992. Morphological classification of a yeast vacuolar protein sorting mutants: evidence for a prevacuolar compartment in class-E vps mutants. *Mol. Biol. Cell* 3:1389–402
94. Rechinger KB, Simpson DJ, Svendsen I, Cameron-Mills V. 1993. A role for γ3 hordein in the transport and targeting of prolamin polypeptides to the vacuole of developing barley endosperm. *Plant J.* 4:841–53
95. Record RD, Griffing LR. 1988. Convergence of the endocytic and lysosomal pathways in soybean protoplasts. *Planta* 176: 425–32
96. Robinson DG, Balusek K, Freundt H. 1989. Legumin antibodies recognize polypeptides in coated vesicles isolated from developing pea cotyledons. *Protoplasma* 150: 79–82
97. Robinson DG, Depta H. 1988. Coated vesicles. *Annu. Rev. Plant Physiol. Plant Mol. Biol.* 39:53–99
98. Robinson DG, Hoh B, Hinz G, Jeong B-K. 1995. One vacuole or two vacuoles: do protein storage vacuoles arise *de novo* during pea cotyledon development? *J. Plant Physiol.* 145:654–64
99. Rose JK, Doms RW. 1988. Regulation of protein export from the endoplasmic reticulum. *Annu. Rev. Cell Biol.* 4:257–88
100. Rosenberg N, Shimoni Y, Altschuler Y, Levanony H, Volokita M, Galili G. 1993. Wheat (*Triticum aestivum* L.) γ-gliadin accumulates in dense protein bodies within the endoplasmic reticulum of yeast. *Plant Physiol.* 102:61–69
101. Rüneberg-Roos P, Kervinen J, Kovaleva V, Raikhel NV, Gal S. 1994. The aspartic proteinase of barley is a vacuolar enzyme that processes probarley lectin in vitro. *Plant Physiol.* 105:321–29
102. Salpeter M, Farquhar MG. 1981. High resolution autoradiographic analysis of the secretory pathway in mammotrophs of the rat anterior pituitary. *J. Cell Biol.* 91: 240–46
103. Sanders SL, Whitfield KM, Vogel JP, Rose MD, Schekman RW. 1992. Sec61p and BiP directly facilitate polypeptide translocation into the ER. *Cell* 69:353–65
104. Schroeder MR, Borkhsenious ON, Matsuoka K, Nakamura K, Raikhel NV. 1993. Colocalization of barley lectin and sporamin in vacuoles of transgenic tobacco plants. *Plant Physiol.* 101:451–58
105. Shewry PR, Napier NA, Tatham AS. 1995. Seed storage proteins: structures and biosynthesis. *Plant Cell* 7:945–56
106. Shewry PR, Sayanova O, Tatham AS, Tamas L, Turner M, et al. 1995. Structure, assembly and targeting of wheat storage proteins. *J. Plant Physiol.* 145:620–25
107. Shewry PR, Tamas L, Tatham AS, Halford NG, Pratt K, et al. 1992. Development of expression and refolding systems for cereal proteins. In *Protein Engineering,* ed. P Goodenough, II:40–51. Newbury, United Kingdom: CPL Press
108. Shewry PR, Tatham AS. 1990. The prolamin storage proteins of cereal seeds: structure and evolution. *Biochem. J.* 267: 1–12
109. Shimoni Y, Zhu X-Z, Levanony H, Segal G, Galili G. 1995. Purification, characterization, and intracellular localization of glycosylated protein disulfide isomerase

from wheat grain. *Plant Physiol.* 108: 327–35

110. Shotwell M, Larkins BA. 1989. The biochemistry and molecular biology of seed storage proteins. In *The Biochemistry of Plants: A Comprehensive Treatise,* ed. E Marcus, pp. 296–45. Orlando, FL: Academic
111. Simon R, Altschuler Y, Rubin R, Galili G. 1990. Two closely related wheat storage proteins follow a markedly different subcellular route in *Xenopus laevis* oocytes. *Plant Cell* 2:941–50
112. Sitia R, Meldolesi J. 1992. Endoplasmic reticulum: a dynamic patchwork of specialized subregions. *Mol. Biol. Cell* 3:1067–72
113. St-Johnston D. 1995. The intracellular localization of messenger RNAs. *Cell* 81: 161–70
114. Staehelin LA, Moore I. 1995. The plant Golgi apparatus: structure, functional organization and trafficking mechanisms. *Annu. Rev. Plant Physiol. Plant Mol. Biol.* 46:261–88
115. Svoboda KKH. 1991. Intracellular localization of types I and II collagen mRNA and endoplasmic reticulum in embryonic corneal epithelia. *J. Cell Sci.* 100:23–33

115a. Takemoto H, Yoshimori T, Yamamoto A, Miyata Y, Yahara I, et al. 1992. Heavy chain binding protein (BiP/GRP78) and endoplasmin are exported from the endoplasmic reticulum in rat exocrin pancreatic cells, similar to protein disulfide-isomerase. *Arch. Biochem. Biophys.* 296: 129–36

116. Tanaka K, Sugimoto T, Ogawa M, Kasai Z. 1980. Isolation and characterization of two types of protein bodies in the rice endosperm. *Agric. Biol. Chem.* 44:1633–39
117. Tanchak MA, Griffing LR, Mersey BG, Fowke LC. 1984. Endocytosis of cationized ferritin by coated vesicles of soybean protoplasts. *Planta* 162:481–86
118. Tepass U, Theres C, Knust E. 1990. *Crumbs* encodes an EGF-like protein expressed on apical membranes of Drosophila epithelial cells and required for organization of epithelia. *Cell* 61:787–99
119. Tooze J, Hollinshead M, Ludwig T, Howell K, Hoflack B, Kern H. 1990. In exocrine pancreas, the basolateral endocytic pathway converges with the autophagic pathway immediately after the early endosome. *J. Cell Biol.* 111:329–45
120. Tooze J, Kern HF, Fuller SD, Howell KE. 1989. Condensation-sorting events in the rough endoplasmic reticulum of exocrine pancreatic cells. *J. Cell Biol.* 109:35–50
121. Torrent M, Geli MI, Ruiz-Avila L, Canals JM, Puigdomenech P, Ludevid D. 1994. Role of structural domains for maize γ-zein retention in *Xenopus* oocytes. *Planta* 192: 512–18
122. Van Vactor D, Krantz DE, Reinke R, Zipursky SL. 1988. Analysis of mutants in chaoptin, a photoreceptor cell-specific glycoprotein in *Drosophila,* reveals its role in cellular morphogenesis. *Cell* 52:281–90
123. Vida TA, Huyer G, Emr SD. 1993. Yeast vacuolar proenzymes are sorted in the late Golgi complex and transported to the vacuole via a prevacuolar endosome-like compartment. *J. Cell Biol.* 121:1245–56
124. Vitale A, Ceriotti A, Denecke J. 1993. The role of the endoplasmic reticulum in protein synthesis, modification and intracellular transport. *J. Exp. Bot.* 44:1417–44
125. Vitale A, Chrispeels MJ. 1992. Sorting of proteins to the vacuoles of plant cells. *BioEssays* 14:151–60
126. Vogel JP, Misra LM, Rose MD. 1990. Loss of Bip/GRP78 function blocks translocation of secretory proteins in yeast. *J. Cell Biol.* 110:1885–95
127. Wallace JC, Galili G, Kawata EE, Cuellar RE, Shotwell MA, Larkins BA. 1988. Aggregation of lysine-containing zeins into protein bodies in *Xenopus* oocytes. *Science* 240:662 64
128. Wandelt CI, Khan RI, Craig S, Schroeder HE, Spencer D, Higgins TJV. 1992. Vicilin with carboxy-terminal KDEL is retained in the endoplasmic reticulum and accumulates to high levels in the leaves of transgenic plants. *Plant J.* 2:181–92
129. Wilhelm JE, Vale RD. 1993. RNA on the move: the mRNA localization pathway. *J. Cell Biol.* 123:269–74
130. Wilkins TA, Tiwari SC. 1995. Biogenesis of distinct vacuole-types during cell differentiation of seed trichomes. *J. Cell. Biochem.* 19A(Suppl.):155
131. Wink M. 1993. The plant vacuole: a multifunctional compartment. *J. Exp. Bot.* 44 (Suppl.):231–46
132. Yamagata H, Tamura K, Tanaka K, Kasai Z. 1986. Cell-free synthesis of rice prolamin. *Plant Cell Physiol.* 27:1419–22
133. Yamagata H, Tanaka K. 1986. The site of synthesis and accumulation of rice storage proteins. *Plant Cell Physiol.* 27:135–45
134. Zhang F, Boston RS. 1992. Increases in binding protein (BiP) accompany changes in protein body morphology in three high-lysine mutants of maize. *Protoplasma* 171: 142–52
135. zur Nieden U, Manteuffel R, Weber E, Neumann D. 1984. Dictysomes participate in the intracellular pathway of storage proteins in developing *Vicia faba* cotyledons. *Eur. J. Cell Biol.* 34:9–17

Annu. Rev. Plant Physiol. Plant Mol. Biol. 1996. 47:351–76

WHAT CHIMERAS CAN TELL US ABOUT PLANT DEVELOPMENT

Eugene J. Szymkowiak

Department of Biological Sciences, The University of Iowa, Iowa City, Iowa 52242; gszymkow@blue.weeg.uiowa.edu

Ian M. Sussex

Department of Plant Biology, University of California, Berkeley, California 94720

KEY WORDS: chimera, genetic mosaic, plant development, cell interactions, meristem layer

ABSTRACT

The generation and analysis of plant chimeras and other genetic mosaics have been used to deduce patterns of cell division and cell fate during plant development and to demonstrate the existence of clonally distinct cell lineages in the shoot meristems of higher plants. Cells derived from these lineages do not have fixed developmental fates but rely on positional information to determine their patterns of division and differentiation. Chimeras with cells that differ genetically for specific developmental processes have been experimentally generated by a variety of methods. This review focuses on studies of intercellular interactions during plant development as well as of the coordination of cells during meristem function and organogenesis. Recent experiments combining mosaic analysis with molecular analysis of developmental mutants have begun to shed light on the nature of the signals involved in these processes and the mechanisms by which they are transmitted and received among cells.

CONTENTS

1040-2519/96/0601-0351$08.00

INTRODUCTION

The analysis of genetic mosaics in which phenotypically marked cells are followed over the course of a plant's development has been invaluable in determining cell lineage patterns in plants. A common conclusion is that, although the fates of cells are generally predictable for any given period in development—i.e. planes of division, duration of division, and elongation and differentiation patterns—these fates are not rigidly fixed. How a cell divides and differentiates is determined by its response to its current position and not by its past lineage. Mosaics have also been used to demonstrate that groups of cells, often from separate lineages, coordinate their activities throughout development. This implies that cells transmit and receive signals, but the mechanisms by which cells determine and respond to position and how they coordinate development with neighboring cells is not well understood. Genetic mosaics provide a unique tool for investigating these questions.

Historical Background

The term "chimera" is applied to plant genetic mosaics in which mosaicism is persistent because cells with two or more different genotypes coexist in a meristem. However, the term has also been used to describe mosaics in which cells with different genotypes are located anywhere in the plant (88). The history of the application of the term "chimera" to mosaics is described in other reviews of plant chimeras (45, 84, 118, 128, 129). We discuss some of the major points of the history that are relevant to the use of chimeras to study development.

Plant mosaics have long been recognized by nurserymen when arose as bud sports with novel phenotypes (28). For example, a shoot that arose from a graft of *Cytisus purpureus* on a stock of *Laburnum vulgare* produced flowers that had intermediate characteristics of the two originally grafted plants (94). This plant produced branches that formed flowers identical to either of the originally grafted plants, i.e. revertant sectors, at a high rate. Such a plant having characteristics of both of the graft partners from which it developed was thought to be the result of somatic fusion of cells of the two plants, and the revertant sectors were the result of the segregation of the characters back to uniformity. Such plants were named graft hybrids (28, 118).

Winkler grafted tomato and nightshade (*Solanum nigrum*) in an attempt to generate such a graft hybrid (136). From grafts of the two plants, several types of shoots were regenerated from cells located at the graft junction. Most of the shoots were identical to either tomato or nightshade, but some showed a blending of nightshade and tomato characters, which Winkler first thought were graft hybrids (137, 138). Winkler also obtained shoots in which the main axis was composed of tomato tissue in a sector adjacent to nightshade tissue (136). Unlike the graft hybrids, which showed a blending of characters, in this shoot both parental types coexisted independently but within a single axis. He called this a "graft chimera," after the "Chimera" of Greek mythology.

The true nature of graft hybrids was revealed in Baur's analysis of variegated pelargoniums (*Pelargonium zonale* var. "albomarginata hort."), which had white leaf margins and green centers (4). Baur observed that some shoots were sectored longitudinally with green and white tissue, reminiscent of Winkler's chimeras. Baur called these "sectorial chimeras." He postulated that plants having the white-margined, variegated leaves were also chimeras, but that in this case the green and white cells must have been in different layers, parallel to the surface of the meristem. A layered arrangement of cells in the meristem had already been observed by histologists (52). Baur suggested that the graft hybrids were in fact periclinal chimeras: One meristem layer was composed of cells from one graft partner, and the other layers were composed of cells from the other graft partner (5).

Although the variegation in the leaves of Baur's pelargoniums suggested a meristem that was mosaic for cells having green or white plastids, this could not be directly demonstrated because green and white plastids cannot be distinguished in meristem cells. Satina, Blakeslee, & Avery (103) approached this problem by generating plants that were chimeric not for plastid phenotypes but for ploidy levels, in which differences could be detected in histological preparations of meristems (cytochimeras). Winkler had used the fact that tomato and nightshade differed in chromosome number in concurring with Baur that Winkler's supposed graft hybrids were actually periclinal chimeras (139). Satina and coworkers treated seeds of *Datura* with colchicine, which induces polyploidy (12), and found that the outer three layers of cells in the meristem (L1, L2, L3) were affected independently of one another. They obtained a number of different stable periclinal cytochimeras, e.g. one in which L1 was octoploid overlying diploid L2 and L3 (8N 2N 2N). The outer two layers, L1 and L2, are maintained as separate lineages because their cells undergo only anticlinal divisions within the meristem, which allows for the existence of stable periclinal chimeras.

As a result of their increased nuclear size, polyploid cells of the *Datura* cytochimeras can be observed both in a meristem layer and in its derivatives in sectioned preparations of mature organs of the plant (99–102). Derivatives of all three layers were found to participate in organ formation. The pattern of tissue contribution to a mature organ from a polyploid meristem layer was found to correspond to the pattern of green or white cells from that same meristem layer in plastid chimeras (30, 112). The plastid chimeras present the advantage that large amounts of tissue can be readily observed without sectioning. Thus, even though plastid phenotype is not discernible in meristems, it was possible to use the patterns of green vs white tissues in leaves of plastid chimeras to construct fate maps that trace the contribution of cells derived from each meristem layer to mature organs of the shoot. In addition, the pattern of green and white tissues in chimeras of other species could be used to deduce the number of stable cell layers in the meristems of those species (114). In general, the L1 forms epidermal tissue on all shoot parts, L2 forms subepidermal tissue, and L3 forms the internal tissue. These fates vary, however, among organs of a plant of a given species and among species for a given organ type, and they may also be influenced by other genetic factors affecting development (31, 129).

COORDINATION OF CELLS FROM DIFFERENT LINEAGES DURING SHOOT DEVELOPMENT

The analysis of shoot meristem layer fates using periclinal chimeras has several important implications concerning cell interactions during development. Similar analyses of root development have not been as extensively pursued; those experiments are discussed in the section on Mosaic Analysis of Root Development. Because shoot meristems of many plants are composed of several separate lineages, as indicated by the occurrence of periclinal chimeras, it is obvious that cells of different lineages must interact in the meristem. Because cells from multiple layers contribute to mature organs, there must also be interactions among cells of separate lineages during organ initiation. Interactions among cells during leaf morphogenesis are even more dramatic. Analysis of leaf development in plastid chimeras of both monocots (114) and dicots showed that the final contribution of cell layers to the mature organs, although predictable, was highly variable for a given location on the leaf. For example, in a periclinal chimera of tobacco, having genetically green L1, white L2, and green L3 (**GWG**), the contributions of L2 and L3 to the leaf blade were readily observable (113). L1-derived cells form the epidermis, but epidermal cells develop chloroplasts only in guard cells, so a genetically green epidermis appears unpigmented. In the central region of the blade, near the midrib, L3-derived cells made up the middle mesophyll, and L2-derived cells gave rise to palisade and lower mesophyll. At the margins, all of the internal cells were derived from L2. The position of the interface between L2-derived mesophyll and L3-derived mesophyll was variable. In regions where there was a large L3 contribution, there was a corresponding smaller L2 contribution, and vice versa. This resulted in leaves with uniform thickness and shape, regardless of the manner in which they were formed by the component cell-layer derivatives. This appears to be a regular feature of leaf development. In fact, such observations have been used to support the view that a plant's morphology is a product of processes that occur at an organismic level and that cell division patterns play a minor role in determining overall morphology (65, 66).

In addition to the typical variation in layer contributions to mature organs, plastid chimeras also allow the visualization of less typical contributions (111). For example, during leaf development in privet, L1 derivatives regularly undergo anticlinal divisions to form the epidermis. Occasional small green sectors at the leaf margins in a **GWG** chimera have been observed (113). These sectors are the result of periclinal divisions of L1. This results in the displacement of L1-derived cells into subepidermal leaf tissue usually formed by L2 or L3 derivatives. These cells, although of L1 lineage, respond to their new position and differentiate as internal, mesophyll cells, developing normal green chloroplasts.

These results depend upon fortuitous alterations in cell position. However, direct experimental evidence that cell fate is altered when position is changed has recently been obtained in *Arabidopsis* root development (130). Cortical initial cells each undergo an asymmetric periclinal division to produce cortical parenchyma and endodermal cells. If a cortical initial cell is laser ablated, adjacent pericycle cells protrude into the vacant site. Here they divide periclinally to maintain the pericycle cell files, and they then divide by an asymmetric periclinal division to form cortical and endodermal cells. Thus, when cells transgress the pericycle-cortical clonal boundary, cell fate is switched to correspond to the new position. In similar ablation studies of early development in *Fucus,* contact by the protoplast of one cell type with the cell wall of the other cell type caused the cell to switch its fate (9), which suggests the cell wall was the source of positional information. Whether the cell wall plays a similar role in higher plant development is not known.

Clonal analyses designed to generate fate maps of cells from any region of the plant, including meristems and organ primordia, have provided further evidence of cell interactions. Cells can be genetically marked for use in clonal analysis by a variety of methods, some of which rely on spontaneous events (21, 73) or involve mutagenesis (90). Clonal analysis in plants has recently been reviewed (29, 88, 89). Clonal analysis of the developing tobacco leaf has shown that this organ is derived from a total of approximately 50–100 founder cells that originate from at least three layers in the meristem (91). Similarly, many cells from both of the two cell layers of maize meristems were found to contribute to leaf initiation in that monocot (87). Fate maps of the shoot apical meristem at the mature seed stage have been generated for several species by clonal analysis (42, 56, 60, 83). A common observation in these studies was that, although a cell having a specific location in the meristem typically gives rise to a clone of cells that will lie in a predictable portion of the plant body, there are no restrictions other than the geometry of cell division on its fate. Because of this, it has been suggested that in plants fate maps be called probability maps (56). Although cell lineages are not constant, the morphology of the mature organs is remarkably constant. Such accommodation (62, 116) is especially apparent in chimeras in which one cell layer is at a developmental disadvantage.

Mutations that mark cells in the plastid chimeras and in clonal analyses must be selectively neutral in somatic growth compared with the rest of the cells in the meristem or developing organ in order to be useful for fate mapping (81). However, plastid mutations that put cells at a developmental disadvantage in chimeras have provided even stronger evidence for accommodation and cell interactions during development. Plastid chimeras of pelargonium having the defective chloroplast mutation *Dpl-W1* in either L2 or L3 showed similar contributions from these two layers in mature leaves, irrespective of which layer carried the mutation. That is, the relative contributions of L2- and L3-derived cells were not dependent on whether the cells were mutant or wild type. In contrast, a different defective chloroplast mutation, *Dpl-W2,* reduced the ability of cells carrying it to contribute to the leaf in chimeras. This resulted in a smaller proportion of the leaf composed of cells having the mutation and a corresponding larger proportion composed of wild-type cells (116). Leaves of normal morphology were nonetheless formed in the two chimeras, from very different developmental pathways, depending on which meristem layer was composed of mutant cells. Similar observations were made by Pohlheim for a variety of species (93).

GENERATION OF DEVELOPMENTAL MOSAICS

The analyses of genetic mosaics to map cell lineages and determine cell fates, in which the cells of a mosaic differed only for some visible marker, clearly demonstrated that cell interactions occur during developmental processes in the meristem and subsequent organ maturation and cellular differentiation. As illustrated by the *Dpl-W2* pelargonium chimeras, mosaics in which cells differ—in addition to being visibly marked—in their relative developmental potential offer a powerful tool for further investigations of the extent and nature of intercellular interactions. Interspecific or intergeneric graft chimeras offer a clear example (64). When chimeras are generated from two different species or genera, there usually is a great number of morphological differences such as leaf and flower size and shape. By analyzing the development of the organs in such chimeras, it can be determined whether cells in different positions of the developing organ contribute equally to determine the final form, or whether cells in certain positions have a greater role in determining the final morphology. The question is whether a certain cell in a chimera develops strictly according to its own genotype (autonomously) or is influenced by the genotype of neighboring cells (nonautonomous development).

Mosaics that address such questions can be generated by a variety of methods, which fall into two main categories: those generated by combining different individual plants and those generated during the development of a single individual (110). When mosaics are generated between two different individuals, the range of possible genetic differences is great. The component cells may differ at a single genetic locus or at multiple loci. They may have different cytoplasms and different ploidies, or they may be of different species or even different genera (129). The range of possibilities is limited only by the ability of two types of cells to form a single meristem coordinately. The factors that govern the ability of genetically dissimilar cells to form a meristem coordinately have not been explicitly examined, but other studies have investigated similar interactions between cells of different species during compatible vs incompatible graft formation (126).

Generation of mosaics from two individual plants requires that cells from both plants be brought into intimate contact. This has been accomplished by grafting tissue explants both in vivo (63, 80) and in vitro (85), and with mixed tissue (17, 79) or protoplast cultures (11). The mixed tissue is then induced to form meristems and shoots. For some species, this can be accomplished in vivo by cutting the graft to expose both tissues at a wound site and allowing spontaneous formation of meristems at that site (23). For other species, such as citrus, plant growth regulators must be applied to the cut graft junction in order for shoots to form (86). From any of these methods, most shoots that regenerate are composed purely of one or the other cell type. A fraction of the shoots that develop are the result of both cell types becoming incorporated into a single meristem. These meristems have the potential to form different types of mosaic shoots.

Mixed callus cultures have yielded variable results in the formation of chimeric meristems (79). The interactions required for a mixture of cells originating from two individual plants to form a meristem cooperatively are poorly understood. Grafting techniques have in general yielded chimeric shoots more reliably, which indicates that some event occurs in the grafted region that is sufficient for cells to be able to form chimeric meristems. Analysis of the events that lead to graft formation has shown that the process occurs through a series of fairly definable steps accompanied by very specific cellular changes (59). No study has explored the stage in the grafting process at which cells become competent to interact to form mosaic meristems. An important step may be the de novo formation of plasmodesmata between the two graft partners. The importance of plasmodesmatal connections among cells in order for those cells to be able to respond to a developmental cue is illustrated by an experiment with *Eucalyptus* parenchyma cells. Nodules of cells interconnected by plasmodesmata responded to addition of IAA by forming tracheary elements, while similar cells that were closely packed but not interconnected by plasmodesmata did not respond to IAA treatment (117). It may be that in order for groups of unrelated cells to respond to a shoot-inducing signal they must also be interconnected in a specific manner.

Mosaics can also arise at any time during ontogeny of a plant when a genetic change occurs in a cell. The clonal derivatives of the cell form a sector, which results in a mosaic plant. Mosaics in which the change affects an aspect of the cell's genotype that is known to be important in development are useful for investigating developmentally important cell interactions. The genetic changes can be either spontaneous or induced, e.g. X-ray-induced chromosome arm loss, and result in the uncovering of a recessive allele (88), insertion or excision of a transposable element (19), transformation (22, 104), somatic elimination of a whole chromosome (97), cytoplasmic changes (72, 109), and even epigenetic changes (77, 82).

In mosaics that arise both from the mixing of cells from different individuals and from a genetic alteration in a cell during the development of a single plant, the type of chimera that forms depends on a number of factors. Regardless of the origin of mosaicism, the relative positions of the genetically different cells in a developing shoot affect the type of mosaic formed and how persistent the mosaicism will be. If the genetically different cells arise in an organ primordium, mosaicism is limited to that organ. The clonal derivatives of a cell on the flank of the meristem persist for a short time, forming only the tissue and organs normally formed from cells in that region. This mosaicism can persist in the plant if some of its derivative cells contribute to the formation of an axillary meristem, which results in a mosaic branch. In order for mosaicism to persist, the population of apical initials in the meristem must include cells of both genetic types (68).

Cellular organization of the meristem (107) also plays a role in the types of chimeras that can form. In plants with a single apical cell, such as ferns, persistent mosaicism is not possible (10). Most higher plant meristems have multiple initials. As has been described above in the section on Historical Background, there can be restrictions in the planes of divisions of some of the initials and their derivatives, depending on their position, which result in the formation of layered meristems. Most angiosperms have a three-layered meristem, though two-layered meristems do occur. Gymnosperm meristems may be layered or unlayered. If an initial cell of one layer is genetically different from the other initials in that and any other layers, a mericlinal chimera results in which in the circumference of the shoot produced from that meristem there are regions that are not periclinally chimeric adjacent to a sector of periclinally chimeric tissue. The proportion of the shoot that is sectored is typically the reciprocal of the number of initials in the meristem layer that are chimeric. The permanence of apical initials has been debated (96, 98, 112), but it appears that whether a cell continues to function as an initial depends only upon its location in the meristem. Which meristem cells are functioning as initials can shift within the layer. In mericlinal chimeras, this results either in the loss of the chimerism or in the formation of a periclinal chimera, in which all of the initials are identical within the layer and genetically different from the other layers. If neither of the two genetic cell types in a mericlinal chimera has a selective advantage in the meristem over the other type, a mericlinal chimera can be persistent. Over time, however, stochastic events result in the fixation of either of the cell types. In contrast, if one cell type has a selective advantage in the meristem, diplontic selection more quickly results in the fixation of the cell type with selective advantage in that layer, which results again in either loss of chimerism or establishment of a periclinal chimera (68, 98). A periclinally chimeric meristem produces a shoot that is uniformly chimeric across its circumference, in contrast with mericlinal chimeras [67 (Figure 16.1)]. Periclinal chimeras can be quite stable. The stability depends upon how precisely the patterns of cell divisions in each meristem layer are maintained. For example, a periclinal division in L1 could displace an initial cell in L2 to form a new mericlinal chimera. In order to identify chimeras unambiguously and detect subsequent rearrangements, it is advantageous to incorporate cell-autonomous markers (those whose phenotype is not affected by that of neighboring cells) when experimentally synthesizing mosaics (44, 88).

When using chimeras to analyze interactions during a specific developmental process, the most information can be gained when the cells of different genotypes are juxtaposed in many different arrangements. In plants with a three-layered meristem, there are six possible periclinal chimeric arrangements of the layers [67 (Figure 16.11)], whereas plants with a two-layered meristem can have only two periclinal chimeric arrangements. Periclinal chimeras are particularly useful when examining interactions between cells occupying different meristem layers and their derivatives in organs. Mericlinal chimerism provides an opportunity to examine lateral interactions. Mosaics that are induced or that occur spontaneously in a single plant often consist of many separate sectors. These sectors are the result of independent events and thus can present a variety of arrangements of the different cell types. Because axillary meristems of periclinal chimeras typically maintain the chimeric arrangement, propagation from axillary buds facilitates experimentation with the same chimera (124). Similarly, a mericlinal chimera can be stabilized as a periclinal chimera if an axillary bud from the periclinally chimeric sector is propagated. Obviously, a specific chimera cannot be propagated sexually because gametes are single cells. Chimeras can be easily propagated if they can produce roots adventitiously. Adventitious roots are typically derived from phloem-associated parenchyma and are therefore usually initiated in cells that are derived from the internal layer of the shoot apical meristem. Thus, roots on a propagated chimeric plant are not themselves chimeric. In two different chimeric plants, e.g. ABB and BAA, adventitious roots of the former would be of the B genotype and those of the latter the A genotype. If there is a root effect on shoot development, it is then possible that the two different root systems would affect shoot development differently. This could be determined by grafting chimeric shoots of each type onto roots carrying one or the other of the genotypes.

ANALYSIS OF DEVELOPMENTAL MOSAICS

Although early in this century chimeras were generated to understand the nature of chimeras themselves, published descriptions from this period provide a great deal of information about cell interactions during development. For example, Jorgenson & Crane (63) generated periclinal chimeras between various pairs of solanaceous species that varied for many developmental processes. One set of chimeras was generated between *Solanum luteum* and tomato. The leaves of *S. luteum* had a simple shape, whereas tomato leaves are compound and composed of several leaflets. The periclinal chimera **TLL** had leaves that were simple in shape like those of *S. luteum,* and the reciprocal chimera **LTT** had compound leaves like those of tomato. [In this review, we use a uniform method to describe the chimeric arrangement of cell layers that may differ from the usage in the original reports. For example, a chimera that has *S. luteum* (L) cells in L1 and tomato (T) cells in L2 and L3 is indicated as **LTT**, whereas Jorgenson & Crane referred to that chimera as *L. esculentum-luteum* (i), meaning that a single layer of *S. luteum* cells overlays a core of tomato cells in the meristem.] The **LTT** leaves did, however, show a reduction in the total number of leaflets. The types and patterns of trichomes in the chimeras were identical to those of the plants that were the source of L1 cells in the chimeras and were not influenced by genetically different underlying tissue, i.e. were cell-layer autonomous. The trichomes could thus be used to establish the genotype of epidermal cells. Therefore, in the chimeras, L1 cells developed in a nonautonomous manner with respect to simple vs compound leaf shape. In chimera **LLT**, the leaves were generally simple. Although there were no cell-layer-autonomous markers that could be used to determine the extent of contribution from the internal meristem layers to the leaf, this set of three chimeras showed concordance between the identity of L2 cells and simple vs compound shape. These chimeras illustrate the value of cell-layer-autonomous markers and of multiple chimeric combinations when using chimeras to investigate cell-cell interactions.

Flower development in the graft chimera *Camellia* "Daisy Eagleson" provides evidence suggesting that nonautonomous differentiation of cells occurs during the acquisition of floral organ identity (115). This chimera arose spontaneously from a graft of *Camellia sasanqua* (S) and *Camellia japonica* (J). The particular line of *C. sasanqua* produces normal, fertile flowers with stamens and pistils, while the particular line of *C. japonica* used produces double flowers with no stamens or pistils but additional petals, which renders the flowers sterile. The chimera, which had the meristem layer arrangement **SJJ**, formed flowers with stamens and pistils. This suggests that L1-derived cells determined the manner in which L2- and L3-derived cells divided and differentiated during floral organ development. Whether *C. sasanqua* cells are needed specifically in L1 or must simply be present in the chimera to induce *C. japonica* cells to differentiate to form stamens and pistils cannot be determined because no reciprocal chimeras were obtained. In addition, conclusive evidence that *C. japonica* cells were present in the internal tissue of stamens and pistils in the chimeric flowers is lacking. It is possible that these organs were entirely L1-derived. Although clonal analysis has shown that in general all three meristem layers contribute to lateral organs, this cannot always be assumed to be true. An analysis of a complete set of graft chimeras—i.e. all six combinations—between *Nicotiana tabacum,* carrying appropriate cell-layer markers, and *N. glauca* demonstrated that expected contributions to organs could be affected by the specific cell-layer arrangement of the chimera (78). The contributions of the three meristem layers to leaves in all six chimeras were found to be typical of what would be predicted from fate maps of tobacco leaves. However, it was found that in certain chimeric arrangements the gametes were not only of the genotype expected for an L2 origin but must have come from L1 or L3. Although a clonal analysis of reproductive organs in *N. glauca* has not been reported, these data suggest that interactions between cells of the two species when in certain chimeric combinations affected the cell lineage patterns.

It is useful to note that when a correspondence is found in a set of chimeras between the genotype of cells in a particular meristem layer and a morphological trait, it does not automatically follow that cells derived from the other layers must have been induced to develop nonautonomously. In an analysis of floral organ morphology (76) in the same *Nicotiana* chimeras described above, it was found that the presence of *N. tabacum* epidermis was associated with wider corollas, whereas *N. glauca* epidermis was associated with narrower corollas. *N. glauca* has narrow corollas relative to those of *N. tabacum.* In such situations, it is important to know the normal extent of the contributions from the three meristem layers to the organs being studied. This may be accomplished with a clonal analysis (34). For example, L1 cells may be found to contribute to all layers of the petal at the margins. It could then be determined whether L1-derived cells in the chimeras influenced the extent of growth of underlying tissues or whether L1-derived cells simply developed as they would have in nonchimeric plants to determine the width of the corolla by contributing more or fewer cells to all layers of the petal.

Combining Developmental Mutants with Mosaic Analysis

Mutations that affect developmental processes can be incorporated into mosaics. *Lateral suppressor* (*ls*) is a mutation in tomato resulting in the failure both to form axillary buds on the shoot and to initiate petals in the flowers. *ls* is not a homeotic mutation, i.e. one that transforms petals into some other floral organ, but instead results in the floral meristem failing to initiate any second whorl organs in the positions where petals would normally develop. In a graft chimera generated between wild-type (+) and *ls* plants having the meristem cell-layer arrangement ***ls*++**, floral meristems initiated second whorl organs that developed into normal petals (121). *ls* cells formed the epidermis of the petals in the chimera, in response to wild-type cells in L2 and L3, as indicated by the presence of the genetically marked *ls* cells in the epidermis. Thus, petals did not form only from cells that had a functional *ls* gene (L2 and L3) but also from the nonautonomous development of cells that normally do not form petals. Axillary bud initiation and development in the chimera was also normal. No meristems formed in axils of leaves in mericlinal sectors having *ls* in internal tissue. Similar effects of the determining role of internal tissue on axillary meristem formation, in this case secondary axillary buds in *Nicotiana,* have also been observed (125).

The *ls* chimeras enabled investigation of interactions between meristem cell layers that occur when an organ is initiated. Chimeras were also used to address whether interactions occur to determine the number of organs initiated by the meristem (120). Two sets of chimeras were generated: one between wild-type tomato that carried cell-layer markers and tomato with the mutation *fasciated* (*f*), which causes an increase in the number of organs in each of the floral whorls, and the other between a line of tomato (*Lycopersicon esculentum,* **E**) that forms more than two carpels (fourth whorl organs) and a line of *L. peruvianum* (Pr) that always forms two carpels. Analysis of these chimeras demonstrated that cells located in a specific location in the meristem, L3, were the major determinant of floral organ number. For example, chimeras **PrEE** and **PrPrE** always formed more than two carpels, in contrast to **PrEPr**, **EEPr**, and **EPrPr**, which always produced two carpels. Similarly, chimeric tomato plants with wild-type cells in L1 and L2 and *fasciated* cells in L3 developed fasciated flowers. It appeared that in the chimeric determination of floral organ number, determination by the L3 was mediated through the effect of cells in that layer on meristem size prior to carpel initiation (120). Cells in the outer floral meristem layers, L1 and L2, underwent a pattern of cell divisions and organ initiation that was dictated not by their own genotypes but by that of L3 cells, i.e. the development of L1 and L2 cells was non-cell-layer autonomous. The ability of L3 cells to determine the pattern of development of the meristem was independent of the layer of origin of the cells and depended only on their position at the time of determination. For example, **+++** sectors on **++*f*** plants and **PrEE** sectors on **PrEPr** plants produced flowers with organ numbers determined by cells in L3, even though those L3 cells originated from L2. Therefore, a cell that originally was in the L2 layer and responded to information from L3 cells could itself, when moved into L3, determine the pattern of development of the floral meristem.

Developmental mutations in combination with autonomous markers can be used to generate mosaics during the development of an individual plant. Periclinal chimeras generated from two individuals provide the advantage that both plants can carry multiple genetic markers; these markers aid in the identification of the chimera and in determining the relative contributions of each layer to mature organs. In addition, the two combined plants can differ for a variety of developmental processes; any of these can be analyzed in the chimera. However, if one is interested in a particular process or the function of a specific gene in that process, chimeras induced in individual plants provide the advantage that there may be far fewer differences between the genetically different cells under comparison. The dominant *Teopod* (*Tp*) mutations of maize all result in the constitutive expression of juvenile traits throughout the development of the shoot. For example, epicuticular wax, usually found only on juvenile leaves (at basal nodes of the plant) and not on adult leaves, is present on adult *Tp* leaves. In addition, vegetative organs—leaves—occur where reproductive structures—tassel branches and flowers—usually form in the tassel. Wild-type sectors were generated on *Tp1* plants by the irradiation-induced loss of a chromosome arm in heterozygous *Tp1*/+ plants (92). These sectors developed in a manner identical to that of the adjacent, mutant tissue, independent of whether the sector was mericlinal, i.e. mutant L1 over wild-type L2 or wild-type L1 over mutant L2, or sectorial, in which both layers were wild type. This indicated that *Tp1* acted in a non-cell-autonomous manner, possibly through the action of a diffusible substance.

When cell-autonomous markers are not proximally linked to a gene of interest, it is nonetheless possible in maize to generate marked sectors for mosaic analysis of that gene by taking advantage of the genetic manipulations that are possible (35). For example, it was of interest to determine whether other *Teopod* mutations, such as *Tp2,* act similarly to *Tp1. Tp2,* however, is located on a chromosome arm that lacks proximal-cell-autonomous markers. A translocation was used to link *Tp2* to a marker, *lemon white 2,* in repulsion

$$\frac{Tp2 \quad +}{+ \quad lw2}$$

in order to visualize tissues in which the dominant allele, *Tp2,* had been lost after irradiation. Large sectors, which would allow one to determine whether diffusion of a substance occurs over many cells, normally require irradiation early in development, such as during embryogenesis. Translocations of the chromosome arm carrying the gene of interest to a supernumerary B chromosome of maize (18), which is frequently lost somatically early in development, provides an alternative to irradiation of developing embryos. Large sectors in which *Tp1,* which had been linked to a B chromosome, was lost were generated in this manner. The sectors that were generated indicated that both *Tp1* and *Tp2* act nonautonomously, because wild-type tissue showed a teopod phenotype. Large mericlinal sectors caused by the loss of the B chromosome translocations also developed teopod characteristics. However, leaves in positions associated with adult differentiation in normal plants having large sectors in which both L1 and L2 derivatives were wild type (had lost *Tp1*) were very wide, typical of adult leaves. Thus, from mosaic analysis it was concluded that both *Tp1* and *Tp2* mutations act in a non-cell-autonomous manner. That is, the presence of mutant cells anywhere in a developing organ resulted in teopod-type differentiation of nearby wild-type cells. The observation that very large wild-type sectors had wild-type differentiation indicated that the range of *Tp1* and *Tp2* activity is limited.

A very different result was obtained from a similar mosaic analysis of the recessive mutation *liguleless-1* (*lg1*) of maize. All leaves of *liguleless-1* mutants lack both a ligule and auricle, structures that normally develop at the junction of leaf blade and sheath (119). The ligule is an outgrowth of epidermal tissue that runs across the adaxial surface of the leaf. The auricle is a wedge-shaped region of tissue that lies between the ligule and the blade. X-rays were used to induce hemizygous *lg1*/– sectors on otherwise wild-type (+/*lg1*) tissue (6). The phenotypes of the sectors obtained in this mosaic analysis indicated that wild-type *Liguleless-1* was required in specific and different tissues for the normal development of the two structures. Normal ligules formed only where wild-type *Liguleless-1* was present in epidermal tissue. Auricles formed only when wild-type *Liguleless-1* was present in mesophyll tissue. However, in some sectors in which mutant epidermis overlay wild-type mesophyll, rudimentary ligules were formed. In contrast, the epidermis in the region of the auricle in sectors of the same arrangement developed like blade epidermis [6 (Figure 11)]. Unlike the results of *Teopod* mosaic analyses, the normal development of ligule and auricle required that wild-type *Liguleless-1* be present in specific tissues. Unlike *Teopod* sectors, laterally adjacent wild-type and *lg1* tissues developed autonomously, which indicated no lateral interactions between the sectors. However, observations that disruption of the ligular region by an *lg1* sector resulted in a basipetal shift in the position of the ligule in wild-type tissue at the margin suggested that wild-type *Liguleless-1* is involved in perpetuating an inductive signal that organizes ligule formation across the leaf (7).

The analysis of various genetic mosaics indicates that cells can develop either autonomously—that is, according to their own genotype—or nonautonomously in response to other cells. It has been observed that for some developmental processes, a cell influences its neighbors because it is in a particular position in the plant, whereas in other processes the ability to influence neighboring cells is independent of position. This latter case is seen in plants mosaic for cells expressing the dominant *Teopod* mutations during phase change. It is likely that autonomous and nonautonomous processes are coordinated during the development of nonmosaic plants as well. In order to understand the role and nature of nonautonomous signaling, an analysis of genes possibly involved in signaling or responding to developmental signals must be combined with analysis of genetic mosaics.

Combining Molecular and Mosaic Analysis of Development

Dominant mutations at the *Knotted-1* locus (*Kn1*) of maize affect leaf development. This mutation causes abnormal cell divisions in all cell layers of the blade, which results in the formation of bumps or knots (15). Abnormal divisions in the epidermis are the first detectable signs of knot formation (43). Ectopic ligule development in the epidermis can also be caused by *Knotted-1* (40). Mosaics of *Knotted-1* were generated by X-rays to produce sectors of wild-type tissue on *Kn1*/+ leaves (50). Despite the apparent importance of the epidermis for the Knotted phenotype, it was found that sectors in which *Kn1* epidermis overlay wild-type mesophyll did not form knots. *Kn1* in mesophyll induced the development of knots in the epidermis, regardless of its genotype. There was no observed effect of *Kn1* tissue on laterally adjacent wild-type tissue, which indicates that induction took place only in an outward direction and not laterally. Additional mosaic analysis determined that *Kn1* was necessary in only a fraction of the mesophyll, the middle mesophyll, and bundle sheath cells in order to induce the *Kn* phenotype (105).

Knotted-1 has been cloned using *Ds2* as a transposon tag and was found to encode a protein with a homeodomain (51, 133). Expression studies of *Kn1* protein showed that it is found ectopically in mutant leaves in regions where knots form, but it is expressed primarily in wild-type plants in the vegetative and reproductive shoot meristems but not in determinate lateral organs such as leaves (106), nor in root meristems. This pattern is consistent with the hypothesis that the normal function of *Knotted-1* is maintenance of indeterminate shoot growth. The most intriguing result is from a comparison of mRNA and protein localizations in the shoot meristem (57). The protein can be detected in all cells of the meristem; however, the mRNA is not detected in L1 cells but is restricted to the internal cells of the meristem. This suggests that Knotted protein may be a signaling molecule that moves between cells. This function would fit the patterns of cell interactions seen in the mosaic analysis of *Kn1* (49).

Floricaula (*flo*) of *Antirrhinum* is another gene that has been subjected to both mosaic and molecular analysis. *flo* mutants are unable to form normal, determinate flowers and instead produce indeterminate inflorescence shoots (27). An unstable allele of *flo,* caused by the insertion of *Tam3,* a transposable element of *Antirrhinum,* was isolated. This insertion allele permitted the cloning of the gene using *Tam3* as a tag (27) and the recovery of mosaic sectors of revertant cells resulting from the excision of the element. Stable periclinal chimeras containing cells that were mutant for *flo* or heterozygous wild type were obtained (19). The various chimeras were all able to form determinate flowers in their inflorescences, but the flowers developed abnormally and were described as having near-wild-type, intermediate, or extreme phenotypes. Wild-type *Flo* is expressed in all meristem cells in wild-type plants. In situ analysis of the chimeras revealed that *flo* was expressed in a different subset of meristem cells in plants of each of the floral phenotypic classes (53). In plants with near-wild-type flowers, *flo* was expressed in L1 cells; in intermediate plants, *flo* was expressed only in L2; and in plants with flowers of the extreme phenotype, *flo* was expressed only in L3. Although the chimeras carried no cell-layer-autonomous markers, the *flo* expression patterns provide good evidence that the three classes were indeed periclinal chimeras with wild-type *Flo* in a single meristem layer. The development of flowers in the three different chimeras indicated that wild-type *Flo* need only be present in any of the meristem cell layers in order for floral development to proceed, although there must also be required functions of wild-type *Flo* activity that are cell autonomous. An additional finding from these chimeras is that the expression of wild-type *Flo* in one meristem layer was sufficient to activate the expression of two downstream, floral organ identity genes, *deficiens* and *plena,* in all meristem layers, although their expressions were slightly altered. Neither of these genes is expressed in stable *flo* mutant meristems. These studies demonstrate an example of expression of a gene in some meristem cells being involved in the coordination of gene expression in all meristem cells.

MOSAIC ANALYSIS OF ROOT DEVELOPMENT

Chimeras and mosaics have not been used extensively to analyze the organization of the root apical meristem or how cells from the meristem contribute to the formation of the differentiated tissues and organs. This neglect of a methodology that has been so powerful in the analysis of shoot development results partly from the general absence of easily observable pigmentation markers to detect cell clones in roots and partly, perhaps, from a perceived lack of need for such analysis in an organ in which cell lineages are so obvious, and where cell files are not disrupted by organ initiation close to the meristem—as occurs in shoots. One example of a detailed cell lineage analysis in a root that did not involve chimeric or mosaic analysis is in *Azolla,* where the complete lineage pattern from divisions of the single apical cell to the production of all the differentiated cells of the organ was described (47).

The roots of angiosperms frequently exhibit precisely aligned cell files that converge at the tip to what are presumed to be initial cells of the cell files. Two types of meristem organization have been described (26). Closed meristems are those in which cell files converge on two to four layers of initial cells; the number of layers differs in different species [38 (Figure 5.10)]. Open meristems are those with a less obvious cell file convergence, and it is presumed that in these roots there is a single layer of initials that generates all the cells of the root axis and the root cap. The organization of the meristem need not be constant, and changes in the number of initial cell layers and between open and closed organization during development have been described (3).

Mosaic analysis of root meristem organization has yielded results that are not consistent with the histological interpretation presented above. Periclinal chimeras, which are frequent in shoots, are almost unknown in roots. They have been described only occasionally (74). Persistent mericlinal chimeras have been reported infrequently (14). Most frequently observed are sectorial chimeras that extend from the epidermis to the vascular tissue and span approximately one third to one half of the root circumference. This observation has led to the conclusion that there is a small number, perhaps two to four, of initial cells in the root meristem and that these lie in a single layer even in roots that histologically appear to possess several discrete meristem layers (33).

The root chimeras described above were generated by treatment of seedlings with X rays or colchicine to produce cytochimeras in which mutant cells were identified by chromosomal abnormalities or by ploidy. It is possible that these mutagens caused mitotic reactivation of cells that are normally in a state of near-mitotic quiescence during root development, resulting in the production of atypical cell lineages (24). It is also possible that X-ray treatment resulted in cell death and regeneration of a reorganized meristem in which the original cell lineages were not preserved. The tip of the root apical meristem contains the quiescent center (25) that includes the presumed initial cells. Within this zone, cell division can be stimulated by a variety of natural and artificial conditions (134). For these reasons, methods that do not disturb the normal patterns of division in the root tip may be required to analyze the organization of the root meristem. Recently, root mosaics have been generated in transgenic plants in which a colored cell autonomous marker (gus) (58) is expressed following excision of a transposable element from the *uid* gene (32).

Fate maps of the root meristem that are based on markers such as this that would not be expected to affect the natural lineage patterns should provide more reliable information than has been obtained using other potentially destructive marking methods.

MECHANISMS BY WHICH CELLS MAY BE DEVELOPMENTALLY COORDINATED

In light of the evidence for nonautonomous interactions during development, which suggests that signals move from one cell to other cells, we next consider possible mechanisms by which signaling could occur. Plasmodesmata are complex structures of membranes and proteins that facilitate the intercellular diffusion of metabolites and small molecules by providing cytoplasmic connections through cell walls (for a review, see 75). Although it has long been recognized that plasmodesmata form during cell division in the newly formed wall (48), they also form de novo and have been observed between genetically different cells in chimeras generated between different plants (11, 13, 16) and between epidermal walls during the fusion of carpels in *Catharanthus roseus* (131). Dye-injection studies indicate that the size of molecules that can pass through plasmodesmata can be regulated. This results in the formation of transient symplastic domains, in which particular molecules can move freely among groups of interconnected cells but not among all cells of a tissue (37). Investigations stemming from studies of movement of tobacco mosaic virus between cells have revealed that macromolecules such as RNA and protein (41) can move through modified plasmodesmata (140, 141). The detection of Knotted protein in cells in which *Kn* mRNA is not detectable (49) implicates plasmodesmata as conduits for intercellular communication. The observations that when and where plasmodesmata form between cells in the course of plant development as well as that sizes of molecules that can pass through them at various developmental stages are regulated strongly suggests an important role for plasmodesmata in the coordination of cells during plant development (75). Chimeras may be useful in elucidating factors that determine density of plasmodesma formation between meristem layers. Preliminary studies of a chimera between *Solanum nigrum* and *Lycopersicon pimpinellifolium* suggested that plasmodesmata between L1 and L2 were preferentially initiated by L1 cells (13).

Biophysical forces provide another mechanism that may coordinate groups of cells during morphogenetic processes (46). Developing sunflower inflorescences were used to examine a mechanism that produces pattern by buckling, which was observed by scanning electron microscopy (54). When the growth of a developing inflorescence was restrained such that it developed as an oval rather than as a disc, predicted modifications of the patterns of bracts and florets was observed, and in addition, organ characters were modified in such a way that they mimicked known mutants. Thus, biophysical constraints may impose a certain pattern of coordination among a group of cells that results in a distinct pattern of development. The growth of cells in particular regions of a developing organ or meristem may produce greater biophysical constraints than do cells in other regions.

GENETIC MOSAICS AND CELL DEATH

Although the genetically different cells in the mosaics discussed up to this point interacted in such a way that growth and development were coordinated, this does not always occur. As previously described, chimeras between *Solanum luteum* and tomato could complete all aspects of vegetative development (63). However, upon the transition to floral development, floral meristems of chimeras with certain cell layer combinations became arrested so that no floral organs matured. In addition, epidermal cells at the bases of the abortive inflorescences died, resulting in the formation of wound tissue. Similar abortive inflorescences and cell death resulted from interactions between genetically different cells of periclinal chimeras between *S. nigrum* and tomato (122). At the bases of the abortive inflorescences where the epidermal cells died, epidermis redifferentiated but was derived from internal tissue, as indicated by the expression of cell layer markers carried by the internal tissue. These observations suggest that the mechanisms coordinating development of cells in different meristem layers may be different in vegetative and floral development between the two plants used to generate the chimeras. The failure of interactions among cell layers that normally occurs in order for floral organs to mature resulted in the death of epidermal cells.

Another interesting example of cellular interactions leading to cell death occurred during leaf development in chimeras generated between tobacco and *Solanum laciniatum* (64). At a specific stage of leaf maturation, small necrotic lesions, normally not exhibited by either species, developed at locations where tobacco and *S. laciniatum* cells were in contact. An example of nonautonomous development of an organelle is found in the pelargonium chimera "A Happy Thought." In this chimera, a nuclear mutation (127) disrupts the development of chloroplasts in cells having the mutation as well as in adjacent wild-type cells [129 (Figure 7.13)]. Whether other cases of cell death that occur in nonchimeric plants are autonomous or nonautonomous was investigated by mosaic analysis. The dominant *Rp1* gene of maize confers resistance to the pathogen *Puccinia sorghi* (8) via local necrosis of plant tissue (hypersensitive response) at the site of infection. Mosaic analysis of *Rp1* clearly demonstrated that *Rp1* is cell-autonomous in its action; that is, necrosis does not spread to adjacent, wild-type tissue. Similarly, analysis of mosaics generated from a semigametic line of cotton (108) revealed that the phenomenon of hybrid necrosis, in which certain hybrid combinations fail to produce viable seed because of the death of the developing F1 hybrid embryo, occurred cell-autonomously. That is, cell death was the result not of intercellular interactions but of intracellular interactions between the two combined genomes in the F1 hybrid cells.

ARE MERISTEM LAYERS DEVELOPMENTALLY SIGNIFICANT?

The specific types of plant mosaics, such as periclinal chimeras, that can be generated to study cell interactions during development, depend on the structure of the meristem. The number of initials, whether they are arranged in layers, and the number of layers determine the patterns in which cells can be arranged in mosaics. Do layered meristems have adaptive or developmental significance? Recent analysis of patterns of gene expression in the meristem has clearly shown that certain genes may be expressed only in L1 cells or may be expressed throughout the meristem except for L1 (39), as is *Knotted* (50). Such patterns have also been observed to change in the meristem during the transition from vegetative to floral development (95). These patterns suggest that meristem layers, which are the result of restrictions in the planes of cell division, may be important in development. It may be that in order for certain processes to take place, gene expression must be limited to certain layers. Alternatively, developmental processes may have taken advantage of the existence of layers. However, mosaic analysis has demonstrated that lineage is not critical in determining how a cell differentiates. Therefore, the arrangement of several lineages into different cell layers may not have developmental significance. In addition, cell lineage analysis revealed that in plants with a nonlayered meristem and a single apical initial, development is still more dependent on position than on lineage (61).

An alternative possibility is that various meristem structures may have been selected for as a response to somatic mutation. In plants with layered meristems, it is generally observed that gametes arise from L2 cells, although they can also be formed by L1 and L3 cells. Analysis of reversion of an unstable mutation showed that reversion in L1-derived cells was about fourfold greater than in L2-derived cells (20). The implication is that if this is typical of somatic mutations, then the majority are not seed-transmissible. Certain mutations may increase the fitness of a mosaic plant (36). It has been suggested that different branches on plants with a long lifespan or that grow in a clonal manner may evolve independently through the independent accumulation of mutations. The structure of the meristem may play an important role in this process (135). Klekowski and coworkers have discussed the genetic consequences of having stratified vs nonstratified meristems. They showed by mathematical models that whether a meristem had a single fixed apical cell or numerous initials had an effect on the retention of neutral (69) or disadvantageous (70) mutations. This analysis showed that stratified meristems promote the long-term retention of somatic mutations, whether advantageous, disadvantageous, or neutral, and that mutations will persist in periclinally chimeric meristems (71).

Plant varieties arising from somatic mutation have been selected for their value in horticulture or agriculture. Because somatic mutations tend to be retained in plants having stratified meristems, vegetatively propagated varieties may be chimeras. For example, many cultivars of banana resulted from somatic mutations that affected inflorescence and leaf morphology or plant height (135). Similarly, cultivars resulting from somatic mutations have been selected in potato [129:111–21]. These are potentially an untapped resource for investigations of cell interactions in plant development by separating these spontaneously arising chimeras into their component parts (1, 132) and analyzing their development.

NATURAL CHIMERAS

The mosaics generated from two individuals that have been used to analyze developmental interactions either arose spontaneously from grafts generated for horticultural purposes or were experimentally synthesized. Examples do exist in which two individuals join to form a mosaic in nature. Individual trees of six different species of strangler figs (*Ficus* spp.) were found to be mosaics, presumably the result of fusion of individual trees during their growth (123). It is not known whether the fused trees maintain physiological independence. It is possible that an adventitious shoot arising from fused individuals could be a spontaneous periclinal chimera.

Lichens offer a clear example of a symbiosis between two different organisms that develop interdependently to determine shape and form. Lichens usually combine one type of fungus and one type of photosynthetic partner in a morphologically homogeneous thallus. However, recent molecular evidence has demonstrated that a single fungus can associate with two different photosynthetic partners—in one case a green alga and in another a cyanobacterium—and that each association results in a distinct thallus morphology (2). Another interesting naturally occurring cell interaction occurs between certain parasitic plants and their hosts. Members of the Balanophoraceae, such as *Langsdorffia* and *Balanophora,* form vegetative tubers on host roots (55). The host-parasite interface forms as a result of the action of a meristem-like structure, composed of both host and parasite, which forms composite vascular tissue within the parasitic tuber. The host and parasitic cells coordinately generate a chimeric structure. The analysis of mechanisms by which cells of different species interact during developmental processes in such "natural chimeras" may offer insights into the mechanisms of cellular interactions that occur during plant development.

ACKNOWLEDGMENTS

Work on chimeras in the authors' labs is supported by National Science Foundation grant IBN 93-17381. We are grateful to Erin Irish for critical comments on the manuscript.

Literature Cited

1. Abu-Qaoud H, Skirvin RM, Chevreau E. 1990. In vitro separation of chimeral pears into their component genotypes. *Euphytica* 48:189–96
2. Armaleo D, Clerc P. 1991. Lichen chimeras: DNA analysis suggests that one fungus forms two morphotypes. *Exp. Mycol.* 15: 1–10
3. Armstrong JE, Heimsch C. 1976. Ontogenetic reorganization of the root meristem in the Compositae. *Am. J. Bot.* 63:212–19
4. Baur E. 1909. Das wesen und die erblichkeitsverhaltnisse der 'Varietates albomarinatae hort' von *Perlargonium zonale.* See Ref. 129, pp. 330–51
5. Baur E. 1910. Pfropfbastarde. *Biol. Zentralbl.* 30:497–514. See Ref. 118
6. Becraft PW, Bongard-Pierce DK, Sylvester AW, Poethig RS, Freeling M. 1990. The *liguleless-1* gene acts tissue specifically in maize leaf development. *Dev. Biol.* 141: 220–32
7. Becraft PW, Freeling M. 1991. Sectors of *liguleless-1* tissue interrupt an inductive signal during maize leaf development. *Plant Cell* 3:801–7
8. Bennetzen JL, Blevins WE, Ellingboe AH. 1988. Cell-autonomous recognition of the rust pathogen determines *Rp1*-specific resistance in maize. *Science* 241:208–10
9. Berger F, Taylor A, Brownlee C. 1994. Cell fate determination by the cell wall in early *Fucus* development. *Science* 263:1421–23
10. Bierhorst DH. 1977. On the stem apex, leaf initiation and early leaf ontogeny in filicalean ferns. *Am. J. Bot.* 64:125–52
11. Binding H, Witt D, Mordhorst G, Kollmann R. 1987. Plant cell graft chimeras obtained by co-cultivation of isolated protoplasts. *Protoplasma* 141:64–73
12. Blakeslee AF, Avery AG. 1937. Methods of inducing doubling of chromosomes in plants. *J. Hered.* 28:393–411
13. Brabec F, Roper S. 1988. On the effect of chimerical composition and of polyploidy on the formation of plasmodesmata. *Mitt. Inst. Allg. Bot. Hamburg* 22:53–61
14. Brumfield RT. 1943. Cell lineage studies in root meristems by means of rearrangements induced by x-rays. *Am. J. Bot.* 31: 101–10
15. Bryan AA, Sass JE. 1941. Heritable characters in maize. *J. Hered.* 32:343–46
16. Burgess J. 1972. The occurrence of plasmodesmata-like structures in a non-division wall. *Protoplasma* 74:449–58
17. Carlson PS, Chaleff RS. 1974. Heterogeneous associations of cells formed in vitro. In *Genetic Manipulations with Plant Materials,* ed. L Ledoux, pp. 245–61. Plenum: New York. 601 pp.
18. Carlson W. 1986. The B chromosome of maize. *CRC Crit. Rev. Plant Sci.* 3:201–26
19. Carpenter R, Coen ES. 1995. Transposon induced chimeras show that *floricaula,* a meristem identity gene, acts non-autonomously between cell layers. *Development* 121:19–26
20. Chapatto JX, Werner DJ, Whetten RW, O'Malley DM. 1995. Characterization of an unstable anthocyanin phenotype and estimation of somatic mutation rates in peach. *J. Hered.* 86:186–93
21. Christianson ML. 1986. Fate map of the organizing shoot apex in *Gossypium. Am. J. Bot.* 73:945–56
22. Christou P. 1990. Morphological description of transgenic soybean chimeras created by the delivery, integration and expression of foreign DNA using electric discharge particle acceleration. *Ann. Bot.* 66: 379–86

23. Clayberg CD. 1975. Insect resistance in a graft-induced periclinal chimera of tomato. *HortScience* 10:13–14
24. Clowes FAL. 1970. The immediate response of the quiescent centre to x-rays. *New Phytol.* 69:1–18
25. Clowes FAL. 1970. The quiescent centre. *Phytomorphology* 17:132–40
26. Clowes FAL. 1981. The difference between open and closed meristems. *Ann. Bot.* 48: 761–67
27. Coen ES, Romero JM, Doyle S, Carpenter R. 1990. *floricaula:* a homeotic gene required for flower development in *Antirrhinum majus. Cell* 63:1311–22
28. Darwin C. 1868. *Variation of Animals and Plants under Domestication,* Parts I, II. London: Murray
29. Dawe RK, Freeling M. 1991. Cell lineage and its consequences in higher plants. *Plant J.* 1:3–8
30. Dermen H. 1947. Histogenesis of some bud sports and variegations. *Proc. Am. Soc. Hort. Sci.* 50:51–73
31. Dermen H, Stewart RN. 1973. Ontogenetic study of floral organs of peach (*Prunus persica*) utilizing cytochimeral plants. *Am. J. Bot.* 60:283–91
32. Dolan L, Duckett CM, Grierson C, Linstead P, Schneider K, et al. 1994. Clonal relationships and cell patterning in the root epidermis of *Arabidopsis. Development* 120:2465–74
33. Dolan L, Linstead P, Poethig RS, Roberts K. 1994. The contribution of chimeras to the understanding of root meristem organization. In *Shape and Form in Plants and Fungi,* ed. DS Ingram, A Hudson, pp. 196–207. London: Academic
34. Dudley M, Poethig RS. 1991. The effect of a heterochronic mutations, Teopod2, on the cell lineage of the maize shoot. *Development* 111:733–39
35. Dudley M, Poethig RS. 1993. The heterochronic *Teopod1* and *Teopod2* mutations of maize are expressed non-cell-autonomously. *Genetics* 133:389–99
36. Edwards PB, Wanjura WJ, Brown WV, Dearn JM. 1990. Mosaic resistance in plants. *Nature* 347:434
37. Erwee MG, Goodwin PB. 1985. Symplastic domains in extrastelar tissues of *Egeria densa* Planch. *Planta* 163:9–19
38. Esau K. 1953. *Plant Anatomy.* New York: Wiley. 2nd ed.
39. Fleming AJ, Mandel T, Roth I, Kuhlemeier C. 1993. The patterns of gene expression in the tomato shoot apical meristem. *Plant Cell* 5:297–309
40. Freeling M, Hake S. 1985. Developmental genetics of mutants that specify *Knotted* leaves in maize. *Genetics* 111:617–34
41. Fujiwara T, Giesman-Cookmeyer D, Ding B, Lommel SA, Lucas W. 1993. Cell-to-cell trafficking of macromolecules through plasmodesmata potentiated by the red clover necrotic mosaic virus movement protein. *Plant Cell* 5:1783–94
42. Furner IJ, Pumfrey JE. 1992. Cell fate in the shoot apical meristem of *Arabidopsis thaliana. Development* 115:755–64
43. Galinas D, Postlethwait SN, Nelson OE. 1969. Characterization of development in maize through the use of mutants. II. The abnormal growth conditioned by the knotted mutant. *Am. J. Bot.* 56:671–78
44. Goffreda J, Szymkowiak EJ, Sussex IM, Mutchler MA. 1990. Chimeric tomato plants show that aphid resistance and triacylglucose production are epidermal autonomous characters. *Plant Cell* 2:643–49
45. Grant V. 1975. Mosaicism. In *Genetics of Flowering Plants,* pp. 278–99. New York: Columbia Univ. Press
46. Green PB. 1994. Connecting gene and hormone action to form, patttern and organogenesis: biophysical transductions. *J. Exp. Bot.* 45:1775–88
47. Gunning BES. 1982. The root of the water fern *Azolla:* cellular basis of development and multiple roles for cortical microtubules. In *Developmental Organization: Its Origin and Regulation. 40th Symp. Soc. Dev. Biol.,* ed. S Subtelny, PB Green, pp. 379–421. New York: Liss
48. Gunning BES, Robard AW, eds. 1976. *Intercellular Communication in Plants: Studies on Plasmodesmata.* Berlin/Heidelberg: Springer-Verlag
49. Hake S. 1992. Unraveling the knots in plant development. *Trends Genet.* 8:109–14
50. Hake S, Freeling M. 1986. Analysis of genetic mosaics shows that the extra epidermal cell divisions in *Knotted* mutant maize plants are induced by adjacent mesophyll cells. *Nature* 320:621–23
51. Hake S, Vollbrecht E, Freeling M. 1989. Cloning *Knotted,* the dominant morphological mutant in maize using Ds2 as a transposon tag. *EMBO J.* 8:15–22
52. Hanstein J. 1868. Die Scheitelzellgroupe in vegetationspunkt der phanerogamen. *Festschr. Neiderrhein. Gesell. Natur und Heilkunde* 109–34. See Ref. 103
53. Hantke SS, Carpenter R, Coen ES. 1995. Expression of *floricaula* in single cell layers of periclinal chimeras activates downstream homeotic genes in all layers of floral meristems. *Development* 121: 27–35
54. Hernandez L, Green PB. 1993. Transduction for the expression of structural pattern: analysis in sunflower. *Plant Cell* 5: 1725–38
55. Hsiao SC, Mauseth JD, Peng CI. 1995. Composite bundles, the host/parasite interface in the holoparasitic angiosperms

Langsdorffia and *Balanophora* (Balanophoraceae). *Am. J. Bot.* 82:81–91
56. Irish VF, Sussex IM. 1992. A fate map of the *Arabidopsis* embryonic shoot apical meristem. *Development* 115:745–53
57. Jackson D, Veit B, Hake S. 1994. Expression of maize *Knotted1* related homeobox genes in the shoot apical meristem predicts patterns of morphogenesis in the vegetative shoot. *Development* 120:405–13
58. Jefferson RA. 1987. Assaying chimeric genes in plants: the GUS gene fusion system. *Plant Mol. Biol. Rep.* 5:387–405
59. Jeffree CE, Yeoman MM. 1983. Development of intercellular connections between opposing cells in a graft union. *New Phytol.* 93:491–509
60. Jegla DE, Sussex IM. 1989. Cell lineage patterns in the shoot meristem of the sunflower embryo in the dry seed. *Dev. Biol.* 131:215–25
61. Jernstedt JA, Cutter EG, Lu P. 1994. Independence of organogenesis and cell pattern in developing angle shoots of *Selaginella martensii*. *Ann. Bot.* 74:343–55
62. Johri MM, Coe EH Jr. 1983. Clonal analysis of corn plant development. I. The development of the tassel and the ear shoot. *Dev. Biol.* 97:154–72
63. Jorgenson CA, Crane MB. 1927. Formation and morphology of *Solanum* chimeras. *J. Genet.* 18:247–73
64. Kaddoura RL, Mantell SH. 1991. Synthesis and characterization of *Nicotiana-Solanum* graft chimeras. *Ann. Bot.* 68:547–56
65. Kaplan DR. 1992. The relationship of cells to organisms in plants: problems and implications of an organismal perspective. *Int. J. Plant Sci.* 153:S28–37
66. Kaplan DR, Hagemann W. 1991. The relationship of cell and organism in vascular plants. *BioScience* 41:693–703
67. Kirk JTO, Tilney-Bassett RAE. 1978. Chimeras. In *The Plastids: Their Chemsitry, Structure, Growth and Inheritance*, pp. 329–49. New York: Elsevier
68. Klekowski EJ. 1988. *Mutation, Developmental Selection, and Plant Evolution*. New York: Columbia Univ. Press. 373 pp.
69. Klekowski EJ, Kazarinova-Fukshansky N. 1984. Shoot apical meristems and mutation: fixation of selectively neutral cell genotypes. *Am. J. Bot.* 71:22–27
70. Klekowski EJ, Kazarinova-Fukshansky N. 1984. Shoot apical meristems and mutation: loss of disadvantageous cell genotypes. *Am. J. Bot.* 71:28–34
71. Klekowski EJ, Kazarinova-Fukshansky N, Mohr H. 1985. Shoot apical meristems and mutation: stratified meristems and angiosperm evolution. *Am. J. Bot.* 72:1788–800
72. Kuehnle AR, Earle ED. 1989. In vitro selection for methomyl resistance in CMS-T maize. *Theor. Appl. Genet.* 78:672–82
73. Langdale JA, Lane B, Freeling M, Nelson T. 1989. Cell lineage analysis of maize bundle sheath and mesophyll cells. *Dev. Biol.* 133:128–39
74. Langlet O. 1927. Zur Kenntnis der Polysomatischen Zellkerne im Wurzelmeristem. *Svens. Bot. Tidskr.* 21:397–422
75. Lucas WJ, Ding B, van der Schoot C. 1993. Plasmodesmata and the supracellular nature of plants. *New Phytol.* 125:435–76
76. Marcotrigiano M. 1986. Experimentally synthesized plant chimeras 3. Qualitative and quantitative characteristics of the flowers of interspecific *Nicotiana* chimeras. *Ann. Bot.* 57:435–42
77. Marcotrigiano M. 1990. Genetic mosaics and chimeras: implications in biotechnology. In *Biotechnology in Agriculture and Forestry: Somaclonal Variation and Crop Improvement I*, ed. YPS Baja, 11:85–111. Berlin: Springer-Verlag
78. Marcotrigiano M, Bernatzky R. 1995. Arrangement of cell layers in the shoot apical meristems of periclinal chimeras influences cell fate. *Plant J.* 7:193–202
79. Marcotrigiano M, Gouin FR. 1984. Experimentally synthesized plant chimeras 1. In vitro recovery of *Nicotiana tabacum* L. chimeras from mixed callus cultures. *Ann. Bot.* 54:503–11
80. Marcotrigiano M, Gouin FR. 1984. Experimentally synthesized plant chimeras 2. A comparison of *in vitro* and *in vivo* techniques for the production of interspecific *Nicotiana* chimeras. *Ann. Bot.* 54:513–21
81. Marcotrigiano M, Morgan PA. 1988. Chlorophyll-deficient cell lines which are genetically uncharacterized can be inappropriate for use as phenotypic markers in developmental studies. *Am. J. Bot.* 75: 985–89
82. Martienssen R, Barkan A, Taylor W, Freeling M. 1990. Somatically heritable switches in the DNA modification of Mu transposable elements monitored with a suppressible mutant in maize. *Genes Dev.* 4:331–43
83. McDaniel C, Poethig RS. 1988. Cell-lineage patterns in the shoot apical meristem of the germinating maize embryo. *Planta* 175: 13–22
84. Nielson-Jones W. 1969. *Plant Chimeras*. London: Methuen. 2nd ed.
85. Noguchi T, Hirata Y, Yagishita N. 1992. Intervarietal and interspecific chimera formation by in vitro graft-culture method in Brassica. *Theor. Appl. Genet.* 83:727–32
86. Ohtsu Y. 1994. Efficient production of a synthetic periclinal chimera of citrus 'NF-5' for introduction of disease resistance. *Ann. Phytopathol. Soc. Jpn.* 60:82–88
87. Poethig RS. 1984. Cellular parameters of leaf morphogenesis in maize and tobacco. In *Contemporary Problems in Plant Anat-*

omy, ed. RA White, WC Dickison, pp. 235–59. New York: Academic
88. Poethig RS. 1987. Clonal analysis of cell lineage patterns in plant development. *Am. J. Bot.* 74:581–94
89. Poethig RS. 1989. Genetic mosaics and cell lineage patterns in plant development. *Trends Genet.* 5:273–77
90. Poethig RS, Coe EH, Johri MM. 1986. Cell lineage patterns in maize embryogenesis: a clonal analysis. *Dev. Biol.* 117:392–404
91. Poethig RS, Sussex IM. 1985. The cellular parameters of leaf development in tobacco: a clonal analysis. *Planta* 165:170–84
92. Poethig S. 1988. A non-cell-autonomous mutation regulating juvenility in maize. *Nature* 336:82–83
93. Pohlheim F. 1983. Vergleichende untersuchungen zur anderung der richtung von Zellteilungen in Blattepidermen. *Biol. Zentralbl.* 102:323–36
94. Poiteau A. 1830. *Cytisus adami. Ann. Soc. Hortic. Paris* 7:95–96
95. Pri-Hadash A, Hareven D, Lifschitz E. 1992. A meristem-related gene from tomato encodes a dUTPase: analysis of expression in vegetative and floral meristems. *Plant Cell* 4:149–59
96. Rogers SO, Bonnett HT. 1989. Evidence for apical initial cells in the vegetative shoot apices of *Hedera helix* cv. Goldheart. *Am. J. Bot.* 76:739–45
97. Rooney WL, Steely DM. 1991. Preferential transmission and somatic elimination of a *Gossypium sturtianum* chromosome in *G. hirsutum. J. Hered.* 82:151–55
98. Ruth JE, Klekowski EJ, Stein OL. 1985. Impermanent initials of the shoot apex and diplontic selection in a juniper chimera. *Am. J. Bot.* 72:1127–35
99. Satina S. 1944. Periclinal chimeras in *Datura* in relation to development and structure a) of the style and stigma, b) of calyx and corolla. *Am. J. Bot.* 31:493–502
100. Satina S. 1945. Periclinal chimeras in *Datura* in relation to the development and structure of the ovule. *Am. J. Bot.* 32:72–81
101. Satina S, Blakeslee AF. 1941. Periclinal chimeras in *Datura stramonium* in relation to development of leaf and flower. *Am. J. Bot.* 28:862–71
102. Satina S, Blakeslee AF. 1943. Periclinal chimeras in *Datura* in relation to the development of the carpel. *Am. J. Bot.* 30:453–62
103. Satina S, Blakeslee AF, Avery AG. 1940. Demonstration of the three germ layers in the shoot apex of *Datura* by means of induced polyploidy in periclinal chimeras. *Am. J. Bot.* 27:895–905
104. Schmulling T, Schell J. 1993. Transgenic tobacco plants regenerated from leaf disks can be periclinal chimeras. *Plant Mol. Biol.* 21:705–8
105. Sinha N, Hake S. 1990. Mutant characters of *Knotted* maize leaves are determined in the intermost tissue layers. *Dev. Biol.* 141:203–10
106. Smith LG, Greene B, Veit B, Hake S. 1992. A dominant mutation in the maize homeobox gene, *Knotted-1,* causes its ectopic expression in leaf cells with altered fates. *Development* 116:21–30
107. Steeves TA, Sussex IM. 1989. *Patterns in Plant Development,* pp. 46–61. New York: Cambridge Univ. Press. 388 pp.
108. Stelly DM, Rooney WL. 1989. Delimitation of the Le_2^{dav} complementary lethality system of *Gossypium* to intracellular interaction. *J. Hered.* 80:100–3
109. Stephens PA, Barwale-Zehr UB, Nickell CD, Widholm JM. 1991. A cytoplasmically inherited, wrinkled-leaf mutant in soybean. *J. Hered.* 82:71–73
110. Stewart RN. 1978. Ontogeny of the primary body in chimeral forms of higher plants. In *The Clonal Basis of Development,* ed. IM Sussex, S Subtelny, pp. 131–59. New York: Academic
111. Stewart RN, Burk LG. 1970. Independence of tissue derived from apical layers in ontogeny of the tobacco leaf and ovary. *Am. J. Bot.* 57:1010–16
112. Stewart RN, Dermen H. 1970. Determination of number and mitotic activity of shoot apical initial cells by analysis of mericlinal chimeras. *Am. J. Bot.* 61:54–67
113. Stewart RN, Dermen H. 1975. Flexibility in ontogeny as shown by the contribution of the shoot apical layers to leaves of periclinal chimeras. *Am. J. Bot.* 62:935–47
114. Stewart RN, Dermen H. 1979. Ontogeny in monocotyledons as revealed by studies of the developmental anatomy of periclinal chloroplast chimeras. *Am. J. Bot.* 66:47–58
115. Stewart RN, Meyer FG, Dermen H. 1972. Camellia + 'Daisy Eagleson,' a graft chimera of *Camellia sasanqua* and *C. japonica. Am. J. Bot.* 59:515–24
116. Stewart RN, Semeniuk P, Dermen H. 1974. Competition and accommodation between apical layers and their derivatives in the ontogeny of chimeral shoots of *Pelargonium X horteum. Am. J. Bot.* 61:54–67
117. Sussex IM, Clutter ME. 1967. Differentiation in tissues, free cells, and reaggregating plant cells. *In Vitro* 3:3–12
118. Swingle CF. 1927. Graft hybrids in plants. *J. Hered.* 18:73–94
119. Sylvester AW, Cande WZ, Freeling M. 1990. Division and differentiation during normal and *liguleless-1* maize leaf development. *Development* 110:985–1000
120. Szymkowiak EJ, Sussex IM. 1992. The internal layer (L3) determines floral meristem size and carpel number in tomato periclinal chimeras. *Plant Cell* 4: 1089–100
121. Szymkowiak EJ, Sussex IM. 1993. Effect

of *lateral suppressor* on petal initiation in tomato. *Plant J.* 4:1–7
122. Szymkowiak G. 1993. Arrested floral development in *Solanum nigrum* tomato chimeras. *Flowering Newsl.* 15:14–21
123. Thompson JD, Herre EA, Hamrick JL, Stone JL. 1991. Genetic mosaics in strangler fig trees: implication for tropical conservation. *Science* 254:1214–16
124. Tian HC, Marcotrigiano M. 1993. Origin and development of adventitious shoot meristems initiated on plant chimeras. *Dev. Biol.* 155:259–69
125. Tian HC, Marcotrigiano M. 1994. Cell-layer interactions influence the number and position of lateral shoot meristems in Nicotiana. *Dev. Biol.* 162:579–89
126. Tiedemann R. 1989. Graft union development and symplastic phloem contact in the heterograft *Cucumis sativus* on *Cucurbita ficifolia. J. Plant Physiol.* 134:427–40
127. Tilney-Bassett RAE. 1963. Genetics and plastid physiology in *Pelargonium. Heredity* 18:485–504
128. Tilney-Bassett RAE. 1963. The structure of periclinal chimeras. *Heredity* 18:265–85
129. Tilney-Bassett RAE. 1986. *Plant Chimeras.* London: Arnold
130. van den Berg C, Willemsen V, Hage W, Weisbeek P, Scheres B. 1995. Directional signals determine cell fate in the *Arabidopsis thaliana* root meristem. *Nature* 378: 62–65
131. van der Schoot C, Dietrich M, Storms M, Verbeke JA, Lucas WJ. 1995. Establishment of a cell-to cell communication pathway between separate carpels during gynoecium development. *Planta* 195: 450–55
132. van Harten AM, Bouter H, van Ommeren A. 1972. Preventing chimerisms in potato (*Solanum tuberosum* L.) *Euphytica* 21: 11–21
133. Vollbrecht E, Veit B, Sinha N, Hake S. 1991. The developmental gene *Knotted-1* is a member of a maize homoebox gene family. *Nature* 350:241–43
134. Webster PL, Langenaur HD. 1973. Experimental control of the activity of the quiescent center in excised root tips of *Zea mays. Planta* 112:91–100
135. Whitham TG, Slobodchikoff CN. 1981. Evolution by individuals, plant-herbivore interactions, and mosaics of genetic variability: the adaptive significance of somatic mutation in plants. *Oecologia* 49:287–92
136. Winkler H. 1907. Uber Pfropfbastarde und pflanzliche Chimaren. *Ber. Dtsch. Bot. Ges.* 25:568–76. See Ref. 129
137. Winkler H. 1908. Solanum tubingense, ein echter Pfropfbastarde zwischen Tomate and Nachtschatten. *Ber. Dtsch. Bot. Ges.* 26:595–608. See Ref. 129
138. Winkler H. 1909. Weitere mitteilungen uber pfropfbastarde. *Z. Bot.* 1:315–45. See Ref. 129
139. Winkler H. 1910. Ueber das Wesen der Pfropfbastarde. *Ber. Dtsch. Bot. Ges.* 28: 116–18. See Ref. 118
140. Wolf S, Deom CM, Beachy RN, Lucas WJ. 1989. Movement protein of tobacco mosaic virus modifies plasmodesmatal size exclusion limit. *Science* 246:377–79
141. Wolf S, Deom CM, Beachy R, Lucas WJ. 1991. Plasmodesmatal function is probed using transgenic tobacco plants that express a virus movement protein. *Plant Cell* 3:593–4

Annu. Rev. Plant Physiol. Plant Mol. Biol. 1996. 47:377–403

THE MOLECULAR BASIS OF DEHYDRATION TOLERANCE IN PLANTS

J. Ingram and D. Bartels

Max-Planck-Institut für Züchtungsforschung, Carl-von-Linné-Weg 10, 50829 Köln, Germany

KEY WORDS: dehydration stress, desiccation tolerance, late-embryogenesis-abundant (LEA) proteins, osmolytes, ABA responsiveness

ABSTRACT

Molecular studies of drought stress in plants use a variety of strategies and include different species subjected to a wide range of water deficits. Initial research has by necessity been largely descriptive, and relevant genes have been identified either by reference to physiological evidence or by differential screening. A large number of genes with a potential role in drought tolerance have been described, and major themes in the molecular response have been established. Particular areas of importance are sugar metabolism and late-embryogenesis-abundant (LEA) proteins. Studies have begun to examine mechanisms that control the gene expression, and putative regulatory pathways have been established. Recent attempts to understand gene function have utilized transgenic plants. These efforts are of clear agronomic importance.

CONTENTS

1040-2519/96/0601-0377$08.00

INTRODUCTION

This review considers molecular mechanisms involved in dehydration tolerance in plants. Most plants encounter at least transient decreases in relative water content at some stage of their life, and many also produce highly desiccation-tolerant structures such as seeds, spores, or pollen. Indeed, physiological drought also occurs during cold and salt stresses, when the main damage caused to the living cell can be related to water deficit (84, 124). Although we are still far from a complete understanding of the damage caused by drought, or the plant's tolerance mechanisms, much molecular data has been collected over the past few years. Current knowledge of the regulatory network governing the drought-stress responses is also fragmentary, with almost no information on signal perception. However, signal transduction, via ABA at least, and the promoter modules of several response genes, are starting to be elucidated.

Some of the most recent efforts to understand gene function have used transgenic plants, and these studies have significant implications for crop development. Plant breeding has already provided an enormous improvement in the drought tolerance of crop plants (1), with selection often allowing desired traits to be transferred from close wild relatives. However, most of the traits are complex, and their molecular basis is frequently not understood. With our rapidly expanding knowledge of the underlying molecular processes involved in dehydration tolerance, together with the technology of gene manipulation, crop improvement can now also be based on genetic material transferred from any organism and used in a directed manner.

RESEARCH STRATEGIES

Dehydration tolerance has been investigated using three main approaches in plants: (*a*) examining tolerant systems, such as seeds and resurrection plants; (*b*) analyzing mutants from genetic model species; and (*c*) analyzing the effects of stress on agriculturally relevant plants.

Tolerant Systems

One approach of physiological research in dehydration tolerance has been to use specific structures or species that can withstand severe desiccation. Most prominent in this category are certain seeds (73, 82), but desiccation-tolerant species such as resurrection plants (angiosperms) (8), mosses (particularly *Tortula ruralis*), and ferns (98) are also included. Both seeds and the resurrection plant *Craterostigma plantagineum* survive severe dehydration; therefore, the detailed molecular analyses of these systems should reveal expressed genes that contain the genetic information for desiccation tolerance.

SEEDS The final maturation stage of the development of seeds is characterized by desiccation, and as much as 90% of the original water is removed in attaining a state of dormancy with unmeasurable metabolism (73). This desiccated state allows survival under extreme environmental conditions and favors wide dispersal. The embryo cannot withstand desiccation at all developmental stages; tolerance is usually acquired well before maturation drying but is lost as germination progresses. The seeds of many species have been used to isolate the mRNA and proteins related to the desiccation-tolerance response, including, in particular, those of *Arabidopsis thaliana* (100) and of crop species such as cotton (*Gossypium* spp.) (6), barley (*Hordeum vulgare*) (9), maize (*Zea mays*) (99), and rice (*Oryza sativa*) (91). However, a significant complication with these studies is the difficulty of separating the pathways leading to desiccation tolerance from those involved with other aspects of development.

The main achievement of molecular studies with seeds has been the identification and characterization of the late-embryogenesis-abundant (LEA) proteins. LEA-protein mRNAs first appear at the onset of desiccation, dominate the mRNA population in dehydrated tissues (111), and gradually fall several hours after embryos begin to imbibe water (see section on Late-Embryogenesis-Abundant Proteins).

RESURRECTION PLANTS Resurrection plants are unique among angiosperms in their ability to survive during drought, when protoplastic desiccation can leave <2% relative water content in the leaves (8). When water is withheld from mature individuals of *C. plantagineum,* changes rapidly occur at the mRNA and protein levels (8), eventually leading to the tolerant state. A particular advantage of these plants in studies at the molecular level is that desiccation tolerance can be investigated in both whole plants and undifferentiated callus cultures (Tolerant callus of *C. plantagineum* is obtained by pretreatment with ABA) (8). In the callus tissue, and to a certain extent in whole *C. plantagineum* plants, the transition to the tolerant state is largely free of the complications of development or other adjustments inherent in seeds or other plant systems. One of the most striking features of the desiccation-induced genes characterized from vegetative tissues of *C. plantagineum* has been their similarity to the genes expressed in seeds of other species.

Genetic Model Systems

Genetic model systems are a second major approach to the examination of dehydration tolerance. These systems take advantage of detailed genetic information, a wide range of mutants, and the feasibility of positional gene cloning. Progress in understanding the role of ABA in desiccation tolerance has been achieved by characterizing mutants, such as the ABA-deficient mutants *flacca* (tomato, *Lycopersicon esculentum*) (22) and *droopy* (potato, *Solanum tuberosum*) (108). A number of mutations related to ABA action are also available in *A. thaliana,* and their analysis has provided many insights into ABA-mediated drought responses. *A. thaliana* lines that are less sensitive to ABA than the wild-type have mutations at the *abi* loci [43; see also the maize *vp1* mutant (82)]. The detailed genetic information available for *A. thaliana* facilitated the isolation of the *ABI1* and *ABI3* genes by positional cloning (42, 74, 86). *ABI3* is specifically expressed in seeds and probably encodes a transcription factor able to activate *lea*-type genes (100), and *ABI1* encodes a calcium-regulated phosphatase.

Crop Plants

A third approach in researching dehydration tolerance has been to use species important to agriculture to analyze the plant response after drought stress. This type of study is useful because, through intensive breeding or in vitro selection, lines are available with differing degrees of tolerance. Thus, correlative evidence can be sought for genes putatively involved in the drought response. The transient and moderate drought stress represented in studies of crop species probably describes the most common form of dehydration that most plants are likely to encounter. The intensity of research has thus enabled a much more complete picture of the possible factors involved in drought tolerance to emerge.

GENES WITH UPREGULATED EXPRESSION IN RESPONSE TO DEHYDRATION

To establish the basic responses of plants to drought, two of the approaches already outlined—examination of tolerant systems and crop plants—have been most productive. One type of analysis involves targeting genes thought to be important, such as those for the many enzymes in drought-induced metabolic pathways. A second approach uses differential screening to isolate upregulated genes. These experiments have been successful in describing many genes encoding proteins of known function associated with desiccation (Table 1). Differential screening has also revealed many genes of unknown function, which are included in Tables 1 and 2; the largest group is the array of LEA-protein-related genes (Table 3). Some of the genes may be involved in secondary problems of drought-stressed plants, such as increased susceptibility to pathogens, e.g. *pcht28* (encoding an acidic endochitinase) (Table 1; 17) and SC514 (encoding lipoxygenase) (Table 1; 10). Genes involved in signaling

Table 1 Genes upregulated by drought stress[a] and encoding polypeptides of known function

cDNA	Source	Encoded polypeptide	Ref
GapC-Crat	*Craterostigma plantagineum*	Cytosolic glyceraldehyde 3-phosphate dehydrogenase	129
pSPS1	*C. plantagineum*	Sucrose-phosphate synthase	b
pSS1; pSS2	*C. plantagineum*	Sucrose synthases	36
pPPC1	*Mesembryanthemum crystallinum*	Phosphoenolpyruvate carboxylase	130
pBAD	*Hordeum vulgare* (barley)	Betaine aldehyde dehydrogenase	54
cAtP5CS	*Arabidopsis thaliana*	δ^1-pyrroline-5-carboxylate synthetase	145
RD28	*A. thaliana*	Water channel	141
SAM1; SAM3	*Lycopersicon esculentum* (tomato)	*S*-adenosyl-L-methionine synthetases	37
rd19A; rd21A	*A. thaliana*	Cysteine proteases	67
UBQ1	*A. thaliana*	Ubiquitin extension protein	66
pMBM1	*Triticum aestivum* (wheat)	L-isoaspartyl methyltransferase	90
SC514	*Glycine max* (soybean)	Lipoxygenase	10
cATCDPK1; cATCDPK2	*A. thaliana*	Ca^{2+}-dependent, calmodulin-independent protein kinases	127
PKABA1	*T. aestivum*	Protein kinase	4
cAtPLC1	*A. thaliana*	Phosphatidylinositol-specific phospholipase C	53
Apx1 gene	*Pisum sativum* (pea)	Cytosolic ascorbate peroxidase	89
Sod 2 gene	*P. sativum*	Cytosolic copper/zinc superoxide dismutase	135
P31	*L. esculentum*	Cytosolic copper/zinc superoxide dismutase	102
pcht28	*L. chilense*	Acidic endochitinase	17
Atmyb2	*A. thaliana*	MYB-protein-related transcription factor	128
ERD11; ERD13	*A. thaliana*	Glutathione *S*-transferases	63
cAtsEH	*A. thaliana*	Soluble epoxide hydrolase	61

[a]The best-characterized plant genes from which cDNA clones have been demonstrated to show increased mRNA expression levels in response to drought stress have been included. Drought stress has been taken to include quite diverse treatments, ideally where water has been withheld from the plant, but also for example by applying osmotic stress with mannitol solutions or by detaching plant organs.

[b]Ingram & Bartels, unpublished data.

and control processes are considered in the section on Second Messengers and Signaling Molecules.

Metabolism

Changes in primary metabolism are a general response to stress in plants. For example, a cDNA-encoding glyceraldehyde-3-phosphate dehydrogenase, isolated from the resurrection plant *C. plantagineum* (Table 1; 129), shows increased expression during drought and upon ABA treatment. However, increased levels of the enzyme are also associated with other environmental stresses in plants, possibly reflecting increased energy demand. Proteases may also be an important feature of stress metabolism, dispensing with redundant proteins and depolymerizing vacuolar storage polypeptides, thereby releasing amino acids for the massive synthesis of new proteins (Tables 1 and 2; 50).

Enzymes of sugar metabolism are probably critical in desiccation tolerance. It has been demonstrated that certain sugars may be central to the protection of a wide range of organisms against drought (see section on Sugars). In *C. plantagineum,* the overall transcript levels of sucrose-phosphate synthase and sucrose synthase increase immediately in response to drought (36; J Ingram & D Bartels, unpublished data). The expression pattern is complex if the kinetics of individual transcript types are followed over the entire course of dehydration.

Enzymes involved in the synthesis of other compounds that can act as compatible solutes—and whose transcript levels are clearly upregulated during drought—include Δ^1-pyrroline-5-carboxylate synthetase (proline biosynthesis) (Table 1; 145) and betaine aldehyde dehydrogenase (glycine betaine biosynthesis) (Table 1; 54).

The induction of the mRNA encoding phospho*enol*pyruvate carboxylase in *Mesembryanthemum crystallinum* (Table 1; 130) highlights the importance of Crassulacean acid metabolism in enabling carbon fixation with minimal water loss. Such metabolism is a major response in a wide variety of plants to growth in dry conditions (139).

Osmotic Adjustment

Total water potential can be maintained during mild drought by osmotic adjustment, which involves utilizing sugars or other compatible solutes (12). Both ion and water channels are likely to be important in regulating water flux, and the relevance of these channels to drought-stress has been supported by the isolation of channel protein genes expressed in response to water deficit. The 7a cDNA from pea (*Pisum sativum*) (Table 2; 50) encodes a polypeptide with characteristic features of ion channels, while the RD28 cDNA (*A. thaliana*) (Table 1; 141) and probably also the H2-5 cDNA (*C. plantagineum*)

Table 2 Genes upregulated by drought stress[a] but encoding polypeptides of unknown function

cDNA	Source	Features of encoded polypeptide	Ref
26g	*Pisum sativum* (pea)	Some similarity to aldehyde dehydrogenase	50
7a	*P. sativum*	Similar to channel proteins	50
kin2	*Arabidopsis thaliana*	Similarity to animal antifreeze proteins	69
pcC 37-31	*Craterostigma plantagineum*	Similar to early-light-inducible proteins	7
TSW12	*Lycopersicon esculentum* (tomato)	A lipid transfer protein	125
pLE16	*L. esculentum*	Similar to lipid transfer proteins	107
15a	*P. sativum*	Similarity to proteases	50
pA1494	*A. thaliana*	Similarity to proteases	136
ERD1	*A. thaliana*	Similar to a Clp ATP-dependent protease subunit	64
Ha hsp17.6; Ha hsp17.9	*Helianthus annuus* (sunflower)	Low-molecular-weight heat-shock proteins	21
Athsp70-1	*A. thaliana*	Similar to the HSP70 heat-shock-protein family	66
Athsp81-2	*A. thaliana*	Similar to the HSP81 heat-shock-protein family	66
BLT4	*Hordeum vulgare* (barley)	Similar to protease inhibitors	32
P22	*Raphanus sativus* (radish)	Similar to protease inhibitors	77
BnD22	*Brassica napus* (rape)	Similar to protease inhibitors	31
pMAH9	*Zea mays* (maize)	Similar to RNA-binding proteins	47
MsaciA	*Medicago sativa* (alfalfa)	Similar to pUM90-1 and pSM2075 polypeptides	70
pUM90-1	*M. sativa*	Similar to MsaciA and pSM2075 polypeptides	80
pSM2075	*M. sativa*	Similar to MsaciA and pUM90-1 polypeptides	79
pBN115	*B. napus*	Similar to polypeptides encoded by pBN19 and pBN26 (*B. napus*), and COR15 (*A. thaliana*)	134
RD22	*A. thaliana*	Similar to an unidentified seed protein from *Vicia faba*	56
salT	*Oryza sativa* (rice)		18
lti65 gene; *lti78* gene	*A. thaliana*		95
pcC 13-62	*C. plantagineum*		104

[a]See Footnote a in Table 1.

(J-B Mariaux & D Bartels, unpublished data) encode putative water-channel proteins (28).

Structural Adjustment

Drought stress has been shown to cause alterations in the chemical composition and physical properties of the cell wall (e.g. wall extensibility), and such changes may involve the genes encoding *S*-adenosylmethionine synthetase (Table 1; 37). Under nonstressful conditions, increased expression of *S*-adenosyl-L-methionine synthetase genes correlates with areas where lignification is occurring (101). Thus, the increased expression in drought-stressed tissue could thus also be due to lignification in the cell wall. Cell elongation stops under prolonged drought stress, and then lignification processes seem to begin (94a). Espartero et al (37) also noted that fungal elicitors cause the coinduction of *S*-adenosyl-L-methionine synthetase transcript with those of other enzymes, e.g. *S*-adenosyl-L-homocystein hydrolase or a methyltransferase, required for cell wall formation.

The *C. plantagineum pcC37-31* cDNA (Table 2; 7) encodes the dsp-22 protein, whose mRNA levels increase in response to various stresses. The cDNA shows significant homology to early light-inducible protein (ELIP) genes (1a). Light is involved in the regulation of the gene expression, and the encoded dsp-22 protein is chloroplastic. ELIPs may play a role in the assembly of the photosystem (1a). During desiccation, *C. plantagineum* chloroplasts undergo morphological changes, and thus the dsp-22 protein could bind pigments or help maintain assembled photosynthetic structures essential for resuming active photosynthesis during resurrection.

Degradation and Repair

Genes encoding proteins with sequence similarity to proteases, and which are induced by drought, have been isolated from both pea (Table 2; 50) and *A. thaliana* (Tables 1 and 2; 64, 67, 136). One of the functions of these enzymes could be to degrade proteins irreparably damaged by the effects of drought (50). During early drought in *A. thaliana,* there is an increase in levels of mRNA encoding ubiquitin extension protein (66), a fusion protein from which active ubiquitin is derived by proteolytic processing. This increase may be significant in terms of protein degradation, because ubiquitin has a role in tagging proteins for destruction. During drought stress, protein residues may be modified by chemical processes such as deamination, isomerization, or oxidation, and it is thus likely that enzymes with functions in protein repair are upregulated in response to drought. Indeed, the response to desiccation in mosses may largely be repair based (98). An example of such repair processes is the observation that L-isoaspartyl methyltransferases may convert modified L-isoaspartyl residues in damaged proteins back to L-aspartyl residues (Table 1; 90).

Mudgett & Clarke (90) have argued that such repair mechanisms could be particularly important during desiccation, when protein turnover rates are low. Although *Escherichia coli* mutants lacking the enzyme grow normally in the logarithmic phase when there is high protein turnover, they survive poorly in the stationary phase when turnover is much lower (75).

The products of two drought-induced genes isolated by differential screening have sequence similarity to heat-shock proteins (Table 2; 66). These encoded proteins are probably chaperonins, involved in protein repair by helping other proteins to recover their native conformation after denaturation or misfolding during water stress. The low-molecular-weight heat-shock proteins (Table 2; 21) may also be chaperonins. This function has been demonstrated for a mammalian low-molecular-weight heat-shock protein (58). An alternative function may be in the sequestration of specific mRNAs in cells subjected to drought (96).

Removal of Toxins

Enzymes concerned with removing toxic intermediates produced during oxygenic metabolism, such as glutathione reductase and superoxide dismutase, increase in response to drought stress and are probably very important in tolerance (89). Decreasing leaf water content and consequent stomatal closure result in reduced CO_2 availability and the production of active oxygen species such as superoxide radicals (117). Increased photorespiratory activity during drought is also accompanied by elevated levels of glycolate-oxidase activity, resulting in H_2O_2 production (89). This could explain why genes encoding enzymes that detoxify active oxygen species such as ascorbate peroxidase (Table 1; 89) and superoxide dismutase (Table 1; 102, 135) have been found upregulated in response to drought.

Late-Embryogenesis-Abundant Proteins

The genes encoding late-embryogenesis-abundant (LEA) proteins are consistently represented in differential screens for transcripts with increased levels during drought. LEA proteins were first described from research into genes abundantly expressed during the final desiccation stage of seed development (see above). Circumstantial evidence for their involvement in dehydration tolerance is strong: The genes are similar to many of those expressed in vegetative tissues of drought-stressed plants (Table 3), and desiccation treatments can often induce precocious expression in seeds. ABA can also induce the *lea* genes in seeds and vegetative tissues.

GENERAL FEATURES Groupings for dividing the LEA proteins originate from a dot matrix analysis with proteins from cotton. A group was assigned on the basis of one cotton LEA protein showing regions of significant homology with

Table 3 Genes upregulated by drought stress[a] that encode polypeptides related to late-embryogenesis-abundant LEA proteins

cDNA	Source	Relationship of encoded polypeptide to LEA proteins	Ref
Ha ds10	*Helianthus annuus* (sunflower)	D19-LEA-protein related	3
Em	*Triticum aestivum* (wheat)	D19-LEA-protein related	76
B19.1; B19.3; B19.4	*Hordeum vulgare* (barley)	D19-LEA-protein related	39
pLE25	*Lycopersicon esculentum* (tomato)	D113-LEA-protein related	23
Ha ds11	*H. annuus*	D113-LEA-protein related	3
pRABAT1	*Arabidopsis thaliana*	D11-LEA-protein related	72
pcC 27-04	*Craterostigma plantagineum*	D11-LEA-protein related	104
M3 (RAB-17)	*Zea mays* (maize)	D11-LEA-protein related	20
B8; B9; B17	*H. vulgare*	D11-LEA-protein related	20
pLE4	*L. esculentum*	D11-LEA-protein related	23
pcC 6-19	*C. plantagineum*	D11-LEA-protein related	104
TAS14	*L. esculentum*	D11-LEA-protein related	46
pLC30-15	*L. chilense*	D11-LEA-protein related	16
H26	*Stellaria longipes*	D11-LEA-protein related	110
pRAB 16A	*Oryza sativa* (rice)	D11-LEA-protein related	91
pcECP40	*Daucus carota* (carrot)	D11-LEA-protein related	62
ERD10; ERD14	*A. thaliana*	D11-LEA-protein related	65
pMA2005	*T. aestivum*	D7-LEA-protein related	26
pMA1949	*T. aestivum*	D7-LEA-protein related	27
pcC 3-06	*C. plantagineum*	D7-LEA-protein related	104
pcC 27-45	*C. plantagineum*	D95-LEA-protein-related	104

[a]See Footnote a in Table 1.

at least one protein from another species (33). The "type" of cotton proteins used for these groupings were LEA D19 (Group 1), LEA D11 [Group 2 (also termed dehydrins)], and LEA D7 (Group 3). The cotton proteins LEA D113 (34, 35) and LEA D95 (40) now define two additional classes. This system will remain useful until clear functions can be assigned.

LEA proteins appear to be located in many cell types and at variable concentrations (19, 34, 35, 45), and within the cell they appear to be predominantly—but not exclusively—cytosolic (19, 45, 91, 114). The concentrations in the cell are characteristically very high. For example, in mature cotton embryo cells, the D7 LEA proteins represent about 4% of nonorganellar cytosolic protein (about 0.34 mM) (111).

A general structural feature of the LEA proteins is their biased amino acid composition, which results in highly hydrophilic polypeptides, with just a few residues providing 20–30% of their total complement. For example, a deduced

D19 protein from cotton contains 13% glycine and 11% glutamic acid (6). Furthermore, most LEA proteins lack cysteine and tryptophan residues.

ROLES We await direct experimental evidence that LEA proteins can protect specific cellular structures or ameliorate the effects of drought stress. Because they are highly hydrophilic, it appears unlikely that they occur in specific cellular structures. Also, their high concentrations in the cell and biased amino acid compositions suggest that they do not function as enzymes (6).

The randomly coiled moieties of some LEA proteins are consistent with a role in binding water. Total desiccation is probably lethal, and therefore such proteins could help maintain the minimum cellular water requirement. McCubbin & Kay (83) have found that the Em protein (D19-group) (Table 3; 76) from wheat is considerably more hydrated than most globular polypeptides because it is over 70% random coil in normal physiological conditions. The random coil tails of the D113 proteins could also bind considerable amounts of water, although the long *N*-terminal helical domain would not share this property (34, 35).

A major problem under severe dehydration is that the loss of water leads to crystallization of cellular components, which in consequence damages cellular structures. This may be counteracted by LEA proteins, and some of the LEA proteins could essentially be considered compatible solutes, which supports the likely role of sugars in maintaining the structure of the cytoplasm in the absence of water. Baker et al (6) have suggested that LEA proteins D11 and D113 could be involved in the "solvation" of cytosolic structures. The random coiling would permit their shape to conform to that of other structures and provide a cohesive layer with possibly greater stability than would be formed by sugars. Their hydroxylated groups would solvate structural surfaces. Furthermore, they could be superior to sucrose as protectants in being less likely to crystallize. However, for the D11-related protein RAB-17, a regulatory role has been postulated (see below).

Baker et al (6) have hypothesized that the 11-amino-acid motif (T/A A/T Q/E A/T A/T K/R Q/ED K/R A/T X ED/Q) (34) of LEA protein D-29 (which is also present in D7 LEA proteins) could counteract the irreversibly damaging effects of increasing ionic strength in the cytosol during desiccation. Such problems could be mitigated by the formation of salt bridges with amino acid residues of highly charged proteins. The repeating elements most likely exist as amphiphilic helices (34), which means that hydrophobic and hydrophilic amino acids are contained in particular sectors of the helix. The helices probably form intramolecular bundles, which would present a surface capable of binding both anions and cations. Further analyses of the D7-group molecules have allowed precise structural predictions to be made: The intersurface edges of the interacting helical regions of the (putative) dimer reveal periodically spaced binding sites for suitably charged ions.

SUGARS

The involvement of soluble sugars in desiccation tolerance in plants is suggested by studies in which the presence of particular soluble sugars can be correlated with the acquisition of desiccation tolerance (73). Such studies have followed work with animals, fungi, yeast, and bacteria, in which a high level of the disaccharide trehalose has been established as important in surviving desiccation. Trehalose is the most effective osmoprotectant sugar in terms of minimum concentration required (25). Whereas trehalose is extremely rare in plants, sucrose—together with other sugars—appears able to substitute. Although sugar accumulation is not the only way in which plants deal with desiccation (12), it is considered an important factor in tolerance.

Many studies with seeds have demonstrated the accumulation of soluble sugars during the acquisition of desiccation tolerance (73); similar results have been demonstrated in resurrection plants. A common theme has emerged. Various soluble carbohydrates may be present in fully hydrated tissues, but sucrose usually accumulates in the dried state. For example, desiccation in the leaves of *C. plantagineum* is accompanied by conversion of the C8-sugar 2-octulose (90% of the total sugar in hydrated leaves) into sucrose, which then comprises about 40% of the dry weight (11).

Total water potential can be maintained during mild drought by osmotic adjustment. Sugars may serve as compatible solutes permitting such osmotic adjustment, although many other compounds usually associated with salt stress are also active, such as proline, glycine betaine, and pinitol (54, 84, 145). Increasing sucrose synthesis and sucrose-phosphate synthase activity is not only a drought-response of desiccation-tolerant plants such as *C. plantagineum* (36) but also of plants that cannot withstand extreme drying, such as spinach (109).

One way sugars may protect the cell during severe desiccation is by glass formation: Rather than solutes crystallizing, through the presence of sugars a supersaturated liquid is produced with the mechanical properties of a solid (68). Glass formation has been demonstrated in viable maize seeds and has been associated with their viability (137). Differential scanning calorimetry has been used to examine the effect of termperature on glass formation by sugar mixtures; only sugar mixtures equivalent in concentration and composition to those in desiccation-tolerant embryos are able to form glass at ambient temperatures (68). It seems likely that sugar composition, rather than just concentration, is related to glass formation. During desiccation, glass would fill space, thus preventing cellular collapse, and in restricting the molecular diffusion required by chemical reactions would permit a stable quiescent state (68).

Phosphofructokinase is a tetrameric enzyme that usually dissociates irreversibly into inactive dimers during dehydration (14). However, it was found that in vitro the disaccharides sucrose, maltose, and trehalose stabilize the activity of the enzyme during drying.

Crowe et al (24) have shown that, in vitro, drying and rehydration of the model-membrane sarcoplasmic reticulum usually results in the fusion of vesicles and loss of the ability to transport calcium. However, when the sugar trehalose was present at concentrations equivalent to those in desiccation-tolerant organisms, functional vesicles were preserved. Many other studies show that sugars can protect membranes in vitro (25); it is suggested that sugars alter physical properties of dry membranes so that they resemble those of fully hydrated biomolecules.

The mechanism by which proteins are stabilized by sugars is better understood than the situation with membranes. Infrared spectroscopy has shown that trehalose probably forms hydrogen bonds between its hydroxyl groups and polar residues in proteins (25). Hydrogen bonding between the hydroxyl group of trehalose and the phosphate head group of phospholipids can be inferred from comparisons of changes in the infrared spectrum of the molecules during dehydration. Strauss & Hauser (120) used the cation Eu^{3+}, which is known to form a specific ionic bridge to the phosphate of phospholipids, to show that sucrose is probably bound between phosphate sites in dry membranes. This was inferred from experiments in which Eu^{3+} ions were added to preparations of sucrose and phosphatidylcholine vesicles; the stabilization of liposomes by sucrose during freeze drying decreased as the Eu^{3+} ions were added, which suggests competitive binding of sucrose and Eu^{3+} at the phosphate sites of the phospholipids.

REGULATION OF GENE EXPRESSION DURING DEHYDRATION

The machinery leading to the expression of drought-stress genes conforms to the general cellular model, with a complex signal transduction cascade that can be divided into the following basic steps: (*a*) perception of stimulus; (*b*) processing, including amplification and integration of the signal; and (*c*) a response reaction in the form of de novo gene expression. No molecular data are available on the perception of drought stress, although turgor change has been suggested as a possible physical signal. An attractive model for the activation of a transduction pathway by a stress signal has been derived from studying the heat-shock response in yeast (60). Kamada et al (60) suggest that heat-induced activation of a particular pathway is in response to increased membrane fluidity in the cell wall. The cell detects this weakness in the cell wall by sensing stretch in the plasma membrane. Examples such as this from simple systems may provide the conceptual framework for devising experiments in plants.

The drought-activated signal transmission process has begun to be dissected at the molecular level, mostly on the basis of studies of isolated drought-responsive genes. Endogenous ABA levels have been reported to increase as a result of water deficit in many physiological studies, and therefore ABA is thought to be involved in the signal transduction (15, 43). Many of the drought-related genes can be induced by exogenous ABA; however, this does not necessarily imply that all these genes are also regulated by ABA in vivo.

We now discuss promoter studies, signaling molecules, and both posttranscriptional and posttranslational modifications in the context of drought-regulated gene expression.

Promoter Studies

CIS- AND *TRANS*-ACTING ELEMENTS Many of the changes in mRNA levels observed during drought reflect transcriptional activation. Treatment with ABA can also induce these changes, and this treatment has been utilized for setting up experimental systems to define *cis*- and *trans*-acting elements. *cis*- and *trans*-acting elements involved in ABA-induced gene expression have been analyzed extensively (Tables 4 and 5; 43).

Table 4 *cis*-acting promoter elements relevant to ABA or drought

Gene	Element	Sequence[a]	Ref
Rab16A (*Oryza sativa*)	ABRE (Motif I)	GT*ACGT*GGCGC	119
EM (*Triticum aestivum*)	Em1A	GGAC*ACGT*GGC	51
Hex3 (synthetic tetramer) (derived from *Nicotiana tabacum*)		GGTGACGTGGC	71
rab28 (*Zea mays*)	ABRE	CC*ACGT*GG	106
Cat1 (*Zea mays*)		CCAAGAAGTC-*ACGT*GGAGGTGGAAGAG	138
HVA22 (*Hordeum vulgare*)	ABRE3 and CE1	GCC*ACGT*ACA and TGCCACCGG	118
CDeT27-45 (*Craterostigma plantagineum*)		AAGCCCAAATTTCA-CAGCCCGATAACCG	93
rd29 (*Arabidopsis thaliana*)	DRE	TACCGACAT	144

[a]The G-box core elements ACGT are in italic.

The best-characterized *cis*-element in the context of drought stress is the ABA-responsive element (ABRE), which contains the palindromic motif CACGTG with the G-box ACGT core element (44). ACGT elements have been observed in a multitude of plant genes regulated by diverse environmental and physiological factors. Systematic DNA-binding studies have shown that nucleotides flanking the ACGT core specify the DNA-protein interactions and subsequent gene activation (57). G-box-related ABREs have been observed in many ABA-responsive genes, although their functions have not always been proven experimentally. The best-studied examples of these ABRE promoter elements are Em1a from wheat and Motif I from the rice *rab 16A* gene (Table 4; 81, 92). Multiple copies of the elements fused to a minimal 35S promoter confer an ABA response to a reporter gene (51, 119), which supports the hypothesis that ABREs are critical for the ABA induction of relevant genes (although it is difficult to explain why single copies are not

Table 5 Characterization of promoters in transgenic plants

Gene	Native gene activity	Reporter gene activity	Ref
Rab 16B	Embryos of Oryza sativa	*Nicotiana tabacum* embryos	142
Em	Embryos of Triticum aestivum	*Nicotiana tabacum* embryos	81
Rab 17	Embryos of Zea mays	The embryos and endosperm of Arabidopsis thaliana	131
Hex3 (synthetic tetramer) (derived from *Nicotiana tabacum*)		Mature seeds of *N. tabacum;* inducible in seedlings by desiccation, salt, and ABA	71
Rd 22	Dehydrated A. thaliana plants	Constitutive in flowers and stems of *A. thaliana;* inducible in *N. tabacum* by ABA or dehydration	56
Rd 29A	Dehydrated A. thaliana plants	Inducible by dehydration in most vegetative parts of *A. thaliana;* inducible in *N. tabacum* by cold, ABA, and salt	143, 144
CDeT27-45	C. plantagineum dehydrated or ABA-treated vegetative tissues	In embryos and mature pollen of both *A. thaliana* and *N. tabacum*	39a, 88
CDeT6-19	C. plantagineum dehydrated or ABA-treated vegetative tissues	In developing embryos and mature pollen of both *A. thaliana* and *N. tabacum* also inducible in their leaves and guard cells	87, 39a, 123
CDeT11-24	C. plantagineum dehydrated or ABA-treated vegetative tissues	Embryos of both *A. thaliana* and *N. tabacum;* inducible in *A. thaliana* leaves by dehydration	[a]
DC8	Embryos of *Daucus carota*	D. carota seed tissues	49
DC3	Embryos of Daucus carota	N. tabacum seedlings; also inducible in the leaves by either drying or ABA treatment	132

[a]R Velasco, F Salamini & D Bartels, unpublished data.

sufficient for this response). The ABA effect on transcription was orientation independent in both the wheat and rice elements, which suggests that they possibly function as enhancer elements in their native genes. Electrophoretic-mobility-shift assays and methylation-interference footprinting have shown that both Em1a and Motif1 interact with nuclear proteins; these DNA-binding proteins are constitutively expressed in an ABA-independent manner (51, 92). cDNAs encoding ABRE-binding proteins (wheat EMBP-1 and tobacco TAF-1) have been cloned and shown to contain a basic region adjacent to a leucine-zipper motif that is characteristic of transcription factors (51, 97). Despite the fact that both proteins exhibit specific and distinct binding properties, their roles in vivo are not understood. It seems possible that they are not directly involved in ABA-responsive gene expression but that they cooperate with other regulatory factors.

Recently, two different elements have been described that must be present to allow a single copy of the ABRE to mediate transcriptional activation in response to ABA, and thus define an ABA response complex. An ABRE element in the barley *Amy32b* α-amylase promoter has been shown to allow ABA-stimulated transcription to increase only in the presence of an O2S element that interacts with the ABRE within tight positional constraints. A second coupling element has been identified during promoter analysis of the ABA-induced barley *HVA22* promoter (118). The coupling element (CE1) acts together with a G-box-type ABRE (GCCACGTACA) in conferring high ABA induction, whereas the ABRE alone is not sufficient for transcriptional activation. CE1-like elements have been found in many other ABA-regulated promoters, but their function remains to be demonstrated (118). The specific sequence of a coupling element may profoundly affect the specificity of ABA-driven gene expression and may explain differences between functional and nonfunctional ABREs.

In promoters such as *CDeT27–45* or *CDeT6–19,* isolated from *C. plantagineum,* G-box-related ABREs do not appear to be major determinants of the ABA or drought response (87, 88). The *CDeT27–45* promoter contains an element that specifically binds nuclear proteins from ABA-treated tissue; this promoter fragment is essential but not sufficient for conferring a response to ABA on a reporter gene (93).

Besides the ABA-mediated gene expression, the investigation of drought-induced genes in *A. thaliana* has also revealed ABA-independent signal transduction pathways (144). The *A. thaliana* genes *rd29A* and *rd29B* are differentially induced under conditions of dehydration, salt or cold stress, and ABA treatment. The *rd29A* gene has at least two *cis*-acting elements. 1. The 9-bp direct repeat sequence, TACCGACAT, termed the dehydration-responsive element (DRE), functions in the initial rapid response of *rd29A* to drought, salt, or low temperature (144). 2. The slower ABA response is mediated by another fragment that contains an ABRE (143). It will be interesting to see whether the same *cis* elements function in other *A. thaliana* genes that are induced during progressive drought; besides ABA, at least two other different signals are involved in this induction (48). The existence of ABA-dependent and -independent pathways is corroborated by studies on the accumulation of three distinct *Lea* transcripts in barley embryos. Selected transcripts increased in response to osmotic stress without requiring ABA, whereas induction by salt did require ABA (38).

A different class of potential transcription factors with relevance to drought stress is represented by the *A. thaliana* gene *Atmyb2*. This gene encodes an MYB-related protein and is induced by dehydration or salt stress and by ABA (128). Plant *myb*-related genes comprise a large family that may play various roles in gene regulation. The ATMYB2 protein expressed in *E. coli* has been shown to bind the MYB-recognition sequence, PyAACTG, which supports its role as a DNA-binding protein. Another *A. thaliana* drought stress–induced gene, *rd22* (56), has a promoter with no ABRE but with two recognition sites for the transcription factors MYC and MYB. Binding of the ATMYB2 protein appears likely but has not been proven experimentally.

ASSESSMENT OF PROMOTERS IN TRANSGENIC PLANTS Promoter analysis using transient expression assays has resulted in the characterization of several distinct *cis*-acting elements and the cloning of related transcription factors. However, tests with a range of promoters derived from drought- or ABA-inducible structural genes in transgenic plants have shown that the promoter activities defined in transient assays are not always correlated with the expression patterns of their corresponding structural genes. A summary of results is given in Table 5. A problem with the approach could be the use of heterologous plant expression systems. Although the genes are always active in seeds, expression in vegetative tissues is not always induced upon drought or ABA treatment, which points to an incomplete activation of the transcriptional machinery. It is interesting to note that ectopic expression of the otherwise seed-specific *abi-3* gene product (42) allows the ABA-mediated activation of *Lea* genes in vegetative tissues of *A. thaliana* (100). Similarly, the *CDeT27–45* promoter from *C. plantagineum* was only fully responsive to ABA in *A. thaliana* in the presence of the ABI3 product (39a). These experiments suggest that the ABI3 gene product can functionally interact with different promoters.

Second Messengers and Signaling Molecules

Protein phosphorylation and dephosphorylation (via kinases and phosphorylases, respectively) are major mechanisms of signal integration in eukaryotic cells. Two *A. thaliana* genes encoding calcium-dependent kinases are induced by dehydration (Table 1; 127), which suggests that they may participate in phosphorylation processes occurring in response to drought. A serine-threonine-type protein kinase has also been isolated from wheat and shows accumulation in ABA-treated embryos and in dehydrated shoots (Table 1; 4). However, the phosphorylation targets of these kinases are not yet known, and their exact roles are obscure.

A role for protein phosphorylation in the drought-stress response is also suggested on the basis of functional studies of the ABA-responsive RAB17 protein from maize (45). This protein is highly phosphorylated in vivo, probably via catalysis by casein kinase 2. The RAB17 protein has been found to be distributed between the cytoplasm and the nucleus of maize embryos, in different states of phosphorylation (5, 45). Biochemical studies showed that RAB17 binds peptides with nuclear localization signals and that the binding is dependent on phosphorylation. It has been suggested that RAB17 mediates the transport of specific nuclear-targeted proteins during stress (45).

Cytoplasmic calcium acts as a second messenger in many cellular processes and may also be involved in the signaling pathways mediating the expression of drought-related genes (13). Stomatal closure is an early plant response to drought, and increases in the cytosolic concentration of free calcium, together with pH changes, are considered to be primary events in the ABA-mediated reduction of stomatal turgor (115). However, it is likely that calcium, together with phosphorylation processes, plays a more general role in the mechanisms associated with drought-stress perception. For example, the *A. thaliana ABI1* gene product is thought to be a calcium-activated phosphoprotein phosphatase (74, 86). Furthermore, a transcript encoding a phosphatidylinositol-specific phospholipase C, an enzyme involved in catalyzing the synthesis of inositol 1,4,5-triphosphate, increases during dehydration (Table 1; 53); inositol-triphosphate stimulates the release of Ca^{2+} from intracellular stores.

Posttranscriptional Control

Much of the effort to understand gene regulation during drought has been devoted to transcriptional mechanisms, but it has become clear that other potential control points include mRNA processing, transcript stability, translation efficiency, and protein modification or turnover. General posttranscriptional mechanisms in plants have recently been reviewed (41, 121). Evidence is emerging that these mechanisms also play a role during stress responses. In *C. plantagineum,* drought stress induces some proteins that are synthesized in a light-dependent manner (see above); for some of these proteins the levels of the mRNA do not parallel those of the proteins, which suggests posttranscriptional regulation (2). A more detailed analysis of alfalfa (*Medicago sativa*) suggests that increased mRNA stability is involved in the accumulation of the MsPRP2 transcript (30). The maize *pMAH9* cDNA clone encodes a transcript that is upregulated by drought. The corresponding protein has RNA-binding characteristics, which suggests that it may play a role in the selective stabilization of mRNAs (Table 2; 78).

A second major control point appears to be the posttranslational modification of proteins, in which phosphorylation is a key mechanism. For example, phosphorylation is involved in the modification of the fructose-1,6-bisphosphatase in drought-stressed leaves of sugar beet (*Beta vulgaris*) (52). Some of the proteases induced by drought stress (Tables 1 and 2) may also have a function in posttranslational modification. Schaffer & Fischer (113) have hypothesized that a thiol protease, the mRNA of which is cold-induced in tomato, could proteolytically activate certain proteins. This mechanism could also operate during drought stress. It has also been suggested that putative protease inhibitors induced during drought (Table 2) have a role in controlling the activity of endogenous proteases (31).

Downregulation of Genes

Until now, most research has focused on understanding how relevant genes are upregulated during drought stress. However, the response to drought also involves the downregulation of several genes. For example, studies of *C. plantagineum* have revealed that transcripts encoding proteins relevant to photosynthesis are downregulated during the dehydration process and thus possibly reduce photooxidative stress (C Bockel & D Bartels, unpublished data). Jiang et al (59) have also shown that the promoter regions of storage protein genes contain the information for their downregulation during seed desiccation. Furthermore, it has recently been reported that histone H1 transcripts accumulate in response to drought stress in vegetative tissues of tomato, and it was suggested that H1 histones are implicated in the repression of gene expression (E Bray, personal communication).

TRANSGENIC PLANTS ASSESSING GENE FUNCTION

Transgenic plants allow the targeted expression of drought-related genes in vivo and are therefore an excellent system to assess the function and tolerance conferred by the encoded proteins. With ectopic expression of genes involved in controlling ABA biosynthesis, it should also be possible to alter the hormonal balance in vivo and thus to clarify the role of ABA in the drought response. Another purpose for using transgenic plants is to improve drought tolerance in agronomically valuable plants. However, despite extensive research, examples of transgenic plants with improved stress tolerance are scarce (see also 12). A reason for this is that stress tolerance is likely to involve the expression of gene products from several pathways.

The accumulation of low–molecular weight metabolites that act as osmoprotectants is a widespread adaptation to dry, saline, and low-temperature conditions in many organisms. In engineering plants that synthesize protective osmolytes, microorganisms appear to be useful sources for genes. Transgenic tobacco plants that synthesize and accumulate the sugar alcohol mannitol have been obtained by introducing a bacterial gene that encodes mannitol 1-phosphate dehydrogenase. Plants producing mannitol showed increased salt tolerance (122). Similarly, a freshwater cyanobacterium that was transformed with *E. coli bet* genes produced significant amounts of glycine betaine; this stabilized photosynthetic activity in the presence of sodium chloride, allowing improved growth (94). Tobacco plants that accumulate the polyfructose molecule fructan have been engineered using microbial (*Bacillus subtilis* or *Streptococcus mutans*) fructosyltransferase genes. These plants showed improved growth under polyethylene-mediated drought stress (105), with a positive correlation observed between the level of accumulated fructans and degree of tolerance. The mechanism by which fructans confer tolerance is not known, although a mere osmotic effect seems unlikely.

One consequence of drought and many other stresses is the production of activated oxygen molecules that cause cellular injury, and therefore plants with increased concentrations of oxygen scavengers should show improved performances under nonlethal stress conditions. When tobacco Mn-superoxide dismutase was overexpressed in alfalfa, the plants showed an increased growth rate after freezing stress (85).

Although *Lea*-related genes are upregulated abundantly in most plants during all types of osmotic stress, separate ectopic expression of three different representatives in tobacco did not yield an obvious drought-tolerant phenotype (55). However, this result is perhaps less surprising considering that drought stress does induce an array of different LEA-related proteins in plants. It is also likely that other factors are required for the expression of tolerance where LEA-type proteins are involved.

FUTURE PERSPECTIVES

Despite the many genes that have been identified in association with drought stress, much of the data is descriptive, with the functions of only a few of the encoded proteins established. The production of mutants using an antisense-RNA approach is a powerful technique that should continue to elucidate certain aspects of stress tolerance, but it has been most successful only with well-characterized areas of plant metabolism. It is also difficult to devise screening procedures for useful dehydration-tolerance mutants, because of the array of processes simultaneously affected by drought. Resurrection plants would be an excellent source for mutants with decreased tolerance, but *C. plantagineum,* as well as many other resurrection-plant species, has a polyploid genome and is thus unsuitable. Mutant analyses so far exploited for drought stress have been with ABA-related mutations, and the power of the approach is shown in the cloning of *Abi1* and *Abi3* (43), which has provided new perspectives. Another valuable approach may be to identify those metabolic steps that are most sensitive to drought stress (a technique used to genetically dissect salt stress in yeast) (116). Such an approach can at least begin to elucidate which gene products are of primary importance.

The plant hormone ABA regulates different aspects of the drought-stress response, and thus the synthesis of pure active ABA analogues (103) may help in the development of probes for ABA-binding proteins, which could then shed some light on primary signals. In contrast with the situation with signal perception, some information is available on *cis*- and *trans*-regulatory factors. Several elements in a promoter need to cooperate with multiple DNA-binding proteins to mediate gene expression. The recently described coupling elements (118) are probably only a beginning in resolving the regulatory network. Little progress has been made with the cloning and analysis of drought-related transcription factors, although a biochemical approach and use of the recently established yeast one- and two-hybrid systems (133) should produce new insights. Regulation at stages beyond transcription must also be further considered, because this could make a major contribution to the final gene expression pattern.

The complexity of drought tolerance apparent throughout this review points to control by multiple genes, and thus the identification of quantitative-trait-loci (QTLs) for drought resistance may well be an effective analytical tool. The approach has just begun to be applied to the environmental-stress responses of plants (126) and is particularly promising considering that saturated DNA–marker maps are now available for both genetic model plants and crop plants.

The molecular analysis of the drought response has arrived at a stage where research can build upon a large collection of characterized genes. The use of novel approaches combining genetic, biochemical, and molecular techniques should provide exciting results in the near future.

ACKNOWLEDGMENTS

The authors wish to thank F. Salamini for his support and for comments on the manuscript, M. Pasemann for help in preparing the manuscript, and the EU PTP project for financial support. We apologize to all colleagues who have contributed to this research area whose work has not been cited because of limited space.

Literature Cited

1. Acevedo E, Fereres E. 1993. Resistance to abiotic stresses. In *Plant Breeding,* ed. MD Hayward, NO Bosemark, I Romagosa, pp. 406-21. London: Chapman & Hall

1a. Adamska I, Kloppstech K. 1994. The role of early light-induced proteins (ELIPs) during light stress. In *Environmental Plant Biology Series,* ed. WJ Davies, *Photoinhibition of Photosynthesis: from Molecular Mechanisms to the Field,* ed. NR Baker, JR Bowyer, pp. 205–19. Oxford: BIOS Sci.

2. Alamillo JM, Bartels D. 1996. Light and stage of development influence the expression of desiccation-induced genes in the resurrection plant *Craterostigma plantagineum. Plant Cell Environ.* 19:In press

3. Almoguera C, Jordano J. 1992. Developmental and environmental concurrent expression of sunflower dry-seed-stored low-molecular-weight heat-shock protein and Lea mRNAs. *Plant Mol. Biol.* 19:781–92

4. Anderberg RJ, Walker-Simmons MK. 1992. Isolation of a wheat cDNA clone for an abscisic acid–inducible transcript with homology to protein kinases. *Proc. Natl. Acad. Sci. USA* 89:10183–87

5. Asghar R, Fenton RD, DeMason DA, Close TJ. 1994. Nuclear and cytoplasmic localization of maize embryo and aleurone dehydrin. *Protoplasma* 177:87–94

6. Baker J, Steele C, Dure L III. 1988. Sequence and characterization of 6 *Lea* proteins and their genes from cotton. *Plant Mol. Biol.* 11:277–91

7. Bartels D, Hanke C, Schneider K, Michel D, Salamini F. 1992. A desiccation-related *Elip*-like gene from the resurrection plant *Craterostigma plantagineum* is regulated by light and ABA. *EMBO J.* 11(8):2771–78

8. Bartels D, Schneider K, Terstappen G, Piatkowski D, Salamini F. 1990. Molecular cloning of abscisic acid–modulated genes which are induced during desiccation of the resurrection plant *Craterostigma plantagineum. Planta* 181:27–34

9. Bartels D, Singh M, Salamini F. 1988. Onset of desiccation tolerance during development of the barley embryo. *Planta* 175: 485–92

10. Bell E, Mullet JE. 1991. Lipoxygenase gene expression is modulated in plants by water deficit, wounding, and methyl jasmonate. *Mol. Gen. Genet.* 230:456–62

11. Bianchi G, Gamba A, Murelli C, Salamini F, Bartels D. 1991. Novel carbohydrate metabolism in the resurrection plant *Craterostigma plantagineum. Plant J.* 1(3): 355–59

12. Bohnert HJ, Nelson DE, Jensen RG. 1995. Adaptations to environmental stresses. *Plant Cell* 7:1099–111

13. Bush DS. 1995. Calcium regulation in plant cells and its role in signaling. *Annu. Rev. Plant Physiol. Plant Mol. Biol.* 46:95–122

14. Carpenter JF, Crowe LM, Crowe JH. 1987. Stabilization of phosphofructokinase with sugars during freeze-drying: characterization of enhanced protection in the presence of divalent cations. *Biochim. Biophys. Acta* 923:109–15

15. Chandler PM, Robertson M. 1994. Gene expression regulated by abscisic acid and its relation to stress tolerance. *Annu. Rev. Plant Physiol. Plant Mol. Biol.* 45: 113–41

16. Chen R-D, Campeau N, Greer AF, Bellemare G, Tabaeizadeh Z. 1993. Sequence of a novel abscisic acid– and drought-induced cDNA from wild tomato (*Lycopersicon chilense*). *Plant Physiol.* 103:301

17. Chen R-D, Yu L-X, Greer AF, Cheriti H, Tabaeizadeh Z. 1994. Isolation of an osmotic stress- and abscisic acid-induced gene encoding an acidic endochitinase from *Lycopersicon chilense. Mol. Gen. Genet.* 245:195–202

18. Claes B, Dekeyser R, Villarroel R, Van den Bulcke M, Bauw G, et al. 1990. Characterization of a rice gene showing organ-specific expression in response to salt stress and drought. *Plant Cell* 2:19–27

18a. Close TJ, Bray EA, eds. 1993. *Current Topics in Plant Physiology: An American Society of Plant Physiologists Series,* Vol. 10, *Plant Responses to Cellular Dehydration During Environmental Stress.* Rock ville, MD: Am. Soc. Plant Physiol.

19. Close TJ, Fenton RD, Yang A, Asghar R, DeMason DA, et al. 1993. Dehydrin: the protein. See Ref. 18a, pp. 104–18

20. Close TJ, Kortt AA, Chandler PM. 1989. A cDNA-based comparison of dehydration-induced proteins (dehydrins) in barley and corn. *Plant Mol. Biol.* 13:95–108

21. Coca MA, Almoguera C, Jordano J. 1994. Expression of sunflower low-molecular-weight heat-shock proteins during embryogenesis and persistence after germination: localization and possible functional implications. *Plant Mol. Biol.* 25:479–92

22. Cohen A, Bray EA. 1990. Characterization of three mRNAs that accumulate in wilted tomato leaves in response to elevated levels of endogenous abscisic acid. *Planta* 182: 27–33

23. Cohen A, Plant ÁL, Moses MS, Bray EA. 1991. Organ-specific and environmentally regulated expression of two abscisic acid-induced genes of tomato. *Plant Physiol.* 97:1367–74

24. Crowe JH, Crowe LM, Jackson SA. 1983. Preservation of structural and functional

activity in lyophilized sarcoplasmic reticulum. *Arch. Biochem. Biophys.* 220(2): 477–84
25. Crowe JH, Hoekstra FA, Crowe LM. 1992. Anhydrobiosis. *Annu. Rev. Physiol.* 54: 579–99
26. Curry J, Morris CF, Walker-Simmons MK. 1991. Sequence analysis of a cDNA encoding a Group 3 LEA mRNA inducible by ABA or dehydration stress in wheat. *Plant Mol. Biol.* 16:1073–76
27. Curry J, Walker-Simmons MK. 1993. Unusual sequence of group 3 LEA (II) mRNA inducible by dehydration stress in wheat. *Plant Mol. Biol.* 21:907–12
28. Daniels MJ, Mirkov TE, Chrispeels MJ. 1994. The plasma membrane of *Arabidopsis thaliana* contains a mercury-insensitive aquaporin that is a homolog of the tonoplast water channel protein TIP. *Plant Physiol.* 106:1325–33
29. Deleted in proof
30. Deutsch CE, Winicov I. 1995. Post-transcriptional regulation of a salt-inducible alfalfa gene encoding a putative chimeric proline-rich cell wall protein. *Plant Mol. Biol.* 27:411–18
31. Downing WL, Mauxion F, Fauvarque M-O, Reviron M-P, de Vienne D, et al. 1992. A *Brassica napus* transcript encoding a protein related to the Künitz protease inhibitor family accumulates upon water stress in leaves, not in seeds. *Plant J.* 2(5): 685–93
32. Dunn MA, Hughes MA, Zhang L, Pearce RS, Quigley AS, Jack PL. 1991. Nucleotide sequence and molecular analysis of the low temperature induced cereal gene, BLT4. *Mol. Gen. Genet.* 229:389–94
33. Dure L III, Crouch M, Harada J, Ho T-HD, Mundy J, et al. 1989. Common amino acid sequence domains among the LEA proteins of higher plants. *Plant Mol. Biol.* 12: 475–86
34. Dure L III. 1993. A repeating 11-mer amino acid motif and plant desiccation. *Plant J.* 3(3):363–69
35. Dure L III. 1993. Structural motifs in Lea proteins. See Ref. 18a, pp. 91–103
36. Elster R. 1994. *Physiologische und molek ulare Charakterisierung des Saccharosestoffwechsels der trockentoleranten Wiederauferstehungspflanze* Craterostigma plantagineum *Hochst.* PhD thesis. Univ. Köln
37. Espartero J, Pintor-Toro JA, Pardo JM. 1994. Differential accumulation of *S*-adenosylmethionine synthetase transcripts in response to salt stress. *Plant Mol. Biol.* 25:217–27
38. Espelund M, De Bedout JA, Outlaw WH Jr, Jakobsen KS. 1995. Environmental and hormonal regulation of barley late-embryogenesis-abundant (Lea) mRNAs is via different signal transduction pathways. *Plant Cell Environ.* 18:943–49
39. Espelund M, Sæboe-Larssen S, Hughes DW, Galau GA, Larsen F, Jakobsen KS. 1992. Late embryogenesis-abundant genes encoding proteins with different numbers of hydrophilic repeats are regulated differentially by abscisic acid and osmotic stress. *Plant J.* 2(2):241–52
39a. Furini A, Parcy F, Salamini F, Bartels D. 1996. Differential regulation of two ABA-inducible genes from *Craterostigma plantagineum* in transgenic *Arabidopsis* plants. *Plant Mol. Biol.* In press
40. Galau GA, Wang HY-C, Hughes DW. 1993. Cotton *Lea5* and *Lea14* encode atypical late embryogenesis-abundant proteins. *Plant Physiol.* 101:695–96
41. Gallie DR. 1993. Posttranscriptional regulation of gene expression in plants. *Annu. Rev. Plant Physiol. Plant Mol. Biol.* 44: 77–105
42. Giraudat J, Hauge BM, Valon C, Smalle J, Parcy F, Goodman HM. 1992. Isolation of the Arabidopsis *ABI3* gene by positional cloning. *Plant Cell* 4:1251–61
43. Giraudat J, Parcy F, Bertauche N, Gosti F, Leung J, et al. 1994. Current advances in abscisic acid action and signalling. *Plant Mol. Biol.* 26:1557–77
44. Giuliano G, Pichersky E, Malik VS, Timko MP, Scolnick PA, Cashmore AR. 1988. An evolutionarily conserved protein binding sequence upstream of a plant light-regulated gene. *Proc. Natl. Acad. Sci. USA* 85: 7089–93
45. Goday A, Jensen AB, Culiáñez-Macià FA, Albà MM, Figueras M, et al. 1994. The maize abscisic acid–responsive protein Rab17 is located in the nucleus and interacts with nuclear localization signals. *Plant Cell* 6:351–60
46. Godoy JA, Pardo JM, Pintor-Toro JA. 1990. A tomato cDNA inducible by salt stress and abscisic acid: nucleotide sequence and expression pattern. *Plant Mol. Biol.* 15:695–705
47. Gómez J, Sánchez-Martínez D, Stiefel V, Rigau J, Puigdomènech P, Pagès M. 1988. A gene induced by the plant hormone abscisic acid in response to water stress encodes a glycine-rich protein. *Nature* 334: 262–64
48. Gosti F, Bertauche N, Vartanian N, Giraudat J. 1995. Abscisic acid–dependent and –independent regulation of gene expression by progressive drought in *Arabidopsis thaliana. Mol. Gen. Genet.* 246: 10–18
49. Goupil P, Hatzopoulos P, Franz G, Hempel FD, You R, Sung ZR. 1992. Transcriptional regulation of a seed-specific carrot gene, DC8. *Plant Mol. Biol.* 18:1049–63
50. Guerrero FD, Jones JT, Mullet JE. 1990.

Turgor-responsive gene transcription and RNA levels increase rapidly when pea shoots are wilted: sequence and expression of three inducible genes. *Plant Mol. Biol.* 15:11–26
51. Guiltinan MJ, Marcotte WR Jr, Quatrano RS. 1990. A plant leucine zipper protein that recognizes an abscisic acid response element. *Science* 250:267–71
52. Harn C, Daie J. 1992. Regulation of the cytosolic fructose-1,6-bisphosphatase by post-translational modification and protein level in drought-stressed leaves of sugarbeet. *Plant Cell Physiol.* 33(6):763–70
53. Hirayama T, Ohto C, Mizoguchi T, Shinozaki K. 1995. A gene encoding a phosphatidylinositol-specific phospholipase C is induced by dehydration and salt stress in *Arabidopsis thaliana. Proc. Natl. Acad. Sci. USA* 92:3903–7
54. Ishitani M, Nakamura T, Han SY, Takabe T. 1995. Expression of the betaine aldehyde dehydrogenase gene in barley in response to osmotic stress and abscisic acid. *Plant Mol. Biol.* 27:307–15
55. Iturriaga G, Schneider K, Salamini F, Bartels D. 1992. Expression of desiccation-related proteins from the resurrection plant *Craterostigma plantagineum* in transgenic tobacco. *Plant Mol. Biol.* 20:555–58
56. Iwasaki T, Yamaguchi-Shinozaki K, Shinozaki K. 1995. Identification of a *cis*-regulatory region of a gene in *Arabidopsis thaliana* whose induction by dehydration is mediated by abscisic acid and requires protein synthesis. *Mol. Gen. Genet.* 247(4): 391–98
57. Izawa T, Foster R, Chua N-H. 1993. Plant bZIP protein DNA binding specificity. *J. Mol. Biol.* 230:1131–44
58. Jakob U, Gaestel M, Engel K, Buchner J. 1993. Small heat shock proteins are molecular chaperones. *J. Biol. Chem.* 268(3): 1517–20
59. Jiang L, Downing WL, Baszczynski CL, Kermode AR. 1995. The 5' flanking regions of vicilin and napin storage protein genes are down-regulated by desiccation in transgenic tobacco. *Plant Physiol.* 107: 1439–49
60. Kamada Y, Jung US, Piotrowski R, Levin DE. 1995. The protein kinase C-activated MAP kinase pathway of *Saccharomyces cerevisiae* mediates a novel aspect of the heat shock response. *Genes Dev.* 9: 1559–71
61. Kiyosue T, Beetham JK, Pinot F, Hammock BD, Yamaguchi-Shinozaki K, Shinozaki K. 1994. Characterization of an *Arabidopsis* cDNA for a soluble epoxide hydrolase gene that is inducible by auxin and water stress. *Plant J.* 6(2):259–69
62. Kiyosue T, Yamaguchi-Shinozaki K, Shinozaki K, Kamada H, Harada H. 1993. cDNA cloning of ECP40, an embryogenic-cell protein in carrot, and its expression during somatic and zygotic embryogenesis. *Plant Mol. Biol.* 21:1053–68
63. Kiyosue T, Yamaguchi-Shinozaki K, Shinozaki K. 1993. Characterization of two cDNAs (ERD11 and ERD13) for dehydration-inducible genes that encode putative glutathione *S*-transferases in *Arabidopsis thaliana* L. *FEBS Lett.* 335(2):189–92
64. Kiyosue T, Yamaguchi-Shinozaki K, Shinozaki K. 1993. Characterization of cDNA for a dehydration-inducible gene that encodes a CLP A, B-like protein in *Arabidopsis thaliana* L. *Biochem. Biophys. Res. Comm.* 196(3):1214–20
65. Kiyosue T, Yamaguchi-Shinozaki K, Shinozaki K. 1994. Characterization of two cDNAs (ERD10 and ERD14) corresponding to genes that respond rapidly to dehydration stress in *Arabidopsis thaliana. Plant Cell Physiol.* 35(2):225–31
66. Kiyosue T, Yamaguchi-Shinozaki K, Shinozaki K. 1994. Cloning of cDNAs for genes that are early-responsive to dehydration stress (ERDs) in *Arabidopsis thaliana* L.: identification of three ERDs as HSP cognate genes. *Plant Mol. Biol.* 25:791–98
67. Koizumi M, Yamaguchi-Shinozaki K, Tsuji H, Shinozaki K. 1993. Structure and expression of two genes that encode distinct drought-inducible cysteine proteinases in *Arabidopsis thaliana. Gene* 129:175–82
68. Koster KL. 1991. Glass formation and desiccation tolerance in seeds. *Plant Physiol.* 96:302–4
69. Kurkela S, Borg-Franck M. 1992. Structure and expression of *kin2,* one of two cold- and ABA-induced genes of *Arabidopsis thaliana. Plant Mol. Biol.* 19:689–92
70. Laberge S, Castonguay Y, Vézina L-P. 1993. New cold- and drought-regulated gene from *Medicago sativa. Plant Physiol.* 101:1411–12
71. Lam E, Chua N-H. 1991. Tetramer of a 21-base pair synthetic element confers seed expression and transcriptional enhancement in response to water stress and abscisic acid. *J. Biol. Chem.* 266(26): 17131–35
72. Lång V, Palva ET. 1992. The expression of a *rab*-related gene, *rab18,* is induced by abscisic acid during the cold acclimation process of *Arabidopsis thaliana* (L.) Heynh. *Plant Mol. Biol.* 20:951–62
73. Leprince O, Hendry GAF, McKersie BD. 1993. The mechanisms of desiccation tolerance in developing seeds. *Seed Sci. Res.* 3:231–46
74. Leung J, Bouvier-Durand M, Morris P-C, Guerrier D, Chefdor F, Giraudat J. 1994. *Arabidopsis* ABA response gene *ABI1:* features of a calcium-modulated protein phosphatase. *Science* 264:1448–52

75. Li C, Clarke S. 1992. A protein methyltransferase specific for altered aspartyl residues is important in *Escherichia coli* stationary-phase survival and heat-shock resistance. *Proc. Natl. Acad. Sci. USA* 89: 9885–89
76. Litts JC, Colwell GW, Chakerian RL, Quatrano RS. 1987. The nucleotide sequence of a cDNA clone encoding the wheat E_m protein. *Nucleic Acids Res.* 15(8):3607–18
77. Lopez F, Vansuyt G, Fourcroy P, Casse-Delbart F. 1994. Accumulation of a 22-kDa protein and its mRNA in the leaves of *Raphanus sativus* in response to salt stress or water deficit. *Physiol. Plant.* 91:605–14
78. Ludevid MD, Freire MA, Gómez J, Burd CG, Albericio F, et al. 1992. RNA binding characteristics of a 16-kDa glycine-rich protein from maize. *Plant J.* 2(6):999–1003
79. Luo M, Lin L, Hill RD, Mohapatra SS. 1991. Primary structure of an environmental stress and abscisic acid–inducible alfalfa protein. *Plant Mol. Biol.* 17: 1267–69
80. Luo M, Liu J-H, Mohapatra S, Hill RD, Mohapatra SS. 1992. Characterization of a gene family encoding abscisic acid– and environmental stress–inducible proteins of alfalfa. *J. Biol. Chem.* 267(22):15367–74
81. Marcotte WR Jr, Russell SH, Quatrano RS. 1989. Abscisic acid–responsive sequences from the Em gene of wheat. *Plant Cell* 1:969–76
82. McCarty DR. 1995. Genetic control and integration of maturation and germination pathways in seed development. *Annu. Rev. Plant Physiol. Plant Mol. Biol.* 46:71–93
83. McCubbin WD, Kay CM. 1985. Hydrodynamic and optical properties of the wheat Em protein. *Can. J. Biochem.* 63:803–10
84. McCue KF, Hanson AD. 1990. Drought and salt tolerance: towards understanding and application. *Trends Biotech.* 8:358–62
85. McKersie BD, Chen Y, de Beus M, Bowley SR, Bowler C, et al. 1993. Superoxide dismutase enhances tolerance of freezing stress in transgenic alfalfa (*Medicago sativa* L.). *Plant Physiol.* 103:1155–63
86. Meyer K, Leube M, Grill E. 1994. A protein phosphatase in ABA signal transduction in *Arabidopsis thaliana. Science* 264: 1452–55
87. Michel D, Furini A, Salamini F, Bartels D. 1994. Structure and regulation of an ABA- and desiccation-responsive gene from the resurrection plant *Craterostigma plantagineum. Plant Mol. Biol.* 24:549–60
88. Michel D, Salamini F, Bartels D, Dale P, Baga M, Szalay A. 1993. Analysis of a desiccation and ABA-responsive promoter isolated from the resurrection plant *Craterostigma plantagineum. Plant J.* 4(1): 29–40
89. Mittler R, Zilinskas BA. 1994. Regulation of pea cytosolic ascorbate peroxidase and other antioxidant enzymes during the progression of drought stress and following recovery from drought. *Plant J.* 5(3): 397–405
90. Mudgett MB, Clarke S. 1994. Hormonal and environmental responsiveness of a developmentally regulated protein repair L-isoaspartyl methyltransferase in wheat. *J. Biol. Chem.* 269(41):25605–12
91. Mundy J, Chua N-H. 1988. Abscisic acid and water-stress induce the expression of a novel rice gene. *EMBO J.* 7(8):2279–86
92. Mundy J, Yamaguchi-Shinozaki K, Chua N-H. 1990. Nuclear proteins bind conserved elements in the abscisic acid–responsive promoter of a rice *rab* gene. *Proc. Natl. Acad. Sci. USA* 87:1406–10
93. Nelson D, Salamini F, Bartels D. 1994. Abscisic acid promotes novel DNA-binding activity to a desiccation-related promoter of *Craterostigma plantagineum. Plant J.* 5(4):451–58
94. Nomura M, Ishitani M, Takabe T, Rai AK, Takabe T. 1995. *Synechococcos* sp. PCC7942 transformed with *Escherichia coli bet* genes produces glycine betaine from choline and acquires resistance to salt stress. *Plant Physiol.* 107:703–8
94a. Nonami H, Boyer JS. 1990. Wall extensibility and cell hydraulic conductivity decrease in enlarging stem tissues at low water potentials. *Plant Physiol.* 93:1610-19
95. Nordin K, Vahala T, Palva ET. 1993. Differential expression of two related, low-temperature-induced genes in *Arabidopsis thaliana* (L.) Heynh. *Plant Mol. Biol.* 21: 641–53
96. Nover L, Scharf K-D, Neumann D. 1989. Cytoplasmic heat shock granules are formed from precursor particles and are associated with a specific set of mRNAs. *Mol. Cell. Biol.* 9(3):1298–308
97. Oeda K, Salinas J, Chua N-H. 1991. A tobacco bZIP transcription activator (TAF-1) binds to a G-box-like motif conserved in plant genes. *EMBO J.* 10(7):1793–802
98. Oliver MJ, Bewley JD. 1996. Desiccation-tolerance of plant tissues: a mechanistic overview. *Hort. Rev.* In press
99. Pagès M, Vilardell J, Jensen AB, Albà MM, Torrent M, Goday A. 1993. Molecular biological responses to drought in maize. In *Global Environmental Change,* NATO Adv. Sci. Inst. Ser., Vol. I 16, *Interacting Stresses on Plants in a Changing Climate,* ed. MB Jackson, CR Black, pp. 583–91. Berlin/Heidelberg: Springer-Verlag
100. Parcy F, Valon C, Raynal M, Gaubier-Comella P, Delseny M, Giraudat J. 1994. Regulation of gene expression programs during *Arabidopsis* seed development: roles of the *ABI3* locus and of endogenous abscisic acid. *Plant Cell* 6:1567–82

101. Peleman J, Boerjan W, Engler G, Seurinck J, Botterman J, et al. 1989. Strong cellular preference in the expression of a housekeeping gene of *Arabidopsis thaliana* encoding *S*-adenosylmethionine synthetase. *Plant Cell* 1:81–93
102. Perl-Treves R, Galun E. 1991. The tomato Cu,Zn superoxide dismutase genes are developmentally regulated and respond to light and stress. *Plant Mol. Biol.* 17:745–60
103. Perras MR, Abrams SR, Balsevich JJ. 1994. Characterization of an abscisic acid carrier in suspension-cultured barley cells. *J. Exp. Bot.* 45(280):1565–73
104. Piatkowski D, Schneider K, Salamini F, Bartels D. 1990. Characterization of five abscisic acid–responsive cDNA clones isolated from the desiccation-tolerant plant *Craterostigma plantagineum* and their relationship to other water-stress genes. *Plant Physiol.* 94:1682–88
105. Pilon-Smits EAH, Ebskamp MJM, Paul MJ, Jeuken MJW, Weisbeek PJ, Smeekens SCM. 1995. Improved performance of transgenic fructan-accumulating tobacco under drought stress. *Plant Physiol.* 107: 125–30
106. Pla M, Vilardell J, Guiltinan MJ, Marcotte WR, Niogret MF, et al. 1993. The *cis*-regulatory element CCACGTGG is involved in ABA and water-stress responses of the maize gene *rab28. Plant Mol. Biol.* 21: 259–66
107. Plant ÁL, Cohen A, Moses MS, Bray EA. 1991. Nucleotide sequence and spatial expression pattern of a drought- and abscisic acid–induced gene of tomato. *Plant Physiol.* 97:900–6
108. Quarrie SA. 1982. Droopy: a wilty mutant of potato deficient in abscisic acid. *Plant Cell Environ.* 5:23–6
109. Quick P, Siegl G, Neuhaus E, Feil R, Stitt M. 1989. Short-term water stress leads to a stimulation of sucrose synthesis by activating sucrose-phosphate synthase. *Planta* 177:535–46
110. Robertson M, Chandler PM. 1992. Pea dehydrins: identification, characterisation and expression. *Plant Mol. Biol.* 19: 1031–44
111. Roberts JK, DeSimone NA, Lingle WL, Dure L III. 1993. Cellular concentrations and uniformity of cell-type accumulation of two Lea proteins in cotton embryos. *Plant Cell* 5:769–80
112. Rogers JC, Rogers S. 1992. Definition and functional implications of gibberellin and abscisic acid *cis*-acting hormone response complexes. *Plant Cell* 4:1443–51
113. Schaffer MA, Fischer RL. 1988. Analysis of mRNAs that accumulate in response to low temperature identifies a thiol protease gene in tomato. *Plant Physiol.* 87:431–36
114. Schneider K, Wells B, Schmelzer E, Salamini F, Bartels D. 1993. Desiccation leads to the rapid accumulation of both cytosolic and chloroplastic proteins in the resurrection plant *Craterostigma plantagineum* Hochst. *Planta* 189:120–31
115. Schroeder JI. 1995. Anion channels as central mechanisms for signal transduction in guard cells and putative functions in roots for plant-soil interactions. *Plant Mol. Biol.* 28:353–61
116. Serrano R. 1995. Salt tolerance in plants and microorganisms: toxicity targets and defense responses. *Intern. Rev. Cytol.* 165: In press
117. Sgherri CLM, Pinzino C, Navari-Izzo F. 1993. Chemical changes and O_2^- production in thylakoid membranes under water stress. *Physiol. Plant.* 87:211–16
118. Shen Q, Ho T-HD. 1995. Functional dissection of an abscisic acid (ABA)–inducible gene reveals two independent ABA-responsive complexes each containing a G-box and a novel *cis*-acting element. *Plant Cell* 7:295–307
119. Skriver K, Olsen PL, Rogers JC, Mundy J. 1991. *Cis*-acting DNA elements responsive to gibberellin and its antagonist abscisic acid. *Proc. Natl. Acad. Sci. USA* 88: 7266–70
120. Strauss G, Hauser H. 1986. Stabilization of lipid bilayer vesicles by sucrose during freezing. *Proc. Natl. Acad. Sci. USA* 83: 2422–26
121. Sullivan ML, Green PJ. 1993. Post-transcriptional regulation of nuclear-encoded genes in higher plants: the roles of mRNA stability and translation. *Plant Mol. Biol.* 23:1091–104
122. Tarczynski MC, Jensen RG, Bohnert H. 1993. Stress protection of transgenic tobacco by production of the osmolyte mannitol. *Science* 259:508–10
123. Taylor JE, Renwick KF, Webb AAR, McAinsh MR, Furini A, et al. 1995. ABA-regulated promoter activity in stomatal guard cells. *Plant J.* 7(1):129–34
124. Thomashow MF. 1993. Characterization of genes induced during cold acclimation in *Arabidopsis thaliana.* See Ref. 18a, pp. 137–43
125. Torres-Schumann S, Godoy JA, Pintor-Toro JA. 1992. A probable lipid transfer protein gene is induced by NaCl in stems of tomato plants. *Plant Mol. Biol.* 18: 749–57
126. Touzet P, Winkler RG, Helentjaris T. 1995. Combined genetic and physiological analysis of a locus contributing to quantitative variation. *Theor. Appl. Genet.* 91: 200–5
127. Urao T, Katagiri T, Mizoguchi T, Yamaguchi-Shinozaki K, Hayashida N, Shinozaki K. 1994. Two genes that encode Ca^{2+}-dependent protein kinases are in-

duced by drought and high-salt stresses in *Arabidopsis thaliana. Mol. Gen. Genet.* 244:331–40

128. Urao T, Yamaguchi-Shinozaki K, Urao S, Shinozaki K. 1993. An *Arabidopsis myb* homolog is induced by dehydration stress and its gene product binds to the conserved MYB recognition sequence. *Plant Cell* 5: 1529–39
129. Velasco R, Salamini F, Bartels D. 1994. Dehydration and ABA increase mRNA levels and enzyme activity of cytosolic GAPDH in the resurrection plant *Craterostigma plantagineum. Plant Mol. Biol.* 26: 541–46
130. Vernon DM, Ostrem JA, Bohnert HJ. 1993. Stress perception and response in a facultative halophyte: the regulation of salinity-induced genes in *Mesembryanthemum crystallinum. Plant Cell Environ.* 16:437–44
131. Vilardell J, Martínez-Zapater JM, Goday A, Arenas C, Pagès M. 1994. Regulation of the *rab*17 gene promoter in transgenic *Arabidopsis* wild-type, ABA-deficient and ABA-insensitive mutants. *Plant Mol. Biol.* 24:561–69
132. Vivekananda J, Drew MC, Thomas TL. 1992. Hormonal and environmental regulation of the carrot lea-class gene *DC3. Plant Physiol.* 100:576–81
133. Wang MM, Reed RR. 1993. Molecular cloning of the olfactory neuronal transcription factor Olf-1 by genetic selection in yeast. *Nature* 364:121–26
134. Weretilnyk E, Orr W, White TC, Iu B, Singh J. 1993. Characterization of three related low-temperature-regulated cDNAs from winter *Brassica napus. Plant Physiol.* 101: 171–77
135. White DA, Zilinskas BA. 1991. Nucleotide sequence of a complementary DNA encoding pea cytosolic copper/zinc superoxide dismutase. *Plant Physiol.* 96:1391–92
136. Williams J, Bulman M, Huttly A, Phillips A, Neill S. 1994. Characterization of a cDNA from *Arabidopsis thaliana* encoding a potential thiol protease whose expression is induced independently by wilting and abscisic acid. *Plant Mol. Biol.* 25:259–70
137. Williams RJ, Leopold AC. 1989. The glassy state in corn embryos. *Plant Physiol.* 89:977–81
138. Williamson JD, Scandalios JG. 1994. The maize (*Zea mays* L.) *Cat1* catalase promoter displays differential binding of nuclear proteins isolated from germinated and developing embryos and from embryos grown in the presence and absence of abscisic acid. *Plant Physiol.* 106:1373–80
139. Winter K, Smith JAC, eds. 1996. *Crassulacean Acid Metabolism: Biochemistry, Ecophysiology and Evolution.* Berlin: Springer-Verlag. In press
140. Deleted in proof
141. Yamaguchi-Shinozaki K, Koizumi M, Urao S, Shinozaki K. 1992. Molecular cloning and characterization of 9 cDNAs for genes that are responsive to desiccation in *Arabidopsis thaliana:* sequence analysis of one cDNA clone that encodes a putative transmembrane channel protein. *Plant Cell Physiol.* 33(3):217–24
142. Yamaguchi-Shinozaki K, Mino M, Mundy J, Chua N-H. 1990. Analysis of an ABA-responsive rice gene promoter in transgenic tobacco. *Plant Mol. Biol.* 15:905–12
143. Yamaguchi-Shinozaki K, Shinozaki K. 1993. Characterization of the expression of a desiccation-responsive *rd29* gene of *Arabidopsis thaliana* and analysis of its promoter in transgenic plants. *Mol. Gen. Genet.* 236:331–40
144. Yamaguchi-Shinozaki K, Shinozaki K. 1994. A novel *cis*-acting element in an *Arabidopsis* gene is involved in responsiveness to drought, low-temperature, or high-salt stress. *Plant Cell* 6:251–64
145. Yoshiba Y, Kiyosue T, Katagiri T, Ueda H, Mizoguchi T, et al. 1995. Correlation between the induction of a gene for δ^1-pyrroline-5-carboxylate synthetase and the accumulation of proline in *Arabidopsis thaliana* under osmotic stress. *Plant J.* 7(5): 751–60
146. Zhang X-H, Moloney MM, Chinnappa CC. 1993. Nucleotide sequence of a cDNA clone encoding a dehydrin-like protein from *Stellaria longipes. Plant Physiol.* 103: 1029–30

Annu. Rev. Plant Physiol. Plant Mol. Biol. 1996. 47:405–30

BIOCHEMISTRY AND MOLECULAR BIOLOGY OF WAX PRODUCTION IN PLANTS

Dusty Post-Beittenmiller

Plant Biology Division, The Samuel Roberts Noble Foundation, Ardmore, Oklahoma 73402-2180

KEY WORDS: very long chain fatty acid elongation, glossy, elongase, condensing enzme, cuticular lipids

ABSTRACT

The aerial surfaces of plants are covered with a wax layer that is primarily a waterproof barrier but that also provides protection against environmental stresses. The ubiquitous presence of cuticular wax is testimony to its essential function. Genetic and environmental factors influence wax quantity and composition, which suggests that it is an actively regulated process. The basic biochemistry of wax production has been elucidated over the past three decades; however, we still know very little about its regulation. This review presents a discussion along with new perspectives on the regulatory aspects of wax biosynthesis. Among the topics discussed are the partitioning of fatty acid precursors into wax biosynthesis and the elongation of fatty acids with particular emphasis on the nature of the acyl primer, and the role of ATP in fatty acid elongation. The recent cloning of wax biosynthetic genes and the transport of wax to plant surfaces are also discussed.

CONTENTS

1040-2519/96/0601-0405$08.00

INTRODUCTION

Waxes are the waterproofing component of the plant cuticle and are therefore essential for life in an aerial environment. They are embedded within cutin or suberin polymers and continue as an amorphous layer on the outer surface of the plant. In many plants, epicuticular wax crystalline structures overlay this layer, which gives the plant surface a glaucous or gray appearance. Plants deficient or altered in surface waxes appear shiny, glossy, or bloomless. This distinct and easily observed phenotype has been widely used in the isolation of mutants defective in wax production.

The cuticle provides the first line of defense between the plant and its environment; the cuticular waxes shed rainwater from the plant surface and limit nonstomatal water loss. In addition, waxes may protect plants from bacterial and fungal pathogens (52) and play a role in plant-insect interactions (26). Its hydrophobicity makes wax a good solvent for organic pollutants and impedes the uptake of aqueous foliar sprays without the addition of surfactants. In addition, the reflective nature of waxes offers some protection against damaging UV radiation.

Several excellent reviews over the past fifteen years have covered wax biosynthesis and composition, wax-deficient mutants, and wax function (19, 100, 100a). Therefore, this review focuses primarily on recent developments and their importance to our understanding of plant wax biosynthesis and its regulation. Emphasis is on the major wax biosynthetic pathways (studied most extensively by biochemical and molecular biological approaches) : decarbonylation, acyl-reduction, and β-ketoacyl elongation.

BIOSYNTHESIS OF WAXES

Composition of Plant Waxes

By definition, cuticular waxes are the hydrophobic compounds on the surface of the plant that are removed by a brief immersion in an organic solvent such as chloroform or hexane. They are complex mixtures of primarily very long chain (VLC, >C18) fatty acids, hydrocarbons, alcohols, aldehydes, ketones, esters, triterpenes, sterols, and flavonoids (51, 105). The proportions of the major classes vary among plant species (Table 1). For example, alkanes and ketones are major components of leek leaf, *Arabidopsis* stem, and *Brassica* leaf waxes but are very low or undetectable in barley and maize leaves. Peanut and alfalfa are rich in alkanes but have no ketones. The major components in alfalfa leaf wax are primary alcohols, compared with peanut leaf, in which the major wax components are fatty acids. In *Arabidopsis*, the proportion of ketones in the leaf wax is 30-fold lower than the proportion of ketones in the stem wax.

Each lipid class of the cuticular wax may be present as a homologous series, or one particular chain length may predominate. When the major lipid class has a predominating homologue, characteristic wax crystals form on the plant's aerial surfaces. The shape and appearance of these crystals are due to the physical-chemical properties of the wax composition. For example, lobed plates are associated with a high proportion of the C24 species in the primary alcohols, which is the major class of cuticular wax on *Quercus robur* (oak) leaves (33). Long thin tubes are characteristic of large amounts of β-diketones on *Hordeum vulgare* (barley) lemma (100), and transversally ridged rodlets are associated with high levels of hentriacontan-16-one on leaf surfaces of

Table 1 Major epicuticular lipid classes among several plant species

Classes	Leek [1]	Barley [2]	Maize [3]	*Arabidopsis*[4] stem	*Arabidopsis*[5] leaf	*Brassica*[6]	Peanut[7]	Alfalfa[8]
Fatty acids	6.4 [9]	10.3	tr [10]	3.2	3.6	1.9	38.1	N.D. [11]
Aldehydes	10.8	1.7	20.0	5.9	2.1	3.9	2.4	20.5
Alkanes	31.0	tr	1	38.0	73.6	40.3	35.7	20.0
2° alcohols	N.D.	N.D.	N.D.	10.3	0.7	11.9	N.D.	N.D.
Ketones	51.8	N.D.	N.D.	30.4	1.2	36.1	N.D.	N.D.
1° alcohols	N.D.	83.0	63.0	11.8	18.5	1.9	23.8	48.4
Wax esters	N.D.	4.7	16.0	0.8	0.2	3.9	N.D.	11.0

1. *Allium porrum,* Y Rhee & D Post-Beittenmiller, unpublished manuscript.
2. *Hordeum vulgare,* primary leaves (32a).
3. *Zea mays* seedlings (6).
4. *A. thaliana* (Landsberg *erecta,* ecotype) stem. Data are averages of two reports (37, 46).
5. *A. thaliana* (Landsberg *erecta,* ecotype) leaf (46).
6. *Brassica oleracea* leaf (91).
7. *Arachis hypogaea,* branch base leaves (111).
8. *Medicago sativa,* seven-day old leaves (6a).
9. % total
10. tr, trace (≤0.5%). 11. N.D., not detected.

members of the subclass Magnoliidae (33). There are also reports of similar crystalline structures from differing compositions (68).

Cuticular wax composition varies among and within species. The same plant may show organ-to-organ differences, tissue-to-tissue differences, and developmental differences. GC and scanning electron microscopy analyses have revealed organ-specific differences in wax composition and content on *Arabidopsis* stems (37), leaves (46), pollen (82), and siliques (53) (Figure 1); between barley spikes and leaf blades (100); and between *Zea mays* (maize) leaf and pollen (12). Tissue-specific differences are exemplified by the adaxial and abaxial leaf surfaces of *Pisum sativa* (pea) that differ by 10-fold in the

Figure 1 Scanning electron micrographs of wild-type and mutant *Arabidopsis* stems (*top*) and siliques (*bottom*). Wild-type stems and siliques (*left*) are characterized by a predominance of varied tubes and lobed plates, respectively. Stems and siliques of *cer7* mutants (*center*) have both reduced numbers of crystals and altered crystalline structures. Siliques of *cer17* mutants differ little from wild-type siliques, whereas *cer17* stems have few tube structures characteristic of wild type (*right*) (J Chen & D Post-Beittenmiller, unpublished information).

level of alkanes (99). Developmental changes in wax production are observed in the leaves of *Tilia tomtentosa* (silver lime trees). Approximately 15 days after unfolding, the leaves begin to synthesize large amounts of waxes, in particular β-amyrenyl acetate and long chain aldehydes (35). Another example of developmental changes in wax production is found in juvenile and adult maize leaves. Young leaves have predominantly primary alcohols (63%), and older leaves have predominantly wax esters (42%) (4, 11). Finally, wax esters are not restricted to the cuticle but also occur as the major storage lipids in jojoba (80) and are minor components of olive oil (13).

The ubiquitous presence of cuticular wax is testimony to its essential function to waterproof the plant, and its immense diversity is evidence of its many successful adaptations to an aerial environment while enhancing its usefulness as protection for plants against biotic and abiotic stresses. Underlying this diversity are sophisticated controls on gene expression at the organ and tissue levels and under a wide range of developmental cues.

De Novo Synthesis and Very Long Chain Fatty Acid Elongation

The precursors of wax biosynthesis are fatty acids that are likely derived from de novo synthesis in plastids. In plants, de novo fatty acid biosynthesis is catalyzed by a series of enzymatic steps, collectively referred to as fatty acid synthase (FAS) (78). The initiation of fatty acid synthesis is the condensation of malonyl-acyl carrier protein (ACP) with acetyl-CoA (42), followed by the sequential reduction of 3-ketoacyl-ACP, the dehydration of 3-hydroxyacyl-ACP, and the reduction of *trans*-Δ^2-enoyl-ACP. The fatty acyl primer remains esterified to the ACP cofactor and is further extended—two carbons at a time—by the donor, malonyl-ACP. For each two-carbon addition, there is a sequential round of condensation, reduction, dehydration, and second-reduction steps. NAD(P)H serve as reducing equivalents for the two reductases. The long chain products (C16, C18) are subsequently processed by one or more enzymes, including stearoyl-ACP desaturase, plastidial acyl-transferases, and acyl-ACP thioesterases (hydrolases). Fatty acids are then utilized for glycerolipids, waxes, or cutin and suberin biosynthesis, depending on the tissue type and developmental stage.

Although de novo fatty acid synthesis occurs ubiquitously, cuticular wax biosynthesis occurs almost exclusively in epidermal tissues (48, 59). The primary enzyme activity characterizing wax biosynthesis is fatty acid elongation. Among membrane lipids, only relatively minor amounts of fatty acids longer than C18 are found (16). In contrast, the majority of wax components are derived from very long chain fatty acids (VLCFA) that are 20–32 carbons in length; fatty acids esterified to alcohols may be 40–60 carbons in length. These VLCFAs are produced from fatty acid precursors (C16 or C18) that are elongated extraplastidially by microsomal enzymes, in a manner biochemically analogous to de novo fatty acid synthesis. The acyl chains undergo the same four basic reactions of condensation, reduction, dehydration, and a second reduction, for each two-carbon elongation, and these four activities are collectively termed elongases (98). Similar to FAS, NAD(P)H serve as the reducing equivalents for the elongase reducing activities (51). However, there are some notable differences between FASs and elongases. First, VLCFA elongation does not occur on ACP, and malonyl-CoA, rather than malonyl-ACP, serves as the two-carbon donor (2, 3). Second, elongases are extraplastidial and membrane-associated rather than stromal and soluble (17, 107). Third, elongases have an apparent but as yet undefined requirement for ATP (17, 28). Therefore, on the basis of these differences, de novo synthesis refers to the FAS-catalyzed plastidial extension of acyl-ACPs, and elongation refers here to the microsomal extensions of the fatty acyl chain by elongases. This distinction is made for clarity because of inconsistencies in the literature in which the plastidial C16:0 to C18:0 extension is referred to as an "elongation reaction."

Because of their essential role in wax production, elongases have been one of the most studied of the wax biosynthetic steps (100). In addition, the corresponding seed elongases have also been studied extensively because the seed oil storage components of the Brassicaceae and jojoba (triacylglycerols and liquid wax, respectively) contain significant levels of the agronomically important VLCFA, erucic acid (C22:1). Seed elongase components have been cloned only recently (41, 54a).

Multiple Elongation Systems

Over the past three decades, ample evidence has been obtained demonstrating multiple elongation systems involved in wax biosynthesis, which are both sequential (generating a homologous series) and parallel reactions (generating different lipid classes) (98). A single elongase catalyzing sequential reactions is exemplified by seed elongases. The condensing enzyme component of a seed elongase has been cloned from jojoba and catalyzes three elongation steps from C18:1 to C24:1 (54a). The corresponding gene, *FAE1,* has been cloned from *Arabidopsis* (41). Although the cloned sequences have not been expressed nor the enzyme activity assayed, the *fae1* mutant is defective in the two elongation steps from C18:1 to C22:1 (54, 55). In contrast, parallel elongases catalyze extensions leading to the production of different wax classes, as shown in Figure 2. In this simplified scheme, the decarbonylation (A), acyl-reduction (B), and β-ketoacyl-elongation (C) are shown as distinct and parallel pathways. All three pathways are found in the epidermal tissue of most plants, but their relative contributions to the cuticular wax composition vary from organ to organ and species to species.

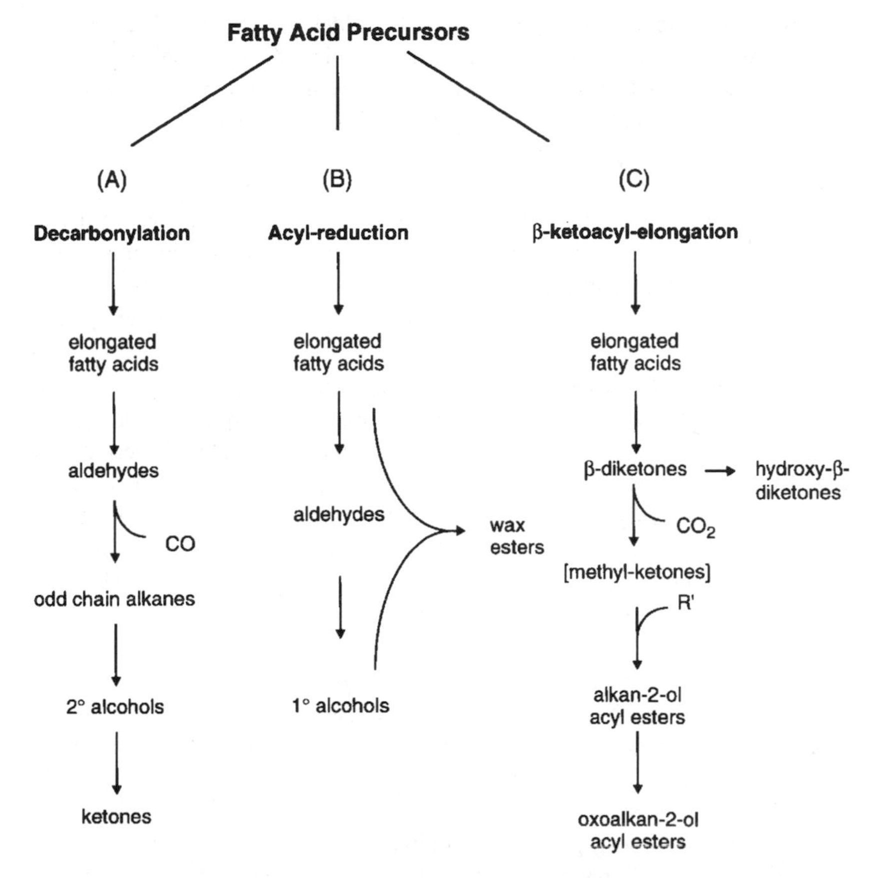

Figure 2 Three primary wax biosynthetic pathways. Pathways depict only the major steps and are not intended to provide biochemical detail. Although pathways are shown as distinct and separate, some intermediates (e.g. aldehydes) may be shared among the pathways.

The decarbonylation pathway results in the production of aldehydes, odd chain alkanes, secondary alcohols, and ketones. Early studies referred to this pathway as the decarboxylation pathway because it was thought that odd chain (n-1) compounds were generated by a decarboxylation of the corresponding even chain (n) fatty acid. It was demonstrated, however, that an aldehyde intermediate is generated and that carbon monoxide, not carbon dioxide, is released (21). Although decarbonylation occurs in plants and microorganisms (23), insects use a cytochrome P450-dependent decarboxylation to generate the odd chain hydrocarbon (Z) 9-tricosene (85).

The acyl-reduction pathway produces aldehydes, primary alcohols, and wax esters derived from the esterification of fatty acids and primary alcohols. In developing jojoba seed, the two-step reduction from fatty acid to primary alcohol is catalyzed by one enzyme, and the aldehyde intermediate does not accumulate (80). In *Brassica oleracea* leaf, the aldehyde intermediate accumulates, and the two reduction steps are catalyzed by separate enzymes (49).

The β-ketoacyl-elongation pathway results in the production of β-diketones and their derivatives. β-diketones are a major component of the cuticular wax of barley spike, uppermost leaf sheath, and internode (103) and are found as minor components of other plant species such as Brassicaceae and carnation (105). Elegant analyses of barley *cer-cqu* mutants elucidated this third pathway and clearly demonstrated that β-ketoacyl elongation was catalyzed by an enzyme system separate from the acyl-reduction and decarbonylation systems (69, 70, 102, 103).

In some plants, the decarbonylation and acyl-reduction pathways may share some or all of the fatty acid elongation reactions and may differ only in the modifying enzymes that act on the elongated fatty acids and their derivatives. The β-ketoacyl-elongation pathway is clearly a separate and parallel elongation pathway in barley (102). Much of the data that distinguish sequential and parallel pathways are derived from inhibitor studies and the characterization of wax-deficient mutants. Studies include photoperiod and chemical inhibition (97). These studies demonstrate that an individual elongation step or a single pathway may be affected without affecting the other parallel pathways. For example, dithiothreitol and 3-mecaptoethanol inhibit synthesis of hydrocarbons but do not affect β-diketone synthesis. In contrast, cyanide affects β-diketone synthesis but not hydrocarbon synthesis (70). Similarly, the sequential condensation steps of de novo fatty acid biosynthesis are differentially inhibited by some of the same chemicals. Cerulenin, a well-characterized inhibitor of 3-ketoacyl synthases, specifically inhibits the C20 to C22 elongation step in leek epidermis (2) but has no effect on the sequential elongation of C22 (28). In contrast, cerulenin does not inhibit the seed elongases of *Brassica napus* or *Arabidopsis* (A Hlousek Radojcic, H Imai & J Jaworski, personal communication). Individual mutants in *Arabidopsis* and maize specifically block C28 and C30 elongation steps (6, 37). These studies together suggest that there may be many elongase systems involving seven sequential reactions for two or three parallel pathways. In the case of the condensing enzymes for de novo fatty acid synthesis, the chain-length specificities are not rigid and show some overlapping activities in vitro. This may also be true for the elongase condensing enzyme reactions. Evidence indicates that the modifying reactions (e.g. aldehyde reductase) do not have chain length specificities (52). However, these conclusions are based on measurements of in vitro activities and should be viewed with caution. In vivo activities may be more selective than in vitro activities.

The parallel pathways are responsible for the wide diversity of cuticular wax composition. Each parallel pathway consists of sequential elongation reactions that terminate with chain modifications or with the release of free elongated fatty acids. However, the extent to which the intermediates are shared among pathways is unclear. For example, aldehydes are intermediates of both the decarbonylation and the acyl-reduction pathways, but it is not known whether one fatty acid reductase generates aldehydes for both the decarbonylase and the aldehyde reductase (such that these enzymes compete for the same substrate pool) or whether there are two fatty acid reductases and channeling of substrates limits competition. *Arabidopsis cer4* mutants are postulated to be defective in aldehyde reductase (37, 46). In the mutant plants, there is a large increase in aldehydes and a decrease in primary alcohols (acyl-reduction pathway) but no change in the products (alkanes, secondary alcohols, and ketones) of the decarbonylation pathway (37, 46). The C28 primary alcohol predominates in wild-type *CER4,* whereas the C30 alcohol predominates in the mutant *cer4* (37). This surprising result suggests that the C28 fatty acid, in the absence of a C28 reductase activity, is elongated to the C30 fatty acid and reduced to the corresponding alcohol (presumably through an aldehyde intermediate) but apparently does not enter the decarbonylation pathway, because there is no increase in the C29 alkane (46a). The lack of an increase in alkanes indicates that aldehyde intermediates may not be readily shared between the pathways, such as might occur when parallel pathways exist as organized complexes.

BRANCH POINTS WITH OTHER LIPID BIOSYNTHETIC PATHWAYS

Fatty acids produced in the plastid from de novo synthesis are utilized by at least three biosynthetic pathways that lead to the production of glycerolipids, waxes, and cutin or suberin. The partitioning of fatty acyl precursors among these pathways is likely to be a key regulatory point that controls both the quantity and quality of cuticular lipids. In most vegetative tissues, de novo synthesized C16:0 and C18:1 are the fatty acid precursors for glycerolipid biosynthesis, whereas the precursors for cuticular wax biosynthesis are primarily saturated fatty acids, probably derived from C18:0 as discussed below, although alkenes are minor components of some cuticular waxes (105). Cutin and suberin are polymers of C16:0 and C18:1 hydroxy fatty acids (14, 50). Thus, following de novo fatty acid biosynthesis, a partitioning occurs that delivers C16:0 and C18:1 fatty acids to glycerolipid or cutin/suberin biosyntheses and C18:0 to wax biosynthesis. Presently, the mechanisms that regulate this partitioning are not understood. Partitioning of precursors into these pathways may be accomplished by enzyme specificities or by substrate availabilities (compartmentalization and/or metabolic channeling). For example, storage and membrane glycerolipids exhibit strong fatty acid bias because of the substrate specificities of acyl transferases (31). Such specificity, however, may only be necessary for branch point enzymes, and enzymes later in the branch pathway may not exhibit narrow substrate specificities. An alternative is that fatty acid partitioning may arise from the supply of specific fatty acids in individual cell types. For example, saturated fatty acids may be elongated in epidermal cells because they are the most abundant in those cells. Similarly, monounsaturated fatty acids may be elongated in seeds because they are most abundant in seeds. Partitioning may therefore be due to the supply of fatty acids rather than to the specificities of the enzymes utilizing the fatty acids.

Termination of Plastidial Fatty Acid Biosynthesis

A tissue in which substantial partitioning must occur (among glycerolipid, wax, and cutin biosyntheses) is leek epidermis. In leek, epicuticular waxes comprise >15% of the total leaf lipid (exclusive of cutin) of which >90% is derived from saturated precursors (61; D Post-Beittenmiller, unpublished data) Only the epidermal cells, which constitute <4% of the leaf fresh weight, contribute to the wax layer. Unlike other cell types in plant vegetative organs (where lipid synthesis utilizes about 10% of the carbon), lipid synthesis in epidermal cells is probably the most dominant metabolic pathway to which carbon is allocated. Potentially significant steps in the regulation of wax production in the epidermis are first, the generation of sufficient fatty acid pools and second, the partitioning of fatty acid precursors among wax biosynthesis, membrane glycerolipid, and cutin or suberin biosynthesis. Therefore, it is important to understand how epidermal cells generate sufficiently large pools of saturated fatty acids (probably C18:0) to support wax biosynthesis while limiting their entry into glycerolipid and cutin or suberin biosynthesis.

When total leaf extracts are assayed for acyl-ACP hydrolyzing activity (acyl-ACP thioesterases), the major activity is directed against oleoyl-ACP (18:1-ACP) (79), which reflects the abundance of unsaturated fatty acids in total leaf glycerolipids. However, when epidermal tissue is removed and assayed separately, acyl-ACP hydrolyzing activities toward saturated acyl-ACPs are significantly higher compared with total leaf extracts (61). Both palmitoyl-ACP (16:0-ACP) and stearoyl-ACP (18:0-ACP) hydrolyzing activities are present in leek (*Allium porrum*) epidermis (61; D Liu & D Post-Beittenmiller, unpublished manuscript). The stearoyl-ACP thioesterase (STE) activity is highly, if not exclusively, expressed in epidermis and has a high specificity for 18:0-ACP (61). Although additional studies directly linking STE activity with wax production are required, the relative activities of stearoyl-ACP desaturase and STE are presumed to be responsible for generating the extraplastidial stearate pool in the epidermis.

Either palmitate (C16:0) or stearate (C18:0) could be the initial saturated substrate for wax biosynthesis. The relative levels of these two fatty acids are determined in part by the activity of 3-ketoacyl-ACP synthase II (KAS II), which catalyzes the two-carbon extension from C16:0 to C18:0 (88). It was previously reported that KAS II activity is not detectable in leek epidermis, which suggested that extraplastidial elongation begins with C16:0 (60). However, in rapidly expanding leek epidermis, KAS II activity is detected (81). Substrate specificity studies of microsomal elongases (prepared from rapidly expanding leek epidermis) also indicate that the rate of C16:0 elongation is low compared with the rate of C18:0 elongation (KJ Evenson & D Post-Beittenmiller, unpublished manuscript). Similarly, jojoba elongase utilizes C16:0 approximately 10- to 20-fold less efficiently than C18:0 (54a). The occurrence of an epidermally expressed STE also implies that there is a pool of stearoyl-ACP in leek epidermis on which this enzyme acts. These data together suggest that stearoyl-ACP is generated in the epidermis by KAS II, the stearoyl-ACP is hydrolyzed by STE, and extraplastidial elongation begins with a C18:0 primer. It should be noted that there are also palmitoyl-ACP thioesterases (24; D Liu & D Post-Beittenmiller, unpublished manuscript) that may generate extraplastidial palmitate. This palmitate, however, may be a precursor for glycerolipid and cutin biosynthesis rather than a substrate for elongation.

An alternative mechanism for partitioning fatty acids into wax, glycerolipid, or cutin or suberin biosynthesis may be based, in part, on the substrate specificity of the elongase. Agrawal and coworkers demonstrated that leek microsomal elongases utilize only saturated endogenous primers (2). However, when longer monounsaturated primers (≥C20:1) are supplied, these are elongated (3). The researchers postulated that the specificity for saturated primers lay with the first elongation step (C18:0 to C20:0) and that monounsaturated primers (C20:1) could be elongated as well as the corresponding saturated primers. In contrast, the partially purified leek epidermal elongase does not show a strict substrate specificity; it will elongate C18:1 as well as C18:0 (57). Recent studies with microsomes prepared from rapidly expanding leek show a preference for C18:0 > C16:0 > C18:1 (KJ Evenson & D Post-Beittenmiller, unpublished manuscript). In addition, seed extracts from *Arabidopsis* (54) and jojoba (54a) will elongate saturated primers (C18:0 to C24:0) at rates comparable to the corresponding monounsaturated primers, although the seed oils contain VLC monounsaturated fatty acids. One simple explanation for these apparently contrasting results in seed is that the in vivo supply of fatty acids is predominantly monounsaturated and that elongases do not have a strict substrate specificity; rather, they elongate whatever fatty acid is supplied.

What Are the Substrates for Elongase Condensing Enzymes?

As discussed above, following termination of plastidial fatty acid synthesis and export of the fatty acid out of the plastid, fatty acids are extended by membrane-associated elongases. In both plants and animals, in vitro fatty acid elongation occurs when microsomes from appropriate tissues are incubated with labeled malonyl-CoA, NAD(P)H, and ATP. This activity is independent of supplied fatty acyl primer; therefore, a pool of endogenous primer is present in microsomal preparations. In leek epidermal microsomes, this pool is substantial because elongation in the absence of primer is 80% of the activity in the presence of supplied primer (28). When supplied, both acyl-CoAs and free fatty acids are effectively elongated by microsomes, but the elongation of fatty acids has an absolute requirement for ATP. Elongation rates in the presence of acyl-CoA but in the absence of ATP are 10–20% of the rates in the presence of ATP. The ATP requirement has often been attributed to its involvement in acyl-CoA synthetase (ligase) activity that generates and/or maintains the acyl-CoA primer pool (58). In a recent review of animal elongation systems, Cinti concluded that the ligase activity under standard elongation conditions (in the absence of supplied CoASH and fatty acids) is not adequate to provide sufficient levels of acyl-CoA primer to achieve the microsomal elongation rates routinely observed (22). Similar comparisons can be made with respect to plant fatty acid elongation. First, the maximum rates of elongation (in the absence of supplied CoASH) by plant microsomes range from 12 to 20 nmol/mg/h (18, 28) compared with the maximal rates (<2 nmol/mg/h) of microsomal acyl-CoA synthesis when no CoASH was supplied (58). The apparent K_m for CoASH in developing safflower microsomes is 24 μM (40). Under standard elongation conditions (in which CoASH is not supplied), the CoASH required for acyl-CoA synthetase activity must either be endogenous or result from the addition of malonyl-CoA. Commercial preparations of [^{14}C]malonyl-CoA contain <1% CoASH. Even if 10% of the added malonyl-CoA (typically 100 μM is used) were hydrolyzed because of microsomal hydrolases or improper storage conditions, the CoASH concentration would only be 10 μM, well below the K_m for CoASH. In addition, if acyl-CoA synthetase is required for elongation of fatty acids, addition of CoASH should stimulate elongase activity. However, fatty acid elongation in microsomes from rapidly expanding leek is inhibited by CoASH (28). Thus, this kinetic analysis indicates that it is not possible for the microsomal acyl-CoA synthetase to generate sufficient acyl-CoA primer for the elongation reaction (40, 58; KJ Evenson & D Post-Beittenmiller, unpublished manuscript).

For reasons cited above, it seems unlikely that ATP is required for acyl-CoA synthesis. If this is true, then why is ATP required, and how are free fatty acids elongated by microsomal elongases? There may be another activated form of the fatty acid that serves as the primer for elongation. For example, ATP may activate free fatty acids to an adenylated intermediate, similar to the acyl-AMP intermediate generated in the synthesis of acyl-CoA:

In the absence of an activated precursor, ATP hydrolysis may enable the free fatty acid to bind directly to an active site cysteine of the condensing enzyme, or it may provide energy for the condensation reaction.

Further evidence that acyl-CoA may not be the immediate substrate for elongation comes from isotope dilution studies using microsomes from developing *Brassica napus* seeds (39a). In these studies, [^{14}C]18:1-CoA was used as the supplied primer and the percent ^{14}C of the substrates and the elongated products were analyzed by GC/MS. These studies revealed that the endogenous pool of C18:1-CoA was quite small and that the pools of C20:1-CoA and C22:1-CoA were undetectable in freshly prepared microsomes, whereas newly synthesized pools of C20:1- and C22:1-CoAs were readily detected after 10 min of incubation with ATP and malonyl-CoA. The percent ^{14}C in the elongated C20:1-CoA product was found to be 2–3-fold less than the percent ^{14}C of the starting C18:1-CoA substrate, which indicates that the [^{14}C]18:1 moiety from the [^{14}C]18:1-CoA was diluted by an endogenous pool other than C18:1-CoA. Preliminary results with microsomes prepared from leek epidermis were similar (A Hlousek-Radojcic, KJ Evenson, D Post-Beittenmiller & JG Jaworski, unpublished manuscript). Clearly, in plants as well as animals, the ATP requirement and the nature of the primer are not adequately understood.

Partitioning Between Wax and Cutin or Suberin

Cutin is a major component of a plant's aerial surfaces, whereas suberin is a polymer associated with roots and wound sites. Cutin or suberin form the insoluble polymeric matrices that are impregnated with wax. They are biochemically and biosynthetically related to wax because they are biopolymers of oxygenated fatty acids (14, 50, 51). They are functionally related to wax in that they are part of the cuticle and therefore serve as the initial barrier to the environment. Cutin and suberin monomers are derived from C16:0 and C18:1 fatty acids synthesized in the plastid (51). They are generated by a family of mixed-function oxidases with activities similar to those of the decarbonylation and β-diketoacyl-elongation pathways for wax biosynthesis (14, 100), although the cutin/suberin oxidases probably do not utilize VLC precursors.

Cutin is made exclusively in the epidermis and therefore is another branch point for partitioning fatty acid precursors. The contribution of saturated fatty acids (as palmitate) is significant for C16:0-derived monomer biosynthesis; therefore, partitioning between wax and cutin biosynthesis cannot be based solely on saturation or unsaturation but must also be based on chain length. Similarly, the monounsaturated fatty acid (C18:1) must be partitioned between glycerolipids and cutin. Glycerolipids do contain some saturated C16:0 (16%) and C18:0 (1%) fatty acids (15a), and limiting rather than excluding saturated fatty acids is an important part of partitioning. In Figure 3, a simple schematic illustrating the complexity of partitioning C16 and C18 fatty acids among the three major lipid synthetic pathways in epidermis in shown.

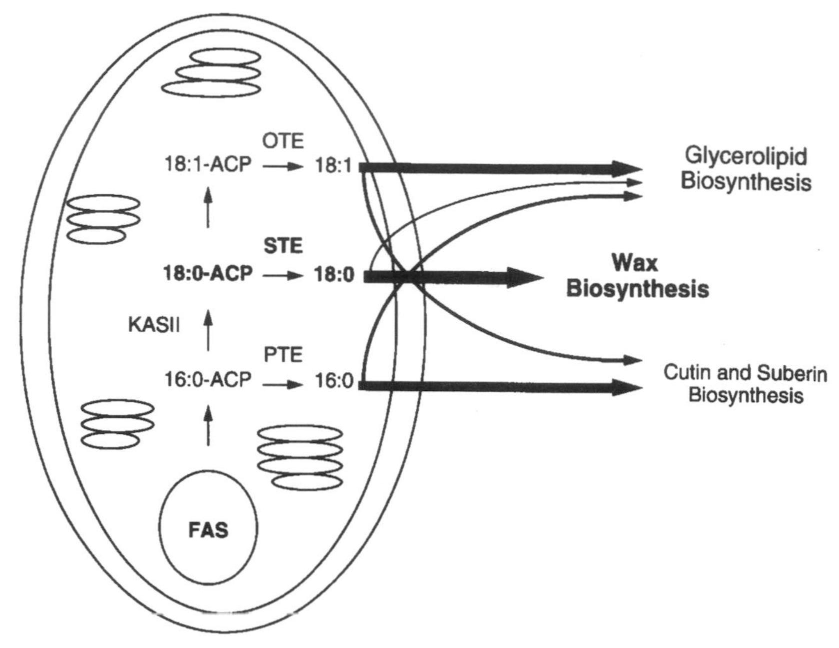

Figure 3 Schematic showing export of fatty acids from the plastid and partitioning between the three lipid biosynthetic pathways in epidermis. FAS, fatty acid synthase; KAS II, 3-ketoacyl-ACP synthase II; OTE, oleoyl-ACP thioesterase; PTE, palmitoyl-ACP thioesterase; STE, stearoyl-ACP thioesterase.

MOVEMENT TO THE OUTER SURFACES

One of the still mysterious aspects of cuticular wax production is the movement of hydrophobic wax components to the plant surface. Early SEM studies led to speculation about two modes of movement: passage through well-defined pores and general diffusion through microscopic spaces in the cell wall (43). The former hypothesis is based on the uneven distribution of wax epicuticular structures and the presence of putative pores on the surfaces of some plants (100). The latter hypothesis is based on the ubiquitous and amorphous coating of wax throughout the cuticle and over the plant surface, and it suggests that cuticular wax is exported at continuous or very frequent intervals. The epicuticular crystalline structures most likely result from the physical-chemical properties of the wax, i.e. the concentrations and types of the specific components, rather than as a result of a transport process. Correlations between wax composition and wax structure, as discussed above, support this hypothesis. Recrystallization studies of wax on spruce needles (that were mechanically injured) show that wax tubes re-form to their original crystalline structure even on nonviable needles, i.e. in the absence of new synthesis or export (7). SEM studies of *Sorghum* sheath tissues demonstrate that wax filaments are secreted at sites around papillae (raised surfaces) on the cork cells. These filaments are associated with intracellular vesicles whose density increases as wax production is induced by light (45). There is no evidence for channel or pore involvement. Although the composition of the vesicles is unknown, they may contain proteins involved specifically in the transport of wax components to the cell surface.

Lipid transfer proteins (LTP) are small, abundant basic proteins that were first identified in animals as possible intracellular carriers of lipids (between membranes or between organelles). They have also been found in microorganisms and in plants (110). The lipid transfer function of plant LTPs has been reported for spinach leaves (86), maize seedlings (25), barley seeds (15, 66), and castor beans (106). They are expressed in the epidermal tissues of a variety of organs including barley aleurone cells (47), leaves and coleoptiles (32), apical meristems (30), and flower buds, but not in roots (83). Immunogold localization studies indicate LTPs are cell wall proteins (94). They have also been isolated from the surface waxes of plants (84). The discovery that LTPs move vectorally into the lumen of the endoplasmic reticulum (8, 67) and are secreted to the cell walls and outer surfaces led to speculation that they may be involved in transporting cuticular lipid components (waxes and cutin or suberin monomers) (89, 90). Although it has been postulated that LTPs transport lipids through the ER and deposit them on the outer surfaces, the abundance of lipid molecules compared with LTPs is a strong stoichiometric argument against such intracellular transfer. Furthermore, the LTPs do not return to the ER lumen for additional lipid transfer, which limits the number of lipid molecules that could be transferred. It is more likely that the movement of LTPs into and through the ER lumen is part of their own secretion process and that association with lipids in the lumen is due to the hydrophobic nature of LTPs. After LTPs are secreted to the outside of the cell wall and cuticle, an environment in which LTPs are likely to move freely, they may be part of a facilitated diffusion mechanism that moves cuticular lipids from the outer layer of the plasma membrane, through the cell wall and cuticle, to the plant surface.

GLOSSY MUTANTS AND THE CLONING OF WAX-RELATED GENES

Mutants with deficient or altered wax coatings have been identified by their nonglaucous or glossy appearance. There are no reports of mutants that are completely deficient in cuticular wax, probably because of the vital function that waxes play in waterproofing the plant cuticle. In many plants, the mutant phenotypes affecting the major wax classes are easily observed by visual

Table 2 Mutants defective or altered in wax production[a]

Species	Locus	Biochemical and genetic phenotype[b]	Cloned	References
Barley	*cer-cqu*[c]	a. loss of β-ketoacyl elongation pathway b. defective in 3 enzymatic functions c. multifunctional protein	no	102, 103
	cer-yy[c]	a. dominant (all 17 alleles) b. regulatory c. affects two pathways d. changes spike wax into leaf wax	no	62
	cer-n[c]	a. both dominant and recessive alleles b. affects two pathways c. regulatory	no	63
Maize	*gl1*[c]	a. blocks early steps or interferes with precursor supply b. epistatic relationships: *gl8* > *gl1* > *gl7*	transposon tagged, gene product unknown	6, 39, 65, 96
	gl2, gl4, gl16	a. structural genes b. blocks C30 elongation	*gl2* and *gl4* tagged	6, 96
	gl3	a. structural gene b. blocks C28 elongation	transposon tagged	6, 65
	gl5	a. increased aldehydes, block in aldehyde reductase	transposon tagged	65
	gl6		transposon tagged	65
	gl7	a. regulatory gene b. blocks early steps c. epistatic relationships: *gl8* > *gl1* > *gl7*	no	6
	gl8[c]	a. blocks early steps or interferes with precursor supply b. epistatic relationships: *gl8* > *gl1* > *gl7*	β-ketoacyl reductase[d]	6, 39, 65, 96
	gl9[c]	a. altered leaf morphology	no	87
	gl11	a. structural gene for EDIa component b. loss of aldehydes with no loss of 1° alcohols c. possible fatty acid reductase	no	6
	gl12		transposon tagged	6, 65
	gl15[c]	a. mutant precociously develops adult leaves b. glossy phenotype is secondary	homologue of *APETALA2*[d]	27, 73, 96
Sorghum	*bm22*	a. cuticle mutant with reduced wax load	no	44
Arabidopsis	cer1[c,d]	a. structural gene for decarbonylase	transposon and T-DNA tagged	1, 37, 46, 53, 56
	cer2[c]	a. regulatory b. blocks C28 elongation and accumulates products of reductive pathway c. altered stem wax but normal leaf wax	T-DNA tagged and cloned by walking, gene product unknown	37, 39, 46, 53, 75
	cer3	a. release of fatty acid from elongase complex	T-DNA tagged	37, 38, 46, 53
	cer4	a. block in aldehyde reductase	T-DNA tagged	36, 37, 46, 53, 56

Table 2 *(continued)*[a]

Species	Locus	Biochemical and genetic phenotype[b]	Cloned	References
	cer6[c]	a. blocks C28 elongation b. male sterile	T-DNA tagged	46, 53, 56, 82
	cer7	a. altered stem wax but normal leaf wax	walking	29, 46, 53
	cer8, cer9, cer19	a. block in substrate transfer		46, 53
	cer10		T-DNA tagged	53, 56
	cer21		T-DNA tagged	56

[a]Not intended to be a comprehensive list of wax defective mutants but rather illustrative. Putative gene product based on phenotype only.

[b]All mutants exhibit a glossy (nonglaucous) phenotype unless indicated otherwise.

[c]See text for more information.

[d]Gene identity based only on sequence similarities, not on functional assays.

examination. Therefore, this phenotype is used extensively as a marker for genetic studies and in breeding programs. Changes in a minor wax component are less likely to have a discernible effect and therefore have not been found. Similarly, plants that overproduce wax are not readily detected by a visual screen. Table 2 is a partial listing of wax mutants in barley, maize, *Sorghum,* and *Arabidopsis,* their phenotypes, and reported progress toward the isolation of corresponding genes. Wax-deficient mutants have also been found in other species, such as pea and broccoli, but are not listed here.

Barley

Some of the most elegant characterizations of *glossy* or *eceriferum* (*cer*) mutants have been conducted in barley by von Wettstein-Knowles and coworkers (97, 98, 100). More than 1560 *cer* mutants have been characterized representing 85 complementation groups (100). Most of these complementation groups probably represent loci that are involved in the production of cuticular wax (biosynthesis, transport, and regulation), although developmental genes (as discussed below for maize *gl15*) cannot be ruled out at this time. More than 30 enzymatic steps are required for the biosynthesis of the major wax components in most plants. This estimate is undoubtedly low because it does not account for differences in organ-specific or developmental expression, differences in substrate chain length specificities, transport to the plant surface, or biosynthesis of minor components. Differences in substrate specificities would presumably result from isozymes (i.e. different loci), whereas differences in expression may be due to a single gene that is differentially expressed under a variety of developmental stages; or differences in expression may be due to several genes, each expressed only during specific developmental stages or in specific organs. These levels of regulatory complexity would further increase the number of loci that could potentially result in mutant phenotypes. Furthermore, the majority of mutant phenotypes do not suggest specific blocks in the pathways, possibly because the pathways share intermediates or because they represent mutations in regulatory genes.

Genetic studies have contributed greatly to our understanding of wax biosynthesis and, in particular, to the biosynthesis of β-diketones and their derivatives. The most illustrative studies are of the *cer-c, -q,* and *-u* mutants. These three complementation groups are tightly linked, and double and even triple mutants were isolated relatively frequently (13 times out of 1252 *cer* mutants), whereas multiple mutants of other *cer* loci were never seen (102). Reversion studies of the double mutants show that single mutational events revert both phenotypes simultaneously. This indicates that the three loci are part of a multifunctional gene, *cer-cqu,* and that each complementation group represents a separate functional domain (103). Mutants defective in one or more of these loci are blocked in the β-ketoacyl-elongation pathway, but the acyl-reduction and decarbonylation pathways are relatively unaffected. Defects in *cer-q* block the synthesis of all β-ketoacyl-derived lipids; therefore, *cer-q* acts early in the pathway or affects the first branch point. The *cer-c* mutant accumulates alkan-2-ols, which indicates a block in β-diketone-derived lipids. The *cer-u* mutant has a decreased level of hydroxy-β-diketones and accumulates β-diketones, which suggests a block in a hydroxylase.

Some mutants (*cer-yy* and *cer-n*) are candidates for regulatory genes. All 17 *cer-yy* alleles are dominant and change the primary alcohol composition of barley spike wax into a composition indistinguishable from that of wild-type barley leaf blade wax (62). To account for this phenotype, the authors hypothesized that the mutant *cer-yy* allele activates the acyl-reduction pathway in leaf blades and suppresses the acyl-reduction and β-ketoacyl-reduction pathways in barley spikes. The dominant nature of this mutant and the fact that it affects three pathways are supportive evidence that it is regulatory. The *cer-n* gene is also potentially regulatory (63). This locus affects all lipid classes and has both dominant and recessive alleles, a property more frequently found among regulatory genes.

Maize and Sorghum

Because maize and *Sorghum* seedlings have a glaucous appearance and adult leaves are glossy, visual screens for wax mutants are done at the seedling stage. This method of screening led to the isolation of some *glossy* mutants not involved in wax biosynthesis. Rather, they are pleiotropic; the wax defect is secondary. Probably the best example of this is the *gl15* mutant in maize. This mutant was originally characterized as defective in the elongation-decarboxylation II pathway (ED-II) resulting in the production of wax esters that are characteristic of adult leaves (10). Subsequent characterization and cloning of

Gl15 demonstrated that this gene is involved in epidermal differentiation and that the mutant defect is a premature transition from juvenile (glaucous leaves) to adult (glossy leaves) (27, 73). Sequencing of the *Gl15* gene revealed that it is a homologue of *APETALA2,* a homeotic flower gene in *Arabidopsis* (73). Similar to *gl15,* the glossy phenotype of *gl9* may be secondary to a developmental lesion because it has an abnormal leaf morphology (87).

Many of the maize *glossy* genes have been tagged (Table 2) using a variety of transposons. One of the most striking observations to come from the characterization of tagged alleles is that wax production is clearly cell autonomous. In many cases where a transposon has excised, individual cells and sectors are covered with wax crystals, whereas adjacent cells are devoid of wax structures (20, 64, 87).

Maddaloni and coworkers reported that *gl1* either blocks an early step in the ED-I pathway or interferes with the supply of precursors to ED-I (64). This mutant has a dramatic reduction of alkanes, aldehydes, and alcohols, but the level of esters is similar to wild type (note that the percentage of esters is higher, whereas the actual level is comparable). The recently cloned gene, *Gl8,* has sequence similarity to β-ketoacyl reductases, which suggests *Gl8* may encode part of an analogue (39, 109). The *gl8* mutant is epistatic to *gl1* because the aldehyde content of the *gl1,gl8* double mutant is more similar to *gl8* than to *gl1* (5). However, the *gl1,gl8* double mutant has an increased level of shorter chain aldehydes, greater than that of either parent, and suggests a synergistic effect (to this reviewer). Consequently, the relationship between *gl1* and *gl8* is unclear.

Arabidopsis

In the past decade, *Arabidopsis* has become the model plant system for genetics and molecular biology. Therefore, it is not surprising that it should also become a model system for genetics and molecular biology of epicuticular wax production. More than 21 complementation groups resulting in a glossy phenotype have been identified in *Arabidopsis*. In many cases, multiple alleles have been isolated, but for some, only one allele has been found (37, 46, 53). Because of the limited number of alleles and for reasons cited above, it is unlikely that the wax biosynthetic pathway has been saturated with mutations in *Arabidopsis*. Compositional analyses of wild-type and *cer* mutants have been conducted on stems (37, 46) and leaves (46), and possible gene products or blocked steps have been proposed (Table 2). There are also significant differences between *Arabidopsis* ecotypes (46).

Hannoufa and coworkers postulated that *cer2* blocks C28 elongation because of the almost complete loss of the C29 alkane, the single major component of stem waxes, and an increase in the shorter chain homologues. Jenks and coworkers see a similar decrease in the C29 alkane, but they postulate that *cer2* may have regulatory functions, based on differences in the wax composition between leaf and stem. Such organ differences, however, may only indicate that *CER2* is differentially expressed because the leaf wax of *cer2* is virtually identical with that of wild-type leaf. *CER2* has been cloned and encodes a 47-kDa novel protein. The transcript level is high, as might be expected for a structural gene (75). If *CER2* encodes a component of the elongase complex, it is surprising that it shows no apparent homology to any of the fatty acid synthase proteins. It will be very interesting to see how this story unfolds with the further characterization of the cloned *CER2*.

Several of the *cer* mutants also show reduced fertility (53). Recently, a new mutant allele of *CER6* was isolated by a screen for male sterile plants (82). The pollen of the *cer6* mutant is viable as assessed by in vitro germination, but it is desiccation-intolerant under normal growing conditions because of reduced wax levels on pollen surfaces. Similar to other *cer6* isolates, fertility is rescued by increasing humidity around the flowers. Interactions between the mutant pollen and the wild-type stigma are aberrant, however, because callose was formed and pollen germination was apparently random. Because of these incompatibility responses, the authors suggest that pollen wax plays a role in pollen-stigma interaction.

Tagging and Walking

Tremendous advances in map-based cloning techniques make it possible to clone many genes that were previously known only by phenotype. In virtually none of the wax mutants characterized in any plant species has a gene function been unequivocally identified on the basis of the mutant phenotype. No loss of a specific enzyme activity has been correlated with any *cer* or *glossy* mutant. To date, the identity of only one wax gene (*gl8* in maize, cloned by transposon tagging) has been revealed by sequence similarity to 3-ketoacyl reductase, but its role in wax biosynthesis has yet to be demonstrated. The *CER1* gene from *Arabidopsis* has also been cloned by transposon tagging (1). The protein encoded by *CER1* has His residues that are thought to be metal-binding sites. The decarbonylase that leads to the production of odd chain alkanes in the alga, *Botryococcus braunii*, is a cobalt-requiring enzyme (23). This information, combined with wax compositional data showing a reduction in alkanes and an increase in aldehydes (37), indicates that *CER1* encodes a decarbonylase (1). Three additional *Arabidopsis* genes (*CER2*, *3*, and *4*) have been reportedly cloned (36, 38, 75), but their gene products have not been identified at the time of this review. There are at least seven tagged *CER* loci in *Arabidopsis*, one of which, *CER2,* has also been isolated by chromosome walking (Table 2). T-DNA tagging has been used for most, but there are increasing reports of *CER* loci tagged by transposons (1, 39, 64, 65, 96). Clearly, the tagging and cloning of genes will open up new areas of research.

Expressed Sequence Tags

With the availability of more than 20,000 expressed sequence tags (EST) from *Arabidopsis* (76), it is now easier to identify clones on the basis of their sequence similarities to known genes. The identification of wax biosynthetic genes is no exception. Given that most of the enzymes are membrane associated, this approach has advantages over protein purification strategies. Because of the sequence similarities between the cloned jojoba condensing enzyme and previously cloned KAS IIIs, the identity of the jojoba clone was determined (39a). Similarly, the jojoba and *FAE1* (41) sequences have been used to identify six *Arabidopsis* ESTs that are putative condensing enzyme components of elongases (95). The *FAE1,* jojoba, and *Arabidopsis* cDNAs encode 56–60-kDa proteins, have membrane-spanning domains, and have two completely conserved Cys residues; one is potentially in the active site. Preliminary data indicate that some of the *Arabidopsis* ESTs are expressed in all organs except seed and root (J Todd, D Post-Beittenmiller & J Jaworski, unpublished manuscript), a pattern that is consistent with the expected expression of an elongase involved in wax biosynthesis. In contrast, *FAE1* is expressed only in seeds, which is expected for a condensing enzyme involved in erucic acid biosynthesis (41).

ENVIRONMENTAL FACTORS

In addition to developmental controls, cuticular lipids are synthesized in response to environmental signals such as light intensity, photoperiod (101), humidity (91), chilling (77), and seasonal variation (34). One dramatic response of wax production to environmental stimuli is observed during tissue culture, where the relative humidity is high and wax production is low. When tissue-culture-grown plants are moved to a less humid environment (i.e. greenhouse or growth chamber), wax production is stimulated, and within days the plant synthesizes a complete and protective coating of wax. *Brassica oleracea* greenhouse-grown plants have 10-fold more wax than tissue-culture-grown plants, and the decarbonylation pathway is stimulated preferentially (91). During the hardening-off period, more wax is produced on new leaves than on leaves that emerged during culturing, which suggests that rapidly expanding tissues have a greater capacity for wax production than nonexpanding tissues (92).

CONCLUSIONS AND FUTURE PROSPECTS

The ubiquitous presence of cuticular wax is testimony to its essential role in a plant's adaptation to an aerial environment. The fact that environmental factors influence wax composition and quantity is evidence that wax production is an actively regulated process. The diversity of wax components is evidence of the wealth of genes devoted to wax production. The differential regulation of cuticular wax production by tissue, organ, and developmental stage provides sophisticated controls on the expression of a complex biosynthetic pathway. The biosynthesis of plant cuticular wax has been studied for over three decades, with pioneering work done in several laboratories using elegant biochemical and genetic approaches. In spite of these efforts, we still know little about the factors that regulate the partitioning of fatty acid precursors and coordinately regulate the synthesis of waxes with the synthesis of cutin and glycerolipids. The cloning of wax biosynthetic genes has just begun, and it promises to open many new paths and bring exciting discoveries.

Literature Cited

1. Aarts MG, Keijzer CJ, Stiekema WJ, Pereira A. 1995. Molecular characterization of the *CER1* gene of *Arabidopsis* involved in epicuticular wax biosynthesis and pollen fertility. *Plant Cell* 7:2115–27
2. Agrawal VP, Lessire R, Stumpf PK. 1984. Biosynthesis of very long chain fatty acids in microsomes from epidermal cells of *Allium porrum* L. *Arch. Biochem. Biophys.* 230:580–89
3. Agrawal VP, Stumpf PK. 1985. Characterization and solubilization of an acyl chain elongation system in microsomes of leek epidermal cells. *Arch. Biochem. Biophys.* 240:154–65
4. Avato P, Bianchi G, Pogna N. 1990. Chemosystematics of surface lipids from maize and some related species. *Phytochemistry* 29:1571–76
5. Avato P, Bianchi G, Salamini F. 1984. Genetic control of epicuticular lipids in maize (*Zea mays* L.). In *Structure, Function and Metabolism of Plant Lipids,* ed. PA Siegenthaler, W Eichenberg, pp. 179–98. Amsterdam: Elsevier/North Holland Biomed. Press
6. Avato P, Bianchi G, Salamini F. 1985. Absence of long chain aldehydes in the wax of the *glossy11* mutant of maize. *Phytochemistry* 24:1995–97

6a. Bergman DK, Dillwith JW, Zarrabi AA, Caddel JL, Berberet RC. 1991. Epicuticular lipids of alfalfa relative to its susceptibility to spotted alfalfa aphids (*Homoptera aphididae*). *Environ. Entomol.* 20:781-85

7. Bermadinger-Stabentheiner E. 1995. Physical injury, re-crystallization of wax tubes and artefacts: identifying some causes of structural alteration to spruce needle wax. *New Phytol.* 130:67–74
8. Bernhard WR, Thoma S, Botella J, Somerville CR. 1991. Isolation of a cDNA clone for spinach lipid transfer protein and evidence that the protein is synthesized by the secretory pathway. *Plant Physiol.* 95: 164–70
9. Deleted in proof
10. Bianchi G, Avato P, Salamini F. 1979. Glossy mutants of maize. IX. Chemistry of *glossy4, glossy8, glossy15,* and *glossy18* surface waxes. *Heredity* 42:391–95
11. Bianchi G, Avato P, Scarpa O, Murelli C, Audisio G, Rossini A. 1989. Composition and structure of maize epicuticular wax esters. *Phytochemistry* 28:165–71
12. Bianchi G, Murelli C, Ottaviano E. 1990. Maize pollen lipids. *Phytochemistry* 29: 739–44
13. Bianchi G, Tava A, Vlahov G, Pozzi N. 1994. Chemical structure of long-chain esters from "Sansa" olive oil. *J. Am. Oil Chem. Soc.* 71:365–69
14. Blée E. 1995. Oxygenated fatty acids and plant defenses. *Inform* 6:852–61
15. Breu V, Guerbette F, Kader J-C, Kannangara CG, Svensson B, von Wettstein-Knowles PM. 1989. A 10-kD barley basic protein transfers phosphatidylcholine from liposomes to mitochondria. *Carlsberg Res. Commun.* 54:81–84

15a. Browse J, McCourt P, Somerville C. 1986. A mutant of *Arabidopsis* deficient in C18:3 and C16:3 leaf lipids. *Plant Physiol.* 81: 859-64

16. Cahoon E, Lynch D. 1991. Analysis of glucocerebrosides of rye (*Secale cerale* L. cv *Puma*) leaf and plasma membrane. *Plant Physiol.* 95:58–68
17. Cassagne C, Lessire R. 1978. Biosynthesis of saturated very long chain fatty acids by purified membrane fractions from leek epidermal cells. *Arch. Biochem. Biophys.* 191: 146–52
18. Cassagne C, Lessire R, Bessoule J-J, Moreau P. 1987. Plant elongases. In *The Metabolism, Structure, and Function of Plant Lipids,* ed. PK Stumpf, JB Mudd, WD Ness, pp. 481–88. New York: Plenum
19. Cassagne C, Lessire R, Bessoule J-J, Moreau P, Créach A, et al. 1994. Biosynthesis of very long chain fatty acids in higher plants. *Prog. Lipid Res.* 33:55–69
20. Cerioli S, Marocco A, Maddaloni M, Motto M, Salamini F. 1994. Early event in maize leaf epidermis formation as revealed by cell lineage studies. *Development* 120:2113–20
21. Cheesbrough TM, Kolattukudy PE. 1984. Alkane biosynthesis by decarbonylation of aldehydes catalyzed by a particulate preparation from *Pisum sativum. Proc. Natl. Acad. Sci. USA* 81:6613–17
22. Cinti DL, Cook L, Nagi MN, Suneja SK. 1992. The fatty acid chain elongation system of mammalian endoplasmic reticulum. *Prog. Lipid Res.* 31:1–51
23. Dennis MW, Kolattukudy PE. 1991. Alkane biosynthesis by decarbonylation of aldehyde catalyzed by a microsomal preparation from *Botryococus braunii. Arch. Biochem. Biophys.* 287:268–75
24. Dörmann P, Voelker T, Ohlrogge J. 1995. Cloning and expression in *E. coli* of a novel thioesterase from *Arabidopsis thaliana* specific for long chain acyl-acyl carrier proteins. *Arch. Biochem. Biophys. Res. Commun.* 316:612–18
25. Douady D, Grosbois M, Guerbette F, Kader J-C. 1986. Phospholipid transfer protein from maize seedlings is partly membrane-bound. *Plant Science* 45:151–56
26. Eigenbrode SD, Espelie KE. 1995. Effects of plant epicuticular lipids on insect herbivores. *Annu. Rev. Entomol.* 40:171–94
27. Evans M, Passas HJ, Poethig R. 1994. Heterochronic effects of *glossy15* mutations on the epidermal identity in maize. *Development* 120:1971–81
28. Evenson KJ, Post-Beittenmiller D. 1995. Fatty acid elongating activity in rapidly-expanding leek epidermis. *Plant Physiol.* 109: 707–16
29. Fahleson J, Kraft M, Dunn P, Post-Beittenmiller D. 1995. *Towards map-based cloning of* CER7, *a gene involved in epicuticular wax production in* Arabidopsis thaliana. Presented at Biochem. Mol. Biol. Plant Fatty Acids Glycerolipids Symp., South Lake Tahoe, Calif.
30. Fleming AJ, Mandel T, Hofmann S, Sterk P, de Vries SC, Kuhlemeier C. 1992. Expression pattern of a tobacco lipid transfer protein gene within the shoot apex. *Plant J.* 2:855–62
31. Frentzen M. 1993. Acyl transferases and triacylglycerols. See Ref. 72a, pp. 195–230
32. Gausing K. 1994. Lipid transfer protein genes specifically expressed in barley leaves and coleoptiles. *Planta* 192:574–80

32a. Giese BN. 1976. Roles of the *cer-j* and *cer-p* loci in determining the epicuticular wax composition on barley seedling leaves. *Hereditas* 82:137-48

33. Gülz P-G. 1994. Epicuticular leaf waxes in the evolution of the plant kingdom. *J. Plant Physiol.* 143:453–64
34. Gülz P-G, Müller E. 1992. Seasonal variation in the composition of epicuticular waxes of *Quercus robur* leaves. *Z. Naturforsch.* 47:800–6
35. Gülz P-G, Prasad RBN, Müller E. 1991. Surface structure and chemical composition of epicuticular waxes during leaf development of *Tilia tomentosa* Moench. *Z. Naturforsch.* 46:743–49
36. Halfter U, Rashotte A, Jenks M, Eigenbrode S, Feldmann KA. 1995. *Characterization and cloning of the* Arabidopsis cer4 (eceriferum) *mutant.* Presented at Int. Conf. *Arabidopsis* Res., 6th, Madison, Wis.
37. Hannoufa A, Mcnevin J, Lemieux B. 1993. Epicuticular waxes of eceriferum mutants of *Arabidopsis thaliana. Phytochemistry* 33:851–55
38. Hannoufa A, Negruk V, Lemieux B. 1995. *Molecular cloning of the* CER3 *gene of* Arabidopsis thaliana. Presented at Int. Conf. *Arabidopsis* Res., 6th, Madison, Wis.
39. Hansen JD, Xu XJ, Xia YJ, Dietrich CR, Delledonne M, et al. 1995. *Molecular cloning and characterization of plant cuticular wax genes.* Presented at Maize Genet. Conf., 37th, Asilomar, Calif.

39a. Hlousek-Radojcic A, Imai H, Jaworski JG. 1995. Oleoyl-CoA is not an immediate substrate for fatty acid elongation in developing seeds of *Brassica napus. Plant J.* 8: 803–9

40. Ichihara K, Nakagawa M, Tanaka K. 1993. Acyl-CoA synthetase in maturing safflower seeds. *Plant Cell Physiol.* 34: 557–66
41. James DW Jr, Lim E, Keller J, Plooy I, Ralston E, Dooner HK. 1995. Directed tagging of the *Arabidopsis* FATTY ACID ELONGATION 1 (*FAE1*) gene with the maize transposon *Activator. Plant Cell* 7: 309–19
42. Jaworski JG, Post-Beittenmiller D, Ohlrogge JB. 1993. Acetyl-acyl carrier protein is not a major intermediate in fatty acid biosynthesis in spinach. *Eur. J. Biochem.* 213:981–87

43. Jeffree C, Baker E, Holloway P. 1976. Origins of the fine structure of plant epicuticular waxes. In *Microbiology of Aerial Plant Surfaces,* ed. C Dickinson, T Preece, pp. 118–58. London: Academic
44. Jenks MA, Joly RJ, Peters PJ, Rich PJ, Axtell JD, Ashworth EN. 1994. Chemically induced cuticle mutation affecting epidermal conductance to water vapor and disease susceptibility in *Sorghum bicolor* (L.) Moench. *Plant Physiol.* 105:1239–45
45. Jenks MA, Rich PJ, Ashworth EN. 1994. Involvement of cork cells in the secretion of epicuticular wax filaments on *Sorghum bicolor* (L.) Moench. *Int. J. Plant Sci.* 155: 506–18
46. Jenks MA, Tuttle HA, Eigenbrode SD, Feldmann KA. 1995. Leaf epicuticular waxes of the *eceriferum* mutants in *Arabidopsis. Plant Physiol.* 108:369–77
46a. Jenks MA, Tuttle HA, Feldmann KA. 1996. Changes in epicuticular waxes on wildtype and *eceriferium* mutants in *Arabidopsis* during development. *Phytochemistry.* In press
47. Kalla R, Shimamoto K, Potter R, Nielsen PS, Linnestad C, Olsen O-A. 1994. The promoter of the barley aleurone-specific gene encoding a putative 7-kDa lipid transfer protein confers aleurone cell-specific expression in transgenic rice. *Plant J.* 6: 849–60
48. Kolattukudy PE. 1968. Further evidence for an elongation-decarboxylation mechanism in the biosynthesis of paraffins in leaves. *Plant Physiol.* 43:375–83
49. Kolattukudy PE. 1971. Enzymatic synthesis of fatty alcohols in *Brassica oleracea. Arch. Biochem. Biophys.* 142:701–9
50. Kolattukudy PE. 1980. Biopolyester membranes of plants: cutin and suberin. *Science* 208:990–1000
51. Kolattukudy PE. 1980. Cutin, suberin, and waxes. In *The Biochemistry of Plants,* ed. PK Stumpf, EE Conn, 4:571–645. New York: Academic
52. Kolattukudy PE. 1987. Lipid-derived defensive polymers and waxes and their role in plant-microbe interaction. In *The Biochemistry of Plants,* ed. PK Stumpf, EE Conn, 9:291–314. New York: Academic
53. Koornneef M, Hanhart CJ, Thiel F. 1989. A genetic and phenotypic description of *eceriferum* (*cer*) mutants in *Arabidopsis thaliana. J. Heredity* 80:118–22
54. Kunst L, Taylor DC, Underhill EW. 1992. Fatty acid elongation in developing seeds of *Arabidopsis thaliana. Plant Physiol. Biochem.* 30:425–34
54a. Lassner MW, Lardizabal K, Metz JG. 1996. A jojoba β-ketoacyl-CoA synthase cDNA complements the canola fatty acid elongation mutation in transgenic plants. *Plant Cell.* In press
55. Lemieux B, Miquel M, Somerville CR, Browse J. 1990. Mutants of *Arabidopsis* with alterations in seed lipid fatty acid composition. *Theor. Appl. Genet.* 80:234–40
56. Lemieux B, Yang P, Hannoufa A, Negruk V, Subramanian M, Deng M-d. 1995. *Molecular genetics of wax biosynthesis in* Arabidopsis. Presented at Int. Conf. *Arabidopsis* Res., 6th, Madison, Wis.
57. Lessire R, Bessoule J-J, Cassagne C. 1989. Involvement of a β-ketoacyl-CoA intermediate in acyl-CoA elongation by an acyl-CoA elongase purified from leek epidermal cells. *Biochim. Biophys. Acta* 1006:35–40
58. Lessire R, Cassagne C. 1979. Long chain fatty acid CoA-activation by microsomes from *Allium porrum* epidermal cells. *Plant Sci. Lett.* 16:31–39
59. Lessire R, Hartmann-Bouillon MA, Cassagne C. 1982. Very long chain fatty acids: occurrence and biosynthesis in membrane fractions from etiolated maize coleoptiles. *Phytochemistry* 21:55–59
60. Lessire R, Stumpf PK. 1982. Nature of the fatty acid synthetase systems in parenchymal and epidermal cells of *Allium porrum* L. leaves. *Plant Physiol.* 73:614–18
61. Liu D, Post-Beittenmiller D. 1995. Discovery of an epidermal stearoyl-acyl carrier protein thioesterase: its potential role in wax biosynthesis. *J. Biol. Chem.* 270: 16962–69
62. Lundqvist U, von Wettstein-Knowles PM. 1982. Dominant mutations at *cer-yy* change barley spike wax into leaf blade wax. *Carlsberg Res. Commun.* 47:29–43
63. Lundqvist U, von Wettstein-Knowles PM. 1983. Phenotypic diversity of barley spike waxes resulting from mutations at locus *cer-n. Carlsberg Res. Commun.* 48:321–44
64. Maddaloni M, Bossinger G, DiFonzo N, Motto M, Salamini F, Bianchi A. 1990. Unstable alleles of the *GLOSSY-1* locus of maize show a light-dependent variation in the pattern of somatic reversion. *Maydica* 35:409–20
65. Maddaloni M, Albano M, Motto M, Salamini F. 1991. Unstable alleles generated at various glossy loci. *Maize Genet. Coop Newsl.* 65:25–26
66. Madrid S, Weber E, von Wettstein-Knowles PM. 1990. *Characterization and targeting of a barley phosphatidylcholine transfer protein (LTP).* Presented at Plant Lipid Biochem. Struct. Util., Wye College, Kent, Engl.
67. Madrid SM. 1991. The barley lipid transfer protein is targeted into the lumen of the endoplasmic reticulum. *Plant Physiol. Biochem.* 29:695–703
68. Meusel I, Leistner E, Barthlott W. 1994. Chemistry and micromorphology of compound epicuticular wax crystalloids (*Strelitzia* type). *Plant Syst. Evol.* 193:115–23

69. Mikkelsen JD. 1979. Structure and biosynthesis of β-diketones in barley spike epicuticular wax. *Carlsberg Res. Commun.* 44: 133–47
70. Mikkelsen JD, von Wettstein-Knowles PM. 1978. Biosynthesis of β-diketones and hydrocarbons in barley spike epicuticular wax. *Arch. Biochem. Biophys.* 1288: 172–81
71. Deleted in proof
72. Deleted in proof
72a. Moore TS, ed. 1993. *Lipid Metabolism in Plants.* Boca Raton, FL: CRC Press
73. Moose SP, Sisco PH. 1995. The maize homeotic gene *GLOSSY15* is a member of the *APETALA2* gene family. *J. Cell. Biochem.* 21A:458
74. Deleted in proof
75. Negruk V, Yang P, Subramanian M, McNevin JP, Lemieux B. 1995. *A novel protein regulates very long chain fatty acid elongation in* Arabidopsis thaliana. Presented at Int. Conf. *Arabidopsis* Res., 6th, Madison, Wis.
76. Newman T, de Bruijn FJ, Green P, Keegstra K, Kende H, et al. 1994. Genes galore: a summary of methods for accessing results from large-scale partial sequencing of anonymous *Arabidopsis* cDNA clones. *Plant Physiol.* 106:1241–55
77. Nordby HE, McDonald RE. 1991. Relationship of epicuticular wax composition of grapefruit to chilling injury. *J. Agric. Food Chem.* 39:957–62
78. Ohlrogge JB, Jaworski JG, Post-Beittenmiller D. 1993. *De novo* fatty acid biosynthesis. See Ref. 72a, pp. 3–32
79. Ohlrogge JB, Shine WE, Stumpf PK. 1978. Fat metabolism in higher plants: characterization of plant acyl-ACP and acyl-CoA hydrolases. *Arch. Biochem. Biophys.* 189: 382–91
80. Pollard M, McKeon T, Gupta L, Stumpf P. 1979. Studies on biosynthesis of waxes by developing jojoba seed. II. The demonstration of wax biosynthesis by cell-free homogenates. *Lipids* 14:651–62
81. Post-Beittenmiller D, Evenson K, Liu D, Jaworski J. 1994. *Examination of the terminal steps of fatty acid biosynthesis in leek epidermis and their relevance to wax biosynthesis.* Presented at Int. Meet. Plant Lipids, 11th, Paris
82. Preuss D, Lemieux B, Yen G, Davis RW. 1993. A conditional sterile mutation eliminates surface components from *Arabidopsis* pollen and disrupts cell signaling during fertilization. *Genes Dev.* 7:974–85
83. Pyee J, Kolattukudy PE. 1995. The gene for the major cuticular wax-associated protein and three homologous genes from broccoli (*Brassica oleracea*) and their expression patterns. *Plant J.* 7:49–59
84. Pyee J, Yu H, Kolattukudy PE. 1994. Identification of a lipid transfer protein as the major protein in the surface wax of broccoli (*Brassica oleracea*) leaves. *Arch. Biochem. Biophys.* 311:460–68
85. Reed JR, Vanderwel D, Choi S, Pomonis JG, Reitz RC, Blomquist GJ. 1994. Unusual mechanism of hydrocarbon formation in the housefly: cytochrome P450 converts aldehyde to the sex pheromone component (Z)-9-tricosene and CO_2. *Proc. Natl. Acad. Sci. USA* 91:10000–4
86. Rickers J, Spener F, Kader J-C. 1985. A phospholipid transfer protein that binds long-chain fatty acids. *FEBS Lett.* 180: 29–32
87. Schnable PS, Stinard PS, Wen T-J, Heinen S, Weber D, et al. 1994. The genetics of cuticular wax biosynthesis. *Maydica* 39: 279–87
88. Shimakata T, Stumpf PK. 1982. Isolation and function of spinach leaf β-ketoacyl-[acyl-carrier-protein] synthases. *Proc. Natl. Acad. Sci. USA* 79:5808–12
89. Sossountzov L, Ruiz-Avila L, Vignols F, Jolliot A, Arondel V, et al. 1991. Spatial and temporal expression of a maize lipid transfer protein gene. *Plant Cell* 3:923–33
90. Sterk P, Booij H, Schellekens GA, Van Kammen A, De Vries SC. 1991. Cell-specific expression of the carrot EP2 lipid transfer protein gene. *Plant Cell* 3:907–21
91. Sutter E. 1984. Chemical composition of epicuticular wax in cabbage plants grown in vitro. *Can. J. Bot.* 62:74–77
92. Sutter E, Langhans RW. 1982. Formation of epicuticular wax and its effect on water loss in cabbage plants regenerated from shoot-tip culture. *Can. J. Bot.* 60:2896–902
93. Deleted in proof
94. Thoma S, Kaneko Y, Somerville CR. 1993. A non-specific lipid transfer protein from *Arabidopsis* is a cell wall protein. *Plant J.* 3:427–36
95. Todd J, Post-Beittenmiller D, Jaworski J. 1995. Characterization and manipulation of fatty acid elongase genes in *Arabidopsis. Presented at Biochem. Mol. Biol. Plant Fatty Acids Glycerolipids Symp.*, South Lake Tahoe, Calif.
96. Vogel JM, Freeling M. 1993. *Molecular genetic analysis of maize glossy genes: role in epicuticular wax production.* Presented at Keystone Symp. Mol. Cell. Biol., Keystone, Colo.
97. von Wettstein-Knowles PM. 1979. Genetics and biosynthesis of plant epicuticular waxes. *Advances in Biochem and Physiol of Plant Lipids,* ed. L-Åppelqvist, C Liljenberg, pp. 1–26. Amsterdam: Elsevier/North Holland Biomed. Press
98. von Wettstein-Knowles PM. 1982. Elongases and epicuticular wax biosynthesis. *Physiol. Vég.* 20:797–809
99. von Wettstein-Knowles PM. 1987. Genes,

elongases and associated enzyme systems in epicuticular wax synthesis. In *The Metabolism, Structure, and Function of Plant Lipids,* ed. PK Stumpf, JB Mudd, WD Ness, pp. 489–98. New York: Plenum

100. von Wettstein-Knowles PM. 1995. Biosynthesis and genetics of waxes. In *Waxes: Chemistry, Molecular Biology and Functions,* ed. RJ Hamilton, 6:91–130. Allowry, Ayr, Scotland: Oily Press

100a. von Wettstein-Knowles PM. 1993. Waxes, cutin, and suberin. See Ref. 72a, pp. 127–66

101. von Wettstein-Knowles PM, Avato P, Mikkelsen JD. 1980. Light promotes synthesis of the very long fatty acyl chains in maize wax. In *Biogenesis and Function of Plant Lipids,* ed. P Mazliak, P Benveniste, C Costes, R Douce, pp. 271–74. Amsterdam: Elsevier/North-Holland Biomed. Press

102. von Wettstein-Knowles PM, Søgaard B. 1980. The *cer-cqu* region in barley: gene cluster or multifunctional gene. *Carlsberg Res. Commun.* 45:125–41

103. von Wettstein-Knowles PM, Søgaard B. 1981. Genetic evidence that *cer-cqu* is a a cluster-gene. *Barley Genetics IV: Proc. 4th Int. Barley Genet. Symp.,* pp. 625–30. Edinburgh: Edinburgh Univ. Press

104. Deleted in proof

105. Walton TJ. 1990. Waxes, cutin and suberin. In *Methods in Plant Biochemistry: Lipids, Membranes and Aspects of Photobiology,* ed. JL Harwood, JR Bowyer, 4:105–58. San Diego: Academic

106. Watanabe S, Yamada M. 1986. Purification and characterization of a non-specific lipid transfer protein from germinated castor bean endosperms which transfers phospholipids and galactolipids. *Biochim. Biophys. Acta* 876:116–23

107. Whitfield HV, Murphy DJ, Hills MJ. 1993. Sub-cellular localization of fatty-acid elongase in developing seeds of *Lunaria annua* and *Brassica napus. Phytochemistry* 32: 255–58

108. Xia Y, Xu X, Hansen JD, Dietrich CR, Delledonne M, et al. 1995. *Molecular cloning and characterization of cuticular wax genes.* Presented at Int. Conf. *Arabidopsis* Res., 6th, Madison, Wis.

109. Xu XJ, Xia YJ, Heinen S, Hansen J, Stinard PS, et al. 1994. *Molecular cloning of cuticular wax genes from maize and* Arabidopsis. Presented at Int. Congr. Plant Mol. Biol., 4th, Amsterdam

110. Yamada M. 1992. Lipid transfer proteins in plants and microorganisms. *Plant Cell Physiol.* 33:1–6

111. Yang G, Espelie KE, Todd JW, Culbreath AK, Pittman RN, Demski JW. 1993. Cuticular lipids from wild and cultivated peanuts and the relative resistance of these peanut species to fall armyworm and thrips. *J. Agr. Food Chem.* 41:814-18

Annu. Rev. Plant Physiol. Plant Mol. Biol. 1996. 47:431–44

ROLE AND REGULATION OF SUCROSE-PHOSPHATE SYNTHASE IN HIGHER PLANTS

Steven C. Huber[1] *and Joan L. Huber*[2]

[1]United States Department of Agriculture/Agricultural Research Service, and Departments of Crop Science and Botany, North Carolina State University, Raleigh, North Carolina 27695-7631

[2]Department of Horticultural Science, North Carolina State University, Raleigh, North Carolina 27695-7631

KEY WORDS: sucrose synthesis, spinach (*Spinacia oleracea* L.), maize (*Zea mays* L.), regulatory protein phosphorylation, protein kinase

ABSTRACT

Sucrose-phosphate synthase (SPS; E.C. 2.4.1.14) is the plant enzyme thought to play a major role in sucrose biosynthesis. In photosynthetic and nonphotosynthetic tissues, SPS is regulated by metabolites and by reversible protein phosphorylation. In leaves, phosphorylation modulates SPS activity in response to light/dark signals and end-product accumulation. SPS is phosphorylated on multiple seryl residues in vivo, and the major regulatory phosphorylation site involved is Ser158 in spinach leaves and Ser162 in maize leaves. Regulation of the enzymatic activity of SPS appears to involve calcium, metabolites, and novel "coarse" control of the protein phosphatase that activates SPS. Activation of SPS also occurs during osmotic stress of leaf tissue in darkness, which may function to facilitate sucrose formation for osmoregulation. Manipulation of SPS expression in vivo confirms the role of this enzyme in the control of sucrose biosynthesis.

CONTENTS

1040-2519/96/0601-0431$08.00

INTRODUCTION

Sucrose plays a pivotal role in plant growth and development because of its function in translocation and storage, and because of the increasing evidence that sucrose (or some metabolite derived from it) may play a nonnutritive role as a regulator of cellular metabolism, possibly by acting at the level of gene expression (19). It is the nonreducing nature of the sucrose molecule that explains its wide distribution and utilization among higher plants. Trehalose, the only other nonreducing disaccharide found in nature, plays a comparable role in insects and fungi.

Sucrose synthesis can be catalyzed by two distinct enzymes in higher plants: sucrose-phosphate synthase (SPS; EC 2.4.1.14):

$$\text{UDP-glucose} + \text{Fru-6-P} \leftrightarrow \text{sucrose-6}'\text{-P} + \text{UDP} + \text{H}^+,$$

and sucrose synthase (SuSy; EC 2.4.1.13):

$$\text{UDP-glucose} + \text{fructose} \leftrightarrow \text{sucrose} + \text{UDP} + \text{H}^+.$$

Although this review focuses on SPS, some comparisons are made with SuSy. Both enzymes are soluble in the cytoplasm and catalyze freely reversible reactions. However, rapid removal of sucrose-6′-P by sucrose phosphatase (SPP; EC 3.1.3.24) keeps the cytosolic [sucrose-P] low and thereby renders the SPS reaction essentially irreversible. In fact, recent evidence suggests that SPS and SPP may actually form a complex in vivo (see section on Possible Occurrence of Enzyme Complexes). Thus, sucrose synthesis is generally considered to be catalyzed by SPS (in conjunction with SPP), whereas sucrose breakdown is largely catalyzed by SuSy. Although a given tissue will tend to have an excess of one activity over the other (depending upon whether it is engaged in net sucrose synthesis or breakdown), many tissues have both enzymes, and it is clear that significant "sugar cycling" can occur (discussed below).

A considerable amount has been learned about SPS since its discovery, but still much remains to be learned. For a general discussion of SPS and its role in sucrose biosynthesis in leaves, see Stitt et al (36) and Huber et al (16). Since about 1990, several new aspects have been elucidated. With respect to localization, it is now clear that SPS is not confined to photosynthetic tissues but also occurs in nonphotosynthetic tissues that are active in sucrose biosynthesis, e.g. ripening fruits. With respect to mechanisms for control, it is now clear that SPS is controlled (*a*) at the level of enzyme protein (e.g. leaf development), (*b*) by allosteric effectors (Glc-6-P and Pi), and (*c*) by reversible seryl phosphorylation. In addition, the gene encoding SPS has been cloned from several species. As a result, the deduced sequence is available, and cDNA probes can be used to monitor changes in the steady-state pool of SPS mRNA. However, despite the progress made to date, we are just beginning to understand the hierarchy of molecular mechanisms that together control SPS enzymatic activity in vivo. This review focuses on recent developments with SPS.

BIOCHEMICAL AND MOLECULAR PROPERTIES

Physical and Regulatory Properties

SPS is a low-abundance protein (<0.1% of leaf soluble protein) and is also relatively unstable. Consequently, progress on the purification and characterization of the enzyme has been slow. In addition, there are some apparent differences among species in some of the properties of the enzyme; not all of our information about SPS has been derived from studies of the same enzyme source. Nonetheless, some general statements can be made. It is now generally accepted that substrate saturation profiles for UDP-Glc and Fru-6-P are hyperbolic rather than sigmoidal and that the enzyme from some species can be allosterically activated by Glc-6-P and inhibited by Pi [see Stitt et al (36) for a review]. These effectors have a large effect on the affinities for both substrates, Fru-6-P and UDP-Glc (29, 33). Alteration of the affinity for substrates and effectors is also involved in the light modulation of SPS that occurs by reversible protein phosphorylation in some species (38). In general, SPS from nonphotosynthetic tissues (e.g. potato tubers) is regulated by metabolites and by protein phosphorylation in an analogous manner to the enzyme from photosynthetic tissues (29).

The native SPS molecule is likely a dimer of 120–138-kDa subunits (15). Anomalous behavior of SPS on gel-filtration chromatography probably accounts for the larger estimates of molecular mass in some studies. The specific activity of the native spinach enzyme is about 150 IU/mg protein. In addition, the reaction catalyzed by SPS is clearly reversible. In the most carefully conducted study to date, Lunn & ap Rees (21) showed that the apparent equilibrium constant (K_{app}) of the pea seed enzyme ranged from 5 to 65 depending upon [Mg^{2+}] and pH. Under assumed in vivo conditions, the K_{app}

for SPS has been estimated to be about 10. The calculated mass-action ratio for the SPS reaction in vivo indicates that the reaction is far from equilibrium, presumably because of rapid removal of sucrose-P by SPP.

Molecular Properties

Cloning of the SPS gene was accomplished first for the enzyme from maize (42) and then for the spinach (20, 35), potato (34), sugarbeet (9), and rice (JJ Valdez-Alarco'n, B Jimenez-Moraila & LR Herrera-Estrella, submitted) enzymes. In general, the N-terminal portions of the ~120-kDa subunit of SPS are highly conserved, and there are also two regions of strong similarity between SPS and SuSy (33). One of the regions that is highly conserved between SPS and SuSy corresponds to residues V176 to S214 of spinach SPS; this region of 39 amino acids exhibits an overall similarity of 64% between SPS and SuSy (Table 1). Salvucci et al (33) noted that within this highly conserved region there is a stretch of 11 amino acids (D197 to E206 of spinach SPS) that resembles the Gly-rich motif of phosphate-binding domains and thus might be involved in binding of Fru-6-P (to SPS) or UDP-Glc. This stretch contains 10 residues that are identical or similar; the only exception is spinach residue 203, which is a conserved Lys residue in SPS but a conserved Val residue at the analogous position in SuSy. It has been suggested that the function of the basic residue in SPS may influence the selectivity for the negatively charged Fru-6-P rather than the neutral Fru molecule.

The portion of the primary sequence in the vicinity of the uridine moiety of the substrate molecule UDP-Glc has been determined by photoaffinity labeling of a recombinant spinach SPS fragment using [β-^{32}P]5-N_3UDP-Glc (32). It was determined that the 5 position of the uridine ring was proximal to the primary sequence Q227 to E239. Note that this sequence is reasonably close to the residues thought to be involved in binding of the other substrate, Fru-6-P (D197 to E206). The uridine-binding region of the SPS molecule is highly conserved among spinach, maize, and potato, but there is relatively little homology with SuSy (33). This is perhaps not surprising because there are no recognized consensus binding motifs for UDP-Glc or other nucleotide diphosphate sugars.

The second region of strong similarity between SPS and SuSy is located toward the C-terminus of the SuSy sequence (residues D587 to P631 of spinach SPS). Of the 44 residues within this region of the SPS molecule, 25 residues are identical and 8 are similar, for an overall similarity of 75%. However, the function of this portion of the molecule is not known.

Another important domain of the SPS molecule that remains to be identified is the effector site involved in Glc-6-P and Pi binding. None of the specific amino acid residues at the effector site has yet been identified. However, the allosteric site contains essential and accessible sulfhydryl group(s),

Table 1 Partial amino acid sequence alignment of SPS from spinach (20, 35), maize (42), potato (34), sugarbeet (9), and rice (JJ Valdez-Alarco'n, B Jimenez-Moraila & LR Herrera-Estrella, submitted) with a homologous region of potato SuSy (30).[a]

Species	Residues	Sequence
		: * : : * * * : * * : – * * – * * * * * * * : : : * :
Spinach	176–212	V V L I S L H G L I R G E N M E L G R D S D T G G Q V K Y V V E L A R A L
Maize	178–214	I . . . . V . . . V . . . . M
Potato	168–204	I . . . . L . . . I L
Sugarbeet	163–199	L . . . . L . . . I L
Rice	179–215	I . . . . L . . . V S . D L
St SUSY	284–318	V . I L . P . . Y F A Q . . V – . . Y P – . . . . . . V . I L D Q V P A L

[a]Within the sequences, identical residues are shown as dots, and gaps are shown as dashes. Residues identical in all six sequences are indicated by an asterisk over the alignment; conserved residues are indicated by a colon. It has been suggested (33) that this highly conserved region may be in the vicinity of the site that binds Fru-6-P (in SPS) or Fru (in SuSy).

whereas the catalytic site does not (2). Comparison of the SPS sequences available to date indicates that there are 10 conserved cysteine residues. Presumably, one or more of these is at the effector site. It is not known whether all the sulfhydryl groups in the molecule are reduced; however, it is known that there are no intersubunit disulfide bonds (J Huber, unpublished data, 1995).

Variation Among Species

There appear to be significant quantitative differences among species in the regulatory properties of SPS in vitro, i.e. the extent of Glc-6-P activation and Pi inhibition (1, 13). There are also differences in the modulation of SPS in vivo. Some species exhibit a marked light activation of SPS (designated as class I and class II species) (13), whereas others do not (class III species). The distinctions among the three classes of plants are quantitative rather than qualitative in nature. For example, our original studies of soybean (a class III species) involved cultivars of maturity group VII, e.g. "Ransom." Although soybean Ransom plants exhibited little, if any, light activation of SPS in vivo, we have recently observed that soybean cultivars of maturity group OOO, e.g. "Maple Presto," show significant light activation (S Huber, unpublished data, 1994). Similarly, we have observed differences among *Nicotiana* species and among cultivars of *Nicotiana tabacum* (S Huber, unpublished data). However, with *N. tabacum,* even when light activation occurs, it is subtle relative to species of classes I and II. Nonetheless, these results suggest that the requisite interconversion enzymes [SPS-kinase (SPSk) and SPS-protein phosphatase (SPS-PP)] may be present at some level in all species.

Further support for this notion has recently been obtained from studies of transgenic tobacco plants expressing the maize SPS gene (18). In control plants, there was a small but significant (~30% increase in the light compared with dark) activation of SPS assayed under selective conditions (limiting substrates plus allosteric inhibitor, Pi). In transgenic tobacco plants expressing the maize SPS gene, V_{max} activity of SPS was increased about 2.5-fold as a result of expression of the transgene, and the maize enzyme was light activated in a very pronounced manner (~150% increase in the light). These results are noteworthy because maize SPS expressed in transgenic tomato plants shows relatively little light modulation (4, 18). The basis for the lack of modulation in this case is not clear but may involve slight differences in quaternary structure, which results in the phosphorylation site becoming less accessible to the endogenous protein kinases.

Possible Occurrence of Enzyme Complexes

There is increasing evidence from other systems that soluble enzymes often occur as complexes with other related enzymes. In the case of SPS, there is evidence for an association with SPSk (12), which may facilitate the phosphorylation of this low-abundance protein. Recent evidence also shows that SPS and SPP may form a complex in vitro. The primary observation is that SPS activity in vitro is reversibly reduced by removal of SPP during purification (GL Salerno, E Echeverria, HG Pontis, submitted); efficient removal of inhibitory sucrose-6′-P via an enzyme complex is speculated to be necessary for maximal SPS activity. It will be interesting to determine whether phosphorylation of SPS (at either regulatory or nonregulatory sites) affects the interaction with SPP.

CONTROL BY REVERSIBLE PROTEIN PHOSPHORYLATION

Light/Dark Modulation of SPS Activity

REGULATORY PHOSPHORYLATION SITE The major (if not sole) regulatory phosphorylation site of spinach SPS has been identified (23) as Ser158 (Table 2). Phosphorylation of Ser158 is both necessary and sufficient for the inactivation of SPS in vitro. Additional lines of evidence consistent with the assignment include: (*a*) labeling of the tryptic phosphopeptide containing Ser158 [previously designated phosphoprotein 7 (11)] in situ correlates with inactivation of SPS; (*b*) a synthetic peptide based on the phosphorylation site sequence is a good substrate for SPSk in vitro and competes with native SPS for phosphorylation/inactivation (22); (*c*) labeling of Ser158 in situ occurs more rapidly than other (nonregulatory) phosphorylation sites on SPS and reflects different turnover rates (17); and (*d*) polyclonal antibodies generated against the phosphorylation site sequence preferentially recognize and immunoprecipitate highly activated dephospho-SPS as opposed to inactivated phospho-SPS (40).

Although the regulatory phosphorylation sequence of spinach SPS is not conserved exactly, all sequences available to date contain a homologous seryl

Table 2 Amino acid sequences surrounding the putative regulatory phosphorylation site of SPS.[a]

Species	Residues	Sequence												
		:		:	:		*		*					:
Spinach	150–162	K	G	R	M	R	R	I	S	S	V	E	M	M
Potato	142–154	R	G	R	L	P	R	I	S	S	V	E	T	M
Sugarbeet	137–149	R	P	R	L	P	R	I	N	S	L	D	A	M
Maize	154–166	K	K	K	F	Q	R	N	F	S	D	V	T	L
Rice	154–166	K	K	K	F	Q	R	N	F	S	E	L	T	V
		−8		−6			−3			0				4

[a]Residues identical in all five sequences are indicated by an asterisk over the alignment; conserved residues are indicated by a colon. Residues are numbered relative to the phosphorylated Ser at position 0. Spinach (20, 35), maize (42), potato (34), sugarbeet (9), and rice (JJ Valdez-Alarco'n, B Jimenez-Moraila & LR Herrera-Estrella, submitted).

residue (Table 2). Evidence consistent with phosphorylation of Ser162 in maize SPS has been obtained in studies of maize leaves as well as transgenic tobacco expressing the maize SPS gene (18). It remains to be determined whether the homologous Ser residue in the other species is phosphorylated and is of regulatory significance, but it seems quite likely. It is important to note that several of the residues surrounding the (putative) phosphorylation site are also conserved among the five species. In particular, there are basic residues at P–3, P–6, and P–8 (numbering relative to the Ser at position 0) and hydrophobic residues at P–5 and P+4 (Table 2). At least several of these conserved residues appear to be important for recognition by protein kinase.

SPS-KINASE Partially purified spinach leaf SPS contains a copurifying protein kinase that can phosphorylate and inactivate SPS with [γ-^{32}P]ATP (12). In vitro, approximately 75–85% of the ^{32}P is incorporated into Ser158, the major regulatory phosphorylation site. Using a synthetic peptide based on the phosphorylation site sequence, two protein kinases with apparent molecular masses of 45 and 150 kDa were resolved chromatographically from spinach leaves (22). The smaller kinase (designated peak I) is most likely a monomer, whereas the larger kinase (peak III) has a subunit molecular mass of ~65 kDa. An important distinction between the two kinases is that the peak I enzyme is strictly Ca^{2+} dependent, whereas the peak III enzyme, which tends to copurify with SPS, is Ca^{2+} independent. The substrate specificity of both kinases has been characterized in vitro using synthetic peptide analogs. The major recognition elements consist of basic residues at P–6 and P–3 (24) and a hydrophobic residue at P–5 (S Huber & D Toroser, unpublished data, 1995). These residues are also conserved among species (Table 2).

Studies with maize leaf SPSk have identified a single form of the enzyme, and there is a clear requirement for peptide substrates with basic residues at P–3/P–6 and a hydrophobic residue at P–5 (R McMichael & S Huber, unpublished data, 1994). Maize leaf SPSk is also strictly Ca^{2+} dependent (18). These observations raise the intriguing possibility that cytosolic [Ca^{2+}] may regulate sucrose biosynthesis, at least in some species. There is evidence that cytosolic [Ca^{2+}] is reduced in the light relative to the dark (26). These changes in cytosolic [Ca^{2+}] could contribute to the light activation of SPS in vivo (Figure 1). Another factor that may be important in vivo is Glc-6-P, which is not only an allosteric activator of SPS but also an inhibitor of SPSk per se (24).

SPS-PROTEIN PHOSPHATASE Phospho-SPS is dephosphorylated/activated by a type 2A protein phosphatase (SPS-PP) that is inhibited by Pi (13). In spinach, there is a distinct light activation of SPS-PP that involves an increase in total extractable activity as well as a decrease in sensitivity to Pi inhibition (41). The light activation of SPS-PP can be blocked by pretreatment of leaves with

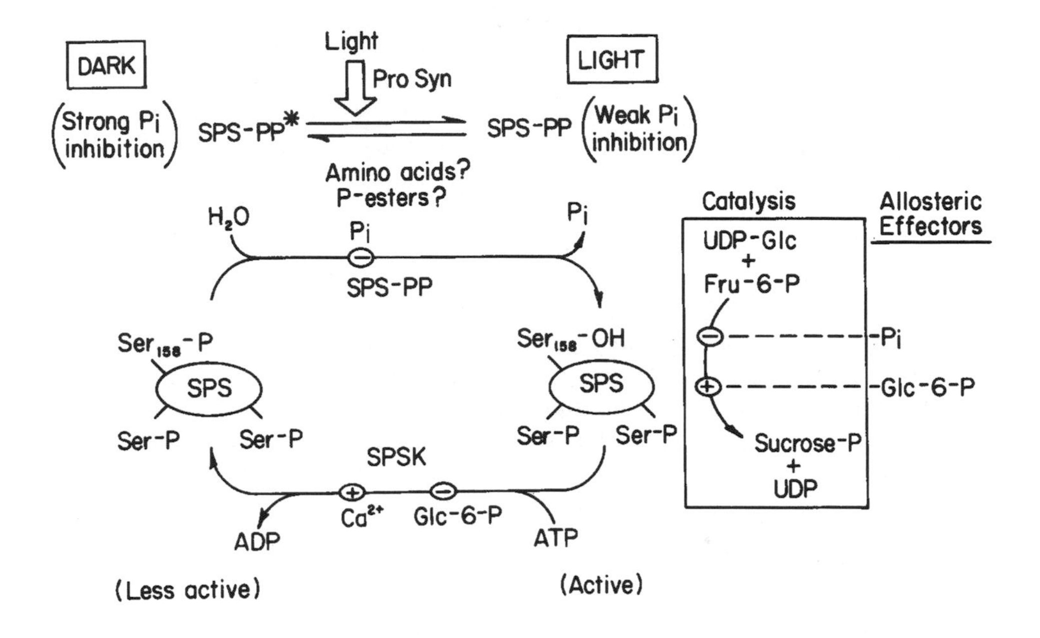

Figure 1 Schematic representation of the regulation of spinach leaf SPS by reversible seryl phosphorylation. Multisite phosphorylation and the identification of the major regulatory site as Ser158 are indicated. An increase in Glc-6-P and a decrease in Pi, as might occur during a dark-to-light transition, would favor dephosphorylation/activation of SPS and would also increase catalytic activity as a result of allosteric regulation. Another important factor may be light modulation of the regulatory properties of SPS-PP, and changes in cytosolic [Ca^{2+}]. Adapted from Reference 13.

cycloheximide (CHX), which suggests a role for cytoplasmic protein synthesis. However, the molecular basis for the light activation remains unclear; it could result from either a covalent modification of existing protein or the synthesis of a target/regulatory subunit or modifying enzyme. Regardless of the mechanism, the light modulation of SPS-PP and its regulation by Pi are thought to play an important role in the activation of SPS after a dark-to-light transition (Figure 1). Other potential effectors of SPS-PP include a variety of P-esters (41) and amino acids (14). The inhibition by amino acids may play an important role in feedback regulation of sucrose synthesis.

Osmotic Stress Activation

Activation of SPS, assayed under selective conditions, occurs in spinach leaves (28) and potato tubers (29) incubated in hyperosmotic solutions of mannitol or sorbitol. The simplest explanation, dephosphorylation of the regulatory site in response to the stress, seems not to be the case. Rather, it appears that a unique site(s) on spinach SPS is phosphorylated during osmotic stress that can partially antagonize the inhibitory effect of phosphorylation of the regulatory site (Ser158 in spinach). Control of this process may occur at the level of gene expression (14). Indeed, stress-induced protein kinases have been demonstrated (10). Whether a stress-induced kinase and the identity of the putative novel phosphorylation site are involved remains to be established.

Nonregulatory Phosphorylation Sites

Spinach leaf SPS appears to be phosphorylated on multiple seryl residues in vivo. Apart from the regulatory site (Ser158 in spinach), the phosphorylation status of the other sites remains relatively constant during light/dark transitions, i.e. the sites may be constitutively phosphorylated. We have tentatively identified two of the nonregulatory sites (17) and are in the process of identifying endogenous protein kinases that might phosphorylate these residues.

SPS IN TRANSGENIC PLANTS

Increased activity of SPS in leaves has been achieved in several species by overexpression of the gene encoding SPS in transgenic plants. Transgenic tomato plants expressing the maize SPS gene had elevated leaf SPS, and the maize enzyme was unregulated with respect to normal light/dark modulation (4, 5, 42). The enhanced SPS activity was associated with an increased light- and CO_2-saturated rate of photosynthesis and, under ambient conditions, with increased ratios of sucrose to starch in leaves (4) and increased partitioning of fixed-C into sucrose (25). Overall growth of the transgenic tomato plants was not increased when the transgene expression was leaf-specific, i.e. expressed from the Rubisco small subunit promoter. However, recent results suggest that growth enhancement may occur when SPS expression is constitutive (i.e. 35S-CaMV promoter) and occurs in both photosynthetic and nonphotosynthetic tissues (3). Micallef et al (25) also found that vegetative growth of transgenic tomato plants (expressing the maize SPS gene driven by the Rubisco SSU promoter) was not increased but noted that reproductive development was enhanced. Total fruit number was increased, the fruit matured earlier, and there was a substantial increase in total fruit dry weight (25). More work needs to be done on the growth response itself and to determine the basis for the enhancement when it is observed, but it is speculated that SPS, by affecting tissue [sucrose] might influence flowering at least in some species. Another intriguing observation by Micallef et al (25) is that SPS transformants did not exhibit the normal acclimation response of leaf photosynthesis to high CO_2. Thus, when grown and measured at high CO_2, the SPS transformants had higher rates of photosynthesis per unit leaf area compared with the control plants.

Spinach SPS has also been expressed in transgenic tobacco and potato plants. However, despite an increase in SPS protein, the additional enzyme was downregulated, apparently by phosphorylation, such that metabolism was not affected (37). In order to effectively upregulate SPS activity and sucrose biosynthesis, it may be necessary to produce plants expressing a genetically modified SPS protein, e.g. with the regulatory phosphorylation site removed.

A substantial reduction in SPS activity in potato leaves and tubers has been achieved with an SPS-antisense construct under control of the 35S-CaMV promoter (7). As a result of the antisense inhibition, sucrose synthesis in leaves was reduced, and starch and amino acid synthesis was increased. A flux control coefficient for SPS was estimated to be 0.30–0.45 in potato leaves. In tubers, the resynthesis of sucrose from starch was reduced in the SPS-antisense transformants. Overall, the results strongly support the notion that SPS is one of the important control points in sucrose biosynthesis.

ROLE OF SPS IN VIVO

Sucrose Synthesis and Sugar Cycling

In addition to the well-recognized role of SPS in sucrose biosynthesis in source leaves, it is becoming clear that some sucrose synthesis occurs even in heterotrophic cells that are engaged in net sucrose degradation. Significant turnover of the endogenous sucrose pool has been identified in a variety of tissues, including potato tubers (7) and germinating *Ricinus* cotyledons (8). Turnover of sucrose is thought to involve a futile cycle of simultaneous synthesis (by SPS and SuSy) and cleavage (by SuSy). Thus, relatively small changes in unidirectional fluxes can occur and produce much larger changes in net flux through the sucrose pool, without large changes in metabolites (7). Changes in the activation state of SPS, presumably as a result of protein phosphorylation, have been shown to contribute to changes in net flux through the sucrose pool in *Ricinus* cotyledons (8) and potato tubers (7).

Factors Affecting SPS Expression

Expression of SPS mRNA and enzyme protein is controlled developmentally, e.g. during leaf development (20), and in mature leaves by a variety of factors, including irradiance (20) and N-nutrition (J Huber, unpublished data, 1994). The responses of SPS to changes in irradiance illustrate the integration of mechanisms for control of SPS activity. Transfer of spinach plants grown at low irradiance to high irradiance results in a rapid increase in net photosynthesis and flux of C into sucrose. Within 3 h of transfer, SPS protein (and V_{max}

activity) remain constant, but activation state of the enzyme is increased presumably by dephosphorylation of Ser158 (15). After longer periods of time at the higher irradiance, there is a gradual increase in SPS protein and mRNA (20). Thus, regulation of enzyme activity by covalent modification and control of SPS gene expression function in an integrated manner to provide short- and long-term control, respectively.

SPS gene expression also responds to sugars. Provision of Glc to excised sugarbeet or potato leaves strongly increased the steady state level of SPS mRNA, whereas exogenous sucrose slightly repressed expression [at least in sugarbeet (9)]. A similar response is seen in potato tubers when starch synthesis is inhibited by antisense repression of ADP-Glc pyrophosphorylase. The transgenic potato tubers accumulated soluble sugars (sucrose and glucose), and there was a tremendous increase in the steady state level of SPS mRNA (27). SPS activity, measured under selective assay conditions, was also increased relative to wild-type tubers (7), but the basis for the increased activity was not determined. The results suggest that hexose sugars, or some related metabolite(s), might be involved in the control of expression of SPS as well as other genes. It is important to note that the sugar effects on SPS expression have been identified both when exogenous sugars are provided (9) and when endogenous sugars are manipulated genetically (27).

CONCLUDING REMARKS

With the cloning of the SPS gene from five species, we are beginning to better understand the SPS molecule with preliminary identification of important domains such as substrate binding sites and phosphorylation sites. In addition to its role in source leaves, SPS is also significant in sink tissues where a futile cycle of simultaneous degradation and resynthesis occurs in a wide range of tissues. Manipulation of SPS activity is now possible and holds promise for impacting on plant growth and resource allocation. SPS is clearly an important factor regulating sucrose biosynthesis, but it is important to recognize that it is not the only factor, and in addition, changes in SPS protein level are often compensated for by adjustments in the activation state of the enzyme as a result of phosphorylation/dephosphorylation. Consequently, future transformation studies need to consider production of plants with SPS protein modified in terms of phosphorylation control.

ACKNOWLEDGMENTS

We thank colleagues for sharing unpublished manuscripts and ideas, especially Mike Salvucci, Christine Foyer, Mark Stitt, Luis Herrera-Estrella, and Lothar Willmitzer, and other members of the laboratory for discussions. Financial support from the US Department of Agriculture/Agricultural Research

Service, the USDA National Research Initiative Competitive Grant Program, and the Department of Energy is gratefully acknowledged.

Literature Cited

1. Crafts-Brandner SJ, Salvucci ME. 1989. Species and environmental variations in the effect of inorganic phosphate on sucrose-phosphate synthase activity. *Plant Physiol.* 91:469–72
2. Doehlert DC, Huber SC. 1985. The role of sulfhydryl groups in the regulation of spinach leaf sucrose-phosphate synthase. *Biochim. Biophys. Acta* 830:267–73
3. Foyer CH, Galtier N, Quick P. 1994. Modifications in carbon assimilation, carbon partitioning and total biomass as a result of over-expression of sucrose phosphate synthase in transgenic tomato plants. *Plant Physiol.* 105:S23
4. Galtier N, Foyer CH, Huber JLA, Voelker TA, Huber SC. 1993. Effects of elevated sucrose-phosphate synthase activity on photosynthesis, assimilate partitioning and growth in tomato (*Lycopersicon esculentum* var. UC 82B). *Plant Physiol.* 101: 535–43
5. Galtier N, Foyer CH, Murchie E, Alred R, Quick P, et al. 1995. Effects of light and atmospheric carbon dioxide enrichment on photosynthesis and carbon partitioning in the leaves of tomato (*Lycopersicon esculentum* L.) plants over-expressing sucrose-phosphate synthase. *J. Exp. Bot.* 46: 1335–44
6. Deleted in proof
7. Geigenberger P, Krause K-P, Hill LM, Reimholz R, MacRae E, et al. 1995. The regulation of sucrose synthesis in leaves and tuber of potato plants. In *Sucrose Metabolism, Biochemistry, Physiology and Molecular Biology,* ed. H Pontis, G Salerno, E Echeverria. Rockville, MD: Am. Soc. Plant Physiol.
8. Geigenberger P, Stitt M. 1991. A "futile" cycle of sucrose synthesis and degradation is involved in regulating partitioning between sucrose, starch and respiration in cotyledons of germinating *Ricinus communis* L. seedlings when phloem transport is inhibited. *Planta* 185:81–90
9. Hesse H, Sonnewald U, Willmitzer L. 1995. Cloning and expression analysis of sucrose-phosphate synthase from sugar beet (*Beta vulgaris*). *Mol. Gen. Genet.* 247: 515–20
10. Holappa LD, Walker-Simmons MK. 1995. The wheat abscisic acid–responsive protein kinase mRNA, PKABA1, is up-regulated by dehydration, cold temperature, and osmotic stress. *Plant Physiol.* 108:1203–10
11. Huber JLA, Huber SC. 1992. Site specific serine phosphorylation of spinach leaf sucrose-phosphate synthase. *Biochem. J.* 283:877–82
12. Huber SC, Huber JL. 1991. In vitro phosphorylation and inactivation of spinach leaf sucrose-phosphate synthase by an endogenous protein kinase. *Biochim. Biophys. Acta* 1091:393–400
13. Huber SC, Huber JLA. 1992. Role of sucrose-phosphate synthase in sucrose metabolism in leaves. *Plant Physiol.* 99: 1275–78
14. Huber SC, Huber JLA, McMichael RW Jr. 1993. The regulation of sucrose synthesis in leaves. In *Carbon Partitioning Within and Between Organisms,* ed. CJ Pollock, JF Farrar, AJ Gordon, pp. 1–26. Oxford: BIOS Sci.
15. Huber SC, Huber JL, McMichael RW Jr. 1994. Control of plant enzyme activity by reversible protein phosphorylation. *Int. Rev. Cytol.* 149:47–98
16. Huber SC, Huber JL, Pharr DM. 1993. Assimilate partitioning and utilization in source and sink tissues. In *International Crop Science,* ed. DR Buxton, I:761–77. Madison, WI: Crop Sci. Soc. Am.
17. Huber SC, McMichael RW Jr, Bachmann M, Huber JL, Shannon JC, et al. 1995. Regulation of leaf sucrose-phosphate synthase and nitrate reductase by reversible protein phosphorylation. In *Protein Phosphorylation in Plants,* ed. PR Shewry. Oxford: Oxford Univ. Press. In press
18. Huber SC, McMichael RW Jr, Huber JL, Bachmann M, Yamamoto YT, Conkling MA. 1995. Light regulation of sucrose synthesis: role of protein phosphorylation and possible involvement of cytosolic [Ca^{2+}]. In *Carbon Partitioning and Source-Sink Interactions in Plants,* ed. MA Madore, W

Lucas, pp. 35–44. Rockville, MD: Am. Soc. Plant Physiol.
19. Jang J-C, Sheen J. 1994. Sugar sensing in higher plants. *Plant Cell* 6:1665–79
20. Klein RR, Crafts-Brandner SJ, Salvucci ME. 1993. Cloning and developmental expression of the sucrose-phosphate-synthase gene from spinach. *Planta* 190: 498–510
21. Lunn JE, ap Rees T. 1990. Apparent equilibrium constant and mass-action ratio for sucrose-phosphate synthase in seeds of *Pisum sativum. Biochem. J.* 267:739–43
22. McMichael RW Jr, Bachmann M, Huber SC. 1995. Spinach leaf sucrose-phosphate synthase and nitrate reductase are phosphorylated/inactivated by multiple protein kinases in vitro. *Plant Physiol.* 108:1077–82
23. McMichael RW Jr, Klein RR, Salvucci ME, Huber SC. 1993. Identification of the major regulatory phosphorylation site in sucrose-phosphate synthase. *Arch. Biochem. Biophys.* 307:248–52
24. McMichael RW Jr, Kochansky J, Klein RR, Huber SC. 1995. Characterization of the substrate specificity of sucrose-phosphate synthase protein kinase. *Arch. Biochem. Biophys.* 321:71–75
25. Micallef BJ, Haskin KA, Vanderveer PJ, Roth K-S, Shewmaker CK, Sharkey TD. 1995. Altered photosynthesis, flowering and fruiting in transgenic tomato plants that have an increased capacity for sucrose synthesis. *Planta* 196:327–34
26. Miller AJ, Sanders D. 1987. Depletion of cytosolic free calcium induced by photosynthesis. *Nature* 326:397–400
27. Müller-Röber BT, Sonnewald U, Willmitzer L. 1992. Inhibition of ADP-glucose pyrophosphorylase leads to sugar storing tubers and influences tuber formation and expression of tuber storage protein genes. *EMBO J.* 11:1229–38
28. Quick P, Siegl G, Neuhaus HE, Feil R, Stitt M. 1989. Short term water stress leads to a stimulation of sucrose synthesis by activating sucrose phosphate-synthase. *Planta* 177:536–46
29. Reimholz R, Geigenberger P, Stitt M. 1994. Sucrose phosphate synthase is regulated, via metabolites and protein phosphorylation in potato tubers, in a manner analogous to the enzyme in leaves. *Planta* 1992: 480–88
30. Salanoubat M, Belliard G. 1987. Molecular cloning and sequencing of sucrose synthase cDNA from potato (*Solanum tuberosum* L.): preliminary characterization of sucrose synthase mRNA distribution. *Gene* 60: 47–56
31. Deleted in proof
32. Salvucci ME, Klein RR. 1993. Identification of the uridine-binding domain of sucrose-phosphate synthase: expression of a region of the protein that photoaffinity labels with 5-azidouridine diphosphate-glucose. *Plant Physiol.* 102:529–36
33. Salvucci ME, van de Loo FJ, Klein RR. 1995. The structure of sucrose-phosphate synthase. In *Sucrose Metabolism, Biochemistry, Physiology and Molecular Biology,* ed. HG Pontis, GL Salerno, E Echeverria. Rockville, MD: Am. Soc. Plant Physiol.
34. Sonnewald U, Basner A. 1993. EMBL Data Library, Accession No. S34172
35. Sonnewald U, Quick WP, MacRae E, Krause KP, Stitt M. 1993. Purification, cloning and expression of spinach leaf sucrose-phosphate synthase in *E. coli. Planta* 189:174–81
36. Stitt M, Huber SC, Kerr P. 1987. Control of photosynthetic sucrose formation. In *Biochemistry of Plants,* ed. MD Hatch, NK Boardman, 8:327–409. New York: Academic
37. Stitt M, Sonnewald U. 1995. Regulation of metabolism in transgenic plants. *Annu. Rev. Plant Physiol. Plant Mol. Biol.* 46:341–68
38. Stitt M, Wilke I, Feil R, Heldt HW. 1988. Coarse control of sucrose-phosphate synthase in leaves: alterations of the kinetic properties in response to the rate of photosynthesis and the accumulation of sucrose. *Planta* 174:217–30
39. Deleted in proof
40. Weiner H. 1995. Antibodies that distinguish between the serine-158 phospho- and dephospho-form of spinach leaf sucrose-phosphate synthase. *Plant Physiol.* 108: 219–25
41. Weiner H, Weiner H, Stitt M. 1993. Sucrose-phosphate synthase phosphatase, a type 2A protein phosphatase, changes its sensitivity towards inhibition by inorganic phosphate in spinach leaves. *FEBS Lett.* 333:159–64
42. Worrell AC, Bruneau J-M, Summerfelt K, Boersig M, Voelker TA. 1991. Expression of a maize sucrose phosphate synthase in tomato alters leaf carbohydrate partitioning. *Plant Cell* 3:1121–30

Annu. Rev. Plant Physiol. Plant Mol. Biol. 1996. 47:445–76

STRUCTURE AND BIOGENESIS OF THE CELL WALLS OF GRASSES

Nicholas C. Carpita

Department of Botany and Plant Pathology, Purdue University, West Lafayette, Indiana 47907

KEY WORDS: cereals, grasses, cell-wall polysaccharides, cell-wall biosynthesis, cell-wall architecture

ABSTRACT

The chemical structures of the primary cell walls of the grasses and their progenitors differ from those of all other flowering plant species. They vary in the complex glycans that interlace and cross-link the cellulose microfibrils to form a strong framework, in the nature of the gel matrix surrounding this framework, and in the types of aromatic substances and structural proteins that covalently cross-link the primary and secondary walls and lock cells into shape. This review focuses on the chemistry of the unique polysaccharides, aromatic substances, and proteins of the grasses and how these structural elements are synthesized and assembled into dynamic and functional cell walls. Despite wide differences in wall composition, the developmental physiology of grasses is similar to that of all flowering plants. Grass cells respond similarly to environmental cues and growth regulators, exhibit the same alterations in physical properties of the wall to allow cell growth, and possess similar patterns of wall biogenesis during the development of specific cell and tissue types. Possible unifying mechanisms of growth are suggested to explain how grasses perform the same wall functions as other plants but with different constituents and architecture.

CONTENTS

1040-2519/96/0601-0445$08.00

INTRODUCTION

In the early 1900s, Walter Norman Haworth and Edmund Langley Hirst, founding fathers of modern carbohydrate chemistry, began studies of the pentose constituents of cell walls of plants. Armed with only rudimentary analytical techniques, they and their colleagues defined the cell walls of esparto grass as composed largely of (1→4)-β-D-xylans (58). By 1970, gas chromatography–mass spectrometry (GC-MS) was employed routinely for unequivocal determination of linkage structure of complex cell-wall polysaccharides. Techniques such as ^{1}H- and ^{13}C-nuclear magnetic resonance (NMR) spectroscopy provided anomeric configurations, linkage structures, and some three-dimensional configurations. Sequence-dependent endoglycanases were used to cleave polysaccharides into oligosaccharides that could be completely sequenced. From characteristic repeating unit structures, the sequence and conformation of very large polymers were deduced (20). By such analyses, the major polysaccharides of the walls of a wide range of flowering plants were defined, and the first models of how cell walls are put together emerged. In subsequent years, the dynamic interactions of individual components were reflected in more current models of the architecture of the primary cell wall of flowering plants—a strong framework of cellulose microfibrils intertwined with xyloglucans that is embedded in a gel of uronic-acid-rich pectins and cross-linked with hydroxyproline-rich glycoproteins (20, 101).

When the first conceptual models were proposed about twenty years ago, the differences in wall compositions between monocots and dicots were just beginning to be catalogued (33). Perhaps because of the socioeconomic importance of the cereals, the vast majority of the monocots studied were grasses. Whistler (164) described grasses as rich sources of xylan, and Aspinall (1), in a review of plant cross-linking glycans, noted the enrichment of xylans and mixed-linked glucans in grasses. Wilkie (165) offered the first comprehensive survey of the cross-linking glycans of grasses. More recent studies of the carbohydrate and aromatic components of cell walls from a broad spectrum of monocots have revealed that the Poales (family Poaceae, formerly the order Graminales, family Gramineae), their progenitors, and related taxonomic orders have primary cell walls completely different from those of other monocots (3, 20, 68). Dahlgren et al (32) proposed phyletical relationships between some two dozen orders of the Monocotyledonae on the basis of several anatomical features and chemical constituents. One of these features, the presence or absence of ferulic acid in the primary walls, is a major distinguishing feature of the Poales and related orders (56). Nonlignified cells of grasses are enriched in aromatic substances, and polymeric forms constitute a second architectural element.

A third type of architectural element is structural protein. The primary walls of the Poales contain substantially less protein than other species, but several classes of proteins are found in elevated amounts in specific cell types during differentiation. Recent reviews (83, 147) note that the grass wall proteins bear reasonable homology to those representing major classes of structural proteins of nongramineous species. In an earlier review (20), structural models for two types of primary walls were provided: the Type I wall, composed of a cellulose-xyloglucan framework embedded in a pectin gel, and the Type II wall, the special wall of the Poales. This review focuses on the Type II wall of the Poales, its composition, architecture, biogenesis, and dynamics during growth.

THE STRUCTURAL ELEMENTS OF THE PRIMARY WALLS OF THE POACEAE

Cellulose

Cellulose microfibrils in all flowering plants are composed of about three dozen linear chains of (1→4)-β-linked D-Glc condensed to form long paracrystalline arrays that spool around each cell (35). Although each chain may be only several thousand units long, they begin and end at different places within a microfibril and make very long microfibrils whose ends are rarely detected.

Glucuronoarabinoxylans

The linkage structure of the grass glucuronoarabinoxylans (GAX) has been known for much of the 20th century. The *t*-α-L-arabinofuranosyl units are attached primarily at the *O*-3 positions along the (1→4)-β-D-xylan backbone, and the *t*-α-D-glucuronic acids are attached to the *O*-2 positions (1, 165). The highly substituted GAXs of the barley aleurone and barley malts contain significant *O*-2- and doubly branched *O*-2, *O*-3-linked arabinosyl units in addition to the abundant *O*-3-linked units (5, 160). Arabinoxylans are widespread in the walls of all flowering plants, but in nongramineous species the polymer is of much lower abundance, and the α-L-arabinosyl units are attached mostly at the *O*-2 rather than the *O*-3 of the xylosyl units (33). Whereas some of the xylem-rich straws may contain exclusively the 4-*O*-methyl derivative (1, 165), underivatized GlcA is the major acidic constituent of the maize primary wall GAXs (22).

A highly substituted GAX (HS-GAX), with six of seven xylosyl units bearing appendant groups, is associated with the maximum growth rate of coleoptiles (22). Smith degradation indicated a unit structure in which the arabinosyl and GlcA units are added to the growing chain in a specific pattern during synthesis (22). Nishitani & Nevins (117) found a sequence-dependent xylanase that requires an appendant GlcA to cleave the neighboring (1→4)-β-D-xylosyl linkage. When maize GAXs are depleted of arabinosyl units by mild-acid hydrolysis, this endo-β-D-xylanase releases a homogeneous group of deca- or undecamers of glucuronoxylan. The molecular structure of the HS-GAX may be a unit structure with GlcA units added to alternate xylan heptamer units (Figure 1A). The arabinosyl units are hydrolyzed after incorporation into the wall to yield the broad range of grass GAXs (46). The xylosyl units are also substituted with acetyl groups at the *O*-2 and *O*-3 position (30); the acetyl content of some ryegrasses has been reported to be almost 10% of the dry mass of the wall (52).

The (1→3),(1→4)-β-D-glucans

Noncellulosic glucans are also found at certain stages in grass development, particularly in the seed brans. These unbranched "mixed-linked" glucans (β-D-glucans) contain both (1→3) and (1→4)-linkages in a ratio of about 1:2 to 1:3. A general sequence structure was unequivocally deduced with a sequence-dependent endoglycanase, a β-D-glucanohydrolase from *Bacillus subtilis* that catalyzes the hydrolysis of a (1→4)-β-D-glucosyl linkage only if preceded by a (1→3)-β-D-linked glucosyl unit on the nonreducing side (151). With oat or barley endosperm walls, this enzyme hydrolyzes about 90% of the β-D-glucan macromolecules into cellobiosyl- and cellotriosyl-(1→3)-β-D-Glc in a ratio of a little over 2:1 (Figure 1B) (151). The remainder of the polymer is made of small amounts of longer runs of the cellodextrin series interrupted by single

Figure 1 (A) (Feruloylated) glucuronoarabinoxylan (GAX). The highly substituted GAX contains α-L-arabinose and α-D-glucuronic acid units on six of every seven (1→4)-β-D-linked xylosyl units (22). Many of the arabinosyl units are cleaved from the GAX during assembly in the wall (46). Ferulic acid esters are attached at the *O*-5 position of a small portion of the arabinosyl units, and these esters may dimerize with other feruloyl groups in several ways to cross-link the GAX into a network (126). (B) Trimer and tetramer unit structure of the mixed-linkage (1→3),(1→4)-β-D-glucans. The *Bacillus subtilis* endo-glucanase cleaves (1→4)-β-D-glucosyl linkages just in front of (1→3) linkages (arrows) to yield cellobiosyl- and cellotriosyl-(1→3)-β-D-glucose oligomers in a ratio of about 2:1 (151).

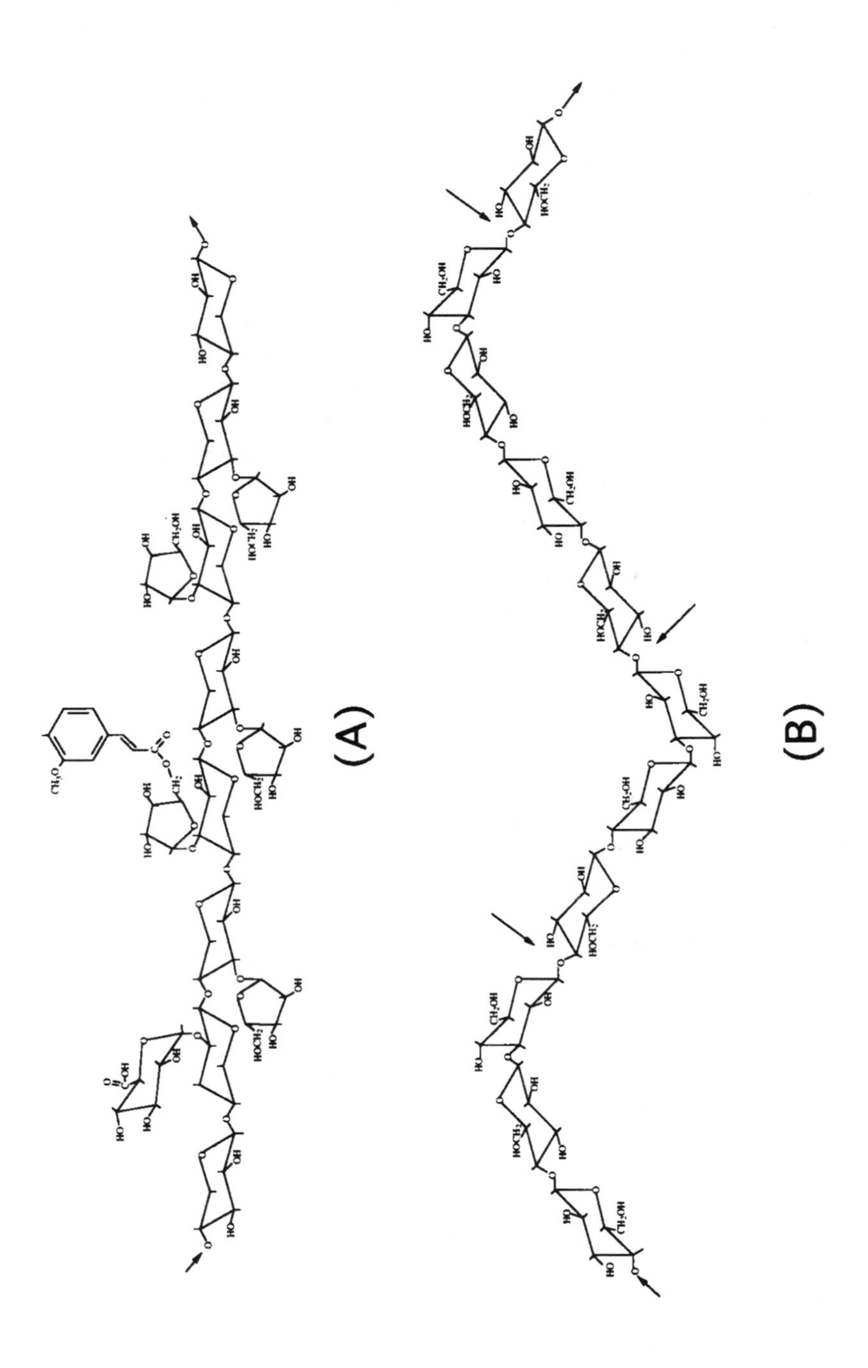
(A)
(B)

(1→3)-linked units (80, 169) These longer runs of cellodextrin units are apparently uniformly distributed throughout the length of the polymer. An endo–β-D-glucanase from cell walls of developing maize seedlings, which can hydrolyze the β-D-glucan macromolecule only at the cellodextrin-rich regions and not the tri- and tetramers, yields a homogeneous fraction of polymers about 50 sugar units long (57, 67). The distribution of sizes of the cellodextrin lengths larger than four are remarkably constant among the cereals (167).

Xyloglucan

Small amounts of xyloglucan (XyG) are found in the grasses, but hydrolysis with 4-glucanohydrolases does not yield the hepta- and nonasaccharides characteristic of the polymer in many Type I walls. The XyGs are enriched in grass meristematic cells before the onset of enhanced β-D-glucan and GAX synthesis during elongation. Enzyme hydrolysis yields the diagnostic disaccharide isoprimeverose, the α-D-Xyl-(1→6)-D-Glc, but the xylosyl units appear on isolated or two contiguously linked glucosyl units of the (1→4)-linked glucan backbone instead of the regular blocks of three typically found in many flowering plants (78). An exception may be the XyG of rice endosperm walls. Shibuya & Misaki (145) found XyG fragments containing both two and three contiguous xylosyl side-groups and a possible attachment of *t*-Gal at the *O*-2 of some of the xylosyl units. The common Type I XyG trisaccharide, *t*-L-Fuc-(1→2)-β-D-Gal-(1→2)-α-D-Xyl-, is absent from the grass XyGs. Consistent with this lack of Fuc and Gal in grass XyG, monoclonal antibodies that recognize this feature of Type I XyGs fail to recognize grass XyG (123). Surprisingly, Fuc is assimilated by suspension cultures of fescue and added to the occasional Gal unit in grass XyG (103).

Other Glycans

Small amounts of glucomannan are also found tightly bound to the cellulose microfibrils. Urea is able to extract selectively a glucomannan from the walls of wheat and barley endosperm walls (5) and the maize coleoptile (15). Although grasses are generally devoid of Fuc, the root cap slime can be considerably enriched in this sugar (24). The maize slime is almost 20% Fuc, in mostly *t*- and 3-linkages, whereas the remainder is pectic-like (4).

Pectic Substances

Two fundamental constituents of all flowering plant pectins are polygalacturonic acid (PGA), which is a homopolymer of (1→4)-α-D-galactosyluronic acid (GalA), and rhamnogalacturonan I (RG I), which is a heteropolymer of repeating (1→2)-α-L-rhamnosyl-(1→4)-α-D-GalA disaccharide units (73). RG I is found in walls of somatic cells of both maize and rice (156). PGA and RG I are found in grasses but in much smaller amounts (17, 146).

Like many flowering plants, grasses contain arabinans, galactans, and highly branched arabinogalactans (AGs) of various configurations and sizes, and they are attached to the *O*-4 of the rhamnosyl residues of RG (17, 146). The arabinans are mostly 5-linked arabinofuranosyl units but can be connected to one another at virtually every free position, the *O*-2, *O*-3, and the *O*-5, to form a diverse group of branched arabinans. The galactans and two classes of AGs are the major side-chains. One class of AGs is associated only with pectins and is composed of (1→4)-β-D-galactan chains with mostly *t*-arabinosyl, and sometimes *t*-galactosyl, units at the *O*-3 of the galactosyl residues of the backbone (3). In maize and rice pectins, the branched 5-linked arabinan and 4-galactan side-chains are attached to the *O*-4 of about two of every three rhamnosyl units (17, 146, 156). The neutral sugar side-chains of the grass RGs are notable only for their lack of Fuc (156). In walls from maize seedlings and rice endosperm, the chelator-soluble pectins are enriched in the HS-GAX as well as RG and arabinogalactan (protein) (AGP). Two fractions of grass pectins are resolved by ion-exchange chromatography (17, 146); one fraction is enriched in HS-GAX and an associated AGP, whereas the other contains mostly GAX and RG I (17).

The second class of AGs constitutes a broad group of short (1→3)- and (1→6)-β-D-galactan chains connected to each other by (1→3,1→6)-linked branch point residues. Most of the remaining *O*-3 or *O*-6 galactosyl positions are filled with *t*-arabinosyl groups (2, 40). These AGs are also associated with proteins (AGPs) whose functions in intracellular, plasma membrane, and cell-wall locations are still unknown.

Aromatic Substances

A major feature of the Poales and their relatives is the enrichment of aromatic substances in nonlignified walls (56, 135). A large portion of the aromatic substances are esters of the hydroxycinnamates, ferulate, and *p*-coumarate (56). The GAXs are cross-linked in walls by both esterified and etherified hydroxycinnamates and by other phenolic substances (68, 138), and the etherified linkages represent complexes of polysaccharide and lignin (69). The ferulate and *p*-coumarate esters are attached to the *O*-5 of arabinosyl units of GAX (54, 79, 114). Markwalder & Neukom (100) suggested that neighboring feruloylated GAX chains are cross-linked by formation of 5,5-diferulate, and this was demonstrated directly in bamboo GAX by Ishii (72). The 5,5-diferulate is only one of a series of dehydrodimers present in grasses (126). These esters are broken by dilute alkali, and release of ferulic acid and diferulic acid is coincident with the release of HS-GAX (16). Polysaccharides may also be cross-linked photochemically by [2 + 2]-homo- and -heterodimerization of the ferulate and *p*-coumarate esters to cyclobutane derivatives called truxillic and truxinic acids (55, 157).

More complex interactions between aromatic substances and polysaccharides involve ester-ether interactions, and the ether bonds are not broken by dilute alkali (70). Scalbert et al (138) suggested that such ester-ether interactions form bridges between polysaccharides and lignin via phenol addition to quinone methide lignin intermediates in several ways. Ferulic acid is the principal component in the ester-ether linkage of carbohydrate and lignin (69, 91, 92). Ferulates are also incorporated into lignins via radical mechanisms to form not only the β-ether linkages but other structural forms that cannot be released by any solvolytic method (125). *p*-Coumaric and ferulic acids form single esters, but only rarely does the *p*-coumaric acid form an ester-ether bridge (91, 92). The *p*-coumaric acid is more heavily associated with lignin, particularly later in cell-wall development, and attached at the γ-positions of the lignin side-chains (124).

The principal monomers of grass lignin are coniferyl and sinapyl alcohols, with some *p*-OH coumaryl alcohol. The latter is often overestimated because the nitrobenzene products from *p*-coumaric acid are attributed to *p*-OH-cinnamyl units of lignin. Syringyl lignin increases in proportion relative to guaiacyl and *p*-hydroxyphenyl lignins during maturation of some grasses.

The formation of the hydroxycinnamate esters of GAX most likely occurs cytosolically, presumably in the Golgi apparatus (115). The cinnamyl alcohol precursors of lignin are also synthesized cytosolically, and several of the key enzymes in the biosynthetic pathway have been characterized in grasses. In addition to phenylalanine ammonia lyase, grasses also possess a tyrosine ammonia lyase that forms *p*-coumaric acid from tyrosine (116). Other hydroxycinnamates are made through oxygenase and methyl transferase reactions. These include *trans*-cinnamate-4-monooxygenase, which forms *p*-coumarate and caffeate from *trans*-cinnamic acid (121); caffeate-*O*-methyl transferase, which catalyzes the conversion of caffeate to ferulate (41); and cinnamyl alcohol dehydrogenase, which catalyzes the conversion of the cinnamyl aldehydes to *p*-coumaryl, coniferyl, and sinapyl alcohols, the direct precursors of lignin (120). The formation of cinnamyl alcohol glucosides may be important in transport, because the unglycosylated precursors are not very soluble in water. Cleavage of the glucosides by cell-wall β-D-glycosidases may generate the active precursors *in muro* (163). The UDP-Glc:coniferyl alcohol transferases have been described in many nongramineous species (163).

The mechanism of the polymerization of the cinnamyl alcohols into lignin is not completely established. For many years, the reactions were thought to be catalyzed solely by peroxidases, but recent evidence has implicated laccase as a participating enzyme (163). Use of ^{13}C-labeling techniques has permitted a convenient monitoring of the synthesis of phenylpropanoid synthesis and poly merization (96, 124). A role of ferulates and diferulates in nucleation of lignification has been suggested (125).

Structural Proteins

In Type I walls, the cell-wall carbohydrates of fully elongated and differentiating cells are cross-linked with structural proteins to produce an inextensible structure. In the primary walls of grasses, this function is carried out largely by phenolic substances. Homologs to several known classes of structural proteins are synthesized in the grasses and are in much larger amounts in specific cell types (83, 147).

In nongramineous plants, the hydroxyproline-rich glycoproteins (HRGPs)—the extensins—are ubiquitous. These rod-shaped proteins owe their structure to two features: a polyproline II helix due to glycosylated Ser-$(Hyp)_4$ repeats and a Tyr-X-Tyr-Lys motif that is the site of formation of a stabilizing intramolecular isodityrosine linkage (83). The grass homolog is a threonine-hydroxyproline-rich glycoprotein (THRGP) found in maize (63, 82, 152), *Sorghum* (127), and rice (12). In the maize and *Sorghum* HRGPs, only one "signature" Ser-$(Hyp)_4$ sequence remains near the carboxy terminus, whereas a majority of the repeats contain a Lys substitution for a Hyp (84, 127). A rice HRGP contains no Ser-Hyp_4 repeats and a repeating Pro-Pro-Thr-Tyr-Lys-Pro in place of Pro-Pro-Thr-Tyr-Thr-Pro of the maize and *Sorghum* proteins (12). This substitution is predicted to interrupt the polyproline II helix, and the increased molecular flexibility should prevent maintenance of a rod-like structure (84). The periderm, a firm, suberized structure, is quite enriched in the THRGP, which supports the idea of a special structural role for these proteins (42, 63). The THRGP is enriched in protoxylem and metaxylem and in the longitudinal radial walls of the epidermis (148). A unique, extensin-like protein with Ser-$(Hyp)_4$ repeats has also been described in the maize pollen grain (27, 134).

Glycine-rich proteins (GRPs) are encoded by a large family of genes and perform a broad spectrum of cytosolic and cell-wall functions (147). GRPs homologous to vascular cell-wall proteins from nongramineous species have been found only in rice (95) and barley (133). Many other GRPs of grasses do not have signal peptides (147), which indicates a possible cytosolic function. Another member of a large gene family, a proline-rich protein (PRP), was also described in maize (76). This protein has repeated Pro-Pro-Tyr-Val and Pro-Pro-Thr-Pro-Arg-Pro-Ser sequences at the N-terminal domain and a cysteine-rich C-terminal domain, features similar to some PRPs from nongramineous species.

The protein portions of AGPs are also from a diverse family of genes encoding Hyp-rich and Hyp-poor polypeptides (40). Ryegrass AGP-peptides are Ala- and Ser-rich and contain Ala-Hyp repeats (51), a motif that is also found in the maize histidine-rich protein (81). van Holst & Fincher (158) showed that the *Lolium* AGP formed a polyproline II helix, but considering the broad diversity of AGP protein structures, this may or may not turn out to be a general feature. The soluble AGPs bind specifically to the Yariv reagent containing phloroglucinol-β-D-linked Glc units (75). In addition to the arabinosylated (1→3,1→6)-β-D-galactan structure (40), small amounts of other neutral and acidic sugars and sugar linkages are found that may give diversity in the carbohydrate domain. The carbohydrates are attached to Hyp residues in some AGPs (2), but it is uncertain whether Ser and Thr residues are the site of attachment in other classes of AGPs. AGPs are found in numerous cellular locations. They are the major polysaccharide in secretory vesicles (46), and distinct AGPs are associated with both the plasma membrane and cell-wall compartments in nongramineous species. Their functions in any location are still speculative.

Nonstructural proteins reside in the grass cell wall. Many are hydrolases, transferases, esterases, peroxidases, and several other enzymes that function in the modification of cell-wall polymers at the various stages of development and differentiation (23). In addition, thionins, which are toxic to fungal pathogens, accumulate in the walls of many grasses (7), and grasses also accumulate pathogenesis-related proteins similar to those found in nongramineous species.

Other Cell-Wall Substances

Silica is particularly abundant in the walls of grasses, mostly as inclusion bodies in the epidermis, periderm, and other specialized cells of the root, rhizome, and aerial shoots (119). Little has been reported on any chemical interaction with other cell-wall constituents. Silica binds uronic acid residues of animal extracellular polysaccharide and peptidoglycans (142), which is an indication of possible complementary interactions with PGA or GAX. A group of related calcium-binding glycoproteins was discovered in specialized silica deposition vesicles involved in the synthesis of the silica-rich walls of marine diatoms (88).

Similar to all flowering plants, grasses possess cutin, suberin, and waxes in specialized cells (3). Osmiophilic granules, apparently derived from the Golgi apparatus, appear in the outer epidermal layers during rapid growth in deepwater rice (89). These granules may carry precursors and enzymes responsible for the synthesis of the cuticle and secretion of waxes (61).

CELL-WALL COMPOSITION AS A TAXONOMIC CHARACTER IN THE MONOCOTYLEDONAE

In a hallmark paper, Chase et al (26) examined the phylogenetics of flowering plants by nucleotide sequence variability of a ribulose bisphosphate carboxylase gene *rbc*L. Duvall et al (36) elaborated on these data in their reevaluation of the Monocotyledonae. From these extensive studies, two major divisions among the Monocotyledonae have been proposed: the "commelinoids," a large clade that includes the grasses, sedges, rushes, palms, and gingers, and the "noncommelinoids," a more basal group of aroids, alismatids, and lilioids (135). Harris & Hartley's (56) detection of the green fluorescence enriched in nonlignified cells is confined, almost without exception, to the commelinoids (Figure 2). Ruddall & Cadick (135) surveyed a broader range of monocots for wall fluorescence, including some of questionable lineage, and established that several genera that had previously been classified as lilioids are actually commelinoids. This classification is largely consistent with the system of Dahlgren et al (32), who showed that several anatomical, developmental, and biochemical traits distinguish the commelinoids from the noncommelinoids. Among these characteristics, the Poales are noted for a starchy endosperm; the presence of silica bodies in the epidermis, pericycle, and other specialized cells; and the absence of calcium oxalate raphide crystals (Figure 2). Silica bodies are found in most commelinoids, with the exceptions of many Commelinales and more basal species. Apart from a few Liliiflorae, silica bodies are absent from the noncommelinoids. Whereas the Poales, Cyperales, Commelinales, and most Zingiberales lack oxalate raphides, the crystals are detected in the Arecales and more primitive commelinoids (32).

Unfortunately, cell-wall polysaccharide composition has not been used extensively enough to confirm some of the phylogenetic relationships, but the data available indicate at least two possible transitional stages in the development of the novel wall structure in the Poales. The first transition is the replacement of XyG with GAX as the principal cross-linking glycan in walls of meristematic cells (Figure 2b). Ferulate, which produces an alkali-induced green fluorescence, and *p*-coumarate are esterified specifically to the *O*-5 position of the arabinofuranosyl units of GAX (54). Because of this chemical feature, all commelinoids should have GAX as a major cell-wall polymer. Consistent with this prediction, GAX is the major cell-wall polysaccharide in the few nongramineous commelinoids that have been examined, including the

Figure 2 Comparison of anatomical and chemical features of the commelinoid and noncommelinoid Monocotyledonae (after 32). (A) The presence of silica bodies (○) is associated with many of the commelinoids, whereas the presence of calcium oxalate raphide crystals (•) is associated with the noncommelinoids (32). (B) Detection of UV-fluorescence (◎) in nonlignified cells is found in most commelinoids and progenitors, whereas there is an absence of fluorescence (◉) in the walls of all noncommelinoids (56, 135). The glucuronoarabinoxylan (❑) is a major cross-linking glycan in many of the commelinoids (3, 10, 149, 165), whereas xyloglucan (■) predominates in the few noncommelinoids that have been examined (3, 128). There is a trend toward low pectin content (✲) in the advanced commelinoids, although some have intermediate amounts (✲), and the noncommelinoids are rich in pectin (✲) (74). The mixed-linkage β-D-glucan is present (○) only in the Poales and absent (•) from all other commelinoid and noncommelinoid species that have been examined (153).

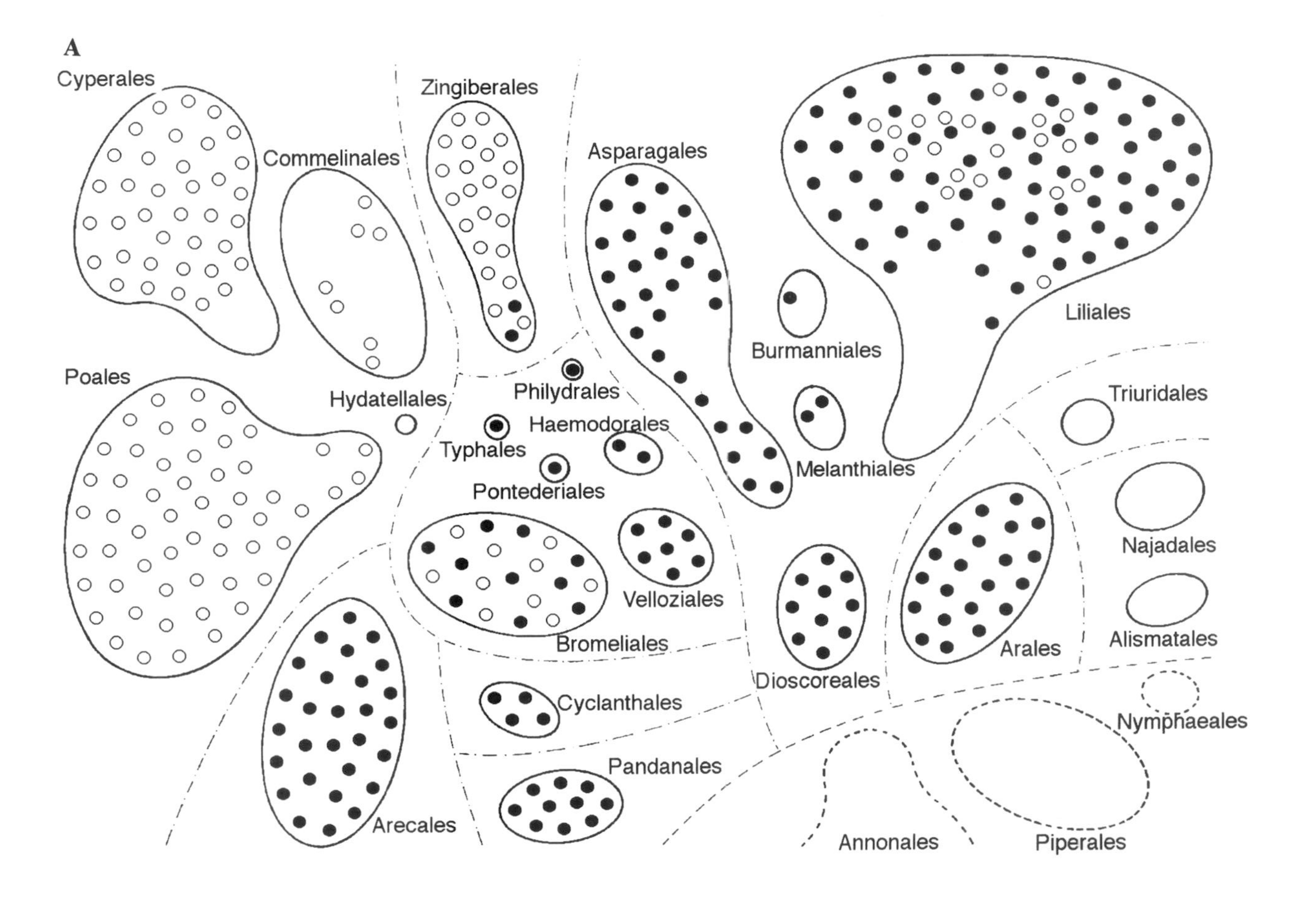
A
Cyperales
Commelinales
Zingiberales
Asparagales
Liliales
Burmanniales
Poales
Hydatellales
Philydrales
Haemodorales
Typhales
Pontederiales
Melanthiales
Triuridales
Velloziales
Bromeliales
Najadales
Alismatales
Arales
Dioscoreales
Cyclanthales
Nymphaeales
Pandanales
Arecales
Annonales
Piperales

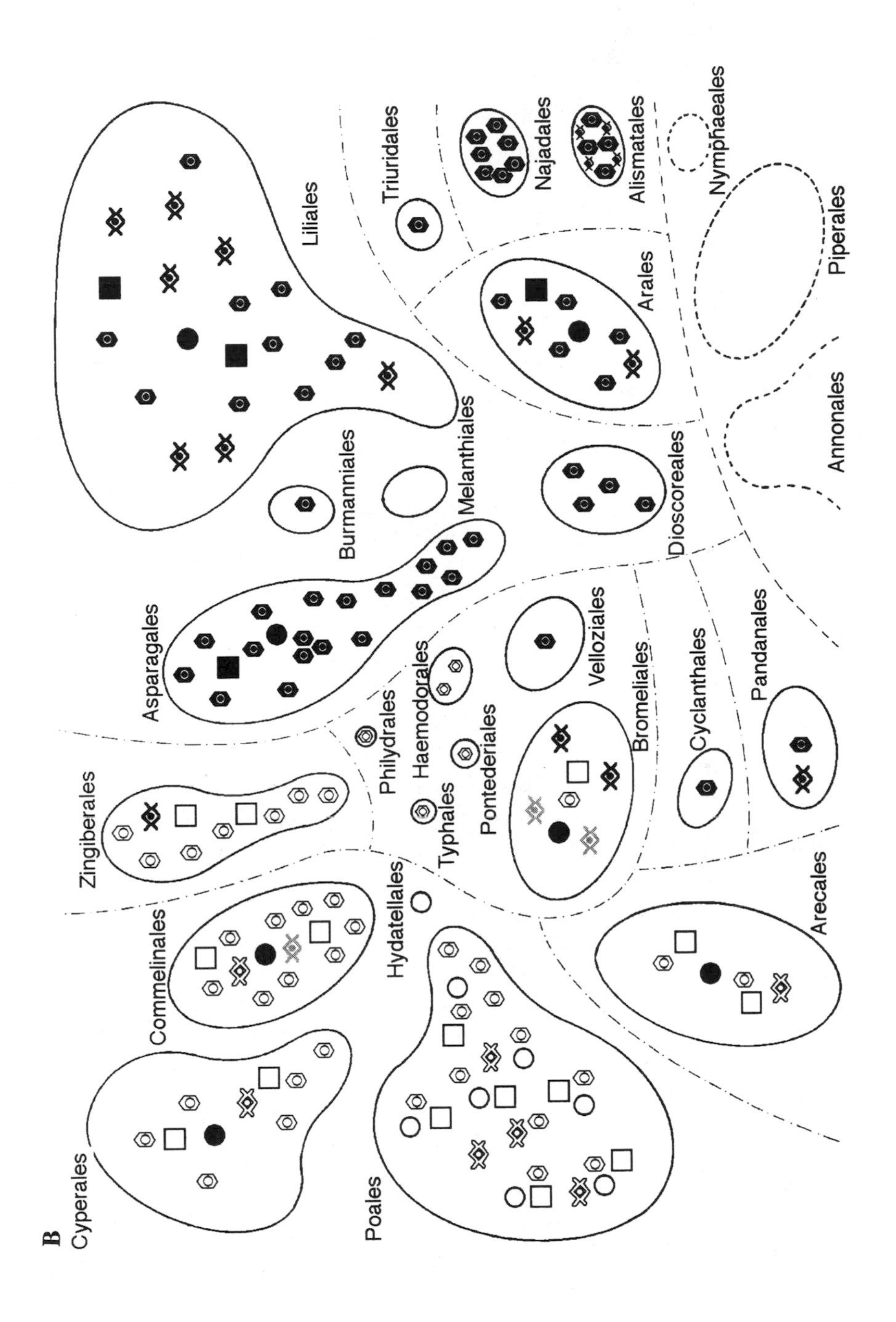
B
Cyperales
Poales
Commelinales
Hydatellales
Zingiberales
Typhales
Philydrales
Haemodorales
Pontederiales
Asparagales
Burmanniales
Melanthiales
Liliales
Triuridales
Najadales
Alismatales
Nymphaeales
Arales
Piperales
Annonales
Dioscoreales
Velloziales
Bromeliales
Cyclanthales
Pandanales
Arecales

Cyperales (10), Zingiberales (3, 165), and Bromeliales (149). Noncommelinoid monocots have XyG- and pectin-rich walls typical of the Type I walls (74, 128). The transition from XyG to GAX also represents an exchange of an acidic polymer for a neutral one, and the tight association of HS-GAX with the small amount of pectic fraction indicates that GAX may have replaced part of the role of the pectins in providing an anionically charged matrix. Jarvis et al (74) found that the content of the pectins among the commelinoids was consistently well below that of noncommelinoids and dicots (Figure 2b). The second transitional stage is marked by the appearance of the (1→3),(1→4)-β-D-glucan (Figure 2). This special glucan is found exclusively in the Poales (149, 153).

ARCHITECTURE

Biochemical studies have provided a good catalog of the wall constituents, but until recently we could only estimate, by virtue of the physical and chemical properties of the polymers in solution and solid-state, how the constituents were assembled, arranged, and cross-linked into a functional matrix. Cryopreservation techniques for electron microscopy have helped in visualizing the fine structure of the wall (102), and antibody and enzyme probes for specific cell-wall epitopes reveal the organization of certain polymers within domains of a single cell wall (87, 136). Fourier transform infrared and FT-Raman spectroscopy are used to detect specific chemical bonds and their orientation within the underivatized cell walls of grasses (143).

In dividing cells, the microfibrils are wound around each cell randomly, and this pattern continues throughout isodiametric expansion. When elongation begins, microfibrils are wound transversely or in a shallow helix around the longitudinal axis. Microfibrils of nongramineous walls are 5–15 nm wide and are spaced 20–40 nm apart (102). Preliminary measurements of mesophyll and epidermal cells of the maize coleoptile showed spacings that were slightly smaller than these dimensions (132), but comparisons of the spacings in growing and nongrowing walls with different architectures still need more thorough evaluation (MC McCann, personal communication). McCann & Roberts (101) suggested that the XyGs not only tether the microfibrils but also establish the spacings between the microfibrils. The ability of the GAX to self–hydrogen bond might result in altered spacings.

Wilkie's review (165) of the gramineous xylans summarized broad variation among cereals in the Ara:Xyl ratios, an estimate of the degree of substitution of the xylan with arabinosyl side-groups. The Ara:Xyl ratio decreases in leaves and stems throughout the development of the oat plant (129), and the number of Ara units along the xylan chain varies markedly during coleoptile elongation—from GAXs whose Xyl units are nearly all branched to those with <10% of the xylosyl units bearing side-groups (14). The side-groups greatly affect the ability of the GAXs to bind to one another and to cellulose. Like XyG, the unbranched (1→4)-linked xylans hydrogen bond to cellulose and to one another, whereas the attachment of arabinosyl and GlcA side-groups to the internal *O*-2 and *O*-3 secondary alcohols of the xylan backbone hinder the formation of hydrogen bonds (13). Because the Ara units markedly alter water solubility and the ability to hydrogen bond, the GAXs may exist in at least two distinct domains within the matrix: an HS-GAX domain that is a structural continuum with the pectin matrix and a relatively unbranched xylan domain that forms tight bonds to the cellulose microfibrils (13, 15).

The distribution of methyl esterified and unesterified PGAs in Type II cell walls was detected by antibodies against these polymers (87). The grasses display a marked developmental preference for accumulating methyl esterified or unesterified pectins in specific cell types. Grasses have mostly esterified PGAs in vascular tissues, whereas walls of the cortical and parenchyma cells have mostly unesterified polymers (87). Neither antibody recognizes PGAs in walls of the root epidermis or root cap cells. As estimated by infrared spectroscopy, the esterified PGAs of grass walls represent a sizeable portion of the total pectin (111) and are easily distinguished by other spectral features from nongramineous species (143). Whereas pectins of the Type II wall are comprised of both PGA and RG, HS-GAX is also a major component closely associated with pectins, particularly in a fraction containing an AGP (17).

The Type II cell-wall model has cellulose microfibrils bound and interlaced with unbranched GAXs (Figure 3a). Additional GAXs with varying degrees of branching may have functionally replaced the predominant pectic substances in the Type I cell wall, and the spacing of the appendant arabinosyl and GlcA units could determine porosity and surface charge. Given that the spacing between cellulose microfibrils measured tangentially is the same as the spacing radially, the lamellate structure is only about four to eight strata thick (Figure 3a).

Figure 3 Architecture of the Type II cell wall of the Poaceae. (A) Representation of four strata of microfibrils and associated polysaccharides and aromatic substances just after cell division. The microfibrils are interlocked by glucurononarabinoxylans. Unlike xyloglucans, the xylans are secreted in a form highly substituted with arabinosyl units that prevent hydrogen bonding. The units are cleaved from a portion of the GAX to yield runs of xylan that can bind on either face to cellulose or to one another (13, 20). Porosity of the GAX domain could be determined by the extent of removal of the appendant units. Some highly substituted GAX remains intercalated in the small amount of pectins that also are found in the primary wall (17). A portion of the noncellulosic polymers are cross-linked to the microfibrils by alkali-resistant phenolic linkages. (B) The expanding Type II cell wall. Absent from the developing wall of dividing cells, the β-D-glucan is synthesized specifically during cell elongation and is the major cross-linking glycan. Long runs of (1→4)-linked glucan could bind to cellulose or to other glucan. Although some tissues accumulate structural proteins, the fundamental cross-linking in the primary wall of recently enlarged cells is by esterified and etherified phenolic compounds that lock the wall into place and halt further stretching of the microfibrils.

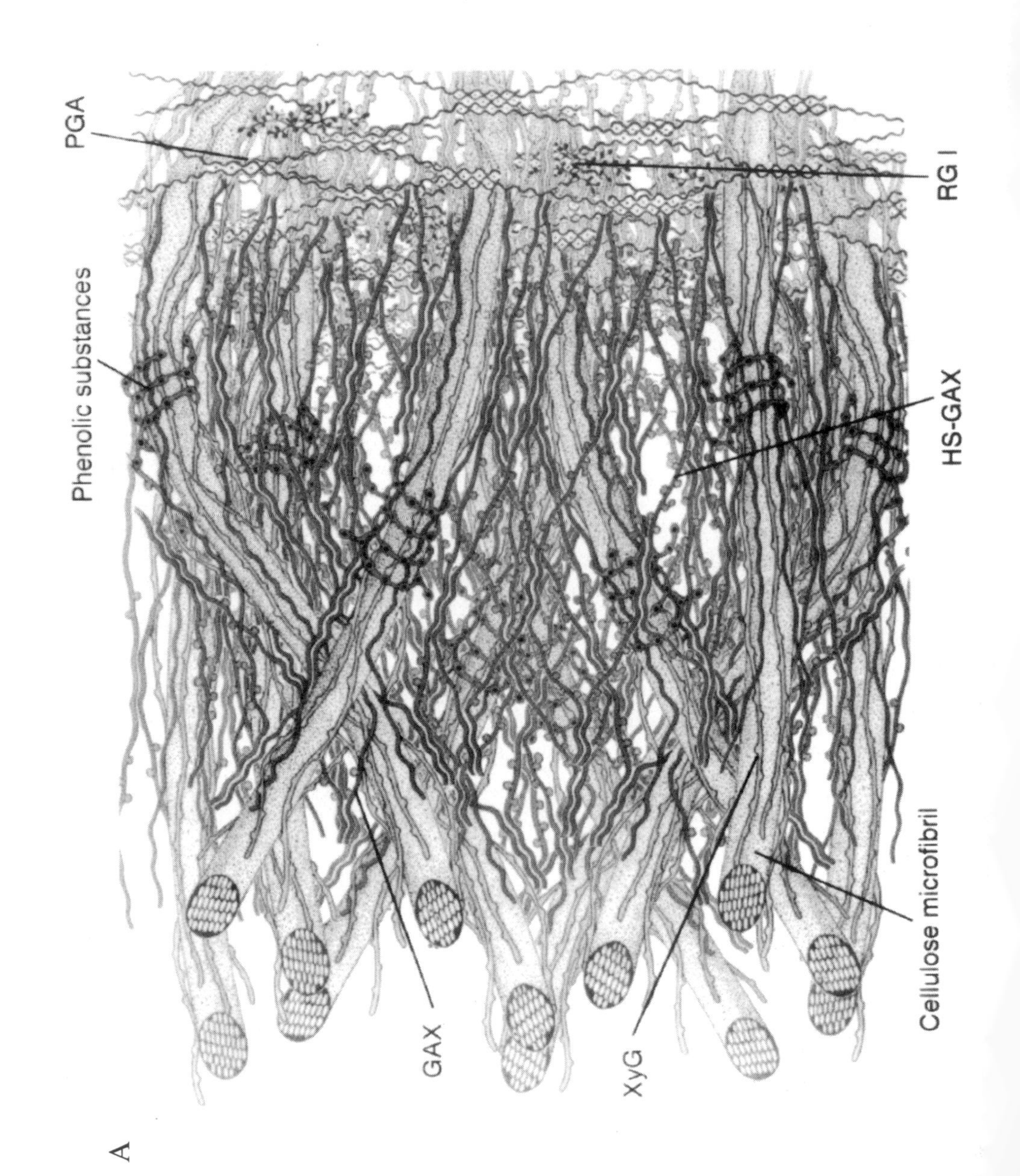
PGA
RG I
Phenolic substances
HS-GAX
GAX
XyG
Cellulose microfibril
A

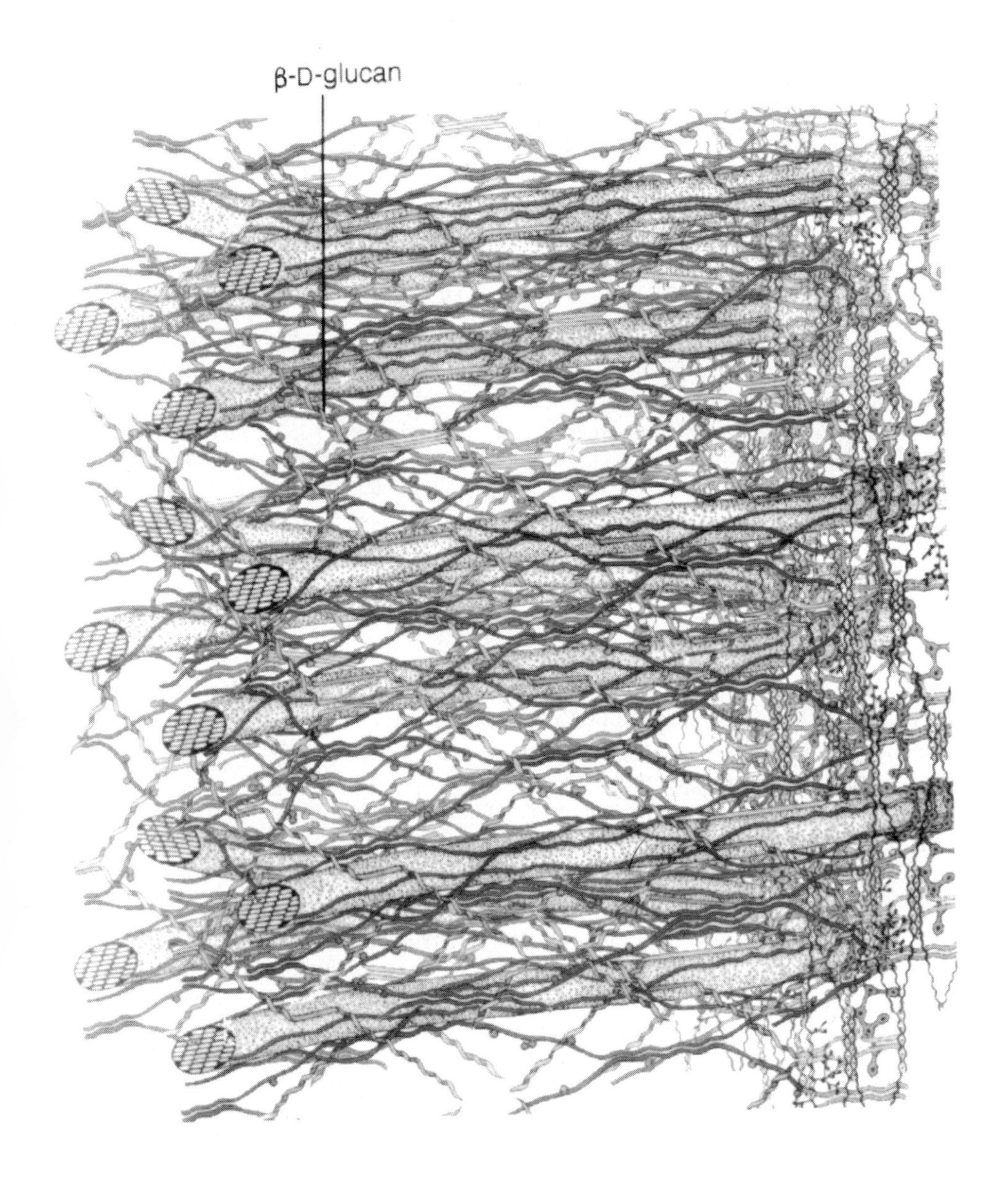
β-D-glucan
B

STRUCTURAL DYNAMICS DURING CELL ELONGATION

Several structural and architectural changes accompany elongation of grass cell walls. Some of the changes involve the synthesis and turnover of developmental-stage-specific cross-linking glycans that are mechanical determinants of growth, whereas subtle alterations in pectin composition and orientation reflect enzymic or physiological control of growth. Undoubtedly, all of these changes reflect either directly or indirectly the biochemical and physical events of wall loosening, expansion, and retightening.

During isodiametric expansion of meristematic cells, the major noncellulosic polysaccharides are GAX and the pectic polysaccharides (14). Liquid cultures of somatic cells, in which little cell elongation is observed, have cell walls that are essentially devoid of β-D-glucan (21). In dividing and elongating cells, highly branched GAXs are abundant, whereas after elongation and differentiation more and more unbranched GAX accumulates (14, 15, 46). Turnover of Ara in coleoptiles has been observed in vitro with isolated coleoptile walls and wall fragments (34, 60) and in vivo by kinetic analysis of the interconversion of labeled Ara into the Xyl moieties of GAX by continuous recycling through the nucleotide-sugar pools (46).

Interactions between the HS-GAX, PGA, and RG may function as a control mechanism of wall metabolism. The PGAs of maize pectins contain nonmethyl esters, whose formation and disappearance are coincident with the most rapid rate of cell elongation (85) and the accumulation of the HS-GAX (14). Maize GalAs are esterified in a greater proportion than can be accounted for by methyl esters alone, and formation of these nonmethyl esters is associated with covalent attachment to the primary wall (85). Brown & Fry (9) discovered an unusual hydrophobic ester of GalA, but they did not identify the alcohol moiety. These nonmethyl esters of PGAs may have escaped detection by both kinds of pectin antibodies (87). Schopfer (141) found Yariv-positive AGPs located at the inner surface of the growth-controlling epidermal wall of the maize coleoptile, a position that suggests an involvement in cell expansion. Schindler et al (140) tested this hypothesis with three monoclonal antibodies directed against different AGP epitopes. Their finding of AGPs in the membranes of developing sclerenchyma and tracheids supports the idea that the AGPs are developmental markers, but the inability of auxin to modulate the amount of soluble AGPs in nonvascular tissues or epidermis is evidence for a lack of involvement of these peptidoglycans in wall loosening (140).

The pectin domain may exert growth-controlling functions, but the interaction of the cross-linking glycans and cellulose is the tensile force-bearing structure that must be loosened. Taiz (154) reviewed much of the early literature on enzymatic activities associated with elongation in grasses and dicots. Subsequent studies focused on the specific polysaccharides in the walls and how they might be modified to permit extension. The synthesis of polysaccharides specific to cell expansion is a special feature of the grasses. Some arabinans, particularly the 5-linked arabinans, are found in the walls of dividing cells but are no longer made during cell expansion (14). Instead, the β-D-glucans are now synthesized along with GAX (Figure 3b). The β-D-glucans are substrate for two enzymes also enriched in growing seedlings: endo-β-D-glucanase, which requires at least four contiguous (1→4)-β-D-Glc units for activity (57), and an exo-glucanase, which yields Glc from the oligosaccharides produced by the endo-β-D-glucanase (67). These types of glycanohydrolases are thought to be secreted to or activated at specific sites within the cell wall to cleave tension-bearing β-D-glucans that interlace the cellulose microfibrils (154). The β-D-glucan is enriched in the outer walls of the epidermal cells (108, 137), the organ that is likely to be the layer that controls elongation rate (89, 90).

The appearance of β-D-glucan during cell expansion, the association of hydrolysis of β-D-glucan *in muro,* and acceleration of its hydrolysis rates by growth regulators all implicate direct physical involvement of the polymer in extension growth. There is some skepticism about this hypothesis because of a lack of correlation between growth and loss of polysaccharide. β-D-glucan is a stage-specific polymer that accumulates only transiently during expansion and disappears when growth is completed (14, 98). Growth in vivo is accompanied by a net accumulation of the β-D-glucan that masks an extremely rapid turnover, and only in excised grass tissues not provided with sugar is the loss of glucan accelerated during elongation (154). Intact grass coleoptiles and deepwater rice accumulate β-D-glucan during the elongation phase of growth (14, 98). In intact plants, the remaining glucan is gradually lost from the wall once growth and synthesis of new β-D-glucan ceases. Hence, the synthesis and hydrolysis of the β-D-glucan are in dynamic equilibrium. The correlation factor for β-D-glucan and growth is not the relative abundance but the rate of turnover, i.e. the balance between synthesis and degradation. Because of the efficiency of sugar recycling mechanisms, turnover in vivo can be measured only indirectly (46).

Enzyme- and substrate-specific antibodies offer an alternative approach to test the role of β-D-glucan hydrolysis in wall expansion. Antibodies against exo- and endoglucanases purified from the walls of growing seedlings partially reverse auxin-induced growth (71). Antibodies against the β-D-glucan itself can also inhibit growth (64, 66). Hoson et al (65) found that antibodies against the XyG nonasaccharide were able to inhibit growth in legumes but not grasses, which is an apparent demonstration of the specificity of the antibody interaction with a specific hydrolase.

New hypotheses have been suggested for how stress relaxation of the wall may result in cell expansion without glucan hydrolysis (44, 105). Two different kinds of wall proteins are implicated in expansion in XyG-cross-linked Type I walls without net depolymerization of a cross-linking glycan. A XyG endotransglycosylase (XET) catalyzes transglycosylation rather than hydrolysis of XyG. In this reaction, one chain of XyG is cleaved but reattached to the nonreducing terminus of another XyG chain (44). By such a mechanism, the microfibrils could undergo a transient slippage without compromising tensile strength. It also provides a mechanism whereby new linkages between polymers can occur when juxtaposed after microfibril slippage. Grass root tips also exhibit substantial XET activity, and this activity is correlated with extension growth (122, 170). A wheat XET gene shows better than 70% sequence homology with several dicot XETs (118). These findings are surprising because XyGs are neither a major structural polymer of the grasses nor possess a unit structure that would be expected to produce effective oligomeric substrates for XET in vivo. No polysaccharide other than XyG has been shown to be substrate for XET.

Cosgrove and his colleagues (106, 107) discovered two proteins, called expansins, that catalyze wall extension in vitro without detectable hydrolysis or transglycosylation. These proteins induce extension of paper in stress relaxation assays, which indicates that they probably catalyze breakage of hydrogen bonds (104). Homologous grass expansins could disrupt the tethering of cellulose by β-D-glucans and xylans in the Type II walls. Expansins induce extension in heat-killed abraded coleoptile sections (31), and oat expansin induces extension of cucumber walls and is similar in activity to the cucumber enzymes (97). Although the oat expansin is abundant in sections of the coleoptile exhibiting substantial growth, expansin is still present in cells that are not growing, which suggests that other mechanisms modify the wall to make it insensitive to expansin action. The substrate(s) for expansin in grasses are not yet established, but expansin binding to cellulose is enhanced by coating the alkali-extracted walls or fibrous cellulose with β-D-glucan (105). A 40-kDa glycoprotein was localized specifically in the walls of elongating cells of several gramineous and nongramineous plant species (130). The protein is found in the walls of coleoptile tissues and is induced in expanding leaves in deep-water rice. Although the appearance of this cell-wall protein is tightly coupled with elongation, the function is unknown.

To summarize, enzymes and proteins implicated directly and indirectly in wall extension and growth include those involved in pectin assembly and gel formation, the turnover of AGPs, the assembly of GAX, the breaking of hydrogen bonds between cross-linking glycans, and the turnover of the β-D-glucans. Once the genes of these enzymes are cloned, molecular experiments could be used to assess the specific role of each of these proteins in growth processes.

Both esterified and etherified cinnamic acid constituents accumulate at rates inversely correlated with elongation rates (16, 110, 155). The cessation of growth is correlated with the appearance of cinnamyl alcohol dehydrogenase (112, 137) and specific peroxidases (86, 99) required for cinnamate synthesis and polymerization in the grasses. The accumulation of the phenolic esters and ethers is an important event not only for locking the cells into their final shape but also for providing tensile and compression strength. Barley cells in liquid culture grown in the presence of dichlorobenzonitrile compensate for the lack of cellulose by increasing cross-linking of polymers by phenolic compounds, an alteration that increases the tensile strength of the wall despite cellulose levels only 30% of normal (144).

CELL-WALL BIOGENESIS IN GRASSES

The chemical structure and architecture of the cell walls of grasses change markedly during the stages of cytokinesis, elongation, and maturation. Like all flowering plants, the developing cell plate—or phragmoplast—is derived from vesicles of the Golgi apparatus and grows outward until it fuses with the existing primary wall (150). As in animal cells, the plant Golgi apparatus is a factory for the synthesis, processing, and targeting of glycoproteins (150). The Golgi apparatus has been established by autoradiography and immunocytochemistry as the site of synthesis and export of all the cross-linking glycans and pectins.

Cellulose and callose, an unbranched (1→3)-linked glucan, are the only polymers known to be made at the plasma membrane surface in any plant (35). Cellulose synthesis in higher plants is likely to occur at six-membered hexagonal terminal complexes called rosettes. The first rosettes observed in freeze fracture replicas of plasma membranes were those of the maize root (113). The quest for cellulose synthesis in vitro has been a long one. Recent attempts to identify polypeptides involved in the synthesis have focused on design of photoaffinity and other labels and the use of product entrapment (35, 48, 68). Meikle et al (109) precipitated callose synthase from *Lolium* membranes with monoclonal antibodies and detected several affinity-labeled polypeptides that were similar to those labeled in nongramineous species. The identity and function of these proteins are unknown.

Pulse-labeling studies with intact seedlings and excised tissues have revealed many dynamic features of polysaccharide synthesis and turnover in vivo. In excised maize coleoptiles, Ara is incorporated into UDP-Ara and UDP-Xyl, but not into any hexose, whereas Xyl is incorporated into the hexose and pentose nucleotide sugars with distribution similar to that of Glc (19). These data reflect differing strategies that plants possess in managing the sugars hydrolyzed from polymers during growth. Plants possess C-1 kinases for Ara, Gal, Man, GalA, GlcA, Rha, and Fuc. Subsequently, NDP pyrophosphorylases catalyze the formation of the UDP or GDP sugars (37). All of these sugars are synthesized de novo from UDP-Glc, or in some instances from GDP-Glc, via nucleotide-sugar interconversion pathways. The existence of these salvage pathways is considered evidence that turnover of polysaccharides involves hydrolysis of the sugars from specific polysaccharides or proteins (46). Whereas stress relaxation of the wall is the physical event, synthesis is the biological event (132); both are vital and, to an extent, biochemically integrated during sustained growth. Even though substantial growth can occur without detectable secretion of new polysaccharide, inhibition by monensin or brefeldin A of secretory vesicle transport to the plasma membrane quickly stops growth (139).

Biosynthesis of (1→3),(1→4)-β-D-Glucan

We know very little about how cell-wall polysaccharides are organized for secretion and then modified for assembly into a functional matrix, but we are beginning to learn some of the features of the polymerization from studies of synthesis in vitro. Some special features of polysaccharide synthesis in the grasses are noteworthy. The grass β-D-glucan differs from other (1→4)-β-D-glucans and callose because the (1→4)- and (1→3)-β-D-glucosyl linkages are ordered within the polymer. Synthesis of authentic β-D-glucan in vitro is determined by hydrolysis of the radioactive products formed with Golgi membranes and UDP-Glc with the sequence-dependent *B. subtilis* endoglucanase, which yields diagnostic cellobiosyl- and cellotriosyl-(1→3)-β-D-Glc. These oligomers are separated by HPLC and assayed for incorporation of radioactivity (47). Synthesis of β-D-glucan with cellular membranes was shown by comparable techniques (59, 109). UDP-Glc is substrate and Mg^{2+} or Mn^{2+} is required as cofactor. Improvement in the recovery assay of polymerase activity was provided by a flotation centrifugation method for isolating Golgi membranes and by improved procedures for separating and identifying, by GLC mass spectrometry and GLC gas proportional counting, diagnostic linkage derivatives of the products (45). The combination of gel permeation chromatography, linkage analysis, and enzymic digestion confirmed that entire tri- and tetrasaccharide units were synthesized and that the macromolecular β-D-glucan synthesized in vitro in the Golgi apparatus was identical to the cell-wall polysaccharide (47).

Unlike the Golgi apparatus from plants with Type I walls, the maize Golgi membranes also possess callose synthases. Because the Golgi synthase activity was stimulated only twofold by $CaCl_2$ as compared with sevenfold in plasma membrane, the activity could not be attributed solely to contamination with plasma membrane (49). Disruption of membrane integrity with detergents and ionophores abolished β-D-glucan synthase activity but increased the synthesis of callose. The Golgi-specific callose synthase may represent the default synthase, and similar to that of the plasma membrane, its activity is stable to solubilization with digitonin or CHAPS. Meikle et al (109) have demonstrated that a polypeptide transiently associated with the synthesis of β-D-glucan synthesizes only (1→3)-linked glucosyl units when isolated with detergents. After removal of inhibitors of activity with a Ca^{2+}-precipitation technique, microfibrillar callose was synthesized in copious amounts by a *Lolium* callose synthase enriched by product entrapment (11). Like no other polysaccharide from higher plants, the grass β-D-glucan contains an ordered arrangement of cellulosic (1→4)-β-D-linkages and callosic (1→3)-β-D-linkages. Because of the kinds of linkages formed by the β-D-glucan synthase, the divalent cation requirements, the substrate and apparent K_m, and the appearance of callose synthase upon damage of Golgi membranes, the grass synthase genes may have evolved from ancestral cellulose synthase genes.

Biosynthesis of Other Cell-Wall Polysaccharides

Other cell-wall polysaccharides are made in the Golgi apparatus, and some of these activities have also been studied in vitro. Incorporation of Ara from UDP-Ara and Xyl from UDP-Xyl into ethanol-insoluble products thought to represent arabinan and xylan, respectively, have been demonstrated in membrane preparations from grass tissues (162). For complex xylans such as GAX, complementary additions of UDP-Ara and UDP-Xyl should result in branched polymers containing increasing amounts of 2,4- and 3,4-linked xylosyl units and a corresponding amount of nonreducing terminal arabinofuranosyl units. Xylans and glucuronoxylans can be made in vitro, although the radioactive products have not been characterized as extensively. One of the more intriguing questions is how arabinofuranosyl units are made. L-Arabinose is in the furanose ring conformation in most plant polymers containing this sugar, including GAX, 5-linked arabinans, AGP, and extensin, whereas UDP-Ara is exclusively in the pyranose form (43). Arabinosyl transferase must be distinct from other glycosyl transferases in its ability to permit ring rearrangement before addition of the sugar to the polymer.

GENETIC MODELS OF CELL-WALL DEVELOPMENT IN THE GRASSES

The work of Reiter et al (131) and Chapple et al (25) illustrates how cell-wall mutants in *Arabidopsis* can be used to understand polysaccharide and lignin biosynthetic pathways and functions. A similar genetic model is needed for the grasses. There are many genera within the Poales that could potentially serve as pure genetic models. For example, *Eragrostis* and *Chloris* are small annual grasses with short life cycles, are 2N, and possess genome sizes only two to three times that of *Arabidopsis* (6). The genome sizes of most of the cereal crop species are enormous by comparison with other flowering plants, which makes them rather unwieldy systems to generate and screen mutants for specific traits and to isolate genes by chromosome walking. Rice possesses the smallest genome of the major cereal crops but is difficult to manage in large numbers within a greenhouse environment. These problems are offset by genetic maps well-populated with markers, and current map-based cloning strategies make it possible to move at a reasonable pace from a phenotypic mutation to a cloned gene.

Although not a single synthase involved in cell-wall polysaccharide synthesis has been purified and its gene knowingly cloned, many of the enzymes that function in the depolymerization of the cell wall are well characterized. In grasses, three separate enzyme activities are associated with β-D-glucan metabolism. They are encoded by genes within several related gene families, and many representative members of the families have been sequenced (62). The activities include (1→3)-β-D-glucanases, which are present in the developing grains, persist throughout germination, and are homologous with the pathogenesis-related glucanases (38); the (1→3,1→4)-β-D-glucanases, which appear exclusively in the grains during germination and depolymerize the cereal β-D-glucan (39); and the (1→4)-β-D-glucanases found in developing seedlings (57). Two exoglucanases have also been described: a 60-kDa enzyme that appears in the grains (94) and a 72-kDa enzyme that is associated with the walls of elongating cells (71). These enzymes hydrolyze nonreducing *t*-Glc(1→3)- or *t*-Glc-(1→4)-linked units from the oligosaccharides produced by endoglucanases.

The cereal (1→3,1→4)-β-D-glucanases, similar to the *B. subtilis* endoglucanase, cleave a (1→4)-linkage only if the adjacent Glc on the nonreducing side is (1→3)-linked (151). This property makes the grass β-D-glucan a good substrate for the enzyme, but the enzyme is unable to hydrolyze glucans composed of solely (1→3)- or (1→4)-β-D-glucosyl units. It is interesting to note that the barley (1→3,1→4)-β-D-glucanase genes are homologous to the (1→3)-β-D-glucanase genes but are completely unrelated to bacterial enzymes with the same carbohydrate linkage specificity (39, 171). The barley glucanase has been crystallized and its three-dimensional structure determined (28, 159). From mapping of the substrate-binding cleft, strategies to alter substrate specificity and other properties of the enzyme by site-directed mutagenesis are possible (62).

Many of the genes that encode enzymes in the synthetic pathways of the lignin precursors have been cloned. Research is now focused on characterizing the regulatory elements as a strategy to modify lignin biosynthesis (163). Several "brown-midrib" (*bmr*) mutants of maize, *Sorghum,* and millet have impaired ability to synthesize lignin, and, in some instances, this factor increases digestibility (29). The maize *bmr3* corresponds to an impaired gene encoding caffeate *O*-methyl transferase (161), an enzyme activity shown to be reduced in maize and millet *bmr* mutants (29, 53). Because lignin is a compression-strength component of the xylem and lignin-like substances can form a line of defense against pathogens, a simple reduction of phenylpropanoid and lignin content in plants is probably not a viable strategy to increase digestibility. Lignin content and structure are not the only factors in digestibility of the grasses. The extensive networks of ester- and ether-linked hydroxycinnamic acids in nonlignified cells and the pectin and other polysaccharide interactions with the phenylpropanoid network undoubtedly are contributing factors in decreased digestibility (77).

The starchy endosperm, a special trait of the Poales, is the fundamental reason that cereals are of central importance in human nutrition (93). The world harvests over one billion tons of cereal grains annually. Rice and wheat alone provide at least half of the calories that humans ingest. The cell walls of the grasses also figure heavily in the nutrition equation, from the β-D-glucans that constitute up to 70% of the endosperm walls, to the vast amounts of xylan- and cellulose-rich walls that are consumed by grazing animals. The β-D-glucans can be both a benefit and a problem. As a benefit, they are the wall constituents implicated in the ability of barley and oat brans to reduce serum cholesterol in hypercholesterolemic individuals (8) and to modulate glucoregulation in diabetics (166). The composition of the endosperm cell walls in flours is a contributing factor to bread quality (50). Incomplete hydrolysis of the viscous β-D-glucans during the brewing process is a major production problem and contributes to "hazing" of beers upon storage (168). Lignin, esterified and etherified aromatic substances, and other chemical modifications of the primary wall polysaccharides greatly reduce the nutritional quality of grasses for ruminants (77). Considering the social and economic impact of the cereals, it is not surprising that there is a considerable knowledge base in the genetics of the cereal crops. This knowledge base could be more widely used to provide information on the structure, biogenesis, and turnover of the special cell wall of the grasses and in the process reveal ways that the cell wall can be modified for agronomic benefit.

ACKNOWLEDGMENTS

I thank Tony Bacic, University of Melbourne; Larry Dunkle, Purdue University; Maureen McCann, John Innes Centre; and John Ralph, University of Wisconsin–Madison for valuable discussions and their critical review of this manuscript. I thank Debby Sherman, Purdue University Electron Microscopy Facility, for her help in the digital imaging of the figures. My work on cell-wall biogenesis in the grasses has been supported largely by grant DE-FG02-88ER13903 of the Division of Biological Energy Sciences, US Department of

Energy. This is publication number 14,910 of the Purdue University Agriculture Experiment Station.

Literature Cited

1. Aspinall GO. 1959. Structural chemistry of the hemicelluloses. *Adv. Carbohydr. Chem.* 14:429–68
2. Bacic A, Churms SC, Stephen AM, Cohen PB, Fincher GB. 1987. Fine structure of the arabinogalactan-protein from *Lolium multiflorum. Carbohydr. Res.* 162:85–93
3. Bacic A, Harris PJ, Stone BA. 1988. Structure and function of plant cell walls. In *The Biochemistry of Plants,* ed. J Priess, 14: 297–371. New York: Academic
4. Bacic A, Moody SF, Clarke AE. 1986. Structural analysis of secreted root slime from maize (*Zea mays* L.). *Plant Physiol.* 80:771–77
5. Bacic A, Stone BA. 1981. Chemistry and organization of aleurone cell wall components from wheat and barley. *Aust. J. Plant Physiol.* 8:475–95
6. Bennett MD, Smith JB. 1976. Nuclear DNA amounts in angiosperms. *Philos. Trans. R. Soc. Lond. Ser. B* 274:228–74
7. Bohlmann H, Apel K. 1991. Thionins. *Annu. Rev. Plant Physiol. Plant Mol. Biol.* 42:227–40
8. Braaten JT, Wood PJ, Scott FW, Wolynetz MS, Lowe MK, et al. 1994. Oat β-glucan reduces blood cholesterol concentration in hypercholesterolemic subjects. *Eur. J. Clin. Nutr.* 48:465–74
9. Brown JA, Fry SC. 1993. Novel *O*-D-galacturonoyl esters in the pectic polysaccharides of suspension-cultured plant cells. *Plant Physiol.* 103:993–99
10. Buchala AJ, Meier H. 1972. Hemicelluloses from the stalk of *Cyperus papyrus. Phytochemistry* 11:3275–78
11. Bulone V, Fincher GB, Stone BA. 1995. *In vitro* synthesis of a microfibrillar (1→3)-β-glucan by a ryegrass (*Lolium multiflorum*) endosperm (1→3)-β-glucan synthase enriched by product entrapment. *Plant J.* 8: 213–25
12. Caelles C, Delseny M, Puigdomènech P. 1992. The hydroxyproline-rich glycoprotein gene from *Oryza sativa. Plant Mol. Biol.* 18:617–19
13. Carpita NC. 1983. Hemicellulosic polymers of cell walls of *Zea* coleoptiles. *Plant Physiol.* 72:515–21
14. Carpita NC. 1984. Cell wall development in maize coleoptiles. *Plant Physiol.* 76: 205–12
15. Carpita NC. 1984. Fractionation of hemicelluloses from maize cell walls with increasing concentrations of alkali. *Phytochemistry* 23:1089–93
16. Carpita NC. 1986. Incorporation of proline and aromatic amino acids into cell walls of maize coleoptiles. *Plant Physiol.* 80: 660–66
17. Carpita NC. 1989. Pectic polysaccharides of maize coleoptiles and proso millet cells in liquid culture. *Phytochemistry* 28: 121–25
18. Carpita NC. 1996. Structure and biogenesis of plant cell walls. In *Plant Metabolism,* ed. DT Dennis, DH Turpin, D LeFebre, D Layzell. London: Longman. In press
19. Carpita NC, Brown RA, Weller KM. 1982. Uptake and metabolic fate of glucose, arabinose, and xylose by *Zea mays* coleoptiles in relation to cell wall synthesis. *Plant Physiol.* 69:1173–80
20. Carpita NC, Gibeaut DM. 1993. Structural models of the primary cell walls in flowering plants: consistency of molecular structure with the physical properties of the walls during growth. *Plant J.* 3:1–30
21. Carpita NC, Mulligan JA, Heyser JW. 1985. Hemicelluloses of cell walls of a proso millet cell suspension culture. *Plant Physiol.* 79:480–84
22. Carpita NC, Whittern D. 1986. A highly substituted glucuronoarabinoxylan from developing maize coleoptiles. *Carbohydr. Res.* 146:129–40
23. Cassab GI, Varner JE. 1988. Cell wall proteins. *Annu. Rev. Plant Physiol. Plant Mol. Biol.* 39:321–53
24. Chaboud A, Rougier M. 1984. Identification and localization of sugar components of rice (*Oryza sativa* L.) root cap mucilage. *J. Plant Physiol.* 116:323–30
25. Chapple CCS, Vogt T, Ellis BE, Somerville CR. 1992. An *Arabidopsis* mutant defec-

tive in the general phenylpropanoid pathway. *Plant Cell* 4:1413–24
26. Chase MW, Soltis DE, Olmstead RG, Morgan D, Les DH, et al. 1993. Phylogenetics of seed plants: an analysis of nucleotide sequences from the plastid gene *rbc*L. *Ann. Missouri Bot. Gard.* 80:528–80
27. Chay CH, Buehler EG, Thorn JM, Whelan TM, Bedinger PA. 1992. Purification of maize pollen exines and analysis of associated proteins. *Plant Physiol.* 100:756–61
28. Chen L, Garrett TPJ, Varghese JN, Fincher GB, Høj PB. 1993. Crystallization and preliminary X-ray analysis of (1,3)- and (1,3;1,4)-β-D-glucanases from germinating barley. *J. Mol. Biol.* 234:888–89
29. Cherney JH, Cherney JR, Akin DE, Axtell JD. 1991. Potential of brown-midrib, low-lignin mutants for improving forage quality. *Adv. Agron.* 46:157–98
30. Chesson A, Gordon AH, Lomax JA. 1983. Substituent groups linked by alkali-labile bonds to arabinose and xylose residues of legume, grass, and cereal straw cell walls and their fate during digestion by rumen microorganisms. *J. Sci. Food Agric.* 34: 1330–40
31. Cosgrove DJ, Li ZC. 1993. Role of expansin in cell enlargement of oat coleoptiles: analysis of developmental gradients and photocontrol. *Plant Physiol.* 103:1321–28
32. Dahlgren RMT, Clifford HT, Yeo PF. 1985. *The Families of the Monocotyledons: Structure, Evolution, and Taxonomy.* New York: Springer-Verlag
33. Darvill A, McNeil M, Albersheim P, Delmer DP. 1980. The primary cell walls of flowering plants. In *The Biochemistry of Plants,* ed. NE Tolbert, 1:91–162. New York: Academic
34. Darvill AG, Smith CJ, Hall MA. 1978. Cell wall structure and elongation growth in *Zea mays* coleoptile tissue. *New Phytol.* 80: 503–16
35. Delmer DP, Amor Y. 1995. Cellulose biosynthesis. *Plant Cell* 7:987–1000
36. Duvall MR, Clegg MT, Chase MW, Clark WD, Kress WJ, et al. 1993. Phylogenetic hypotheses for the monocotyledons constructed from *rbc*L sequence data. *Ann. Missouri Bot. Gard.* 80:607–19
37. Feingold DS. 1982. Aldo (and keto) hexoses and uronic acids. In *The Encyclopedia of Plant Physiology, New Series,* ed. FA Loewus, W Tanner, 13A:3–76. Berlin: Springer-Verlag
38. Fincher GB. 1989. Molecular and cellular biology associated with endosperm mobilization in germinating cereal grains. *Annu. Rev. Plant Physiol. Plant Mol. Biol.* 40: 305–46
39. Fincher GB, Lock PA, Morgan MM, Lingelbach K, Wettenhall REH, et al. 1986. Primary structure of the (1→3,1→4)-β-D-glucan 4-glucohydrolase from barley aleurone. *Proc. Natl. Acad. Sci. USA* 83: 2081–85
40. Fincher GB, Stone BA, Clarke AE. 1983. Arabinogalactan-proteins: structure, biosynthesis, and function. *Annu. Rev. Plant Physiol.* 34:47–70
41. Finkle BJ, Masri MS. 1964. Methylation of polyhydroxyaromatic compounds by pampas grass *O*-methyltransferase. *Biochim. Biophys. Acta* 85:167–69
42. Fritz SE, Hood KR, Hood EE. 1991. Localization of soluble and insoluble fractions of hydroxyproline-rich glycoproteins during maize kernel development. *J. Cell Sci.* 98: 545–50
43. Fry SC, Northcote DH. 1983. Sugar-nucleotide precursors of arabinopyranosyl, arabinofuranosyl, and xylopyranosyl residues in spinach polysaccharides. *Plant Physiol.* 73:1055–61
44. Fry SC, Smith RC, Renwick KF, Martin DJ, Hodge SK, Matthews KJ. 1992. Xyloglucan endotransglycosylase, a new wall-loosening enzyme activity from plants. *Biochem. J.* 282:821–28
45. Gibeaut DM, Carpita NC. 1990. Separation of membranes by flotation centrifugation for *in vitro* synthesis of plant cell wall polysaccharides. *Protoplasma* 156:82–93
46. Gibeaut DM, Carpita NC. 1991. Tracing cell wall biogenesis in intact cells and plants: selective turnover and alteration of soluble and cell wall polysaccharides in grasses. *Plant Physiol.* 97:551–61
47. Gibeaut DM, Carpita NC. 1993. Synthesis of (1→3),(1→4)-β-D-glucan in the Golgi apparatus of maize coleoptiles. *Proc. Natl. Acad. Sci. USA* 90:3850–54
48. Gibeaut DM, Carpita NC. 1994. Biosynthesis of plant cell wall polysaccharides. *FASEB J.* 8:904–15
49. Gibeaut DM, Carpita NC. 1994. Improved recovery of (1→3),(1→4)-β-D-glucan synthase activity from Golgi apparatus of *Zea mays* (L.) using differential flotation centrifugation. *Protoplasma* 180:92–97
50. Girhammerar U, Nakamura M, Nair BM. 1986. Water-soluble pentosans from wheat and rye: chemical composition and some physical properties in solution. In *Gums and Stabilisers for the Food Industry,* ed. GO Phillips, PA Wedlock, DJ Williams, pp. 123–34. London: Elsevier
51. Gleeson PA, McNamara M, Wettenhall REH, Stone BA, Fincher GB. 1989. Characterization of the hydroxyproline-rich protein core of an arabinogalactan-protein secreted from suspension-cultured *Lolium multiflorum* (Italian ryegrass) endosperm cells. *Biochem. J.* 264:857–62
52. Gordon AH, Lomax JA, Dalgarno K, Chesson A. 1985. Preparation and composition of mesophyll, epidermis, and fibre cell

walls from leaves of perennial ryegrass (*Lolium perenne*) and Italian ryegrass (*Lolium multiflorum*). *J. Food Sci. Agric.* 36:509–19
53. Grand C, Parmentier P, Boudet A, Boudet AM. 1985. Comparison of lignins and of enzymes involved in lignification in normal and brown midrib (*bm3*) mutant corn seedlings. *Physiol. Vég.* 23:905–11
54. Gubler F, Ashford AE, Bacic A, Blakeney AB, Stone BA. 1985. Release of ferulic acid esters from barley aleurone. II. Characterisation of the feruloyl compounds released in response to GA3. *Aust. J. Plant Physiol.* 12:307–17
55. Hanley AB, Russell WR, Chesson A. 1993. Formation of substituted truxillic and truxinic acids in plant cell walls—a rationale. *Phytochemistry* 33:957–60
56. Harris PJ, Hartley RD. 1980. Phenolic constituents of the cell walls of monocotyledons. *Biochem. System. Ecol.* 8:153–60
57. Hatfield RD, Nevins DJ. 1987. Hydrolytic activity and substrate specificity of an endoglucanase from *Zea mays* seedling cell walls. *Plant Physiol.* 83:203–7
58. Haworth WN, Hirst EL, Oliver E. 1934. Polysaccharides. Part XVIII. The constitution of xylan. *J. Chem. Soc.* 1917–23
59. Henry RJ, Stone BA. 1982. Factors influencing β-glucan synthesis by particulate enzymes from suspension-cultured *Lolium multiflorum* endosperm cells. *Plant Physiol.* 69:632–36
60. Heyn ANJ. 1986. A gas chromatographic analysis of the release of arabinose from coleoptile cell walls under the influence of auxin and dextranase preparations. *Plant Sci.* 45:77–82
61. Hoffmann-Benning S, Klomparens KL, Kende H. 1994. Characterization of growth-related osmiophilic particles in corn coleoptiles and deepwater rice internodes. *Ann. Bot.* 74:563–72
62. Høj PB, Fincher GB. 1995. Molecular evolution of plant β-glucan endohydrolases. *Plant J.* 7:367–79
63. Hood KR, Baasiri RA, Fritz SE, Hood EE. 1991. Biochemical and tissue print analyses of hydroxyproline-rich glycoproteins in cell walls of sporophytic maize tissues. *Plant Physiol.* 96:1214–19
64. Hoson T, Masuda Y, Nevins DJ. 1992. Comparison of the inner and outer epidermis. Inhibition of auxin-induced elongation of maize coleoptiles by glucan antibodies. *Plant Physiol.* 98:1298–1303
65. Hoson T, Masuda Y, Sone Y, Misaki A. 1991. Xyloglucan antibodies inhibit auxin-induced elongation and cell wall loosening of azuki bean epicotyls but not of oat coleoptiles. *Plant Physiol.* 96:551–57
66. Hoson T, Nevins DJ. 1989. β-D-glucan antibodies inhibit auxin-induced cell elongation and changes in the cell wall of *Zea* coleoptile segments. *Plant Physiol.* 90: 1353–58
67. Huber DJ, Nevins DJ. 1981. Partial purification of endo- and exo-β-D-glucanase enzymes from *Zea mays* L. seedlings and their involvement in cell wall autohydrolysis. *Planta* 151:206–14
68. Iiyama K, Lam TBT, Meikle PJ, Ng K, Rhodes DI, Stone BA. 1993. Cell wall biosynthesis and its regulation. In *Forage Cell Wall Structure and Digestibility*, ed. HG Jung, DR Buxton, RD Hatfield, J Ralph, pp. 621–83. Madison, WI: Am. Soc. Agron.
69. Iiyama K, Lam TBT, Stone BA. 1990. Phenolic acid bridges between polysaccharides and lignin in wheat internodes. *Phytochemistry* 29:733–37
70. Iiyama K, Lam TBT, Stone BA. 1994. Covalent cross-links in the cell wall. *Plant Physiol.* 104:315–20
71. Inouhe M, Nevins DJ. 1991. Inhibition of auxin-induced cell elongation of maize coleoptiles by antibodies specific for cell wall glucanases. *Plant Physiol.* 96:426–31
72. Ishii T. 1991. Isolation and characterization of a diferuloyl arabinoxylan hexasaccharide from bamboo shoot cell walls. *Carbohydr. Res.* 219:15–22
73. Jarvis MC. 1984. Structure and properties of pectin gels in plant cell walls. *Plant Cell Environ.* 7:153–64
74. Jarvis MC, Forsyth W, Duncan HJ. 1988. A survey of the pectic content of nonlignified monocot cell walls. *Plant Physiol.* 88: 309–14
75. Jermyn MA, Yeow YM. 1975. A class of lectins present in the tissues of seed plants. *Aust. J. Plant Physiol.* 2:501–31
76. Josè-Estanyol M, Ruiz-Avila L, Puigdomènech P. 1992. A maize embryo-specific gene encodes a proline-rich and hydrophobic protein. *Plant Cell* 4:413–23
77. Jung HJG, Buxton DR. 1994. Forage quality variation among maize inbreds: relationships of cell-wall composition and in-vitro degradability for stem internodes. *J. Sci. Food Agric.* 66:313–22
78. Kato Y, Iki K, Matsuda K. 1981. Cell-wall polysaccharides of immature barley plants. II. Characterization of a xyloglucan. *Agric. Biol. Chem.* 45:2745–53
79. Kato Y, Nevins DJ. 1985. Isolation and identification of *O*-(5-*O*-feruloyl-α-L-arabinosyl) - (1→3) - *O*-β-D-xylopyranosyl- (1 →4)-D-xylopyranose as a component of *Zea* shoot cell-walls. *Carbohydr. Res.* 137: 139–50
80. Kato Y, Nevins DJ. 1986. Fine structure of (1→3),(1→4)-β-D-glucan from *Zea* shoot cell-walls. *Carbohydr. Res.* 147:69–85
81. Kieliszewski MJ, Kamyab A, Leykam JF, Lamport DTA. 1992. A histidine-rich ex-

tensin from *Zea mays* is an arabinogalactan protein. *Plant Physiol.* 99:538–47
82. Kieliszewski M, Lamport DTA. 1987. Purification and partial characterization of a hydroxyproline-rich glycoprotein in a graminaceous monocot, *Zea mays. Plant Physiol.* 85:823–27
83. Kieliszewski MJ, Lamport DTA. 1994. Extensin: repetitive motifs, functional sites, post-translocational codes, and phylogeny. *Plant J.* 5:157–72
84. Kieliszewski MJ, Leykam JF, Lamport DTA. 1990. Structure of the threonine-rich extensin from *Zea mays. Plant Physiol.* 92: 316–26
85. Kim JB, Carpita NC. 1992. Changes in esterification of the uronic acid groups of cell wall polysaccharides during elongation of maize coleoptiles. *Plant Physiol.* 98: 646–53
86. Kim SH, Shinkle JR, Roux SJ. 1989. Phytochrome induces changes in the immunodetectable level of a wall peroxidase that precede growth changes in maize seedlings. *Proc. Natl. Acad. Sci. USA* 86: 9866–70
87. Knox JP, Linstead PJ, King J, Cooper C, Roberts K. 1990. Pectin esterification is spatially regulated both within cell walls and between developing tissues of root apices. *Planta* 181:512–21
88. Kröger N, Bergsdorf C, Sumper M. 1994. A new calcium binding glycoprotein family constitutes a major diatom cell wall component. *EMBO J.* 13:4676–83
89. Kutschera U, Bergfeld R, Schopfer P. 1987. Cooperation of epidermis and inner tissues in auxin-mediated growth of maize coleoptiles. *Planta* 170:168–80
90. Kutschera U, Kende H. 1988. The biophysical basis of elongation growth in internodes of deepwater rice. *Plant Physiol.* 88:361–66
91. Lam TBT, Iiyama K, Stone BA. 1992. Cinnamic acid bridges between cell wall polymers in wheat and phalaris internodes. *Phytochemistry* 32:1179–83
92. Lam TBT, Iiyama K, Stone BA. 1994. An approach to the estimation of ferulic acid bridges in unfractionated cell walls of wheat internodes. *Phytochemistry* 37: 327–33
93. Langenheim JH, Thimann KV. 1982. *Botany: Plant Biology and its Relation to Human Affairs.* New York: Wiley
94. Leah R, Kigel J, Svendsen I, Mundy J. 1995. Biochemical and molecular characterization of a barley seed β-glucosidase. *J. Biol. Chem.* 270:15789–97
95. Lei M, Wu R. 1991. A novel glycine-rich cell wall protein gene in rice. *Plant Mol. Biol.* 16:187–98
96. Lewis NG, Yamamoto E, Wooten JB, Just G, Ohashi H, Towers GHN. 1987. Monitoring biosynthesis of wheat cell-wall phenylpropanoids *in situ. Science* 237:1344–46
97. Li ZC, Durachko DM, Cosgrove DJ. 1993. An oat coleoptile wall protein that induces wall extension *in vitro* and that is antigenically related to a similar protein from cucumber hypocotyls. *Planta* 191:349–56
98. Luttenegger DG, Nevins DJ. 1985. Transient nature of a (1→3),(1→4)-β-D-glucan in *Zea mays* coleoptile walls. *Plant Physiol.* 77:175–78
99. MacAdam JW, Sharp RE, Nelson CJ. 1992. Peroxidase activity in the leaf elongation zone of tall fescue. II. Spatial distribution of apoplastic peroxidase activity in genotypes differing in length of the elongation zone. *Plant Physiol.* 99:879–85
100. Markwalder HU, Neukom H. 1976. Diferulic acid as a possible crosslink in hemicelluloses from wheat germ. *Phytochemistry* 15:836–37
101. McCann MC, Roberts K. 1992. Architecture of the primary cell wall. In *The Cytoskeletal Basis of Plant Growth and Form,* ed. CW Lloyd, pp. 109–29. London: Academic
102. McCann MC, Wells B, Roberts K. 1990. Direct visualization of cross-links in the primary plant cell wall. *J. Cell Sci.* 96: 323–34
103. McDougall GJ, Fry SC. 1994. Fucosylated xyloglucan in suspension-cultured cells of the gramineous monocotyledon, *Festuca arundinacea. J. Plant Physiol.* 143:591–95
104. McQueen-Mason S, Cosgrove DJ. 1994. Disruption of hydrogen bonding between plant cell wall polymers by proteins that induce plant wall extension. *Proc. Natl. Acad. Sci. USA* 91:6574–78
105. McQueen-Mason SJ, Cosgrove DJ. 1995. Expansin mode of action on cell walls: analysis of wall hydrolysis, stress relaxation, and binding. *Plant Physiol.* 107: 87–100
106. McQueen-Mason S, Durachko DM, Cosgrove DJ. 1992. Two endogenous proteins that induce cell wall extension in plants. *Plant Cell* 4:1425–33
107. McQueen-Mason SJ, Fry SC, Durachko DM, Cosgrove DJ. 1993. The relationship between xyloglucan endotransglycosylase and in-vitro cell wall extension in cucumber hypocotyls. *Planta* 190:327–31
108. Meikle PJ, Hoogenraad NJ, Bonig I, Clarke AE, Stone BA. 1994. A (1→3,1→4)-β-glucan-specific monoclonal antibody and its use in the quantitation and immunocytochemical location of (1→3),(1→4)-β-glucans. *Plant J.* 5:1–9
109. Meikle PJ, Ng KF, Johnson E, Hoogenraad NJ, Stone BA. 1991. The β-glucan synthase from *Lolium multiflorum:* detergent solubilization, purification using monoclonal antibodies, and photoaffinity labeling with a

novel photoreactive pyrimidine analogue of uridine 5′-diphosphoglucose. *J. Biol. Chem.* 266:22569–81
110. Miyamoto K, Ueda J, Takeda S, Ida K, Hoson T, et al. 1994. Light-induced increase in the contents of ferulic and diferulic acids in cell walls of *Avena* coleoptiles: its relationship to growth inhibition by light. *Physiol. Plant.* 92:350–55
111. Morikawa H, Senda M. 1978. Infrared analysis of oat coleoptile cell walls and oriented structure of matrix polysaccharides in the walls. *Plant Cell Physiol.* 19: 327–36
112. Morrison TA, Kessler JR, Hatfield RD, Buxton DR. 1994. Activity of two lignin biosynthesis enzymes during development of a maize internode. *J. Sci. Food Agric.* 65:133–39
113. Mueller SC, Brown RM Jr. 1980. Evidence for an intramembrane component associated with a cellulose microfibril-synthesizing complex in higher plants. *J. Cell Biol.* 84:315–26
114. Mueller-Harvey I, Hartley RD, Harris PJ, Curzon EH. 1986. Linkage of *p*-coumaroyl and feruloyl groups to cell-wall polysaccharides of barley straw. *Carbohydr. Res.* 148:71–85
115. Myton KE, Fry SC. 1994. Intraprotoplasmic feruloylation of arabinoxylans in *Festuca arundinacea* cell cultures. *Planta* 193: 326–30
116. Neish AC. 1961. Formation of *m*- and *p*-coumaric acids by enzymatic deamination of the corresponding isomers of tyrosine. *Phytochemistry* 1:1–24
117. Nishitani K, Nevins DJ. 1991. Glucuro noxylan xylanohydrolase: a unique xylanase with the requirement for appendant glucuronosyl units. *J. Biol. Chem.* 266: 6539–43
118. Okazawa K, Sato Y, Nakagawa T, Asada K, Kato I, et al. 1993. Molecular cloning and cDNA sequencing of endoxyloglucan transferase, a novel class of glycosyltransferase that mediates molecular grafting between matrix polysaccharides in plant cell walls. *J. Biol. Chem.* 268:25364–68
119. Parry DW, Hodson MJ, Sangster AG. 1984. Some recent advances in studies of silicon in higher plants. *Philos. Trans. R. Soc. London Ser. B* 304:537–49
120. Pillonel C, Hunziker P, Binder A. 1992. Multiple forms of the constitutive wheat cinnamyl alcohol dehydrogenase. *J. Exp. Bot.* 43:299–305
121. Potts JRM, Weklych R, Conn EE. 1974. The 4-hydroxylation of cinnamic acid by sorghum microsomes and the requirement of cytochrome P–450. *J. Biol. Chem.* 249: 5019–26
122. Pritchard J, Hetherington PR, Fry SC, Tomos AD. 1993. Xyloglucan endotransglycosylase activity, microfibril orientation and the profiles of cell wall properties along growing regions of maize roots. *J. Exp. Bot.* 44:1281–89
123. Puhlmann J, Bucheli E, Swain MJ, Dunning N, Albersheim P, et al. 1994. Generation of monoclonal antibodies against plant cell-wall polysaccharides. I. Characterization of a monoclonal antibody to a terminal α-(1→2)-linked fucosyl-containing epitope. *Plant Physiol.* 104:699–710
124. Ralph J, Hatfield RD, Quideau S, Helm RF, Grabber JH, Jung HJG. 1994. Pathway of *p*-coumaric acid incorporation into maize lignin as revealed by NMR. *J. Am. Chem. Soc.* 116:9448–56
125. Ralph J, Helm RF, Quideau S, Hatfield RD. 1992. Lignin-feruloyl ester cross-links in grasses. I. Incorporation of feruloyl esters into coniferyl alcohol dehydrogenation polymers. *J. Chem. Soc. Perkin Trans.* 1, pp. 2961–69
126. Ralph J, Quideau S, Grabber JH, Hatfield RD. 1994. Identification and synthesis of new ferulic acid dehydrodimers present in grass cell walls. *J. Chem. Soc. Perkin Trans.* 1, pp. 3485–98
127. Raz R, Crétin C, Puigdomènech P, Martínez-Izquierdo JA. 1991. The sequence of a hydroxyproline-rich glycoprotein gene from *Sorghum vulgare. Plant Mol. Biol.* 16:365–67
128. Redgwell RJ, Selvendran RR. 1986. Structural features of cell-wall polysaccharides in onion *Allium cepa. Carbohydr. Res.* 157: 183–99
129. Reid JSG, Wilkie KCB. 1969. Total hemicelluloses from oat plants at different stages of growth. *Phytochemistry* 8: 2059–65
130. Reinard T, Sprunck S, Altherr S, Jacobsen HJ. 1994. Biochemical properties of a novel cell wall protein associated with elongation growth in higher plants. *J. Exp. Bot.* 45:1593–601
131. Reiter WD, Chapple CCS, Somerville CR. 1993. Altered growth and cell walls in a fucose-deficient mutant of *Arabidopsis. Science* 261:1032–35
132. Roberts K. 1994. The plant extracellular matrix: in a new expansive mood. *Curr. Opin. Cell Biol.* 6:688–94
133. Rohde W, Rosch K, Kröger K, Salamini F. 1990. Nucleotide sequence of a *Hordeum vulgare* gene encoding a glycine-rich protein with homology to vertebrate cytokeratins. *Plant Mol. Biol.* 14:1057–59
134. Rubinstein AL, Broadwater AH, Lowrey KB, Bedinger PA. 1995. *Pex1,* a pollen-specific gene with an extensin-like domain. *Proc. Natl. Acad. Sci. USA* 92:3086–90
135. Rudall PJ, Caddick LR. 1994. Investigation of the presence of phenolic compounds in monocotyledonous cell walls, using UV

fluorescence microscopy. *Ann. Bot.* 74: 483–91
136. Ruel K, Joseleau JP. 1984. Use of enzyme-gold complexes for the ultrastructural localization of hemicelluloses in the plant cell wall. *Histochemistry* 81:573–80
137. Sauter M, Kende H. 1992. Levels of β-glucan and lignin in elongating internodes of deepwater rice. *Plant Cell Physiol.* 33: 1089–97
138. Scalbert A, Monties B, Lallemand J-Y, Guittet E, Rolando C. 1985. Ether linkage between phenolic acids and lignin fractions from wheat straw. *Phytochemistry* 24: 1359–62
139. Schindler T, Bergfeld R, Hohl M, Schopfer P. 1994. Inhibition of Golgi-apparatus function by brefeldin A in maize coleoptiles and its consequences on auxin-mediated growth, cell-wall extensibility and secretion of cell-wall proteins. *Planta* 192: 404–13
140. Schindler T, Bergfeld R, Schopfer P. 1995. Arabinogalactan proteins in maize coleoptiles: developmental relationship to cell death during xylem differentiation but not to extension growth. *Plant J.* 7:25–36
141. Schopfer P. 1990. Cytochemical identification of arabinogalactan protein in the outer epidermal wall of maize coleoptiles. *Planta* 183:139–42
142. Schwarz K. 1973. A bound form of silicon in glycosaminoglycans and polyuronides. *Proc. Natl. Acad. Sci. USA* 70:1608–12
143. Séné CFB, McCann MC, Wilson RH, Grinter R. 1994. Fourier-transform Raman and Fourier-transform infrared spectroscopy. An investigation of five higher plant cell walls and their components. *Plant Physiol.* 106:1623–31
144. Shedletzky E, Shmuel M, Trainin T, Kalman S, Delmer D. 1992. Cell wall structure in cells adapted to growth on the cellulose-synthesis inhibitor 2,6-dichlorobenzonitrile. A comparison between two dicotyledonous plants and a graminaceous monocot. *Plant Physiol.* 100:120–30
145. Shibuya N, Misaki A. 1978. Structure of hemicellulose isolated from rice endosperm cell wall: Mode of linkages and sequences in xyloglucan, β-glucan and arabinoxylan. *Agric. Biol. Chem.* 42:2267–74
146. Shibuya N, Nakane R. 1984. Pectic polysaccharides of rice endosperm cell walls. *Phytochemistry* 23:1425–29
147. Showalter AM 1993. Structure and function of plant cell wall proteins. *Plant Cell* 5:9–23
148. Smallwood M, Martin H, Knox JP. 1995. An epitope of rice threonine- and hydroxyproline-rich glycoprotein is common to cell wall and hydrophobic plasma-membrane glycoproteins. *Planta* 196:510–522.
149. Smith BG, Harris PJ. 1995. Polysaccharide composition of unlignified cell walls of pineapple [*Ananas comosus* (L.) Merr.] fruit. *Plant Physiol.* 107:1399–1409
150. Staehelin LA, Moore I. 1995. The plant Golgi apparatus: structure, functional organization and trafficking mechanisms. *Annu. Rev. Plant Physiol. Plant Mol. Biol.* 46:261–88
151. Staudte RG, Woodward JR, Fincher GB, Stone BA. 1983. Water-soluble (1→3), (1→4)-β-D-glucans from barley (*Hordeum vulgare*) endosperm. III. Distribution of cellotriosyl and cellotetraosyl residues. *Carbohydr. Polym.* 3:299–312
152. Stiefel V, Ruiz-Avila L, Raz R, Vallés MP, Gómez J, et al. 1990. Expression of a maize cell wall hydroxyproline-rich glycoprotein gene in early leaf and root vascular differentiation. *Plant Cell* 2:785–93
153. Stinard PS, Nevins DJ. 1980. Distribution of noncellulosic β-D-glucans in grasses and other monocots. *Phytochemistry* 19: 1467–68
154. Taiz L. 1984. Plant cell expansion: regulation of cell-wall mechanical properties. *Annu. Rev. Plant Physiol.* 35:585–657
155. Tan KS, Hoson T, Masuda Y, Kamisaka S. 1992. Involvement of cell wall-bound diferulic acid in light-induced decrease in growth rate and cell wall extensibility of *Oryza sativa* coleoptiles. *Plant Cell Physiol.* 33:103–08
156. Thomas JR, Darvill AG, Albersheim P. 1989. Rhamnogalacturonan I, a pectic polysaccharide that is a component of monocot cell-walls. *Carbohydr. Res.* 185: 279–305
157. Turner LB, Mueller-Harvey I, McAllan AB. 1993. Light-induced isomerization and dimerization of cinnamic acid derivatives in cell walls. *Phytochemistry* 33: 791–96
158. van Holst GJ, Fincher GB. 1984. Polyproline II conformation in the protein component of arabinogalactan-protein from *Lolium multiflorum. Plant Physiol.* 75:1163–64
159. Varghese JN, Garrett TPJ, Colman PM, Chen L, Høj PB, Fincher GB. 1994. Three-dimensional structures of two plant β-glucan endohydrolases with distinct substrate specificities. *Proc. Natl. Acad. Sci. USA* 91:2785–89
160. Viëtor RJ, Kormelink FJM, Angelino SAGF, Voragen AGJ. 1994. Substitution patterns of water-unextractable arabinoxylans from barley and malt. *Carbohydr. Polym.* 24:113–18
161. Vignols F, Rigau J, Torres MA, Capellades M, Puigdomènech P. 1995. The *brown midrib3* (*bm3*) mutation in maize occurs in the gene encoding caffeic acid *O*-methyltransferase. *Plant Cell* 7:407–16

162. Waldron KW, Brett CT. 1985. Interaction of enzymes involved in cell-wall heteropolysaccharide biosynthesis. In *Biochemistry of Plant Cell Walls, SEB Semin. Ser. 28,* ed. CT Brett, JR Hillman, pp. 79–97. Cambridge: Cambridge Univ. Press
163. Whetten R, Sederoff R. 1995. Lignin biosynthesis. *Plant Cell* 7:1001–13
164. Whistler RL. 1950. Xylan. *Adv. Carbohydr. Chem.* 5:269–90
165. Wilkie KCB. 1979. The hemicelluloses of grasses and cereals. *Adv. Carbohydr. Chem. Biochem.* 36:215–64
166. Wood PJ, Braaten JT, Scott FW, Riedel D, Poste LM. 1990. Comparisons of viscous properties of oat and guar gum and the effects of these and oat bran on glycemic index. *J. Agric. Food Chem.* 38:753–57
167. Wood PJ, Weisz J, Blackwell BA. 1994. Structural studies of (1→3),(1→4)-β-D-glucans by ^{13}C-nuclear magnetic resonance spectroscopy and by rapid analysis of cellulose-like regions using high-performance anion-exchange chromatography of oligosaccharides released by lichenase. *Cereal Chem.* 71:301–7
168. Woodward JR, Fincher GB. 1983. Water soluble barley β-glucans. *Brew. Digest,* May, pp. 28–32
169. Woodward JR, Fincher GB, Stone BA. 1983. Water soluble (1→3),(1→4)-β-D-glucans from barley (*Hordeum vulgare*) endosperm. II. Fine structure. *Carbohydr. Polym.* 3:207–25
170. Wu Y, Spollen WG, Sharp RE, Hetherington PR, Fry SC. 1994. Root growth maintenance at low water potentials. Increased activity of xyloglucan endotransglycosylase and its possible regulation by abscisic acid. *Plant Physiol.* 106:607–15
171. Xu P, Wang J, Fincher GB. 1992. Evolution and differential expression of the (1→3)-β-glucan endohydrolase-encoding gene fam-

Annu. Rev. Plant Physiol. Plant Mol. Biol. 1996. 47:477–508

SOME NEW STRUCTURAL ASPECTS AND OLD CONTROVERSIES CONCERNING THE CYTOCHROME *b6f* COMPLEX OF OXYGENIC PHOTOSYNTHESIS

W. A. Cramer, G. M. Soriano, M. Ponomarev, D. Huang, H. Zhang, S. E. Martinez, and J. L. Smith

Department of Biological Sciences, Purdue University, West Lafayette, Indiana 49707-1392

KEY WORDS: chloroplast, cytochrome *f*, electron transfer, energy transducing membrane, proton translocation

ABSTRACT

The cytochrome b_6f complex functions in oxygenic photosynthetic membranes as the redox link between the photosynthetic reaction center complexes II and I and also functions in proton translocation. It is an ideal integral membrane protein complex in which to study structure and function because of the existence of a large amount of primary sequence data, purified complex, the emergence of structures, and the ability of flash kinetic spectroscopy to assay function in a readily accessible ms–100 μs time domain. The redox active polypeptides are cytochromes *f* and b_6 (organelle encoded) and the Rieske iron-sulfur protein (nuclear encoded) in a mol wt = 210,000 dimeric complex that is believed to contain 22–24 transmembrane helices. The high resolution structure of the lumen-side domain of cytochrome *f* shows it to be an elongate (75 Å long) mostly β-strand, two-domain protein, with the N-terminal α-amino group as orthogonal heme ligand and an internal linear 11-Å bound water chain. An unusual electron transfer event, the oxidant-induced reduction of a significant fraction of the *p* (lumen)-side cytochrome *b* heme by plastosemiquinone indicates that the electron transfer pathway in the b_6f complex can be described by a version of the Q-cycle mechanism, originally proposed to describe similar processes in the mitochondrial and bacterial bc_1 complexes.

1040-2519/96/0601-0477$08.00

CONTENTS

INTRODUCTION[1]

The cytochrome b_6f complex is one of three multimeric, large (mol wt = 210,000 as a dimer) integral membrane protein complexes that is responsible in oxygenic membranes for electron transport from water to $NADP^+$ coupled to proton translocation and generation of the transmembrane proton electrochemical potential. The complex contains four well-studied polypeptides of mol wt = 18,000–31,000, of which three are organelle encoded and one (the mol wt = 19,000 Rieske iron-sulfur protein) is nuclear encoded. Electron transfer through the cyt b_6f complex is catalyzed by redox prosthetic groups on three of these polypeptides—cytochrome *f*, cytochrome b_6, and the Rieske protein. There are at least three additional small (mol wt = 5000) hydrophobic polypeptides. The complex as a dimer, which is believed to be the functional form, contains 22–24 transmembrane helices. An overview of the function of the b_6f complex is that it functions as the intermediate complex in the transfer of electrons from the photosystem II (water-splitting) to the photosystem I ($NADP^+$-reducing) complex and in the process also functions to translocate protons across the membrane. Although these processes can be studied by precise spectrophotometric techniques and with site-directed mutants, studies

[1] *Abbreviations* chl, chlorophyll; chr, chromatophore; $\Delta\psi$, transmembrane potential; $\Delta\tilde{\mu}_{H^+}$, electrochemical potential; cyt, cytochrome; *n*, *p*, electrochemically negative and positive sides of the membrane, respectively; EDC, *N*-ethyl-3-[3-(dimethylamino)propyl] carbodiimide; E_m, midpoint oxidation-reduction potential; EPR, electron paramagnetic resonance; K_{eq}, equilibrium constant; LHC, light-harvesting chlorophyll protein; MOA-stilbene, (E,E)-methyl-3-methoxy-2-(styrylphenyl)propenoate; mol wt, molecular weight; M_r, relative molecular weight; NQNO, 2-n-4-hydroxyquinoline-N-oxide; PC, plastocyanin; PQ, plastoquinone; PS, photosystem; RFeS, Rieske iron-sulfur protein; SQ, semiquinone; suIV, subunit four; TM, transmembrane.

of the details of the electron transfer pathway as it traverses the complex—and the coupled proton translocation—are complicated by the branching of the electron transport pathway within the complex. New structure data on cytochrome *f* also introduce the ideas of long-distance intraprotein electron transfer and the role of internal bound water into the problem. The following review addresses the experimental basis of the concepts that describe the function of the cytochrome b_6f complex and of the present level of understanding and uncertainty.

The cytochrome bc_1 (ubiquinol:cytochrome c_1 reductase) and b_6f (plastoquinol:plastocyanin oxidoreductase) complexes participate in electron transport, H+ translocation, and generation of the electrochemical potential ($\Delta\tilde{\mu}_{H^+}$) in the respiratory and photosynthetic bacterial electron transport chains and in oxygenic photosynthesis. They also occupy a central position with respect to the redox potential scale and pathway of the respective electron transport chains (e.g. respiration, oxygenic photosynthesis). In addition to these overall functional similarities, there is significant sequence identity between the cytochrome *b* polypeptides (17, 100, 115) and an overall membrane topographical similarity in the arrangement of the redox prosthetic groups, i.e. *b* hemes, *f* and *c*1 hemes, and the Rieske [2Fe-2S] center. The cyt *b* polypeptide of the complexes from mitochondria (mi) and photosynthetic bacteria (ca 400 residues) is approximately twice as long as cyt b_6 (ca 215 residues), but the N-terminal half of the former appears to contain the heme-binding domain that defines the latter. The stoichiometry of heme *b*:heme *c* (cytochrome *f* or c_1):[2Fe-2S] center is 2:1:1 in b_6f and bc_1 complexes. The *bis*-histidine heme ligation of the two *b* hemes occurs through four histidines, absolutely conserved in all of the approximately 110 bc_1 and b_6f sequences that are available in the data bank (Most cyt b_6f sequences are collected in References 17, 29). The *bis*-histidine coordination of the two *b* hemes results in one heme near each side of the membrane bilayer. The His pair on each side of the membrane consists of one His on transmembrane (TM) helix B and one on helix D [the four TM helices of cyt b_6 are labeled A–D], resulting in one heme on the electrochemically positive (*p*) side and one heme on the negative (*n*) side of the membrane. The two hemes are oriented perpendicularly to the plane of the membrane, with their edges separated by ~12 Å [14, 17 (Figure 1C), 100, 115]. The structural motif of heme(s) bridging TM α-helices is common not only to other integral membrane cyt *b* polypeptides (21) but also to the binding of the *a*-type heme in the bacterial (44) and bovine mitochondrial (108) cytochrome oxidase, as shown by crystallographic analysis. We direct the reader to recent reviews on this subject that contain, among other aspects, a more complete presentation of sequence data (16, 17, 37, 53). Reviews on cyclic electron transport (4) and on the bc_1 complex (106, 107) are also pertinent.

The change between the eight-helix cyt *b* found in photosynthetic bacteria and mitochondria into the four-helix cyt b_6 and three-helix subunit four (suIV) that define the equivalent structural units as they appear in Widger et al (115) coincides approximately with the appearance in evolution of oxygenic photosynthetic membranes (29). A second diagnostic feature of cyt b_6, in comparison to cyt $b(bc_1)$, is that the two His residues on helix D are separated by 14 residues in cyt b_6 and 13 residues in cyt $b(bc_1)$ (115). This structural difference may underlie the relative homogeneity of the visible and electron paramagnetic resonance (EPR) spectra of the two hemes of cyt b_6 relative to cyt $b(bc_1)$ (14). Cyt b_6 and suIV have a high degree of sequence identity. The degree of pseudo-identity of thirteen cyt b_6 and sixteen suIV polypeptides is 79% and 63%, respectively (17). Sequences for cyt b_6 and suIV from *Chlorella* (91a) and *Chlorobium* (101a) have subsequently appeared in the data bases. The family of bc_1–b_6f cyt *b* polypeptides is the most thoroughly sequenced of all integral membrane polypeptides, with more than 800 different polypeptides having been at least partly sequenced (21), and ~110 complete sequences from the total family (including b_6) now available through accession numbers in the data banks.

Polypeptide Subunits

The polypeptide content of the b_6f complex, which is simpler than that of the eukaryotic mitochondrial bc_1 complex, contains four polypeptides of relative molecular weight (M_r) ≥ 17,000 (Figure 1A), three of which contain redox prosthetic groups, cyt *f* (mol wt = 32,038 in spinach chloroplasts), cyt b_6 (mol wt = 24,166), and the Rieske [2Fe-2S] protein (mol wt = 19,116). Cytochromes b_6 and *f* and suIV (mol wt = 17,445) are chloroplast encoded; the organelle origin of cyt *f* contrasts with the nuclear origin of the functionally analogous cyt c_1. Additional polypeptides in this size range are M_r = 15,000 and M_r = 19,500 components in complexes isolated from the chloroplast grana compartment (99) and *Chlamydomonas reinhardtii* (66), respectively. One or more bands are often found in the M_r = 64,000 region, which is a molecular size of interest because of an identification of a polypeptide of this size with the light-harvesting chloroplast (LHC) II kinase. A Western blot using antibody to the putative M_r = 64,000 kinase showed its presence in cyt b_6f complex isolated from the unicellular green alga, *Acetabularia acetabulum* (27). The presence of the putative kinase is of interest in connection with the hypothesis that the cyt b_6f complex is a redox sensor for the activation of the kinase (1, 27, 111). However, the significance of the M_r = 64,000 band in the b_6f preparation is now unclear because of the retraction of its proposed association with the kinase (39). In any case, an alternative explanation of the meaning of bands in this M_r region of the b_6f complex can be based on Western blot reactions (27) with antibodies to cyt *f*, suIV, and the Rieske

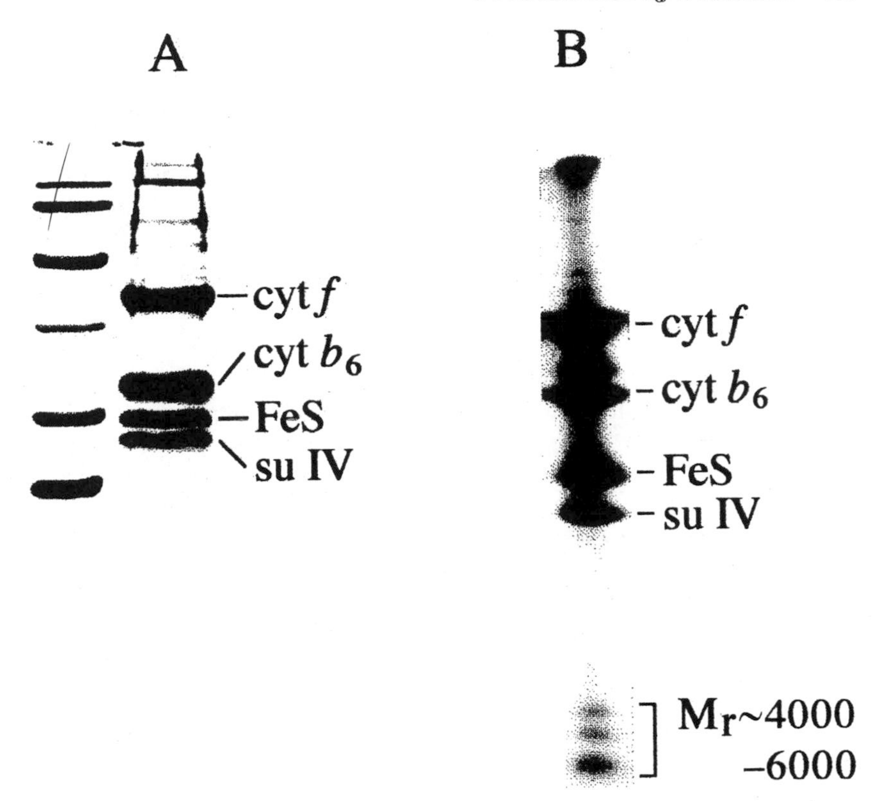

Figure 1 (A) First-dimension SDS-PAGE of four $M_r \geq 17,000$ subunits of the b_6f complex from spinach chloroplasts; (B) second-dimension SDS-PAGE from first-dimension native gel according to Schägger et al (101) showing four large and three small hydrophobic subunits. Gel scanned in color at high contrast and the program NIH image used to subtract background and amplify density by a factor of two.

protein. The Westerns suggest that the M_r 50,000–60,000 bands could be attributed to subcomplexes, e.g. the cyt *f*–Rieske protein subcomplex with M_r = 50,000–55,000.

The complex contains at least three small (mol wt ≈ 4) hydrophobic subunits—suIV–VII—the organelle-encoded 37-residue *Pet*G (36), nuclear-encoded 39-residue *Pet*X (23), and a M_r = 3400 *Pet*L (88) gene products, as shown in a second-dimension SDS-PAGE (see bottom, Figure 1B). The function of the small subunits is not known. Small hydrophobic polypeptides are also found in the photosystem (PS) II and I reaction center complexes. The structural requirement for such polypeptides—to provide low dielectric or hydrophobic "fill"—may be analogous to the role of the small amphiphile heptane-1,2,3-triol as an additive to the detergent environment required for the crystallization of reaction center membrane protein complexes (73a).

The stoichiometry of the subunits in the b_6f complex purified from *C. reinhardtii*, determined by [^{14}C]-acetate labeling, is approximately 1:1:1. With unity stoichiometry of the major polypeptides, the mol wt of the seven polypeptide (four relatively large, three small) components of a monomeric complex is ~105 kDa. It has been proposed that the subunit components of the complex are different in grana and stromal membranes, with the stromal preparation containing more bound plastocyanin (PC) and M_r = 4000 petG hydrophobic polypeptide (99).

Other Components: Chlorophyll a

The b_6f complex isolated from spinach chloroplasts was found to contain one tightly bound chlorophyll (chl) molecule, inferred to be chl *a* from its spectral peak at 669 nm (17, 42). The presence of one tightly bound chl *a* molecule has been confirmed in the *C. reinhardtii* preparation, where it was found not to exchange with [^{3}H]-labeled chl *a* in detergent micelles (88). Its presence in the original benchmark preparation of the b_6f (43) is now realized not to have been adventitious. One aesthetic consequence of the presence of this chl *a* is that the purified cyt b_6f complex in solution and in crystals, instead of having the striking red color that is characteristic of the heme absorption and of crystals of the mitochondrial cyt bc_1 complex (7, 120), has a (more dull) brown color (D Huang & WA Cramer, unpublished data). The function of the bound chl is unknown. Given the structure of the LHCII protein (61), the chl in the b_6f complex may reside at the level of the membrane interfacial layer. A position of the special chl at the *p*-side interfacial layer of the thylakoid membrane is implied by its participation in a local electrochromic event whose time course correlates with the reduction of cyt *f* and the oxidation of plastoquinol and, by inference, local H^+ transfer (51). This binding is relevant to the nature of the lumen-side exit pathway for H^+ in the b_6f complex and possibly to the origin of the slow electrochromic phase (see section on Origin of Slow Electrochromic Band Shift).

Dimeric Structure and Activity

The complex is dimeric as isolated from spinach chloroplast thylakoid membranes in detergent with mol wt $\approx 2.3 \times 10^5$ (42, 101). The true dimer mol wt, including three small ($M_r \approx 4000$) subunits (Figure 1A), bound lipid, but not detergent, would be mol wt $\approx 2.2 \times 10^5$. The in vitro electron transport activity ($PQ_2H_2 \rightarrow PC$) of the separated dimer was four- to fivefold greater than that of the monomer preparation, whose activity could be attributed to residual dimer. The most active cyt b_6f preparation is that from *C. reinhardtii*, also found to be dimeric, whose turnover, 250–300 s^{-1} (86), is approximately three times that measured in the spinach preparation (42). The presence of the active dimeric form at high levels in the detergent-extracted b_6f complex and the absence of activity in the monomer suggest that the dimer is the active complex in the membrane. Monomeric b_6f complex has been isolated from cyanobacteria (3; G Tsiotis, personal communication) but without the Rieske protein (M Rögner & G Tsiotis, personal communication). Conversion or interconversion of dimer to monomer was inferred from the observation of two M_r forms (presumably dimer and monomer) on a sucrose gradient. The larger converted to the smaller with time, and the conversion was inhibited by crosslinking reagents (11).

The major suggestions about the functional significance of a structurally dimeric complex, all of which concern cyt bc_1, have been summarized by Huang et al (42) and relate to proton translocation activity. An additional function of the dimer would be to facilitate docking of peripheral proteins through its increased surface area. Two candidates for such peripheral proteins that may dock on the *n* (stromal) side of the complex are the "G" cytochrome-like component, whose interaction with the complex in *Chlorella* was inferred from spectrophotometric studies (48), and the LHCII kinase or phosphatase system that may be regulated by redox signaling from the b_6f complex (1).

Transmembrane α-helices

The number of TM α-helices in the monomeric seven-subunit b_6f complex, inferred from nucleotide sequence and biochemical topography and binding studies is:

1 [petA (103, 118)] + 4 [petB (103, 115)] + 0–1 [petC (115)] + 3 [PetD (115)] + 1 (PetG) + 1 (PetX) + 1 [PetL (according to Reference 88)] = 11–12.

For a dimeric structure (see below), the number of TM helices is 22–24. The small uncertainty results from the mode of membrane anchoring of the PetC Rieske protein. The Rieske protein in plant and *C. reinhardtii* chloroplasts contains a hydrophobic segment of ~25 residues near the N-terminus that is followed by a glycine-rich nonpolar region. It has recently been found that the 179-residue Rieske protein in the complex can be cleaved four residues before the first Gly in the Gly-enriched segment (87). The cleavage by proteolysis of the b_6f complex is into a C-terminal 139-residue fragment that retains the [2Fe-2S] center. Thus, the question of the anchor concerns the ca 20-residue hydrophobic domain that is upstream from the cleavage site. A 20-residue hydrophobic sequence is generally used to imply a TM α-helix. However, extraction of the Rieske protein from chloroplast membranes with chaotropic agents, along with sedimentation behavior characteristic of a soluble protein, led to the conclusion by Breyton et al (10) that it is an extrinsic protein. However, the Rieske protein could not be extracted by high ionic strength (2M NaCl) from the membranes (10), which would be expected for an extrinsic protein. Alternatively, there is a structure-based precedent in the enzyme prostaglandin synthase for a hydrophobic association of the protein in the intermediate dielectric constant interfacial layer of the membrane (85). Considering that (*a*) the extraction with chaotropic agents but not with high ionic strength suggests a hydrophobic association with the membrane, and that (*b*) channel-forming toxin-like or colicin proteins with a long terminal hydrophobic segment can become water-soluble when the segment in the protein core is buried (84), the Rieske protein may also be bound hydrophobically in the interfacial layer. This association would occur as it does in the prostaglandin synthase, through a hydrophobic 20–25-residue segment, in this case near the Rieske N-terminus. Such a mode of hydrophobic association, in contrast with a TM α-helix, would be consistent with the inability to cleave the Rieske protein with protease probes added to the stromal side of thylakoid membranes (103a). Thus, with regard to the anticipated mode of membrane association of the b_6f complex, to be tested by high-resolution structure analysis, the dimeric structure of the cyt b_6f complex would have 22 TM- and 24 membrane-embedded α-helices.

Dissociation of the subunits at high pH in spinach thylakoids suggests that the hydrophobic and nonelectrostatic helix-helix interactions between the subunits of the complex are unusually weak (104). This may explain why it has been difficult thus far to obtain well-defined three-dimensional crystals of the b_6f complex in a detergent environment.

HIGH-RESOLUTION STRUCTURE INFORMATION; CYTOCHROME *f*

Structure of the Lumen-Side Domain of Cytochrome f at 1.96-Å Resolution

The crystal structure of the reduced redox-active 252-residue lumen-side domain of the 285-residue cyt *f* has been solved by multiple isomorphous replacement to a resolution of 2.3 Å (70) and recently to 1.96 Å (SE Martinez, D Huang, M Ponomarev, WA Cramer & JL Smith, unpublished manuscript). The protein that was solved was the soluble active 252-residue lumen-side domain, cleaved after Gln252, of the 285-residue (mol wt = 32,038 in turnip chloroplasts) mature polypeptide that spans the membrane once (Table 1). Preparations of cyt *f* from cruciferous plants, however, were reported to be soluble and monomeric with M_r = 27,000 instead of M_r = 32,000–33,000 (32). Turnip cyt *f* (mol wt = 27,500) was truncated near the C-terminus during purification, as shown by C-terminal sequencing of the turnip cyt polypeptide with carboxypeptidases Y and A and electrospray mass spectrometry (34, 71). The occurrence of proteolytic cleavage at a site two residues into the 20-residue putative hydrophobic TM α-helix (Table 1) suggests that the protease might be related to a processing or leader peptidase. The thylakoid processing

Table 1 Amino acid sequences of cytochrome *f* of the b_6f complex from turnip (*Brassica campestris*), *Chlamydomonas reinhardtii,* and three cyanobacteria. The following features are indicated: N-terminal amino acid, Tyr1 (bold), the heme-binding domain (italics), and the single transmembrane helix region (underlined).

B. campestris	————————	————————	————————-**Y**	PIFAQQNY-ENPREATGRIV
C. reinhardtii	MSNQV—FTTLRAATL——	————————AVILGM	AGGLAV—————SPAQA**Y**	PVFAQQNY-ANPREANGRIV
Synechocystis 6803	MRN—PDTLGLWTKTMVA—	——————-LRRFTVLAI	ATVSVFLITDLGLPQAASA**Y**	PFWAQETAPLTPREATGRIV
A. quadrupl	MKT—PELMAIWQR———	——————-LKTACLVAI	ATFGLFFASDVLFPQAAAA**Y**	PFWAQQTAPETPREATGRIV
Nostoc 7906	MRN—ASVTARLTRSVRA—	——————-IVKTLLIAI	ATVTFYFSCDLALPQSAAA**Y**	PFWAQQTYPETPREPTGRIV
B. campestris	CANCHLASKPVDIEVPQAVL	PDTVFEAVVKIPYDMQLKQV	LANGKKGALNVGAVLILPEG	FELAPPDRISPEMKEKIGNL
C. reinhardtii	*CANCH*LAQKAVEIEVPQAVL	PDTVFEAVIELPYDKQVKQV	LANGKKGDLNVGMVLILPEG	FELAPPDRVPAEIKEKVGNL
Synechocystis 6803	CANCHLAQKAAEVEIPQAVL	PDTVFEAVVKIPYDLDSQQV	LGDGSKGGLNVGAVLMLPEG	FKIAPPDRLSEGLKEKVGGT
A. quadrupl	*CANCH*LAAKEAEVEIPQSVL	PDQVFEAVVKIPYDHSQQQV	LGDGSKGGLNVGAVLMLPDG	FKIAPADRLSDELKEKTEGL
Nostoc 7906	CANCHLAAKPTEVEVPQSVL	PDTVFKAVVKIPYDTSAQQV	GADGSKVGLNVGAVLMLPEG	FKIAPEDRISEELQEEIGDT
B. campestris	SFQNYRPNKKNILVIGPVPG	QKYSEITFPILAPDPATNKD	VHFLKYPIYVGGNRGRGQIY	PDGSKSNNTVYNATAGGIIS
C. reinhardtii	YYQPYSPEQKNILVVGPVPG	KKYSEMVVPILSPDPAKNKN	VSYLKYPIYFGGNRGRGQVY	PDGKKSNNTIYNASAAGKIV
Synechocystis 6803	YFQPYREDMENVVIVGPLPG	EQYQEIVFPVLSPDPAKDKS	INYGKFAVHLGANRGRGQIY	PTGLLSNNNAFKAPNAGTIS
A. quadrupl	YFQSYAPDQENVVIIGPISG	DQYEEIVFPVLSPDPKTDKN	INYGKYAVHLGANRGRGQVY	PTGELSNNNQFKASATGTIT
Nostoc 7906	YFQPYSEDKENIVIVGPLPG	EQYQEIVFPVLSPNPATDKN	IHFGKYSVHVGGNRGRGQVY	PTGEKSNNNLYNASATGTIA
				↓
B. campestris	KI—LRKEKGG—YEITIVD	ASNERQVIDIIPRGLELLVS	EGESIKLDQPLTSNPNVGGF	GQGDAEIVLQDPLRVQGLLF
C. reinhardtii	AITALSEKKGG—FEVSI-E	KANGEVVVDKIPAGPDLIVK	EGQTVQADQPLTNNPNVGGF	GQAETEIVLQNPARIQGLLV
Synechocystis 6803	EVNALEAGG———YQLIL-T	TADGTETVD-IPAGPELIVS	AGQTVEAGEFLTNNPNVGGF	GQKDTEVVLQNPTRIKFLVL
A. quadrupl	NIAVNEAAG———TDITI-S	TEAGEVIDT-IPAGPEVIVS	EGQAIAAGEALTNNPNVGGF	GQKDTEVVLQNPARIYGYMA
Nostoc 7906	KIAKEEDEDGNVKYQVNI-Q	PESGDVVVDTVPAGPELIVS	EGQAVKAGDALTNNPNVGGF	GQRDAEIVLQDAGRVKGLIA
B. campestris	FLGSVVLAQIFLVLKKKQFE	KVQLSEM-NF		
C. reinhardtii	FFSFVLLTQVLLVLKKKQFE	KVQLAEM-NF		
Synechocystis 6803	FLAGIMLSQILLVLKKKQIE	KVQAAEL-NF		
A. quadrupl	FVAGIMLTQIFLVLKKKQVE	RVQAAGNCDF		
Nostoc 7906	FVALVMLAQVMLVLKKKQVE	RVQAAEM-NF		

peptidase of cyt *f* is membrane bound (46), and the active site of the processing enzyme for PC is on the lumen side of the membrane (55).

In addition to being the first polypeptide of the cyt b_6f or bc_1 complexes to be solved crystallographically, cyt *f* has several unique aspects. Unlike all other c-type cytochrome structures that consist of one predominantly α-helical domain, cyt *f* contains, first, two domains in an elongate (ca 75 Å × 35 Å × 25 Å) geometry and, second, a secondary structure that is mostly β sheet (Figure 2). The covalently bound heme lies within the larger domain near the interface between the two domains; its Fe is 45 Å from the Arg250 near the C-terminus that is adjacent or close to the single TM α-helix. Third, the axial sixth heme ligand is the α-amino group of the N-terminal tyrosine residue (Figure 3A). A number of spectroscopic studies had noted similarities between spectra of cyt *f* and those of soluble eukaryotic cyt *c* at alkaline pH where the axial methionine sulfur ligand is exchanged for a lysine ε-amino group as an axial heme ligand, and a conserved Lys145 was proposed as the axial sixth ligand (discussed in 70). However, Lys145 is 33 Å from the heme iron (Figure 4). The N-terminal amino group as a heme ligand (Figure 3A) is unprecedented for a cytochrome or any heme protein. Fourth, the cytochrome contains an internal H_2O chain. The 1.96-Å structure revealed an L-shaped array of five buried water molecules, which extends in two directions from the Nδ1 of the heme ligand His25

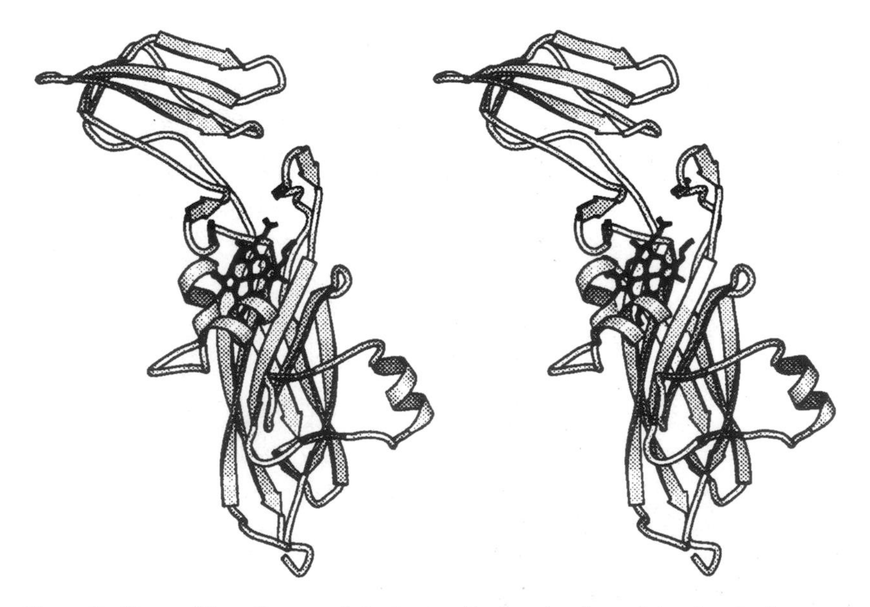

Figure 2 Stereo ribbon diagram of the lumen-side domain of cyt *f* showing the heme and predominantly β-strand secondary structure, in the N→C directional ribbon format. The C-terminus that enters the membrane is at the bottom of the figure. The α-carbon of Arg250 close to the membrane surface is 45 Å from the heme Fe. Drawing made with the MOLSCRIPT program (59).

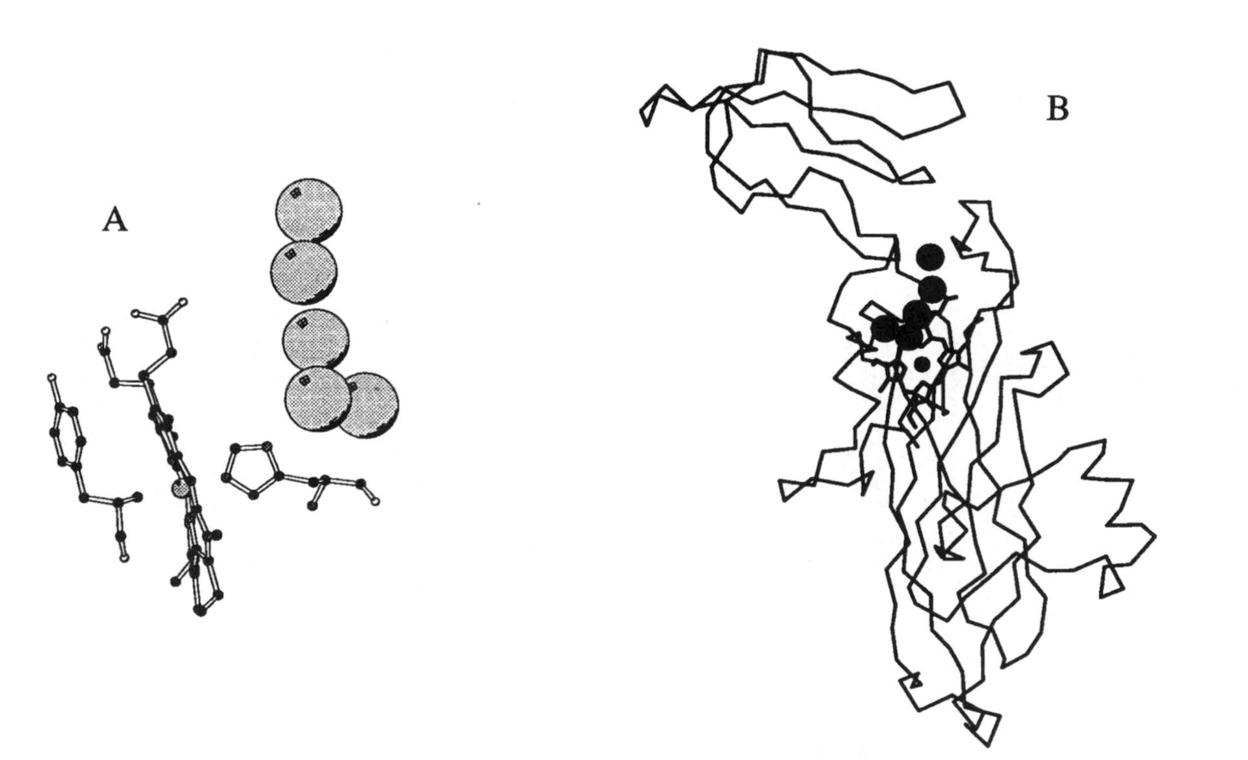

Figure 3 Bound water chain of cyt *f* (A) in a "ball and stick" profile of the region of the heme and its ligands showing a side view of the heme and of the axial heme ligands, His25 (right) and the amino group of N-terminal residue, Tyr1 (left); and (B) in the perspective of the entire lumen-side domain of the cyt *f* molecule. In A and B the oxygens of the water chain are shown with a size equal to the average Van der Waals radius of 1.4 Å.

(Figure 3). The longer branch of the L that contains four H_2O molecules extends toward Lys66 of the putative basic PC docking domain. The water sites are highly occupied, and their temperature factors are comparable to those of protein atoms (see section on The Cyt *f* Internal H_2O Chain as Part of a *p*-Side Exit Port for H^+ from the b_6f Complex).

The Interaction of Cytochrome f with Plastocyanin

An additional feature of the cyt *f* structure is a prominent positively charged region at the interface between the large and small domains that contains the basic residues Lys58, Lys65, and Lys66 (large domain) and Lys187 (small domain) (Figure 4). The latter residue of cyt *f* was crosslinked with the water-soluble linker *N*-ethyl-3-[3-(dimethylamino) propyl] carbodiimide (EDC) to Asp44 of PC (78).

The Cu protein PC (97–104 residues) (91), or cyt c_6 (54) in the absence of Cu in *C. reinhardtii* or cyanobacteria, catalyzes electron transfer from cyt *f* to P700 in the photosystem (PS) I reaction center. The structure of oxidized PC from poplar revealed the protein as an eight-stranded antiparallel β barrel (predominantly β strand, similar to cyt *f*) with a single Cu atom at one end ("northern" face) liganded by two His, one Cys, and one Met and surrounded by a hydrophobic surface and two negative regions, consisting of carboxylates 42–44 and 59–61 or 59–60, on the "east" side (12). In addition to the crosslinking of PC Asp44 to cyt *f* Lys187, Glu59/60 in the second acidic patch region was also crosslinked with EDC to an unknown residue on cyt *f* (78). Two paths of electron transfer to and from the Cu atom were identified in the original work (12): an "adjacent" and hydrophobic site connected to His87 (the only solvent-accessible copper ligand) and a "remote" site via Tyr83 on the "east" side of the protein. The two surface acidic patches are located within the immediate vicinity of the Tyr83 (91).

Five high-resolution PC structures are now known: poplar, *Scenedesmus, Enteromorpha, Chlamydomonas,* and French Bean (91). Before the cyt *f* structure was known, it was appreciated that the PC–cyt interaction is electrostatic in nature (2, 78, 89, 98). The electrostatic character of the PC–cyt *f* interaction was confirmed by the in vitro ionic strength dependence of the rate constant for electron transfer. In laser flash photolysis measurements of the electron transfer kinetics between cyt *f* and PC, a small increase in transfer rate was observed when the ionic strength was initially increased, which suggests that the proteins are not optimally oriented for electron transfer. A major role of the electrostatic interaction in forming the PC–cyt *f* complex was inferred by the monotonic decrease in rate with increasing ionic strength at higher values of ionic strength (73). The in vivo rate constant for electron transfer between cyt *f* and PC is 5×10^3 s^{-1} (9).

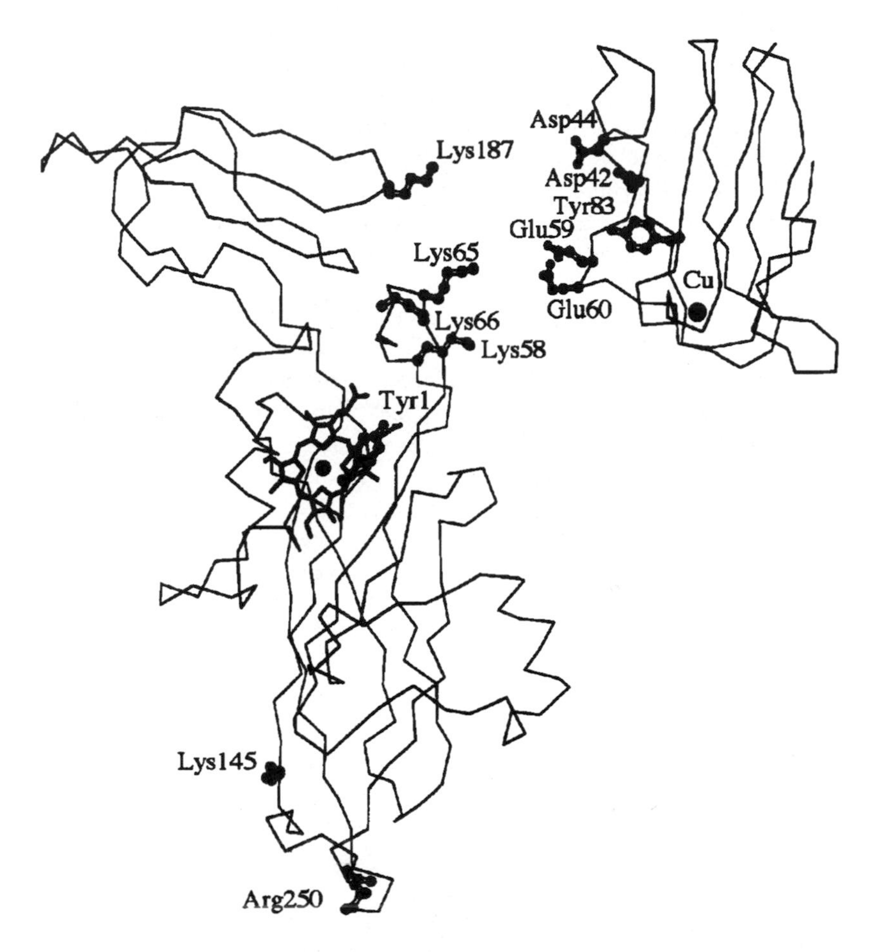

Figure 4 Cα trace of cyt *f* and poplar plastocyanin (PC) in a "pre-docking" state (right) that shows side chains of basic residues near Lys187, which can be crosslinked to Asp44 of PC (78). Lys187 and Arg209 in the "small domain" are adjacent to Lys65, Lys66, and Lys58 on the "large domain" that form a cluster having the most pronounced positive surface potential in the protein. The role of the acidic residues of PC and Tyr83 is discussed in the text.

DOCKING SITES It was inferred from experiments on cyt *f* reactions with negatively charged semiquinones (SQ) that turnip cyt *f* displays a localized positive charge that affects the electron transfer rate (73, 89). As noted above, the turnip cyt *f* structure reveals an extended patch of positively charged residues (Figure 4): Lys187, Arg209, Lys66, Lys65, and Lys58 (70). Except for Arg209, these residues are conserved in the higher plant cyt *f*, as is the negatively charged patch on the surface of higher plant PCs. The basic patch is mostly missing, however, from the cyanobacterial cyt *f* (Lys66 is preserved), and there are 1–2 negative charges in the region of Lys187 (33, 113). Cyt *f* of *Nostoc PCC 7906*, which binds a basic plastocyanin, has the pronounced acidic sequence Glu185-Glu186-Asp187-Glu188-Asp189 in the region of Lys187 of the higher plant cyt *f* basic patch. Changes in the complementary PCs of the cyanobacterial cyt *f* are consistent with a role of an electrostatic interaction in the docking as discussed in the context of *Phormidium* (113). For *Synechocystis PCC 6803* (91), the PC has two neutral and one acidic residue at positions 42–44, and His-Lys substituted at two of the three positions 59–61.

CONFORMATIONAL FLEXIBILITY The electrostatic interaction and strong coupling alone appear insufficient for competent electron transfer. The PC–cyt *f* crosslinked complex was found to be incompetent in electron transfer (upper limit of rate constant, 0.1 s^{-1}), which suggests that efficient electron transfer also requires conformational rearrangement after the initial docking (90). Therefore, the "eastern" and northern faces of PC may interact sequentially with cyt *f*. The initial docking of PC would use one or both of its "eastern" acidic patches followed by contact of the northern hydrophobic patch with a more hydrophobic region of the cyt *f* surface that is closer to the heme.

SITE-DIRECTED MUTAGENESIS *Plastocyanin* Several PC residues implicated in interaction with cyt *f* have been probed by site-directed mutagenesis using either transgenic tobacco plants or *Escherichia coli* as the expression system (38, 75, 76, 80). 1. Mutation of Tyr83 to Phe or Leu decreased the binding constant of PC for cyt *f* by a factor of 8 to 10. The rate constant for electron transfer was decreased by 15-fold in the *Y83L* mutant (38), which indicated that the Tyr83 participates both in cyt *f* binding and electron transfer. The Tyr-OH group and aromatic ring appear to be important for binding and for electron transfer, respectively. 2. Asp42 and Leu12 in the negative and hydrophobic patches, respectively, have been mutagenized to Asn42 and Glu12, Ala12, or Asn12. The *D42N* mutant did not affect cyt *f* binding, either because of alternative interaction with neighboring carboxylates or because of lack of direct contact between D42 and cyt *f*. The *L12E* and *L12A* mutants bound cyt *f* less strongly than wild type, which suggests a role of a specific hydrophobic interaction, but the binding of L12N by cyt *f* was four times stronger than wild type (76).

Cytochrome f Site-directed mutagenesis of cyt *f* was carried out in *C. reinhardtii* to test residues involved in PC docking. A mutant with Lys58, Lys65, and Lys66—changed to Gln, Ser, and Glu, respectively—showed no decrease in the amplitude or kinetics of flash-induced cyt *f* oxidation and of the 515-nm slow (millisecond) electrochromic band shift when grown under phototrophic conditions (87). Double mutants that removed the positive charge on Lys187 in the small domain and one of Lys58, Lys65, or Lys66 in the large domain displayed a decrease in the half time of cyt *f* photooxidation by a factor of 10 to 20 (J Zhou, J Fernandez-Velasco & R Malkin, unpublished data). These data are consistent with a role of the basic surface region of cyt *f*, utilizing either the large and small domain, in PC docking. On the other hand, the heterotrophically grown triple Lys small domain mutant was impaired in cyt *f* oxidation and assembly (87).

THE CYT *f* INTERNAL H_2O CHAIN AS PART OF A *p*-SIDE EXIT PORT FOR H^+ FROM THE b_6f COMPLEX The internal buried five-water chain of cyt *f* is a conserved feature of the molecule. The residues participating in side-chain H bonds are five Asn and Gln residues: Q59, N153, Q158, N168, and N232. It appears that all residues forming H bonds with the water chain, either through the side chain or backbone, are conserved (SE Martinez et al, submitted). The one exception is in *Vicia fabia,* where an ACC codon was reported at position 153 instead of Asn (possible codon AAC). The high degree of identity of the residues H-bonding to the H_2O chain and the high degree of occupancy of the water sites, as well as the similarity of the temperature factors of the water oxygens and neighboring amino acid atoms, suggest that one should think about the H_2O chain in the same way as the amino acids—as a conserved intrinsic aspect of the structure of the protein. Therefore, the H_2O chain is likely to have an important role in the function of the protein as well as in its structure. This water chain may define the exit port for H^+ from the *p* side of the cyt b_6f complex (Figure 5). This hypothesis is based on the following: 1. Four of the five H_2Os in the long branch of the L-shaped water chain extend 11 Å toward Lys66 of the putative docking site for PC, which could act through its carboxylates (X^- in Figure 5) as an H^+, as well as an e^- acceptor. 2. Protonation of PC carboxylates could provide a mechanism for its release from cyt *f* after reduction. 3. The water chain has many features that optimize it as a "proton wire," including insulation from the protein. 4. Direct deposition into the bulk aqueous phase of H^+ generated by quinol oxidation, often depicted as the mechanism of H^+ transfer in traditional "cartoons" on this subject, seems unlikely if the membrane surfaces are coated with peripheral protein. Without a channel through the surface protein "jungle," protons might not be able to enter or escape from the membrane. 5. There is a specific precedent for H^+ uptake from the *n*-side bulk aqueous phase through an extended H_2O chain to the quinone binding site (Q_B) in the bacterial photosynthetic reaction center (25). 6. It has already been proposed that oxidation of ubiquinol at the *p*-side interface is linked to protonation of residues with defined pK values (Y^- in Figure 5) in the mitochondrial Rieske protein (67). The obvious gap in the model in Figure 5 is that there is at present no structural data for the Rieske protein and therefore no structural basis for a H^+ connection between the quinol at the *p*-side interface and the H_2O chain of cyt *f*. The proposed H^+ pathway requires (*a*) a proton wire through the Rieske protein and part of the cyt *f* for the long-distance H^+ transfer; and (*b*) one or more amino acid residues

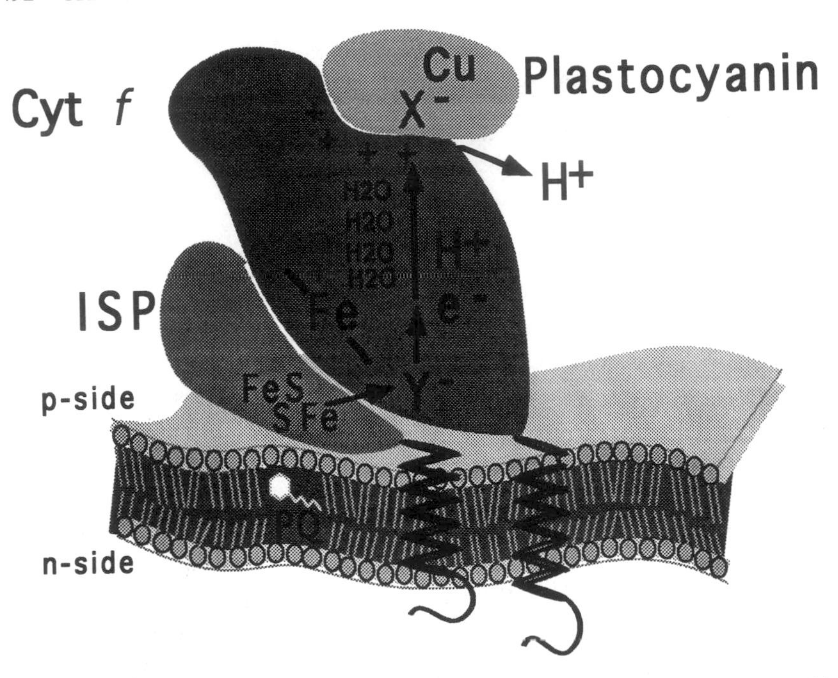

Figure 5 Model for the electron transport exit pathway from the b_6f complex on the lumen-side (p-side) of the membrane from PQH_2 in the bilayer to plastocyanin (PC) and the coupled exit port for H^+ translocation from the bulk p-side aqueous phase. Y^- and X^- represent residues (e.g. carboxylates) in the Rieske iron-sulfur protein (ISP)-cyt f interface and PC, respectively, that can serve as H^+ sinks. Protonation of such a group in PC, in the region of PC Asp 44 or Glu59, may provide a mechanism for the release of bound PC. The PQH_2 binding niche is thought to include the RFeS protein and the peripheral "cd" and "ef" surface helices of cyt b_6 and subunit four (suIV) (17).

(Y^-, e.g. Asp, Glu, Tyr, or His in Figure 5) that can undergo redox-linked pK changes in the appropriate pH range (acidic inside thylakoids), accept H^+ from quinol, and donate to the H_2O chain. Given the bound water chains in the reaction center and cyt f, such chains may turn out to be an emerging structural motif for long-distance H^+ transfer in energy-transducing membranes (e.g. in bacteriorhodopsin).

LONG-DISTANCE CYT $f \rightarrow$ PC ELECTRON TRANSFER The H_2O chain also presents an interesting problem for electron transfer within the protein and to PC. Because of the effectiveness of donor-acceptor coupling across H bonds (22, 109), nature may utilize this water chain for long-distance (10–20-Å) e^- and H^+ transfer. Long-distance e^- transfer depends on the driving force $\Delta G°$ and reorganization energy of the reaction (68), distance between donor and acceptor sites (26), secondary and/or tertiary structures of the intervening protein medium (20, 65), type of coupling between donor and acceptor (i.e. through covalent bonds, H bonds, or through space) (5, 22, 109), and orientation of donor and acceptor sites (110). The "Pathways" model (5, 6, 110) can be used as an approach to calculate and consider the most efficient paths for e^- transfer within cyt *f*.

GENERALITY OF THE CYTOCHROME *f* STRUCTURE The primary sequence including the region around the heme domain and putative PC-binding region of the 285-residue cyt *f* is highly conserved (ca 90%) in 10 cytochrome sequences from higher plants (33) and *Brassica rapa* (34). The sequence identity between the solved domain of turnip cyt *f* and that of the green alga *C. reinhardtii* used for mutagenesis studies is 82%. All but five of the residue differences between the latter are predicted to be on the surface of the protein, and the one charge change in the PC-binding region is compensated, so that it is likely that the structure of higher plant and *C. reinhardtii* is very similar. The sequences of the three cyanobacterial sequences compiled by Gray (33) differ more markedly—and also differ from one another—so that one might expect larger differences in structure. The pI of the cyanobacterial cyt *f* is more acidic—by 2 to 4 pH units—than the plant chloroplast cyt *f* (114). Although two of the three cyanobacterial cytochromes utilize an acidic PC, all are missing three of the four basic residues—at positions 187, 65, and 58 (Figure 5)—that define the major surface positive patch in the turnip cyt *f* structure. The cyanobacterial *Nostoc PCC 7906,* which binds a basic PC, has the pronounced acidic residue sequence Glu185-Glu186-Asp187-Glu188-Asp189 around residue 187, which is known to be involved in PC binding from its crosslinking to Asp44 of PC (78). Studies of the PC–cyt *f* interaction have been extended in detail to *Phormidium laminosum,* which has also lost its "basic patch" (113).

Relation to cytochrome c_1 A major difference in the coding and assembly of the redox-active polypeptides between the b_6f and mitochondrial bc_1 complex is that cyt *f* is organelle encoded and cyt c_1 is nuclear encoded. The different cellular origin of cytochromes *f* and c_1 is one of three reasons for suggesting that, although the b_6f and bc_1 complexes have functional similarities and although there is an evolutionary relation between cyt b_6-suIV and cyt $b(bc_1)$, cytochromes *f* and c_1 may be present in the two complexes as a result of convergent functional evolution (29, 70). The other two reasons are as follows: 1. The crystal structure of cyt *f* (70) indicates that the distal axial heme ligand of cyt *f*, the N-terminal amino group, is different from the internal methionine utilized for cyt c_1 (35). 2. The degree of amino acid identity is very low except for the ubiquitous Cys-X-Y-Cys-His heme binding signature, as noted previously in a comparison of five cyt c_1 and one cyt *f* sequences (77).

PROSPECTS FOR STRUCTURES OF OTHER SUBUNITS A soluble 139-residue fragment of the Rieske protein was obtained by proteolysis (67) of the b_6f complex purified as previously described (42). This fragment has a homogeneous N-terminus with the N-terminal sequence NH_2-Phe-Val-Pro-Pro-Gly-Gly-, consists of residues 41–179 of the mature Rieske protein, and contains an approximately fully populated and native [2Fe-2S] cluster; crystals with reasonable diffraction have been obtained (87). The Rieske fragment appears suitable for crystallization. A high-resolution data set has been obtained from x-ray structure analysis of a soluble fragment of the Rieske protein from the bovine mitochondrial bc_1 complex (H Michel, personal communication).

ASPECTS OF ASSEMBLY

Mutants in the cyt b_6f complex of *C. reinhardtii* missing any one of the major subunits (63, 119), as well as the PetG hydrophobic subunit (8), appear to exert a pleiotropic effect on the stability of the other subunits. In addition, the other subunits of the complex are unstable in the presence of a soluble truncated mutant (cyt *f*s of cyt *f* that, similar to the turnip cyt *f* fragment discussed above, is missing the C-terminal 35 residues and the hydrophobic anchor region (64, 119). The absence of the C-terminal 35-residue segment also results in a threefold increase in the rate of synthesis of cyt *f*s (64). Thus, the 20-residue TM helix of cyt *f* and/or the 15-residue stromal-side peripheral C-terminal peptide of the cytochrome affects the stability of the other subunits and the rate of synthesis of cyt *f* itself.

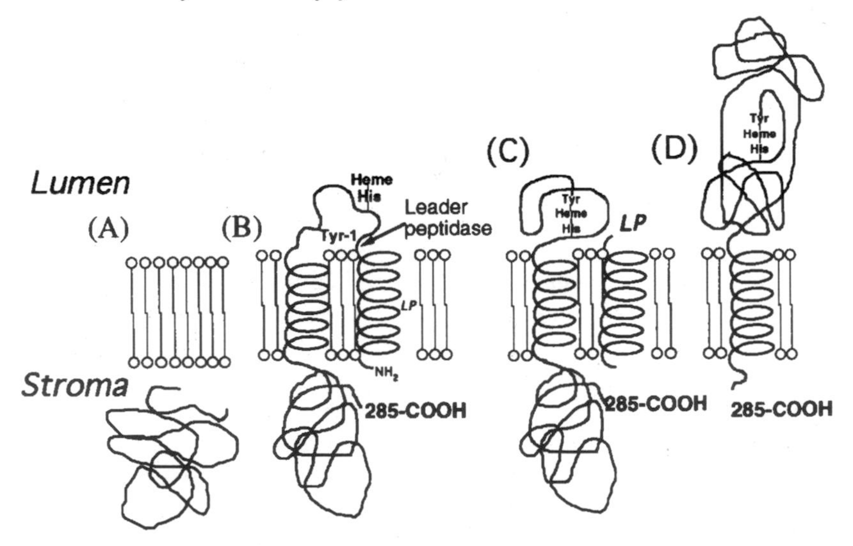

Figure 6 Proposed role of processing in the assembly of the translocated cyt b_6f complex. (A) Unfolded cyt *f* polypeptide approaching membrane from stromal side; (B) signal peptide, exposed at its C-terminus to leader peptidase, and TM helix inserted into membrane bilayer; (C) N-terminal α-amino group of Tyr1 available after cleavage by leader peptidase; and (D) formation of six-coordinate heme through ligation of N-terminal α-amino group.

Translocation and Folding of Cytochrome f

The coordination of the heme by the N-terminal α-amino group of the protein implied that processing of the cyt *f* signal peptide would be necessary for completion of the heme coordination and the final folding and assembly of the protein (17, 70). This provides some knowledge of the sequence of events in the translocation of the cytochrome across the membrane, its processing by the lumen-side leader peptidase, and its subsequent folding. The protein would not fully assemble until translocation across the membrane had proceeded sufficiently to allow cleavage of the cytochrome signal peptide and liberation of Tyr1 of the mature protein (Figure 6A–D). This hypothesis for assembly has been tested with cyt *f* processing mutants in which the consensus processing site—AQA—was changed to LQL (62). This change only partly prevented processing. The amount of heme binding to the unprocessed protein, detected by heme stain, appeared to be quite small (62), so there is either a small amount of heme with a different sixth ligand or adventitiously bound five-coordinate heme.

PROTON TRANSLOCATION FUNCTION OF THE COMPLEX: Q CYCLE

The quinone- or Q-cycle model (74) has been used to describe e^- transfer and H^+ translocation events in the bc_1 complexes of chromatophores (chr) and mitochondria and in the b_6f complex (see below).

Briefly, the Q-cycle mechanism involves:

1. The oxidation of a quinol bound on the *p* side of the complex to an SQ by the iron-sulfur center (Figure 7A, B; steps 1, 2). This 1-e^- oxidation may cause the transfer of 2 H^+ directly to the *p*-side bulk lumenal space (92, 106), or possibly via protonatable amino acids and H^+-conducting bound water chains in the *p*-side polar phase of the Rieske iron-sulfur protein and cyt f/c_1 (Figure 5):

$$Q_p\text{–}H_2 + RFeS_p(ox) \rightarrow Q_p^{\bullet-} + RFeS_p(red) + 2H^+ \ (p \text{ side}). \qquad 1.$$

The data do not exclude the possibility that fewer than 2 H^+/e^- are transferred in this step with the formation of some QH, and then additional H^+ transferred upon oxidation of QH.

2. The SQ formed by the 1-e^- quinol oxidation, in concert with the transfer of the first electron to $RFeS_p$, transfers its electron to the cyt *b* or b_6 heme on the *p* side of the membrane (heme b_p):

$$Q_p^{\bullet-} + b_p(ox) \rightarrow Q_p + b_p(red), \qquad 2.$$

or when written as a concerted reaction (56),

$$Q_pH_2 + RFeS_p(ox) + b_p(ox) \rightarrow Q_p + b_p(red) + RFeS_p(red) + 2H^+ \ (p \text{ side}).$$

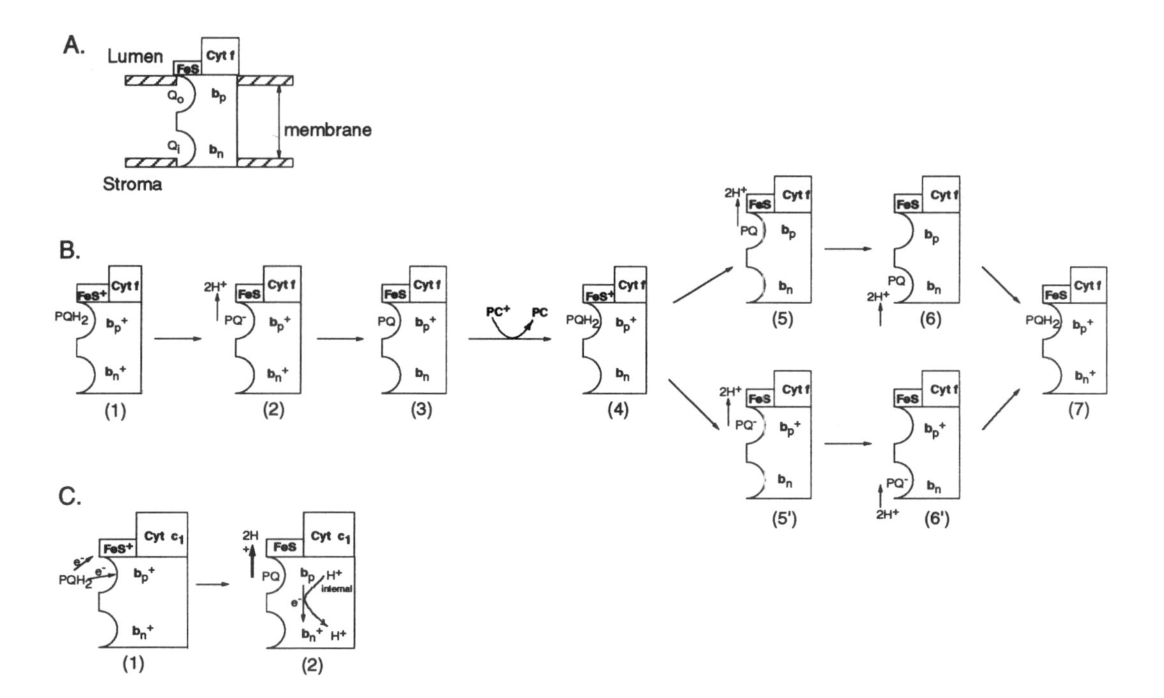

Figure 7 Possible mechanisms of charge translocation in the cyt *b6f* complex. (A) Definition of Q-cycle scheme, (B) Q (steps 1–7) and semiquinone (SQ) (steps 5′, 6′, 7) cycles as described in text. Charge translocation and generation of associated membrane potential according to Mulkidjanian & Junge (79). Models (A, B) are based on those in Joliot & Joliot (50).

The electron is rapidly transferred to heme b_n, although reduced heme b_p is not observed through its reduction because of the kinetics of its turnover (Figure 7B, step 3):

$$b_n(red) + b_p(ox) \rightarrow b_n(ox) + b_p(red). \quad 3.$$

The rate constant for this reaction in chromatophores (chr) has been estimated to be $>10^5\ s^{-1}$ (19).

Formation of reduced heme b_n and the basic phenomenon of oxidant-induced reduction are readily observed only in the presence of antimycin A or related inhibitors that also define antimycin A as a specific inhibitor of the rapid oxidation of heme b_n. The transfer to heme b_n in chromatophores and mitochondria can be observed because its reduced α-band maximum is at 561–562 nm, compared with 565–566 nm for heme b_p. The transfer from heme b_p to heme b_n occurs spontaneously, and it is consistent with the higher potential (more positive in mitochondria and chr by approximately 100 mV and 140 mV, respectively) measured for heme b_n compared with b_p. This midpoint potential difference has led to a common notation for the hemes b_p and b_n that is commonly written as b_l (b low potential) and b_h (b high potential).

3. The next step in the Q cycle proposed for the bc_1 complexes (e.g. see 106) is a 1-e^- reduction by heme b_n of a quinone, Q_n, bound in a nearby niche:

$$b_n(red) + Q_n \rightarrow b_n(ox) + Q_n^{\overset{\bullet}{-}}. \quad 4.$$

4. The Q cycle is completed by oxidation of a second quinol at the p site, Q_pH_2, and a second set of reactions (1–2) above. The second cyt b_n oxidation reaction reduces the semiquinone $Q_n^{\overset{\bullet}{-}}$ to Q_nH_2. That is,

$$b_n(red) + Q_n^{\overset{\bullet}{-}} + 2H^+ \rightarrow b_n(ox) + Q_nH_2. \quad 5.$$

This quinol then diffuses to the p site via the quinol pool,

$$Q_nH_2 \rightarrow Q_pH_2 \quad 6.$$

from which it can again be oxidized by the RFeS center, thus restarting the cycle. Thus, the cytochrome complex must turn over twice for every reaction of $Q_n \rightarrow Q_nH_2$. This implies that the cytochrome complex must interact with at least two quinone molecules. Two tightly bound quinone molecules have been implicated by EPR spectra to interact with the mitochondrial bc_1 complex (24). The reduction of Q_n is shown in Figure 7B (steps 6, 7) as a concerted 2-e^- oxidation of hemes b_p and b_n according to a modification of the Q cycle used to explain data for the b_6f complex (see section on One Major Difference: The Q_n Site).

Thermodynamics of the Q Cycle

The $K_{eq} = 1.5$ for the 1-e^- transfer from PQH_2 to heme b_p, with E_m values of +90 mV, +290 mV, and −90 mV for the PQ/PQH_2, RFeS, and b_p couples is barely favorable. Thus, a favorable K_{eq} requires that the reaction be coupled to a favorable ΔE_m for the reaction between hemes b_n and b_p. Every 60-mV increase in the E_m of heme b_n is correlated with a 10-fold increase in the K_{eq}. The somewhat favorable ΔE_m between the Rieske protein and cyt c_1 would also increase the effective operating potential of the RFeS and increase the K_{eq} of the Q-cycle transfer to heme b_n.

The Q cycle has been successful in providing an explanation for the following experimental observations regarding H^+ and/or e^- transfer in the bc_1 complexes: (*a*) H^+/e^- ratios ≈ 2, (*b*) the oxidant-induced reduction of cyt *b*, (*c*) formation of a TM potential associated with e^- transfer through the two hemes of cyt *b*, and (*d*) a function for cyt *b*.

In more detail: 1. For every e^- that is transferred through the complex, ~2 H^+ are observed on the *p* side. A H^+/e^- stoichiometry of 1–2 for H^+ translocation in the bc_1 complex is required for the energetics of ATP synthesis in mitochondria. The ATP-synthesis requirements in chromatophores cannot be used to set a lower limit on the stoichiometry of H^+ translocation by bc_1 complex because the bacterial cyclic electron transport system does not allow determination of the ATP/2e^- ratio. The H^+/e^- stoichiometry argument does not by itself, however, strongly imply a Q-cycle mechanism because the cytochrome oxidase complex of the respiratory chain can also pump H^+ with a $H^+/e^- \approx 2$ (117). The ability of membrane protein systems such as cytochrome oxidase to generate an electrogenic H^+ pump with a stoichiometry of 2 became known after the original formulation of the Q cycle.

2. The experiment that uniquely and strongly implies a Q-cycle mechanism is the oxidant-induced reduction of cyt *b* (117). When oxygen or ferricyanide is added to oxidize cyt c_1 in the presence of the *n*-site inhibitor antimycin A, one observes reduction of cyt *b*. Antimycin was added to prevent the reoxidation of the high-potential heme by binding to the *n* site. This oxidant-induced reduction of cyt *b* in the presence of antimycin occurs only in the presence of the oxidized RFeS protein (105). A simple thermodynamic explanation for this seemingly anomalous reduction is that the SQ anion, $Q^{\bullet-}$, formed by 1-e^- oxidation of the quinol is a stronger reductant than the reduced quinol, QH_2 [for a review of quinone thermodynamics, see Cramer & Knaff (15, Chapter 2)]. The $E_m \approx -50$ mV– –100 mV of heme b_p in chr and mitochondria, which is 120–180 mV more negative than the ubiquinone/ubiquinol couple, would require a ubisemiquinone dismutation constant of 10^{-4}–10^{-6} at equilibrium. The concentration of SQ at equilibrium is then 10^{-2}–10^{-3} times the concentration of UQ and UQH_2. The experimental detection of SQ species in mitochondria and chr (e.g. 72) is part of the proof of the existence of the Q cycle.

3. The interheme $b_p \rightarrow b_n$ electron transfer is electrogenic and generates a membrane potential, $\Delta\psi$. This membrane potential has been measured in chr by the slow (ms) component of the carotenoid electrochromic bandshift in the

515–520-nm region (45, 47), which is antimycin insensitive and therefore reflects interheme electron transfer.

4. In addition to explaining oxidant-induced reduction and H^+/e^- ratios of ~2, the Q-cycle model provides a mechanism for a two-step oxidation of the quinol and a function for the two hemes of the cyt *b* and for the cyt *b* subunit polypeptide.

Application of the Q Cycle to the Cytochrome b_6f Complex

SOME QUESTIONS AND ALTERNATIVE MECHANISMS Because the Q cycle is almost universally accepted as a description of the pathway of electron transfer in the cyt bc_1 systems of mitochondria and chromatophores, it is a natural inference that a Q-cycle pathway is also obligatory in the cyt b_6f system of oxygenic photosynthesis. This biochemical analogy is supported by the occurrence of H^+/e^- ratios of 2 as discussed above. On the other hand, there is some dispute about (*a*) whether this H^+/e^- is closer to 2 (93) or 1 for b_6f in an energized thylakoid membrane (data summarized in 14), and (*b*) whether a sufficient $\Delta\tilde{\mu}_{H^+}$ is generated in a noncyclic electron transport pathway—or whether additional $\Delta\tilde{\mu}_{H^+}$ must be provided by a cyclic electron transport pathway (4) to provide enough ATP for carbon fixation. The noncyclic electron transport pathway from water to $NADP^+$, with $H^+/e^- = 1$ for electron transfer through the b_6f complex, can theoretically support the synthesis of almost all (8/9) of the ATP needed for CO_2 fixation by the C_3 pathway, assuming the proton requirement for ATP synthesis (H^+/ATP) = 3. An H^+/e^- ratio of 2 for e^- transfer through the b_6f complex would support synthesis of more than enough (4/3) of the ATP needed for C_3-pathway CO_2 fixation. For carbon fixation by the C_4 pathway, not even $H^+/e^- = 2$ in the noncyclic pathway would be sufficient, and cyclic electron transport is required (see section on Cyclic Electron Transport) (4). However, the arguments about the absolute value of the H^+/e^- may not be critical for the Q cycle in any case because, as noted above, the mammalian cytochrome oxidase pumps 2 H^+ per electron without a Q cycle.

2. Additional arguments that support the existence of an obligatory Q-cycle in the cytochrome b_6f complex are similar to those for the bc_1 complex discussed above: The Q cycle mechanism provides an explanation for (*a*) oxidant (light)-induced reduction of cyt b_6; it also provides mechanisms for (*b*) the slow (ms) electrochromic phase that is an indicator of a membrane potential, (*c*) oxidation of plastosemiquinone presumably generated in the oxidant-reduced, and (*d*) a function for the cyt b_6 polypeptide. This rationale is perhaps sufficient to conclude that the Q cycle must operate in the b_6f complex of oxygenic photosynthesis by mechanisms similar to those used by the bc_1 complex discussed above. However, although there is disagreement as to the magnitude and depth of differences between the two systems, some properties of the b_6f complex differ from those of the bc_1 complexes.

ONE MAJOR DIFFERENCE: THE Q_n SITE *Inhibitors* There is no specific high-affinity Q_n-site inhibitor, such as antimycin, that causes major inhibition of the main chain electron transfer. Antimycin at levels greater than stoichiometric concentrations has no obvious effect on the cyt b_6 oxidation rate, and this insensitivity to antimycin can be explained in terms of sequence differences in cyt b_6 (41). The two inhibitors that mimic antimycin by causing an increase in the amplitude of cyt b_6 reduction by a short light flash are 2-n-4-hydroxyquinoline-N-oxide (NQNO) (102) and (E,E)-methyl-3-methoxy-2-(styrylphenyl) propenoate (MOA-stilbene) (95). However, at the same concentrations that cause an increase in cyt b_6 reduction, NQNO has no effect on noncyclic or linear e^- transport (52). There is partial inhibition by MOA-stilbene— ~40% relative to its effect on PSII (95). Thus, the Q cycle may not be obligatory or may occur only in a fraction of the membranes. Alternatively, the basic Q cycle may be obligatory, although the Q_n site is different and no inhibitor equivalent to antimycin has been found. For the latter case, NQNO effects can be explained by assuming: (*a*) that the rate constant for b_n oxidation in the absence of NQNO is approximately threefold greater than that for b_n reduction via $b_p \rightarrow b_n$ interheme transfer, thus explaining the small amplitude—ca 0.2–0.3 heme—of flash-induced reduction under reducing conditions of intermediate strength; (*b*) the rate constant for heme b_n oxidation was proposed to decrease in the presence of NQNO by an amount, approximately a factor of 9, to make it three times smaller than that for b_n reduction, thus explaining the roughly threefold increase in amplitude of flash-induced reduction of heme b_n with NQNO (102). The absence of inhibition by NQNO of noncyclic electron transport would be explained by the decreased rate constant of b_n oxidation not affecting the rate-limiting step of electron transport (96). However, the rate of cyt *b* reoxidation in the absence of NQNO following reduction by a single flash under many conditions is ≥50 ms (28, 57, 94, 96), which is substantially greater than the rate-limiting step of linear electron transport. This latter observation led to a modification of the Q cycle (57, 95) in which the slow oxidation after the single flash is explained by the stable formation of heme b_n(red). Reduction of the Q_n would occur by a concerted 2-e^- oxidation of both hemes after formation of both b_p(red) and b_n(red) by a second flash (Figure 7B, steps 5–7). The occurrence of a modified Q cycle with b_6f compared with the bc_1 complexes could be a consequence of a decreased E_m of the $Q_n/Q_n^{\bullet}$ couple in the b_6f compared with the bc_1 complex. It would prevent Q_n from accepting one electron from b_n(red). The lowered E_m would also explain why a convincing EPR signal corresponding to a $Q_n^{\bullet}$ species has not been obtained (82).

Redox potentials of the b hemes The modified Q cycle also requires a substantial positive ΔE_m for the electron transfer from heme b_p to b_n. A detectable (>50 mV) ΔE_m has always been found in the isolated b_6f complex, ranging from values for the b_l and b_h components in the complex from spinach chloroplasts of −50 mV and −170 mV, ΔE_m = +120 mV (36a); and −84 mV and −158 mV, ΔE_m = +74 mV, in the active preparation recently obtained from *C. reinhardtii* (86). The E_m of cyt *f* in the *C. reinhardtii* complex (+330 mV) is ca 40 mV more negative than the E_m of +370 mV for the spinach or turnip cyt *f*. A thermodynamic problem with the above E_m values of −150– −170 mV for heme b_l is that, according to Equation 2, the K_{eq} and $\Delta G°$ would be quite unfavorable for transfer of the first electron to heme b_l if the E_m(RFeS) = +290 mV and marginally favorable if this E_m = +370 mV (40, 94).

For E_m determination of the two *b* hemes in situ, single E_{m7} values of −30 (31), −50 (28), and −70 (96) mV have been measured for the hemes. It was not possible to resolve different E_m values in situ in thylakoid membranes. In contrast, from different rates of dark reduction of the two hemes, a ΔE_m of +140 mV was calculated (48), and a ΔE_m = +105 mV for E_{m7} values of −45 mV and −150 mV was obtained from spectral deconvolution of redox titrations (58). An additional perspective on these disparate measurements of the heme E_m values is provided by the concept of significant heme-heme interaction energies across the 12-Å edge-edge distance caused by a change on oxidation state of one heme relative to the other (113). The data of References 58 and 28 could be fit by the model of Wagner & Walz (112), assuming single E_m values for both sets of data and a larger interaction energy for the data of Reference 58. The larger interaction energy would cause a larger splitting between heme E_m values, thus explaining the larger ΔE_m in Reference 58. Interaction between the hemes is also implied by splitting of the Soret band circular dichroism spectra of the two hemes measured in bc_1 complex from yeast mitochondria (83). The absolute value of the heme E_m values in References 58 and 28 could be different because of changes in the effective dielectric constant of the interfacial layer that would cause different packing of the peripheral protein layer caused by different ionic conditions. A more polar environment will cause a shift to more negative values (60). A consequence of the analysis of Wagner & Walz (112) for the Q cycle is that the ΔE_m values could be smaller than required. However, the physiological meaning of different values of heme interaction energy in different preparations is unclear. Therefore, it is not possible to dissociate the problem from the question of unexplained experimental differences in the nature of the preparations used for determination of the midpoint potentials.

With regard to other manifestations of heme heterogeneity, the reduced α-band peaks of the two hemes are both at 564 nm and not distinguishable in the *C. reinhardtii* complex and are different by 1 nm in the main peaks measured at 77 K, where the difference will be greater than at room temperature (43). The difference in α-band maxima of the two hemes in thylakoids was reported to be ≈0.6 nm (28) and 0.6 nm in the green alga *Chlorella*, where the peaks of hemes b_l and b_h measured in situ are 563.4 nm and 564.0 nm (48). The largest difference was found in a study of spinach thylakoids (58), where the peaks of hemes b_e and b_n were 565.0 and 563.2 nm, respectively. Most of the data indicate that the α band peaks of the two hemes are more similar in cyt b_6 compared with cyt *b* of the bc_1 complex. Much of the redox data discussed above also indicates less heterogeneity. The resolution of these differences may await molecular analysis found in high resolution structures of the two complexes.

Origin of slow electrochromic band shift Measurement of a concerted (i.e. same kinetics and amplitude) reduction of high (cyt *f*) and low (cyt b_6) potential chains implies existence of a Q-cycle mechanism (56), with the time course of the cyt *b* reduction matching that of the slow (ms) electrogenic phase attributed to interheme transfer (58). However, a recent study in chromatophores suggests a different origin of the slow electrogenic phase (79). Measurement of the relative time course of flash-induced heme b_h reduction, slow phase, and H^+ transfer indicated that the H^+ transfer ($t_{1/2}$ = 10 ms) and not e^- transfer ($t_{1/2}$ = 3 ms) is the main source of the slow electrogenic phase ($t_{1/2}$ = 10 ms). It is proposed in the latter study (79) that the interheme transfer is neutralized by initial cotransfer of H^+ and that the transfer of H^+ to the *p* side of the membrane is electrogenic (Figure 7C). This model is of interest in light of the extended *p*-side intraprotein pathway proposed for the H^+ exit port from the b_6f complex (Figure 5), the measurement discussed above of the local field sensed by a chl *a* band shift on the *p* side of the b_6f complex (51), and discussions on the origin of the slow component of the 515-nm absorbance change (81).

Reactions of the b_6f Complex Under Reducing Conditions: The Semiquinone Cycle

Approximately full amplitude of the slow electrogenic phase with a $t_{1/2}$ = 4 ms was measured at low ambient redox potential (–200 mV), conditions under which the cyt *b* hemes and PQ pool were initially poised in the reduced state, according to their E_m value (–30– –75 mV) measured under the same conditions (31). The slow phase could not be explained by interheme electron transfer because of the much slower oxidation of cyt *b,* and it was proposed that the origin of the slow phase might be movement of anionic plastosemiquinone formed in the flash ["Q Non-Cycle" (30)], as proposed in the "semiquinone cycle" (97, 116). A somewhat similar conclusion was reached in studies on *Chlorella* under reducing conditions (49, 50), which suggested that the semiquinone, $PQ^{\bullet}$, formed at the Q_p site would diffuse to the Q_n site at least when heme b_n is reduced (Figure 7B; steps 5′, 6′), and that because oxidized PQ appeared not to have a high affinity for the Q_n site, the SQ cycle may operate under oxidizing as well as reducing conditions. On the other hand, it has been argued that the SQ cycle implies slippage in the Q-cycle mechanism, the concerted reduction of cyt b_6 and cyt f implies no such slippage, and therefore the semiquinone cycle should not occur (56).

Cyclic Electron Transport

In terms of elementary structure considerations, the cyclic electron transport pathway provides a priori an obvious function in which to seek other functions in which cyt b_6 might be involved. This pathway has been reviewed recently (4), and it has been pointed out that the existence of a cyclic phosphorylation pathway in vivo is well established (see above).

As discussed below, ATP input from cyclic phosphorylation is essential to provide the ATP needed for C_4 photosynthesis. The transfer of electrons from PSI on the stromal side of the membrane (e.g. from reduced ferredoxin) to the quinone pool in the center of the bilayer implies the need for a redox prosthetic group with an appropriate E_m at the *n*-side membrane interface. Heme b_n of cyt b_6 fulfills these requirements. Thus, a mechanism of cyclic electron transport has been proposed (15) that would use the well-established "oxidant-induced reduction" to transfer an electron from Q_pH_2 to heme b_p (Equations 1, 2); concomitantly, an electron would be transferred from PSI via ferredoxin and possibly other peripheral redox docking components such as cyt "G" (see above) to heme b_n (14, 15). The two hemes would be oxidized by bound PQ in a concerted reaction similar to that proposed in the context of the "modified" Q cycle (58, 95). On the one hand, this mechanism is supported by the four- to fivefold increase in oxidation rate of cyt b_6 after a flash or flash train that occurs in the presence of NADPH-ferredoxin used to prereduce heme b_n before a flash (28). On the other hand, although antimycin is the major inhibitor of cyclic phosphorylation and, by analogy with bc_1 complexes, might be expected to interact with cyt b_6f, its binding activity fractionates with PSI and not with the b_6f complex (4). In addition, ferredoxin does not stimulate the slow rate of cyt b_6 by dithionite (13), as discussed by Bendall & Manasse (4). The latter data imply the existence of a redox carrier alternative to cyt b_6, "ferredoxin-plastoquinone reductase," which is presently a hypothetical component, to close the cycle.

CONCLUSION

Cytochrome b_6f is a fascinating membrane protein complex because of the combination of precise spectrophotometric assays of function, active purified preparations, and for the integral cytochrome *b*, the largest amount of primary sequence data for any known membrane protein. The proton translocation function of the complex is clear. The consensus mechanism for the H^+ translocation, partly based on analogy with mitochondrial and bacterial systems, is the Q cycle. There is a high resolution structure of cytochrome *f*, which has allowed the first view of the details of the *p*-side structure of the complex.

Resolution of some of the detailed questions about the electron transfer pathway, the mechanism of H^+ translocation, and possible differences between the bc_1 and b_6f complexes will be facilitated in the near future by the impending availability of high resolution structures of the complexes and the Rieske protein extrinsic domain.

ACKNOWLEDGMENTS

The authors' research included in this review was supported by grants from the NIH (GM-38323) and USDA (9501148). The authors are indebted to P. N. Furbacher, J. B. Heymann, S. L. Schendel, and W. J. Vredenberg for helpful discussions, as well as to many colleagues: D. S. Bendall, G. Hauska, A. Hope, A. Joliot, P. Joliot, J.-L. Popot, C. Wagner, D. Walz, and F.-A. Wollman who kindly shared manuscripts with us prior to their publication.

Literature Cited

1. Allen JF. 1992. Protein phosphorylation in regulation of photosynthesis. *Biochim. Biophys. Acta* 1098:275–335
2. Bagby S, Driscoll PC, Goodall KG, Redfield C, Hill H. 1990. The complex formed between plastocyanin and cytochrome c: investigation by NMR spectroscopy. *Eur. J. Biochem.* 188:411–20
3. Bald D, Kruip J, Boekema EJ, Rogner M. 1992. Structural investigations of cyt. b_6/f-complex and PSI-complex from the cyanobacterium *Synechocystis PCC6803*. See Ref. 79a, 1:629–32
4. Bendall DS, Manasse RS. 1995. Cyclic phosphorylation and electron transport. *Biochim. Biophys. Acta* 1229:23–38
5. Beratan DN, Betts JN, Onuchic JN. 1991. Protein electron transfer rates set by the bridging secondary and tertiary structure. *Science* 252:1285–88
6. Beratan DN, Onuchic JN, Hopfield JJ. 1987. Electron tunneling through covalent and noncovalent pathways in proteins. *J. Chem. Phys.* 86:4488–98
7. Berry EA, Huang L-S, Earnest TN, Jap BK. 1992. X-ray diffraction by crystals of beef heart ubiquinol:cytochrome c oxidoreductase. *J. Mol. Biol.* 224:1161–66
8. Berthold D, Schmidt CL, Malkin R. 1995. The deletion of PetG in *C. reinhardtii* disrupts the cytochrome bf complex. In *Photosynth. Res. 10th Int. Photosynth. Congr.*, 103rd, Montpellier, France, 2:571–74
9. Bouges-Bocquet B. 1977. Cytochrome *f* and plastocyanin kinetics in *C. pyrenoidosa. Biochim. Biophys. Acta* 462:362–70
10. Breyton C, de Vitry C, Popot J-L. 1994. Membrane association of cytochrome b6f subunits: the Rieske iron-sulfur protein from *C. reinhardtii* is an extrinsic protein. *J. Biol. Chem.* 269:7597–602
11. Chain R, Malkin R. 1991. The chloroplast *b6f* complex can exist in monomeric and dimeric states. *Photosynth. Res.* 28:59–68
12. Coleman PM, Freeman HC, Guss JM, Muarata M, Norris VA. 1978. X-ray crystal analysis of plastocyanin at 2.7 Å resolution. *Nature* 272:319–24
13. Cox RP. 1979. Chloroplast cytochrome b-563. Hydrophobic environment and lack of direct reaction with ferredoxin. *Biochem. J.* 184:39–44
14. Cramer WA, Black MT, Widger WR, Girvin ME. 1987. Comparative structure and function of the b cytochromes of bc_1 and b_6f complexes. In *The Light Reactions*, ed. J. Barber, pp. 446–93. Amsterdam: Elsevier
15. Cramer WA, Knaff DB. 1991. *Energy Transduction in Biological Membranes.* New York: Springer-Verlag
16. Cramer WA, Martinez SE, Huang D, Furbacher PN, Smith JL. 1994. The cytochrome b_6f complex. *Curr. Opin. Struct. Biol.* 4:536–44
17. Cramer WA, Martinez SE, Huang D, Tae G-S, Everly RM, et al. 1994. Structural aspects of the cytochrome *b6f* complex;

structure of the lumen-side domain of cytochrome *f*. *J. Bioenerg. Biomembr.* 26:31–47
18. Crofts AR. 1985. The mechanism of the ubiquinol:cytochrome c oxidoreductases of mitochondria and of *Rb. sphaeroides.* In *The Enzymes of Biological Membranes,* ed. AN Martonosi, 4:347–82. New York: Plenum
19. Crofts AR, Wang Z. 1989. How rapid are the internal reactions of the ubiquinol:cytochrome c2 oxidoreductase? *Photosynth. Res.* 22:69–87
20. Curry WB, Grabe MD, Kurnikov IV, Skourtis SS, Beratan DN, et al. 1995. Pathways, pathway tubes, pathway docking, and propagators in electron transfer proteins. *J. Bioenerg. Biomembr.* 27:285–93
21. Degli Esposti M, De Vries S, Crimi M, Ghelli A, Patarnello T, Meyer A. 1993. Mitochondrial cytochrome b: evolution and structure of the protein. *Biochim. Biophys. Acta* 1143:243–71
22. de Rege PJF, Williams SA, Therein MJ. 1995. Direct evaluation of electronic coupling mediated by hydrogen bonds: implications for biological electron transfer. *Science* 269:1409–13
23. de Vitry C, Breyton C, Pierre Y, Popot J-L. 1995. The 4-kD chloroplast polypeptide of cytochrome b_6f complex encoded by the nuclear *PetX* gene: nucleic and protein sequences, targeting signals and membrane topology. *Proc. 10th Int. Congr. Photosynth.,* Montpellier, France, pp. 595–98
24. Ding H, Robertson DE, Daldal F, Dutton PL. 1992. Cytochrome bc_1 complex [2Fe-2S] cluster and its interaction with ubiquinone and ubihydroquinone at the Q_o site: a double-occupancy Q_o site model. *Biochemistry* 31:3144–58
25. Ermler UG, Fritsch SK, Buchanan SK, Michel H. 1994. Structure of the photosynthetic reaction centre from *Rb. sphaeroides* at 2.65 Å resolution: cofactors and protein-cofactor interactions. *Structure* 2:925–36
26. Farid RS, Moser CS, Dutton PL. 1993. Electron transfer in proteins. *Curr. Opin. Struct. Biol.* 3:225–33
27. Frid D, Gal A, Oettmeier W, Hauska G, Berger S, Ohad I. 1992. The redox-controlled light-harvesting chlorophyll a/b protein kinase: deactivation by substituted quinones. *J. Biol. Chem.* 267:25908–15
28. Furbacher PN, Girvin ME, Cramer WA. 1989. On the question of interheme electron transfer in the chloroplast cytochrome b6 in situ. *Biochemistry* 28:8990–98
29. Furbacher PN, Tae G-S, Cramer WA. 1996. Evolution and origins of cytochrome bc_1 and b_6f complexes. In *Origin and Evolution of Biological Energy Conservation,* ed. H Baltscheffsky. New York: VCH. In press
30. Girvin ME. 1985. *Electron and proton transfer in the quinone-cytochrome b/f region of chloroplasts.* PhD thesis. Purdue Univ., W. Lafayette, IN
31. Girvin ME, Cramer WA. 1984. A redox study of the electron transport pathway responsible for generation of the slow electrochromic phase in chloroplasts. *Biochim. Biophys. Acta* 767:29–38
32. Gray JC. 1978. Purification and properties of a monomeric cytochrome f from Charlock. *Eur. J. Biochem.* 82:133–41
33. Gray JC. 1992. Cytochrome *f,* structure, function, and biosynthesis. *Photosynth. Res.* 34:359–74
34. Gray JC, Rochford RJ, Packman LC. 1994. Proteolytic removal of the C-terminal trans-membrane region of cytochrome f during extraction from turnip and charlock leaves generates a water-soluble monomeric form of the protein. *Eur. J. Biochem.* 223:481–88
35. Gray KA, Davidson E, Daldal F. 1992. Mutagenesis of Met183 drastically affects the physicochemical properties of cytochrome c_1 of *Rb. capsulatus. Biochemistry* 31:11864–73
36. Haley J, Bogorad L. 1989. A 4-kDa chloroplast polypeptide associated with the cytochrome *b6f* complex. *Proc. Natl. Acad. Sci. USA* 86:1534–38
36a. Hauska G, Hurt E, Gabellini N, Lockau W. 1983. Comparative aspects of quinol-cytochrome *c*/plastocyanin oxidoreductases. *Biochim. Biophys. Acta* 726:97–133
37. Hauska G, Schütz M, Büttner M. 1996. The cytochrome b_6f complex-composition, structure, and function. In *Oxygenic Photosynthesis: The Light Reactions,* ed. DR Ort, CF Yocum. Amsterdam: Kluwer. In press
38. He S, Modi S, Bendall DS, Gray JC. 1991. The surface-exposed tyrosine residue Tyr-83 of pea plastocyanin is involved in both binding and electron transfer reactions with cytochrome *f. EMBO J.* 10:4011–16
39. Hind G, Marshak DR, Coughlan SJ. 1995. Spinach thylakoid polyphenol oxidase: cloning, characterization, and relation to a putative protein kinase. *Biochemistry* 34: 8157–64
40. Hope AB, Huilgol RR, Panizza M, Thompson M, Matthews DB. 1992. The flash-induced turnover of cytochrome b-563, cytochrome *f* and plastocyanin in chloroplasts: models and estimation of kinetic parameters. *Biochim. Biophys. Acta* 1100: 15–26
41. Howell N, Gilbert K. 1988. Mutational analysis of the mouse mitochondrial cytochrome *b* gene. *J. Mol. Biol.* 203:607–18
42. Huang D, Everly RM, Cheng RH, Heymann JB, Schägger H, et al. 1994. Characterization of the chloroplast cytochrome b_6f as a structural and functional dimer. *Biochemistry* 33:4401–9
43. Hurt EC, Hauska G. 1981. A cytochrome

f/b$_6$ complex of five polypeptides with plastoquinol-plastocyanin-oxidoreductase activity from spinach chloroplasts. *Eur. J. Biochem.* 117:591–99
44. Iwata S, Ostermeier C, Ludwig B, Michel H. 1995. Structure at 2.8 Å resolution of cytochrome *c* oxidase from *P. denitrificans*. *Nature* 376:660–69
45. Jackson JB, Dutton PL. 1973. The kinetic and redox potentiometric resolution of the carotenoid shifts in *Rb. sphaeroides* chromatophores: their relationship to electric field alterations in electron transport and energy coupling. *Biochim. Biophys. Acta* 325:102–13
46. Johnson EM, Schabelrauch LS, Sears BB. 1991. A plastome mutation affects processing of both chloroplast and nuclear DNA-encoded plastid proteins. *Mol. Gen. Genet.* 225:106–12
47. Joliot P, Delosme R. 1974. Flash-induced 519 nm absorption change in green algae. *Biochim. Biophys. Acta* 357:267–84
48. Joliot P, Joliot A. 1988. The low potential electron transfer chain in the cytochrome b/f complex. *Biochim. Biophys. Acta* 933: 319–33
49. Joliot P, Joliot A. 1992. Electron transfer between photosystem II and the cytochrome *b/f*: mechanistic and structural implications. *Biochim. Biophys. Acta* 1102: 53–61
50. Joliot P, Joliot A. 1994. Mechanism of electron transfer in the cytochrome b/f complex of algae: evidence for a semiquinone cycle. *Proc. Natl. Acad. Sci. USA* 91:1034–38
51. Joliot P, Joliot A. 1995. A shift in chlorophyll spectrum associated with electron transfer within cytochrome *bf* complex. *Proc. 10th Int. Photosynth. Congr.*, pp. 615–18
52. Jones RW, Whitmarsh J. 1985. Origin of the electrogenic reaction in the chloroplast cytochrome *b/f* complex. *Photochem. Photobiol.* 9:119–27
53. Kallas T. 1994. The cytochrome *b*$_6$*f* complex. In *The Molecular Biology of Cyanobacteria,* ed. DA Bryant, pp. 259–317. Dordrecht: Kluwer
54. Kerfeld CA, Anwar HP, Interrante R, Merchant S, Yeates TO. 1995. The structure of chloroplast cytochrome c_6 at 1.9 Å resolution: evidence of functional oligomerization. *J. Mol. Biol.* 250:627–47
55. Kirwin PM, Elderfield PD, Williams RS, Robinson C. 1991. Transport of protein into chloroplasts: organization, orientation, and lateral distribution of the plastocyanin processing peptidase. *J. Biol. Chem.* 263: 18128–33
56. Kramer DM, Crofts AR. 1993. The concerted reduction of the high and low potential chains of the bf complex by plastocyanin. *Biochim. Biophys. Acta* 1183:72–84
57. Kramer DM, Crofts AR. 1992. A Q cycle type model for turnover of the *bf* complex under a wide range of redox conditions. See Ref. 79a, 2:491–94
58. Kramer DM, Crofts AR. 1994. Re-examination of the properties and function of the *b* cytochromes of the thylakoid cytochrome *bf* complex. *Biochim. Biophys. Acta* 1184: 193–201
59. Kraulis PJ. 1991. MOLSCRIPT: a program to produce both detailed and schematic plots of protein structure. *J. Appl. Cryst.* 24:946–50
60. Krishtalik L, Tae G-S, Cherepanov DA, Cramer WA. 1993. The redox properties of cytochromes *b* imposed by the membrane electrostatic environment. *Biophys. J.* 65: 184–95
61. Kühlbrandt W, Wang DN, Fujiyoshi Y. 1994. 3-D structure of plant light harvesting complex determined by electron crystallography. *Nature* 350:130–34
62. Kuras R, Büschlen S, Wollman F-A. 1995. Maturation of pre-apocytochrome *f* in vivo. *J. Biol. Chem.* 270:27797–930
63. Kuras R, Wollman F-A. 1994. The assembly of cytochrome *b*$_6$*f* complexes: an approach using genetic transformation of the green alga, *C. reinhardtii. EMBO J.* 13: 1019–27
64. Kuras R, Wollman F-A, Joliot P. 1995. Conversion of cytochrome *f* to a soluble form in vivo in *Chlamydomonas reinhardtii. Biochemistry* 34:7468–75
65. Langen R, Chang IJ, Germanas JP, Richards JH, Winkler JR, Gray HB. 1995. Electron tunneling in proteins: coupling through a β strand. *Science* 268: 1733–35
66. Lemaire C, Girard-Bascou J, Wollman F-A, Bennoun P. 1986. Studies on the cytochrome *b*$_6$*f* complex. *Biochim. Biophys. Acta* 851:229–38
67. Link TA, Hagen WR, Pierik AJ, Assmann C, von Jagow G. 1992. Determination of the redox properties of the Rieske [2Fe–2S] cluster of bovine heart bc_1 complex by direct electrochemistry of a water-soluble fragment. *Eur. J. Biochem.* 208:685–91
68. Marcus R, Sutin N. 1985. Electron transfers in chemistry and biology. *Biochim. Biophys. Acta* 811:265–322
69. Deleted in proof
70. Martinez SE, Huang D, Szczepaniak A, Cramer WA, Smith JL. 1994. Crystal structure of the chloroplast cytochrome *f* reveals a novel cytochrome fold and unexpected heme ligation. *Structure* 2:95–105
71. Martinez SE, Smith JL, Huang D, Szczepaniak A, Cramer WA. 1992. Crystallographic studies of the lumen-side domain of turnip cytochrome *f*. See Ref. 79a, 2: 495–98
72. Meinhardt SW, Yang XH, Trumpower BL, Ohnishi T. 1987. Identification of a stable

ubisemiquinone and characterization of the effects of ubiquinone on oxidation-reduction status of the Rieske iron-sulfur protein in three subunit ubiquinol-cytochrome *c* oxidoreductase complexes of *P. denitrificans. J. Biol. Chem.* 262:8702–6
73. Meyer TE, Zhao ZG, Cusanovich MA, Tollin G. 1993. Transient kinetics of electron transfer from a variety of c-type cytochromes to plastocyanin. *Biochemistry* 32: 4552–59
73a. Michel H. 1991. In *Crystallization of Membrane Proteins,* ed. H Michel, pp. 73–86. Boca Raton, FL: CRC Press
74. Mitchell P. 1976. Possible molecular mechanisms of the protonmotive function of cytochrome systems. *J. Theor. Biol.* 62: 327–67
75. Modi S, He S, Gray J, Bendall D. 1992. The role of surface-exposed Tyr-83 of plastocyanin in electron transfer from cytochrome *c. Biochim. Biophys. Acta* 1101:64–68
76. Modi S, Nordling M, Lundberg LG, Örjan H, Bendall DS. 1992. Reactivity of cytochromes *c* and *f* with mutant forms of plastocyanin. *Biochim. Biophys. Acta* 1102: 85–90
77. Moore GR, Pettigrew GW. 1990. *Cytochromes c: Evolutionary, Structural and Physicochemical Aspects.* Heidelberg: Springer-Verlag
78. Morand L, Frame MK, Colvert KK, Johnson DA, Krogmann DW, Davis DJ. 1989. Plastocyanin-cytochrome *f* interaction. *Biochemistry* 28:8039–47
79. Mulkidjanian AY, Junge W. 1994. Calibration and time resolution of lumenal pH-transients in chromatophores of *Rb. capsulatus* following a single turnover flash of light: proton release by the cytochrome bc_1 complex is strongly electrogenic. *FEBS Lett.* 353:189–93
79a. Murata N, ed. 1992. *Research in Photosynthesis.* Dordrecht: Kluwer
80. Nordling M, Olausson T, Lundberg LG. 1990. Expression of plastocyanin in *E. coli. FEBS Lett.* 276:98–102
81. Ooms JJJ, Versluis W, van Vliet PH, Vredenberg WJ. 1991. The flash-induced P515 shift in relation to ATPase activity in chloroplasts. *Biochim. Biophys. Acta* 1056: 293–300
82. Pace RJ, Hope AB, Smith P. 1992. Detection of flash-induced quinone radicals in spinach chloroplasts. *Biochim. Biophys. Acta* 1098:209–16
83. Palmer G, Degli Esposti M. 1994. Application of exciton coupling theory to the structure of mitochondrial cytochrome *b. Biochemistry* 33:176–85
84. Parker MW, Postma JPM, Pattus F, Tucker AD, Tsernoglou D. 1992. Refined structure of the pore-forming domain of colicin A at 2.4 Å resolution. *J. Mol. Biol.* 224:639–57
85. Picot D, Loll PJ, Garavito MR. 1994. The x-ray crystal structure of the membrane protein prostaglandin H2 synthase-1. *Nature* 367:243–49
86. Pierre Y, Breyton C, Tribet C, Kramer D, Olive J, Popot JL. 1995. Purification and characterization of the cytochrom b_6f complex from *Chlamydomonas reinhardtii. J. Biol. Chem.* 270:29342–49
87. Ponomarev M, Soriano GM, Huang D, Zhang H, Carrell C, et al. 1996. *Biophys. J.* In press
88. Popot J-L, Pierre Y, Breyton C, Lemoine Y, Takahashi Y, Rochaix JD. 1995. Purification and composition of cyt b_6f complex from *Chlamydomonas reinhardtii. Proc. 10th Int. Congr. Photosynth.,* Montpellier, France, pp. 507–12
89. Qin L, Kostić NM. 1992. Electron-transfer reactions of cytochrome *f* with flavin semiquinones and with plastocyanin: importance of protein-protein electrostatic interactions and of donor-acceptor coupling. *Biochemistry* 31:5145–50
90. Qin L, Kostić NM. 1993. Importance of protein rearrangement in the electron-transfer reaction between the physiological partners cytochrome *f* and plastocyanin. *Biochemistry* 32:6073–80
91. Redinbo MR, Yeates TO, Merchant S. 1994. Plastocyanin: structural and functional analysis. *J. Bioenerg. Biomembr.* 26: 49–66
91a. Reimann A, Kuck U. 1989. Nucleotide sequence of the plastid genes for apocytochrome b_6 (petB) and subunit IV of the cytochrome b_6f complex (petD) from the green alga *Chlorella protothecoides*: lack of introns. *Plant Mol. Biol.* 136:255–56
92. Rich PR. 1986. A perspective on Q-cycles. *J. Bioenerg. Biomembr.* 18:145–56
93. Rich PR. 1988. A critical examination of the supposed variable proton stoichiometry of the chloroplast cytochrome *bf* complex. *Biochim. Biophys. Acta* 932:33–42
94. Rich PR, Heathcote P, Moss DA. 1987. Kinetic studies of electron transfer in a hybrid system constructed from the cytochrome *bf* complexes & Photosystem I. *Biochim. Biophys. Acta* 892:138–51
95. Rich PR, Madgwick SA, Brown S, Von Jagow G, Brandt U. 1992. MOA-stilbene, a new tool for investigation of the chloroplast *bf* complex. *Photosynth. Res.* 34: 465–77
96. Rich PR, Madgwick SA, Moss DA. 1991. The interactions of duroquinol, DBMIB, and NQNO with the chloroplast cytochrome *bf* complex. *Biochim. Biophys. Acta* 1058:312–28
97. Rich PR, Wikström M. 1986. Evidence for a mobile semiquinone in the redox cycle of the mammalian cytochrome bc_1 complex. *FEBS Lett.* 194:176–81

98. Roberts VA, Freeman HC, Olson AJ, Tainer JA, Getzoff ED. 1991. Electrostatic orientation of the electron-transfer complex between plastocyanin and cytochrome *f*. *J. Biol. Chem.* 266:13431–41
99. Romanowska E, Albertsson P-Å. 1994. Isolation and characterization of the cytochrome *bf* complex from whole thylakoids, grana, and stroma lamellae vesicles from spinach chloroplasts. *Plant Cell Physiol.* 35:557–68
100. Saraste M. 1984. Location of heme binding sites in the mitochondrial cytochrome *b*. *FEBS Lett.* 166:367–72
101. Schägger H, Cramer WA, von Jagow G. 1994. Analysis of molecular masses and oligomeric states of protein complexes by blue native electrophoresis and isolation of membrane protein complexes by two-dimensional native electrophoresis. *Anal. Biochem.* 217:220–30
101a. Schütz M, Zirngibl S, le Coutre J, Büttner M, Xie D-L, Nelson N, et al. 1994. A transcription unit for the Rieske re-S protein and cytochrome b in *Chlorobium limicola*. *Photosynth. Res.* 39:163–74
102. Selak MA, Whitmarsh J. 1982. Kinetics of the electrogenic step and cytochrome b_6 and *f* redox changes in chloroplasts. *FEBS Lett.* 150:286–92
103. Szczepaniak A, Cramer WA. 1990. Thylakoid membrane topography. *J. Biol. Chem.* 265:17720–26
103a. Szczepeniak A, Frank K, Rybka J. 1995. Membrane association of the Rieske iron-sulfur protein. *Z. Naturforsch.* 50C:535–42
104. Szczepaniak A, Huang D, Keenan TW, Cramer WA. 1991. Electrostatic destabilization of the cytochrome b_6f complex in the thylakoid membrane. *EMBO J.* 10: 2757–64
105. Trumpower BL. 1981. Function of the iron-sulfur protein of the cytochrome b-c_1 segment in electron transfer and energy-conserving reactions of the mitochondrial respiratory chain. *Biochim. Biophys. Acta* 639: 129–55
106. Trumpower BL. 1990. The protonmotive Q cycle: energy transduction by coupling of proton translocation to electron transfer by the cytochrome bc_1 complex. *J. Biol. Chem.* 265:11409–12
107. Trumpower BL, Gennis RB. 1994. Energy transduction by cytochrome complexes in mitochondrial and bacterial respiration: the enzymology of coupling electron transfer reactions to transmembrane proton translocation. *Annu. Rev. Biochem.* 63:675–716
108. Tsukihara T, Aoyama H, Yamashita E, Tomizaki T, Yamaguchi H, et al. 1995. Structures of metal sites of oxidized bovine heart cytochrome *c* oxidase at 2.8 Å. *Science* 269:1069–74
109. Turro C, Chang CK, Leroi GE, Cukier RI, Nocera DG. 1992. Photoinduced electron transfer mediated by a hydrogen-bonded surface. *J. Am. Chem. Soc.* 114:4013–15
110. Ullmann GM, Kostić NM. 1995. Electron-tunneling paths in various electrostatic complexes between cytochrome c and plastocyanin: anisotropy of the copper-ligand interactions and dependence of the iron-copper electronic coupling on the metalloprotein orientation. *J. Am. Chem. Soc.* 117: 4766–74
111. Vener AV, van Kan PJM, Gal A, Andersson B, Ohad I. 1995. Activation/deactivation cycle of redox-controlled thylakoid protein phosphorylation: role of plastoquinol bound to the reduced cytochrome b_6f complex. *J. Biol. Chem.* 270:25225–32
112. Wagner C, Walz D. 1995. Analysis of b-cytochrome titrations in terms of interacting redox couples. *Proc. 10th Int. Congr. Photosynth*, pp. 781–84
113. Wagner MJ, Packer JC, Howe CJ, Bendall DS. 1995. Some properties of cytochrome *f* in the cyanobacterium *Pharmidium laminosum*. *Proc. 10th Int. Congr. Photosynth*, pp. 745–48
114. Widger WR, Cramer WA. 1991. The cytochrome b_6f complex. In *Molecular Biology of Plastids and the Photosynthetic Apparatus*, ed. IK Vasil, L Bogorad, pp. 149–76. Orlando: Academic
115. Widger WR, Cramer WA, Herrmann R, Trebst A. 1984. Sequence homology and structural similarity between the b cytochrome of mitochondrial complex III and the chloroplast b_6f complex: position of the cytochrome *b* hemes in the membrane. *Proc. Natl. Acad. Sci. USA* 81:674–78
116. Wikström M, Krab K. 1986. The semiquinone cycle: a hypothesis of electron transfer and proton translocation in cytochrome bc-type complexes. *J. Bioenerg. Biomembr.* 18:181–93
117. Wikström MK, Berden JA. 1972. Oxidoreduction of cytochrome *b* in the presence of antimycin. *Biochim. Biophys. Acta* 283:403–20
118. Willey DL, Auffret AD, Gray JC. 1984. Structure and topology of cytochrome *f* in pea chloroplast membranes. *Cell* 36: 555–62
119. Wollman F-A, Kuras R, Choquet Y. 1996. Epistatic effects in thylakoid protein synthesis: the example of cytochrome *f*. 1995. *Proc. 10th Int. Congr. Photosynth.*, Montpellier, France, 3:737–42
120. Yue W-H, Zou Y-P, Yu L, Yu C-A. 1991. Crystallization of mitochondrial ubiquinol-cytochrome c reductase. *Biochemistry* 30: 2303–6

Annu. Rev. Plant Physiol. Plant Mol. Biol. 1996. 47:509–40

CARBOHYDRATE-MODULATED GENE EXPRESSION IN PLANTS

K. E. Koch

Plant Molecular and Cellular Biology Program, Horticultural Sciences Department, University of Florida, Gainesville, Florida 32611

KEY WORDS: catabolite repression, metabolite-regulated genes, photosynthate partitioning, source/sink relations, sucrose

ABSTRACT

Plant gene responses to changing carbohydrate status can vary markedly. Some genes are induced, some are repressed, and others are minimally affected. As in microorganisms, sugar-sensitive plant genes are part of an ancient system of cellular adjustment to critical nutrient availability. However, in multicellular plants, sugar-regulated expression also provides a mechanism for control of resource distribution among tissues and organs. Carbohydrate depletion upregulates genes for photosynthesis, remobilization, and export, while decreasing mRNAs for storage and utilization. Abundant sugar levels exert opposite effects through a combination of gene repression and induction. Long-term changes in metabolic activity, resource partitioning, and plant form result. Sensitivity of carbohydrate-responsive gene expression to environmental and developmental signals further enhances its potential to aid acclimation. The review addresses the above from molecular to whole-plant levels and considers emerging models for sensing and transducing carbohydrate signals to responsive genes.

CONTENTS

1040-2519/96/0601-0509$08.00

INTRODUCTION

In plants and microorganisms, sugars not only function as substrates for growth but affect sugar-sensing systems that initiate changes in gene expression. Both abundance and depletion of carbohydrates can enhance or repress expression of genes. Responses vary depending on the carbohydrate, though metabolic flux may be more important than actual levels of carbon resources. Many sugar-modulated genes have direct and indirect roles in sugar metabolism, which suggests that their altered expression may have adaptive value. Not only do collective, long-term changes in metabolism result, but patterns of carbohydrate allocation among plant parts can also be altered.

The existence and potential importance of sugar-regulated gene expression in plants has become apparent only in the past few years. Previous evidence indicated that sugar supplies could alter enzyme activities, metabolism, and development, but these data and their significance were generally not viewed in the context of gene expression. Initial work on photosynthetic genes and their metabolic effectors is reviewed by Sheen (161) and discussed by Stitt et al (175). Thomas & Rodriguez (187) summarize metabolite regulation in cereal seedlings and further appraise the germinating cereal seed as a model system (188). Koch & Nolte (84) relate advances in sugar-modulated gene expression to effects on transport paths. Classical aspects of altered carbohydrate availability on whole-plant and organ processes are appraised by Farrar & Williams (34) and Wardlaw (199), with updates by Geiger et al (43), Quick & Schaffer (139), and Pollock & Farrar (134). Information on sugar-responsive gene expression is also available for microbial (15, 40, 150, 158, 192) and animal systems (195).

CARBOHYDRATES AS SUBSTRATES AND SIGNALS

Biological Significance

In microbes, carbohydrate signals to sugar-responsive genes provide a way for these organisms to adjust to changes in availability of essential nutrients. This capacity is vital to their survival and/or effective competition. Classic examples include control of the lactose operon in *Escherichia coli* and the glucose responsive genes for sugar metabolism in *Saccharomyces cerevisiae* (15, 40, 150, 158, 192). Similar responses have more recently been identified in unicellular algae (9, 93, 155, 172).

In multicellular organisms, however, acclimation to altered carbohydrate availability occurs within a complex structure. Sugar-regulated genes provide a means not only for integrating cellular responses to transport sugars (carrying information on carbohydrate status of the whole) but also for coordinating changes in resource utilization and allocation among parts. In addition, carbohydrate-responsive genes can effect changes in organismal development.

For plants in particular, carbohydrate-regulated genes represent an especially valuable mechanism for adjusting to environmental change. Plants are extremely sensitive and responsive to their surroundings because immobility leaves them few options for survival other than acclimation. Sugar concentrations vary over a wide range in plant tissues. This range typically exceeds that found in more homeostatic systems (such as the mammalian blood stream) and provides plants with both a broader range of signals and a greater challenge to adjustment. Sugar-mediated changes in gene expression are also unique in plants because changes in carbohydrate allocation can ultimately modulate form through processes affecting import/export balance (photosynthesis vs utilization).

Effects of carbohydrate availability on expression of specific genes may complement and amplify the influence of more immediate metabolic controls. Although gene-level responses are slower, they provide a magnitude and duration of change that cannot be accommodated by other means of regulation. The signals and regulatory mechanisms controlling the two processes appear to be quite different.

"FEAST AND FAMINE" RESPONSES AT THE GENE EXPRESSION LEVEL

"Feast and famine" is used here in a relative context and is not necessarily based on absolute levels of carbohydrate (see section on Carbohydrate-Sensing Systems). In the same way, "sugar-modulated," "carbohydrate-responsive," and "metabolite-regulated" gene expression are broadly inclusive. Transcriptional regulation is usually implied and is substantiated in many instances (77, 89, 160, 161), but message stability and turnover can also be involved (162). In any case, the ultimate effects of altered mRNA levels depend on the efficacy of translation, turnover and/or modification of protein products, and the metabolic context into which such changes are introduced.

The overall theme of Tables 1 and 2, together with discussion of salient features in this section, is that of carbohydrate-responsive gene expression as a mechanism for plant adjustment to altered availability of this essential resource. Known examples of sugar-responsive gene expression are organized by carbon-exporting and -importing tissues to help clarify the potential of their collective relevance to each. In general, carbohydrate depletion enhances expression of genes for photosynthesis, reserve mobilization, and export processes (Table 1), whereas abundant carbon resources favor genes for storage and utilization (Table 2). These effects, summarized schematically in Figure 1, reinforce the suggestion that sugar-responsive genes provide a means of adjusting whole-plant resource allocation and may ultimately contribute to adaptive changes in form.

Table 1 "Famine" genes: enhanced by sugar depletion

Genes/enzymes (function)	Evidence: plant, tissue	Effectors tested	Refs
Photosynthesis			
Rubisco S–subunit [rbcS]	*Zea* protoplast, trans expr	S, acet	160, 161
	tomato lvs	S, G	194
	Chenopodium cell cult +	G	89
	tobacco and potato plants	S, G + girdl	89
	tobacco leaf protoplasts	G	25
Rubisco L–subunit [rbcL]	*Chlorogonium* cell cult	acet	9, 172
chl a/b–binding protein (cab, Lhcb)	*Zea* protoplasts, trans expr	S, G, acet, oth	160, 161
	Chenopodium cell cult	G	89
	tobacco leaf protoplasts	G, S	25
	rape cell cult	G	55
	Chlamydomonas cults	acet	79
atp-δ thylakoid ATPase	*Chenopodium* cell cult	G	89
malic enzyme, C4 [Me1]	*Zea* protoplasts, trans expr	S, G, acet, oth	160, 161
PEP carboxylase, C4 [Pepc1]	*Zea* protoplasts, trans expr	S, G, acet, oth	160, 161
triose-phosphate translocator	tobacco lvs	S	82
pyruvate PPdikin [Ppdk1]	*Zea* protoplasts, trans expr	S, G, acet, oth	160, 161
C4-pyruvate phosphodikinase	*Zea* protoplasts, trans expr		160, 161
(C4 psynth)	*Chenopodium* cell cult		70
			89
Remobilization (starch, lipid, and protein breakdown)			
Amy3D, Amy3E α-amylase	rice cell cults	S, G, F, Mal	61, 180
α-amylase	rice, cult embryo, and scutel	S, G, F, endo extract	74, 187, 214
	barley aleurone	Na-butyrate	94
plastid starch phosphorylase	*Chenopodium* cells, lvs	G	91
phosphoglucose mutase	*Chenopodium* cells, lvs	G	91
isocitrate lyase [Icl] (glyox cycle)	cucumber cotyledons	S, G, F, 2dG,	50, 51
	Chlorogonium cells	M + acet	155
malate synth (glyox cycle)	cucumber cotyledons	S, G, F, 2dG, M	50, 51
proteases	maize root tips	G	13
asparagine synthetase (N cycling)	*Arabidopsis* shoot tissues	S	95
Sucrose and mannitol metabolism (synthesis and breakdown)			
acid invertase	maize root tips [*Ivr1*]	G, S, F, oth	86, 209, 210
S synth	maize root tips [*Sh1*]	S, G, F, oth	85
	maize protoplasts [*Sh1*]	S	104
	carrot, whole plant	pruning	179
	Vicia seeds, cotyledons	F, G	57
	Arabidopsis [*ASus1*]	sink manip	106
SPS	sugar beet rts, lvs	S	60
Mtol dehydrogenase	celery cell cult	S, Mtol	133

Abbreviations: 2dG, 2-deoxy-glucose; acet, acetate; cult, culture; endo, endosperm; F, fructose, G, glucose, Lhcb, light-harvesting chlorophyll-binding protein (also cab); lvs, leaves; M, mannose; Mal, maltose; Mtol, mannitol; PEP, phosphoenolpyruvate; PPdikin, phosphodikinase (cytosolic); rts, roots; scutel, scutellum; *Sh1*, *Shrunken1*; S, sucrose; SPS, sucrose phosphate synthase; trans expr, transient expression; synth, synthase.

Carbohydrate Depletion and Sugar-Responsive Genes

Plant and microbial gene responses to carbohydrate depletion have important similarities but differ as well. In both plants and microbes, sugar and acetate effects favor uptake of preferred substrates requiring the least metabolic cost and promote heterotrophic growth over photosynthesis when possible. How-

Table 2 "Feast" genes: enhanced by sugar abundance

Genes/enzymes (function)	Evidence: plant, tissue	Effectors tested	Refs
Polysaccharide biosynthesis (starch and other)			
AGPase [*Sh2*] (starch)	*Chenopodium* cell cult + spinach,		91
	transgenic potato lvs		92, 116
	potato/detached lvs in dark		87
starch phosphorylase	potato tuber	S	171
	potato/detached lvs in dark	S	87
starch synth [GBSS]	potato/detached lvs in dark	S, G, F	87
branching enzyme [BE]	potato/detached lvs in dark	S, G, F	87
	cassava stems and lvs	S, G, F	152
Storage proteins			
sporamin, A & B types	sweet potato/cult plts/stems	S	56
	sweet potato/lvs and petiole	S, G, F	65, 122
		S, pgal a	
β-amylase (storage protein?)	sweet potatoes and lvs	S, G, F	122
	sweet potatoes and lvs	pgal a	127
patatin class I	transgenic potato	S	52
	promoter	S	36, 143
	potato tubers and lvs	S	78
	transgenic potato lvs and tuber	S, starch	99
	potato tuber/transgenic tobacco	S	202
	potato leaf and stem explants	S	202
	potato leaf and stem explants	S, Gln, dk	131
	transgenic potato tubers	sol. sugs.	117
proteinase inhibitor II [Pin2]	transgenic potato tubers	sol. sugs.	117
	potato lvs/transgenic tobacco	S, G, F, Mal	70
	detached potato lvs	ABA, MeJA	131
	transgenic tobacco	S, G, F	78
lipoxygenase (storage protein)	soybean *Lox-NR*	depodding	53
	soybean *VspB*	S, Mal	149
	soybean *Vsps*	S, G, F, MeJA	107
Pigments and defense			
chalcone synth (pigment/path.)	petunia in *Arabidopsis*, alfalfa	S, G, F	193
	protoplasts, *Camelia sinensis*	p-coumar	101
		sugars	183
RT locus (pigment synth)	petunia/petal, anther	G + light	14
dihydroflavonol-reductase	ivy lvs and stems	sugars	120
Mn-superoxide dismutase	rubber tree/all tissues	S	110
hrp (pathology)	*Xanthomonas campestris*	S + Met	157
chaperonin 60*B* (protein synth)	*Arabidopsis* lvs	S	215
Respiration			
PGAL-dehydrog. (*GapC*) cyto	*Arabidopsis* lvs	S	211
β-isopropylmalate dehydrog.	potato, tomato, *Arabidopsis*	S, AA	66
apocytochrome 6 (*co6*)	*Chlorogonium* cell cult		93
PP-F-6-P phosphotransferase	*Chenopodium* cell cult, tobacco,	G	92
(cytosolic enzyme)	and spinach lvs	G	92

Table 2 (*Continued*)

Genes/enzymes (function)	Evidence: plant, tissue	Effectors tested	Refs
Sucrose metabolism			
invertase	maize root tips [*Ivr2*]	S, G, F	87, 209
	Chenopodium rubrum	S, G, F, 6dG	144
	carrot, whole plant	manip.	179
S synth	maize [*Sus1*]	S, G, F	86
	rice embryos	S, G, F	74
	Vicia faba cotyledons	S	57
	potato tubers, lvs, stems	S	151
	potato plants, throughout	S	37, 38
	Chenopodium cell cults	S, G, F	46
SPS	sugar beet petioles	G	60
	transgenic potato	sol. sugs.	117
Other			
nitrate reductase	*Arabidopsis* lvs light/dark	S, G, F	196
	Arabidopsis plants light/dark	G	18
	Chenopodium cells/spinach lvs		91
SAM synth	*Lolium* lvs	S	208
rolC gene of Ri plamid	transgenic tobacco/phloem	S	213
30-kD Rubisco-assoc. protein	soybean lvs	pod removal	171

Abbreviations: 6dG, 6-deoxy-glucose; AA, amino acids; cult plts, cultured plants; F, fructose; G, glucose; GapC, PGAL-dehydrogenase (cytoplasmic); Gln, glutamine; Glu, glutamate; lvs, leaves; Mal, maltose; MeJA, methyl jasmonate; Met, methionine; p-coumar, p-coumaric acid; PGAL, glyceraldehyde-3-phosphate dehydrogenase; pgal a., polygalacturonic acid; PP-F-6-P phosphotransferase, pyrophosphate:fructose-6-phosphate-phototransferase; S, sucrose; SPS, sucrose phosphate synthase

ever, in complex plant systems, carbohydrate-regulated genes also provide a means for optimizing investment of C, N, P, etc among different plant parts and processes. Localized expression of starvation-induced genes may also aid survival of key cells and tissues under stress (84, 85).

In carbon-exporting or other autotrophic cells, photosynthetic genes are typically upregulated by sugar depletion. These include genes for the primary CO_2 fixation enzymes of both C3 and C4 plants (18, 160, 161) and other genes critical to photosynthesis (113, 160). Both nuclear (Table 1; 89) and plastid genes (25, 89, 160, 161) are affected, though the latter may respond more slowly to altered carbohydrate levels (25, 163). Enhanced expression results largely from derepression of sugar and acetate controls on transcription (89), though longevity of mRNA can also contribute to sugar modulation in vivo (162). Photosynthetic genes are repressed most by acetate (160) and often more strongly by hexoses than sucrose (69, 160, 161). Acetate effects are observed in cotyledons and in unicellular algae (Table 1; 50, 51).

The physiological consequences of sugar-induced changes in gene expression are discussed further in the section on "Implications at the Cell and Organism Level." Coordinated but often contrasting responses to sugar depletion are also evident at the enzyme level (90, 153, 175, 194). Plastid proteins and enzymes can differ in their responses depending on whether they are encoded by nuclear or plastid genes (10, 190, 194) and whether they are involved in photosynthesis or other processes (91).

Genes for remobilization of sugars and other small molecules from polymers and/or vacuoles are also induced in exporting cells by carbohydrate depletion. In photosynthetic leaves, those genes associated with starch breakdown can be upregulated by carbohydrate depletion and repressed by glucose (91). A similar response is observed in source tissues of germinating monocot seeds, where starch hydrolysis in endosperm provides the bulk of exported sugars (61, 74, 187, 214). In addition, carbohydrate depletion in cotyledons of germinating dicot seeds upregulates genes for remobilization of lipid reserves via the glyoxylate cycle (50, 51). β-amylase genes are not enhanced by carbohydrate depletion (125, 127); however, their in vivo function remains unclear.

The extent of protein remobilization (and associated gene expression) can vary markedly with the degree of carbohydrate depletion (33, 68, 173). Leaf storage proteins are broken down under these conditions (53, 170, 173), though typically nonphotosynthetic cells are involved (53). Starvation effects (see below) occur only if photosynthetic capacity is severely compromised.

In carbon-importing cells, transitions to net carbon export are favored by "famine-induced" changes in gene expression. This cellular altruism in higher plants is distinct from responses of microbes and unicellular algae. Genes related to carbohydrate, lipid, and protein remobilization (Table 2 and below), amino acid synthesis (95), and sucrose formation [sucrose phosphate synthase (SPS; 60, 81)] are upregulated. Responses of SPS genes to sugar availability may be complex. In sugarbeet, a taproot-specific form of this enzyme is upregulated and downregulated by glucose and sucrose, respectively (60), whereas a spinach gene is regulated in synchrony with the sink-to-source transition (81).

Starvation and carbon conservation responses at the level of gene expression initially affect genes related to reserve remobilization (see above) and respiration (see below) that preserve structural constituents of the cell (12, 13, 68). Prolonged stress may induce genes related to breakdown, shuttling, and scavenging of cellular resources (P Ramond, unpublished data). Detoxification of nitrogenous compounds may be facilitated by upregulation of asparagine synthase (68) and sucrose synthase (Table 1). These changes are accompanied by sugar-repressible increases in activity of endoglycosidases [implicated in glycoprotein breakdown (100)], endopeptidases (49), and unidentified starvation proteins (5, 200). Under extreme starvation, the activity of enzymes involved in β-oxidation (although not a fully operative glyoxylate cycle) increases and may be associated with metabolism of membrane lipids (12, 13, 31, 68). Specific subgroups of apparently immature mitochondria and their associated proteins also disappear from starved cells under these conditions (23, 73).

Starvation stress and transport sugars can also affect accumulation of osmoprotectants. Upregulation of a mannitol dehydrogenase gene by carbohydrate depletion allows use of this sugar alcohol as a carbohydrate source (133), whereas sugar repression favors accumulation and salt tolerance (133, 185, 186). A starvation-tolerant class of genes for sucrose synthase and for invertase is also induced (85, 86); these proteins are expressed in key tissues under stress (84, 85) and at specific stages in development (209; see below).

Carbohydrate Abundance and Sugar-Responsive Genes

A large but specific set of genes is positively regulated by sugars. The majority of identified genes that are induced by elevated sugar levels encode products that help set capacity for carbon storage, utilization, and import. Other important classes include defense genes, secondary product pathways, and storage proteins.

In carbon-exporting and/or autotrophic cells, the transitions to import and storage programs are typically initiated at elevated sugar levels. The decreased expression of photosynthetic genes described above allows reallocation of the C and N (otherwise utilized in photosynthetic proteins) to other processes more advantageous under the prevailing carbohydrate environment. In this context, the concurrent upregulation of genes for nitrate reductase and a putative SAM synthase in leaves (205; Table 2) could facilitate amino acid synthesis and turnover of other N sources. Such genes may also contribute to synthesis of leaf storage proteins (Table 2; 170) and other signals including polyamines that may enhance the positive effects of sugars (180). Other aspects of gene expression related to storage and carbon use also change in *Lolium* leaves as sugar levels rise (206).

Genes related to storage reserve synthesis can be upregulated by sugars. These genes are similarly affected in both photosynthetic and nonphotosynthetic organs (84, 122, 131, 188). These changes may be associated with conversion of chloroplasts to either amyloplasts (154) or chromoplasts as sugar levels rise (64).

Genes for sucrose metabolism can be upregulated in photosynthetic tissues following manipulations that cause sugars to accumulate (151). These changes often result in elevated starch levels (167, 168) especially in cells alongside major veins (124).

In carbon-importing cells, genes for starch biosynthesis have received the greatest attention. Those encoding ADPG-pyrophosphorylase (AGPase), a key step in starch biosynthesis, are markedly sugar responsive in potato (87, 89, 116). AGPase expression is also strongly enhanced by sugars in transgenic potato cell cultures (116) and in other species (175). Corresponding increases in activity of the AGPase enzyme are not necessarily observed but may occur more slowly (175). Starch synthase and branching enzyme are also induced and/or expressed at elevated levels when sugars are plentiful (87, 152). Carbohydrate regulation of sucrose synthase genes is complex and not necessarily directly related to starch synthesis. The *Shrunken1 (Sh1)* gene for sucrose synthase in maize can directly affect synthesis of starch and/or cell wall materials (20, 123, 124); however, a second gene, *Sus1,* responds more strongly to elevated sugars (85; see section on Contrasting Response Classes Among Genes for Sucrose Metabolism).

Genes encoding storage proteins were among the first sugar-responsive genes identified. A gene for sporamin storage protein in sweet potatoes (56) is upregulated in situ, and it is ectopically expressed in plantlets treated with high sugars. The patatin storage protein genes of potato also respond positively to high sugar levels (78, 99, 131, 202). An additional group of sugar-modulated genes includes vegetative storage proteins, which are expressed at elevated sugar levels in several species (29, 107, 149). Many of these proteins have enzyme activity in addition to a storage function. The *Vsp* gene groups A and B of tobacco encode proteins with phosphatase activity (149); a soybean vegetative storage protein has lipoxygenase activity (53); the patatins are lipid acyl hydrolases (52, 202); and the WIN and Pin-II storage protein genes of *Solanum* spp. are proteinase inhibitors (70). Studies of sugar responsiveness in these genes have revealed important interactions between carbohydrate supply and other signals (e.g. N, P, auxin, etc).

A number of pigment and defense genes are positively modulated by carbohydrates (77; Table 2). The products of these genes mediate plant interactions with other organisms, either as pathogens, pollinators, or fruit dispersal agents. Often these interactions involve enhanced carbon use by the plant; however, effects on biosynthesis of pigments, proteins, and chaperonins can be distinct.

Respiratory genes are affected to varying degrees by sugars (Table 2). Both nuclear- and plastid-encoded genes can show positive responses, with the latter upregulated through both mRNA abundance and gene copy number (93). Mitochondrial ubiquinone mRNA is also strongly affected (B Collins, P Raymond, R Brouquisse, CJ Pollock & JF Farrar, unpublished data), as are levels of cytochrome oxidase and activity of fumarase (91). As in yeast, however, carbohydrates do not globally upregulate respiratory genes. For example, mRNA levels may remain constant for glycolytic genes [often used as controls (89, 92)] even though other respiratory genes respond to elevated glucose levels (91, 92; see Table 2).

Genes for sucrose metabolism can be strongly affected by high sugar levels in importing as well as exporting cells. The complex carbohydrate regulation of the invertases and sucrose synthase genes that control the two known paths for sucrose breakdown is discussed in the following section.

Contrasting Response Classes Among Genes for Sucrose Metabolism

Genes for sucrose metabolism occupy a central position not only in carbon flow but also in the production of alternate potential effectors of the sugar-sensing system. This altered expression of genes in sucrose metabolism could affect whole-plant adjustment to changes in carbohydrate supplies at several levels. Shifts in resource allocation are often directly correlated with activity of the respective enzymes, and indirect effects on signaling systems could further amplify changes in expression of these genes as well as genes affecting developmental programs. Collectively, these changes could also lead to changes in plant form that fine-tune acclimation.

Sucrose metabolism is the first step in carbon use by the majority of importing cells in plants (21, 115, 181). Two recently appreciated features of sucrose metabolism are particularly interesting. First, the genes for invertase as well as sucrose synthase are sugar-modulated (84, 144, 179, 209). Second, isozyme forms of each enzyme show contrasting carbohydrate responses (84, 85, 209). (In each instance, one isozyme is upregulated while one or more others are repressed.) Sugar-modulation of genes for both known paths for sucrose metabolism provides a potential mechanism for coarse control of this process.

The presence of isozyme forms with contrasting carbohydrate responsiveness was an unexpected finding. Reciprocal expression was first observed for sucrose synthases (85) and subsequently for invertases (209, 210). Initial studies of sugar-modulated gene expression were perplexing because of contradictory results. Sucrose synthase was reportedly both repressed (83, 104, 169) and enhanced in the presence of abundant carbohydrate supplies (74, 151). The reciprocal sugar responsiveness of genes encoding distinct isozymes is likely to have been responsible (85). Differentially responsive genes could also explain the contrasting effects of light on expression of sucrose synthases of wheat (105).

The surprising similarity between differential sugar-modulation of different genes for invertases (209, 210) and sucrose synthases in maize indicates that there are two sugar-response classes among genes for sucrose metabolism. Both the *Sh1* gene for sucrose synthase and the *Ivr1* gene for invertase (210) are expressed maximally when supplies of metabolizable sugars are limited [e.g. ca 10 mM glucose (0.2% w/v)] (85, 209). Both types of mRNA persist during carbohydrate starvation stress in root tips, and they are enhanced at key sites and times during reproductive development (84, 85, 209, 210). In contrast, the *Sus1* gene for sucrose synthase and the *Ivr2* gene for invertase both respond positively to abundant carbohydrate supplies [e.g. ca 100 mM glucose (2.0% w/v)] and are expressed in a broad range of importing tissues (209, 210). A "feast"-responsive set of isozymes for both paths of sucrose breakdown could aid adjustment of import and metabolism relative to photosynthate availability. The potential value of up- or downregulating sucrose utilization in balance with its supply is consistent with the broad distribution of this isozyme form among importing tissues. In contrast, the potential physiological significance is less clear for the isozyme genes expressed when carbohydrate supplies are limited. To date, little is known about differences between properties of the invertase isozymes, although the sucrose synthases appear to be enzymatically similar (32). It is possible that more recent work showing phosphorylation of this enzyme (63, 86, 159) may be related to differences in properties of the sucrose synthase isoforms (63).

Additional clues to the biological importance of sucrose synthase and invertase isozymes whose genes are upregulated under "famine" conditions may lie in the altered protein localization and reproductive timing of expression. Under starvation stress, sucrose synthase protein in maize root tips is localized to epidermis and vascular strands while being markedly depleted from the cortex (56a, 85). Cortical cells are often sacrificed during various stresses, including low oxygen, N, or P availability (56a, 84, 108), whereas vascular and epidermal tissues are preserved. Living epidermis appears to be essential for nutrient and water uptake in many species and is very often associated with a hypodermis (having endodermal-like functions) and a rhizosheath of soil particles bound to the root surface by polysaccharide secretions [also sugar modulated in their extent (114)] (132, 108). It is possible that import priority could be conferred on essential cells and tissues during periods of limited resources by localized upregulation of special isoforms of sucrose metabolizing enzymes (84, 85). This would be consistent with upregulation of the same starvation-tolerant isoforms in specific reproductive tissues and in their additional localization in apices of roots and shoots preserved at the expense of other tissues in a starving plant (6). Our current knowledge of the *Sh1* sucrose synthase and *Ivr1* invertases of maize is consistent with this hypothesis (86, 209).

IMPLICATIONS AT THE CELL AND ORGANISM LEVEL

Long-Term Metabolic Changes

The relatively slow kinetics of the carbohydrate-induced changes in gene expression (85, 175) and enzyme activity (84, 86) are consistent with the time frame often required for source/sink adjustments at the whole-plant level (41). The physiological changes parallel the altered expression patterns of individual genes. Photosynthesis and C-conservation are generally enhanced when sugar supplies are limited, and utilization usually predominates when sugars are abundant.

Changes in photosynthetic processes resulting from sugar-modulated gene expression generally occur over an extended time period (ca 3–7 days) (84,

86, 175) and may amplify shorter-term effects of direct metabolic control (35, 62). Altered transcript abundance can occur within a few hours or less in some systems (144, 161), although initial changes are often not evident until ca 12–24 h and progress slowly thereafter. Consistent with this time scale, early work by Geiger (41) showed that changes in photosynthetic capacity in response to altered source-sink balance takes 3–4 days. Also, as plants acclimate to elevated CO_2, photosynthetic rates decline over a period of days after an initial increase (27), possibly because of the repressive effects of accumulated sugars in leaves (194).

Invertase activity in leaves may indirectly affect repression of photosynthetic genes by accumulated sugars because hexoses may affect the sugar-sensing system more directly than sucrose (69, 160, 161; see section on Carbohydrate-Sensing Systems). Hexoses in particular have been implicated in long-term repression of photosynthetic genes (59, 175). Species with high levels of leaf invertase show a greater degree of photosynthetic inhibition in instances of reduced sucrose export (47). This effect is substantiated in transgenic invertase overexpressors (30, 58, 166, 168, 177, 198), where sucrose export from mature leaves is inhibited and hexose production enhanced. Pollock et al (136) also found that elevated leaf sucrose had little inhibitory effect on photosynthesis while fructans were being actively synthesized and stored in vacuoles (presumably removing hexoses from the cytosol). Long-term influences of hexoses on photosynthesis are likely to involve other factors as well (35, 42, 62, 176). Krapp et al (89) suggested that, as in yeast, hexose metabolism may be required for repressive effects of sugars on gene expression; they found an imperfect correlation between hexose levels and photosynthetic repression.

Leaf senescence may also be enhanced by long-term hexose effects on gene expression. In this context it is interesting that invertase activity is elevated during aging (135). Hexoses in particular may favor expression of genes involved in remobilization of photosynthetic machinery and altered pigment synthesis (Tables 1 and 2). Acetate effects are still more pronounced (160), which indicates that lipid breakdown and mobilization may accelerate senescence. The putative advantage of sugar repression of photosynthetic genes is that valuable resources need not be committed to this process if carbohydrate supplies are already sufficient.

Pigment changes and chloroplast-to-chromoplast conversions during fruit ripening and senescence may be affected by carbohydrate-sensitive genes (201). In ivy, sugar-sensitive gene expression has been directly related to induction of enzyme activity leading to pigment accumulation (119, 120). In citrus peel, sugars mediate interconversions between chloroplasts and chromoplasts (64). Regreening occurs in late-ripening oranges as peel sugar levels drop in spring. Rising sugar levels are consistently associated with the chloroplast-to-chromoplast conversion in autumn. Sugars can also stimulate these changes in chlorophyll, carotenoid, and plastid characters in vitro (48). Chloroplast-to-chromoplast conversions are also enhanced by the induction of invertase at low temperature in citrus (138). Many of the nuclear-encoded plastid genes are also readily responsive to sugar levels (Tables 1 and 2), whereas chloroplast genes appear to be less so (194). As noted above, however, the responsiveness of chloroplast genes to acetate suggests that this metabolite could be important in plastid conversions.

Changes in storage processes are closely related to carbohydrate-responsive gene expression. Sugar effects on storage organ formation are discussed below (see section on Interaction with Developmental and Environmental Signals). In addition, most instances of enhanced gene expression cited in Table 2 are accompanied by respective increases in enzyme activity and storage of carbohydrate and/or proteins (e.g. 146, 197). Further, sugar effects on an α-amylase gene family and on sucrose synthase correlate with the balance between endosperm remobilization and the demands of the growing seedling (188).

Respiratory changes related to sugar-modulated gene expression are less clear. The hypothesis for "coarse control" described by Farrar & Williams (34) indicates that long-term respiratory responses follow extended changes in carbohydrate availability and probably require altered gene expression. Although evidence supports this view, the relationship between carbohydrates and gene expression is complex, and aspects of the story remain unresolved. Respiration typically rises in response to increasing levels of sugars (8, 34, 96, 109), and it decreases with starvation (12, 68, 140)—as do expression levels of many genes related to respiratory processes (Table 2). Concurrent increases in levels of key mRNAs and of associated respiratory activity have been observed as sugar content rises in maturing leaves of transgenic plants over-expressing invertase (175). Similar correlations have been made in several species (89, 93). Kroymann et al (93) suggested coordination through a signal related to cellular energy charge. The ATP/ADP ratio rises along with respiration and associated transcript levels (89). However, the evidence gathered thus far does not necessarily support adenylate charge as a direct signal for carbohydrate-responsive genes (175, 161; see section on Key Metabolites as Direct Signals). Other issues not yet fully resolved include the role of changes in organelle number (140) and genome copy number (93), the importance of the alternate oxidase (96), significance of Pi and adenylates to gene expression (161, 175), and the potential impact of acetate on the carbohydrate signaling.

Carbohydrate-Responsive Genes, Assimilate Partitioning, and Development

The ultimate significance of sugar-modulated gene expression may be induction of changes in whole-plant morphology. Taken together, the trends in gene expression, subsequent metabolic changes, and shifts in resource allocation are consistent with this suggestion. Sugar modulation of developmental genes is implied by responses such as potato tuber induction (165); however, the specific genes involved have not been identified. The pronounced interactions between carbohydrate levels and plant growth regulators (especially auxin/sugar antagonisms) and other essential nutrients such as N or P suggest that sugars may affect development at the cell, organ, and whole-plant levels. The type of carbohydrate supplied to cells and callus cultures can also affect morphological change not attributable to osmotic effects.

Figure 1 summarizes the developmental trends implied by the gene expression changes known to occur in carbon-importing and -exporting tissues under feast and famine conditions (see Figure 1). A number of studies support the notion that C-availability can affect C-allocation through altered gene expression (see discussion on partition in section on Contrasting Response Classes Among Genes for Sucrose Metabolism). This in turn can affect partitioning between root and shoot structures (43, 134, 167, 179, 204). Other seemingly contradictory results obtained when different methods are used to manipulate sugar availability might also be inter- preted within this simplified framework. Both root and shoot growth are inhibited in transgenic plants with excess invertase, presumably because trans- location to the root system is disrupted (175) and high hexose levels simultaneously repress photosynthetic genes in leaves. In contrast, shoot growth may be indirectly enhanced when sugars are supplied to whole plants via the root system (56, 88), if increased root growth leads to increased capacity for cytokinin synthesis, which in turn may stimulate shoot growth and photosynthetic processes (112).

Carbohydrate-induced changes in vegetative morphology often involve an altered balance of growth regulators and mineral nutrients (130, 207). Sugars can repress auxin-mediated processes including apical dominance and upright stem growth (negative geotropism) resulting in more spreading, procumbent growth forms (203). Expression of genes now known to be sugar responsive may also have a role in gravitropism (75, 76). In addition, sugar induction of storage organs in potato and sweet potato can be distinguished from regulation of the genes associated with storage processes per se. Although some of the latter processes are coordinately regulated (197), the morphological program remains separately sugar responsive (117).

Cell differentiation and the cell cycle can also be strongly affected by sugar availability. Development of tracheid vs phloem cells can be controlled by sugar/auxin balance (39), and other effects of specific sugars on differentiation have been reevaluated in cultured cells (191). The cell cycle within a given tissue can be synchronized by withholding and resupplying sugars (200). In addition, cell divisions can be induced in nongrowing buds of sunflower by elevated sugar levels (2).

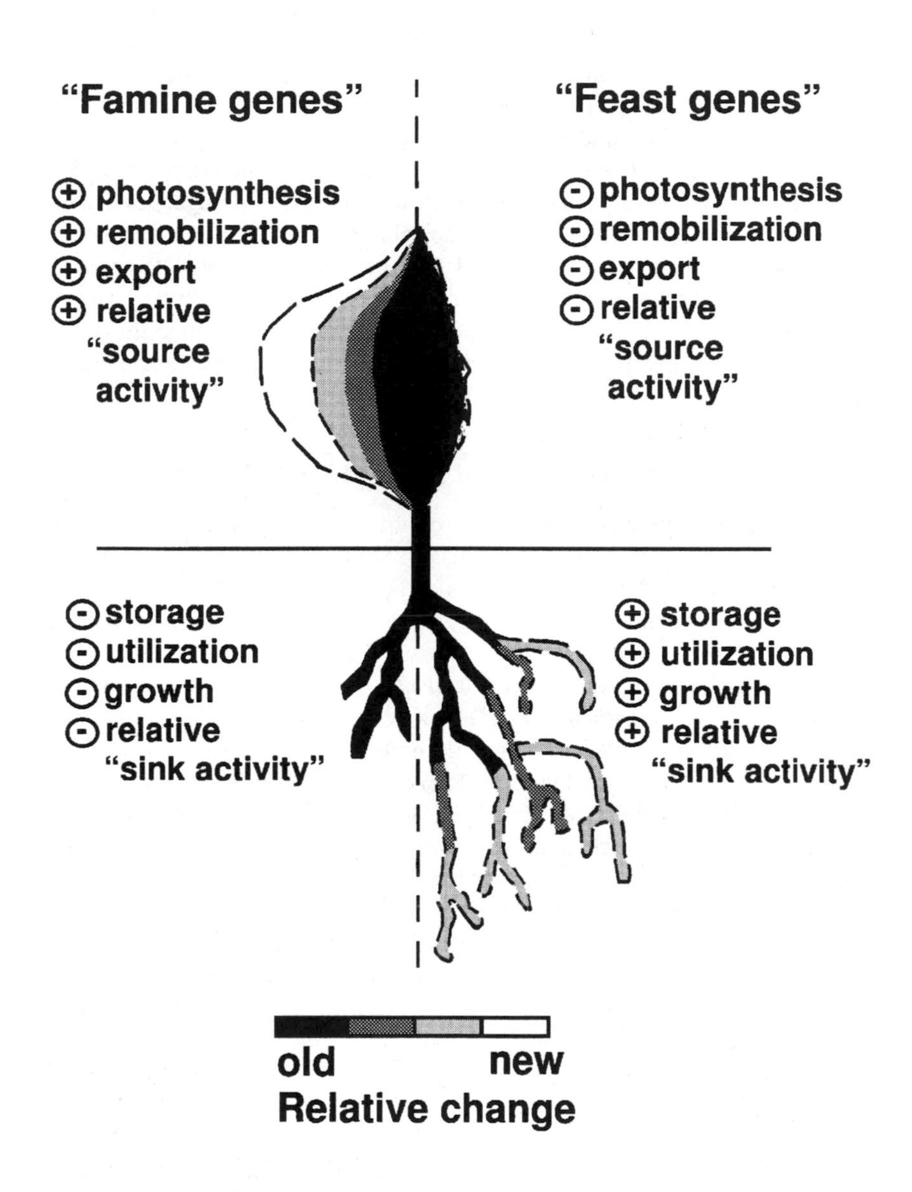

Figure 1 The impact of sugar-modulated gene expression on overall activity and resource allocation is diagrammed figuratively at the whole-plant level (changes do not necessarily represent actual morphological changes per se). "Feast" and "Famine" genes are those upregulated and downregulated under conditions of limited and abundant carbohydrate supplies, respectively, in either exporting (upper half) or importing tissues (lower half). Processes favored by these changes in gene expression are designated with a (+), and processes that are diminished are designated with a (–).

Sugar-induced changes to reproductive programs may be closely related to effects on the cell cycle. Sugar pulses to apical meristems can initiate synchronous cell divisions that precede other aspects of floral meristem differentiation (7, 97). High levels of apical sugars can also amplify photoperiod effects on floral evocation in *Lolium temulentum* and replace it fully in *Sinapis alba* (80, 137). In the latter case, concomitant with increased apical sugar levels, increases in invertase are observed in the apical meristem (137). Some invertase genes are sugar-regulated (84, 144, 210) and have the potential to enhance sugar perception by hexose sensing systems (see below). In addition, sugars supplied through roots can suppress phenotypes of early- and late-flowering mutants in dark-grown *Arabidopsis* plants (J Salinas, personal communication).

Interactions with Developmental and Environmental Signals

Effects of carbohydrate availability on fruit and seed set may mediate responses to certain environmental stresses. Studies of stress-induced kernel abortion in maize show that exogenous carbohydrate supply (11) and short-term reserves in young ovules (141) are crucial to kernel set in conditions of low water (216) or high temperature stress (17). Sugar-feeding studies have implicated the final phase of import and use of these substrates within the developing ovule as critical (11, 17; J Boyer, unpublished data). Effective sugar utilization in vivo is strongly dependent on the activity of sucrose metabolizing enzymes (21, 111, 115, 181) that are encoded by sugar-responsive genes (Tables 1 and 2; 85, 209). Soluble invertase occupies a conspicuous position during the earliest phases of fruit and seed set (181, 209), and that activity is selectively affected by abortion-inducing stresses such as low water potential (216). Although the evidence is thus far largely correlative, the sugar responsiveness of soluble invertase genes could provide a mechanism for integrating and transducing information on the C-resources available to the fertilized ovule.

Developmental signals mediated by growth regulators can have marked effects on carbohydrate-modulated genes. The nature of this interface is still poorly defined. However, work by Mullet and coworkers (29) indicates that one of two different sugar-responsive promotors they studied is sensitive to auxin/sugar antagonism. Similar response elements could explain the repression of sugar responses (179) by auxin analogs in cell cultures (178) and auxin modification of sugar effects on invertase expression at the whole-plant level (45, 148, 203). Gibberellin interaction with sugar signaling is apparent in germinating grain seeds (187, 188) and stolen starch metabolism (4). Cytokinin and sugar signals overlap in transcriptional regulation of nitrate reductase (18, 196), invertase (209), and other genes (26). They can also affect respiration

(121), the cell cycle (67, 98), auxin antagonism (112), kernel abortion (17, 71), and an array of morphological changes (112).

Interactions between environmental signals and sugar-responsive genes do not necessarily involve developmental programs. Elevated sugar supply, osmotic stress, and pathogen invasion all upregulate the *mtd* gene for mannitol metabolism and enhance synthesis of this osmotic protectant (133). Mannitol in turn imparts enhanced salt tolerance in transgenic plants (184, 185). Osmotic adjustment can also be affected by sugar-regulation of invertases (212), which could in turn sensitize cells to sugar supplies by the elevation of hexoses (161, 177). Similar to *mtd* and some other sugar-responsive genes (70), invertase is induced by wounding (178).

Recent progress has also been made in defining the interface between sugar-sensing systems and the transduction of various light signals (19). Sheen pointed out (161) that the carbohydrate-repression of photosynthetic genes supersedes many of the light effects. Chory (19) proposed that light signals are partly filtered through a sugar-regulated segment of the transduction pathway. Parks & Hangarter (128) also found that blue light effects can differ depending on tissue sugar status. The effects of sugars on photoperiodic responses were discussed above. In addition, high irradiance responses (HIR) such as anthocyanin biosynthesis in fruit skins overlap sugar effects.

Influence of essential mineral nutrients such as N (24) and P (186) on gene expression and morphology is often strongly linked to carbohydrate status (130). Several possible avenues for C/N interactions or P effects on C-signaling have thus far tested negative (175). Sadka et al (149), however, found that P availability altered sugar-regulated transcription of a carbohydrate-sensitive promoter element.

CARBOHYDRATE-SENSING SYSTEMS

Several lines of evidence indicate that sugar effects on gene expression involve specific signaling mechanisms and do not simply result from their nonspecific effects as substrates for plant growth. First, the effects of sugars on gene expression are highly selective; many genes are not affected. Second, sugars can repress as well as activate responsive genes. Third, in many cases, sugar-modulated gene expression can be mimicked by nonmetabolizable sugar analogs (69, 160, 161) and altered by selective metabolic perturbations (102, 161, 175, 182). Finally, sugars are well-known effectors of gene expression in microbes.

Microbial sugar-sensing mechanisms are an important resource for development of testable hypotheses in plants (50, 51, 69, 160). However, the emerging picture of sugar signaling in plants highlights important differences (40, 150, 192) and intriguing similarities with microbes.

Hexose Phosphorylation and Protein Kinase Cascades

An important hypothesis for sugar sensing stems from the evidence that phosphorylation of hexoses by hexokinase is a well-documented primary source of sugar signaling in yeast (161). The hexokinase itself is proposed to have a dual function as a protein kinase (40) sensitive to the flux of sugars entering metabolism. Because the rate of hexose phosphorylation is more important than the steady-state level of hexose-P produced in yeast, the concentration of hexokinase enzyme-product complex is proposed to be directly involved in signaling carbon flux through the pathway.

As in yeast, sugar concentrations per se are not necessarily correlated with changes in plant gene expression. Analysis of maize mutants with high-sugar kernels (44) and transgenic, sugar-storing potatoes (117) showed little or no change in expression of genes otherwise affected by sugars. Although compartmentalization of sugars was not addressed in these studies, the results may be interpreted as evidence that carbon flux rather than steady-state sugar level is the critical signal. The data that support a corresponding role for hexokinases in plants center largely on responses to sugars (2-deoxyglucose and mannose) that are rapidly phosphorylated by hexokinases but that do not readily undergo subsequent metabolism. Positive responses to these sugars have now been observed in several plants (50, 51, 69, 160, 161, 175). In at least one study, the effects of nonmetabolizable sugars were shown to be blocked by addition of mannoheptulose, an inhibitor of hexokinase (69). By analogy to yeast, these results are interpreted as favoring flux through the hexokinase reaction as the inductive signal (50, 51, 69, 160). Other associated perturbations have not been fully excluded, although changes in Pi levels appear to have little effect (175).

If the hexokinase hypothesis remains viable in plants, the wide variation in specificity of plant hexokinases for glucose as opposed to fructose (156) may add a fascinating layer of complexity to the regulatory scenario in plants. Recent data from yeast indicate that glucokinases do not have the same sugar-sensing impact as the hexokinases (145).

Extensive studies in yeast have also identified downstream components of a protein kinase cascade involved in transmitting signals to the nucleus (40). Putative homologs of the yeast *snf1* gene have been identified in rye (RKIN1) (1), tobacco (NPK5) (3, 118), and barley (a multigene family) (54). Nakamura and coworkers have shown that calcium levels (125) and a calcium-dependent protein kinase (CDPK) (126) may be involved in sugar induction of sporamin and β-amylase genes in transgenic tobacco. The latter finding suggests that sugar sensing in plants may be distinct from mechanisms in other systems because CDPKs appear to be lacking in other organisms (142). The localization of this CDPK on the plasmamembrane of plant cells suggest a possible association with the sites of membrane transfer. Other recently isolated plant genes are similar to the glucose-regulated proteins (GRP) involved in secretion and ER function in mammals (28, 164). Putative sugar-sensing-related DNA binding proteins have been isolated from several plants (65, 78, 103), but as yet they have no clear relationship to components of signal transduction pathways in other organisms. The functions of these genes are under study.

Plasmamembrane Transfer

In microbes, transfer of sugars across the cell membrane is critical for sugar sensing and may be closely coupled to hexokinase action (150). In plants, the role of membrane transport has been tested using sugar analogs that are nonmetabolizable and nonphosphorylatable (as distinct from the analogs used to implicate hexokinases) but actively taken up by plant cells. Data from W Frommer (unpublished) and Roitsch et al (144) suggest that transfer across the plasmamembrane alone (or the configuration of the sugar analog per se) can initiate a signal. The studies of Jang & Sheen (69) indicate that transfer across the membrane was necessary but not sufficient to initiate a response. In plants, a direct involvement of membrane transport has the added implication that sugars entering the cell via plasmodesmata (symplastic transfer) might be perceived differently from sugars taken up from the apoplast (see Figure 2).

Depending on the tissue, sugars may enter a plant cell via any of three routes: (*a*) through plasmodesmata (symplastic transfer), (*b*) across the plasmamembrane as sucrose (from the apoplast), and/or (*c*) across the plasmamembrane as hexoses (again from the apoplast). As illustrated in Figure 2, each path has the potential to transmit different signals to a sugar-sensing mechanism. If plasmamembrane transfer is directly involved in signaling, then hexose uptake from the apoplast would potentially exert a greater effect per unit C than sucrose. Sucrose arriving via plasmodesmatal connections would not exert a similar membrane signal, although plasmodesmata might have an as yet undefined role in sugar sensing. Preliminary evidence indicates that altered photosynthate availability may affect size exclusion limits in plasmodesmata and promote pathway switching (i.e. extent of apoplastic vs symplastic transfer) (129, 147).

The alternative pathways by which sugars enter cellular metabolism may also impact sugar-sensing mechanisms (Figure 3). Hydrolysis of sucrose by invertase generates twice as much substrate for a hexokinase-based sensor as does sucrose synthase. Sucrose synthase, on the other hand, generates UDPG, which may feed into other signaling pathways. Vacuolar compartmentalization and hydrolysis by invertase may affect the timing of hexose-signaling events.

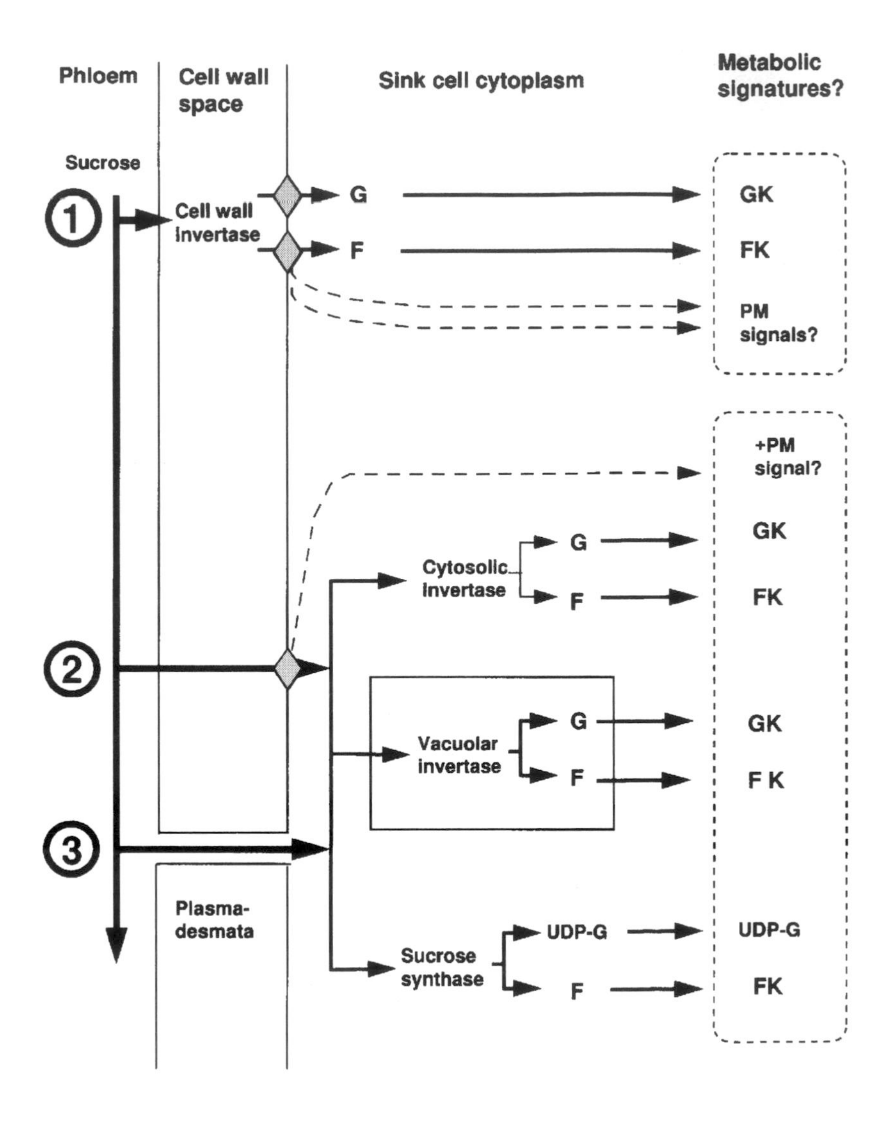

Figure 2 Potential differences in metabolic signatures of sugars entering plant cells and their relevance to the carbohydrate-sensing system. Three different physical paths of sucrose import are shown, with potential signals from plasmodesmata identified with diamonds and dashed lines. Within importing cells, potential input into the hexokinase aspect of the signaling system is designated by FK or GK. Differences in possible metabolic signatures depending on the entry path and initial sucrose cleavage reaction are shown in dashed boxes at the far right.

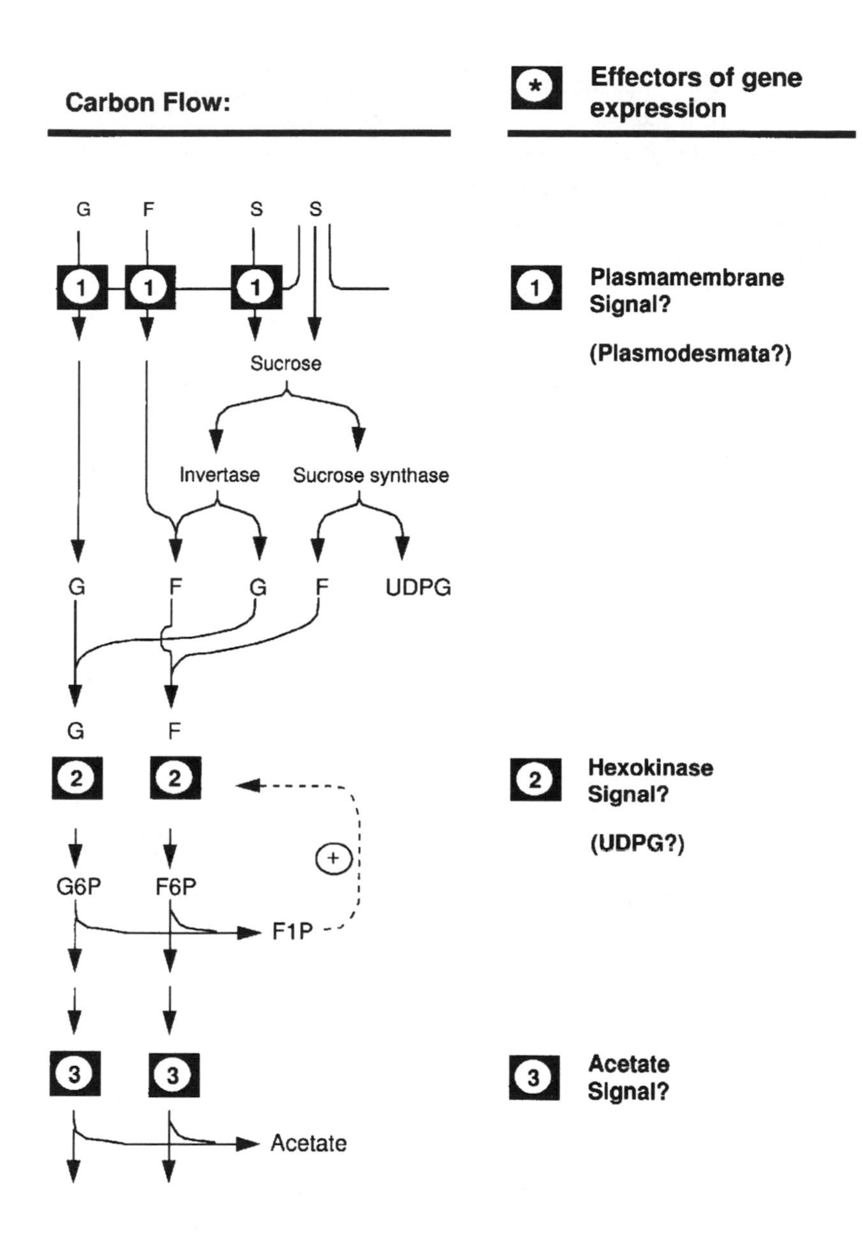

Figure 3 Potential points of signal input into the carbohydrate-sensing system of plants. A simplified path of C-flow is shown at the left with corresponding sites of signal input at the right.

Key Metabolites as Direct Signals

A second source of hypotheses for sugar signaling has come from analogies to pathways for regulation of metabolism (175). The possibility that the same metabolites could mediate regulation of both metabolism and genes in these pathways would provide an attractive mechanism for coordinating rapid metabolic changes with longer-term adjustments in gene expression (34, 134, 175). However, key metabolites such as F-2,6-BP, sugar-P, adenylates, and the Pi/PPi ratio, which collectively regulate carbohydrate metabolism, appear to have little or no direct involvement in sugar regulation of gene expression in the systems studied thus far (161, 175). Levels of F-2,6-BP have been altered using inhibitors and in transgenic plants expressing an antisense gene for PPi-dependent phosphofructokinase without affecting carbohydrate-responsive genes (175; A Krapp and M Stitt, unpublished data). Likewise, various manipulations of Pi levels altered metabolism in predictable ways but did not change expression of sugar-modulated genes (175). Introduction of various phosphorylated sugars, adenylated sugars, and ATP into cells by electroporation failed to affect sugar-responsive genes (161).

Although sugar levels can vary widely in plants, maintenance of "energy homeostasis" is one proposed function of carbohydrate-regulated gene expression (69). The search for a link between respiratory metabolism and gene expression continues. One possible link to a mitochondrial function involves its role as a calcium reservoir sensitive to changes in respiratory substrates (72). Changes in redox potential are also a possibility, as observed for chloroplast mRNAs (28a). Adenylate balance may also be involved through an influence on a hexokinase-based sensing mechanism. In this respect, it seems not widely appreciated that the concentration of enzyme-product complex will be directly proportional to the net forward flux through the reaction only under initial velocity conditions, and these conditions are not likely to apply in vivo. Nearer to equilibrium, the intermediate enzyme complexes will be affected by concentrations of all substrates and products including hexose, hexose-P, ADP, and ATP. It might be more accurate to view yeast hexokinase as a sensor of some ratio of these metabolites rather than flux per se.

The profound effects of acetate on yeast and algal cultures are well known (9, 93, 172), yet the significance of this metabolite to higher plants has been explored only recently in leaf protoplasts (160) and intact cotyledons (50, 51). Acetate appears to be the strongest input into the carbohydrate-sensing system of maize leaf protoplasts (160). It is not clear whether there are points of overlap between this signal and sugar inputs. Signal initiation from these two metabolites appears to occur differently. In this context, it is also intriguing that lipid acyl hydrolases have been recruited as carbohydrate-responsive storage proteins in potato (52, 202).

CLOSING COMMENTS

In multicellular organisms, carbohydrate-responsive gene expression acquires a functional significance beyond that observed in microorganisms. As in yeast and bacteria, specific groups of genes show dramatically different responses to changes in the carbohydrate environment and include both up- and downregulation of gene expression. In multicellular structures, however, individual cells respond to changes in the internal carbohydrate environment of the organism, thus allowing coordinated long-term adjustments for the benefit of the whole.

Plants in particular appear to have successfully employed this mechanism for meeting the adaptive demands of their sessile existence. The sites, timing, and extent of sugar-modulated gene expression described here indicate that these processes may contribute to the dynamic allocation of carbon resources and the continuous adaptive adjustment of form so characteristic of multicellular plants. Sugars in vascular plants are thus long-distance messengers of whole-organism carbohydrate status as well as substrates for both cellular metabolism and local carbohydrate-sensing systems. Although the primary source of carbohydrate signals is currently unclear, hexokinase action and acetate levels remain a common theme shared by mammalian and microorganism sensing systems. The pathways for transduction of sugar signals overlap with other environmental and developmental signals affecting gene expression.

ACKNOWLEDGMENTS

Sincere appreciation is extended to Tami Spurlin, Thea Edwards, Wayne Avigne, and Kalavathy Padmanabhan for their invaluable help in compilation of the work presented here. Research contributions from the author's laboratory were made possible by funding through the National Science Foundation, the USDA-NRI, and Florida Agricultural Experiment Station. This paper is Journal Series No. R-04999

Literature Cited

1. Alderson A, Sabelli PA, Dickinson JR, Cole D, Richardson M, et al. 1991. Complementation of *snf1,* a mutation affecting global regulation of carbon metabolism in yeast, by a plant protein kinase cDNA. *Proc. Natl. Acad. Sci. USA* 88:8602–5
2. Ballard LAT, Wildman SG. 1963. Induction of mitosis by sucrose in excised and attached dormant buds of sunflower (*Helianthus annuus* L.). *Aust. J. Biol. Sci.* 17:36–43
3. Banno H, Muranaka T, Ito Y, Moribe T, Usami S, et al. 1993. Isolation and

characterization of cDNA clones that encode protein kinases of *Nicotiana tabacum*. *J. Plant Res.* 3:181–92
4. Baur-Höch B, Mächler F, Nösberger J. 1990. Effect of carbohydrate demand on the remobilization of starch in stolons and roots of white clover (*Trifolium repens* L.) after defoliation. *J. Exp. Bot.* 41:573–78
5. Baysdorfer C, Van Der Woude WJ. 1988. Carbohydrate responsive proteins in the roots of *Pennisetum americanum*. *Plant Physiol.* 87:566–70
6. Baysdorfer C, Warmbrodt RD, Van Der Woude WJ. 1988. Mechanisms of starvation tolerance in pearl millet. *Plant Physiol.* 88:1381–87
7. Bernier G, Havelange A, Houssa C, Petitjean A, Lejeune P. 1993. Physiological signals that induce flowering. *Plant Cell* 5:1147–55
8. Bingham IJ, Stevenson EA. 1993. Control of root growth: effects of carbohydrates on the extension, branching and rate of respiration of different fractions of wheat roots. *Physiol. Plant.* 88:149–58
9. Böege F, Westhoff P, Zimmerman K, Zetsche K. 1981. Regulation of the synthesis of ribulose-1,5-bisphosphate carboxylase and its subunits in the flagellate *Chlorogonium elongatum:* the effect of light and acetate on the synthesis and the degradation of the enzyme. *Eur. J. Biochem.* 113:581–86
10. Börner T, Hess WR. 1993. Altered nuclear, mitochondrial and plastid gene expression in white barley cells containing ribosome-deficient plastids. In *Plant Mitochondria,* ed. A Brennicke, U Kück, pp. 207–20. New York: VCH Publishers Inc.
11. Boyle MG, Boyer JS, Morgan PW. 1991. Stem infusion of liquid culture medium prevents reproductive failure of maize at low water potential. *Crop Sci.* 31:1246–52
12. Brouquisse R, James F, Pradet A, Raymond P. 1992. Asparagine metabolism and nitrogen distribution during protein degradation in sugar-starved maize root tips. *Planta* 188:384–95
13. Brouquisse R, James F, Raymond P, Pradet A. 1991. Study of glucose starvation in excised maize root tips. *Plant Physiol.* 96:619–26
14. Brugliera F, Holton TA, Stevenson TW, Farcy E, Lu CY, Cornish EC. 1994. Isolation and characterization of a cDNA clone corresponding to the *Rt* locus of *Petunia hybrida. Plant J.* 5:81–92
15. Carlson M. 1987. Regulation of sugar utilization in *Saccharomyces* species. *J. Bacteriol.* 169:4873–77
16. Cheikh N, Jones RJ. 1994. Disruption of maize kernel growth and development by heat stress: role of cytokinin/abscisic acid balance. *Plant Physiol.* 106: 45–51
17. Cheikh N, Jones RJ. 1995. Heat stress effects on sink activity of developing maize kernels grown in vitro. *Physiol. Plant.* 94:59–66
18. Cheng CL, Acedo GN, Cristinsin M, Conkling MA. 1992. Sucrose mimics the light induction of *Arabidopsis* nitrate reductase gene transcription. *Proc. Natl. Acad. Sci. USA* 89:1861–64
19. Chory J. 1993. Out of darkness: mutants reveal pathways controlling light-regulated development in plants. *Trends Genet.* 9:167–72
20. Chourey PS, Nelson OE. 1976. The enzymatic deficiency conditioned by the *shrunken-1* mutations in maize. *Biochem. Genet.* 14:1041–55
20a. He C-J, Drew ME, Morgan P. 1994. Induction of enzymes associated with lysigenous aerenchyma formation in roots of *Zea mays* during hypoxia or nitrogen starvation. *Plant Physiol.* 105: 861–65
21. Claussen W. 1983. Investigations on the relationship between the distribution of assimilates and sucrose synthetase activity in *Solanum melongena* L. II. Distribution of assimilates and sucrose synthetase activity. *Z. Pflanzen Physiol.* 110:175–82
22. Claussen W, Hawker JS, Loveys BR. 1985. Sucrose synthase activity, invertase activity, net photosynthetic rates and carbohydrate content of detached leaves of eggplants as affected by attached stems and shoots (sinks). *Plant Physiol.* 119:123–31
23. Couée I, Murielle J, Carde JP, Brouquisse R, Raymond P, Pradet A. 1992. Effects of glucose starvation on mitochondrial subpopulations in the meristematic and submeristematic regions of maize root. *Plant Physiol.* 100:1891–900
24. Crawford NM. 1995. Nitrate: nutrient and signal for plant growth. *Plant Cell* 7:859–68
25. Criqui MC, Durr A, Parmentier Y, Marbach J, Fleck J, Jamet E. 1992. How are photosynthetic genes repressed in freshly-isolated mesophyll protoplasts of *Nicotiana sylvestris*? *Plant Physiol. Biochem.* 30:597–601
26. Crowell DN, Amasino RM. 1994. Cytokinins and plant gene expresson. See Ref. 112a, pp. 233–42

27. Curtis PS, Vogel C, Pregitzer KS, Zak DR, Teeri JA. 1995. Interacting effects of soil fertility and atmospheric CO_2 on leaf area growth and carbon gain physiology in *Populus* X *euroamericana* (Dode) Guinier. *New Phytol.* 129:253–63
28. D'Amico L, Valsasina B, Daminati MG, Fabbrini MS, Nitti G, et al. 1992. Bean homologs of the mammalian glucose-regulated proteins: induction by tunicamycin and interaction with newly synthesized seed storage proteins in the endoplasmic reticulum. *Plant J.* 2:443–55
28a. Danon A, Mayfield SP. 1994. Light-regulated translation of chloroplast messenger RNAs through redox potential. *Science* 266:1717-19
29. DeWald DB, Sadka A, Mullet JE. 1994. Sucrose modulation of soybean *Vsp* gene expression is inhibited by auxin. *Plant Physiol.* 104:439–44
30. Dickinson CD, Altabella T, Chrispeels MJ. 1991. Slow-growth phenotype of transgenic tomato expressing apoplastic invertase. *Plant Physiol.* 95:420–25
31. Dieuaide M, Brouquisse R, Pradet A, Raymond P. 1991. Increased fatty acid β-oxidation after glucose starvation in maize root tips. *Plant Physiol.* 99:595–600
32. Echt CS, Chourey PS. 1985. A comparison of two sucrose synthetase isozymes from normal and *Shrunken-1* maize. *Plant Physiol.* 79:530–36
33. Elamrani A, Gaudillère JP, Raymond P. 1994. Carbohydrate starvation is a major determinant of the loss of greening capacity in cotyledons of dark-grown sugar beet seedlings. *Physiol. Plant* 91: 56–64
34. Farrar JF, Williams JHH. 1991. Control of the rate of respiration in roots: compartmentation, demand, and the supply of substrate. In *Compartmentation of Metabolism,* ed. M Emms, pp. 167–88. London: Butterworths
35. Foyer CH. 1988. Feedback inhibition of photosynthesis through source-sink regulation in leaves. *Plant Physiol. Biochem.* 26(4):483–92
36. Frommer WB, Mielchen C, Martin T. 1994. Metabolic control of patatin promoters from potato in transgenic tobacco and tomato plants. *Life Sci. Adv.* 13:329–34
37. Fu H, Kim SY, Park WD. 1995. High-level tuber expression and sucrose inducibility of a potato *Sus4* sucrose synthase gene require 5′ and 3′ flanking sequences and the leader intron. *Plant Cell* 7:1387–94
38. Fu H, Park WD. 1995. Sink- and vascular-associated sucrose synthase functions are encoded by different gene classes in potato. *Plant Cell* 7: 1369–85
39. Fukuda H, Komamine A. 1985. Cytodifferentiation. In *Cell Culture and Somatic Cell Genetics of Plants: Cell Growth, Nutrition, Cytodifferentiation, and Cryopreservation,* ed. IK Vasil, pp. 150–212. New York: Academic
40. Gancedo JM. 1992. Carbon catabolite repression in yeast. *Eur. J. Biochem.* 206:297–313
41. Geiger DR. 1976. Effect of translocation and assimilate demand of photosynthesis. *Can. J. Bot.* 54:2337–45
42. Geiger DR, Servaites JC. 1994. Diurnal regulation of photosynthetic carbon metabolism in C_3 plants. *Annu. Rev. Plant Physiol. Plant Mol. Biol.* 45:235–56
43. Geiger DR, Koch KE, Shieh WJ. 1996. Effect of environmental factors on whole plant assimilate partitioning and associated gene expression in the whole plant. *J. Exp. Bot.* In press
44. Giroux MJ, Boyer C, Feix G, Hannah LC. 1994. Coordinated transcriptional regulation of storage product genes in the maize endosperm. *Plant Physiol.* 106:713–22
45. Glasziou KT, Waldron JC, Bull TA. 1966. Control of invertase synthesis in sugar cane: loci of auxin and glucose effects. *Plant Physiol.* 41:282–88
46. Godf DE, Reigel A, Roitsch T. 1995. Regulation of sucrose synthase expression in *Chenopodium rubrum:* characterization of sugar induced expression in photoautotrophic suspension cultures and sink tissue specific expression in plants. *J. Plant Physiol.* 146:231–38
47. Goldschmidt EE, Huber SC. 1992. Regulation of photosynthesis by end-product accumulation in leaves of plants storing starch, sucrose, and hexose sugars. *Plant Physiol.* 99:1443–48
48. Goldschmidt EE, Koch KE. 1996. Source sink relations and carbohydrate economy of citrus. See Ref. 215a, 797–824
49. Gordon AJ, Kessler W. 1990. Defoliation-induced stress in nodules of white clover. II. Immunological and enzymic measurements of key proteins. *J. Exp. Bot.* 41:1255–62
50. Graham IA, Baker CJ, Leaver CJ. 1994. Analysis of the cucumber malate synthase gene promotor by transient expression and gel retardation assays. *Plant J.* 6:893–902
51. Graham IA, Denby KJ, Leaver CJ. 1994. Carbon catabolite repression regulates

glyoxylate cycle gene expression in cucumber. *Plant Cell* 6:761–72
52. Grierson C, Du JS, Zabala MD, Beggs K, Smith C, et al. 1994. Separate *cis* sequences and *trans* factors direct metabolic and developmental regulation of a potato-tuber storage protein gene. *Plant J.* 5:815–26
53. Grimes HD, Tranbarger TJ, Franceschi VR. 1993. Expression and accumulation patterns of nitrogen-responsive lipoxygenase in soybeans. *Plant Physiol.* 103: 457–66
54. Halford NG, Vincente-Carbajosa J, Sabelli PA, Shewry PR, Hannappel U, Kreis M. 1992. Molecular analyses of a barley multigene family homologous to the yeast protein kinase gene *SNF1*. *Plant J.* 2:791–97
55. Harter K, Talke-Messerer C, Barz W, Schäfer E. 1993. Light- and sucrose-dependent gene expression in photomixotrophic cell suspension cultures and protoplasts of rape (*Brassica napus* L.). *Plant J.* 4:507–16
56. Hattori T, Nakagawa S, Nakamura K. 1990. High-level expression of tuberous root storage protein genes of sweet potato in stems of plantlets grown in vitro on sucrose medium. *Plant Mol. Biol.* 14:595–604
57. Heim U, Weber H, Bäumlein H, Wobus U. 1993. A sucrose-synthase gene of *Vicia faba* L.: expression pattern in developing seeds in relation to starch synthesis and metabolic regulation. *Planta* 191:394–401
58. Heineke D, Sonnewald U, Büssis D, Günter G, Leidreiter K, et al. 1992. Apoplastic expression of yeast-derived invertase in potato: effects on photosynthesis, leaf solute composition, water relations, and tuber composition. *Plant Physiol.* 100:301–8
59. Heineke D, Wildenberger K, Sonnewald U, Willmitzer L, Heldt HW. 1994. Accumulation of hexoses in leaf vacuoles: studies with transgenic tobacco plants expressing yeast-derived invertase in the cytosol, vacuole or apoplasm. *Planta* 194:29–33
60. Hesse H, Sonnewald U, Willmitzer L. 1995. Cloning and expression analysis of sucrose-phosphate-synthase from sugar beet (*Beta vulgaris* L.). *Mol. Gen. Genet.* 247:515–20
61. Huang N, Chandler J, Thomas BR, Koizumi N, Rodriguez RL. 1993. Metabolic regulation of α-amylase gene expression in transgenic cell cultures of rice. *Plant Mol. Biol.* 23:737–47
62. Huber SC. 1989. Biochemical mechanism for regulation of sucrose accumulation in leaves during photosynthesis. *Plant Physiol.* 91:656–62
63. Huber SC, Huber JL, Chourey PS, Hannah LC, Koch KE. 1995. Sucrose synthase is phosphorylated in vivo in maize leaves. *Plant Physiol.* 108:142
64. Huff A. 1984. Sugar regulation of plastid interconversions in epicarp of citrus fruit. *Plant Physiol.* 76:307–12
65. Ishiguro S, Nakamura K. 1994. Characterization of a cDNA encoding a novel DNA binding protein, SPF1, that recognizes SP8 sequences in the 5′ upstream regions of genes coding for sporamin and beta-amylase from sweet potato. *Mol. Gen. Genet.* 244:563–71
66. Jackson SD, Sonnewald U, Willmitzer L. 1993. Cloning and expression analysis of β-isopropylmalate dehydrogenase from potato. *Mol. Gen. Genet.* 236:309–14
67. Jacqmard A, Houssa C, Bernier G. 1994. Regulation of the cell cycle by cytokinins. See Ref. 112a, pp. 197–215
68. James F, Brouquisse R, Pradet A, Raymond P. 1993. Changes in proteolytic activities in glucose-starved maize roottips: regulation by sugars. *Plant Physiol. Biochem.* 31:845–56
69. Jang J-C, Sheen J. 1994. Sugar sensing in higher plants. *Plant Cell* 6:1665–79
70. Johnson R, Ryan CA. 1990. Wound-inducible potato inhibitor II genes: enhancement of expression by sucrose. *Plant Mol. Biol.* 14:527–36
71. Jones JR, Schreiber BM, McNeil K, Brenner ML, Foxon G. 1992. Cytokinin levels and oxidase activity during maize kernel development. In *Physiology and Biochemistry of Cytokinins in Plants,* ed. M Kaminek, DWS Mok, E Zazimalova, pp. 235–39. The Hague: Academic
72. Jouaville LS, Ichas F, Holmuhamedov EL, Camacho P, Lechleiter JD. 1995. Synchronization of calcium waves by mitochondrial substrates in *Xenopus laevis* oocytes. *Nature* 377:438–41
73. Journet EP, Bligny R, Doucr R. 1986. Biochemical changes during sucrose deprivation in higher plant cells. *J. Biol. Chem.* 261:3193–99
74. Karrer EE, Rodriguez RL. 1992. Metabolic regulation of rice α-amylase and sucrose synthase genes in planta. *Plant J.* 2:517–23
75. Kaufman PB, Ghosheh NS, LaCroix JD, Soni SL, Ikuma H. 1973. Regulation of invertase levels in *Avena* stem segments by gibberellic acid, sucrose, glucose, and fructose. *Plant Physiol.* 52:221–28
76. Kaufman PB, Song I. 1987. Hormones and the orientation of growth. In *Plant*

Hormones and Their Role in Plant Growth and Development, ed. PF Davies, pp. 375–92. Dordrecht: Nijhoff
77. Kim SR, Costa MA, An G. 1991. Sugar response element enhances wound response of potato proteinase inhibitor II promotor in transgenic tobacco. *Plant Mol. Biol.* 17:973–83
78. Kim SY, May GD, Park WD. 1994. Nuclear-protein factors binding to a class-I patatin promoter region are tuber-specific and sucrose-inducible. *Plant Mol. Biol.* 26:603–15
79. Kindle KL. 1987. Expression of a gene for a light harvesting chlorophyll a/b-binding protein in *Chlamydomonas reinhardtii:* effect of light and acetate. *Plant Mol. Biol.* 9:547–63
80. King RW, Evans LT. 1991. Shoot apex sugars in relation to long-day induction of flowering in *Lolium temulentum* L. *Aust. J. Plant Physiol.* 18:121–35
81. Klein RR, Crafts-Brandner SJ, Salvucci ME. 1993. Cloning and developmental expression of the sucrose-phosphate-synthase gene from spinach. *Planta* 190: 498–510
82. Knight JS, Gray JC. 1994. Expression of genes encoding the tobacco chloroplast phosphate translocator is not light-regulated and is repressed by sucrose. *Mol. Gen. Genet.* 242:586–94
83. Koch KE, McCarty DR. 1988. Induction of sucrose synthase by sucrose depletion in maize root tips. *Plant Physiol.* 86:35
84. Koch KE, Nolte KD. 1995. Sugar modulated expression of genes for sucrose metabolism and their relationship to transport pathways. See 104a, pp. 68–77
85. Koch KE, Nolte KD, Duke ER, McCarty DR, Avigne WT. 1992. Sugar levels modulate differential expression of maize sucrose synthase genes. *Plant Cell* 4:59–69
86. Koch KE, Xu J, Duke ER, McCarty DR, Yuan CX, et al. 1995. Sucrose provides a long distance signal for coarse control of genes affecting its metabolism. In *Sucrose Metabolism, Biochemistry, and Molecular Biology,* ed. HG Pontis, G Salerno, E Echeverria. Rockville, MD: Am. Soc. Plant Physiol. 14:266–77
87. Koßmann J, Visser RGF, Müller-Röber BT, Willmitzer L, Sonnewald U. 1991. Cloning and expression analysis of a potato cDNA that encodes branching enzyme: evidence for coexpression of starch biosynthetic genes. *Mol. Gen. Genet.* 230:39–44
88. Kovtun Y, Daie J. 1995. End-product control of carbon metabolism in culture-grown sugar beet plants: molecular and physiological evidence on accelerated leaf development and enhanced gene expression. *Plant Physiol.* 108:1647–56
89. Krapp A, Hofmann B, Schäfer C, Stitt M. 1993. Regulation of the expression of *rbcS* and other photosynthetic genes by carbohydrates: a mechanism for the "sink regulation" of photosynthesis? *Plant J.* 3:817–28
90. Krapp A, Quick WP, Stitt M. 1991. Ribulose-1,5-bisphosphate carboxylase-oxygenas, other Calvin-cycle enzymes, and chlorophyll decrease when glucose is supplied to mature spinach leaves via the transpiration stream. *Planta* 186:58–69
91. Krapp A, Stitt M. 1994. Influence of high-carbohydrate content on the activity of plastidic and cytosolic isoenzyme pairs in photosynthetic tissues. *Plant Cell Environ.* 17:861–66
92. Krapp A, Stitt M. 1995. An evaluation of direct and indirect mechanisms for the "sink regulation" of photosynthesis of spinach: changes in gas exchange, carbohydrates, metabolites, enzyme activities and steady-state transcript levels after cold girdling source leaves. *Planta* 195:313–23
93. Kroymann J, Schneider W, Zetsche K. 1995. Opposite regulation of copy number and the expression of plastid and mitochondrial genes by light and acetate in the green flagellate *Chlorogonium. Plant Physiol.* 108:1641–46
94. Kumar S, Chandra GR, Albaugh GP, Muthukrishnan S. 1985. Regulation of the expression of α-amylase gene by sodium butyrate. *Plant Mol. Biol.* 5: 269–79
95. Lam H-M, Peng SS-Y, Coruzzi GM. 1994. Metabolic regulation of the gene encoding glutamine-dependent asparagine synthetase in *Arabidopsis thaliana. Plant Physiol.* 106:1347–57
96. Lambers H, Atkin O. 1995. Regulation of carbon metabolism in roots. See 104a, pp. 226–38
97. Lejeune P, Bernier G, Requier MC, Kinet JM. 1993. Sucrose increase during floral induction in the phloem sap collected at the apical part of the shoot of the long-day plant *Sinapis alba* L. *Planta* 190:71–74
98. Lindsey K, Yeoman MM. 1985. Dynamics of plant cell cultures. In *Cell Culture and Somatic Cell Genetics of Plants: Cell Growth, Nutrition, Cytodifferentiation, and Cryopreservation,* ed. IK Vasil, pp. 61–101. New York: CRC Academic
99. Liu XY, Rocha-Sosa M, Hummel S, Willmitzer L, Frommer WB. 1991. A

detailed study of the regulation and evolution of the two classes of patatin genes in *Solanum tuberosum* L. *Plant Mol. Biol.* 17:1139–54

100. Llernould S, Karamanos Y, Priem B, Morvan H. 1994. Carbon starvation increases endoglycosidase activities and production of "unconjugated N-glycans" in *Silene alba* cell-suspension cultures. *Plant Physiol.* 106:779–84
101. Loake GJ, Choudhary AD, Harrison MJ, Mavandad M, Lamb CJ, Dixon RA. 1991. Phenylpropanoid pathway intermediates regulate transient expression of a chalcone synthase gene promoter. *Plant Cell* 3:829–40
102. Lue MY, Lee HT. 1994. Protein phosphatase inhibitors enhance the expression of an α-amylase gene, α*Amy3,* in cultured rice cells. *Biochem. Biophys. Res. Comm.* 205:807–16
103. Lugert T, Werr W. 1994. A novel DNA-binding domain in the *Shrunken* initiator-binding protein (ibp1). *Plant Mol. Biol.* 25:493–506
104. Maas C, Schaal S, Werr W. 1990. A feedback control element near the transcription start site of the maize *Shrunken* gene determines promoter activity. *EMBO J.* 9:3447–52

104a. Madore MM, Lucas WL, eds. 1995. *Carbon Partitioning and Source Sink Interactions in Plants.* Rockville, MD: Am. Soc. Plant Physiol.

105. Maraña C, Garcia-Olmedo F, Carbonero P. 1990. Differential expression of two types of sucrose synthase-encoding genes in wheat in response to anaerobiosis, cold shock, and light. *Gene* 88:167–72
106. Martin T, Frommer WB, Salanoubat M, Willmitzer L. 1993. Expression of an *Arabidopsis* sucrose synthase gene indicates a role in metabolization of sucrose both during phloem loading and in sink organs. *Plant J.* 4:367–77
107. Mason H, DeWald DB, Creelman RA, Mullet JE. 1992. Coregulation of soybean vegetative storage protein gene expression by methyl jasmonate and soluble sugars. *Plant Physiol.* 98:859–67
108. McCully ME, Canny MJ. 1995. How do real roots work: some new views of root structure. *Plant Physiol.* 109:1–6
109. McDonnell E, Farrar JF. 1992. Substrate supply and its effect on mitochondrial and whole tissue respiration in barley roots. In *Molecular, Biochemical and Physiological Aspects of Plant Respiration,* ed. H Lambers, LHW van der Plas, pp. 455–62. The Hague: Academic
110. Miao Z, Gaynor JJ. 1993. Molecular cloning, characterization and expression of Mn-superoxide dismutase from the rubber tree (*Hevea brasiliensis*). *Plant Mol. Biol.* 23:267–77
111. Miller ME, Chourey PS. 1992. The maize invertase-deficient *minature-1* seed mutation is associated with aberrant pedicel and endosperm development. *Plant Cell* 4:297–305
112. Mok MC. 1994. Cytokinins and plant development: an overview. See Ref. 112a, pp. 155–66

112a. Mok WS, Mok MC, eds. 1994. *Cytokinins: Chemistry, Activity, and Function.* Boca Raton: CRC Press

113. Monroy AF, Schwartzbach SD. 1984. Catabolite repression of chloroplast development in *Euglena. Proc. Natl. Acad. Sci. USA* 81:2786–90
114. Morré DJ, Jones DD, Mollenhauer HH. 1967. Golgi apparatus mediated polysaccharide secretion by outer root cap cells of *Zea mays.* I. Kinetics and secretory pathway. *Planta* 74:286–301
115. Morris DA, Arthur ED. 1984. An association between acid invertase activity and cell growth during leaf expansion in *Phaseolus vulgaris* L. *J. Exp. Bot.* 35:1369–79
116. Müller-Röber BT, Koßmann J, Hannah LC, Willmitzer L, Sonnewald U. 1990. One of two different ADP-glucose pyrophosphorylase genes from potato responds strongly to elevated levels of sucrose. *Mol. Gen. Genet.* 224:136–46
117. Müller-Röber BT, Sonnewald U, Willmitzer L. 1992. Inhibition of the ADP-glucose pyrophosphorylase in transgenic potatoes leads to sugar-storing tubers and influences tuber formation and expression of tuber storage protein genes. *EMBO J.* 11: 1229–38
118. Muranaka T, Banno H, Machida Y. 1994. Characterization of tobacco protein-kinase npk5, a homolog of *Saccharomyces-cerevisiae snf1* that constitutively activates expression of the glucose-repressible *suc2* gene for secreted invertase of *Saccharomyces cerevisiae. Mol. Cell. Biol.* 14:2958–65
119. Murray JR, Hackett WP. 1991. Dihydroflavonol and reductase activity in relation to differential anthocyanin accumulation in juvenile and mature phase *Hedera helix* L. *Plant Physiol.* 97:343–51
120. Murray JR, Smith AG, Hackett WP. 1994. Differential dihydroflavonol reductase transcription and anthocyanin pigmentation in the juvenile and mature phases of ivy (*Hedera helix* L.). *Planta* 194:102–9

121. Musgrave ME. 1994. Cytokinins and oxidative processes. See Ref. 112a, pp. 167–78
122. Nakamura K, Ohto M, Yoshida N, Nakamura K. 1991. Sucrose-induced accumulation of β-amylase occurs concomitant with the accumulation of starch and sporamin in leaf-petiole cuttings of sweet potato. *Plant Physiol.* 96:902–9
123. Nolte KD, Hendrix DL, Radin JW, Koch KE. 1995. Sucrose synthase localization during initiation of seed development and trichome differentiation in cotton ovules. *Plant Physiol.* 109:1285–93
124. Nolte KD, Koch KE. 1993. Companion-cell specific localization of sucrose synthase in zones of phloem loading and unloading. *Plant Physiol.* 101:899–905
125. Ohto MA, Hayashi K, Isobe M, Nakamura K. 1995. Involvement of Ca^{2+} signaling in the sugar-inducible expression of genes coding for sporamin and β-amylase of sweet potato. *Plant J.* 7: 297–307
126. Ohto M, Nakamura K. 1996. Sugar-induced increase of calcium-dependendent protein kinases associated with the plasmamembrane in leaf tissues of tobacco. *Plant Physiol.* 109:In press
127. Ohto M, Nakamura-Kito K, Nakamura K. 1992. Induction of expression of genes coding for sporamin and β-amylase by polygalacturonic acid in leaf-petiole cuttings of sweet potato. *Plant Physiol.* 99:422–27
128. Parks BM, Hangarter RP. 1994. Blue light sensory systems in plants. *Cell Biol.* 5:347–53
129. Patrick JW, Offler CE. 1996. Switching of the cellular pathway of post-sieve element transport in roots and stems. *J. Exp. Bot.* In press
130. Paul MJ, Stitt M. 1993. Effects of nitrogen and phosphorus deficiencies on levels of carbohydrates, respiratory enzymes and metabolites in seedlings of tobacco and their response to exogenous sucrose. *Plant Cell Environ.* 16:1047–57
131. Peña-Cortes H, Liu X, Sanchez-Serrano J, Schmid R, Willmitzer L. 1992. Factors affecting gene expression of patatin and proteinase-inhibitor-II gene families in detached potato leaves: implications for their co-expression in developing tubers. *Planta* 186:495–502
132. Peterson CA. 1988. Exodermal Casparian bands: their significance for ion uptake by roots. *Physiol. Plant* 72: 204–8
133. Pharr DM, Stoop JMH, Studer-Feusi ME, Williamson JD, Massel MO, Conkling MA. 1995. Mannitol catabolism in plant sink tissues. See 104a, pp. 180–93
134. Pollock CJ, Farrar J. 1996. Source-sink relations: the role of sucrose. In *Environmental Stress and Photosynthesis,* ed. NR Baker. The Netherlands: Kluwer. In press
135. Pollock CJ, Lloyd EJ. 1978. Acid invertase activity during senescence of excised leaf tissue of *Lolium temulentum. Z. Pflanzenphysiol. Bd.* 90:79–84
136. Pollock CJ, Winters AL, Gallagher J, Cairns AJ. 1996. Sucrose and the regulation of fructan metabolism in leaves of temperate gramineae. In *Sucrose Metabolism, Biochemistry, Physiology, and Molecular Biology,* ed. HG Pontis, GL Salerno, E Echeverria, 14:167–78. Rockville, MD: Am. Soc. Plant Physiol.
137. Pryke JA, Bernier G. 1978. Acid invertase activity in the apex of *Sinapis alba* during transition to flowering. *Ann. Bot.* 42:747–49
138. Purvis AC, Rice JD. 1983. Low temperature induction of invertase activity in grapefruit flavedo tissue. *Phytochemistry* 22:831–34
139. Quick PW, Schaffer AA. 1996. Sucrose metabolism in sources and sinks. See Ref. 215a
140. Rebeille F, Bligny R, Martin JB, Douce R. 1985. Effect of sucrose starvation on sycamore (*Acer pseudoplatanus*) cell carbohydrate and Pi status. *Biochem. J.* 226:679–84
141. Reed AJ, Singletary GW. 1989. Roles of carbohydrate supply and phytohormones in maize kernel abortion. *Plant Physiol.* 91:986–92
142. Roberts DM, Harmon AC. 1992. Calcium-modulated proteins: targets of intracellular calcium signals in higher plants. *Annu. Rev. Plant Physiol. Plant Mol. Biol.* 43:375–414
143. Rocha-Sosa M, Sonnewald U, Frommer W, Stratmann M, Schell J, Willmitzer L. 1989. Both developmental and metabolic signals activate the promoter of a class-I patatin gene. *EMBO J.* 8(1):23–29
144. Roitsch T, Bittner M, Godt DE. 1995. Induction of apoplastic invertase of *Chenopodium rubrum* by D-glucose and a glucose analog and tissue-specific expression suggest a role in sink-source regulation. *Plant Physiol.* 108:285–94
145. Rose M, Albig W, Entian KD. 1991. Glucose repression in *Saccharomyces cerevisiae* is directly associated with hexose phosphorylation by hexokinases PI and PII. *Eur. J. Biochem.* 199:511–18
146. Ross HA, Davies HV. 1992. Sucrose metabolism in tubers of potato (*Solanum*

tuberosum L.): effects of sink removal and sucrose flux on sucrose-degrading enzymes. *Plant Physiol.* 98:287–93

147. Ruan YL, Patrick JW. 1995. The cellular pathway of postphloem sugar transport in developing tomato fruit. *Planta* 196: 434–44
148. Sacher JA, Hatch MD, Glasziou KT. 1963. Regulation of invertase synthesis in sugar cane by an auxin- and sugar-mediated control system. *Physiol. Plant.* 16:836–42
149. Sadka A, Dewald DB, May GD, Park WD, Mullet JE. 1994. Phosphate modulates transcription of soybean vspB and other sugar-inducible genes. *Plant Cell.* 6:737–49
150. Saier MH. 1989. Protein phosphorylation and allosteric control of inducer exclusion and catabolite repression by the bacterial phosphoenol pyruvate: sugar phosphotransferase system. *Microbiol. Rev.* 53:109–20
151. Salanoubat M, Belliard G. 1989. The steady-state level of potato sucrose synthase mRNA is dependent on wounding, anaerobiosis and sucrose concentration. *Gene* 84:181–85
152. Salehuzzaman SNIM, Jacobsen E, Visser RGF. 1994. Expression patterns of two starch biosynthetic genes in in vitro cultured cassava plants and their induction by sugars. *Plant Sci.* 98:53–62
153. Schäfer C, Simper H, Hofmann B. 1992. Glucose feeding results in coordinated changes of chlorophyll content, ribulose-1,5-bisphosphate carboxylase-oxygenase activity and photosynthetic potential in photoautrophic suspension cultured cells of *Chenopodium rubrum. Plant Cell Environ.* 15:343–50
154. Schaffer AA, Liu KC, Goldschmidt EE, Boyer CD, Goren R. 1986. *Citrus* leaf chlorosis induced by sink removal: starch, nitrogen, and chloroplast ultrastructure. *J. Plant Physiol.* 124:111–21
155. Schmidt A, Zetsche K. 1990. Regulation of the synthesis of isocitrate lyase and the corresponding mRNA in the green alga *Chlorogonium. Bot. Acta.* 103:48–53
156. Schnarrenberger C. 1990. Characterization and compartmentalization in green leaves of hexokinases with different specificities for glucose, fructose and mannose and for nucleoside triphosphates. *Planta* 181:249–55
157. Schulte R, Bonas U. 1992. A *Xanthomonas* pathogenicity locus is induced by sucrose and sulfur-containing amino acids. *Plant Cell* 4:79–86
158. Schuster JR. 1989. Regulated transcriptional systems for the production of proteins in yeast: regulation by carbon source. In *Yeast Genetic Engineering,* ed. PJ Barr, AJ Brike, P Valenzuela, pp. 83–108. London: Butterworths
159. Shaw JR, Ferl RJ, Baier J, St Clair D, Carson C, et al. 1994. Structural features of the maize *sus1* gene and protein. *Plant Physiol.* 106:1659–65
160. Sheen J. 1990. Metabolic repression of transcription in higher plants. *Plant Cell* 2:1027–38
161. Sheen J. 1994. Feedback-control of gene-expression. *Photosynth. Res.* 39: 427–38
162. Sheu JJ, Jan SP, Lee HT, Yu SM. 1994. Control of transcription and mRNA turnover as mechanisms of metabolic repression of α-amylase gene expression. *Plant J.* 5:655–64
163. Shih MC, Goodman HM. 1988. Differential light regulated expression of nuclear genes encoding chloroplast and cytosolic glyceraldehyde-3-phosphate dehydrogenase in *Nicotiana tabacum. EMBO J.* 7:893–98
164. Shorrosh BS, Dixon RA. 1992. Molecular characterization and expression of an alfalfa protein with sequence similarity to mammalian ER_p72, a glucose-regulated endoplasmic reticulum protein containing active site sequences of protein disulphide isomerase. *Plant J.* 2:51–58
165. Simko I. 1994. Sucrose application causes hormonal changes associated with potato-tuber induction. *J. Plant Growth Regul.* 13:73–77
166. Sonnewald U, Brauer M, von Schaewen A, Stitt M, Willmitzer L. 1991. Transgenic tobacco plants expressing yeast-derived invertase in either the cytosol, vacuole or apoplast: a powerful tool for studying sucrose metabolism and sink/source interactions. *Plant J.* 1(1): 95–106
167. Sonnewald U, Lerchl J, Zrenner R, Frommer W. 1994. Manipulation of sink-source relations in transgenic plants. *Plant Cell Environ.* 17:649–58
168. Sonnewald W, Willmitzer L. 1992. Molecular approaches to sink-source interactions. *Plant Physiol.* 99:1267–70
169. Springer B, Werr W, Starlinger P, Bennett DC, Zokolica M et al. 1986. The *Shrunken* gene on chromosome 9 of *Zea mays* L. is expressed in various plant tissues and encodes an anareobic protein. *Mol. Gen. Genet.* 205:461–68
170. Staswick PE. 1994. Storage proteins of vegetative plant tissues. *Annu. Rev. Plant Physiol. Plant Mol. Biol.* 45:303–22
171. St. Pierre B, Brisson N. 1995. Induction

of the plastidic starch-phosphorylase gene in potato storage sink tissue: effect of sucrose and evidence for coordinated regulation of phosphorylase and starch biosynthetic genes. *Planta* 195:339–44
172. Steinbiß HJ, Zetsche K. 1986. Light and metabolite regulation of the synthesis of ribulase-1,5-bisphosphate carboxylase/oxygenase and the corresponding mRNAs in the unicellular alga *Chlorogonium. Planta* 167:575–81
173. Stepien V, Sauter JJ, Martin F. 1994. Vegetative storage proteins in woody plants. *Plant Physiol. Biochem.* 32:185–92
174. Stitt M. 1990. Fructose-2,6-bisphosphate as a regulatory molecule in plants. *Annu. Rev. Plant Physiol. Plant Mol. Biol.* 44:153–85
175. Stitt M, Krapp A, Klein D, Röper-Schwarz U, Paul M. 1995. Do carbohydrates regulate photosynthesis and allocation by altering gene expression? See 104a, pp. 68–77
176. Stitt M, Sonnewald U. 1995. Regulation of metabolism in transgenic plants. *Annu. Rev. Plant Physiol. Plant Mol. Biol.* 46:341–68
177. Stitt M, von Schaewen A, Willmitzer L. 1990. "Sink" regulation of photosynthetic metabolism in transgenic tobacco plants expressing yeast invertase in their cell wall involves a decrease of the Calvin-cycle enzymes and an increase of glycolytic enzymes. *Planta* 183:40–50
178. Sturm A, Chrispeels MJ. 1990. cDNA cloning of carrot extracellular β-fructosidase and its expression in response to wounding and bacterial infection. *Plant Cell* 2:1107–19
179. Sturm A, Sebkova V, Lorenz K, Hardegger M, Leinhard S, Unger C. 1995. Development- and organ-specific expression of the genes for sucrose synthase and three isozymes of acid β-fructofuranosidase in carrot. *Planta* 195: 601–10
180. Sung HI, Liu LF, Kao C. 1994. The induction of α-amylase activity by sucrose starvation in suspension-cultured rice cells is regulated by polyamines. *Physiol. Plant.* 91:137–40
181. Sung SJ, Xu DP, Black CC. 1988. Identification of actively filling sucrose sinks. *Plant Physiol.* 89:1117–21
182. Takeda S, Mano S, Ohto M, Nakamura K. 1994. Inhibitors of protein phosphatases 1 and 2a block the sugar inducible gene expression in plants. *Plant Physiol.* 106:567–74
183. Takeuchi A, Matsumoto S, Hayatsu M. 1994. Chalcone synthase from *Camellia sinensis:* isolation of the cDNAs and the organ-specific and sugar-responsive expression of the genes. *Plant Cell Physiol.* 35:1011–18
184. Tarczynski MC, Jenson RG, Bohnert HJ. 1992. Expression of a bacterial mtlD gene in transgenic tobacco leads to production and accumulation of mannitol. *Proc. Natl. Acad. Sci. USA* 89:2600–4
185. Tarczynski MC, Jenson RG, Bohnert HJ. 1993. Stress protection of transgenic tobacco by the osmolyte mannitol. *Science* 259:508–10
186. Theodorou ME, Plaxton WC. 1993. Metabolic adaptations of plant respiration to nutritional phosphate deprivation. *Plant Physiol.* 101:339–44
187. Thomas BR, Rodriguez RL. 1994. Metabolite signals regulate gene expression and source/sink relations in cereal seedlings. *Plant Physiol.* 106:1235–39
188. Thomas BR, Terashima M, Katoh S, Stoltz T, Rodriguez RL. 1995. Metabolic regulation of source-sink relations in cereal seedlings. See Ref. 104a, pp. 78–90
189. Deleted in proof
190. Topping JF, Leaver CJ. 1990. Mitochondrial gene expression during wheat leaf development. *Planta* 182:399–407
191. Tran Thanh Van K. 1980. Control of morphogenesis by inherent and exogenously applied factors in thin cell layers. In *International Review of Cytology 11A: Perspectives in Plant Cell and Tissue Culture,* ed. IK Vasil, pp. 175–94. New York: CRC Academic
192. Trumbly RJ. 1992. Glucose repression in the yeast *Saccharomyces cerevisiae. Mol. Microbiol.* 6:15–21
193. Tsukaya H, Ohshima T, Naito S, Chino M, Komeda Y. 1991. Sugar-dependent expression of the CHS-A gene for chalcone synthase from petunia in transgenic *Arabidopsis. Plant Physiol.* 97:1414–21
194. Van Oosten JJ, Besford RT. 1994. Sugar feeding mimics effect of acclimation to high CO_2: rapid downregulation of RuBisCO small subunit transcripts but not of the large subunit transcripts. *J. Plant Physiol.* 143:306–12
195. Vaulont S, Kahn A. 1994. Transcriptional control of metabolic regulation genes by carbohydrates. *FASEB J.* 8:28–35
196. Vincentz M, Moureaux T, Leydecker MT, Vaucheret H, Caboche M. 1993. Regulation of nitrate and nitrite reductase expression in *Nicotiana plumbaginifolia* leaves by nitrogen and carbon metabolites. *Plant J.* 3:315–24
197. Visser RGF, Vreugdenhil D, Hendriks

T, Jacobsen E. 1994. Gene-expression and carbohydrate content during stolon to tuber transition in potatoes (*Solanum tuberosum*). *Physiol. Plant.* 90:285–92

198. von Schaewen A, Stitt M, Schmidt R, Sonnewald U, Willmitzer L. 1990. Expression of yeast-derived invertase in the cell wall of tobacco and *Arabidopsis* plants leads to accumulation of carbohydrate and inhibition of photosynthesis and strongly influences growth and phenotype of transgenic tobacco plants. *EMBO J.* 9:3033–44
199. Wardlaw IF. 1990. The control of carbon partitioning in plants. *New Phytol.* 116:341–81
200. Webster PL, Henry M. 1987. Sucrose regulation of protein synthesis in pea root meristem cells. *Environ. Exp. Bot.* 27:253–62
201. Wen IC, Sherman WB, Koch KE. 1995. Heritable pleiotropic effects of the nectarine mutant from peach. *J. Am. Soc. Hort. Sci.* 120:721–25
202. Wenzler HC, Mignery GA, Fisher LM, Park WD. 1989. Analysis of a chimeric class-I patatin-GUS gene in transgenic potato plants: high-level expression in tubers and sucrose-inducible expression in cultured leaf and stem explants. *Plant Mol. Biol.* 12:41–50
203. Willemoës JG, Beltrano J, Montaldi ER. 1988. Diagravitropic growth promoted by high sucrose contents in *Paspalum vaginatum* and its reversion by GA. *Can. J. Bot.* 66:2035–37
204. Williams JHH, Winters AL, Farrar JF. 1992. Sucrose: a novel plant growth regulator. In *The Molecular, Biochemical and Physiological Aspects of Plant Respiration,* ed. H Lambers, LHW van der Plas, pp. 463–69. The Hague: Academic
205. Winters AL, Gallagher J, Pollock CJ, Farrar JF. 1995. Isolation of a gene expressed during sucrose accumulation in leaves of *Lolium temulentum* L. *J. Exp. Bot.* 46:1345–50
206. Winters AL, Williams JHH, Thomas DS, Pollock CJ. 1994. Changes in gene expression in response to sucrose accumulation in leaf tissue of *Lolium temulentum* L. *New Phytol.* 128:591–600
207. Wu LL, Mitchell JP, Cohn NS, Kaufman PB. 1993. Gibberellin (GA_3) enhances cell wall invertase activity and mRNA levels in elongating dwarf pea (*Pisum sativum*) shoots. *Int. J. Plant Sci.* 154(2): 280–89
208. Wu LL, Song I, Karuppiah N, Kaufman PB. 1993. Kinetic induction of oat shoot pulvinus invertase mRNA by gravistimulation and partial cDNA cloning by the polymerase chain reaction. *Plant Mol. Biol.* 21(6):1175–79
209. Xu J. 1995. *Cloning and molecular characterization of sugar and developmental regulation in a maize invertase gene family.* Ph.D. thesis, Univ. Florida. 123 pp.
210. Xu J, Pemberton GH, Almira EC, McCarty DR, Koch KE. 1995. The *Ivr1* gene for invertase in maize. *Plant Physiol.* 108:1293–94
211. Yang Y, Kwon HB, Peng HP, Shih MC. 1993. Stress responses and metabolic regulation of glyceraldehyde-3-phosphate dehydrogenase genes in *Arabidopsis. Plant Physiol.* 101:209–16
212. Yelenosky G, Guy CL. 1977. Carbohydrate accumulation in leaves and stems of 'Valencia' orange at progressively colder temperatures. *Bot. Gaz.* 138:13–17
213. Yokoyama R, Hirose T, Fujii N, Aspuria ET, Kato A, Uchimiya H. 1994. The rolC promoter of *Agrobacterium rhizogenes* Ri plasmid is activated by sucrose in transgenic tobacco plants. *Mol. Gen. Genet.* 244:15–22
214. Yu SM, Kuo YH, Sheu G, Sheu YJ, Liu LF. 1991. Metabolic derepression of α-amylase gene expression in suspension-cultured cells of rice. *J. Biol. Chem.* 266:21131–37
215. Zabaleta E, Oropeza A, Jiménez B, Salerno G, Crespi M, Herrera-Estrella L. 1992. Isolation and characterization of genes encoding chaperonin 60β from *Arabidopsis thaliana. Gene* 111:175–81

215a. Zamski E, Schaffer AA, eds. 1996. *Photoassimilate Distribution in Plants and Crops: Source-Sink Relationships.* New York: Dekker

216. Zinselmeier C, Westgate ME, Schussler JR, Jones RJ. 1995. Low water potential disrupts carbohydrate metabolism in maize (*Zea mays* L.) ovaries. *Plant Physiol.* 107:385–91

Annu. Rev. Plant Physiol. Plant Mol. Biol. 1996. 47:541–68

CHILLING SENSITIVITY IN PLANTS AND CYANOBACTERIA: The Crucial Contribution of Membrane Lipids

I. Nishida and N. Murata

National Institute for Basic Biology, Okazaki, 444 Japan

KEY WORDS: acyl-lipid desaturase, genetic manipulation, glycerol-3-phosphate acyltransferase, lipid molecular species, photoinhibition

ABSTRACT

The contribution of membrane lipids, particularly the level of unsaturation of fatty acids, to chilling sensitivity of plants has been intensively discussed for many years. We have demonstrated that the chilling sensitivity can be manipulated by modulating levels of unsaturation of fatty acids of membrane lipids by the action of acyl-lipid desaturases and glycerol-3-phosphate acyltransferase. This review covers recent studies on genetic manipulation of these enzymes in transgenic tobacco and cyanobacteria with special emphasis on the crucial importance of the unsaturation of membrane lipids in protecting the photosynthetic machinery from photoinhibition under cold conditions. Furthermore, we review the molecular mechanism of temperature-induced desaturation of fatty acids and introduce our hypothesis that changes in the membrane fluidity is the initial event of the expression of desaturase genes.

CONTENTS

1040-2519/96/0601-0541$08.00

INTRODUCTION

Chilling sensitivity is manifested by certain plants whose germination, growth, development of reproductive organs, and postharvest longevity are restricted within a range of chilling temperatures from 0°C (nonfreezing) to about 15°C (68, 148). Certain stages in the life cycle of a plant are more sensitive to chilling than are other stages of development. For example, seedlings appear to be more susceptible to chilling than plants at advanced stages of development (68), and the maturation of pollen is the process that is the most sensitive to chilling temperatures during the entire life cycle of chilling-sensitive plants (97, 112).

In the past several decades, extensive attention has been paid to the molecular mechanisms of the chilling sensitivity of plants (68, 148) because of the agricultural demands for improvements in the chilling tolerance of horticultural crops. In particular, the role of the unsaturation of membrane lipids in tolerance of plants to chilling has been discussed for many years. In 1973, Lyons (68) and Raison (100) first proposed the hypothesis that the thermotropic phase transition of membrane lipids might play an initiative role in the chilling sensitivity of plants. With further exposure to chilling, the phase-separated biomembranes become incapable of maintaining ionic gradients and cellular metabolism becomes disrupted, with the death of plant cells being the final result (77). The occurrence of phase separation as the initial event in chilling injury has been demonstrated in the unicellular cyanobacterium *Anacystis nidulans* (80). However, because plant cells contain high levels of polyunsaturated fatty acids in their membranes, it has been argued that the chilling-induced phase transition would not occur in lipids in intact membranes. In our previous work (76, 82), we found a positive correlation between the chilling sensitivity of herbaceous plants and the level of saturated and *trans*-monounsaturated molecular species of phosphatidylglycerol (PG), which are also termed "high-melting point molecular species" in thylakoid membranes. A similar correlation was observed in other herbaceous plants (6, 64, 105), alpine plants (18), and woody plants (137). However, such studies do not indicate more than a correlation, and there remains the question of whether these high-melting point molecular species are directly related to the chilling sensitivity of plants.

In recent years, there have been extensive advances in our understanding of the regulation of the unsaturation of the membrane lipids of plants and cyanobacteria. The successful genetic manipulation of enzymes responsible for the biosynthesis (126) and desaturation (83) of fatty acids has been achieved, as well as the isolation of mutants with defective enzymes (8, 129). In this context, the roles of the unsaturation of membrane lipids in tolerance and response to low temperature have been reexamined using transgenic and mutant strains. In this review, we briefly summarize recent progress in the genetic manipulation of acyltransferases and acyl-lipid desaturases in transgenic plants and cyanobacteria and effects of lipid changes caused by transformation on the chilling sensitivity. It appears that the unsaturation of lipids in thylakoid membranes protects the photosynthetic machinery from photoinhibition at low temperature. We also review the molecular mechanisms of desaturation of fatty acids in response to low temperature and introduce our hypothesis that changes in membrane fluidity are the initial events that lead to the expression of desaturases.

UNSATURATION OF FATTY ACIDS AND CHILLING SENSITIVITY

It has been well-established in a wide range of organisms, such as microorganisms, plants, and poikilotherms, that the level of unsaturated fatty acids in the glycerolipids of membranes changes with changes in growth temperature (14). A downward shift in growth temperature generally increases the degree of unsaturation of membrane lipids. By measuring a number of physical parameters that define the thermal motility of lipid molecules, such an increase in the degree of unsaturation of membrane lipids can be demonstrated to compensate for the decrease in the fluidity of membrane lipids that is brought about by the downward shift in temperature (15, 101). The increase in the degree of unsaturation of membrane lipids is also correlated with the sustained activity of membrane-bound enzymes at lower temperature (100, 101). Thus, the unsaturation of membrane lipids is considered to be one of the most critical parameters for the functioning of biological membranes and, therefore, for the survival of organisms at lower temperatures (15).

The above argument might, however, be invalid because low temperature induces a number of other changes in cellular metabolism, e.g. in protein synthesis (13, 34, 72) and in the levels of expression of a number of novel genes, the so-called cold-inducible genes or cold-responsive genes (10, 11, 27, 28, 44, 45, 58, 60, 61, 72a,b, 86, 92, 93, 96, 108, 113, 114, 150, 151, 153,

154). Therefore, to evaluate the role of unsaturated fatty acids in cold tolerance, it is essential to manipulate the level of unsaturated fatty acids independently of low temperature, as has recently been achieved both by manipulating the genes in question (57, 79, 83, 155) and also by creating mutant strains that are partially defective in cellular lipid metabolism (8).

ENZYMES INVOLVED IN THE REGULATION OF FATTY ACID UNSATURATION

Biosynthesis of Lipids

Before attempts can be made to manipulate the unsaturation of fatty acids, it is necessary to understand the biosynthetic pathway and the enzymes involved in the biosynthesis and desaturation of fatty acids. In plant cells, saturated fatty acids are synthesized within plastids (133, 135). The major reactions are catalyzed by members of the so-called type II fatty acid synthase system (36), which consists of multiple enzymes that are required for fatty acid biosynthesis and which requires a protein cofactor, acyl-carrier protein (ACP). In contrast with the bacterial type II fatty acid synthase system in *Escherichia coli,* the fatty acid synthase system of plastids cannot produce unsaturated fatty acids as products of reactions that lead to fatty acid synthesis de novo (69). Palmitoyl-ACP (16:0-ACP) and stearoyl-ACP (18:0-ACP) are synthesized by this system as two major and key substrates for further metabolism.

The biosynthesis of unsaturated fatty acids in plant cells has been reviewed elsewhere (8, 36, 50). In the first committed step toward the biosynthesis of C18-unsaturated fatty acids, plant cells convert 18:0-ACP to oleoyl-ACP (18:1-ACP) in a reaction catalyzed by stearoyl-ACP desaturase in plastids (50, 133). By contrast, 16:0-ACP is not a good substrate for the stearoyl-ACP desaturase. Further desaturation of 18:1(9) into polyunsaturated fatty acids is catalyzed by acyl-lipid desaturases once the fatty acids have been esterified on glycerolipids (83, 117). Therefore, in plant cells, the level of polyunsaturated fatty acids in the membrane lipids depends to a considerable extent on the activities of acyl-lipid desaturases.

Fatty acids are incorporated into glycerolipids via the stepwise esterification of glycerol 3-phosphate (106). The reaction is catalyzed by glycerol-3-phosphate acyltransferase (GPAT) and 1-acylglycerol-3-phosphate acyltransferase (106). The phosphatidic acid produced is a common precursor to cellular glycerolipids. Plant cells contain separate sets of acyltransferases in the endoplasmic reticulum (ER), chloroplasts (or plastids), and mitochondria. Once fatty acids have been esterified to the *sn*-1 and *sn*-2 positions in phosphatidic acid, they are hardly ever reorganized by the further metabolism of glycerolipids in plant cells (106). Furthermore, in plant cells, saturated fatty acids esterified on the glycerol backbone of most lipids cannot be desaturated (81). Thus, the specificities of the acyltransferases to fatty acids and to the *sn*-positions of the glycerol backbone become key factors in the regulation of the levels of unsaturated molecular species of glycerolipids. In plastids, the *sn*-2 position of glycerol 3-phosphate is exclusively esterified with 16:0 (106), and levels of high-melting-point molecular species of PG are regulated by the specificity to fatty acids of the GPAT in the plastids (23, 24).

In cyanobacterial cells, fatty acids are synthesized via a mechanism similar to that in the cells of higher plants (80). 16:0-ACP and 18:0-ACP are synthesized as end-products by the fatty acid synthase system (62, 63, 130, 131). Neither acyl-ACP desaturase nor acyl-ACP hydrolase has been found in cyanobacterial cells. The acyl groups of acyl-ACPs are directly incorporated into glycerolipids by acyltransferases and then converted to unsaturated fatty acids by acyl-lipid desaturases (83). Thus, the levels of unsaturated fatty acids in the membrane lipids of cyanobacterial cells are determined exclusively by the acyl-lipid desaturases.

Acyltransferases

GLYCEROL-3-PHOSPHATE ACYLTRANSFERASE IN THE PLASTIDS The GPAT in plastids is a nucleus-encoded stromal protein that utilizes acyl-ACP as a substrate in vivo. This enzyme in chilling-resistant plants, such as pea and spinach, is specific to 18:1-ACP (23, 24). By contrast, the enzyme in chilling-sensitive plants, such as squash, utilizes both 18:1-ACP and 16:0-ACP as its substrate with a preference for unsaturated acyl-ACP. The enzyme was first purified to homogeneity from squash cotyledons (88), and the cDNA for the precursor to this GPAT was isolated immunochemically from a cDNA expression library of squash cotyledons using antibodies raised against the purified protein (48). The gene and the cDNA of *Arabidopsis thaliana* for the precursor to GPAT were then isolated by heterologous hybridization (89). To date, cDNAs for GPAT have also been isolated from cucumber (51) and French bean (24a), two chilling-sensitive plants, and from pea (149) and spinach (49), two chilling-resistant plants.

GLYCEROL-3-PHOSPHATE ACYLTRANSFERASE FROM *ESCHERICHIA COLI* This enzyme, encoded by the *plsB* gene (65), is bound to the cytoplasmic membrane and utilizes both acyl-ACP and acyl-CoA as substrates (33). It is selective for saturated acyl-ACP as compared to unsaturated acyl-ACP. As we discuss below, the level of unsaturated molecular species of PG can be successfully manipulated when the product of the *plsB* gene is targeted to the chloroplasts of *Arabidopsis* (155).

Desaturases

Fatty acid desaturases play a central role in regulating the level of unsaturation of fatty acids in membrane lipids. The desaturases can be classified into three groups according to their specificities to the biochemical form of their fatty acid substrates, namely, acyl-lipid desaturases, acyl-ACP desaturases, and acyl-CoA desaturases (83). Acyl-lipid desaturases, which are specific to fatty acids esterified to glycerolipids (50, 117, 147), are most common in plants and cyanobacteria. They are transmembrane proteins in the ER and chloroplast membranes of plant cells (50). The desaturase in the ER uses cytochrome b_5 as the electron donor (54, 128), whereas the desaturases in chloroplasts and cyanobacteria use ferredoxin as the electron donor (119, 120, 147). We demonstrated recently by immunogold labeling and electron microscopy that acyl-lipid desaturases are located in both the plasma and thylakoid membranes of the cells of *Synechocystis* sp. PCC6803 (L Mustardy, D Los, Z Gombos & N Murata, unpublished information). Acyl-ACP desaturases are present in the stroma of plastids (70) but have not been found in cyanobacteria (62, 131). Neither plants nor cyanobacteria contains acyl-CoA desaturases (42).

Each desaturase introduces a double bond into fatty acids specifically at the specific Δ (counting from the carboxyl group) or ω (counting from the methyl end) position of the fatty-acyl chain and at specific *sn* positions of glycerides. The positional specificity of the desaturation of fatty acids in *Synechocystis* sp. PCC6803, as determined by feeding of cells with heptanoic acid (39), revealed that this cyanobacterium contains four desaturases, which are designated Δ6, Δ9, Δ12, and ω3 acyl-lipid desaturases (39).

ACYL-LIPID DESATURASES FROM CYANOBACTERIA The *desA* gene, which encodes a Δ12 acyl-lipid desaturase, was first isolated from a genomic library of *Synechocystis* sp. PCC6803 (142) by complementation (20) of a mutant, designated *Fad12* (145), that was unable to desaturate the Δ12 position of C18 fatty acids esterified at the *sn*-1 position of all classes of glycerolipids. The *desA* gene has also been isolated from other cyanobacterial strains, namely, *Synechococcus* sp. PCC7002 (111), *Synechocystis* sp. PCC6714 (111), *Anabaena variabilis* (111), and *Spirulina platensis* (78).

The *desC* gene, which encodes a Δ9 acyl-lipid desaturase, was found in the 5′ upstream region of the *desA* gene on the chromosome of *A. variabilis* (110). The *desC* gene of *Synechocystis* sp. PCC6803 was then cloned by heterologous hybridization (110). The deduced amino acid sequences of the products of those *desC* genes are about 25% identical to those of Δ9 acyl-CoA desaturases from rat (138), mouse (52, 94), and yeast (132).

The *desB* gene, which encodes an ω3 acyl-lipid desaturase, has been cloned from a *desA*-depleted strain of *Synechocystis* sp. PCC6803 by heterologous

Table 1 Genes for acyl-lipid desaturases that have been cloned from cyanobacteria

Strain	Amino acids encoded by the open reading frame	Reference
Δ12 Desaturase (*desA*)		
Synechocystis sp. PCC6803	351	Wada et al (142)
Synechocystis sp. PCC6714	349	Sakamoto et al (111)
Synechococcus sp. PCC7002	347	Sakamoto et al (111)
Anabaena variabilis	350	Sakamoto et al (111)
Spirulina platensis	351	Murata et al (78)
ω3 Desaturase (*desB*)		
Synechocystis sp. PCC6803	359	Sakamoto et al (109)
Δ9 Desaturase (*desC*)		
Synechocystis sp. PCC6803	318	Sakamoto et al (110)
Anabaena variabilis	272	Sakamoto et al (110)
Δ6 Desaturase (*desD*)		
Synechocystis sp. PCC6803	359	Reddy et al (104)
Spirulina platensis	368	Murata et al (78)

hybridization with a probe derived from the *desA* gene (109). The *desD* gene, which encodes a Δ6 desaturase from *Synechocystis* sp. PCC6803, has been cloned (104) by a gain-of-function method using *Anabaena* sp. PCC7120, which does not contain a Δ6 desaturase. Table 1 summarizes the desaturases that have been cloned from cyanobacteria.

ACYL-LIPID DESATURASES FROM HIGHER PLANTS Acyl-lipid desaturases have also been cloned from several plant sources by different techniques, such as a cloning method based on the polymerase chain reaction (PCR) (118) with a primer designed by reference to the amino acid sequence of the purified protein (121), chromosome walking (3, 47), differential screening (158, 159), and T-DNA tagging (21, 95, 157) (Table 2). In contrast with these successful results, the molecular cloning of enzymes responsible for either the Δ3-*trans*-unsaturation of 16:0 esterified at the *sn*-2 position of PG or the Δ7 (ω9) unsaturation of 16:0 esterified at the *sn*-2 position of monogalactosyldiacylglycerol has not been successful.

ACYL-ACP DESATURASES FROM HIGHER PLANTS A Δ9 acyl-ACP desaturase (or stearoyl-ACP desaturase) was partially purified from maturing safflower seeds (70) and was purified to homogeneity from avocado mesocarp (123) and developing embryos of safflower seeds (140). cDNAs encoding stearoyl-ACP desaturases have been isolated from castor bean (55, 123), saf-

Table 2 cDNAs and genes for acyl-lipid desaturases that have been cloned from higher plants

Strain	cDNA or gene	Amino acids encoded by the open reading frame	Reference
Δ12 Desaturase (ER-located, *FAD2*)			
Arabidopsis thaliana	cDNA	383	Okuley et al (95)
Glycine max	cDNA-1	387	Heppard et al (37)
	cDNA-2	387	Heppard et al (37)
Δ12 Desaturase (chloroplast-located, *FAD6*)			
Arabidopsis thaliana	cDNA	418	Falcone et al (20a)
Spinacia oleracea	cDNA	447	Schmidt et al (118)
Glycine max	cDNA	424	Hitz et al (41)
Brassica napus	cDNA	443	Hitz et al (41)
Δ15 (ω3) Desaturase (ER-located, *FAD3*)			
Vigna radiata (*ARG1*)	cDNA	380	Yamamoto (158)
Arabidopsis thaliana	gene	386	Yadav et al (157)
	cDNA	386	Yadav et al (157)
Brassica napus	cDNA-1	383	Arondel et al (3)
	cDNA-2	377	Yadav et al (157)
Glycine soja	cDNA	380	Yadav et al (157)
Δ15 (ω3) Desaturase (chloroplast-located, *FAD7* or *8*)			
Arabidopsis thaliana	gene *FAD7*	446	Iba et al (47)
	cDNA *FAD7*	446	Yadav et al (157)
	cDNA *FAD8*	435	Gibson et al (26)
Brassica napus	cDNA	>404	Yadav et al (157)
Glycine soja	cDNA	453	Yadav et al (157)

flower seeds (140), cucumber (122), spinach (87), *Brassica rapa* (56), and *Brassica napus* (127). The stearoyl-ACP desaturase of castor bean is detected as an active homodimer that contains four iron atoms when its cDNA is overexpressed in *E. coli* (22).

GENETIC MANIPULATION OF THE UNSATURATION OF MEMBRANE LIPIDS AND CHILLING SENSITIVITY

Overexpression of cDNA for Glycerol-3-Phosphate Acyltransferases in Tobacco Plants

CHANGES IN LEVEL OF UNSATURATED PG MOLECULAR SPECIES

The isolation of cDNA for the acyltransferase allows the selective manipulation of the unsaturation of fatty acids of PG. Tobacco is one of the most useful plants for the establishment of transgenic plants (38). In addition, both the chilling sensitivity and the extent of fatty acid saturation of PG in tobacco

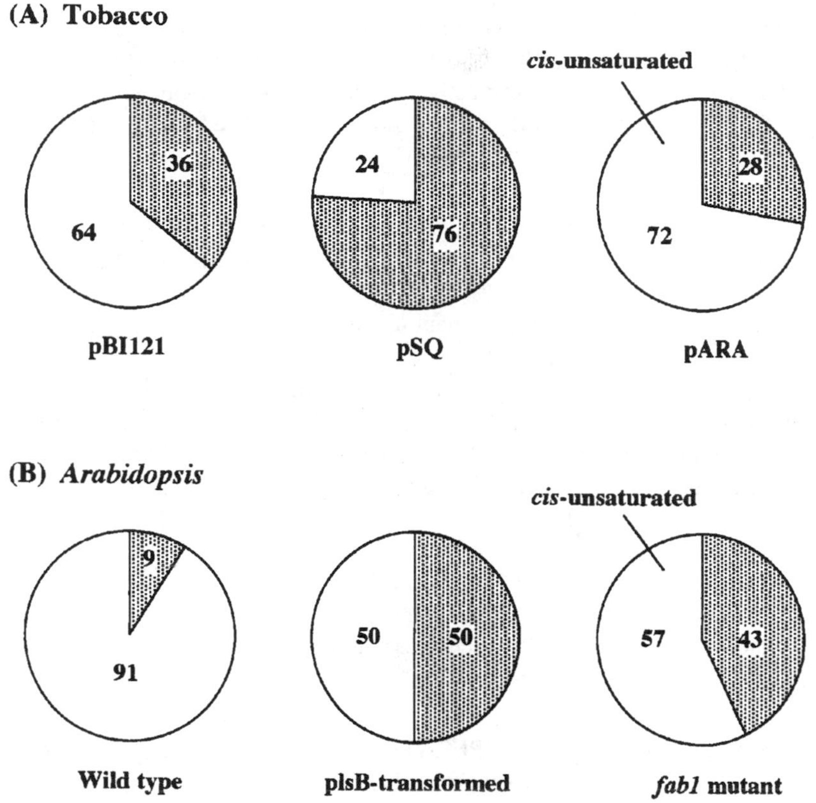

(A) Tobacco

cis-unsaturated

36
64

24
76

28
72

pBI121 pSQ pARA

(B) ***Arabidopsis***

cis-unsaturated

9
91

50
50

57
43

Wild type plsB-transformed *fab1* mutant

Figure 1 Changes in relative levels of *cis*-unsaturated molecular species of PG. (A) Transgenic tobacco plants transformed with pBI121 (control), with pSQ (cDNA for squash GPAT), and with pAR (cDNA for *Arabidopsis* GPAT). Values were calculated from the data in Reference 79. (B) *Arabidopsis* plants: wild type, a transgenic plant transformed with the *plsB* gene, and the *fab1* mutant. Values were calculated from the data in References 155 and 156. Open areas correspond to the sum of all the *cis*-unsaturated molecular species, and the shaded areas correspond to the sum of the saturated and *trans*-monounsaturated molecular species.

plants are intermediate between those of squash and *Arabidopsis*. Therefore, we transformed tobacco plants with respect to these traits by inducing the overexpression of cDNAs for the plastid-located GPATs from squash and *Arabidopsis* (79). The transformation of tobacco plants with cDNAs for GPATs from squash and *Arabidopsis* significantly altered the molecular species composition of PG (Figure 1A), but it did not significantly affect the relative composition of lipid classes or the fatty acid composition of the other glycerolipids. The level of *cis*-unsaturated molecular species of PG in tobacco leaves was 64% of the total molecular species of PG, whereas in leaves of transgenic tobacco plants transformed with pBI121 (the control plasmid), the cDNA for squash GPAT (pSQ), and the cDNA for *Arabidopsis* GPAT (pARA), it was 64%, 24%, and 72%, respectively. These results support our hypothesis that the GPAT in the plastids plays a major role in determining the level of *cis*-unsaturated molecular species of PG (76).

CHILLING SENSITIVITY In transgenic tobacco plants that had been transformed with pBI121, pSQ, and pARA, the photosynthesis of leaf discs was inhibited by 25 ±11%, 88 ±12% and 7 ±3%, respectively, during incubation for 4 h at 1°C under strong illumination. Under similar conditions, photosynthesis of spinach and squash leaf discs was inhibited by 7% and 82%, respectively. These findings demonstrate that the chilling tolerance of tobacco with respect to leaf disc photosynthesis was enhanced by transformation with pARA while it was diminished by transformation with pSQ. The chilling sensitivity of intact plants has also been estimated directly by exposing plants in a plant box to a temperature of 1°C for 10 days under continuous illumination and then incubating them at 25°C for another two days. Chlorosis was observed in leaves from control (pBI121) and pSQ-transformed tobacco plants. These results demonstrate that the chilling sensitivity of tobacco leaves can be manipulated by use of GPAT, which alters the level of unsaturated molecular species of PG in chloroplasts.

SENSITIVITY TO HIGH TEMPERATURE The effect of suppressed levels of *cis*-unsaturated molecular species of PG on the stability of the photosystem II (PSII) complex was investigated with tobacco plants that had been transformed with pBI121 and pSQ (75). No significant differences were found in the sensitivity to high temperature of the oxygen-evolving activity between the transformed plants. This result stands in marked contrast to the inferences made from results of some previous studies (98, 102), in which the heat stability of the photosynthetic machinery increased in parallel with the saturation of fatty acids in the thylakoid membrane lipids. However, because such changes in the composition of membrane lipids in these studies were achieved by altering the growth temperature, which might also affect the levels of other cellular metabolites and enzymes, the proposal of a direct link between the changes in lipids and the thermal stability of photosynthetic activities seems unjustified.

Overexpression of the plsB Gene and Mutation of Fatty Acid Elongase in Arabidopsis

Wolter et al (155) succeeded in reducing the level of *cis*-unsaturated molecular species of PG by targeting the product of the *plsB* gene of *E. coli* to the chloroplasts of *Arabidopsis.* The level of *cis*-unsaturated molecular species in PG fell from 90% in the wild-type plants to <50% in the transgenic plants (Figure 1B). In this experiment, changes were also observed in the fatty acid composition—to a minor extent—in all the other glycerolipids. *Arabidopsis* plants transformed with the *plsB* gene exhibited a chilling-sensitive phenotype when they were incubated in darkness for seven days at 4°C and then returned to 20°C. Their older leaves wilted and became brown and necrotic after two days at 20°C. The chilling-sensitive phenotype was enhanced by light, and the plants under illumination started to wilt during incubation for three to four days at 4°C and eventually died. These observations provide further evidence for the contribution of the *cis*-unsaturated molecular species of PG in chloroplasts to the chilling tolerance of higher plants.

In contrast with the result of Wolter et al (155), Wu & Browse (156) reported that, in a *fab1* mutant of *Arabidopsis* (defective in palmitoyl-ACP elongase and having, therefore, an elevated level of 16:0 in its cellular lipids), the level of *cis*-unsaturated molecular species of PG was greatly reduced from 91% in the wild-type plant to 57% in the mutant plant (Figure 1B). The wild-type plant and the *fab1* mutant showed no differences in growth and in chilling injury when they were incubated at 2°C for seven days under continuous, moderately intense illumination (156). Because chilling-sensitive plants, such as cucumber or mung bean, died under the same conditions, they implied that the level of unsaturated molecular species of PG would not be a major factor for defining the chilling resistance of plants, at least in *Arabidopsis.* However, their experiments also showed that the *fab1* mutant plant displayed retardation of growth after incubation at 2°C under illumination for more than two weeks, and that the mutant plant eventually died after incubation for more than four weeks. Under these conditions the wild-type plant survived normally. As is discussed in a later section (Molecular Mechanism of Low-Temperature Photoinhibition Regulated by the Unsaturation of Membrane Lipids), we recently found that increased levels of *cis*-unsaturated molecular species of PG in transgenic tobacco relieve the photoinhibition of the photosystem II comlex at low temperature by accelerating a recovery process from the low-temperature-induced photoinactivated state. In such experiments, the intensity of light during incubation at low temperature plays a key role. When the intensity of light is relatively low as in experiments by Wu & Browse (156), photoinhibition in vivo is not apparent.

Disruption of Genes for Desaturases in Synechocystis sp. PCC6803

DECREASE IN LEVELS OF POLYUNSATURATED FATTY ACIDS The cyanobacterium, *Synechocystis* sp. PCC6803, contains four desaturases that control the levels of polyunsaturated fatty acids in membrane lipids (84). A great advantage to using this organism is that a specific function of a gene can be defined by use of the method known as insertional mutation (152). Using this method, we created cyanobacterial strains with different levels of unsaturated fatty acids in their membrane lipids (109, 142, 144). These strains allow us to examine the roles of the unsaturation of membrane lipids in various phenomena, such as growth, chilling tolerance, and temperature acclimation (29–32, 143). Figure 2A shows a typical example of changes in levels of unsaturated molecular species of membrane lipids. In this experiment, the *desA* and *desD* genes of the wild-type strain of *Synechocystis* sp. PCC6803 were disrupted by inserting antibiotic-resistance gene cassettes (e.g. Km^r, a kanamycin-resistance gene cassette) into the genome. The results demonstrated that individual unsaturated bonds of membrane lipids in this *Synechocystis* strain can be specifically deleted by the stepwise disruption of the genes for desaturases.

GROWTH RATE The wild-type, Fad6 (defective in the expression of the *desD* gene) and Fad6/desA::Km^r cells of *Synechocystis* sp. PCC6803 grow at similar rates during the exponential phase at 34°C. At 22°C, the wild-type and Fad6 cells grow at about the same rate, but the Fad6/desA::Km^r cells grow at a markedly lower rate. These observations suggest that disruption of the *desA* gene or the elimination of the Δ12 double bond from the membrane lipids has a deleterious effect on the growth of this cyanobacterium at the low temperature (144).

CHILLING SENSITIVITY When wild-type Fad6 and Fad6/desA::Km^r cells that had been grown at 34°C were exposed to a low temperature, the extent of photoinhibition of photosynthesis, which was assayed at 34°C (31, 32, 144), depended on the extent of unsaturation of membrane lipids (Figure 3). At 10°C, the Fad6/desA::Km^r cells, which contained only monounsaturated and saturated lipid molecular species, suffered the most severe photoinhibition among the three types of cell, whereas the wild-type and Fad6 cells were indistinguishable, and levels of photoinhibition were insignificant. At room temperature, the photoinhibition was less significant than at low temperatures in all the strains, but the Fad6/desA::Km^r cells showed the highest degree of photoinhibition among the three types of cell (Figure 3) (31, 32, 144). Photosynthetic electron-transport activities were, by contrast, unaffected by the changes in the extent of unsaturation of membrane lipids (31). These results suggest that introduction of the second double bond into the membrane lipids protects the cells against photoinhibition, in particular at low temperature.

HIGH-TEMPERATURE SENSITIVITY In contrast with the severe damage

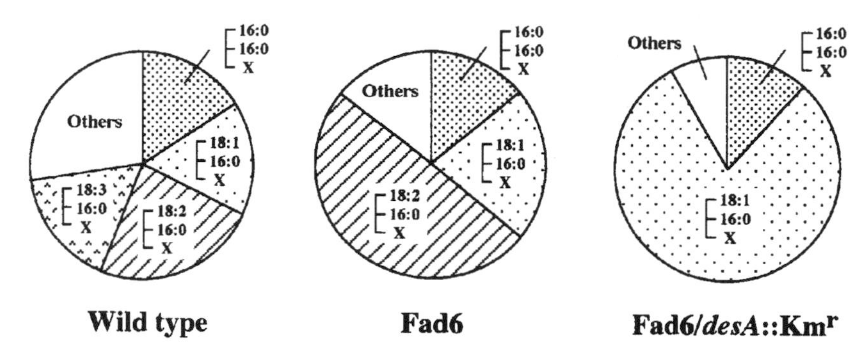

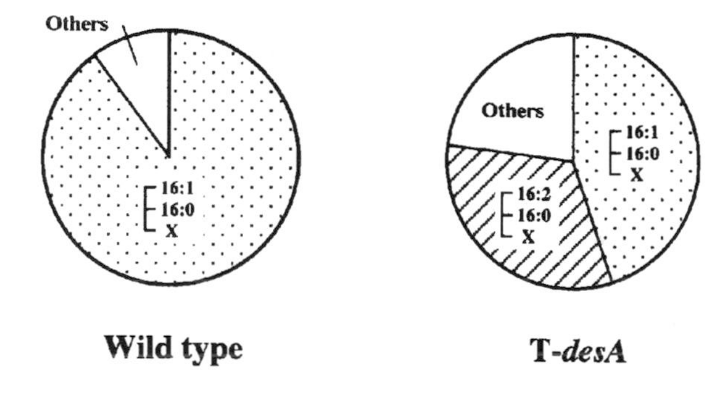

Figure 2 Changes in molecular species composition after manipulation of genes for desaturases in cyanobacteria. Reproduced from Murata & Wada (83). (A) Mutation of the Δ6 and Δ12 desaturases in *Synechocystis* sp. PCC6803 (144). (B) Introduction of the Δ12 desaturase by transformation with the *desA* gene from *Synechocystis* sp. PCC6803 (143). T-*desA,* a strain transformed with the *desA* gene.

to the photosynthetic machinery at low temperature with decreases in the unsaturation of membrane lipids, the sensitivity to high temperature of the photosynthetic machinery was barely affected or was only slightly reduced by a decrease in the unsaturation of membrane lipids (29, 143). Again, our results clearly refute the possibility that in cyanobacteria unsaturated membrane lipids might be unfavorable for high-temperature tolerance (98, 102). We have found

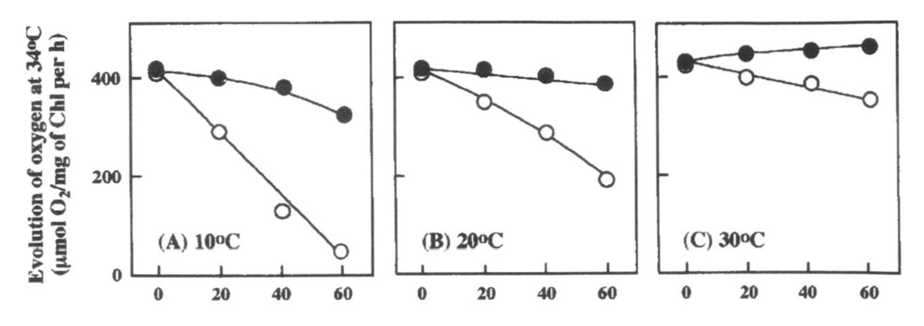

Figure 3 Photoinhibition of photosynthesis in wild-type and Fad6/desA::Km^r cells of *Synechocystis* sp. PCC6803 grown at 34°C. •–• , Wild type; ○–○, Fad6/desA::Km^r. Reproduced from Murata & Wada (83) and based on results of Gombos et al (32).

recently that the high-temperature tolerance of photosynthesis that is acquired by a change in the growth temperature is caused by protein factors (90, 91).

Expression of the Gene for a Desaturase in Synechococcus sp. PCC7942

INCREASES IN LEVELS OF POLYUNSATURATED FATTY ACIDS *Synechococcus* sp. PCC7942 (*Anacystis nidulans* strain R2) is a Group 1 strain of cyanobacteria (84), and it is unable to introduce a second double bond into monounsaturated fatty acids (84). We transformed wild-type cells of this strain with the *desA* gene of *Synechocystis* sp. PCC6803 in an attempt to increase the extent of unsaturation of membrane lipids (109, 142). Figure 2B shows the changes in the composition of lipid molecular species in cells of the wild-type strain and in cells that had been transformed with the *desA* gene. The *desA* gene endowed the *Synechococcus* strain with the novel ability to synthesize 16:2(9, 12) and 18:2(9, 12) at the *sn*-1 position.

CHILLING SENSITIVITY When wild-type cells and cells that had been transformed with a vector plasmid were exposed to temperatures below 10°C, the photosynthetic activity decreased irreversibly; more than 50% of the activity was lost during incubation at 5°C for 60 min (142, 143). The cells transformed with the *desA* gene, by contrast, did not show any significant changes in photosynthetic activity under the same conditions (142). These observations demonstrate that an elevated level of unsaturation of membrane lipids enhances the low-temperature tolerance of *Synechococcus* sp. PCC7942. This result is consistent with the observations in *Synechocystis* sp. PCC6803, in which the extent of unsaturation of membrane lipids was decreased, with a loss of low-temperature tolerance, by the disruption of genes for the desaturases.

HIGH-TEMPERATURE SENSITIVITY We have also compared wild-type and transformed cells of *Synechococcus* with respect to the high-temperature tolerance of the photosynthetic machinery and photosynthetic electron transport (143). No detectable changes accompanied the increases in the extent of unsaturation of membrane lipids.

MOLECULAR MECHANISM OF LOW-TEMPERATURE PHOTOINHIBITION REGULATED BY THE UNSATURATION OF MEMBRANE LIPIDS

For oxygenic photosynthetic organisms, light is an absolute prerequisite for photosynthesis, but, at the same time, it is harmful to the photosynthetic apparatus (2). There is a general consensus that the light-induced damage to the photosynthetic machinery is particularly serious in photosystem II (PSII) (2, 99). The photoinhibition of photosynthesis is such a major stress that the rate of light-induced inactivation of the reaction center exceeds the rate of the recovery process from the light-induced inactivated state (2). Furthermore, photoinhibition is aggravated if strong light is combined with other stresses, such as low temperature (2). Then the question arises as to how unsaturated membrane lipids might alleviate the low-temperature-induced photoinhibition. Do they suppress the light-induced inactivation process, or do they accelerate the recovery process?

Photoinhibition in Transgenic Tobacco Plants

The level of *cis*-unsaturated molecular species of PG does not affect the process of light-induced inactivation of photoinhibition in transgenic tobacco. This conclusion is based on the observation that the photoinhibition at high and low temperatures of leaf discs from both pSQ- and pBI121-transformed plants occurs at about the same rate if the recovery from the photoinhibited state is blocked by administration of lincomycin, an inhibitor of protein synthesis (75). In contrast, the recovery from the photoinhibited state does occur under weak illumination in leaf discs that have been photoinhibited to a certain level (about 80%). During the recovery period, wild-type and pBI121-transformed tobacco plants resume photosynthesis at a rate equal to 50–60% of the original rate, whereas the recovery of the pSQ-transformed plants is much slower than that of the pBI121-transformed plants (Figure 4). These results indicate that *cis*-unsaturated molecular species of PG accelerate the recovery of the photosyn-

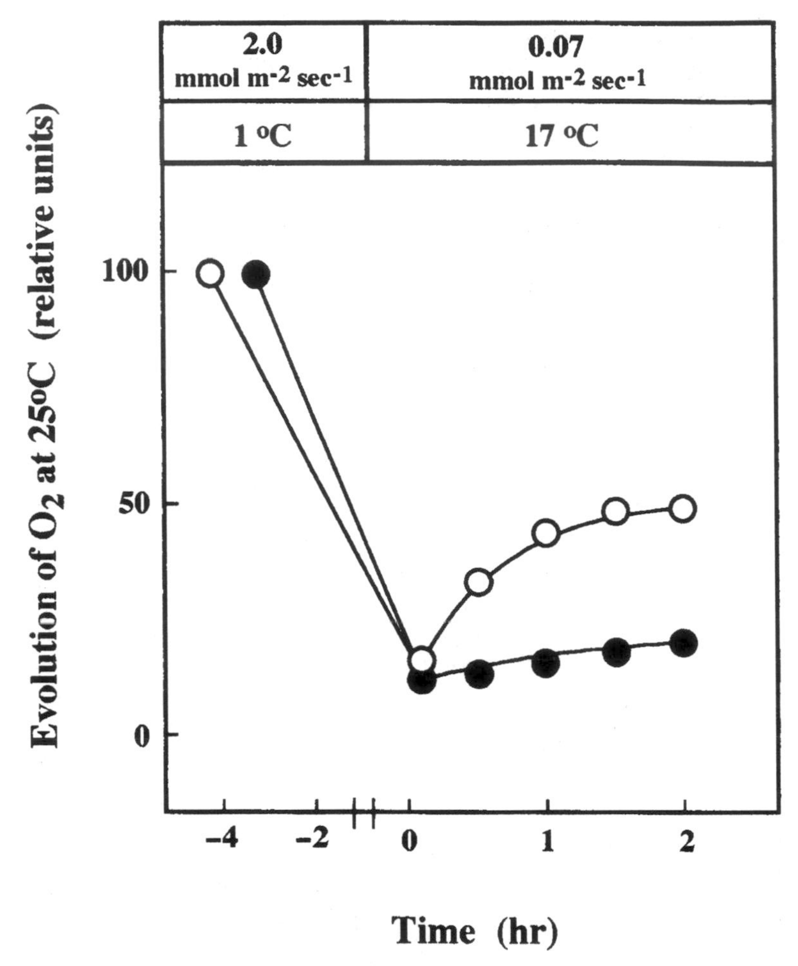

Figure 4 The recovery of the photosynthetic machinery from photoinhibition in leaves of pBI121-transformed (○–○) and pSQ-transformed plants (●–●). Reproduced from Moon et al (75).

thetic machinery from chilling-induced photoinhibition. Recently, it was reported (9) that there are no significant differences in the fluorescence parameters of photosynthesis between *Arabidopsis* plants transformed with the *plsB* gene and wild-type plants. However, in the cited study, the chilling-induced photoinhibition was not examined by dissecting the photoinactivation process and the recovery process.

Photoinhibition in Synechocystis sp. PCC6803

Results similar to those described above have been obtained with *Synechocystis* sp. PCC6803 strains with different levels of unsaturation of membrane lipids (32). Under photoinactivation conditions with recovery blocked by lincomycin, there is no significant difference among the rates of photoinhibition among different strains with different levels of unsaturation of membrane lipids (32). By contrast, under recovery conditions when the photoinactivation process is negligible, an elevated level of unsaturation of membrane lipids considerably enhances the rate of recovery from photoinhibition. Therefore, it seems very likely that the apparent enhancement of the photoinhibition of photosynthesis in intact cells at low temperature might be caused by disruption of the recovery process (32). This conclusion is consistent with that drawn from transgenic tobacco plants, and it suggests that the unsaturation of membrane lipids is definitely important for the survival of oxygenic photosynthetic organisms at low temperature.

Turnover of the D_1 Protein

According to a current hypothetical molecular mechanism of photoinhibition in vivo (1, 2), the photoinactivation process is caused by the light-induced damage to the D_1 protein of the PSII reaction center (85), and the recovery process involves several steps, as follows: (*a*) removal of the light-inactivated D_1 protein from the PSII complex by proteolysis, (*b*) synthesis de novo of the precursor to the D_1 protein (pre-D_1 protein), (*c*) sorting of the pre-D_1 protein into the PSII complex, (*d*) maturation of the pre-D_1 protein by processing, and (*e*) reconstitution of the oxygen-evolving complex and recovery of the fully active PSII complex. Therefore, we can assume that one of the above recovery steps is accelerated by the unsaturation of membrane lipids.

The proteolytic digestion of the damaged D_1 protein was investigated in wild-type, Fad6, and Fad6/desA::Km^r cells of *Synechocystis* sp. PCC6803 by immunoblotting (53). It was clear that the unsaturation of membrane lipids did not limit the removal of the damaged D_1 protein. Transcription of the *psbA* gene (the structural gene for the pre-D_1 protein) was not affected by the unsaturation of membrane lipids. It is likely that either reassembly of the pre-D_1 protein with the PSII complex or the processing of the pre-D_1 protein that yields the mature D_1 protein is accelerated by the unsaturation of membrane lipids.

CHILLING SENSITIVITY OF *ARABIDOPSIS* MUTANTS AND OTHER TRANSGENIC PLANTS

Several *Arabidopsis* mutants that are defective in lipid-synthesizing activities have been shown to exhibit an altered growth phenotype at low temperature. The *fad5* (previously designated *fadB;* defective in chloroplast *sn*-2-palmitoyl-

MGDG desaturase) and *fad6* (previously designated *fadC;* defective in chloroplast Δ12 desaturase) mutant plants are characterized, at low temperatures, by leaf chlorosis, a reduced growth rate, and changes in chloroplast morphology, such as a decrease in size and appressed regions of thylakoid membranes (46). The chilling-induced phenotypes of these mutants are similar to those of cyanobacterial strains with decreased levels of the Δ12 double bond. The *fad2* mutant (defective in microsomal Δ12 desaturase) had a greatly reduced rate of stem elongation at 12°C and died at 6°C (71).

Kodama et al (57) reported that chilling-induced damage to very young tobacco seedlings could be alleviated by a slight increase in levels of 18:3 as a result of overexpression of the *Fad7* gene of *Arabidopsis* that encodes a plastid-located ω3 desaturase. Because the changes in the extent of the unsaturation are very minor, it is very unlikely that the change in chilling sensitivity in this case is caused by a membrane-related mechanism.

TEMPERATURE ACCLIMATION

Desaturation of Membrane Lipids

It is important for living organisms to maintain the fluidity of their cell membranes at a certain level, and this requirement is achieved by changes in the levels of unsaturated fatty acids in membrane lipids that are catalyzed by fatty acid desaturases (15, 107). Several mechanisms for the regulation of low-temperature-induced desaturation of fatty acids have been reviewed, as follows (15, 25, 139). 1. The concentration of oxygen, a substrate of desaturases, limits the rate of desaturation of fatty acids in some plant systems, e.g. nonphotosynthetic plant tissues (35) and cultured sycamore cells (103). 2. The importance of the synthesis de novo of desaturases in response to low temperature was first demonstrated by Fulco and coworkers (25), who observed the low-temperature-induced synthesis of an enzyme responsible for the Δ5 desaturation of 16:0 in *Bacillus megaterium.* A similar response in terms of the synthesis of fatty acid desaturases has been found in *Tetrahymena* (139). 3. Although the second mechanism is considered the most likely in the regulation of the unsaturation of membrane lipids in *Tetrahymena,* there remains the possibility that at an early stage of the low-temperature response, the desaturase itself is activated by the decrease in the fluidity of the membrane lipids that is caused by low temperature (43, 125, 139). 4. It has also been proposed that the relative rates of biosynthesis of saturated vs unsaturated fatty acids control the levels of unsaturated fatty acids in safflower seeds (7).

Cyanobacterial cells respond to low temperature by an increase in the level of unsaturated fatty acids in membrane lipids (77, 80, 115, 146). Therefore, the regulation of the activities of acyl-lipid desaturases has been examined in

Synechocystis sp. PCC6803, from which all four acyl-lipid desaturases have been cloned. The crucial step in the regulation of the unsaturation of membrane lipids in cyanobacterial cells is the synthesis de novo of acyl-lipid desaturases. The chilling-induced desaturation of fatty acids can be inhibited by rifampicin and chloramphenicol but not by cerulenin, an inhibitor of the synthesis of fatty acids (67, 116, 146). Northern blot analyses (67) have indicated that the level of the mRNA for the Δ12 desaturase (*desA*) increases tenfold upon a decrease in ambient temperature from 36°C to 22°C (67), which suggests that the chilling-induced desaturation of membrane lipids might be caused by upregulation of the expression of the gene for the desaturase. Further studies suggested that the increase in the level of *desA* mRNA at low temperature is caused by the stabilization of the mRNA and by the enhanced transcription of the *desA* gene (D Los, M Ray & N Murata, unpublished information).

The effect of low temperature on the upregulation of expression of the *desA* gene can be mimicked by the catalytic hydrogenation of only a small portion of the plasma membrane lipids in cells of *Synechocystis* sp. PCC6803 (141). After this intervention, the level of *desA* mRNA increased tenfold within 1 h (141). This time course was similar to that observed after a shift from 36°C to 22°C. Therefore, it appears that changes in membrane fluidity could be involved in the initial event that leads to the expression of genes for desaturases in cyanobacterial cells.

The effect of temperature on the activity of desaturases has been investigated in extracts of *E. coli* cells that overexpress the product of a *desA* gene (S Pampoon, D Los & N Murata, unpublished information). The desaturase exhibited lower activity at lower temperatures, as is observed with most other enzymes.

The Signal Transduction Pathway to the Expression of Cold-Inducible Genes

Low temperature induces a number of changes in cellular metabolism. These changes include increases in the extent of unsaturation of fatty acids and phospholipid content, changes in protein composition, and induction of cold-inducible genes. Cold-inducible genes have been cloned from numerous plant species (10, 11, 27, 28, 44, 45, 58, 60, 61, 72a,b, 86, 92, 93, 96, 108, 150, 151, 153, 154) and from cyanobacteria (113, 114). The genes can be classified as *dhn-*, *lea-*, and *rab-*relabled genes [dehydrin- (5, 12), late embryogenesis-abundant- (4), and responsive to abscisic acid- (124) related genes]. Other cold-inducible genes encode polypeptides that are homologous to the product of a human tumor gene, *bbc1* (108), to a lipid transfer protein (45), and to the translation elongation factor 1α (19). In addition, RNA-binding proteins and the ribosomal protein S21 have been identified as products of cold-inducible genes in cyanobacteria (113, 114). The expression of these genes at low temperature is controlled by various mechanisms, and some of the genes are also induced by environmental stresses other than low temperature. However, all attempts to improve chilling tolerance by inducing overexpression of any one of these genes have been unsuccessful, which perhaps indicates that cooperation of products of cold-inducible genes is necessary for expression of a chilling-tolerant phenotype.

Much current attention is being focused on the way in which a low temperature is perceived by the cell and the way in which the signal is transduced to the nucleus to induce the expression of cold-inducible genes. To understand how the temperature signal might regulate the levels of transcripts of genes for the desaturases, we have to take into consideration the following results. Monroy & Dhindsa (73) demonstrated that an influx of Ca^{2+} ions from the extracellular apoplasmic space into the cytoplasm is involved in the expression of the *cas* (cold-acclimation specific) genes in alfalfa cells in suspension culture. They also examined the possible involvement of protein phosphorylation in cold acclimation (74) and have cloned several Ca^{2+}-dependent protein kinases from alfalfa by a consensus sequence-based amplification method (73). Shinozaki and his coworkers (40) succeeded in amplifying the gene for phosphatidylinositol-specific phospholipase C from *A. thaliana* by PCR. This gene is induced by cold, drought, salt, and abscisic acid.

CONCLUSIONS

1. The weight of evidence indicates that the unsaturation of membrane lipids is correlated with the chilling sensitivity of plants and cyanobacteria. The unsaturation apparently protects the photosystem II complex from low-temperature photoinhibition by accelerating the recovery of the complex from the photoinhibited state. The processing of the pre-D_1 protein to the mature D_1 protein and the sorting of the pre-D_1 protein to the thylakoid membranes are possible sites of the acceleration of the turnover cycle of the D_1 protein. It is noteworthy in this regard that acidic lipids, such as PG, are involved in the sorting of proteins to membranes in microbial systems (16, 17, 59, 66).
2. Membrane lipids are not the sole factors that regulate the chilling sensitivity of plants. At low temperatures, most plants accumulate polyols and amino acids (or amino acid derivatives) as compatible solutes, which might contribute to the chilling sensitivity of these plants. It is also possible that some other factors, such as specific proteins, are responsible for chilling tolerance.
3. The degree of chilling sensitivity of a plant depends on the stage of its life cycle. Maturation of pollen is the most chilling-sensitive of all stages. The

imbibition of seeds is also a very chilling-sensitive process. It is unclear as yet whether the unsaturation of membrane lipids is related to the chilling sensitivity of plants at these stages.

4. We have suggested that the temperature sensor of cyanobacterial cell is located in the plasma membrane. However, the sensor, the signal transducers, the details of the regulation of expression of genes for desaturases, and a role for Ca^{2+} ions in the signaling all remain mysteries that remain to be resolved.
5. Most crops of tropical origin are sensitive to chilling temperatures. This sensitivity limits the worldwide production of agricultural materials. A better understanding of the molecular mechanisms of chilling sensitivity is important if we are to improve the chilling tolerance of crops by genetic engineering.

Acknowledgments

This work was supported, in part, by Grants-in-Aid for Scientific Research on Priority Areas (Nos. 04273102 and 04273103) to N. Murata and by a grant for General Scientific Research (No. 06640854) to I. Nishida from the Ministry of Education, Science, Sports and Culture, Japan.

Literature cited

1. Aro E-M, Hundal T, Carlberg I, Andersson B. 1990. In vitro studies on light-induced inhibition of photosystem II and D_1-protein degradation at low temperature. *Biochim. Biophys. Acta* 1019: 269–75
2. Aro E-M, Virgin I, Andersson B. 1993. Photoinhibition of photosystem II: inactivation, protein damage and turnover. *Biochim. Biophys. Acta* 1143:113–34
3. Arondel V, Lemieux B, Hwang I, Gibson S, Goodman HM, Somerville CR. 1992. Map-based cloning of a gene controlling omega-3 fatty acid desaturation in *Arabidopsis*. *Science* 258: 1353–55
4. Baker J, Steele C, Dure L. 1988. Sequence and characterization of 6 *Lea* proteins and their genes from cotton. *Plant Mol. Biol.* 11:277–91
5. Bartels D, Schneider K, Terstappen G, Piatkowski D, Salamini F. 1990. Molecular cloning of abscisic acid–modulated genes which are induced during desiccation of the resurrection plant *Craterostigma plantagineum*. *Planta* 182: 27–34
6. Bishop DG. 1986. Chilling sensitivity in higher plants: the role of phosphatidylglycerol. *Plant Cell Environ.* 9: 613–16
7. Browse J, Slack CR. 1983. The effect of temperature and oxygen on the rates of fatty acid synthesis and oleate desaturation in safflower (*Carthamus tinctorius*) seed. *Biochim. Biophys. Acta* 753: 145–52
8. Browse J, Somerville CR. 1994. Glycerolipids. In *Arabidopsis*, ed. EM Meyerowitz, CR Somerville, pp. 881–912. New York: Cold Spring Harbor Lab. 1300 pp.
9. Brüggemann W, Wolter F-P. 1995. Decrease of energy-dependent quenching,

but no major changes of photosynthesis parameters in *Arabidopsis thaliana* with genetically engineered phosphatidylglycerol composition. *Plant Sci.* 108:13–21

10. Catevelli L, Bartels D. 1990. Molecular cloning and characterization of cold-regulated genes in barley. *Plant Physiol.* 93:1504–10
11. Chauvin L-P, Houde M, Sarhan F. 1993. A leaf-specific gene stimulated by light during wheat acclimation to low temperature. *Plant Mol. Biol.* 23:255–65
12. Close TJ, Kortt AA, Chandler PM. 1989. A cDNA-based comparison of dehydration-induced proteins (dehydrins) in barley and corn. *Plant Mol. Biol.* 13:95–108
13. Cooper P, Ort DR. 1988. Changes in protein synthesis induced in tomato by chilling. *Plant Physiol.* 88:454–61
14. Cossins AR. 1994. Homeoviscous adaptation of biological membranes and its functional significance. See Ref. 15, pp. 63–76
15. Cossins AR, ed. 1994. *Temperature Adaptation of Biological Membranes*. London: Portland. 227 pp.
16. de Vrije GJ, Batenburg AM, Killian JA, de Kruijff B. 1990. Lipid involvement in protein translocation in *Escherichia coli*. *Mol. Microbiol.* 4:143–50
17. de Vrije T, de Swart RL, Dowhan W, Tommassen J, de Kruijff B. 1988. Phosphatidylglycerol is involved in protein translocation across *Escherichia coli* inner membranes. *Nature* 334:173–75
18. Dorne A-J, Cadel G, Douce R. 1986. Polar lipid composition of leaves from nine typical alpine species. *Phytochemistry* 25:65–68
19. Dunn MA, Morris A, Jack PL, Hughes MA. 1993. A low temperature–responsive translation elongation factor 1α from barley (*Hordeum vulgare L.*). *Plant Mol. Biol.* 23:221–25
20. Dzelzkalns VA, Bogorad L. 1988. Molecular analysis of a mutant defective in photosynthetic oxygen evolution and isolation of a complementing clone by a novel screening procedure. *EMBO J.* 7:333–38

20a. Falcone DL, Gibson S, Lemieux B, Somerville C. 1994. Identification of a gene that complements an *Arabidopsis* mutant deficient in chloroplast ω6 desaturase activity. *Plant Physiol.* 106: 1453–38

21. Feldmann KA. 1991. T-DNA insertion mutagenesis in *Arabidopsis*: mutational spectrum. *Plant J.* 1:71–82
22. Fox BG, Shanklin J, Somerville CR, Münck E. 1993. Stearoyl-acyl carrier protein Δ^9 desaturase from *Ricinus communis* is a diiron-oxo protein. *Proc. Natl. Acad. Sci. USA* 90:2486–90
23. Frentzen M, Heinz E, McKeon TA, Stumpf PK. 1983. Specificities and selectivities of glycerol-3-phosphate acyltransferase from pea and spinach chloroplasts. *Eur. J. Biochem.* 129:629–36
24. Frentzen M, Nishida I, Murata N. 1987. Properties of the plastidial acyl-(acyl-carrier-protein):glycerol-3-phosphate acyltransferase from the chilling-sensitive plant squash (*Cucurbita moschata*). *Plant Cell Physiol.* 28:1195–201

24a. Fritz M, Heinz E, Wolter FP. 1995. Cloning and sequencing of a full-length cDNA coding for *sn*-glycerol-3-phosphate acyltransferase from *Phaseolus vulgaris*. *Plant Physiol.* 107:1039–40

25. Fulco AJ. 1969. The biosynthesis of unsaturated fatty acids by bacilli. I. Temperature induction of the desaturation reaction. *J. Biol. Chem.* 244:889–95
26. Gibson S, Arondel V, Iba K, Somerville CR. 1994. Cloning of a temperature-regulated gene encoding a chloroplast omega-3 desaturase from *Arabidopsis thaliana*. *Plant Physiol.* 106:1615–21
27. Gilmour SJ, Artus NN, Thomashow MF. 1992. cDNA sequence analysis and expression of two cold-regulated genes of *Arabidopsis thaliana*. *Plant Mol. Biol.* 18:13–21
28. Goddard NJ, Dunn MA, Zhang L, White AJ, Jack PL, Hughes MA. 1993. Molecular analysis and spatial expression pattern of a low-temperature-specific barley gene, *blt101*. *Plant Mol. Biol.* 23:871–79
29. Gombos Z, Wada H, Hideg E, Murata N. 1994. The unsaturation of membrane lipids stabilizes photosynthesis against heat stress. *Plant Physiol.* 104:563–67
30. Gombos Z, Wada H, Murata N. 1991. Direct evaluation of effects of fatty-acid unsaturation on the thermal properties of photosynthetic activities, as studied by mutation and transformation of *Synechocystis* PCC6803. *Plant Cell Physiol.* 32:205–11
31. Gombos Z, Wada H, Murata N. 1992. Unsaturation of fatty acids in membrane lipids enhances tolerance of the cyanobacterium *Synechocystis* PCC6803 to low-temperature photoinhibition. *Proc. Natl. Acad. Sci. USA* 89:9959–63
32. Gombos Z, Wada H, Murata N. 1994. The recovery of photosynthesis from low-temperature photoinhibition is accelerated by the unsaturation of membrane lipids: a mechanism of chilling tolerance. *Proc. Natl. Acad. Sci. USA* 91:8787–91

33. Green PR, Merrill AH Jr, Bell RM. 1981. Membrane phospholipid synthesis in *Escherichia coli*: purification, reconstitution, and characterization of *sn*-glycerol-3-phosphate acyltransferase. *J. Biol. Chem.* 256:11151–59
34. Guy CL, Niemi KJ, Brambl R. 1985. Altered gene expression during cold acclimation of spinach. *Proc. Natl. Acad. Sci. USA* 82:3673–77
35. Harris P, James AT. 1969. The effect of low temperatures on fatty acid biosynthesis in plants. *Biochem. J.* 112: 325–30
36. Harwood JL. 1988. Fatty acid metabolism. *Annu. Rev. Plant Physiol. Plant Mol. Biol.* 39:101–38
37. Heppard EP, Kinney AJ, Stecca KL, Miao G-H. 1995. Developmental and growth temperature regulation of two different microsomal ω6 desaturase genes in soybeans. *Plant Physiol.* 110: 311–19
38. Herrera-Estrella L, Simpson J. 1988. Foreign gene expression in plants. In *Plant Molecular Biology. A Practical Approach*, ed. CH Shaw, pp. 131–60. Oxford: IRL Press. 313 pp.
39. Higashi S, Murata N. 1993. An in vivo study of substrate specificities of acyl-lipid desaturases and acyltransferases in lipid synthesis in *Synechocystis* PCC6803. *Plant Physiol.* 102:1275–78
40. Hirayama T, Ohto C, Mizoguchi T, Shinozaki K. 1995. A gene encoding a phosphatidylinositol-specific phospholipase C is induced by dehydration and salt stress in *Arabidopsis thaliana*. *Proc. Natl. Acad. Sci. USA* 92:3903–7
41. Hitz WD, Carlson TJ, Booth JR Jr, Kinney AJ, Stecca KL, Yadav NS. 1994. Cloning of a higher-plant plastid ω-6 fatty acid desaturase cDNA and its expression in a cyanobacterium. *Plant Physiol.* 105:635–41
42. Holloway PW. 1983. Fatty acid desaturation. In *The Enzymes*, ed. PD Boyer, 16:63–83. New York: Academic. 783 pp.
43. Horváth I, Török Z, Vigh L, Kates M. 1991. Lipid hydrogenation induces elevated 18:1-CoA desaturase activity in *Candida lipolytica* microsomes. *Biochim. Biophys. Acta* 1085:126–30
44. Houde M, Danyluk J, Laliberte J-F, Rassart E, Dhindsa RS, Sarhan F. 1992. Cloning, characterization, and expression of a cDNA encoding a 50-kilodalton protein specifically induced by cold acclimation in wheat. *Plant Physiol.* 99: 1381–87
45. Hughes MA, Dunn MA, Pearce RS, White AJ, Zhang L. 1992. An abscisic-acid-responsive, low temperature barley gene has homology with a maize phospholipid transfer protein. *Plant Cell Environ.* 15:861–65
46. Hugly S, Somerville CR. 1992. A role for membrane lipid polyunsaturation in chloroplast biogenesis at low temperature. *Plant Physiol.* 99:197–202
47. Iba K, Gibson S, Nishiuchi T, Fuse T, Nishimura M, et al. 1993. A gene encoding a chloroplast ω-3 fatty acid desaturase complements alterations in fatty acid desaturation and chloroplast copy number of the *fad7* mutant of *Arabidopsis thaliana*. *J. Biol. Chem.* 268:24099–105
48. Ishizaki O, Nishida I, Agata K, Eguchi G, Murata N. 1988. Cloning and nucleotide sequence of cDNA for the plastid glycerol-3-phosphate acyltransferase from squash. *FEBS Lett.* 238:424–30
49. Ishizaki-Nishizawa O, Azuma M, Ohtani T, Murata N, Toguri T. 1995. Nucleotide sequence of cDNA from *Spinacia oleracea* encoding plastid glycerol-3-phosphate acyltransferase. *Plant Physiol.* 108:1342
50. Jaworski JG. 1987. Biosynthesis of monoenoic and polyenoic fatty acids. See Ref. 136, pp. 159–74
51. Johnson TC, Schneider JC, Somerville CR. 1992. Nucleotide sequence of acyl-acyl carrier protein: glycerol-3-phosphate acyltransferase from cucumber. *Plant Physiol.* 99:771–72
52. Kaestner KH, Ntambi JM, Kelly TJ Jr, Lane MD. 1989. Differentiation-induced gene expression in 3T3-L1 preadipocytes: a second differentially expressed gene encoding stearoyl-CoA desaturase. *J. Biol. Chem.* 264:14755–61
53. Kanervo E, Aro E-M, Murata N. 1995. Low unsaturation level of thylakoid membrane lipids limits turnover of the D1 protein of photosystem II at high irradiance. *FEBS Lett.* 364:239–42
54. Kearns EV, Hugly S, Somerville CR. 1991. The role of cytochrome b_5 in Δ12 desaturation of oleic acid by microsomes of safflower (*Carthamus tinctorius* L.). *Arch. Biochem. Biophys.* 284: 431–36
55. Knutzon DS, Scherer DE, Schreckengost WE. 1991. Nucleotide sequence of a complementary DNA clone encoding stearoyl-acyl carrier protein desaturase from castor bean, *Ricinus communis*. *Plant Physiol.* 96:344–45
56. Knutzon DS, Thompson GA, Radke SE, Johnson WB, Knauf VC, Kridl JC. 1992. Modification of *Brassica* seed oil by antisense expression of a stearoyl-acyl

carrier protein desaturase gene. *Proc. Natl. Acad. Sci.* USA 89:2624–28

57. Kodama H, Hamada T, Horiguchi G, Nishimura M, Iba K. 1994. Genetic enhancement of cold tolerance by expression of a gene for chloroplast ω-3 fatty acid desaturase in transgenic tobacco. *Plant Physiol.* 105:601–5
58. Kurkela S, Franck M. 1990. Cloning and characterization of a cold- and ABA-inducible *Arabidopsis* gene. *Plant Mol. Biol.* 15:137–44
59. Kusters R, Dowhan W, de Kruijff B. 1991. Negatively charged phospholipids restore prePhoE translocation across phosphatidylglycerol-depleted *Escherichia coli* inner membranes. *J. Biol. Chem.* 266:8659–62
60. Laberge S, Castonguay Y, Vézina L-P. 1993. New cold- and drought-regulated gene from *Medicago sativa. Plant Physiol.* 101:1411–12
61. Lång V, Palva ET. 1992. The expression of a rab-related gene, *rab18*, is induced by abscisic acid during the cold-acclimation process of *Arabidopsis thaliana* (L.) Heynh. *Plant Mol. Biol.* 20:951–62
62. Lem NW, Stumpf PK. 1984. In vitro fatty acid synthesis and complex lipid metabolism in the cyanobacterium *Anabaena variabilis*. I. Some characteristics of fatty acid synthesis. *Plant Physiol.* 74:134–38
63. Lem NW, Stumpf PK. 1984. In vitro fatty acid synthesis and complex lipid metabolism in the cyanobacterium *Anabaena variabilis*. II. Acyltransfer and complex lipid formation. *Plant Physiol.* 75:700–4
64. Li T, Lynch DV, Steponkus PL. 1987. Molecular species composition of phosphatidylglycerols from rice varieties differing in chilling sensitivity. *Cryo-Lett.* 8:314–21
65. Lightner VA, Bell RM, Modrich P. 1983. The DNA sequences encoding *plsB* and *dgk* loci of *Escherichia coli. J. Biol. Chem.* 258:10856–61
66. Lill R, Dowhan W, Wickner W. 1990. The ATPase activity of SecA is regulated by acidic phospholipids, SecY, and the leader and mature domains of precursor proteins. *Cell* 60:271–80
67. Los DA, Horváth I, Vigh L, Murata N. 1993. The temperature-dependent expression of the desaturase gene *desA* in *Synechocystis* PCC6803. *FEBS Lett.* 318: 57–60
68. Lyons JM. 1973. Chilling injury in plants. *Annu. Rev. Plant Physiol.* 24: 445–66
69. Magnuson K, Jackowski S, Rock CO, Cronan JE Jr. 1993. Regulation of fatty acid biosynthesis in *Escherichia coli. Microbiol. Rev.* 57:522–42
70. McKeon TA, Stumpf PK. 1982. Purification and characterization of the stearoyl-acyl carrier protein desaturase and the acyl-acyl carrier protein thioesterase from maturing seeds of safflower. *J. Biol. Chem.* 257:12141–47
71. Miquel M, James D, Dooner H, Browse J. 1993. *Arabidopsis* requires polyunsaturated lipids for low-temperature survival. *Proc. Natl. Acad. Sci. USA* 90: 6208–12
72. Mohapatra SS, Poole RJ, Dhindsa RS. 1987. Changes in protein patterns and translatable messenger RNA populations during cold acclimation of alfalfa. *Plant Physiol.* 84:1172–76

72a. Mohapatra SS, Poole RJ, Dhindsa RS. 1988. Abscisic acid–regulated gene expression in relation to freezing tolerance in alfalfa. *Plant Physiol.* 89:468–74

72b. Mohapatra SS, Wolfraim L, Poole RJ, Dhindsa RS. 1989. Molecular cloning and relationship to freezing tolerance of cold acclimation-specific genes of alfalfa. *Plant Physiol.* 89:375–80

73. Monroy AF, Dhindsa RS. 1995. Low-temperature signal transduction: induction of cold acclimation–specific genes of alfalfa by calcium at 25°C. *Plant Cell* 7:321–31
74. Monroy AF, Sarhan F, Dhindsa RS. 1993. Cold-induced changes in freezing tolerance, protein phosphorylation, and gene expression: evidence for a role of calcium. *Plant Physiol.* 102:1227–35
75. Moon BY, Higashi S, Gombos Z, Murata N. 1995. Unsaturation of the membrane lipids of chloroplasts stabilizes the photosynthetic machinery against low-temperature photoinhibition in transgenic tobacco plants. *Proc. Natl. Acad. Sci. USA* 92:6219–23
76. Murata N. 1983. Molecular species composition of phosphatidylglycerols from chilling-sensitive and chilling-resistant plants. *Plant Cell Physiol.* 24:81–86
77. Murata N. 1989. Low-temperature effects on cyanobacterial membranes. *J. Bioenerg. Biomembr.* 21:61–75
78. Murata N, Deshnium P, Tasaka Y. 1996. Biosynthesis of gamma-linoleic acid in the cyanobacterium *Spirulina platensis*. In *Gamma-Linolenic Acid: Metabolism and Role in Nutrition and Medicine.* Champaign, IL: Am. Oil Chem. Soc. Press. In press
79. Murata N, Ishizaki-Nishizawa O, Higashi S, Hayashi H, Tasaka Y, Nishida I. 1992. Genetically engineered altera-

tion in the chilling sensitivity of plants. *Nature* 356:710–13
80. Murata N, Nishida I. 1987. Lipids of blue-green algae (cyanobacteria). See Ref. 136, pp. 315–47
81. Murata N, Nishida I. 1990. Lipids in relation to chilling sensitivity of plants. See Ref. 148, pp. 181–99
82. Murata N, Sato N, Takahashi N, Hamazaki Y. 1982. Compositions and positional distributions of fatty acids in phospholipids from leaves of chilling-sensitive and chilling-resistant plants. *Plant Cell Physiol.* 23:1071–79
83. Murata N, Wada H. 1995. Acyl-lipid desaturases and their importance in the tolerance and acclimatization to cold of cyanobacteria. *Biochem. J.* 308:1–8
84. Murata N, Wada H, Gombos Z. 1992. Modes of fatty-acid desaturation in cyanobacteria. *Plant Cell Physiol.* 33:933–41
85. Nanba O, Satoh K. 1987. Isolation of a photosystem II reaction center consisting of D-1 and D-2 polypeptides and cytochrome *b*-559. *Proc. Natl. Acad. Sci. USA* 84:109–12
86. Neven L, Haskell DW, Hofig A, Li Q-B, Guy CL. 1993. Characterization of a spinach gene responsive to low temperature and water stress. *Plant Mol. Biol.* 21:291–305
87. Nishida I, Beppu T, Matsuo T, Murata N. 1992. Nucleotide sequence of a cDNA clone encoding a precursor to stearoyl-(acyl-carrier-protein) desaturase from spinach, *Spinacia oleracea*. *Plant Mol. Biol.* 19:711–13
88. Nishida I, Frentzen M, Ishizaki O, Murata N. 1987. Purification of isomeric forms of acyl-[acyl-carrier-protein]: glycerol-3-phosphate acyltransferase from greening squash cotyledons. *Plant Cell Physiol.* 28:1071–79
89. Nishida I, Tasaka Y, Shiraishi H, Murata N. 1993. The gene and the RNA for the precursor to the plastid-located glycerol-3-phosphate acyltransferase of *Arabidopsis thaliana*. *Plant Mol. Biol.* 21:267–77
90. Nishiyama Y, Hayashi H, Watanabe T, Murata N. 1994. Photosynthetic oxygen evolution is stabilized by cytochrome c550 against heat inactivation in *Synechococcus* sp. PCC7002. *Plant Physiol.* 105:1313–19
91. Nishiyama Y, Kovács E, Lee CB, Hayashi H, Watanabe T, Murata N. 1993. Photosynthetic adaptation to high temperature associated with thylakoid membranes of *Synechococcus* PCC7002. *Plant Cell Physiol.* 34:337–43
92. Nordin K, Heino P, Palva ET. 1991. Separate signal pathways regulate the expression of a low-temperature-induced gene in *Arabidopsis thaliana* (L.) Heynh. *Plant Mol. Biol.* 16:1061–71
93. Nordin K, Vahala T, Palva ET. 1993. Differential expression of two related, low-temperature-induced genes in *Arabidopsis thaliana* (L.) Heynh. *Plant Mol. Biol.* 21:641–53
94. Ntambi JM, Buhrow SA, Kaestner KH, Christy RJ, Sibley E, et al. 1988. Differentiation-induced gene expression in 3T3-L1 preadipocytes: characterization of a differentially expressed gene encoding stearoyl-CoA desaturase. *J. Biol. Chem.* 263:17291–300
95. Okuley J, Lightner J, Feldmann K, Yadav N, Lark E, Browse J. 1994. *Arabidopsis FAD2* gene encodes the enzyme that is essential for polyunsaturated lipid synthesis. *Plant Cell* 6:147–58
96. Orr W, Iu B, White TC, Robert LS, Singh J. 1992. Complementary sequence of a low temperature-induced *B. napus* gene with homology to the *A. thaliana kin1* gene. *Plant Physiol.* 98:1532–34
97. Patterson BD, Mutton L, Paull RE, Nguyen VQ. 1987. Tomato pollen development: stages sensitive to chilling and a natural environment for the selection of resistant genotypes. *Plant Cell Environ.* 10:363–68
98. Pearcy RW. 1978. Effect of growth temperature on the fatty acid composition of the leaf lipids in *Atriplex lentiformis* (Torr.) Wats. *Plant Physiol.* 61:484–86
99. Powles SB. 1984. Photoinhibition of photosynthesis induced by visible light. *Annu. Rev. Plant Physiol.* 35:15–44
100. Raison JK. 1973. The influence of temperature-induced phase changes on kinetics of respiratory and other membrane-associated enzymes. *J. Bioenerg.* 4:258–309
101. Raison JK. 1980. Membrane lipids: structure and function. See Ref. 134, pp. 57–83
102. Raison JK, Roberts JKM, Berry JA. 1982. Correlations between the thermal stability of chloroplast (thylakoid) membranes and the composition and fluidity of their polar lipids upon acclimation of the higher plant, *Nerium oleander*, to growth. *Biochim. Biophys. Acta* 688:218–28
103. Rebeille F, Bligny R, Douce R. 1980. Oxygen and temperature effects on the fatty acid composition of sycamore cells (*Acer pseudoplatanus* L.). *Biochim. Biophys. Acta* 620:1–9
104. Reddy AS, Nuccio ML, Gross LM,

Thomas TL. 1993. Isolation of a Δ^6-desaturase gene from the cyanobacterium *Synechocystis* sp. strain PCC6803 by gain-of-function expression in *Anabaena* sp. strain PCC7120. *Plant Mol. Biol.* 22:293–300

105. Roughan PG. 1985. Phosphatidylglycerol and chilling sensitivity in plants. *Plant Physiol.* 77:740–46
106. Roughan PG, Slack CR. 1982. Cellular organization of glycerolipid metabolism. *Annu. Rev. Plant Physiol.* 33:97–132
107. Russell NJ. 1984. Mechanisms of thermal adaptation in bacteria: blueprint for survival. *Trends Biochem. Sci.* 9:108–12
108. Sáez-Vásquez J, Raynal M, Meza-Basso L, Delseny M. 1993. Two related, low-temperature-induced genes from *Brassica napus* are homologous to the human tumor *bbc1* (breast basic conserved) gene. *Plant Mol. Biol.* 23:1211–21
109. Sakamoto T, Los DA, Higashi S, Wada H, Nishida I, et al. 1994. Cloning of ω3 desaturase from cyanobacteria and its use in altering the degree of membrane-lipid unsaturation. *Plant Mol. Biol.* 26:249–64
110. Sakamoto T, Wada H, Nishida I, Ohmori M, Murata N. 1994. Δ9 acyl-lipid desaturases of cyanobacteria: molecular cloning and substrate specificities in terms of fatty acids, *sn*-positions, and polar head groups. *J. Biol. Chem.* 269: 25576–80
111. Sakamoto T, Wada H, Nishida I, Ohmori M, Murata N. 1994. Identification of conserved domains in the Δ12 desaturases of cyanobacteria. *Plant Mol. Biol.* 24:643–50
112. Satake T, Koike S. 1983. Sterility caused by cooling treatment at the flowering stage in rice plants. *Jpn. J. Crop Sci.* 52:207–13
113. Sato N. 1994. A cold-regulated cyanobacterial gene cluster encodes RNA-binding protein and ribosomal protein S21. *Plant Mol. Biol.* 24:819–23
114. Sato N. 1995. A family of cold-regulated RNA-binding protein genes in the cyanobacterium *Anabaena variabilis* M3. *Nucleic Acids Res.* 23:2161–67
115. Sato N, Murata N. 1980. Temperature shift–induced response in lipids in the blue-green alga, *Anabaena variabilis*: the central role of diacylmonogalactosylglycerol in thermo-adaptation. *Biochim. Biophys. Acta* 619:353–66
116. Sato N, Murata N. 1981. Studies on the temperature shift–induced desaturation of fatty acids in monogalactosyl diacylglycerol in the blue-green alga (cyanobacterium), *Anabaena variabilis*. *Plant Cell Physiol.* 22:1043–50
117. Sato N, Seyama Y, Murata N. 1986. Lipid-linked desaturation of palmitic acid in monogalactosyl diacylglycerol in the blue-green alga (cyanobacterium) *Anabaena variabilis* studied in vivo. *Plant Cell Physiol.* 27:819–35
118. Schmidt H, Dresselhaus T, Buck F, Heinz E. 1993. Purification and PCR-based cDNA cloning of a plastidial n-6 desaturase. *Plant Mol. Biol.* 26:631–42
119. Schmidt H, Heinz E. 1990. Desaturation of oleoyl groups in envelope membranes from spinach chloroplasts. *Proc. Natl. Acad. Sci. USA* 87:9477–80
120. Schmidt H, Heinz E. 1990. Involvement of ferredoxin in desaturation of lipid-bound oleate in chloroplasts. *Plant Physiol.* 94:214–20
121. Schmidt H, Heinz E. 1993. Direct desaturation of intact galactolipids by a desaturase solubilized from spinach (*Spinacia oleracea*) chloroplast envelopes. *Biochem. J.* 289:777–82
122. Shanklin J, Mullins C, Somerville CR. 1991. Sequence of a complementary DNA from *Cucumis sativus* L. encoding the stearoyl-acyl-carrier protein desaturase. *Plant Physiol.* 97:467–68
123. Shanklin J, Somerville CR. 1991. Stearoyl-acyl-carrier-protein desaturase from higher plants is structurally unrelated to the animal and fungal homologs. *Proc. Natl. Acad. Sci. USA* 88:2510–14
124. Skriver K, Mundy J. 1990. Gene response to ABA and osmotic stress. *Plant Cell* 2:503–12
125. Skriver L, Thompson GA Jr. 1979. Temperature-induced changes in fatty acid unsaturation of *Tetrahymena* membranes do not require induced fatty acid desaturase synthesis. *Biochim. Biophys. Acta* 572:376–81
126. Slabas AR, Fawcett T, Grithiths G, Stobart K. 1993. Biochemistry and molecular biology of lipid biosynthesis in plants: potential for genetic manipulation. In *Biosynthesis and Manipulation of Plant Products*, ed. D Grierson, 3: 104–38. London: Blackie Acad. Prof. 253 pp.
127. Slocombe SP, Cummins I, Jarvis RP, Murphy DJ. 1992. Nucleotide sequence and temporal regulation of a seed-specific *Brassica napus* cDNA encoding a stearoyl-acyl carrier protein (ACP) desaturase. *Plant Mol. Biol.* 20:151–55
128. Smith MA, Cross AR, Jones OTG, Griffiths WT, Stymne S, Stobart K. 1990. Electron-transport components of the 1-acyl-2-oleoyl-*sn*-glycerol-3-phosphoc holine Δ^{12}-desaturase (Δ^{12}-desaturase)

in microsomal preparations from developing safflower (*Carthamus tinctorius* L.) cotyledons. *Biochem. J.* 272:23–29
129. Somerville CR, Browse J. 1991. Plant lipids: metabolism, mutant, and membranes. *Science* 252:80–87
130. Stapleton SR, Jaworski JG. 1984. Characterization and purification of malonyl-coenzyme A:[acyl-carrier-protein] transacylases from spinach and *Anabaena variabilis*. *Biochim. Biophys. Acta* 794: 240–48
131. Stapleton SR, Jaworski JG. 1984. Characterization of fatty acid biosynthesis in the cyanobacterium *Anabaena variabilis*. *Biochim. Biophys. Acta* 794:249–55
132. Stukey JE, McDonough VM, Martin CE. 1990. The *OLE1* gene of *Saccharomyces cerevisiae* encodes the Δ9 fatty acid desaturase and can be functionally replaced by the rat stearoyl-CoA desaturase gene. *J. Biol. Chem.* 265: 20144–49
133. Stumpf PK. 1980. Biosynthesis of saturated and unsaturated fatty acids. See Ref. 134, pp. 177–204
134. Stumpf PK, ed. 1980. *The Biochemistry of Plants*, Vol. 4. New York: Academic. 693 pp.
135. Stumpf PK. 1987. The biochemistry of saturated fatty acids. See Ref. 136, pp. 121–36
136. Stumpf PK, ed. 1987. *The Biochemistry of Plants*, Vol. 9. Orlando: Academic. 363 pp.
137. Tasaka Y, Nishida I, Higashi S, Beppu T, Murata N. 1990. Fatty acid composition of phosphatidylglycerols in relation to chilling sensitivity of woody plants. *Plant Cell Physiol.* 31:545–50
138. Thiede MA, Ozols J, Strittmatter P. 1986. Construction and sequence of cDNA for rat liver stearoyl coenzyme A desaturase. *J. Biol. Chem.* 261:13230–35
139. Thompson GA Jr, Nozawa Y. 1984. The regulation of membrane fluidity in *Tetrahymena*. In *Biomembranes,* Vol. 12, *Membrane Fluidity*, ed. M Kates, LA Manson, pp. 387–432. New York: Plenum. 693 pp.
140. Thompson GA, Scherer DE, Aken SF-V, Kenny JW, Young HL, et al. 1991. Primary structures of the precursor and mature forms of stearoyl-acyl carrier protein desaturase from safflower embryos and requirement of ferredoxin for enzyme acitivity. *Proc. Natl. Acad. Sci. USA* 88:2578–82
141. Vigh L, Los DA, Horváth I, Murata N. 1993. The primary signal in the biological perception of temperature: Pd-catalyzed hydrogenation of membrane lipids stimulated the expression of the *desA* gene in *Synechocystis* PCC6803. *Proc. Natl. Acad. Sci. USA* 90:9090–94
142. Wada H, Gombos Z, Murata N. 1990. Enhancement of chilling tolerance of a cyanobacterium by genetic manipulation of fatty acid desaturation. *Nature* 347:200–3
143. Wada H, Gombos Z, Murata N. 1994. Contribution of membrane lipids to the ability of the photosynthetic machinery to tolerate temperature stress. *Proc. Natl. Acad. Sci. USA* 91:4273–77
144. Wada H, Gombos Z, Sakamoto T, Murata N. 1992. Genetic manipulation of the extent of desaturation of fatty acids in membrane lipids in the cyanobacterium *Synechocystis* PCC6803. *Plant Cell Physiol.* 33:535–40
145. Wada H, Murata N. 1989. *Synechocystis* PCC6803 mutants defective in desaturation of fatty acids. *Plant Cell Physiol.* 30:971–78
146. Wada H, Murata N. 1990. Temperature-induced changes in the fatty acid composition of the cyanobacterium, *Synechocystis* PCC6803. *Plant Physiol.* 92: 1062–69
147. Wada H, Schmidt H, Heinz E, Murata N. 1993. In vitro ferredoxin-dependent desaturation of fatty acids in cyanobacterial thylakoid membranes. *J. Bacteriol.* 175:544–47
148. Wang CY, ed. 1990. *Chilling Injury of Horticultural Crops*. Boca Raton, FL: CRC. 313 pp.
149. Weber S, Wolter F-P, Buck F, Frentzen M, Heinz E. 1991. Purification and cDNA sequencing of an oleate-selective acyl-ACP:*sn*-glycerol-3-phosphate acyltransferase from pea chloroplasts. *Plant Mol. Biol.* 17:1067–76
150. Weretilnyk E, Orr W, White TC, Iu B, Singh J. 1993. Characterization of three related low temperature-regulated cDNAs from winter *Brassica napus*. *Plant Physiol.* 101:171–77
151. White TC, Simmonds D, Donaldson P, Singh J. 1994. Regulation of *BN115*, a low-temperature-responsive gene from winter *Brassica napus*. *Plant Physiol.* 106:917–28
152. Williams JGK. 1988. Construction of specific mutations in photosystem II photosynthetic reaction center by genetic engineering methods in *Synechocystis* 6803. *Methods Enzymol.* 167: 766–78
153. Wolfraim LA, Dhindsa RS. 1993. Cloning and sequencing of the cDNA for *cas17*, a cold acclimation-specific gene of alfalfa. *Plant Physiol.* 103:667–68
154. Wolfraim LA, Langis R, Tyson H,

Dhindsa RS. 1993. cDNA sequence, expression, and transcript stability of a cold acclimation-specific gene, *cas18*, of alfalfa (*Medicago falcata*) cells. *Plant Physiol.* 101:1275–82

155. Wolter F-P, Schmidt R, Heinz E. 1992. Chilling sensitivity of *Arabidopsis thaliana* with genetically engineered membrane lipids. *EMBO J.* 11: 4685–92
156. Wu J, Browse J. 1995. Elevated levels of high-melting-point phosphatidylglycerols do not induce chilling sensitivity in an *Arabidopsis* mutant. *Plant Cell* 7:17–27
157. Yadav NS, Wierzbicki A, Aegerter M, Caster CS, Pérez-Grau L, et al. 1993. Cloning of higher plant ω-3 fatty acid desaturase. *Plant Physiol.* 103:467–76
158. Yamamoto KT. 1994. Further characterization of auxin-regulated mRNAs in hypocotyl sections of mung bean [*Vigna radiata* (L.) Wilczek]: sequence homology to genes for fatty-acid desaturases and atypical late-embryogenesis-abundant protein, and the mode of expression of the mRNAs. *Planta* 192:359–64
159. Yamamoto KT, Mori H, Imaseki H. 1992. Novel mRNA sequences induced by indole-3-acetic acid in sections of elongating hypocotyls of mung bean (*Vigna radiata*). *Plant Cell Physiol.* 33: 13–20

Annu. Rev. Plant Physiol. Plant Mol. Biol. 1996. 47:569–93

THE MOLECULAR-GENETICS OF NITROGEN ASSIMILATION INTO AMINO ACIDS IN HIGHER PLANTS

H.-M. Lam, K. T. Coschigano, I. C. Oliveira, R. Melo-Oliveira, G. M. Coruzzi

Department of Biology, New York University, New York, NY 10003

KEY WORDS: glutamine, glutamate, aspartate, asparagine, gene regulation

ABSTRACT

Nitrogen assimilation is a vital process controlling plant growth and development. Inorganic nitrogen is assimilated into the amino acids glutamine, glutamate, asparagine, and aspartate, which serve as important nitrogen carriers in plants. The enzymes glutamine synthetase (GS), glutamate synthase (GOGAT), glutamate dehydrogenase (GDH), aspartate aminotransferase (AspAT), and asparagine synthetase (AS) are responsible for the biosynthesis of these nitrogen-carrying amino acids. Biochemical studies have revealed the existence of multiple isoenzymes for each of these enzymes. Recent molecular analyses demonstrate that each enzyme is encoded by a gene family wherein individual members encode distinct isoenzymes that are differentially regulated by environmental stimuli, metabolic control, developmental control, and tissue/cell-type specificity. We review the recent progress in using molecular-genetic approaches to delineate the regulatory mechanisms controlling nitrogen assimilation into amino acids and to define the physiological role of each isoenzyme involved in this metabolic pathway.

CONTENTS

1040-2519/96/0601-0569$08.00

INTRODUCTION

The assimilation of inorganic nitrogen onto carbon skeletons has marked effects on plant productivity, biomass, and crop yield (45, 64). Nitrogen deficiency in plants has been shown to cause a decrease in the levels of photosynthetic structural components such as chlorophyll and ribulose bisphosphate carboxylase (rubisco), with resulting reductions in photosynthetic capacity and carboxylation efficiency (26). Because enzymes involved in the assimilation of nitrogen into organic form in plants are crucial to plant growth, they are also effective targets for herbicide development (19).

A tremendous amount of biochemical and physiological studies have been performed on nitrogen assimilatory enzymes from a variety of plant species. Summaries of these biochemical studies can be found in several comprehensive reviews (40, 69, 70, 84, 91). The biochemical reactions of nitrogen assimilatory enzymes discussed herein are summarized in Table 1. Although these biochemical studies have provided a solid groundwork for the understanding of nitrogen assimilation in plants, a complete picture of the factors controlling and the enzymes involved in this process in a single plant is still lacking. The existence of multiple isoenzymes for each step in nitrogen metabolism has complicated biochemical purification schemes (95). Because the mechanisms controlling intra- and intercellular transport of inorganic and organic nitrogen in plants are presently unknown, it is impossible to predict the in vivo function of nitrogen assimilatory enzymes localized in distinct cells or subcellular compartments based on in vitro biochemistry.

Recently, molecular techniques and the analysis of plant mutants deficient in a particular isoenzyme have been employed to study nitrogen assimilation and metabolism. These studies have shown that the genes involved in nitrogen assimilation are not constitutively expressed "housekeeping" genes but are

Table 1 Biochemical reactions and mutants of plant nitrogen assimilation enzymes

Enzyme	Reaction	Mutant identification
GS1/GS2	glutamate + NH_4^+ + ATP = glutamine + ADP + Pi	barley (GS2) (143)
Fd-GOGAT	glutamine + 2-oxoglutarate + 2 Fd (red) = 2 glutamate + 2 Fd (ox)	Arabidopsis (116), barley (11), pea (8)
NADH-GOGAT	glutamine + 2-oxoglutarate + NADH = 2 glutamate + NAD	none
GDH	glutamate + H_2O + NAD/NADP = NH_4^+ + 2-oxoglutarate + NADH/NADPH	Arabidopsis (62; R Melo-Oliveira, I Oliveira & G Coruzzi, unpublished data)
AspAT	glutamate + oxaloacetate = aspartate + 2-oxoglutarate	Arabidopsis (62, 107)
AS	glutamine + aspartate + ATP = asparagine + glutamate + AMP +PPi	none

Abbreviations: GS1, cytoplasmic glutamine synthetase; GS2, chloroplastic glutamine synthetase; Fd-GOGAT, ferredoxin-dependent glutamate synthase; NADH-GOGAT, NADH-dependent glutamate synthase; GDH, glutamate dehydrogenase; AspAT, aspartate aminotransferase; AS, asparagine synthetase; Fd, ferredoxin; Pi, inorganic phosphate; PPi, pyrophosphate; NH_3, ammonia

carefully regulated by factors such as light, metabolites, and cell type. This review highlights examples where molecular, genetic, and biochemical analyses have begun to define the in vivo roles of individual isoenzymes in plant nitrogen assimilation and to uncover the mechanisms regulating this process. Special attention is paid to the genetically tractable system Arabidopsis, because it enables studies of biochemistry, molecular biology, and genetics of nitrogen assimilation in a single species. We also focus on the metabolism of glutamine, glutamate, aspartate, and asparagine, which are the dominant components in the total free amino acid pool in most legumes and crop plants (69, 88). While most studies of nitrogen metabolism have previously been performed in legumes and crop species, HPLC analyses of Arabidopsis have also demonstrated that these four amino acids can account for 60–64% of the total free amino acids present in leaves and are transported in the vascular tissues (62, 107). Thus, Arabidopsis appears to be a suitable model plant for the study of nitrogen assimilation; the results should have an impact on understanding less genetically tractable plants.

ASSIMILATION OF INORGANIC NITROGEN INTO GLUTAMINE AND GLUTAMATE

In plants, all inorganic nitrogen is first reduced to ammonia before it is incorporated into organic form (21, 50). Ammonia is then assimilated into glutamine and glutamate, which serve to translocate organic nitrogen from sources to sinks in legumes and nonlegumes, including Arabidopsis (62, 69, 88, 107). The major enzymes involved are glutamine synthetase (GS), glutamate synthase (GOGAT, glutamine-2-oxoglutarate aminotransferase), and glutamate dehydrogenase (GDH). Each of these enzymes occurs in multiple isoenzymic forms encoded by distinct genes (see below). The individual isoenzymes of GS, GOGAT, or GDH have been proposed to play roles in three major ammonia assimilation processes: primary nitrogen assimilation, reassimilation of photorespiratory ammonia, and reassimilation of recycled nitrogen.

Primary Nitrogen Assimilation

In legumes, ammonia can be formed by the direct fixation of atmospheric dinitrogen atoms within root nodules (13, 135, 136). In nonlegumes, ammonia is generated by the concerted reactions of nitrate reductase and nitrite reductase (21, 50). In most tropical and subtropical species, nitrate taken up by the roots is largely transported to leaves where it is reduced to ammonia in plastids (3). Because chloroplastic GS2 and ferredoxin-GOGAT (Fd-GOGAT) are the predominant GS/GOGAT isoenzymes in leaves located in plastids, they have been proposed to function in the assimilation of this primary nitrogen into glutamine and glutamate (84). Because the predominant forms of GS and GOGAT in roots are cytosolic GS1 and NADH-GOGAT, these isoenzymes have been proposed to be involved in primary nitrogen assimilation in roots (84). GDH is less likely to be involved in primary nitrogen assimilation because of its K_m for ammonia (119).

The traditional assignments of GS/GOGAT isoenzyme function based on organ-specific distribution have been challenged by the phenotype of plant mutants defective in these enzymes. For example, although chloroplastic GS2 and Fd-GOGAT are proposed to be important for primary nitrogen assimilation in leaves, plant GS2 or Fd-GOGAT-deficient mutants appear to be competent in primary assimilation and specifically defective in the reassimilation of photorespiratory ammonia (9, 117; see sections on GS and GOGAT). No plant mutants yet exist in cytosolic GS1 or NADH-GOGAT to address whether they in fact are the major isoenzymes involved in primary nitrogen assimilation in leaves and/or roots.

Reassimilation of Photorespiratory Ammonia

Photorespiration is thought to be a wasteful process occurring predominantly in C3 plants that is initiated by rubisco oxygenase activity (41, 61). Thus, in plants grown in air, the oxygenation by rubisco results in the diversion of a portion of ribulose bisphosphate from the Calvin cycle and its conversion to two molecules of phosphoglycolate. The photorespiratory enzymes in plants catalyze a series of metabolic conversions of phosphoglycolate that occur sequentially in chloroplasts, peroxisomes, and mitochondria. These reactions lead to the release of carbon dioxide and photorespiratory ammonia. In C3 plants, the ammonia released through photorespiration may exceed primary nitrogen assimilation by 10-fold (61). Therefore, to survive, a plant must be able to reassimilate this photorespiratory ammonia into glutamine or glutamate. Plant mutants defective in enzymes of the photorespiratory pathway have been identified by a conditional lethal phenotype screen (9, 117; also see below). The existence of photorespiratory mutants specifically defective in chloroplastic GS2 or Fd-GOGAT countered the suggestion that GDH, located in mitochondria, played a major role in reassimilation of photorespiratory ammonia (148). Thus, although the biochemical data and subcellular localization studies suggested that GDH played a major role in the reassimilation of photorespiratory ammonia, genetic data suggested otherwise.

Assimilation of Recycled Nitrogen

Ammonia is released during biochemical processes such as protein catabolism, amino acid deamination, and some specific biosynthetic reactions such as those involving methionine, isoleucine, phenylpropanoid, and lignin biosynthesis (66, 84). For plants to efficiently utilize nitrogen assimilated from the soil, they must be able to recycle nitrogen released during various catabolic reactions. While ammonia recycling occurs at all times in a plant, there are two major times when massive amounts of recycled ammonia must be reassimilated into glutamine or glutamate for transport. The first is during germination, when seed storage proteins are broken down and nitrogen is transported as glutamine to the growing seedling (69). Later, proteins in senescing leaves are degraded, and the nitrogen is reassimilated as glutamine for transport to the developing seed (83). Increased activities for cytosolic GS1, NADH-GOGAT, and GDH during these processes have suggested the involvement of these particular isoenzymes (70, 119).

GLUTAMINE SYNTHETASE

Biochemistry Background of Glutamine Synthetase

Two classes of glutamine synthetase (GS: E.C.6.3.1.2) isoenzymes that are located in the cytosol (GS1) or chloroplast (GS2) have been identified by ion-exchange chromatography. Although there are multiple forms of cytosolic GS, we refer to all cytosolic forms of GS as GS1 for simplicity. The distinct physiological roles of GS2 and GS1 have been implicated by their organ-specific distributions. For instance, because GS2 is the predominant isoenzyme in leaves, it has been proposed to function in primary assimilation of ammonia reduced from nitrate in chloroplasts and/or in the reassimilation of photorespiratory ammonia (84). Because cytosolic GS1 is predominant in roots, it has been proposed to function in root nitrogen assimilation, although root plastid GS2 has also been implicated in this process (82). The finding that cytosolic GS1 is the predominant GS isoenzyme expressed during senescence in different plant species suggests that this GS isoenzyme plays a role in the mobilization of nitrogen for translocation and/or storage (56–58). The localization of GS1 in vascular bundles further supports the notion that cytosolic GS functions to generate glutamine for intercellular nitrogen transport (16, 55).

Despite the numerous studies on GS isoenzymes performed at the biochemical level, the exact in vivo role of each GS isoenzyme in plant metabolism is equivocal. The GS isoenzymes are encoded by a gene family in all plant species examined to date. A thorough characterization of the members of the GS gene family found in pea (127, 141), rice (106), Arabidopsis (90), *Phaseolus* (22, 37, 73), maize (72, 104, 115), and soybean (49, 102) showed that each species appears to possess a single nuclear gene for chloroplastic GS2 and multiple genes for cytosolic GS1. These studies have demonstrated that several members of the GS gene families are regulated differently by cell type, light, and metabolites as outlined below.

Molecular and Genetic Studies of Chloroplastic GS2

The in vivo function of chloroplastic GS2 has been elucidated by both molecular studies on the genes and genetic studies of plant GS2 mutants. The GS2 gene is primarily expressed in green tissues in all species examined (20, 30, 72, 104). Indeed, the developmental onset of GS2 gene expression coincides with the maturation of chloroplasts in pea (30, 141) and the development of photosynthetic cotyledons in *Phaseolus* (20). Studies performed in pea (30), maize (104), *Phaseolus* (29), and Arabidopsis (90) demonstrated that GS2 gene expression is tightly regulated by light, and in several cases this has been shown to be mediated at least in part by phytochrome activation (30). GS2 gene expression can also be regulated by metabolic control in response to carbohydrate and amino acid supplementation in tobacco and Arabidopsis (33; I Oliveira & G Coruzzi, unpublished data). In addition, GS2 mRNA accumulation has been reported to increase in leaves of plants cultivated under photorespiratory conditions (20, 30), a finding in line with one of the proposed functions of GS2, the reassimilation of photorespiratory ammonia (82).

Although screens for plant mutants unable to survive in photorespiratory conditions were conducted in Arabidopsis and later in barley, mutants specifically defective in GS2 were identified only in the barley screen (116, 143). The barley GS2 mutants lack the ability to reassimilate ammonia lost during photorespiration. These mutants die not because of a toxic buildup of ammonia but because of the drain on the organic nitrogen pool (8), as the decrease of photosynthetic rate in GS2 mutants can be rescued by supplementation of alanine, asparagine, and glutamine (67). A dramatic result of this mutant study is the finding that a GS isoenzyme located in the chloroplast is essential for the reassimilation of photorespiratory ammonia released in mitochondria. Because the parameters regulating the intra- and intercellular transport of inorganic and organic nitrogen are presently unknown, this is a dramatic example of how a mutant deficient in a particular subcellular isoenzyme can be used to define the true in vivo role of an isoenzyme.

Paradoxically, the barley mutants deficient in chloroplast GS2 were unable to reassimilate photorespiratory ammonia released in the mitochondria even though they contained normal levels of GS1 in the cytosol (143). This apparent paradox has been resolved by studies on the cell-specific expression patterns of genes for chloroplastic GS2 and cytosolic GS1. Studies of GS-promoter-GUS fusions revealed that chloroplastic GS2 is expressed predominantly in leaf mesophyll cells, where photorespiration occurs, whereas cytosolic GS1 is expressed exclusively in the phloem (31, 34). Although this observation contradicts previous biochemical data that suggested that a large portion of cytosolic GS1 activity was located in mesophyll protoplasts of pea (142), these promoter-GUS fusion results were later confirmed by in situ immunolocalization studies of the native cytosolic GS1 proteins in rice and tobacco (16, 55). This vascular-specific expression pattern may explain why cytosolic GS1 cannot compensate for the loss of chloroplastic GS2 in mesophyll cells of the barley GS2 mutants.

One piece of the GS isoenzyme puzzle that is outstanding is the fact that the screens for photorespiratory mutants in Arabidopsis failed to uncover any mutants defective in GS, either chloroplastic GS2 or cytosolic GS1. There are several possible explanations for this finding. 1. The Arabidopsis photorespiratory screen was not saturating. This is unlikely because multiple alleles for many enzymes in the photorespiratory pathway were isolated in that screen, including 58 mutants affecting Fd-GOGAT (4). 2. Both chloroplastic GS2 and cytosolic GS1 are expressed in mesophyll cells, so that a mutation in one gene is masked. Again, this is unlikely, because cytosolic GS1 is not expressed in mesophyll cells, at least in tobacco and rice (16, 55). 3. There is more than one gene for chloroplastic GS2 in Arabidopsis. 4. A mutation in chloroplastic or cytosolic GS is lethal in Arabidopsis and prevents the isolation of mutants.

Molecular Studies of Cytosolic GS1

Because cytosolic GS is an enzyme involved in the assimilation of ammonia fixed by *Rhizobium,* the early studies on genes for cytosolic GS1 were conducted in legumes such as *Phaseolus,* soybean, pea, and alfalfa (126). In each case, multiple genes for cytosolic GS1 existed, certain members of which were expressed at highest levels in nodules. On the basis of the identification of the "nodule-specific" or "nodule-enhanced" expression of certain GS genes, it was proposed that the multiplicity of genes for cytosolic GS1 has evolved in legumes for this purpose (20, 102, 126, 141). The finding that Arabidopsis also contains at least three genes for cytosolic GS1 (90) suggests that the GS gene family has evolved independently of its role in the nitrogen-fixation process in legumes. Genes for all three cytosolic GS1 isoenzymes of Arabidopsis are expressed at high levels in roots (90). The use of gene-specific probes has revealed that the three genes for cytosolic GS1 of Arabidopsis have subtle differences in their expression patterns. Two genes for cytosolic GS1 are both expressed at high levels in germinated seeds, which suggests that these genes may play a role in the synthesis of glutamine for transport of nitrogen out of cotyledons (90). This is consistent with previous data showing that individual genes for cytosolic GS1 of pea and bean are likely to serve this function in legumes (22, 128). Promoter-GUS fusion studies (31, 34) and subsequent immunolocalization analyses (16, 55) also demonstrated that cytosolic GS1 is expressed in phloem in several species. This further supports the role of GS1 in generating glutamine for intercellular transport.

Expression of genes encoding cytosolic GS1 is under different modes of regulation. For instance, ammonia supplementation has been reported to induce mRNA accumulation of a soybean GS1 gene (49). Treatment with either ammonia or nitrate has also been shown to elicit an increase in the steady state levels of one of the maize GS1 isoforms (GS1-1) in roots and another isoform in shoots (GS1-2). In contrast, another isoform of cytosolic GS in maize (GS1-3) was not affected by the same treatment (121). Studies of transgenic plants transformed with a soybean GS1 promoter-GUS fusion construct revealed that ammonia could induce the accumulation of β-glucuronidase in *Lotus* but not in tobacco plants, although the tissue specificity was conserved in both species (81). In addition, studies of soybean GS1 promoter deletion constructs in transgenic *Lotus* plants demonstrated that the elements responsible for tissue specificity and the induction by ammonia are located in two distinct regions of the GS1 promoter (78, 79). Together, the data accumulated on the expression of the different forms of GS enzymes further illustrate the complexity of the biological function of the GS genes as reflected by the diversity and the differential pattern of expression of each isoenzyme.

GLUTAMATE SYNTHASE

Biochemistry Background of Glutamate Synthase

In higher plants, there are two antigenically distinct forms of glutamate synthase (GOGAT) that use NADH (NADH-GOGAT: E.C.1.4.1.14) or ferredoxin

(Fd-GOGAT: E.C.1.4.7.1) as the electron carrier (70, 110, 119, 122). NADH-GOGAT is located primarily in plastids of nonphotosynthetic tissues such as roots (80, 122). In root nodules of legumes, NADH-GOGAT is involved in the assimilation of nitrogen fixed by *Rhizobium* (2, 17). It has been hypothesized that NADH-GOGAT catalyzes the rate-limiting step of ammonia assimilation in these root nodules (43). In nonlegumes, NADH-GOGAT may function in primary assimilation or reassimilation of ammonia released during amino acid catabolism (84).

In contrast with NADH-GOGAT, Fd-GOGAT is located primarily in the leaf chloroplast where light leads to an increase in Fd-GOGAT protein and activity (70, 110). These findings suggested that the physiological role(s) of Fd-GOGAT is related to light-inducible processes in leaves such as photosynthesis and photorespiration. Fd-GOGAT may also play a smaller role in nonphotosynthetic tissues, because some Fd-GOGAT activity is associated with roots (123). The molecular and genetic studies outlined below have helped to clarify the relative in vivo roles of NADH- vs Fd-GOGAT.

Molecular Studies of NADH–Glutamate Synthase

cDNA clones of NADH-GOGAT were successfully isolated from the legume alfalfa (43) and the nonlegume Arabidopsis (H-M Lam & G Coruzzi, unpublished data). Both the alfalfa and Arabidopsis NADH-GOGAT genes encode putative functional domains within the mature protein that are highly homologous to the large and small subunits of *Escherichia coli* NADPH-GOGAT (43; H-M Lam & G Coruzzi, unpublished data). A putative NADH-binding motif, contained in the small subunit of *E. coli* NADPH-GOGAT, is also found in the corresponding C-terminal domain of both the alfalfa and Arabidopsis NADH-GOGAT enzymes (43; H-M Lam & G Coruzzi, unpublished data).

Measurements of mRNA levels and promoter-GUS fusions of the NADH-GOGAT genes in alfalfa and *Lotus* have shown the tight relationship of the regulated expression of NADH-GOGAT to the nodulation process in legumes (137). It was found that the NADH-GOGAT gene is expressed primarily in cells of effective nodules and is maintained at low or undetectable levels in other tissues. In the nonlegume Arabidopsis, mRNA levels of NADH-GOGAT are enhanced in roots as opposed to leaves (H-M Lam & G Coruzzi, unpublished data). Preliminary studies also show that the expression of the Arabidopsis NADH-GOGAT gene increases during the early stages of seed germination (H-M Lam & G Coruzzi, unpublished data). Because the expression patterns of the genes for cytosolic GS1 and NADH-GOGAT appear coordinated, they may function together in processes such as the primary assimilation of nitrate-derived ammonia in root cells, the reassimilation of ammonia released during catabolic reactions, and/or remobilization of ammonia released during germination. Because plant mutants in NADH-GOGAT have not been identified, its true in vivo role remains conjectural.

Molecular and Genetic Studies of Ferredoxin–Glutamate Synthase

Fd-GOGAT is uniquely found in photosynthetic organisms. Fd-GOGAT genes have been cloned from six plant species: maize (105), tobacco (149), barley (5), spinach (85), Scots pine (36), and Arabidopsis (K Coschigano & G Coruzzi, unpublished data). A single gene was identified in every species except Arabidopsis, which has been shown to contain two expressed genes (*GLU1* and *GLU2*), each encoding a distinct form of Fd-GOGAT.

Fd-GOGAT mRNA accumulates primarily in leaf tissue in response to light, as has been shown in maize, tobacco, and Arabidopsis (*GLU1*) (105, 149; K Coschigano & G Coruzzi, unpublished data). Involvement of phytochrome in the light induction of Fd-GOGAT has been demonstrated at the mRNA level in tomato (6) and observed at the protein level in mustard cotyledons and Scots pine seedlings (32, 48). In Arabidopsis, Fd-GOGAT mRNA accumulation (*GLU1*) can also be induced in the absence of light by exogenous sucrose applications (K Coschigano & G Coruzzi, unpublished data).

In addition to the highly expressed *GLU1* gene for Fd-GOGAT in Arabidopsis described above, a second expressed gene encoding Fd-GOGAT (*GLU2*) was isolated in Arabidopsis. The discovery of the second gene encoding a distinct form of Fd-GOGAT is consistent with the observance of two antigenically distinct Fd-GOGAT isoforms in rice (123). In contrast with *GLU1* mRNA, accumulation of *GLU2* mRNA is low in leaves but high in roots. *GLU2* mRNA expression does not appear to be significantly influenced by light or sucrose but instead is observed at constitutive, low levels (K Coschigano & G Coruzzi, unpublished data). The expression pattern of the Arabidopsis *GLU2* gene for Fd-GOGAT is very similar to that seen for the gene encoding NADH-GOGAT (see above).

The roles of the various GOGAT isoenzymes are being elucidated through the isolation of plant mutants. Photorespiratory mutants specifically lacking Fd-GOGAT enzyme activity have been isolated from three plant species: Arabidopsis (116), barley (11), and pea (8). In the three Arabidopsis *gluS* mutants initially characterized, leaf Fd-GOGAT activity was reduced to <5% of wild-type levels, whereas NADH-GOGAT (which contributes about 5% of the total GOGAT activities in normal conditions) remained unchanged (roots were not analyzed) (116). In the photorespiratory barley mutants, both leaf and root Fd-GOGAT activity was reduced to <6% of wild type, which suggests that these activities are under the control of the same gene (11). All of the Fd-GOGAT-deficient mutants isolated from the three plant species were chlorotic and eventually died when grown in atmospheric conditions promoting photorespiration (air), which thus established an essential role for Fd-GOGAT in photorespiration. However, because the Fd-GOGAT-deficient mutants recovered and were viable when grown in conditions where photorespiration was suppressed (high CO_2 or low O_2), Fd-GOGAT appeared at first glance to be dispensable for nonphotorespiratory roles, such as in primary nitrogen assimilation. This conclusion was paradoxical because most primary assimilation probably occurs in leaves, where Fd-GOGAT activity predominates (95% of total GOGAT activity) and NADH-GOGAT is a minor component (5% of total GOGAT activity).

The presence of two expressed Fd-GOGAT genes in Arabidopsis is curious because a single gene mutation affecting Fd-GOGAT activity had been isolated in the phenotypic screen for photorespiratory mutants (116). Thus, although there were two genes for Fd-GOGAT, a mutation in one gene produced a photorespiratory-deficient phenotype. However, it appears that a mutation in the highly expressed *GLU1* gene results in a photorespiratory defect as the *GLU1* gene maps to the region of the *gluS* photorespiratory mutation (K Coschigano & G Coruzzi, unpublished data). The *GLU2* gene, which is expressed at constitutively low levels in leaves and at higher levels in roots, maps to a different chromosome and thus may be involved in the primary assimilation process. Interestingly, Fd-GOGAT was also implicated in playing a role in primary assimilation in maize by observance of a rapid, transient, and cycloheximide-independent accumulation of Fd-GOGAT transcripts in maize roots after treatment with nitrate (98). Arabidopsis mutants null for GLU1 activity would be quite valuable to elucidate the role of the *GLU1* gene, and thus a comprehensive analysis of all of the *gluS* alleles is being performed (K Coschigano & G Coruzzi, unpublished data). These *GLU1* mutants could in turn be used to isolate mutations in *GLU2*. The phenotype of a Fd-GOGAT null mutant (*GLU1*, *GLU2* double mutant) could be used to distinguish between Fd-GOGAT roles and NADH-GOGAT roles.

GLUTAMATE DEHYDROGENASE

Biochemistry Background of Glutamate Dehydrogenase

Two major forms of glutamate dehydrogenase (GDH) have been reported: an NADH-dependent form (NADH-GDH: E.C.1.4.1.2) found in the mitochondria (25, 74) and an NADPH-dependent form (NADPH-GDH: E.C.1.4.1.4) localized to the chloroplast (71). The GDH enzyme is abundant in several plant organs (15, 70, 76). Moreover, the GDH isoenzymatic profile can be influenced by dark stress, natural senescence, or fruit ripening (15, 75, 118). These studies suggest that GDH may play a specific or unique role in assimilating ammonia or catabolizing glutamate during these processes.

Although GDH enzyme activity exists in plant tissues at high levels, there is an ongoing debate about its physiological role in higher plants. Originally, GDH was proposed to be the primary route for the assimilation of ammonia in plants. However, this biosynthetic role of GDH has been challenged by the discovery of an alternative pathway for ammonia assimilation via the GS/GOGAT cycle. Moreover, the fact that the GDH enzyme has a high K_m for ammonia argues against a role in primary nitrogen assimilation (119). Studies have shown that GDH enzyme activity can be induced in plants exposed to high levels of ammonia (15), and as such GDH has been proposed to be important specifically for ammonia-detoxification purposes. Mitochondrial GDH has been proposed to be involved in the assimilation of high levels of photorespiratory ammonia released in mitochondria (148). However, the isolation of photorespiratory mutants defective in chloroplastic GS2 (in barley) (143) or Fd-GOGAT (in barley and in Arabidopsis) (8, 59, 116) suggests that GDH is not important in photorespiration (143). Furthermore, treatment of plants with the GS inhibitor MSO prevents the incorporation of ammonia into glutamate and glutamine, even though both GDH activity and ammonia levels remain high (70). Together, these results may be used to argue against a biosynthetic role for GDH. Instead, a catabolic role for GDH has been invoked, which is supported by the fact that GDH activity is induced during germination and senescence, two periods where amino acid catabolism occurs (70, 119).

Molecular and Genetic Studies of Glutamate Dehydrogenase

Studies of plant GDH genes and mutants have begun to shed some light on the role of GDH in plants. In both Arabidopsis and maize there appear to be two genes for GDH based on Southern analysis and mutant analysis. The predicted peptide sequences encoded by cDNAs for maize and Arabidopsis *GDH1* reveal high identity to the GDH enzymes of other organisms (103; R Melo-Oliveira, I Oliveira & G Coruzzi, unpublished data). Furthermore, the predicted protein sequences of Arabidopsis and maize GDH suggest that they encode NADH-dependent enzymes that are likely to be associated with the mitochondria (103; R Melo-Oliveira, I Oliveira & G Coruzzi, unpublished data).

Studies have also been performed on GDH gene regulation. The transcripts for maize GDH have been shown to be predominant in roots and present in the bundle sheath cells in leaf tissues (103). This evidence agrees with results at the level of NADH-GDH activity in maize. In contrast, the level of *GDH1* mRNA in Arabidopsis, a C3 plant, is higher in leaves than in roots (R Melo-Oliveira, I Oliveira & G Coruzzi, unpublished data). *GDH1* mRNA also accumulates to high levels in dark-adapted plants, and this accumulation is repressed by light or sucrose (R Melo-Oliveira, I Oliveira & G Coruzzi, unpublished data). This observation is consistent with previous biochemical data that showed that GDH activity increased in response to carbon limitation in maize (87). It appears that the *GDH1* and *GLN2* (GS2) genes of Arabidopsis are reciprocally regulated by light and sucrose (R Melo-Oliveira, I Oliveira & G Coruzzi, unpublished data) as shown previously in lupine at the level of enzyme activity (97). These gene expression data suggest that GDH1 and GS2 play nonoverlapping roles in Arabidopsis nitrogen metabolism.

An Arabidopsis mutant deficient in GDH was identified in the M2 generation of EMS-mutagenized Arabidopsis using a GDH activity stain on crude leaf protein extracts following electrophoresis on native gels (62, 145; R Melo-Oliveira, I Oliveira & G Coruzzi, unpublished data). The GDH enzymes of Arabidopsis can be resolved into seven isoenzymes in this manner (14, 62, 107). These seven GDH activity bands are the result of the random association of two types of subunits into a hexameric complex (15). It has been proposed that two nonallelic genes are responsible for the synthesis of the GDH1 and GDH2 subunits (14, 15). A single Arabidopsis GDH mutant, *gdh1-1*, has been identified that has an altered pattern of GDH activity: It possesses a single GDH2 holoenzyme and is missing the GDH1 holoenzyme as well as the heterohexamers (R Melo-Oliveira, I Oliveira & G Coruzzi, unpublished data). The Arabidopsis *gdh1-1* mutant displays an impaired growth phenotype compared with wild type specifically when plants are grown in media containing exogenous inorganic nitrogen. This conditional phenotype suggests a nonredundant role for GDH in the assimilation of ammonia under conditions of inorganic nitrogen excess. A similar GDH-deficient mutant has been previously described in *Zea mays*, a C4 plant, which also appears to be affected in the *GDH1* gene product (92, 93). Preliminary studies showed that the maize GDH mutant displays a growth phenotype only under low night temperatures (94). Moreover, it has been reported that the maize *GDH1* mutant shows a 10- to 15-fold lower total GDH activity when compared with wild-type maize (77). Because the photorespiratory rate is very low or nonexistent in a C4 plant, the maize *GDH1* mutant cannot be used to assess the role of GDH in photorespiration. Therefore, the Arabidopsis *GDH1* mutant will be valuable to assess the function of this enzyme in photorespiration in a C3 plant. It should be noted that neither the maize nor the Arabidopsis *GDH1* mutants are null for GDH, because they each possess a second *GDH2* gene. Isolation of *GDH2* mutants and creation of *GDH1/GDH2* double mutants will be needed to define the role of GDH unequivocally.

DOWNSTREAM METABOLISM OF GLUTAMINE AND GLUTAMATE

Following the assimilation of ammonia into glutamine and glutamate, these two amino acids act as important nitrogen donors in many cellular reactions, including the biosynthesis of aspartate and asparagine (40, 66). Aspartate contributes an integral part of the malate-aspartate shuttle that allows the transfer of reducing equivalents from mitochondria and chloroplast into the cytoplasm (52). In C4 plants, aspartate shuttles carbon between mesophyll cells and bundle sheath cells (47). Asparagine is thought to be an important compound for transport and storage of nitrogen resources because of its relative stability and high nitrogen to carbon ratio. Asparagine is a major nitrogen-transport compound in both legumes and nonleguminous plants. In seeds of *Lupinus albus*, 86.5% of the nitrogen from protein is remobilized into asparagine (69). Radioactive nitrogen feeding experiments in peanut indicate that up to 80% of the label of ^{15}N-[N_2] was recovered as asparagine in the sap of nodules (89). Asparagine also acts as the major constituent of nitrogen transported out of nodules in leguminous plants (69, 109). In nonleguminous plants such as Arabidopsis, asparagine is also a major transported amino acid detected in the phloem exudates (62, 107). In the following sections, we discuss AspAT and AS, which are the two major enzymes involved in the downstream metabolism of assimilated nitrogen into aspartate and asparagine.

ASPARTATE AMINOTRANSFERASE

Biochemistry Background of Aspartate Aminotransferase

Biochemical studies show that aspartate aminotransferase (AspAT: E.C.2. 6.1.1) can exist as distinct isoenzymes (144). The activities of various AspAT isoenzymes have been found in different tissues and different subcellular locations such as the cytosol, mitochondria, chloroplasts, glyoxysomes, or peroxisomes (for examples, see 108, 125, 140, 144). The subcellular compartmentation of AspAT isoenzymes suggests that the different forms of AspAT might serve distinct roles in plant metabolism. It is also important to note that individual AspAT isoenzymes respond differently to environmental conditions and metabolic status such as light treatment or nitrogen starvation, which suggests that they serve distinct roles (101, 125).

Molecular and Genetic Studies of Aspartate Aminotransferase

Molecular and genetic analyses of AspAT genes have begun to elucidate the in vivo function of each AspAT isoenzyme. cDNA clones encoding AspAT have been isolated in both legumes and nonlegumes such as alfalfa, Arabidopsis, *Panicum,* and soybean (108, 125, 132, 140, 147). The regulation of AspAT in legumes is tightly coupled with the symbiotic process. In alfalfa, the levels of AspAT mRNA are induced during effective nodule development (35, 132).

In the C3 plant Arabidopsis, the entire gene family of AspAT isoenzymes has recently been characterized (108, 147). Five different AspAT cDNA clones

[*ASP1–4* and *ASP5* (formally *AAT1*)] were obtained, including those encoding the mitochrondrial, plastidic, peroxisomal, and cytosolic forms of AspAT. Although two of the five *ASP* genes encode cytosolic forms of AspAT (*ASP2* and *ASP4*), only *ASP2* is expressed at high levels, especially in roots (108). The *ASP1* and *ASP3* genes, which encode a mitochondrial and a peroxisomal form of AspAT respectively, are each expressed at relatively high levels in all organs examined (108). In the C4 plant *Panicum miliaceum*, AspAT genes encoding the cytosolic, mitochrondrial, and plastidic AspAT isoenzymes are all expressed at higher levels in green leaves than in mesocotyls and root tissue (125). It has also been reported that nitrogen availability shows a positive effect on the levels of mRNAs for cytosolic AspAT and mitochondrial AspAT genes but not for plastidic AspAT genes in *Panicum miliaceum* (125).

To help determine the in vivo function(s) of each AspAT isoenzyme, a screen for mutants defective in the predominant forms of AspAT in leaf extracts was performed in Arabidopsis. Crude leaf extracts contain two major AspAT isoenzymes, AAT2 (cytosolic) and AAT3 (chloroplastic), as detected by activity staining of native gels (62, 107). M2 seedlings were screened for alterations in AAT profiles using the native gel screen on crude leaf extracts (107). Four classes of AspAT mutants were obtained from a screen of 8000 EMS-mutagenized Arabidopsis seeds: (*a*) loss of cytosolic AAT2 activity, (*b*) loss of chloroplastic AAT3 activity, (*c*) alteration of cytosolic AAT2 gel mobility, and (*d*) alteration of chloroplastic AAT3 gel mobility (107). By analyzing the effects of these mutations on the growth phenotypes and the balance of free amino acid pools in different classes of mutant plants, the in vivo importance of each isoenzyme can be determined. Preliminary analyses of these AAT mutants show that a mutation in the cytosolic *ASP2* gene results in a retarded growth phenotype and a decrease in the pools of free aspartate (C Schultz & G Coruzzi, unpublished data). Thus, despite the presence of two genes for cytosolic AspAT (*ASP2* and *ASP4*), a mutation in the highly expressed *ASP2* gene causes a growth defect and aspartate-deficient phenotype.

ASPARAGINE SYNTHETASE

Biochemistry Background of Asparagine Synthetase

Asparagine was the first amino acid discovered and was isolated in asparagus 190 years ago (138). Despite this historical placement, the mechanism of asparagine biosynthesis in plants has been elucidated only recently. The glutamine-dependent asparagine synthetase enzyme (AS: E.C.6.3.5.4) is now generally accepted as the major route for asparagine biosynthesis in plants (70, 100). However, ammonia is also a possible AS substrate, particularly in the case of maize roots (86). In some cases, asparagine is believed to act as an ammonia detoxification product produced when plants encounter high concentrations of ammonia (39, 113).

The hypothesis that asparagine serves to transport nitrogen in plants is supported by high levels of AS activity detected in nitrogen-fixing root nodules (10, 51, 109) and in cotyledons of germinating seedlings (28, 60, 68). Biochemical studies on partially purified plant AS enzymes have been seriously hampered by the copurification of a heat-stable, dialyzable inhibitor (54, 60), the instability of AS enzyme in vitro (113), and the presence of contaminating asparaginase activity in plant extracts (51). These problems in detection of AS activities have made it difficult to monitor low-level AS activities in certain organs or slight but important changes of AS activity levels resulting from changes in growth conditions.

Molecular Studies of Asparagine Synthetase

The first two cDNA clones encoding plant AS (*AS1* and *AS2*) were obtained from a pea library using a human AS cDNA clone as a heterologous probe (130, 131). Both the pea *AS1* and *AS2* genes are expressed in leaves as well as in roots. Subsequently, studies of AS cDNA clones isolated from Arabidopsis and asparagus have shown that AS genes in these plants are expressed primarily in the leaves or the harvested spears, respectively (24, 63). The AS polypeptides encoded by these cDNA clones each contain a PurF-type glutamine-binding domain (100). This supports the notion that glutamine is the preferred substrate of plant AS. Moreover, studies of these AS cDNA clones, together with the previous biochemical data, have suggested that asparagine metabolism is regulated by the carbon/nitrogen status of a plant (63). The levels of asparagine and AS activities are also controlled by environmental and metabolic signals. Both the asparagine content in phloem exudates and AS activities are induced when light-grown plants are dark adapted (133, 134). Conversely, light and/or sucrose have been shown to result in a decrease in AS activity, as observed in sycamore cell cultures (38) and root tips of corn (12, 120).

The first striking observation of AS gene expression in pea and Arabidopsis was the high level of AS mRNA in dark-grown or dark-adapted plants (63, 130, 131). The light repression of gene expression of *AS1* in pea and *ASN1* in Arabidopsis is at least in part mediated through the action of phytochrome (63, 130). In addition to the direct phytochrome-mediated effects, light appears to exert indirect effects on AS gene expression via associated changes in carbon metabolites. In asparagus spears, it was shown that AS mRNA levels increase in harvested spears, in parallel with the decline of cellular sugar content and independent of light (24). In Arabidopsis, *ASN1* mRNA is high in dark-adapted plants, and treatment with exogenous sucrose represses the steady state level of *ASN1* transcripts (63). These molecular data are consistent with the biochemical data discussed above. Further information concerning the metabolic control of AS gene expression was obtained by demonstrating that the addition of exogenous amino acids (glutamate, glutamine, asparagine) to the growth medium was able to partially relieve the sucrose repression of the *ASN1* gene of Arabidopsis (63). This finding suggests that the ratio of organic nitrogen to carbon in a plant may be the ultimate factor controlling *ASN1* gene expression. Under conditions where levels of carbon skeletons are low relative to organic nitrogen, asparagine synthesis stores the excess nitrogen as an inert nitrogen reserve. Interestingly, in high-protein maize lines and high-protein rye ecotypes, there seems to be a shift in the composition of transported nitrogen from metabolically active glutamine to inertly stored asparagine (27).

Two new cDNA clones for Arabidopsis AS (*ASN2*, *ASN3*) were recently obtained by functional complementation of a yeast mutant lacking all AS activity (H-M Lam & G Coruzzi, unpublished data). The expression of the *ASN2* and *ASN3* genes seems to be at relatively lower levels compared with the Arabidopsis *ASN1* gene (H-M Lam & G Coruzzi, unpublished data). The mRNA levels of *ASN2* gene are regulated in an opposite manner than the *ASN1* gene. The AS enzymes encoded by these new AS genes may function to provide the required asparagine for other physiological processes such as photorespiration (124).

LIGHT AND METABOLIC CONTROL OF NITROGEN ASSIMILATION

Evidence shows that the process of nitrogen assimilation into amino acids is subject to light and metabolic control at the molecular level. Light exerts a positive effect on the expression of genes involved in ammonia assimilation into glutamine/glutamate such as on GS2 and Fd-GOGAT (30, 90, 105, 127, 149; K Coschigano & G Coruzzi, unpublished data). Conversely, light has been shown to repress genes encoding AS and GDH (42, 63, 131; R Melo-Oliveira, I Oliveira & G Coruzzi, unpublished data). The involvement of phytochrome in these light effects has been reported in some experiments (30, 63, 127, 130, 131). Further genetic experiments using the available phytochrome-deficient mutants available in Arabidopsis (99, 146) should provide more clues about which phytochrome regulates nitrogen assimilation. Although phytochrome is known to be the primary light receptor (96), the downstream signal transduction cascade is not understood. Thus, a direct linkage between the expression of genes involved in nitrogen assimilation and the light signal pathway is still lacking. The identification of light-responsive elements in plant promoters of genes encoding enzymes such as GS2 and AS in pea

(129; N Ngai & G Coruzzi, unpublished data) may be important in finding the missing link for light regulation of these nitrogen assimilatory genes.

In Arabidopsis, the reciprocal control of *GLN2* vs *ASN1* by light at the mRNA level has been shown to reflect similar light-induced changes in the levels of glutamine and asparagine. Glutamine levels are higher in light-grown plants, whereas asparagine levels are highest in dark-adapted plants (62, 107). This was also found previously in pea (133, 134). Under light conditions, nitrogen is assimilated into metabolically active glutamine and glutamate and transported as such for use in anabolic reactions in plants. Under dark-growth conditions (low carbon concentration relative to organic nitrogen), the plants direct the assimilated nitrogen into inert asparagine for long-distance transport or long-term storage.

Recently, there has been some discussion about the possible cross-talk between light control of gene expression and metabolic regulation by sugars (53). It is interesting to note that sucrose can mimic the effects of light on the expression of genes related to nitrogen metabolism such as nitrate reductase, nitrite reductase, GS2, Fd-GOGAT, GDH, and AS (18, 63; R Melo-Oliveira, I Oliveira & G Coruzzi, unpublished data; K Coschigano & G Coruzzi, unpublished data). Regulation of nitrogen assimilatory genes by the cellular carbon status reflects the interrelationship between carbon and nitrogen metabolism in plants.

Several lines of studies have focused on the metabolic control by sugars on genes related to photosynthesis and carbon metabolism (53, 111, 112). Hexose kinase is proposed to be the switching enzyme that can sense carbon availability inside the cell (53). On the basis of studies in microorganisms, a plant homologue of the yeast catabolic repression *trans*-acting factor SNF1 has been identified in rye (1). Subsequently, SNF1-related genes were isolated from Arabidopsis and barley (46, 71a). In barley, two SNF1-related protein kinases show differential expression patterns in different tissues (46). It will be important to see whether a SNF1 mutant might alter the balance of carbon and nitrogen metabolism.

In addition to the control by carbon status in the cell, it has been proposed that the relative abundance of nitrogen pools also plays a significant role in regulating nitrogen assimilation. In fact, some reports claim that the ratio of cellular carbon to nitrogen is a major player in the metabolic control of nitrogen assimilation. A homologue of a yeast general nitrogen regulatory protein NIT2 was obtained in tobacco (23). Cross-talk between the regulation of two amino acid pathways has also been reported in plants in which a blockage of histidine biosynthesis leads to a decrease in the mRNA levels of most amino acid biosynthetic enzymes, which suggests that general control of amino acid biosynthesis occurs in plants (44).

CONCLUSION

Molecular and genetic analyses have provided important tools to extend our knowledge of nitrogen assimilation based on biochemical studies. The mechanisms by which light and/or metabolic status regulate nitrogen assimilation are beginning to be dissected using cloned genes. For example, some potential regulatory genes have already been identified. In addition, specific screens for mutants in this process can be conducted in a genetically tractable system such as Arabidopsis. A combined molecular and genetic study on the regulatory network by which a gene responds to the metabolic status will lead to a better understanding of the interaction of genes controlling different carbon and nitrogen metabolic pathways. Basic research studies in these areas of nitrogen metabolism may also make significant contributions to the improvement of nitrogen usage efficiency and crop yield.

ACKNOWLEDGMENTS

This work was supported by National Institutes of Health Grant No. GM32877, United States Department of Energy Grant No. DEFG02-92-ER20071, and National Science Foundation Grant No. MCB 9304913 to G. Coruzzi, and by United States Department of Agriculture Grant No. 93-37306-9285 to K. Coschigano. We acknowledge Dr. Carolyn Schultz (University of Melbourne) and Nora Ngai (New York University) for contributions to the manuscript.

Literature cited

1. Alderson A, Sabelli PA, Dickinson JR, Cole D, Richardson M, et al. 1991. Complementation of *snf1*, a mutation affecting global regulation of carbon metabolism in yeast by a plant protein kinase cDNA. *Proc. Natl. Acad. Sci. USA* 88:8602–5
2. Anderson MP, Vance CP, Heichel GH, Miller SS. 1989. Purification and characterization of NADH-glutamate synthase from alfalfa root nodules. *Plant Physiol.* 90:351–58
3. Andrews M. 1986. The partitioning of nitrate assimilation between root and shoot of higher plants. *Plant Cell Environ.* 9:511–19
4. Artus NN. 1988. *Mutants of* Arabidopsis thaliana *that either require or are sensitive to high atmospheric CO_2 concentrations.* PhD thesis. Michigan State Univ., East Lansing, MI
5. Avila C, Márquez AJ, Pajuelo P, Cannell ME, Wallsgrove RM, Forde BG. 1993. Cloning and sequence analysis of a cDNA for barley ferredoxin-dependent glutamate synthase and molecular analysis of photorespiratory mutants deficient in the enzyme. *Planta* 189:475–83
6. Becker TW, Nef-Campa C, Zehnacker C, Hirel B. 1993. Implication of the phytochrome in light regulation of the tomato gene(s) encoding ferredoxin-dependent glutamate synthase. *Plant Physiol. Biochem.* 31:725–29
7. Blackwell RD, Murray AJS, Lea PJ. 1987. Inhibition of photosynthesis in barley with decreased levels of chloro-

plastic glutamine synthetase activity. *J. Exp. Bot.* 38:1799–809
8. Blackwell RD, Murray AJS, Lea PJ. 1987. The isolation and characterisation of photorespiratory mutants of barley and pea. In *Progress in Photosynthesis Research,* ed. J Biggins, III:625–28. Dordrecht: Nijhoff
9. Blackwell RD, Murray AJS, Lea PJ, Kendall AC, Hall NP, et al. 1988. The value of mutants unable to carry out photorespiration. *Photosynth. Res.* 16: 155–76
10. Boland MJ, Hanks JF, Reynolds PHS, Blevins DG, Tolbert NE, Schubert KR. 1982. Subcellular organization of ureide biogenesis from glycolytic intermediates in nitrogen-fixing soybean nodules. *Planta* 155:45–51
11. Bright SWJ, Lea PJ, Arruda P, Hall NP, Kendall AC, et al. 1984. Manipulation of key pathways in photorespiration and amino acid metabolism by mutation and selection. In *The Genetic Manipulation of Plants and Its Application to Agriculture*, ed. PJ Lea, GR Stewart, pp. 141–69. Oxford: Oxford Univ. Press
12. Brouquisse R, James F, Pradet A, Ray mond P. 1992. Asparagine metabolism and nitrogen distribution during protein degradation in sugar-starved maize root tips. *Planta* 188:384–95
13. Burris RH, Roberts GP. 1993. Biological nitrogen fixation. *Annu. Rev. Nutr.* 13:317–35
14. Cammaerts D, Jacobs M. 1983. A study of the polymorphism and the genetic control of the glutamate dehydrogenase isozymes in *Arabidopsis thaliana. Plant Sci. Lett.* 31:65–73
15. Cammaerts D, Jacobs M. 1985. A study of the role of glutamate dehydrogenase in the nitrogen metabolism of *Arabidopsis thaliana. Planta* 163:517–26
16. Carvalho H, Pereira S, Sunkel C, Salema R. 1992. Detection of cytosolic glutamine synthetase in leaves of *Nicotiana tabacum* L. by immunocytochemical methods. *Plant Physiol.* 100:1591–94
17. Chen FL, Cullimore JV. 1988. Two isozymes of NADH-dependent glutamate synthase in root nodules of *Phaseolus vulgaris* L.: purification, properties and activity changes during nodule development. *Plant Physiol.* 88: 1411–17
18. Cheng C-L, Acedo GN, Cristinsin M, Conkling MA. 1992. Sucrose mimics the light induction of Arabidopsis nitrate reductase gene transcription. *Proc. Natl. Acad. Sci. USA* 89:1861–64
19. Cobb A. 1992. The inhibition of amino acid biosynthesis. In *Herbicides and Plant Physiology*, pp. 126–44. New York: Chapman & Hall
20. Cock JM, Brock IW, Watson AT, Swarup R, Morby AP, Cullimore JV. 1991. Regulation of glutamine synthetase genes in leaves of *Phaseolus vulgaris. Plant Mol. Biol.* 17:761–71
21. Crawford NM, Arst HN Jr. 1993. The molecular genetics of nitrate assimilation in fungi and plants. *Annu. Rev. Genet.* 27:115–46
22. Cullimore JV, Gebhardt C, Saarelainen R, Miflin BJ, Idler KB, Barker RF. 1984. Glutamine synthetase of *Phaseolus vulgaris* L.: organ-specific expression of a multigene family. *J. Mol. Appl. Genet.* 2:589–99
23. Daniel-Vedele F, Dorbe M-F, Godon C, Truong H-N, Caboche M. 1994. Molecular genetics of nitrate assimilation in Solanaceous species. In *7th NATO/ASI on Plant Molecular Biology: Molecular-Genetic Analysis of Plant Development and Metabolism*, ed. P Puigdomenech, G Coruzzi, pp. 129–39. New York: Springer-Verlag
24. Davis KM, King GA. 1993. Isolation and characterization of a cDNA clone for a harvest induced asparagine synthetase from *Asparagus officinalis* L. *Plant. Physiol.* 102:1337–40
25. Day DA, Salom CL, Azcon-Bieto J, Dry IB, Wiskich JT. 1988. Glutamate oxidation by soybean cotyledon and leaf mitochondria. *Plant Cell Physiol.* 29: 1193–200
26. Delgado E, Mitchell RAC, Parry MA, Driscoll SP, Mitchell VJ, Lawlor DW. 1994. Interacting effects of CO_2 concentration, temperature and nitrogen supply on the photosynthesis and composition of winter wheat leaves. *Plant Cell Environ.* 17:1205–13
27. Dembinski E, Bany S. 1991. The amino acid pool of high and low protein rye inbred lines (*Secale cereale* L.). *J. Plant Physiol.* 138:494–96
28. Dilworth MF, Dure L III. 1978. Developmental biochemistry of cotton seed empbryogenesis and germination. X. Nitrogen flow from arginine to asparagine in germination. *Plant Physiol.* 61:698–702
29. Duke SH, Schrader LE, Miller MG, Niece RL. 1978. Low temperature effects on soybean (*Glycine max* (L) Merr. cv. Wells) free amino acid pools during germination. *Plant Physiol.* 62:642–47
30. Edwards JW, Coruzzi GM. 1989. Photorespiration and light act in concert to regulate the expression of the nuclear gene for chloroplast glutamine synthetase. *Plant Cell* 1:241–48

31. Edwards JW, Walker EL, Coruzzi GM. 1990. Cell-specific expression in transgenic plants reveals nonoverlapping roles for chloroplast and cytosolic glutamine synthetase. *Proc. Natl. Acad. Sci. USA* 87:3459–63
32. Elmlinger MW, Mohr H. 1991. Coaction of blue/ultraviolet-A light and light absorbed by phytochrome in controlling the appearance of ferredoxin-dependent glutamate synthase in the Scots pine (*Pinus sylvestris* L.) seedling. *Planta* 183:374–80
33. Faure JD, Jullien M, Caboche M. 1994. *Zea3*: a pleiotropic mutation affecting cotyledon development, cytokinin, resistance and carbon-nitrogen metabolism. *Plant J.* 5:481–91
34. Forde BG, Day HM, Turton JF, Shen W-J, Cullimore JV, Oliver JE. 1989. Two glutamine synthetase genes from *Phaseolus vulgaris* L. display contrasting developmental and spatial patterns of expression in transgenic *Lotus corniculatus* plants. *Plant Cell* 1:391–401
35. Gantt JS, Larson RJ, Farnham MW, Pathirana SM, Miller SS, Vance CP. 1992. Aspartate aminotransferase in effective and ineffective alfalfa nodules. *Plant Physiol.* 98:868–78
36. García-Gutiérrez A, Cantón FR, Gallardo F, Sánchez-Jiménez F, Ceanovas FM. 1995. Expression of ferredoxin-dependent glutamate synthase in dark-grown pine seedlings. *Plant Mol. Biol.* 27:115–28
37. Gebhardt C, Oliver JE, Forde BG, Saarelainen R, Miflin BJ. 1986. Primary structure and differential expression of glutamine synthetase genes in nodules, roots and leaves of *Phaseolus vulgaris*. *EMBO J.* 5:1429–35
38. Genix P, Bligny R, Martin J-B, Douce R. 1994. Transient accumulation of asparagine in sycamore cells after a long period of sucrose starvation. *Plant Physiol.* 94:717–22
39. Givan CV. 1979. Metabolic detoxification of ammonia in tissues of higher plants. *Phytochemistry* 18:375–82
40. Givan CV. 1980. Aminotransferases in higher plants. In *The Biochemistry of Plants: Amino Acids and Derivatives*, ed. BJ Miflin, 5:329–57. New York: Academic
41. Givan CV, Joy KW, Kleczkowski LA. 1988. A decade of photorespiratory nitrogen recycling. *Trends Biol. Sci.* 13: 433–37
42. Goldberg RB, Barker SJ, Perez-Grau L. 1989. Regulation of gene expression during plant embryogenesis. *Cell* 56: 149–60
43. Gregerson RG, Miller SS, Twary SN, Gantt JS, Vance CP. 1993. Molecular characterization of NADH-dependent glutamate synthase from alfalfa nodules. *Plant Cell* 5:215–26
44. Guyer D, Patton D, Ward E. 1995. Evidence for cross-pathway regulation of metabolic gene expression in plants. *Proc. Natl. Acad. Sci. USA* 92:4997–5000
45. Hageman RH, Lambert RJ. 1988. The use of physiological traits for corn improvement. In *Corn and Corn Improvement*, ed. GF Sprague, JW Dudley, pp. 431–61. Madison: Am. Soc. Agron. Soil Soc. Am. 3rd ed.
46. Hannappel U, Vicente-Carbajosa J, Barker JHA, Shewry PR, Halford NG. 1995. Differential expression of two barley SNF-1 related protein kinase genes. *Plant Mol. Biol.* 27:1235–40
47. Hatch MD, Mau S-L. 1973. Activity, location, and role of asparate aminotransferase and alanine aminotransferase isoenzymes in leaves with C4 pathway photosynthesis. *Arch. Biochem. Biophys.* 156:195–206
48. Hecht U, Oelmuller R, Schmidt S, Mohr H. 1988. Action of light, nitrate and ammonium on the levels of NADH- and ferredoxin-dependent glutamate synthases in the cotyledons of mustard seedlings. *Planta* 175:130–38
49. Hirel B, Bouet C, King B, Layzell B, Jacobs F, Verma DPS. 1987. Glutamine synthetase genes are regulated by ammonia provided externally or by symbiotic nitrogen fixation. *EMBO J.* 6: 1167–71
50. Hoff T, Truong H-N, Caboche M. 1994. The use of mutants and transgenic plants to study nitrate assimilation. *Plant Cell Environ.* 17:489–506
51. Huber TA, Streeter JG. 1985. Purification and properties of asparagine synthetase from soybean root nodules. *Plant Sci.* 42:9–17
52. Ireland RJ, Joy KW. 1985. Plant transaminases. In *Transaminases*, ed. P Christen, DE Metzler, pp. 376–84. New York: Wiley
53. Jang J-C, Sheen J. 1994. Sugar sensing in higher plants. *Plant Cell* 6: 1665–79
54. Joy KW, Ireland RJ, Lea PJ. 1983. Asparagine synthesis in pea leaves, and the occurrence of an asparagine synthetase inhibitor. *Plant Physiol.* 73:165–68
55. Kamachi K, Yamaya T, Hayakawa T, Mae T, Ojima K. 1992. Changes in cytosolic glutamine synthetase polypeptide and its mRNA in a leaf blade of

rice plants during natural senesence. *Plant Physiol.* 98:1323–29
56. Kamachi K, Yamaya T, Hayakawa T, Mae T, Ojima K. 1992. Vascular bundle-specific localization of cytosolic glutamine synthetase in rice leaves. *Plant Physiol.* 99:1481–86
57. Kamachi K, Yamaya T, Mae T, Ojima K. 1991. A role for glutamine synthetase in the remobilization of leaf nitrogen during natural senescence in rice leaves. *Plant Physiol.* 96:411–17
58. Kawakami N, Watanable A. 1988. Senescence-specific increase in cytosolic glutamine synthetase and its mRNA in radish cotyledons. *Plant Physiol.* 98: 1323–29
59. Kendall AC, Wallsgrove RM, Hall NP, Turner JC, Lea PJ. 1986. Carbon and nitrogen metabolism in barley (*Hordeum vulgare* L) mutants lacking ferredoxin-dependent glutamate synthase. *Planta* 168:316–23
60. Kern R, Chrispeels MJ. 1978. Influence of the axis in the enzymes of protein and amide metabolism in the cotyledons of mung bean seedlings. *Plant Physiol.* 62:815–19
61. Keys AJ, Bird IF, Cornelius MJ, Lea PJ, Wallsgrove RM, Miflin BJ. 1978. Photorespiratory nitrogen cycle. *Nature* 275:741–43
62. Lam H-M, Coschigano K, Schultz C, Melo-Oliveira R, Tjaden G, et al. 1995. Use of Arabidopsis mutants and genes to study amide amino acid biosynthesis. *Plant Cell* 7:887–98
63. Lam H-M, Peng SS-Y, Coruzzi GM. 1994. Metabolic regulation of the gene encoding glutamine-dependent asparagine synthetase in *Arabidopsis thaliana*. *Plant Physiol.* 106:1347–57
64. Lawlor DW, Kontturi M, Young AT. 1989. Photosynthesis by flag leaves of wheat in relation to protein, ribulose bisphosphate carboxylase activity and nitrogen supply. *J. Exp. Bot.* 40:43–52
65. Deleted in proof
66. Lea PJ. 1993. Nitrogen metabolism. In *Plant Biochemistry and Molecular Biology*, ed. PJ Lea, RC Leegood, pp. 155–80. New York: Wiley
67. Lea PJ, Blackwell RD, Murray AJS, Joy KW. 1988. The use of mutants lacking glutamine synthetase and glutamate synthase to study their role in plant nitrogen metabolism. In *Plant Nitrogen Metabolism Recent Advances in Phytochemistry,* ed. JE Poulton, JT Romeo, EE Conn, 23:157–89. New York/London: Plenum
68. Lea PJ, Fowden L. 1975. The purification and properties of glutamine-dependent asparagine synthetase isolated from *Lupinus albus*. *Proc. R. Soc. London Ser. B* 192:13–26
69. Lea PJ, Miflin B. 1980. Transport and metabolism of asparagine and other nitrogen compounds within the plant. See Ref. 40, pp. 569–607
70. Lea PJ, Robinson SA, Stewart GR. 1990. The enzymology and metabolism of glutamine, glutamate, and asparagine. In *The Biochemistry of Plants,* ed. BJ Miflin, PJ Lea, 16:121–59. New York: Academic
71. Lea PJ, Thurman DA. 1972. Intracellular Location and properties of plant L-glutamate dehydrogenases. *J. Exp. Bot.* 23:440–49
71a. Le Guen L, Thomas M, Bianchi M, Halford NG, Kreis M. 1992. Structure and expression of a gene from *Arabidopsis thaliana* encoding a protein related to SNF1 protein kinase. *Gene* 120: 249–54
72. Li M-G, Villemur R, Hussey PJ, Silflow CD, Gantt JS, Snustad DP. 1993. Differential expression of six glutamine synthetase genes in *Zea mays*. *Plant Mol. Biol.* 23:401–7
73. Lightfoot DA, Green NK, Cullimore JV. 1988. The chloroplast located glutamine synthetase of *Phaseolus vulgaris* L: nucleotide sequence, expression in different organs and uptake into isolated chloroplasts. *Plant Mol. Biol.* 11:191–202
74. Loulakakis CA, Roubelakis-Angelakis KA. 1990. Intracellular localization and properties of NADH-glutamate dehydrogenase from *Vitis vinifera* L.: purification and characterization of the major leaf isoenzyme. *J. Exp. Bot.* 41: 1223–30
75. Loulakakis KA, Roubelakis-Angelakis KA, Kanellis AK. 1994. Regulation of glutamate dehydrogenase and glutamine synthetase in avocado fruit during development and ripening. *Plant Physiol.* 106:217–22
76. Loyola-Vargas VM, Jimenez ES. 1984. Differential role of glutamate dehydrogenase in nitrogen metabolism of maize tissues. *Plant Physiol.* 76:536–40
77. Magalhaes JR, Ju GC, Rich PJ, Rhodes D. 1990. Kinetics of $^{15}NH_4^+$ assimilation in *Zea mays*: preliminary studies with a glutamate dehydrogenase (*GDH1*) null mutant. *Plant Physiol.* 94:647–56
78. Marsolier M-C, Carrayol E, Hirel B. 1993. Multiple functions of promoter sequences involved in organ-specific expression and ammonia regulation of a cytosolic soybean glutamine synthetase

gene in transgenic Lotus. *Plant J.* 3:405–14

79. Marsolier M-C, Hirel B. 1993. Metabolic and developmental control of cytosolic glutamine synthetase genes in soybean. *Physiol. Plant.* 89:613–17
80. Matoh T, Takahashi E. 1982. Changes in the activites of ferredoxin and NADH-glutamate synthase during seedling development of peas. *Planta* 154: 289–94
81. Miao GH, Hitel B, Marsolier MC, Ridge RW, Verma DP. 1991. Ammonia-regulated expression of a soybean gene encoding cytosolic glutamine synthetase in transgenic *Lotus corniculatus*. *Plant Cell* 3:11–22
82. Miflin BJ. 1974. The location of nitrite reductase and other enzymes related to amino acid biosynthesis in the plastids of root and leaves. *Plant Physiol.* 54: 550–55
83. Miflin BJ, Lea PJ. 1976. The pathway of nitrogen assimilation in plants. *Phytochemistry* 15:873–85
84. Miflin BJ, Lea PJ. 1980. Ammonia assimilation. See Ref. 40, pp. 169–202
85. Nalbantoglu B, Hirasawa M, Moomaw C, Nguyen H, Knaff DB, Allen R. 1994. Cloning and sequencing of the gene encoding spinach ferredoxin-dependent glutamate synthase. *Biochim. Biophys. Acta* 1183:557–61
86. Oaks A, Ross DW. 1984. Asparagine synthetase in *Zea mays*. *Can. J. Bot.* 62:68–73
87. Oaks A, Stulen I, Jones K, Winspear MJ, Misra S, Boesel IL. 1980. Enzymes of nitrogen assimilation in maize roots. *Planta* 148:477–84
88. Peoples MB, Gifford RM. 1993. Long-distance transport of carbon and nitrogen from sources to sinks in higher plants. In *Plant Physiology, Biochemistry and Molecular Biology*, ed. DT Dennis, DH Turpin, pp. 434–47. New York: Wiley
89. Peoples MB, Pate JS, Atkins CA, Bergersen FJ. 1986. Nitrogen nutrition and xylem sap composition of peanut (*Arachis hypogaea* L. cv Virginia Bunch). *Plant Physiol.* 82:946–51
90. Peterman TK, Goodman HM. 1991. The glutamine synthetase gene family of *Arabidopsis thaliana*: light-regulation and differential expression in leaves, roots, and seeds. *Mol. Gen. Genet.* 230: 145–54
91. Poulton JE, Romeo JT, Conn CC, eds. 1989. *Plant Nitrogen Metabolism: Recent Advances in Phytochemistry,* 23. New York: Plenum
92. Pryor AJ. 1974. Allelic glutamic dehydrogenase isozymes in maize: a single hybrid isozyme in heterozygotes? *Heredity* 32:397–419
93. Pryor AJ. 1979. Mapping of glutamic dehydrogenase (*Gdh*) on chromosome 1, 20.1 recombination units distal to *Adh 1*. *Maize Genet. Coop. Newsl.* 53:25–26
94. Pryor AJ. 1990. A maize glutamic dehydrogenase null mutant is cold temperature sensitive. *Maydica* 35:367–72
95. Quail PH. 1979. Plant cell fractionation. *Annu. Rev. Plant Physiol.* 30:425–84
96. Quail PH, Boylan MT, Parks BM, Short TW, Xu Y, Wagner D. 1995. Phytochromes: photosensory perception and signal transduction. *Science* 168:675–80
97. Ratajczak L, Ratajczak W, Mazurowa H. 1981. The effect of different carbon and nitrogen sources on the activity of glutamine synthetase and glutamate dehydrogenase in lupine embryonic axes. *Physiol. Plant.* 51:277–80
98. Redinbaugh MG, Campbell WH. 1993. Glutamine synthetase and ferrodoxin-dependent glutamate synthase expression in the maize (*Zea mays*) root primary response to nitrate. *Plant Physiol.* 101:1249–55
99. Reed JW, Chory J. 1994. Mutational analyses of light-controlled seedling development in Arabidopsis. *Semin. Cell Biol.* 5:327–34
100. Richards NGJ, Schuster SM. 1992. An alternative mechanism for the nitrogen transfer reaction in asparagine synthetase. *FEBS Lett.* 313:98–102
101. Robinson DL, Kahn ML, Vance CP. 1994. Cellular localization of nodule-enchanced aspartate aminotransferase in *Medicago sativa* L. *Planta* 192:202–10
102. Roche D, Temple SJ, Sengupta-Gopalan C. 1993. Two classes of differentially regulated glutamine synthetase genes are expressed in the soybean nodule: a nodule-specific and a constitutively expressed class. *Plant Mol. Biol.* 22:971–83
103. Sakakibara H, Fujii K, Sugiyama T. 1995. Isolation and characterization of a cDNA that encodes maize glutamate dehydrogenase. *Plant Cell Physiol.* 36(5):789–97
104. Sakakibara H, Kawabata S, Takahashi H, Hase T, Sugiyama T. 1992. Molecular cloning of the family of glutamine synthetase genes from maize: expression of genes for glutamine synthetase and ferrodoxin-dependent glutamate synthase in photosynthetic and nonphotosynthetic tissues. *Plant Cell Physiol.* 33:49–58
105. Sakakibara H, Matanabe M, Hase T, Sugiyama T. 1991. Molecular cloning

and characterization of complementary DNA encoding for ferredoxin dependent glutamate synthase in maize leaf. *J. Biol. Chem.* 266:2028–35

106. Sakamoto A, Ogawa M, Masumura T, Shibata D, Takeba G, et al. 1989. Three cDNA sequences coding for glutamine synthetase polypeptides in *Oryza sativa* L. *Plant Mol. Biol.* 13:611–14
107. Schultz CJ. 1994. *A molecular and genetic dissection of the aspartate aminotransferase isoenzymes of* Arabidopsis thaliana. PhD thesis. New York Univ., New York
108. Schultz CJ, Coruzzi GM. 1995. The aspartate aminotransferase gene family of Arabidopsis encodes isoenzymes localized to three distinct subcellular compartments. *Plant J.* 7:61–75
109. Scott DB, Farnden KJF, Robertson JG. 1976. Ammonia aasimilation in lupin nudules. *Nature* 263:703–5
110. Sechley KA, Yamaya T, Oaks A. 1992. Compartmentation of nitrogen assimilation in higher plants. *Int. Rev. Cytol.* 134:85–163
111. Sheen J-Y. 1990. Metabolic repression of transcription in higher plants. *Plant Cell* 2:1027–38
112. Sheen J-Y. 1994. Feedback control of gene expression. *Photosynth. Res.* 39: 427–38
113. Sieciechowicz KA, Joy KW, Ireland RJ. 1988. The metabolism of asparagine in plants. *Phytochemistry* 27:663–71
114. Sivasankar S, Oaks A. 1995. Regulation of nitrate reductase during early seedling growth: a role for asparagine and glutamine. *Plant Physiol.* 107:1225–31
115. Snustad DP, Hunsperger JP, Chereskin BM, Messing J. 1988. Maize glutamine synthetase cDNAs: isolation by direct genetic selection in *Escherichia coli*. *Genetics* 120:1111–24
116. Somerville CR, Ogren WL. 1980. Inhibition of photosynthesis in Arabidopsis mutants lacking leaf glutamate synthase activity. *Nature* 286:257–59
117. Somerville CR, Ogren WL. 1982. Genetic modification of photorespiration. *Trends Biochem. Sci.* 7:171–74
118. Srivastava HS, Singh Rana P. 1987. Role and regulation of L-glutamate dehydrogenase activity in higher plants. *Phytochemistry* 26:597–610
119. Stewart GR, Mann AF, Fentem PA. 1980. Enzymes of glutamate formation: glutamate dehydrogenase, glutamine synthetase, glutamate synthase. See Ref. 40, pp. 271–327
120. Stulen I, Oaks A. 1977. Asparagine synthetase in corn roots. *Plant Physiol.* 60:680–83
121. Sukanya R, Li M-G, Snustad DP. 1994. Root- and shoot-specific responses of individual glutamine synthetase genes of maize to nitrate and ammonia. *Plant Mol. Biol.* 26:1935–46
122. Suzuki A, Gadal P. 1984. Glutamate synthase: physicochemical and functional properties of different forms in higher plants and other organisms. *Physiol. Veg.* 22:471–86
123. Suzuki A, Vidal J, Gadal P. 1982. Glutamate synthase isoforms in rice: Immunological studies of enzymes in green leaf, etiolated leaf, and root tissues. *Plant Physiol.* 70:827–32
124. Ta TC, Joy KW. 1986. Metabolism of some amino acids in relation to the photorespiratory nitrogen cycle of pea leaves. *Planta* 169:117–22
125. Taniguchi M, Kobe A, Kato M, Sugiyama T. 1995. Aspartate aminotransferase isoenzymes in *Panicum miliaceum* L., and NAD-malic enzyme–type C_4 plant: comparison of enzymatic properties, primary structures, and expression patterns. *Arch. Biochem. Biophys.* 318:295–306
126. Temple SJ, Heard J, Ganter J, Dunn G, Sengupta-Gopalan C. 1995. Characterization of a nodule-enhanced glutamine synthetase from alfalfa: nucleotide sequence, *in situ* localization, and transcript analysis. *Mol. Plant-Microbe Interact.* 8:218–27
127. Tingey SV, Tsai F-Y, Edwards JW, Walker EL, Coruzzi GM. 1988. Chloroplast and cytosolic glutamine synthetase are encoded by homologous nuclear genes which are differentially expressed *in vivo*. *J. Biol. Chem.* 263: 9651–57
128. Tingey SV, Walker EL, Coruzzi GM. 1987. Glutamine synthetase genes of pea encode distinct polypeptides which are differentially expressed in leaves, roots and nodules. *EMBO J.* 6:1–9
129. Tjaden G, Edwards JW, Coruzzi GM. 1995. *cis* elements and *trans*-acting factors affecting regulation of a non-photosynthetic light-regulated gene for chloroplast glutamine synthetase. *Plant Physiol.* 108:1109–17
130. Tsai F-Y, Coruzzi GM. 1990. Dark-induced and organ-specific expression of two asparagine synthetase genes in *Pisum sativum*. *EMBO J.* 9:323–32
131. Tsai F-Y, Coruzzi GM. 1991. Light represses the transcription of asparagine synthetase genes in photosynthetic and non-photosynthetic organs of plants. *Mol. Cell. Biol.* 11:4966–72
132. Udvardi MK, Kahn ML. 1991. Isolation and analysis of a cDNA clone that en-

codes an alfalfa (*Meticago sativa*) aspartate aminotransferase. *Mol. Gen. Genet.* 231:97–105
133. Urquhart AA, Joy KW. 1981. Use of phloem exudate technique in the study of amino acid transport in pea plants. *Plant Physiol.* 68:750–54
134. Urquhart AA, Joy KW. 1982. Transport, metabolism, and redistribution of Xylem-borne amino acids in developing pea shoots. *Plant Physiol.* 69:1226–32
135. Vance CP. 1990. Symbiotic nitrogen fixation: recent genetic advances. See Ref. 70, pp. 43–88
136. Vance CP, Gantt JS. 1992. Control of nitrogen and carbon metabolism in root nodules. *Physiol. Plant.* 85:266–74
137. Vance CP, Miller SS, Gregerson RG, Samac DA, Robinson DL, Gantt JS. 1995. Alfalfa NADH-dependent glutamate synthase: structure of the gene and importance in symbiotic N_2 fixation. *Plant J.* 8:345–58
138. Vauquelin LN, Robiquet PJ. 1806. The discovery of a new plant principle in *Asparagus sativus. Ann. Chim.* 57:88–93
139. Vincentz M, Moureaux T, Leydecker M-T, Vaucheret H, Caboche M. 1993. Regulation of nitrate and nitrite reductase expression in *Nicotiana plumbaginifolia* leaves by nitrogen and carbon metabolites. *Plant J.* 3:315–24
140. Wadsworth GW, Marmaras SM, Matthews BF. 1993. Isolation and characterization of a soybean cDNA clone encoding the plastid form of aspartate aminotransferase. *Plant Mol. Biol.* 21: 993–1009
141. Walker EL, Coruzzi GM. 1989. Developmentally regulated expression of the gene family for cytosolic glutamine synthetase in *Pisum sativum. Plant Physiol.* 91:702–8
142. Wallsgrove RM, Lea PJ, Miflin BJ. 1979. Distribution of the enzymes of nitrogen assimilation within the pea leaf cell. *Plant Physiol.* 63:232–36
143. Wallsgrove RM, Turner JC, Hall NP, Kendall AC, Bright SWJ. 1987. Barley mutants lacking chloroplast glutamine synthetase-biochemical and genetic analysis. *Plant Physiol.* 83:155–58
144. Weeden NF, Gottlieb LD. 1980. The genetics of chloroplast enzymes. *J. Hered.* 71:392–96
145. Wendel JF, Weeden NF. 1989. Visualization and interpretation of plant isozymes. In *Isoenzymes in Plant Biology*, ed. DE Soltis, PE Soltis, pp. 5–45. Oregon: Dioscorides Press
146. Whitelam GC, Harberd N. 1994. Action and function of phytochrome family members revealed through the study of mutant and transgenic plants. *Plant Cell Environ.* 17:615–25
147. Wilkie SE, Roper J, Smith A, Warren MJ. 1995. Isolation, characterisation and expression of a cDNA clone encoding aspartate aminotransferase from *Arabidopsis thaliana. Plant Mol. Biol.* 27: 1227–33
148. Yamaya T, Oaks A. 1987. Synthesis of glutamate by mitochondria: an anaplerotic function for glutamate dehydrogenase. *Physiol. Plant.* 70:749–56
149. Zehnacker C, Becker TW, Suzuki A, Carrayol E, Caboche M, Hirel B. 1992. Purification and properties of tobacco ferredoxin-dependent glutamate synthase, and isolation of corresponding cDNA clones. *Planta* 187:266–74

Annu. Rev. Plant Physiol. Plant Mol. Biol. 1996. 47:595–626

MEMBRANE TRANSPORT CARRIERS

W. Tanner and T. Caspari
Lehrstuhl für Zellbiologie und Pflanzenphysiologie, Universität Regensburg, 93040 Regensburg, Germany

KEY WORDS: plant carriers, carriers of *Saccharomyces cerevisiae,* membrane topology, helix packing, regulation

ABSTRACT

Plant and fungal membrane proteins catalyzing the transmembrane translocation of small molecules without directly using ATP or acting as channels are discussed in this review. Facilitators, ion-cotransporters, and exchange translocators mainly for sugars, amino acids, and ions that have been cloned and characterized from *Saccharomyces cerevisiae* and from various plant sources have been tabulated. The membrane topology and structure of the most extensively studied carriers (lac permease of *Escherichia coli,* Glut1 of man, HUP1 of *Chlorella*) are discussed in detail as well as the kinetic analysis of specific Na^+ and H^+ cotransporters. Finally, the knowledge concerning regulatory phenomena of carriers—mainly of *S. cerevisiae*—is summarized.

CONTENTS

INTRODUCTION

Plant membrane transport has frequently been reviewed in recent years. These reviews were focused on either ion channels (14, 69, 153, 176), ion pumps

1040-2519/96/0601-0595$08.00

(26, 132, 158), or carriers for specific classes of molecules (19, 23, 56, 131, 149, 158). Progress in the field, however, especially since the advent of molecular biology, has been more rapid than ever before. The emphasis in this review is shifted from previous reviews on carriers. We summarize those carriers that have been cloned and characterized from plant and fungal cells in the past five to six years, and we discuss their more general properties, such as their structure (what little we know of it), regulation, and kinetics. The review is not restricted to membrane proteins of plant cells. Percentage identities and degrees of similarity cannot be confined to DNA sequences. Whatever the rules are for a membrane protein to catalyze a transmembrane transport step, it probably matters little whether this protein is localized in a bacterial, plant, or mammalian membrane. Thus, whenever we can learn from an organism belonging to a neighboring territory, we should certainly consider—at least for an initial working hypothesis—the knowledge already available.

Fungal cells, for example *Saccharomyces cerevisiae,* are discussed in this review. Because bacterial and plant membrane proteins can be functionally expressed in yeasts (73, 143, 144) and because this feature has been elegantly used to clone plant genes by complementation (4, 138, 157), yeasts and their transport properties are of great interest to plant transport physiologists. Yeast as a tool has recently been reviewed (57). Yeast with its own membrane transport properties is partly covered in this review.

NOMENCLATURE

Names and designations given to transport proteins catalyzing the passage of molecules through biological membranes are manifold, and their precise definitions are often overlapping, or even confusing (see discussion in 19). Confusion recently arose over the observation that the ATP–binding cassette superfamily of transporter proteins (ABC transporters), such as multidrug resistance proteins and the famous human cystic fibrosis transmembrane conductance regulator (*CFTR*), act or at least seem to act as channels and to require ATP (66, 177). It is more than unlikely, however, that uphill transport through channels could exist. It seems clear that—except for *CFTR*—whenever transport, ATP requirement, ATP hydrolysis, and stoichiometry of ABC transporters have been measured they qualify as primary active transporters or pumps using ATP for creating a steep uphill gradient of substrate (2, 72). The stoichiometry of the maltose transporter, for example, has been shown to be 1–2 ATP per maltose (72). The alleged Cl^--channel properties of multidrug resistance proteins were recently shown to be a result of the fact that these proteins regulate independent membrane proteins, which are Cl^- channels (71). Finally, for *CFTR,* which acts as a Cl^- channel under physiological conditions, ATP as well as its hydrolysis seems to be required for the opening/closing cycle of the channel (66). Energy is thus required to allow downhill transport in a regulated manner by an unusual ligand-gated ion channel. Whether bifunctional transport molecules really exist and whether in vivo they work in both modes, e.g. as pumps and as channels or as carriers and channels (155), are completely open questions and should not be considered too seriously until hard evidence is available.

In this review, we concentrate on membrane transport proteins generally classified as carriers, i.e. facilitators or uniporters, secondary active transporters such as sym- or antiporters, and exchange translocators. In all these cases, the transport protein is more or less considered an enzyme catalyzing transmembrane translocation without ever directly using ATP. Proteins hydrolyzing ATP during transport generally are pumps such as ion ATPases or ABC transporters—in the area of the latter, important discoveries have lately been made in plants and yeast (10, 76, 109)—and they, as well as channels, are not covered.

A BRIEF HISTORICAL EXCURSION

In July 1995, for the first time, the total genome of an organism was decoded. 1743 genes make a prokaryotic cell of the genus *Haemophilus influenzae* (49). This number may be astonishingly low, but it certainly is a surprise that 12.2% of the genes of an organism code for membrane transporters and assisting proteins.

As little as thirty years ago, most biologists and especially biochemists were still skeptical about the existence of carriers. We present some historical data that show the rapid and exciting progress in the field.

Seventy years ago, Cori (29, 30) laid the experimental groundwork for the carrier concept. Following sugar absorption in the small intestine of rats, the observed specificities, the saturation behavior, and the competition between chemically related substrates (D-glucose and D-galactose) suggested that some transport catalysts exist in cell membranes. Plant physiologists and bacteriologists came to the same conclusion when they tried to explain the enormous intracellular accumulation of ions and amino acids, respectively [see Epstein (43), Cohen & Monod (28), and references therein]. The thorough treatment of potassium uptake, like an enzymatic reaction in the study of Epstein & Hagen (44), made the physicochemical subject of transport more accessible for plant physiologists and plant biochemists.

Forty years ago Rickenberg et al (136) demonstrated that the *Y* gene of the lac-operon of *Escherichia coli* coded for a protein, which they called lac-permease. About ten years later, the gene product was made visible for the first time on a gel as a labeled protein band with an apparent molecular mass of 30 kDa (52). In 1980, the first gene coding for a transport protein—again the β-galactoside transporter of *E. coli*—was cloned (20, 38), and the gene product was shown in vitro to be solely responsible for active lactose transport (116). As mentioned above, we can now estimate that a typical bacterial cell contains approximately 200 genes responsible for its transport endeavors, which code for pumps, channels, and carriers (49). For a eukaryotic cell such as *S. cerevisiae,* the corresponding genes may amount to about 1000. Judging from the chromosomes sequenced to date, about 33% of the ORFs are membrane proteins (estimated total ORFs of *S. cerevisiae* ≈ 6500), which, however, include those that are components of internal membranes (46). These considerations, however, already extrapolate into the future.

The problem of ion and nonelectrolyte accumulation had at least in principle been solved in the 1960s and 1970s, when the Na^+- and the H^+-gradient hypothesis of Crane (32) and Mitchell (114) had been proven experimentally for animals (32), bacteria (185), plants (93, 95), and fungi (156, 159). Mechanistic details, however, are even today hardly understood.

CARRIERS CLASSIFIED ACCORDING TO SUBSTRATES: MOLECULAR CLONING

In the following paragraph, the molecular cloning of *S. cerevisiae* and plant carriers, an activity booming in the field, is briefly summarized.

Amino Acids, Peptides

FUNGI Amino acid transport of fungi has recently been excellently summarized by Grenson (65), who laid the foundation for all we know today about this topic in *S. cerevisiae*. These cells possess a large number of amino acid permeases with considerable overlapping substrate specificity. Prerequisite to the successful molecular cloning of amino acid transporters was careful physiological characterization of the various transporters and the isolation of mutants. The latter has been based either on the resistance to toxic amino acid analogues such as the arginine analogue canavanine or on the lack of growth of auxotrophic strains on certain amino acids. Complementation of such mutants allowed the cloning first of the *CAN1* and the *HIP1* gene, coding for the arginine and histidine transporters (75, 171), and subsequently the genes for proline and the general amino acid transporters (82, 178), as well as a specific transporter for γ-aminobutyric acid (5). It is interesting to note that a mutation in the *S. cerevisiae SHR3* gene affects pleiotropically the secretion of a large number of amino acid permeases, but not other secretory proteins such as invertase (105). In Table 1, additional amino acid transporter genes of *S. cerevisiae* are compiled.

Table 1 Cloned carriers[a]

I. Amino acid and peptide transporters

A. Fungi

Gene	Organism	Reaction	Remarks[b]	Reference
CAN1	*S. cerevisiae*	arginine transport		75
HIP1	*S. cerevisiae*	histidine transport		171
PUT4	*S. cerevisiae*	proline transport	NCR[c] and NCI[c]	178
GAP1	*S. cerevisiae*	amino acid transport	NCR and NCI, protein phosphorylated	82, 164
LYP1	*S. cerevisiae*	lysine transport		169
UGA4	*S. cerevisiae*	GABA transport	GABA induced, NCR and NCI	5
TAT1	*S. cerevisiae*	tyrosine transport	expression confers resistance to FK506	151
TAT2/SCM2	*S. cerevisiae*	tryptophane transport	expression confers resistance to FK506	151
PTR2	*S. cerevisiae*	peptide transport	two further genes, inducible	127

B. Plants

Gene	Organism	Reaction	Remarks	Reference
AAP1/NAT2	*A. thaliana*	amino acid transport	flower and silique	54, 78
AAP2	*A. thaliana*	amino acid transport	stem, silique, flower, root, source leaf	48
AAP3	*A. thaliana*	amino acid transport	root	48
AAP4	*A. thaliana*	amino acid transport	stem, source leaf, flower	48
AAP5	*A. thaliana*	amino acid transport	source leaf, stem, flower, silique, root	48
AAP6	*A. thaliana*	amino acid transport	sink leaf	48
AAT1	*A. thaliana*	basic amino acid transport		55
ProT1	*A. thaliana*	proline transport		134
ProT2	*A. thaliana*	proline transport		134
NTR1	*A. thaliana*	peptide transport	source leaf, silique	53
At PTR2	*A. thaliana*	di- and tripeptide transport	root	165
BAP1	*H. vulgare*	proline transport		101
BAP2	*H. vulgare*	proline transport		101

II. Sugar Transporters

A. Fungi

Gene	Organism	Reaction	Remarks	Reference
SNF3	*S. cerevisiae*		regulator	12, 25
HXT1	*S. cerevisiae*	glucose transport	high glucose induced	104, 124
HXT2	*S. cerevisiae*	glucose transport	low glucose induced, high glucose repressed	
HXT3	*S. cerevisiae*	glucose transport	glucose induced	91, 124
HXT4/LGT1	*S. cerevisiae*	glucose transport	low glucose induced, high glucose repressed	124, 173
HXT5-7	*S. cerevisiae*	glucose transport		133
GAL2	*S. cerevisiae*	galactose transport	galactose induced, glucose repressed	170
ITR1	*S. cerevisiae*	myo-inositol transport	turnover by End3p, End4p, Pep4p	102, 118

Table 1 (*continued*)

Gene	Organism	Reaction	Remarks[b]	Reference
ITR2	*S. cerevisiae*	myo-inositol transport		118
MAL61	*S. cerevisiae*	maltose transport	maltose induced, glucose repressed	27
AGT1	*S. cerevisiae*	maltose transport	maltose induced, glucose repressed	68
B. Plants				
HUP1	*C. kessleri*	monosaccharide transport	glucose induced	148
HUP2	*C. kessleri*	monosaccharide transport	glucose induced	163
HUP3	*C. kessleri*	monosaccharide transport	glucose induced	163
STP1	*A. thaliana*	monosaccharide transport	ovaries	142, 144
STP2	*A. thaliana*		anthers	142
STP3	*A. thaliana*		leaf, sepals	142
STP4	*A. thaliana*		anthers, root tips	142
AtSUC1	*A. thaliana*	sucrose transport		147
AtSUC2	*A. thaliana*	sucrose transport	phloem	147, 174
MST1	*N. tabacum*	monosaccharide transport	root, sink leaf	146
Hex3	*R. communis*	monosaccharide transport	root, sink leaf	184
SoSUT1	*S. oleracea*	sucrose transport		138
StSUT1	*S. tuberosum*	sucrose transport	phloem	137
PmSUC1	*P. major*	sucrose transport	developing seeds	61
PmSUC2	*P. major*	sucrose transport	phloem	62, 162

III. Inorganic Cation and Anion Transporters

A. Fungi

Gene	Organism	Reaction	Remarks	Reference
MEP1	*S. cerevisiae*	ammonium transport	NCR, NCI	108
TRK1	*S. cerevisiae*	potassium transport	high affinity (>0.1 mM)	60
TRK2	*S. cerevisiae*	potassium transport	low affinity (5 mM)	90
PHO84	*S. cerevisiae*	phosphate transport	phosphate repressed,	18
SUL1	*S. cerevisiae*	sulfate transport		161
B. Plants				
AMT1	*A. thaliana*	ammonium transport	root, leaf, stem, seedling	119
HKT1	*T. aestivum*	potassium transport	root	150
CHL1	*A. thaliana*	nitrate transport	nitrate induced	175
NAR3	*C. reinhartii*	nitrate transport	nitrate induced	129
SHST1	*S. hamata*	sulfate transport	root	160
SHST3	*S. hamata*	sulfate transport	leaf	160

Table 1 (*continued*)

Gene	Organism	Reaction	Remarks[b]	Reference
IV. Others				
A. Fungi				
FUR4	*S. cerevisiae*	uracil transport	protein phosphorylated, turnover by End3p, End4p, Pep4p	83, 180, 181
FCY2	*S. cerevisiae*	purine-cytosine transport		183
DAL5	*S. cerevisiae*	allantoate transport	NCR, NCI	130
DUR3	*S. cerevisiae*	urea transport	NCR	39
CTR1	*S. cerevisiae*	choline transport		117
AAC1	*S. cerevisiae*	ADP/ATP exchange	mitochondria	1
AAC2	*S. cerevisiae*	ADP/ATP exchange	mitochondria	103
AAC3	*S. cerevisiae*	ADP/ATP exchange	mitochondria	92
PTP	*S. cerevisiae*	phosphate transport	mitochondria	128
B. Plants				
PTBC3	*S. oleracea*	triosephosphate-phosphate exchange	chloroplast	50
PPT1	*P. sativum*	triosephosphate-phosphate exchange	chloroplast	189
cTPT	*F. trinervia* *F. pringlei*	triosephosphate-phosphate exchange	chloroplast	47
SODiT1	*S. oleracea*	2-oxogluterate-malate exchange	chloroplast, twelve instead of six helices	182
ANT1	*Z. mays*	ADP/ATP exchange	mitochondria,	8
ANT2	*Z. mays*	ADP/ATP exchange	mitochondria,	11
ANT1	*S. tuberosum*	ADP/ATP exchange	mitochondria	41
AAT	*C. kessleri*	ADP/ATP exchange	mitochondria, glucose induced	74

[a]This table is representative and certainly not complete. Of fungal cells, only *S. cerevisiae* is represented.
[b]Gene expression and regulatory features.
[c]NCI, nitrogen catabolite inactivation (see section on Regulation); NCR, nitrogen catabolite repression (see section on Regulation).

PLANTS Whereas amino acid transport phenomena have been intensively studied in unicellular organisms and in animal tissues and cells, comparatively few reports have appeared over the years on plants. This has changed recently because all of the various yeast mutants deficient in amino acid permeases (see above) can be used for complementation by plant genes as well. The first group to show this successfully was Lacroute et al (113). In this way, approximately 10 different genes coding for amino acid transporters have now been cloned, sequenced, and partially characterized (56). Of these, the *AAP1/NAT2* gene product of *Arabidopsis thaliana* has a very broad specificity (54, 78) and likely transports all protein amino acids except asparagine and the basic amino acids (Table 1). This transporter belongs to a gene family with at least five more members, all possessing a broad but not identical specificity (48). When individually expressed in corresponding yeast mutants, transformants containing *AAP4* preferentially transported proline and valine and transformants containing *AAP5* lysine and arginine. The expression in various *A. thaliana* tissues differed considerably (Table 1). The *NTR1/AtPTR1* gene is not related to the *AAP* family (53). Originally thought to be a histidine transporter, *NTR1/AtPTR1* surprisingly shows a high degree of similarity to yeast, mammalian, and *Arabidopsis* peptide transporters (*PTR2, PepT1, AtPTR2*) and to the putative *Arabidopsis* nitrate transporter (*CHL1*) (45, 127, 165, 175). A gene, *AAT1,* has been cloned from *A. thaliana* (55). It codes for a basic amino acid transporter, and its sequence is related to the yeast *GAP1* gene. Further amino acid and peptide transport genes from higher plants are listed in Table 1.

Sugars

FUNGI Yeast sugar transport was extensively reviewed three years ago by Bisson et al (13). D-glucose is transported by facilitators in *S. cerevisiae,* not by H^+ symport. There is a large family of potential hexose transporters. Gene sequences of seven such genes (*HXT1–7*) have been published (13, 133), but the total number in this family may be more than 14 (M Ciriacy, personal communication). The important question of why these transporters exist in such large families is of course even more pertinent in a unicellular organism, where the argument of tissue-specific expression cannot be raised. However, only for a very few members of the family has hexose transport been directly demonstrated. Within the large "major facilitator superfamily" (MFS; 107) to which these yeast carriers also belong, other members transport, e.g. H^+/tetracyclin, H^+/lactose, citrate, and many other substrates. The first gene cloned and, because of homology, thought to be a high-affinity glucose transporter—*SNF3* (25)—is a regulator for the expression of *HXT* genes and thereby for growth on glucose (12). For additional genes, see Table 1.

PLANTS Sugars are synthesized by plants, and therefore sugar uptake from external sources generally is not a relevant process for these organisms. Redistribution, however, of assimilates from photosynthesizing tissues to non-green cells is of major importance. In addition, because water-soluble carbohydrates are the main solutes for long-distance transport, membrane transport of sugars is and has been a central issue in plant physiology. In spite of this, for pragmatic reasons transmembrane transport of sugars has mainly been studied as sugar uptake by unicellular organisms. Being unicellular and green, *Chlorella kessleri* has been an ideal model plant, not the least because sugar uptake can be dramatically induced in these cells (172). Because of this feature, the *HUP1* gene has been cloned by differentially screening a cDNA library of induced cells (148). With the help of *HUP1* cDNA, the *STP1* gene of

A. thaliana has been cloned (144) and, contrary to the opinion stated in a recent review by Sussman (168), both these genes coding for carriers of the plasma membrane were cloned and fully characterized (143, 144) several years before those to which Sussman assigned priority.

Similar to the situation with the yeast *HXT* genes and the *A. thaliana AAP* genes, the large number of *STP* genes with high similarity is astonishing. The family in *A. thaliana* contains at least 12 *STP* genes (142), *Ricinus communis* contains at least 8 (184), and *Chenopodium rubrum* contains 7 (139). In the case of *STP2–4,* the tissue-specific expression in *A. thaliana* has been demonstrated, using β-glucuronidase as a reporter gene under the control of the promoters of the three genes (142). *STP2* is expressed in the anthers, *STP3* in leaves and sepals, and *STP4* in anthers and root tips. Thus, *STP2* and *STP4* seem to be strictly sink specific, which is also true for *MST1,* a transporter of tobacco plants (146).

A disturbing situation existed for a while, because in spite of the great importance of sucrose as the main transport sugar in plants, no corresponding transporter gene was in sight. The riddle was solved with a detour via *S. cerevisiae* by Riesmeier et al (138). After *SoSUT1* was cloned from spinach, the corresponding gene from potatoes was obtained by the same procedure (137). On the basis of the spinach gene, two sucrose transporters have been cloned from both *A. thaliana* and *Plantago major* (see Table 1). Although one sucrose transporter in all plants tested is localized in the phloem (62, 137, 147, 174), an interesting and stimulating dispute has arisen, however, as to whether it is present exclusively in the sieve tubes or in the companion cells (58, 162). The second *PmSUC1* sucrose transporter of *P. major* is expressed only during seed development (61). There is no doubt, on the other hand, that all sucrose as well as hexose transporters that have been tested are H^+ symporters (62, 138, 147). This feature has also been studied in plasma membrane vesicles energized with cytochrome oxidase c (123, 167), a method originally worked out by Driessen & Konings (36).

An interesting application of sugar transporters has recently been reported. *Volvox carteri,* which normally does not take up D-glucose, can be heavily radiolabeled with ^{14}C-D-glucose if transformed with an algal glucose transporter gene (67).

Inorganic Cation and Anion Transporters

FUNGI In section III of Table 1, the cloned carriers responsible for the transport of inorganic ions at the plasmalemma are summarized. The field has been well reviewed up to 1991 by Gaber (59). Gaber also cloned and characterized two yeast transporters responsible for the uptake of potassium, the most prominent cytosolic ion in all organisms (60, 90). A double knockout mutant requires

50–100 mM K^+ for growth, and Ko & Gaber claim that additional K^+ transporters must be present in yeast cells (90). On the other hand, they note that suppressor mutations are located within hexose transporters (91), which suggests that relative minor changes in special integral membrane proteins may significantly increase K^+ permeability. Could the residual K^+ permeability of the double disruptant possibly be caused by the basic K^+ leak of all the various membrane proteins? Only detection of further K^+ carriers or channels could disprove this possibility.

PLANTS A popular but biased conception in general biology is that green plants supply almost all the rest of the living world with organic nutrients. This is only half the truth: For all animals, including humans, plants supply the inorganic as well as the organic nutrients. Epstein has brilliantly advocated this fact (42) by pointing out that nowhere in the universe will intelligent life on solid ground have developed without some sort of organisms that are rooted and therefore nonmotile and able to extract with an immense root surface (34) all the inorganic ingredients of life from the soil. In contrast, our knowledge of the molecular biology of membrane transporters of plant roots is meager. However, although only about half a dozen transporters are currently known in terms of their molecular biology (see Table 1), this part of the plant transport field is developing rapidly, which was apparent at the 10th International Workshop in Plant Membrane Biology in August 1995 in Regensburg.

The main progress in this area to date has been brought about by complementing yeast mutants: the ammonium (119), the high-affinity potassium (150), and three sulfate (160) transporter genes all have been cloned in this way. The potassium transporter *HKT1,* first thought to work as an H^+ symporter (150), seems instead to be a Na^+ symporter. Micromolar amounts of Na^+ were shown to stimulate K^+ uptake and vice versa (140). The question arises as to whether the classical concept that for plants Na^+ is not an essential ion is wrong after all.

Other Carriers

FUNGI In the last group in Table 1, the purine and pyrimidine transporters, the mitochondrial ones, and a few others are compiled. The purine/cytosine and the uracil transporters of *S. cerevisiae* have no equivalent in plants at this time. They are not related to any of the larger carrier families. In their regulation, they show a number of parallels with the amino acid permeases (see below).

The mitochondrial ATP/ADP exchange proteins belong to the most thoroughly studied carriers (89). The one from beef heart was the first membrane transporter purified to homogeneity and functionally reconstituted in vesicles

(98) as well as sequenced via protein sequencing (7). These exchange proteins belong to a large family of related mitochondrial transport proteins (MCF), all of which possess six transmembrane helices (100).

PLANTS The main carbon flow between the chloroplast and the cytosol is thought to proceed via the triose-phosphate/phosphate exchange protein, which has been intensively studied and biochemically characterized in the groups of Heldt and Flügge and which finally gave rise to the first carrier cloned and sequenced from plants (50). The area has been recently reviewed (51). As the mitochondrial ATP/ADP exchange proteins, also the chloroplast ones typically possess six transmembrane helices. The 2-oxoglutarate-malate exchanger of chloroplasts (182), by contrast, is an exception, and has putative 12 transmembrane helices. An ATP/ADP exchanger with 12 transmembrane helices that is not related to the mitochondrial ones has been detected in *A. thaliana* (88). The authors speculate that this carrier exchanges ATP for ADP at the inner envelope membrane of plastids.

TOPOLOGY/STRUCTURE

The majority of sequences of all carriers cloned to date suggest that the corresponding gene products are membrane proteins with 12 putative membrane-spanning helices. The main exceptions are the ATP/ADP and the phosphate/triosephosphate exchange proteins of mitochondria and chloroplasts, respectively (51, 89), which generally possess six transmembrane helices. They form dimers, whereas the 12 transmembrane helix transporters in their active forms—wherever thoroughly tested—seem to be monomers (35). Because no three-dimensional structure of any carrier protein is available to date, all pictures and models of their appearance are hypothetical. In spite of this restriction, or more likely because of it, a large number of intellectually very appealing approaches have been taken to learn as much as possible about the topology and preliminary structural features of carriers.

This topic can be subdivided with these questions:

1. Do the transmembrane helices that have been predicted really exist? Where are the N- and C-termini? Where are the individual loops between the helices located—toward the inside or the outside of the cell or organelle?
2. How are the various transmembrane helices located in relation to one another? Which form the actual substrate-specific pore through the membrane, in case a pore is a good picture for the real situation?
3. What is the role of the individual loops between the helices? Are they simply keeping the correct distance between the transmembrane parts, or do they play a more active role in substrate recognition and translocation?

We now summarize what is known in answer to these questions. Essentially, sugar carriers of *E. coli* (*LacY*), *Homo sapiens* (*Glut1*), and *C. kessleri* (*HUP1* and *HUP2*) are discussed.

First, the number of transmembrane helices is predicted from the hydropathy plot: Each stretch of amino acids that is about 20 amino acids long and that has a high lipophilic index represents a potential transmembrane part of the protein in question. Because such hydropathy plots are generally open to discussion, the number of transmembrane helices is debatable. For the lac-permease, for example, 10, 12, and even 13 such helices have been suggested (86, 179).

Two elegant procedures have recently been devised to tackle the membrane topology of proteins. One is based on the construction of fusion proteins between an alkaline phosphatase (lacking its own signal sequence) at the C-terminus and N-terminal segments of variable length of the membrane protein in question, e.g. the lac-permease (21). In all cases where the N-terminal part of the fusion protein directs the alkaline phosphatase to the outside, i.e. to the periplasmic space of *E. coli,* the enzyme is active, and this activity can easily be measured colorimetrically with intact cells. When the alkaline phosphatase is located in the cytoplasm, the test does not show activity. Thus, depending on the exact point of fusion, the enzymatic test was positive when the phosphatase was fused to an external loop of the permease and inactive when the point of fusion was at an internal loop or at the C-terminal end. In this way, all 11 loop fusions gave the result expected from the topology of the model (5 loops inside the cell, 6 outside; see Figure 1), a result which of course also strongly supports the existence of 12 membrane-spanning helices (21).

The second procedure can only be used for eukaryotic cells; however, it does not suffer from the problem of expressing eukaryotic/prokaryotic fusion proteins in *E. coli,* which sometimes are toxic. The method—glycosylation scanning mutagenesis—first applied to an ER protein (121), in principle consists of screening a variable artificial N-glycosylation site on a membrane protein to see whether it is used. The topology of the Glut1 transporter, facilitating D-glucose diffusion equilibrium in erythrocytes and many other human cells, has been determined with this method (77). The protein has one putative N-glycosylation site within the first loop, which according to the predicted topology is located facing the cell exterior (Figure 1). This site is in fact glycosylated, which supports the prediction for the first loop. When the N-glycosylation site is removed by site-directed mutagenesis and the protein is expressed in mammalian cells, it runs somewhat faster on SDS-PAGE than the product of the wild-type gene. One new N-glycosylation site [-N-X-S/T (N, asparagine; X, any amino acid except proline; S/T, serine or threonine)] was introduced into this mutated gene at each predicted loop, as well as into the C-terminal end predicted to be located intracellularly. The N-glycosylation

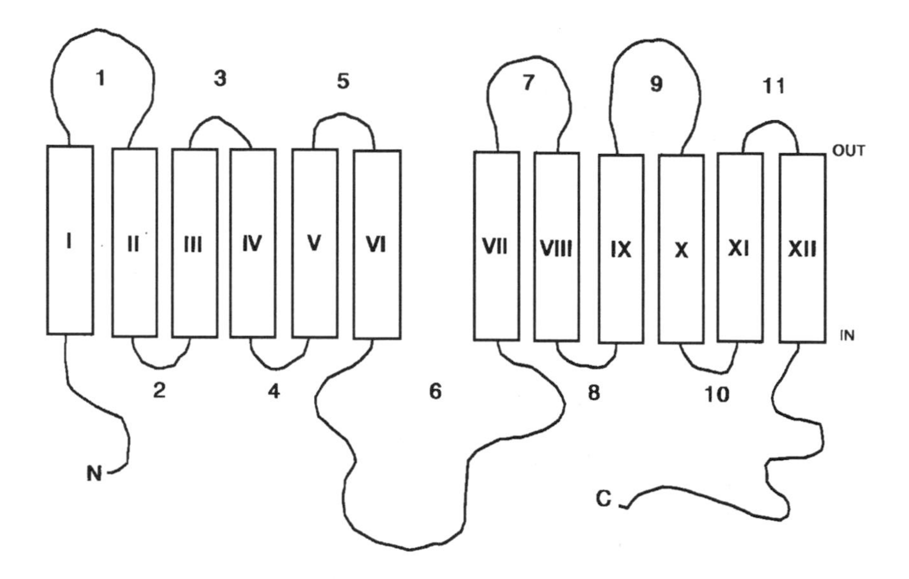

Figure 1 Suggested topology of carriers of the MFS family (107), I–XII, putative transmembrane helices; 1–11, predicted external and internal loops.

sites located in loops 3, 5, 7, 9, and 11 were glycosylated—as was loop 1 of the wild-type gene product—whereas sites in loops 2, 4, 6, 8, 10, and the C-terminus were not used for N-glycosylation. This is in complete accordance with the model, because protein N-glycosylation is restricted to the luminal side of the endoplasmic reticulum (ER), and solely those parts of a membrane protein face the ER lumen, which later end up facing the cell exterior.

The topology presented in Figure 1 can certainly be extrapolated to hold for all members of the MFS family, and scarce funding money should, therefore, not be spent for analogous work with other members of this family of carriers. In addition, all the results obtained in previous work using peptide-specific antibodies, proteases, and impermeable inhibitors also agree with the proposed topology of the 12 membrane spanners (reviewed in 9, 86). Second, because our knowledge of the number of transmembrane helices as well as the orientation of the loops and both the "loose" ends of these carriers is fairly sound, the question of how the 12 helices are located in relation to one another—the so-called helix packing problem—is more pertinent.

Again, lac-permease is the molecule most actively studied (86, 87). The results of various investigations using a number of quite different techniques are summarized in Figure 2A. Thus, evidence for close proximity of helix VII

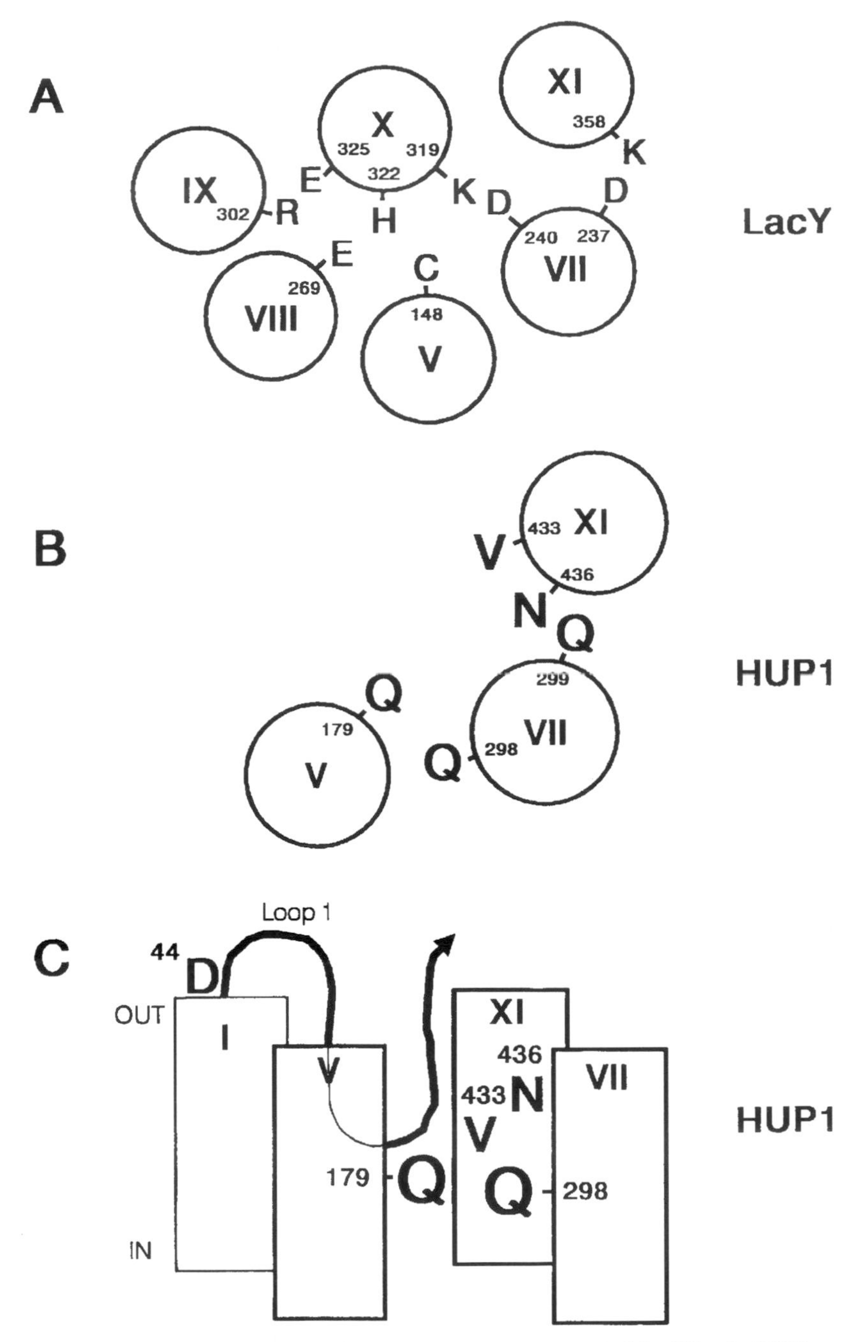

Figure 2 Postulated helix packing of the *E. coli LacY* (87, 191) and of the *Chlorella HUP1* gene product (187,188). (A) Helical wheel model of putative helices V and VII–XI in lac permease; (B) helical wheel model of putative helices V, VII, and XI in the *Chlorella* hexose transporter; (C) schematic representation of the postulated substrate recognition site of the *HUP1* transporter.

both with helix X and helix XI has been obtained from second-site suppression experiments.

The postulated relationship of helix X to helices VIII and IX is based on the observation that a fluorophor coupled to position H322 (helix X) as well as to E269 (helix VIII) forms an excited state dimer (excimer), which only occurs when the distance between the two conjugated ring systems is less than 3.5 Å (84). The same phenomenon was observed with R302 (helix IX) and E325 (helix X).

When 5-(α-bromoacetamido)-1,10-phenanthroline-copper [OP-(Cu)] was coupled to specific cysteines introduced into a C-less lac-permease (191), this modification under certain reducing conditions acts as a site-specific "chemical protease." The site specificity is determined by the amino acid of the C-less permease, which is replaced by cysteine. From the pattern of peptides obtained by this method, the authors concluded that helix V is closely packed to helix VII and VIII in lac permease (Figure 2A). That this helix is part of the substrate "pore" has also been concluded from the observation that C 148 (helix V) is located in the sugar-binding site, although C 148 as such is not essential for transport.

Finally, the idea that two correctly spaced histidyl side chains should result in an excellent chelator for divalent metal ions has recently been successfully applied to study the problem of helix packing (40, 85).

We have investigated helix packing in our laboratory with the *HUP1* gene of *C. kessleri* (188). The basic idea has been that screening for K_m mutants from a set of completely randomly produced mutants of the *HUP1* gene should yield mutations in all those amino acids contributing to the interaction of the substrate with the protein, and possibly only very few additional ones, because of allosteric effects on substrate binding. The randomly mutated carrier DNA was produced by PCR, aiming at an error rate of 1 per 1800 bp (37). The mutated DNA pool was used to transform a *Schizosaccharomyces pombe* mutant, which was unable to take up and to grow on glucose (112). Although most of the transformants selected by a decreased sensitivity toward 2-deoxy-glucose were V_{max} mutants, four were obtained that showed distinctly increased K_m values for D-glucose. The corresponding mutations were all located in transmembrane helices: Q179E (helix V), Q298R (helix VII), and V433L and N436Y (both in helix XI). By site-directed mutagenesis, the three amides were subsequently replaced in the most conservative way possible (Q179N, Q298N, and N436Q), but increased K_m values by a factor of 10 to 20 were retained (188). Assuming that with these mutations the carrier-substrate-binding or recognition site has been hit, it may be suggested that in the *HUP1* protein helices, V, VII, and XI are in close proximity (Figure 2B). Q299, the glutamine neighboring the one found by random mutagenesis (Q298) in helix VII has been shown previously by site-directed mutagenesis to affect the

K_m for glucose (22). It is interesting to note that, of the five contacts glucose has been shown to make when co-crystallized with hexokinase (3), four are amides as is suggested here.

As pointed out in the beginning of the review, as long as no three-dimensional structure is available, all these considerations are speculations. Whether they are good or bad will be revealed by the first crystals. Two things shall be pointed out, however. First, all the random mutations of the *HUP1* K_m mutants fall more or less into the middle of three transmembrane helices. Beforehand, one might have as well—or even moreso—expected that substrate recognition is happening at the outer rim of some pore. The data obtained certainly do not suggest this. It is true, however, that the exact positioning of each helix is not known, and to shift them up or down for a few amino acid residues is possible for all of them. Thus, the precise location of the mutated amino acids in the middle of the bilayer could be accidental after all. Second, significant parallels between Figures 2A and 2B become obvious: For both carriers, helices V, VII, and XI are suggested to be in proximity. It is dangerous, of course, to press similarities too hard, especially because amino acid identities between the two transporters hardly exist (15–27% in helices V, VII, and XI), and in addition, one translocates disaccharides, the other monosaccharides. Nevertheless, the transporters of the MFS superfamily (107) may very well do their jobs along an identical construction plan; if once successfully invented in evolution the construct may have been modified in details but not in principle anymore. Other investigations with the aim to characterize the substrate recognition site have been carried out with *E. coli GalP* (70) and with the *S. cerevisiae Gal2* transporter (120).

Third, what is the function of the loops between the transmembrane helices? Intuitively, their importance may seem secondary for the actual task of a device that catalyzes the translocation of molecules through membranes. This view is strengthened by the finding that fairly large peptides can be inserted into some loops of lac-permease without interfering with its activity (111). On the other hand, the literature on membrane transport proteins contains at least one exciting example that a loop may actually carry out the main job of catalyzing transmembrane transport: the loop between helix V and VI of the best characterized K^+ channel (153). Recently, we have obtained evidence that loop 1 may also be involved in substrate recognition (187). When chimeras were constructed of the *Chlorella HUP1* and *HUP2* genes, the products of which differ significantly in their substrate specificity [*HUP1*:Glc/Gal = 12; *HUP2*: Glc/Gal = 0.7; (163)], it was sufficient to insert part of loop 1 of the *HUP2* gene into the rest of the *HUP1* gene, in order to significantly improve the transport of D-galactose. Even the K_m for D-galactose of the chimera was significantly improved as compared with that of the *HUP1* carrier. In addition, D44 at the beginning of loop 1 is a crucial amino acid: The carrier with D44N is completely inactive, and the protein with D44E, although fully expressed, has 10% of wild-type activity and shows a 15-fold increased K_m for D-glucose (22). These observations suggest that loop 1 is taking part somehow in substrate binding. Whether this means that loop 1 participates in forming the substrate pore, as diagrammed in Figure 2C, is of course speculative. Finally, when possible functions of the various loops are considered in general, it should be pointed out that very likely at least the cytoplasmic ones are involved in regulatory phenomena (see below).

KINETICS

Kinetics of membrane transport not only started this field of research, as mentioned in the historical section of this review, it also has been for many years the queen of all transport physiological approaches. This has changed only in the past decade, when the first transporters were cloned and the fascination with molecular biology did not leave room for classical experimentation. It is because of this cloning of genes and their heterologous expression in *Xenopus* oocytes, for example (152), that studies on carrier kinetics have been revitalized. In particular, the possibility of investigating H^+ and Na^+ symporters, highly expressed in large cells amenable to electrophysiological measurements, has attracted anew a number of plant and animal physiologists to kinetics.

The questions asked today are the same as those asked ten years ago, but the current methods are better suited not to find the correct answer but to approach it with greater chance. How are the three variables—the substrate and ion concentration and the membrane voltage—interdependent? How are they interlinked by a membrane protein that, because of its construction, is able to bridge those 80 Å of a biological membrane?

It is likely that the most complete sets of data related to the above questions have been obtained for the Na^+-alanine symporter of mouse pancreas cells, carefully studied using the method of whole-cell recording with patch pipettes (80, 81), and for the Na^+-glucose symporter, cloned from rabbit intestine, expressed in *Xenopus* oocytes, and studied with the two-electrode voltage-clamp technique (125, 126). In the following paragraphs, the main results obtained for these mammalian transporters are summarized and briefly discussed. Studies with plant cell transporters are then discussed, and parallels and differences are pointed out.

For the mammalian Na^+-alanine transporter, the half-saturation concentration for alanine according to the typical Michaelis-Menten formalism depends on the extracellular Na^+ concentration. The value increases about sixfold when the Na^+ concentration is decreased from 150 to 5 mM (81). Similarly for the intestinal Na^+-glucose symporter, the half-saturation binding constant for α-methyl-D-glucose $K_{0.5}^{sugar}$ increases 36-fold when the Na^+ concentration is reduced from 100 to 2 mM.

Of special interest concerning the half-saturation constants for substrates and cotransported ions is their dependence on the membrane potential $\Delta\psi$. Thus, the Na^+-alanine cotransporter shows a voltage-dependent Na^+-binding behavior: At 0 mV, the $K_{0.5}^{Na^+}$ amounts to 50 mM, and it is decreased to 18 mM at a potential of –40 mV (81). Again, the analogous behavior is seen in the analysis of the Na^+-glucose cotransporter. Changing the membrane potential of the *Xenopus* oocyte under the voltage-clamp condition from –20 to –150 mV leads to a concomitant increase in the affinity for Na^+ from $K_{0.5}^{Na^+}$ 50 to 5 mM. Whereas in the case of the alanine transporter the maximal current observable is nearly voltage independent, maximal currents in the case of the Na^+-glucose transporter are little but significantly affected by membrane voltage (126). This difference resulted in partly different interpretations by the two investigating groups concerning their respective carriers. Jauch & Läuger (80) favor the "Na^+ well" concept, which assumes that a fraction of the transmembrane voltage drops between the ion-binding site and the external medium. Another way of looking at it is that the Na^+-binding site is let down within the membrane and that part of $\Delta\psi$ is used to concentrate Na^+ ions in the space between the bulk phase and the Na^+-binding site. The actual Na^+-binding site would not increase its affinity for the ion because of voltage increase, but rather the lower concentrations of ions in the bulk phase would be compensated for by the increased $\Delta\psi$. However, the possibility cannot be excluded that because of changes in $\Delta\psi$, the Na^+-binding site of the carrier changes its position and thus its dielectric environment in the membrane. The Na^+ well concept dates back to a suggestion by Mitchell (115), who discussed a potential "proton well" in the mitochondrial ATP synthase.

Parent et al (125, 126) assume for the Na^+-glucose cotransporter that membrane voltage affects the ion-binding reaction (Na^+ well) as well as a translocation step (the "empty" negatively charged carrier C^- "moving" from the cell inside to the outside). Parent et al also observed an effect of the membrane potential on the half-saturation constant for sugar, however, only at very low concentrations of Na^+. At physiological concentration of Na^+, this effect was gone. Because at high membrane potential sugar affinity approached the same value as that observed at saturating external Na^+ concentration, negative $\Delta\psi$ partially compensated for the decrease in $K_{0.5}^{sugar}$ as $[Na^+]$ was reduced. This suggests that the voltage dependence of $K_{0.5}^{sugar}$ is because of the voltage dependence of Na^+ binding discussed above, i.e. is again because of a potential Na^+ well effect.

Similar extensive studies with plant transporters are also under way. The hexose H^+ symporter HUP1 from *Chlorella* (6) as well as the hexose H^+ symporter STP1 from *Arabidopsis* (16, 17) have been functionally expressed in *Xenopus* oocytes and characterized with the two-electrode voltage-clamp technique. The currents obtained with STP1-transformed oocytes were 20–40 times larger than those from *HUP1* injected oocytes (6, 17), which most likely indicates that the *HUP1* construct is less well expressed. The first electrophysiological data have also been reported for the H^+–amino acid transporter *AAP1* or *NAT2* (15).

In contrast with the mammalian Na^+ cotransporters, the half-saturation constant of *STP1* for the sugar analogue 3-0-methylglucose increases with increasing H^+ concentration (decreasing pH value) and at pH 5.5 $K_{0.5}^{sugar}$ also increases with increasing $\Delta\psi$. Because this result does not agree with the behavior of the Na^+ cotransporters discussed above, it has been interpreted that *STP1* works by a "sequential" or "consecutive" mechanism. For a sequentially working carrier C, one of the two substrates (either sugar or ion) is bound on the outside, causing a conformational change to C^*, thereby translocating (without the second substrate being bound) to the inside (Figure 3). There the substrate is released to the cytoplasm, and at the same time C^* can take the second substrate from the outside. Reorientation of C^* to C leads to release of the second substrate in the cytoplasm, and C can start all over. For carriers working in this way, it is predicted that increasing the concentration of cosubstrate (ion) decreases the affinity of the carrier toward the substrate (80). The mammalian Na^+-alanine and the Na^+-glucose transporters are assumed to work by the conventional, so-called simultaneous mechanism (80, 81, 125, 126). In this case, a ternary complex of carrier with both substrates, e.g. sugar plus Na^+, is formed and only then is a conformational transition from one state with outward-facing to one with inward-facing binding sites taking place (Figure 3).

The indications that the two mammalian Na^+ cotransporters discussed above and the H^+-hexose cotransporter of *A. thaliana* should be working in such different modes is somewhat disturbing for anyone who prefers to stress biological unity. Neither the alanine nor the Na^+-glucose cotransporters do, however, belong to the evolutionarily related MFS transporters (see above), which could make the discrepancy more easily acceptable. The same argument holds for the *AAP1/NAT2* proton cotransporter, of which preliminary kinetic data indicate that its mode of action is also of the simultaneous type, forming a carrier-H^+–amino acid complex before the transmembrane transition (15). At least this shows that there does not exist a principal borderline between Na^+ and H^+ cotransporters.

The question then remains about whether all the H^+ cotransporters of the MFS group may kinetically behave like the *STP* transporter. The answer seems to be "no," because all the detailed data available on the "old"—not to say classical—*Chlorella* studies do indicate that *HUP1* works according to the simultaneous mechanism (94, 96, 97, 154). As has been mentioned, a critical

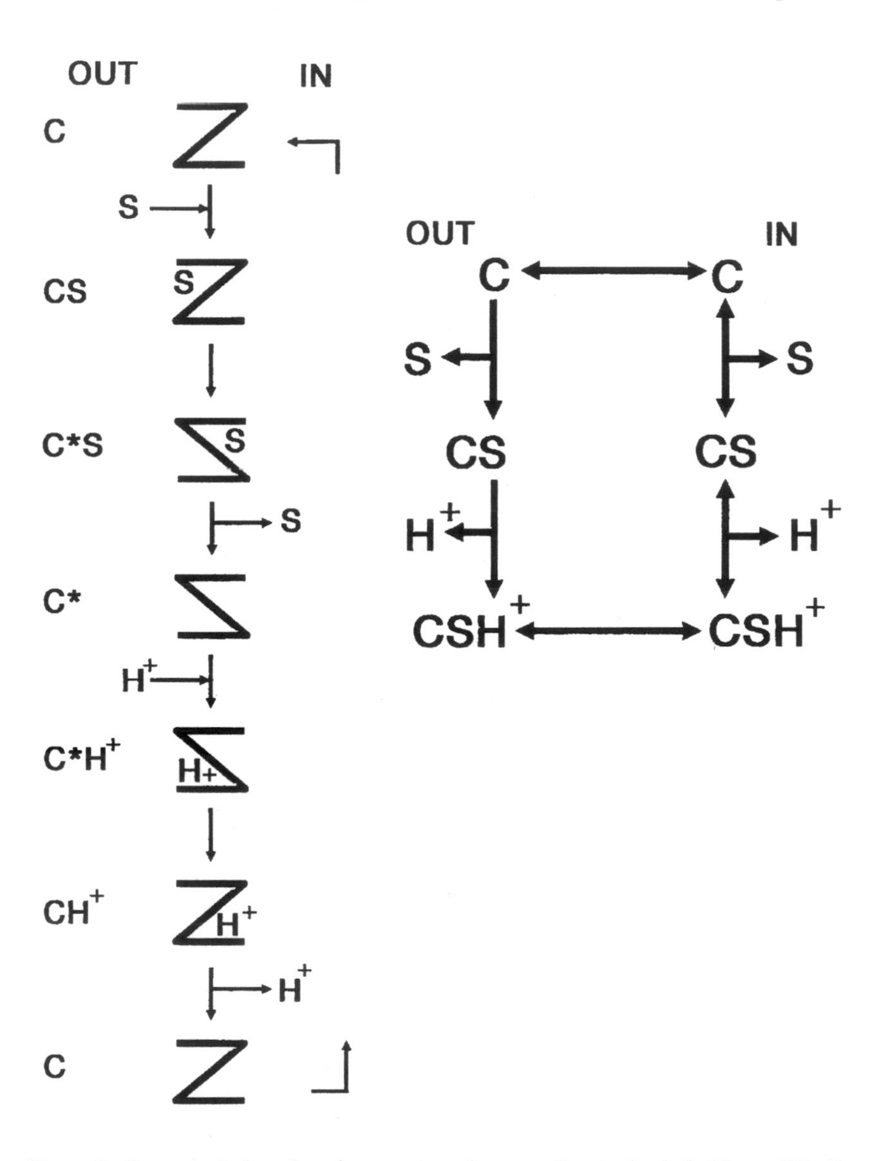

Figure 3 Two principal modes of symporter action according to Jauch & Läuger (80). See discussion in the text.

feature in distinguishing between "simultaneous" and "consecutive" is the dependence of the half-saturation constant for sugar on the concentration of the cotransported ion, here the proton. For the sugar analogue 6-deoxyglucose, this value changes continuously from 0.3 mM at pH 6.0 to 43 mM at pH 8–9 (97). In addition, the voltage dependence of H^+-sugar symport of *Chlorella* could be best explained by either a proton well concept or a change in the pK value of a protonable group of the carrier because of some conformational change (154). Finally, from transport studies with D-glucosamine, the following was concluded: (*a*) that a proton is symported and not a hydroxyl ion antiported, and (*b*) that sugar and proton are bound in close proximity at a single site, because D-glucosamine protonated at the amino group—not the unprotonated form!—was transported by uniport (96).

There is one main reason, however, why these previous data should be interpreted with a grain of salt. It turned out (163) that the glucose uptake activity of *C. kessleri* measured in vivo is the result of three transporters, which are all induced by D-glucose. At least the gene products of *HUP1* and *HUP2*, the ones mainly expressed after induction, are responsible for the kinetic features determined in vivo. The results of Aoshima et al (6), who measured exclusively the *HUP1* protein expressed in *Xenopus* oocytes, differ in part from the results obtained with intact *Chlorella*. Thus the pH dependence, the voltage dependence of V_{max} at pH 6.0 (154), and the substrate specificity did not correspond under the two experimental setups. The latter has been explained in the meantime, because the *HUP2* gene product turned out to preferentially take up D-galactose (163), whereas for the *HUP1* transporter this hexose is a poor substrate. It seems unlikely that the other differences will also be explainable in this way. Possibly the different membrane environment of the oocyte may, after all, change some carrier characteristics. A detailed electrophysiological investigation of the *HUP* transporters should clarify the question, whether within one evolutionarily conserved carrier family (MFS) quite different transport mechanisms are at work.

Finally, it must be pointed out that modeling of the alanine study discussed above (80) has been based on the assumption that the rate constants for binding and release of the cotransported ion and the substrate are large and that the association-dissociation reaction with the carrier is therefore always in equilibrium. This assumption has quite generally been made in transport kinetics and has correctly been criticized by Sanders et al (141). In addition, Sanders et al also criticized assumptions such as "conformational changes," "ion wells," and other "less well-defined explanations." We should not forget, however, that a mathematically better defined model is not automatically closer to the truth. An "ion well," for example, is not simply "out," because a specific phenomenon can be independently modeled with the help of a dozen rate constants (141). An assumed conformational change of a transporter may indeed be less well defined, but is the difference between talking about allosteric and conformational changes on one hand and choosing greatly differing pairs of rate constants for the "out"- and "in"-binding site of a carrier on the other hand really more than semantics?

REGULATION OF CARRIER ACTIVITY AND TURNOVER

The decade to come in membrane transport studies will be concerned with carrier regulation. For unicellular organisms, growth rates are usually determined by the amount and activity of specific carriers. The enzymatic machinery for metabolism within cells is generally active in excess even under saturating external substrate supply (63). With limiting amounts of substrate, the rate of uptake—depending on the carrier properties—directly determines the growth rate of the organism (186). *S. pombe* cells transformed with the *Chlorella HUP1* may serve to illustrate this: Because the *HUP1* carrier possesses a K_m for D-glucose that is 100 times lower than that of the endogenous *S. pombe* carrier (143), the transformants grow 1.6 times faster compared with *S. pombe* wild-type cells at sugar concentrations <2.5 mM (T Caspari, unpublished results). Because membrane transport activities are qualitatively and quantitatively important for a cell, and because cells invest 12% of their genomic information in transport proteins (see the section A Brief Historical Excursion), it is not surprising that membrane transport activities are regulated in a rather sophisticated and strict way. Although there is an extensive literature on the physiological phenomena of transport regulation (e.g. 65), little is known in molecular terms.

A regulatory event discovered early in the history of membrane transport is the observation that amino acid uptake into yeast cells leads to an inhibition of the uptake system by the accumulated amino acid (31). Although this regulatory event has generally been interpreted as a feedback mechanism, not only inhibiting further entry of the amino acid in question but possibly also explaining the phenomenon that amino acid does not efflux via the carrier (65), this feedback mechanism has never been proven at the protein level. Are the cytoplasmic protein loops or termini of a carrier protein required for this kind of regulation, and if so, which ones? The interpretation that the lack of amino acid efflux is due to their storage in the vacuole is unlikely, because the phenomenon can be demonstrated even in vesicles (122). Unidirectional uptake and lack of efflux of amino acids has also been observed in *Chlorella* (145).

Another regulatory phenomenon, called catabolite inactivation by Holzer (110) and generally observed in parallel with catabolite repression, has also been intensively studied in yeast. It has been shown that glucose downregulates the galactose carrier in this way (33, 110). An analogous phenomenon caused by rich nitrogen source is called nitrogen catabolite inactivation (NCI). It takes place, for example, with the general amino acid permease (*GAP1*) and also with various other amino acid carriers of yeast (see Table 1). The *GAP1* activity is completely but reversibly inhibited, when instead of a poor nitrogen source (proline) a good one (NH_4^+) is added to the cells. In the presence of specific mutations, this inactivation is suppressed (the genes *NPI1* and *NPI2*, "nitrogen permease inactivator"; 65). Another group of such mutations is not pleiotropic; they are located within the reading frame of the very permease affected by NCI, which suggests that the site getting covalently modified within the transporter is affected by this mutation. In one such case, that of the GABA permease, the C-terminus was changed by a mutation suppressing nitrogen catabolite inactivation (65). Obviously, one suspects that protein phosphorylation and dephosphorylation are involved in this kind of regulation; in fact, phosphorylated forms of *Gap1p* and *Fur4p* have been observed (164, 180). The clear correlation of protein phosphorylation with carrier activity is still missing, however. In addition to the inhibitory genes (*NPI*s), there are also pleiotropic-activating ones (*NPR1,* nitrogen-permease reactivation). Because the nucleotide sequence of *NPR1* resembles a protein kinase, the phorphorylated form of *GAP1,* for example, may be the active one (65).

In addition to this inactivation of carriers (*NCI*), an independent, genetically separable nitrogen catabolite repression (*NCR*) exists in yeast and acts by affecting carrier gene transcription. In this case, glutamine seems to be the crucial effector, and NH_4^+, which causes *NCR,* does so via glutamine (65).

The amount of carrier working in a plasma membrane depends on the rate of its synthesis as well as the rate of its degradation, its turnover. Again, although hardly anything is known in higher plants about carrier half-life, a wealth of information exists for *S. cerevisiae* and shows that catabolite inactivation can also be irreversible, i.e. proteolytic in nature. In the first place, it is interesting that transport proteins have a very much shorter half-life than bulk yeast proteins. Whereas $t_{1/2}$ of up to 160 h were measured for bulk proteins (64), 1 h was determined for the maltose transporter in the presence of glucose (135). In yeast cells, careful analyses have shown that the degradation of the maltose, inositol, and uracil transporters requires endocytosis and proteolysis in the vacuole (102, 135, 181); proteasomes are not involved (135).

CONCLUSIONS

In spite of fascinating advances in the field of membrane carriers, much remains to be done. To date, no three-dimensional structures of carriers are known, though several laboratories are able to produce milligram amounts of pure proteins—also from plant carriers (106, 166; T Caspari, I Robl & W Tanner, unpublished manuscript). Moreover, new crystallization procedures for membrane proteins (79) raise new hopes. Further kinetic data derived from still more sophisticated experiments [cut-open oocytes (190)] will aid in solving the question about whether principally different mechanisms do exist for symporters. In the future, however, we will mostly see a detailed understanding of carrier regulation, especially in relation to cell and tissue specificity—and this in relation to the total developmental spectrum from embryogenesis and seed germination all the way to the flowering plant. Transport regulation will also be found to be associated with environmental changes. Finally, there will most likely arise interesting applications. It will be fun to watch progress in the years to come!

ACKNOWLEDGMENTS

We are grateful to W. Frommer, M. Hawkesford, R. Kaback, R. Lagunas, E. Martinoia, and N. Sauer for making results available before publication; to E. Komor, N. Sauer, M. Opekarova, and A. Will for helpful suggestions and discussions; and to V. Mrosek and D. Urbanek for writing and helping to organize the review. The work carried out at the University of Regensburg has been supported by grants from the Deutsche Forschungsgemeinschaft (SFB 43) and by the Fonds der Chemischen Industrie.

Literature Cited

1. Adrian GS, McCammon MT, Montgomery DL, Douglas MG. 1986. Sequences required for delivery and localization of the ADP/ATP translocator to the mitochondrial inner membrane. *Mol. Cell. Biol.* 6:626–34
2. Ames GFL, Mimura CS, Holbrook SR, Shyamala V. 1992. Traffic ATPases: a superfamily of transport proteins operating from *Escherichia coli* to humans. *Adv. Enzymol.* 65:1–47
3. Anderson CM, Stenkamp RE, McDonald RC, Steitz TA. 1978. A refined model of the sugar binding site of yeast hexokinase B. *J. Mol. Biol.* 123:207–19
4. Anderson JA, Huprikar SS, Kochian LV, Lucas WJ, Gaber RF. 1992. Functional expression of a probable *A. thaliana* potassium channel in *S. cerevisiae*. *Proc. Natl. Acad. Sci. USA* 89: 3736–40
5. Andre B, Hein B, Grenson M, Jauniaux JC. 1993. Cloning and expression of the UGA4 gene coding for the inducible GABA-specific transport protein of *Saccharomyces cerevisiae*. *Mol. Gen. Genet.* 237:17–25
6. Aoshima H, Yamada M, Sauer N, Komor E, Schobert C. 1993. Heterologous expression of the H^+/hexose cotransporter from *Chlorella* in *Xenopus oocytes* and its characterization with respect to sugar specificity, pH and membrane potential. *J. Plant Physiol.* 141: 293–97
7. Aquila H, Misra D, Eulitz M, Klingenberg M. 1982. Complete amino acid sequence of the ADP/ATP carrier from beaf heart mitochondria. *Hoppe-Seyler's Z. Physiol. Chem.* 363:345–49
8. Baker A, Leaver CJ. 1985. Isolation and sequence analysis of a cDNA encoding the ATP/ADP translocator of *Zea mays* L. *Nucleic Acids Res.* 13: 5857–67
9. Baldwin SA. 1992. Probing the structure

and function of the human erythrocyte glucose transporter. *Biochem. Soc. Trans.* 20:533–37
10. Balzi E, Goffeau A. 1995. Yeast multidrug resistance: the PDR network. *J. Bioenerg. Biomembr.* 27:71–76
11. Bathgate B, Baker A, Leaver CJ. 1989. Two genes encode the adenine nucleotide translocator of maize mitochondria: isolation, characterisation and expression of the structural genes. *Eur. J. Biochem.* 183:303–10
12. Bisson LF, Coons DM, Fong NM, Mazer J, Vagnoli P. 1995. Characterization of the physiological role of the SNF3 gene of *Saccharomyces. Yeast* 11:S27
13. Bisson LF, David MC, Kruckeberg AL, Lewis DA. 1993. Yeast sugar transporters. *Crit. Rev. Biochem. Mol. Biol.* 28:259–308
14. Blatt MR, Thiel G. 1993. Hormone control of ion channel gating. *Annu. Rev. Plant Physiol. Plant Mol. Biol.* 44:543–67
15. Boorer KJ. 1995. The transport mechanism of a plant H^+/amino acid transporter expressed in *Xenopus oocytes. Int. Workshop Plant Membr. Biol., 10th, Regensburg*, L09 (Abstr.)
16. Boorer KJ, Forde BG, Leigh RA, Miller AJ. 1992. Function expression of a plant plasma membrane transporter in *Xenopus* oocytes. *FEBS Lett.* 302:166–68
17. Boorer KJ, Loo DDF, Wright EM. 1994. Steady-state kinetics of the H^+/hexose cotransporter (STP1) from *Arabidopsis thaliana* expressed in *Xenopus oocytes. J. Biol. Chem.* 269:1–8
18. Bun YM, Nishimura M, Harashima S, Oshima Y. 1991. The PHO84 gene of *Saccharomyces cerevisiae* encodes an inorganic phosphate transporter. *Mol. Cell. Biol.* 11:3229–38
19. Bush DR. 1993. Proton-coupled sugar and amino acid transporters in plants. *Annu. Rev. Plant Physiol. Plant Mol. Biol.* 44:513–42
20. Büchel DE, Gronenborn B, Müller-Hill B. 1980. Sequence of the lactose permease gene. *Nature* 283:541–43
21. Calamia J, Manoil C. 1990. Lac permease of *Escherichia coli.* Topology and sequence elements promoting membrane insertion. *Proc. Natl. Acad. Sci. USA* 87:4937–41
22. Caspari T, Stadler R, Sauer N, Tanner W. 1994. Structure/function relationship of the *Chlorella* glucose/H^+ symporter. *J. Biol. Chem.* 269:3498–502
23. Caspari T, Will A, Opekarova M, Sauer N, Tanner W. 1994. Hexose/H^+ symporters in lower and higher plants. *J. Exp. Biol.* 196:483–91
24. Deleted in proof
25. Celenza JL, Marshall-Carlson L, Carlson M. 1988. The yeast SNF3 gene encodes a glucose transporter homologous to the mammalian protein. *Proc. Natl. Acad. Sci. USA* 85:2130–34
26. Chanson A. 1993. Active transport of proton and calcium in higher plant cells. *Plant Physiol. Biochem.* 31:943–55
27. Cheng Q, Michels CA. 1989. The maltose permease encoded by the MAL61 gene of *Saccharomyces cerevisiae* exhibits both sequence and structural homology to other sugar transporters. *Genetics* 123:477–84
28. Cohen GN, Monod J. 1957. Bacterial permeases. *Bacteriol. Rev.* 21:169–94
29. Cori CF. 1925. The rate of absorption of a mixture of glucose and galactose. *Proc. Soc. Exp. Biol. Med.* 23:290–91
30. Cori CF. 1925. The fate of sugar in the animal body. I. The rate of absorption of hexoses and pentoses from the intestinal tract. *J. Biol. Chem.* 66:691–715
31. Crabeel M, Grenson M. 1970. Regulation of histidine uptake by a specific feedback inhibition of two histidine permeases in *Saccharomyces cerevisiae. Eur. J. Biochem.* 14:197–204
32. Crane RK. 1962. Hypothesis for mechanism of intestinal active transport of sugars. *Federation Proc.* 21:891–95
33. DeJuan C, Lagunas R. 1986. Inactivation of the galactose transport system in *Saccharomyces cerevisiae. FEBS Lett.* 207:258–61
34. Dittmer H. 1937. A quantitative study of the roots and root hairs of a winter rye plant (*Secale cereale*). *Ann. J. Bot.* 24:414–20
35. Dornmair K, Corin AF, Wright JK, Jähnig F. 1985. The size of the lactose permease derived from rotational diffusion measurements. *EMBO J.* 4:3633–38
36. Driessen AJM, Konings WN. 1993. Insertion of lipids and proteins into bacterial membranes by fusion with liposomes. *Methods Enzymol.* 221:394–408
37. Eckert KA, Kunkel TA. 1991. The fidelity of DNA polymerase used in the polymerase chain reactions. In *PCR—A Practical Approach*, ed. MJ Mc Pherson, pp. 225–44. Oxford: Oxford Univ. Press
38. Ehring R, Beyreuther K, Wright JK, Overath P. 1980. In vitro and in vivo products of *E. coli* lactose permease gene are identical. *Nature* 283:537–40
39. Elberry HM, Majudmar ML, Cunning-

ham TS, Sumrada RA, Cooper TG. 1993. Regulation of the urea active transporter gene (DUR3) in *Saccharomyces cerevisiae*. *J. Bacteriol.* 175: 4688–98
40. Elling CE, Nielsen SM, Schwartz TW. 1995. Conversion of antagonist-binding site to metal-ion site in the tachykinin NK-1 receptor. *Nature* 374:74–77
41. Emmermann M, Braun HP, Schmitz UK. 1991. The ADP/ATP translocator from potato has a long amino-terminal extension. *Curr. Genet.* 20:405–40
42. Epstein E. 1973. Roots. *Sci. Am.* 228/5: 48–58
43. Epstein E. 1973. Mechanisms of ion transport through plant cell membranes. *Int. Rev. Cytol.* 34:123–68
44. Epstein E, Hagen CE. 1952. A kinetic study of the absorption of alkali cations by barley roots. *Plant Physiol.* 27:457–74
45. Fei YJ, Kanai Y, Nussberger S, Ganapathy V, Leibach FH, et al. 1994. Expression cloning of a mammalian proton-coupled oligopeptide transporter. *Nature* 368:563–66
46. Feldmann H, Aigle M, Aljinovic G, André B, Baclet MC, et al. 1994. Complete DNA sequence of yeast chromosome II. *EMBO J.* 13:5795–809
47. Fischer K, Arbinger B, Kammerer B, Busch C, Brink S, et al. 1994. Cloning and in vivo expression of functional triose phosphate/phosphate translocators from C3- and C4-plants: evidence for the putative participation of specific amino acid residues in the recognition of phosphoenolpyruvate. *Plant. J.* 5: 215–26
48. Fischer WN, Kwart M, Hummel S, Frommer WB. 1995. Substrate specificity and expression profile of amino acid transporters (AAPs) in *Arabidopsis*. *J. Biol. Chem.* 270:16315–20
49. Fleischmann RD, Adams MD, White W, Clayton RA, Kirkness EF, et al. 1995. Whole-genome random sequencing and assembly of *Haemophilus influenzae* Rd. *Science* 269:449–604
50. Flügge UI, Fischer K, Gross A, Sebald W, Lottspeich F, Eckerskorn C. 1989. The triose phosphate-3-phosphoglycerate-phosphate translocator from spinach chloroplasts. *EMBO J.* 8:39–45
51. Flügge UI, Heldt H. 1991. Metabolite translocators of the chloroplast envelope. *Annu. Rev. Plant. Physiol. Plant Mol. Biol.* 42:129–44
52. Fox CF, Kennedy EP. 1965. Specific labeling and partial purification of the M protein, a component of the galactoside transport system of *Escherichia coli*. *Proc. Natl. Acad. Sci. USA* 54:891–99
53. Frommer WB, Hummel S, Rentsch D. 1994. Cloning of an *Arabidopsis* histidine transporting protein related to nitrate and peptide transporters. *FEBS Lett.* 347: 185–89
54. Frommer WB, Hummel S, Riesmeier JW. 1993. Expression cloning in yeast of a cDNA encoding a broad specificity amino acid permease from *Arabidopsis thaliana*. *Proc. Natl. Acad. Sci. USA* 90:5944–48
55. Frommer WB, Hummel S, Unseld M, Ninnemann O. 1995. Seed and vascular expression of high affinity transporter for cationic amino acids in *Arabidopsis*. In Press
56. Frommer WB, Kwart M, Hirner B, Fischer WN, Hummel S, Ninnemann O. 1994. Transporters for nitrogenous compounds in plants. *Plant Mol. Biol.* 26: 1651–70
57. Frommer WB, Ninnemann O. 1995. Heterologous expression of genes in bacterial, fungal, animal, and plant cells. *Annu. Rev. Plant Physiol. Plant Mol. Biol.* 46:419–44
58. Frommer W, Lauter FR, Ninnemann O, Rentsch D. 1995. Molecular approaches to study the function of ion and metabolite transporters of the plant plasma membrane. *Int. Workshop Plant Membr. Biol., 10th, Regensburg,* L06 (Abstr.)
59. Gaber RF. 1992. Molecular genetics of yeast ion transport. *Int. Rev. Cytol.* 137A: 299–353
60. Gaber RF, Styles CA, Fink GR. 1988. TRK1 encodes a plasma membrane protein required for high-affinity potassium transport in *Saccharomyces cerevisiae*. *Mol. Cell. Biol.* 8:2848–59
61. Gahrtz M, Schmelzer E, Stolz J, Sauer N. 1996. Expression of the *PmSUC1* sucrose carrier gene from *Plantago major* L. is induced during seed development. *Plant J.* 9:93–100
62. Gahrtz M, Stolz J, Sauer N. 1994. A phloem-specific sucrose H^+ symporter from *Plantaga major* supports the model of apoplastic phloem loading. *Plant. J.* 6:697–706
63. Gancedo C, Serrano R. 1989. Energy yielding metabolism in yeast. In *The Yeasts*, ed. JS Harrison, AH Rose, 3: 205–59. New York: Academic. 2nd ed.
64. Gancedo JM, Lopez S, Ballesteros F. 1982. Calculation of half-lives of proteins in vivo: heterogeneity in the rate of degradation of yeast proteins. *Mol. Cell. Biochem.* 43:89–95
65. Grenson M. 1992. Amino acid transporters in yeast: structure, function and

regulation. In *Molecular Aspects of Transport Proteins,* ed. JJHHM de Pont, pp. 219–45. Amsterdam: Elsevier Sci.
66. Gunderson KL, Kopito RR. 1995. Conformational states of CFTR associated with channel gating: the role of ATP binding and hydrolysis. *Cell* 82:231–39
67. Hallmann A, Sumper M. 1995. The *Chlorella* hexose/H^+ symporter is a useful selectable marker and biochemical reagent when expressed in *Volvox. Proc. Natl. Acad. Sci. USA.* 91:11562–66
68. Han EK, Michels CA. 1992. Glucose-induced inactivation of maltose permease in *Saccharomyces. Yeast* 8:S512
69. Hedrich R, Schroeder JI. 1989. The physiology of ion channels and electrogenic pumps in higher plants. *Annu. Rev. Plant Physiol.* 40:539–69
70. Henderson PJF, Martin GEM, McDonald TP, Steel A, Walmsley AR. 1994. Dissection of discrete kinetic events in the binding of antibiotics and substrates to the galactose-H^+ symport protein, GalP, of *Escherichia coli. Antonie van Leeuwenhoek* 65:349–58
71. Higgins CF. 1995. The ABC of channel regulation. *Cell* 82:693–96
72. Higgins CF, Gallagher MP, Hyde SC, Mimmack ML, Pearce SR. 1990. Periplasmic binding protein-dependent transport systems: the membrane-associated components. *Philos. Trans. R. Soc. London Ser. B* 326:353–65
73. Hildebrandt V, Ramezani-Rad M, Swida U, Wrede P, Grzesiek S, et al. 1989. Genetic transfer of the pigment bacteriorhodopsin into the eukaryote *Schizosaccharomyces pombe. FEBS Lett.* 243:137–40
74. Hilgarth C, Sauer N, Tanner W. 1991. Glucose increases the expression of the ATP/ADP translocator and the glyceraldehyde-3-phosphate dehydrogenase genes in *Chlorella. J. Biol. Chem.* 266: 24044–47
75. Hoffmann W. 1985. Molecular characterization of the CAN1 locus in *Saccharomyces cerevisiae. J. Biol. Chem.* 260:11831–37
76. Hörtensteiner S, Vogt E, Hagenbuch B, Meier PJ, Amrhein N, Martinoia E. 1993. Direct energization of bile acid transport into plant vacuoles. *J. Biol. Chem.* 268:18446–49
77. Hresko RC, Kruse M, Strube M, Mueckler M. 1994. Topology of the Glut1 glucose transporter deduced from glycosylation scanning mutagenesis. *J. Biol. Chem.* 269:20482–88
78. Hsu LC, Chiou TJ, Chen L, Bush DR. 1993. Cloning a plant amino acid transporter by functional complementation of a yeast amino acid transport mutant. *Proc. Natl. Acad. Sci. USA* 90:7441–45
79. Iwata S, Ostermeier C, Ludwig B, Michel H. 1995. Structure at 2.8 Å resolution of cytochrome oxidase from *Paracoccus denitrificans. Nature* 376: 660–69
80. Jauch P, Läuger P. 1986. Electrogenic properties of the sodium-alanine cotransporter in pancreatic acinar cells. II. Comparison with transport models. *J. Membr. Biol.* 94:117–27
81. Jauch P, Petersen OH, Läuger P. 1986. Electrogenic properties of the sodium-alanine cotransporter in pancreatic acinar cells. I. Tight-seal whole-cell recordings. *J. Membr. Biol.* 94:99–115
82. Jauniaux JC, Grenson M. 1990. GAP1, the general amino acid permease gene of *Saccharomyces cerevisiae*: nucleotide sequence, protein similarity with the other bakers yeast amino acid permeases, and nitrogen catabolite repression. *Eur. J. Biochem.* 190:39–44
83. Jund R, Weber E, Chevallier MR. 1988. Primary structure of the uracil transport protein of *Saccharomyces cerevisiae. Eur. J. Biochem.* 171:417–24
84. Jung K, Jung H, Kaback HR. 1994. Dynamics of lactose permease of *E. coli* determined by site-directed fluorescence labeling. *Biochemistry* 33:3980–85
85 Jung K, Voss J, He M, Hubbell WL, Kaback HR. 1995. Engineering a metal binding site within a polytopic membrane protein, the lactose permease of *Escherichia coli. Biochemistry* 34: 6272–77
86. Kaback HR. 1992. In and out and up and down with lac permease. *Int. Rev. Cytol.* 137A:97–125
87. Kaback HR, Frillingos S, Jung H, Jung K, Prive GG, et al. 1994. The lactose permease meets Frankenstein. *J. Exp. Biol.* 196:183–95
88. Kampfenkel K, Neuhaus E. 1995. Molecular characterization of an *Arabidopsis thaliana* cDNA encoding a novel putative adenylate translocator of higher plants. *Int. Workshop Plant Membr. Biol., 10th, Regensburg,* R47 (Abstr.)
89. Klingenberg M. 1992. Structure-function of the ADP/ATP carrier. *Biochem. Soc. Trans.* 20:547–50
90. Ko CH, Gaber RF. 1991. TRK1 and TRK2 encode structurally related K^+ transporters in *Saccharomyces cerevisiae. Mol. Cell. Biol.* 11:4266–73
91. Ko CH, Liang H, Gaber RF. 1993. Roles of multiple glucose transporters in *Saccharomyces cerevisiae. Mol. Cell. Biol.* 13:638–48

92. Kolarov J, Kolarov N, Nelson N. 1990. A third ADP/ATP translocator gene in yeast. *J. Biol. Chem.* 265:12711–16
93. Komor E. 1973. Proton-coupled hexose transport in *Chlorella vulgaris*. *FEBS Lett.* 38:16–18
94. Komor E, Haass D, Komor B, Tanner W. 1973. The active hexose-uptake system of *Chlorella vulgaris*: K_m values for 6-deoxyglucose influx and efflux and their contribution to sugar accumulation. *Eur. J. Biochem.* 39:193–200
95. Komor E, Rotter M, Tanner W. 1977. A proton-cotransport system in a higher plant: sucrose transport in *Ricinus communis*. *Plant Sci. Lett.* 9:153–62
96. Komor E, Schobert C, Cho BH. 1983. The hexose uptake system of *Chlorella*: Is it a proton symport or a hydroxyl ion antiport system? *FEBS Lett.* 156:6–10
97. Komor E, Tanner W. 1974. The hexose-proton cotransport system of *Chlorella*: pH dependent change in K_m values and translocation constants of the uptake system. *J. Gen. Physiol.* 64:568–81
98. Krämer R, Klingenberg M. 1977. Reconstitution of adenine nucleotide transport with purified ADP/ATP-carrier protein. *FEBS Lett.* 82:363–67
99. Kruckeberg AL, Bisson LF. 1990. The HXT2 gene of *Saccharomyces cerevisiae* is required for high-affinity glucose transport. *Mol. Cell. Biol.* 10: 5903–13
100. Kuan J, Saier MH Jr. 1993. The mitochondrial carrier family of transport proteins: structural, functional, and evolutionary relationships. *Crit. Rev. Biochem. Mol. Biol.* 28:209–33
101. Kuusinen A, Saloheimo A, Jussila M, Tikka L, Penttilä M, Sopanen T. 1995. Cloning and characterization of two amino acid permease cDNA's of *Barley scutellum*. *Int. Workshop Plant Membr. Biol., 10th, Regensburg,* R21 (Abstr.)
102. Lai K, Bolognese CP, Swift S, McGraw P. 1995. Regulation of inositol transport in *Saccharomyces cerevisiae* involves inositol-induced changes in permease stability and endocytic degradation in the vacuole. *J. Biol. Chem.* 270:2525–34
103. Lawson JE, Douglas MG. 1988. Separate genes encode functionally equivalent ADP/ATP carrier proteins in *Saccharomyces cerevisiae*: isolation and analysis of AAC2. *J. Biol. Chem.* 263: 14813–18
104. Lewis DA, Bisson LF. 1991. The HXT1 gene product of *Saccharomyces cerevisiae* is a new member of the family of hexose transporters. *Mol. Cell. Biol.* 11:3804–13
105. Ljungdahl PO, Gimeno CJ, Styles CA, Fink GR. 1992. SHR3: a novel component of the secretory pathway specifically required for localization of amino acid permeases in yeast. *Cell* 71:463–78
106. Loddenkötter B, Kammerer B, Fischer K, Flügge UI. 1993. Expression of the functional mature chloroplast triose phosphate translocator in yeast internal membranes and purification of the histidine-tagged protein by a single metal-affinity chromatography step. *Proc. Natl. Acad. Sci. USA* 90:2155–59
107. Marger MD, Saier MH Jr. 1993. A major superfamily of transmembrane facilitators that catalyse uniport, symport and antiport. *Trends Biol. Sci.* 18:13–20
108. Marini AM, Vissers S, Urrestarazu A, Andre B. 1994. Cloning and expression of the MEP1 gene encoding an ammonium transporter in *Saccharomyces cerevisiae*. *EMBO J.* 13:3456–63
109. Martinoia E, Grill E, Tommasini R, Kreuz K, Amrhein N. 1993. ATP-dependent glutathione S-conjugate "export" pump in the vacuolar membrane of plants. *Nature* 364:247–49
110. Matern H, Holzer H. 1977. Catabolite inactivation of the galactose uptake system in yeast. *J. Biol. Chem.* 252:6399–6402
111. McKenna E, Hardy D, Kaback HR. 1992. Insertional mutagenesis of hydrophilic domains in the lactose permease of *Escherichia coli*. *Proc. Natl. Acad. Sci. USA* 89:11954–58
112. Milbradt B, Höfer M. 1994. Glucose-transport-deficient mutants of *Schizosaccharomyces pombe*: phenotype, genetics and use for genetic complementation. *Microbiology* 140:2617–23
113. Minet M, Dufour M, Lacroute F. 1992. Complementation of *Saccharomyces cerevisiae* auxotrophic mutants by *Arabidopsis thaliana* cDNAs. *Plant J.* 2:417–22
114. Mitchell P. 1963. Molecule, group and electron translocation through natural membranes. *Biochem. Soc. Symp.* 22: 142–68
115. Mitchell P, Moyle J. 1974. The mechanism of proton translocation in reversible proton-translocating adenosine triphosphatase. *Biochem. Soc. Spec. Publ.* 4:91–111
116. Newman MJ, Foster DL, Wilson TH, Kaback HR. 1981. Purification and reconstitution of functional lactose carrier from *Escherichia coli*. *J. Biol. Chem.* 256:11804–8
117. Nikawa J, Hosaka K, Tsukagoshi Y, Yamashita S. 1990. Primary structure of the yeast choline transport gene and

regulation of its expression. *J. Biol. Chem.* 265:15996–6003

118. Nikawa JI, Tsukagoshi Y, Yamashita S. 1991. Isolation and characterisation of two distinct *myo*-inositol transporter genes of *Saccharomyces cerevisiae*. *J. Biol. Chem.* 266:11184–91
119. Ninnemann O, Jauniaux JC, Frommer WB. 1994. Identification of a high affinity NH^+_4 transporter from plants. *EMBO J.* 13:3464–71
120. Nishizawa K, Shimoda E, Kasahara M. 1995. Substrate recognition domain of the Gal2 galactose transporter in yeast *Saccharomyces cerevisiae* as revealed by chimeric galactose-glucose transporters. *J. Biol. Chem.* 270:2423–26
121. Olander EH, Simoni RD. 1992. The intracellular targeting and membrane topology of 3-hydroxy-3-methylglutaryl-CoA reductase. *J. Biol. Chem.* 267:4223–35
122. Opekarova M, Caspari T, Tanner W. 1993. Unidirectional arginine transport in reconstituted plasma-membrane vesicles from yeast overexpressing CAN1. *Eur. J. Biochem.* 211:683–88
123. Opekarova M, Caspari T, Tanner W. 1994. The HUP1 gene product of *Chlorella kessleri*: H^+/glucose symport studied in vitro. *Biochem. Biophys. Acta* 1194:149–54
124. Özcan S, Johnston M. 1995. Three different regulatory mechanisms enable yeast hexose transporter (HXT) genes to be induced by different levels of glucose. *Mol. Cell. Biol.* 15:1564–72
125. Parent L, Supplisson S, Loo DDF, Wright EM. 1992. Electrogenic properties of the cloned Na^+/glucose cotransporter. II. A transport model under nonrapid equilibrium conditions. *J. Membr. Biol.* 125:63–79
126. Parent L, Supplisson S, Loo DDF, Wright EM. 1992. Electrogenic properties of the cloned Na^+/glucose cotransporter. I. Voltage-clamp studies. *J. Membr. Biol.* 125:49–62
127. Perry JR, Basrai MA, Steiner HY, Naider F, Becker JM. 1994. Isolation and characterization of a *Saccharomyces cerevisiae* peptide transport gene. *Mol. Cell. Biol.* 14:104–15
128. Phelps A, Schobert CT, Wohlrab H. 1991. Cloning and characterization of the mitochondrial phosphate transport protein gene from the yeast *Saccharomyces cerevisiae*. *Biochemistry* 30:248–52
129. Quesada A, Galvan A, Fernandes E. 1994. Identification of nitrate transporter genes in *Chlamydomonas reinhardtii*. *Plant. J.* 5:407–19
130. Rai R, Genbauffe FS, Cooper TG. 1988. Structure and transcription of the allantoate permease gene (DAL5) from *Saccharomyces cerevisiae*. *J. Bacteriol.* 170:266–71
131. Rausch T. 1991. The hexose transporters at the plasma membrane and the tonoplast of higher plants. *Physiol. Plant.* 82:134–42
132. Rea PA, Poole RJ. 1993. Vacuolar H^+-translocating pyrophosphatase. *Annu. Rev. Plant Physiol. Plant Mol. Biol.* 44:157–80
133. Reifenberger E, Freidel K, Ciriacy M. 1995. Identification of novel HXT genes in *Saccharomyces cerevisiae* reveals the impact of individual hexose transporters on glycolytic flux. *Mol. Microbiol.* 16:157–67
134. Rentsch D, Laloi M, Frommer WB. 1995. Molecular analysis of peptide and proline transport in *Arabidopsis*. *Int. Workshop Plant Membr. Biol., 10th, Regensburg*, R22 (Abstr.)
135. Riballo E, Herweijer M, Wolf D, Lagunas R. 1995. Catabolite inactivation of the yeast maltose transporter occurs in the vacuole after internalization by endocytosis. *J. Bacteriol.* 177:5622–27
136. Rickenberg HV, Cohen GN, Buttin G, Monod J. 1956. La galactoside-permease d'*Escherichia coli*. *Ann. Inst. Pasteur* 91:829–57
137. Riesmeier JW, Hirner B, Frommer WB. 1993. Potato sucrose transporter expression in minor veins indicates a role in phloem loading. *Plant Cell* 5:1591–98
138. Riesmeier JW, Willmitzer L, Frommer WB. 1992. Isolation and characterization of a sucrose carrier cDNA from spinach by functional expression in yeast. *EMBO J.* 11:4705–13
139. Roitsch T, Tanner W. 1994. Expression of a sugar-transporter gene family in a photoautotrophic suspension culture of *Chenopodium rubrum* L. *Planta* 193:365–71
140. Rubio F, Gassmann W, Schroeder JI. 1995. Na^+-coupled K^+ uptake and functional domains involved in Na^+ tolerance of the high-affinity K^+ uptake transporter, HKT1, from higher plants. *Int. Workshop Plant Membr. Biol., Regensburg*, 10th, R40 (Abstr.)
141. Sanders D, Hansen UP, Gradmann D, Slayman CL. 1984. Generalized kinetic analysis of ion-driven cotransport systems: a unified interpretation of selective ionic effects on Michaelis parameters. *J. Membr. Biol.* 77:123–52

142. Sauer N, Baier K, Gahrtz M, Stadler R, Stolz J, Truernit E. 1994. Sugar transport across the plasma membranes of higher plants. *Plant Mol. Biol.* 26:1671–79
143. Sauer N, Caspari T, Klebl F, Tanner W. 1990. Functional expression of the *Chlorella* hexose transporter in *Schizosaccharomyces pombe*. *Proc. Natl. Acad. Sci. USA* 87:7949–52
144. Sauer N, Friedländer K, Gräml-Wicke U. 1990. Primary structure, genomic organization and heterologous expression of a glucose transporter from *Arabidopsis thaliana*. *EMBO J.* 9:3045–50
145. Sauer N, Komor E, Tanner W. 1983. Regulation and characterization of two inducible amino-acid transport systems in *Chlorella vulgaris*. *Planta* 159:404–10
146. Sauer N, Stadler R. 1993. A sink-specific H^+/monosaccharide co-transporter from *Nicotiana tabacum*: cloning and heterologous expression in baker's yeast. *Plant J.* 4:601–10
147. Sauer N, Stolz J. 1994. SUC1 and SUC2: two sucrose transporters from *Arabidopsis thaliana*, expression and characterization in baker's yeast and identification of the histidine-tagged protein. *Plant J.* 6:67–77
148. Sauer N, Tanner W. 1989. The hexose carrier from *Chlorella*: cDNA cloning of a eucaryotic H^+ cotransporter. *FEBS Lett.* 259:43–46
149. Sauer N, Tanner W. 1993. Molecular biology of sugar transporters in plants. *Bot. Acta* 4:277–86
150. Schachtman DP, Schroeder JI. 1994. Structure and transport mechanism of a high-affinity potassium uptake transporter from higher plants. *Nature* 370: 655–58
151. Schmidt A, Hall MN, Koller A. 1994. Two FK506 resistance-conferring genes in *Saccharomyces cerevisiae*, TAT1 and TAT2, encode amino acid permeases mediating tyrosine and tryptophan uptake. *Mol. Cell. Biol.* 14: 6597–6606
152. Schroeder JI. 1994. Heterologous expression and functional analysis of higher plant transport proteins in *Xenopus* oocytes. *METHODS: A Companion to Methods of Enzymology* 6:70–81
153. Schroeder JI, Ward JM, Gassmann W. 1994. Perspectives on the physiology and structure of inward-rectifying K^+ channels in higher plants: biophysical implications for K^+ uptake. *Annu. Rev. Biophys. Biomol. Struct.* 23:441–71
154. Schwab WGW, Komor E. 1978. A possible mechanistic role of the membrane potential in proton-sugar cotransport of *Chlorella*. *FEBS Lett.* 87:157–60
155. Schwarz M, Gross A, Steinkamp T, Flügge UI, Wagner R. 1994. Ion channel properties of the reconstituted chloroplast triose phosphate/phosphate translocator. *J. Biol. Chem.* 269:29481–89
156. Scaston A, Inkson C, Eddy AA. 1973. The absorption of protons with specific amino acids and carbohydrates by yeast. *Biochem. J.* 134:1031–43
157. Sentenac H, Bonneaud N, Minet M, Lacroute F, Salmon JM, et al. 1992. Cloning and expression in yeast of a plant potassium ion transport system. *Science* 256:663–65
158. Serrano R. 1991. Transport across yeast vacuolar and plasma mebranes. In *The Molecular and Cellular Biology of the Yeast Saccharomyces,* 1:523–85. Cold Spring Harbor, NY: Cold Spring Harbor Lab. Press
159. Slayman CL, Slayman CW. 1974. Depolarisation of the plasma membrane of *Neurospora* during active transport of glucose: evidence for a proton-dependent cotransport system. *Proc. Natl. Acad. Sci. USA* 71:1935–39
160. Smith FW, Ealing PM, Hawkesford MJ, Clarkson DT. 1995. Plant members of a family of sulfate transporters reveal functional subtypes. *Proc. Natl. Acad. Sci. USA* 92:9373–77
161. Smith FW, Hawkesford MJ, Prosser IM, Clarkson DT. 1995. Isolation of a cDNA from *Saccharomyces cerevisiae* that encodes a high affinity sulfate transporter of the plasma membrane. *Mol. Gen. Genet.* 247:709–15
162. Stadler R, Brandner J, Schulz A, Gahrtz M, Sauer N. 1995. Phloem loading by the PmSUC2 sucrose carrier from *Plantago major* occurs into companion cells. *Plant Cell* 7:9545–54
163. Stadler R, Wolf K, Hilgarth C, Tanner W, Sauer N. 1995. Subcellular localization of the inducible *Chlorella* HUP1 monosaccharide-H^+ symporter and cloning of a co-induced galactose-H^+ symporter. *Plant Physiol.* 107: 33–41
164. Stanbrough M, Magasanik B. 1995. Transcriptional and posttranslational regulation of the general amino acid permease of *Saccharomyces cerevisiae*. *J. Bacteriol.* 177:94–102
165. Steiner HY, Song W, Zhang L, Naider F, Becker JM, Stacey G. 1994. An *Arabidopsis* peptide transporter is a member of a new class of membrane

transport proteins. *Plant Cell* 6:1289–99

166. Stolz J, Darnhofer-Demar B, Sauer N. 1995. Rapid purification of a functionally active sucrose carriers from transgenic yeast using a bacterial biotin acceptor domain. *FEBS Lett.* 377:167–71
167. Stolz J, Stadler R, Opekarova M, Sauer N. 1994. Functional reconstitution of the solubilized *Arabidopsis thaliana* STP1 monosaccharide-H^+ symporter in lipid vesicles and purification of the histidine tagged protein from transgenic *Saccharomyces cerevisiae*. *Plant J.* 6: 225–33
168. Sussman MR. 1995. Molecular analysis of proteins in the plant plasma membrane. *Annu. Rev. Plant Phys. Plant Mol. Biol.* 45:211–34
169. Sychrova H, Chevallier MR. 1993. Cloning and sequencing of the *Saccharomyces cerevisiae* gene LYP1 coding for a lysine-specific permease. *Yeast* 9:771–82
170. Szkutnicka K, Tschopp JF, Andrews L, Cirillo VP. 1989. Sequence and structure of the yeast galactose transporter. *J. Bacteriol.* 171:4486–93
171. Tanaka JI, Fink GR. 1985. The histidine permease gene (HIP1) of *Saccharomyces cerevisiae*. *Gene* 38:205–14
172. Tanner W. 1969. Light-driven active uptake of 3-O-methylglucose via an inducible hexose uptake system of *Chlorella*. *Biochem. Biophys. Res. Commun.* 36:278–83
173. Theodoris G, Fong NM, Coons DM, Bisson LF. 1994. High-copy suppression of glucose transport defects by HXT4 and regulatory elements in the promotors of the HXT genes in *Saccharomyces cerevisiae*. *Genetics* 137: 957–66
174. Truernit E, Sauer N. 1995. The promoter of the *Arabidopsis thaliana* SUC2 sucrose-H^+ symporter gene directs expression of β-glucuronidase to the phloem: evidence for phloem loading and unloading by SUC2. *Planta* 196:564–70
175. Tsay Y-F, Schroeder JI, Feldmann KA, Crawford NM. 1993. The herbicide sensitivity gene CHL1 of *Arabidopsis* encodes a nitrate-inducible nitrate transporter. *Cell* 72:705–13
176. Tyerman SD. 1992. Anion channels in plants. *Annu. Rev. Plant. Physiol. Plant Mol. Biol.* 43:351–73
177. Valverde MA, Diaz M, Sepulveda FV, Gill DR, Hyde SC, Higgins CF. 1992. Volume-regulated chloride channels associated with the human multidrug-resistance P-glycoprotein. *Nature* 355: 830–33
178. Vandenbol M, Jauniaux JC, Grenson M. 1989. Nucleotide sequence of the *Saccharomyces cerevisiae* PUT4 proline-permease-encoding gene: similarities between CAN1, HIP1 and PUT4 permeases. *Gene* 83:153–59
179. Vogel H, Wright JK, Jähnig F. 1985. The structure of the lactose permease derived from Raman spectroscopy and prediction methods. *EMBO J.* 5:3625–31
180. Volland C, Garnier C, Haguenauer-Tsapis R. 1992. In vivo phosphorylation of the yeast uracil permease. *J. Biol. Chem.* 267:23767–71
181. Volland C, Urban-Grimal D, Geraud G, Haguenauer-Tsapis R. 1994. Endocytosis and degradation of the yeast uracil permease under adverse conditions. *J. Biol. Chem.* 269:9833–41
182. Weber A, Menzlaff E, Arbinger B, Gutensohn M, Eckerskorn C, et al. 1995. The 3-oxo-glutarate/malate translocator of chloroplast envolope membranes: molecular cloning of a transporter containing a 12-helix motif and expression of the functional protein in yeast cells. *Biochemistry* 34:2621–27
183. Weber E, Rodriguez C, Chevallier MR, Jund R. 1990. The purine-cytosine permease gene of *Saccharomyces cerevisiae*: primary structure and deduced protein sequence of the FCY2 gene product. *Mol. Microbiol.* 4:585–96
184. Weig A, Franz J, Sauer N, Komor E. 1994. Isolation of a family of cDNA clones from *Ricinus communis* L. with close homology to the hexose carriers. *J. Plant. Physiol.* 143:178–83
185. West IC. 1970. Lactose transport coupled to proton movements in *Escherichia coli*. *Biochem. Biophys. Res. Commun.* 41:655–61
186. Weusthuis RA, Pronk JT, van den Broek PJA, van Dijken JP. 1994. Chemostat cultivation as a tool for studies on sugar transport in yeast. *Microbiol. Rev.* 58: 616–30
187. Will A, Tanner W. 1996. Importance of the first external loop for substrate recognition as revealed by chimeric *Chlorella* monosaccharide/H^+ symporters. *FEBS Lett.* In press
188. Will A, Caspari T, Tanner W. 1994. K_m mutants of the *Chlorella* monosaccharide/H^+ cotransporter randomly generated by PCR. *Proc. Natl. Acad. Sci. USA* 91:10163–67
189. Willey DL, Fischer K, Wachter E, Link TA, Flügge UI. 1991. Molecular cloning

and structural analysis of the phosphate translocator from pea chloroplasts and its comparison to the spinach phosphate translocator. *Planta* 183:451–61

190. Wright EM, Loo DDF, Panayotova-Heiermann M, Lostao MP, Hirayama BH, et al. 1994. Active sugar transport in eukaryotes. *J. Exp. Biol.* 196:197–212

191. Wu J, Perrin DM, Sigman DS, Kaback HR. 1995. Helix packing of lactose permease in *Escherichia coli* studied by site-directed chemical cleavage. *Proc. Natl. Acad. Sci. USA 92:9186–90*

Annu. Rev. Plant Physiol. Plant Mol. Biol. 1996. 47:627–54

LIPID-TRANSFER PROTEINS IN PLANTS

Jean-Claude Kader

Laboratoire de Physiologie Cellulaire et Moléculaire, Université Pierre et Marie Curie (Paris 6)(Unité de Recherche Associée au CNRS 1180), 4 place Jussieu, 75252 Paris Cedex 05, France

KEY WORDS: acyl-binding proteins, membrane lipids, cutin synthesis, embryogenesis, plant-pathogen interaction

ABSTRACT

Lipid-transfer proteins (LTP) are basic, 9-kDa proteins present in high amounts (as much as 4% of the total soluble proteins) in higher plants. LTPs can enhance the in vitro transfer of phospholipids between membranes and can bind acyl chains. On the basis of these properties, LTPs were thought to participate in membrane biogenesis and regulation of the intracellular fatty acid pools. However, the isolation of several cDNAs and genes revealed the presence of a signal peptide indicating that LTPs could enter the secretory pathway. They were found to be secreted and located in the cell wall. Thus, novel roles were suggested for plant LTPs: participation in cutin formation, embryogenesis, defense reactions against phytopathogens, symbiosis, and the adaptation of plants to various environmental conditions. The validity of these suggestions needs to be determined, in the hope that they will elucidate the role of this puzzling family of plant proteins.

CONTENTS

1040-2519/96/0601-0627$08.00

INTRODUCTION

Twenty years after their discovery in plants (51), lipid-transfer proteins (LTPs), defined by their ability to facilitate transfer of phospholipids between membranes in vitro, have yet to be assigned a biological role. Their isolation followed a simple idea. Because the enzymes governing the synthesis of phospholipids are not distributed among all membranes (13, 14, 101, 102), membrane biogenesis requires import of newly synthesized phospholipids. Various mechanisms were suggested: flux of vesicles (74), spontaneous movements of phospholipids (52, 123), and involvement of proteins transporting phospholipids (123). The isolation of LTPs from several plants seemed consistent with the latter hypothesis. However, with recent observations showing that LTPs are extracellularly located and are secreted, a possible role for these proteins in intracellular lipid transfer seems unlikely. Other roles in cutin formation or in defense reactions against phytopathogens were recently suggested. This review aims to contribute to the debate existing about this puzzling family of plant proteins.

BIOCHEMICAL PROPERTIES

Assays

The use of proper assays is of major importance for a correct definition of lipid-transfer proteins. The principle of the assays is to monitor the transfer of labeled lipids from donor to acceptor membranes. Acceptor membranes are either natural membranes such as mitochondria, chloroplasts, plasma membranes, and "microsomes" (endoplasmic rich-fraction) or artificial membranes (liposomes or lipid vesicles) (41, 53, 82). Donor membranes are either natural membranes or liposomes. The lipids to be transferred are either radioactive (118, 125), spin labeled (35, 79, 94), or fluorescent (22, 35, 73). The assays involving radioactive phospholipids are organized as follows: The donor membranes (liposomes) are prepared from ^{3}H-phosphatidylcholine (PC) (the phospholipid to be transferred) and ^{14}C-cholesteryl-oleate (or ^{14}C trioleoylglycerol) (nontransferable compounds). The acceptors are routinely plant mitochondria. After incubation in the presence of lipid transfer proteins, the mitochondria are collected by centrifugation. The ^{3}H/^{14}C ratio of the lipids recovered in the mitochondria indicates the extent of the transfer of individual PC molecules from liposomes to mitochondria. The weak ^{14}C label found in acceptor membranes shows that LTP does not provoke a cosedimentation of donor with acceptor membranes (118).

Fluorescent assays do not require a separation of acceptor and donor membranes. In these assays, donor vesicles consist of self-quenching vesicles of nitrobenzoxadiazol-phosphatidylcholine or of pyrenyl-phosphatidylcholine. When the fluorescent phospholipids leave the donor membranes to reach the acceptor liposomes, the fluorescence increases because of a decrease in the quenching in the donor membrane. This assay allows a continuous monitoring of the lipid transfer (22, 73).

These assays have been used to determine the lipid transfer activity of protein extracts prepared from eukaryotic and prokaryotic cells. Among eukaryotic cells, in addition to plants, mammalian cells, yeasts, and fungi have been studied (28, 92, 108, 123, 124). These assays are important to confirm that a protein belongs to the category of LTPs; one example is the barley LTP, which was initially designated as "putative protease amylase inhibitor" (PAPI) (8, 12, 75). The assays should be carried out with all proteins proposed to be LTPs on the basis of their homology with purified proteins as well as with the products of LTP-like genes.

In all these assays, a bidirectional movement of phospholipids occurs; the lipids of the donor membranes are exchanged with those of the acceptor membranes. For this reason, the proteins involved in this process were first named "phospholipid-exchange proteins" (PLEPs). However, the fact that there is not a true one-for-one exchange of phospholipids led to the name "phospholipid-transfer protein" (PLTP) (52, 123, 124) and then to the generic name "lipid-transfer protein" (LTP), accepted now because of the ability of these proteins to promote the movement of lipids other than phospholipids. The term "non specific lipid-transfer protein" (nsLTP) was also used in reference to the apparent lack of specificity for the various phospholipids.

Purification

The procedures used to purify LTPs from plants are based on LTP biochemical properties (small size—9–10 kDa) and basic character (isoelectric point around 9), which were found to be remarkably similar in all higher plants studied. Plant LTPs are generally purified from soluble proteins through a combination of gel filtration, cation exchange chromatography, and reverse-phase HPLC (20, 26, 27, 53, 72, 77, 110). The lipid transfer proteins are monitored in the purification process by the various transfer assays. Several peaks of lipid transfer activity are often detected, suggesting the existence of various isoforms. The purification of LTPs was helped by their relative abundance in plants [about 4% of the supernatant proteins in maize (40)]. A purification factor of about 100 is sufficient for obtaining a homogeneous fraction from various seeds or leaves. Another approach has been followed by Pyee & Kolattukudy (89), who extracted wax 9 LTP with a short surface wash of broccoli leaves and then purified this LTP, which represents 90% of the total proteins from the surface wax, with electrophoresis.

LTPs have been purified to homogeneity from various monocotyledonous and dicotyledonous plants (3, 17, 125). Plant LTPs are remarkably stable and keep their activity after months of storage at 4°C. Moreover, after a 5-min incubation at 90°C, the transfer activity of maize LTP is preserved, perhaps because of the presence of several disulfide bridges.

Their molecular mass varies from 9 to 10 kDa as determined either by gel filtration or by SDS-gel electrophoresis and confirmed by amino acid sequence determinations. However, in barley and wheat, shorter (7 kDa) LTPs have been detected (22, 54). Animal cells contain several categories of proteins with molecular mass varying from 11 to 33 kDa and that transfer either specifically PC, or preferentially PI (phosphatidylinositol) (PI-TP) or nonspecific (nsLTP) (123). Yeasts contain 35-kDa proteins transferring preferentially PI (37).

The major part of the lipid transfer activity of protein extracts from plants is associated with basic proteins (95% in the case of spinach leaves) (53). Only the basic LTPs (isoelectric point varying from 8.8 to 10.0, determined by chromatofocalization or isoelectric focusing) have been purified. Acidic proteins, with an apparent molecular mass of 20 to 30 kDa, have been detected in various plants, particularly in castor bean, but neither purified to homogeneity nor sequenced (125).

The availability of polyclonal antibodies prepared against LTPs purified from various plants or fusion proteins (105, 112) has provided tools to confirm, in some cases, the nature of a putative LTP (105), to determine the LTP contents either by ELISA or Western blot (40, 116), or to localize several LTPs immunohistochemically (104, 112). These antibodies also helped in the isolation of cDNAs encoding plant LTPs (109, 115). However, no monoclonal antibody is available for plant LTPs.

Specificity for Lipids and Acyl-Binding

Plant LTPs are able to transfer not only PC from liposomes to mitochondria but also PI and, to lesser extents, PE or PG (phosphatidylglycerol). They are also able to transfer galactolipids but not triacylglycerols (79). The ability of plant LTPs to transfer sterols has not been studied. LTPs from castor bean, spinach, sunflower, and oilseed rape are able to bind acyl-CoA (4, 81, 93, 114). In contrast, animal cells contain fatty acid–binding proteins or acyl-CoA-binding proteins that are unable to transfer phospholipids (44). The binding of fatty acids or acyl-CoA esters by plant LTPs was determined by separation of the acyl-LTP complex by gel filtration or by temperature-dependent ligand affinity (114). In castor bean, the saturating binding capacities for oleic acid and oleoyl-CoA per mole of LTP were 1:1 (114). In oilseed rape, several isoforms of LTP, separated by cation-exchange chromatography, revealed both ability for transfer of PC and binding of oleoyl-CoA (81). Three of these isoforms have been recently purified and sequenced (80). This ability to bind acyl chains has also been demonstrated with a barley 7-kDa protein that has some features in common with LTPs (16) and with an LTP secreted from carrot embryogenic cultures that was found to be able to bind oleoyl-CoA in a near-equimolar ratio (42, 69).

It can thus be concluded that the same proteins have the dual ability to transfer lipids and to bind acyl chains. Are the same sites involved in these two processes? To answer this question, structural studies of LTPs were performed.

STRUCTURE AND MODE OF ACTION

Amino Acid Sequence

The complete amino acid sequences have been determined for LTPs purified from various plants (11, 22, 75, 106, 109, 128) and were found to correspond to those deduced from the nucleotide sequences of LTP genes with the exception of the presence of a signal sequence at the amino terminal end of the gene products (Figure 1). These sequences thus correspond to mature LTPs. In the case of barley and rice, LTPs have been primarily called putative amylase-protease inhibitors, because of their homology with a protein isolated from Indian finger millet and because they were initially supposed to belong to this category, although it is probably an LTP (15). Plant LTPs have a total number of amino acids varying from 91 to 95 residues. They lack tryptophan, and they have eight cysteine residues located at conserved positions according to the pattern 2/3-C-8-C-12/15-CC-19-C-1-C-21/23-C-13-C-4/8. The cysteine residues are engaged in four disulfide bridges as determined by tryptic digestion of castor bean LTP (Figure 2) (106).

Plant LTPs exhibit strong structural homologies as revealed by sequence alignment and hydrophobic cluster analysis (22, 43, 109). However, no sequence homology was found between LTPs from mammalian and plant LTPs (123). Plant LTPs differ from oleosins, which are hydrophobic proteins associated with oil bodies in seeds (90).

Tertiary Structure

On the basis of the sequences of plant LTPs, a tentative tertiary structure model was proposed that consisted mainly of beta-sheets (65, 109). This conclusion appeared to be wrong in light of the structural studies carried out on various LTPs. Studies of wheat LTP by NMR, infrared, and Raman spectroscopy (22,

```
                                                       41
Maize     MARTQQLAVVATAVVALVLLAAATSEAAIS C GQVASAIAP C
Carrot     MGVLRSSFVAMMVMYMVLATTPNAEAVLT C GQVTGALAP C
Tobacco        MEIAGKIACFVVLCMVVAAPCAEAIT C GQVTSNLAP C
Spinach   MASSAVIKLACAVLLCIVV-AAPYAEAGIT C GMVSSKLAP C
Gerbera       MASMVMNVLCVAVACMVFSASYADAIS C GQVTSGLVP C
Barley    MARAQVL-LMAAALV-LMLTAAPRAAVALN C GQVDSKMKP C
Broccoli     MAGLMKLACLIFACMIVAGPITSNAALS C GTVSGYVAP C
Rice      MARAQLVLVALVAAALLLAGPHTTM-AAIS C GQVNSAVSP C
Castor        MKNVVFSVLLLLSFLFCLANTNEAAVP C STVDMKAAA C
Arabido               MLALGLHDCGRSNTSNAALS C GSVNSNLAA C

                                                    81
Maize     ISYARGQGSGPSA-G CC SGVRSLNNAARTTADRRAA CNC LK
Carrot    LGYLRSQVNVPVPLT CC NVVRGLNNAARTTLDKRTA CGC LK
Tobacco   LAYLRN--TGPLGR- CC GGVKALVNSARTTEDRQIA CTC LK
Spinach   IGYLKG---GPLGGG CC GGIKALNAAAATTPDRKTA CNC LK
Gerbera   FGYL--AAGGPVPPA CC NGVRGLNNAAKTTPDRQTA CGC LK
Barley    LTYVOG-GP-GPSGE CC NGVRDLHNQAQSSGDRQTV CNC LK
Broccoli  IGYLAQNAP-AVPTA CC SGVTSLNNMARTTPDRQQA CRC LV
Rice      LSYARGLRP---SAA CC SGVRSLNSAASTTADRRTA CNC LK
Castor    VGFATGKDSKP-SQA CC TGLQQLAQTVKTVDDKKAI CRC LK
Arabido   IGYVLQGGVIP--PA CC SGVKNLNSIAKTTPDRQQA CNC IQ

                                                      120
Maize     NAAAGVSGLNAGN-AASIPSK C GVSIPYTISTSTD C SRVN
Carrot    QTANAVTGLNLNA-AAGLPAR C GVNIPYKISPTTD C NRVV
Tobacco   SAAGAISGINLGK-AAGLPST C GVNIPYKISPSTD C SKVO
Spinach   SAANAIKGINYGK-AAGLPGM C GVHIPYAISPSTN C NAVH
Gerbera   GILAANTRINLNN-ANSLPGK C GISIGYKITPNID C SKIH
Barley    GIARGIHNLNLNN-AASIPSK C NVNVPYTISPDID C SRIY
Broccoli  GAANALPTINVAR-AAGLPKA C GVNIPYKISKTTN C NSVK
Rice      NVAGSISGLNAGN-AASIPSK C GVSIPYTISPSID C SS
Castor    ASSKSLGIKDQFLSK--IPAA C NIKVGFPVSTNTN C ETIH
Arabido   GAARALGSGLNAGRAAGIPKA C GVNIPYKISTSTN C KTVR
```

Figure 1 Comparison of the amino acid sequences of lipid-transfer proteins deduced from cDNAs or genes encoding these proteins in plants. The putative cleavage site of the signal sequence is indicated by an arrow. The cysteine residues are within boxes. The following sequences are presented: maize (109), carrot (105), tobacco (30), spinach (9), *Gerbera hybrida* (58), barley (75), broccoli (wax 9; 88), rice (119), castor bean (nsLTP C, 115), *Arabidopsis thaliana* (111).

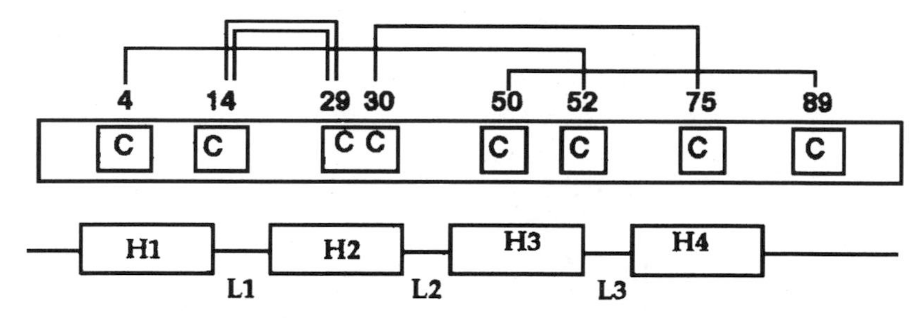

Figure 2 Structural features of plant lipid-transfer proteins. They appear as 91–95 amino acid polypeptides comprising four α-helices joined by loops. The eight conserved cysteine residues form four disulfide bridges (97, 106).

98) showed that the protein is organized mainly as helical segments (40% of total structures) connected by disulfide bridges. This helical structure plays a role in the transfer activity. A model of the three-dimensional structure of a wheat LTP was built on the basis of 1H-NMR data (36). The polypeptide backbone appears to be composed of four helices linked by flexible loops.

In order to check these models, crystallographic studies have been carried out. Although wheat (83) and rice (47) LTPs have been crystallized, the first crystal structure has been recently determined on maize LTP either complexed or not complexed with palmitic acid (97). The protein appears composed of a single compact domain with four α-helices and a long C-terminal domain, which confirms that it is an all-alpha-type structure, contrary to the initial model proposed for LTPs (Figure 3).

Mode of Action

How do LTPs facilitate movements of phospholipids between membranes? Several years after the detection of their activity, their mode of action is not completely understood. A shuttle mechanism has been proposed for the phosphatidylcholine-specific LTP from mammalian cells, which suggests the formation of a phospholipid-LTP complex that interacts with the membrane and exchanges its bound phospholipid with a phospholipid molecule from the membrane (123). A similar sequence of events has been suggested for plant LTPs (50, 52, 53). However, such a complex has never been isolated with a plant LTP (and also with nonspecific LTPs from mammalian cells), which suggests that the binding is too weak to allow formation of a stable complex (52). In contrast, a strong binding of acyl chains (93) or of lyso-phosphatidylcholine (22) has been noted for several plant LTPs. These binding properties are in agreement with the model suggesting that LTPs contain a hydrophobic cavity that can accept one acyl chain but not a whole phospholipid molecule

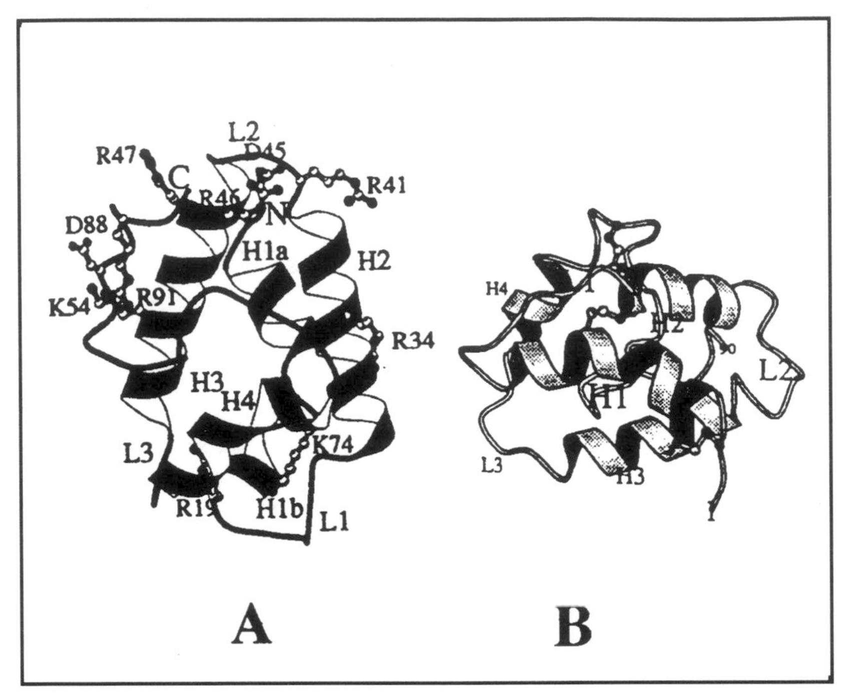

Figure 3 Ribbon diagrams of plant lipid-transfer proteins: (A) diagram of maize LTP established from crystallographic studies by Shin et al (97); (B) diagram of wheat LTP established from NMR studies by Gincel et al (36). The eight conserved cysteines form four disulfide bridges at the same positions suggested by previous chemical studies of castor bean LTP (107). In the absence of a bound ligand, the protein has a tunnel-like hydrophobic cavity that is large enough to accommodate a long fatty acyl chain. When a complex is formed with the fatty acid, most of the acyl chain is buried inside this hydrophobic cavity (97). One end of the tunnel has a wider mouth and is close to a flexible and polar region, which might play a role in the binding and the subsequent release of lipids. The other end of the tunnel has a narrower opening and nonpolar residues. NMR studies carried out on the same protein have recently yielded similar conclusions (85). This cavity is absent from a hydrophobic protein from soybean exhibiting a four-helices structure (6). It may be interesting to study the structure of modified LTPs after expression of the corresponding genes, normal or mutated, in heterologous organisms, as has been done with wheat and castor bean (24, 66). [Reproduced with permission from *Current Biology* (A) and *European Journal of Biochemistry* (B).]

(97). The two acyl chains of the phospholipid may thus interact differently with the protein, with one more weakly bound than the other, which facilitates the extraction of the phospholipid when the LTP interacts with a membrane surface (Figure 4).

It is not known whether the protein undergoes conformational change when it interacts with lipids or with membranes. The reduction of a wheat LTP by dithiothreitol leads to a decrease of α-helix proportion (from 40 to 25%).

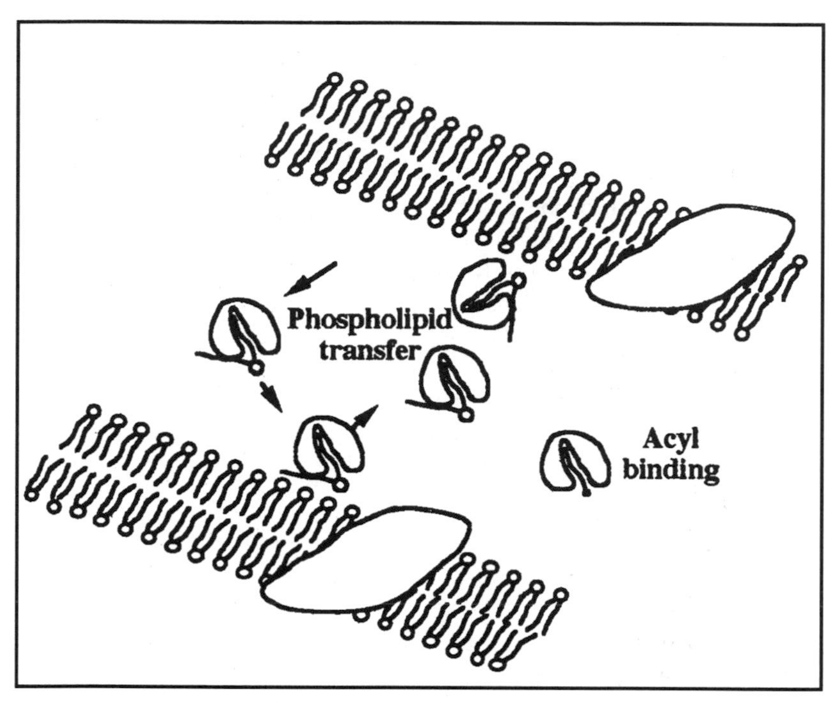

Figure 4 Postulated mode of action of plant lipid-transfer proteins. A reversible complex between LTP and phospholipid is formed. A hydrophobic tunnel recently suggested to be able to accommodate only one acyl chain (97) could be involved in this process. The binding and release of phospholipid molecules, facilitated by polar domains situated around the opening of the cavity, leads to their exchange with phospholipids of the membranes. It is also indicated that LTP can bind fatty acids or acyl-CoA esters.

The reduction of maize LTP was also found to inhibit the lipid transfer activity, which confirms the important role of disulfide bridges (39). The fact that the lipid transfer activity can be disturbed when ions are added or when the content of the acceptor liposomes in acidic phospholipids varies (50) indicates that electrostatic interactions between LTP and membranes are involved in the function of the protein. LTP-like proteins present in wheat, barley, pine, and *Petunia* (76, 77, 87) were found to be the substrates, with various phosphorylation sites, of Ca^{2+}-dependent protein kinases. Moreover, after labeling of *Petunia* petals with ^{32}P phosphate, the major labeled protein has an apparent molecular mass close to that of LTPs (77). The physiological significance of these observations remains to be determined. It is worth noting that plants contain calcium-dependent phospholipid-binding proteins called annexins but that their biological role is still unknown (91). The biotechnological role of LTP in the stabilization of the beer foam has been recently considered, although the mechanism of this action is unknown (103).

Manipulation of Membrane Lipids

In vitro plant LTPs are able to replace the phospholipids of an acceptor membrane with those of the donor membrane, and vice versa. As a consequence, LTPs were used as tools in order to modify the lipid composition of a membrane and to study the consequences of these changes on its properties. This modification has been done with chloroplast envelope membranes (70), leading to changes in their acyl composition. Plant LTPs have also been used to manipulate the lipid composition of erythrocyte membranes (94) and human platelets (7). The possibility that plant LTPs promote a net transfer—i.e. an increase in the amount of lipids in the acceptor membrane—has not been examined enough, in contrast with animal nsLTP (123).

GENE EXPRESSION

Cloning

The first cDNA encoding a plant LTP was isolated from a library constructed from mRNAs extracted from maize seedlings (109). The library was screened with polyclonal antibodies prepared against maize LTP. The amino acid sequence, deduced from the cDNA clone, was in agreement with the sequence determined on the purified protein with the exception of an unexpected leader sequence of 27 amino acids. All the cDNAs, further determined for plant LTPs, exhibited the same features. On the basis of the maize cDNA sequence, several homologous cDNAs or genes have been characterized or, in some cases, renamed. This is the case for the cDNA encoding barley LTP, which was published two years before the maize LTP cDNA as a cDNA clone encoding a putative amylase/protease inhibitor (75). Since then, 37 cDNA and gene sequences encoding plant LTPs have been published (Table 1). More than 30 sequences available from projects devoted to partial sequencing of anonymous cDNA clones and expressed sequence tags (EST) from *A. thaliana,* rice, castor bean, maize, tobacco, oilseed rape, *Senecio odorus,* and *Brassica campestris* have also been deposited in databases. This makes it necessary to clarify the nomenclature of these genes, which are in some cases defined on the basis of a weak sequence homology. It can be easily predicted that the number of LTP gene sequences will considerably increase in the future, owing in part to their general occurrence in plants and to the complexity of this gene family.

The LTP gene family is indeed complex. Several cDNAs encoding LTPs have been characterized in the same plant. This is the case for maize, in which it has been proposed that a mechanism of alternative splicing might exist in

Table 1 Genes encoding lipid-transfer proteins in higher plants

Source	Gene or cDNA	Accession number[a]	Protein[b]	Ref
Arabidopsis thaliana	LTP1 (cDNA)		110	111
	LTP1 (gene)	S73825	111	111
Brassica napus (oilseed rape)	E2(cDNA)		119(24)	32
Brassica oleracea (broccoli)	wax9A to wax 9D (gene)	L33904 to L33907	117 to 120	88
Daucus carota (carrot)	EP2 (cDNA)	M64746	120(26)	105
Gerbera hybrida	gltp1 (cDNA)	Z31588	116(25)	58
Gossypium hirsutum (cotton)	GH3 (cDNA)		120	63
Hordeum vulgare (barley)	LTP1 (cDNA)		117(25)	75
	Ltp1 (gene)	X59253, X60292	117(26)	60 100
	LTP2,LTP3, LTP4 (cDNA)	X68654,X68655 X68656	115	71
	blt 4.1 (cDNA)		131	46
	gblt 4.2,gblt4.6,		115	122
	gbl 4.9 (gene)		115	1
	pKG2316 (cDNA)	Z37114	120	34
	pKG285	Z37115	115	34
	7kDa-LTP2 (cDNA)	X60793	102(35)	54
Lycopersicon esculentum (tomato)	TSW12 (cDNA)	X56040	114(23)	113
Nicotiana tabacum (tobacco)	Ltp1 (cDNA)	D13952		66
	Ltp1 (gene)	X62395	118(23)	30
Oryza sativa (rice)	Ltp (gene)	Z23271	117	119
Pachyphytum	EPI 12 (cDNA)	L14770	116(23)	18
Pinus taeda (loblolly pine)	pPtIF (cDNA)	U10432	123	55
Ricinus communis (castor bean)	nsLTPC (cDNA)	D11077	106(24)	114
	nsLTPC1 (cDNA)	M86353	118(21)	120
	nsLTPC2 (cDNA)	M86354	118(23)	
Sorghum vulgare	Ltp1,LtP2 (cDNA)	X71667,X71668	93	84
Spinacia oleracea (Spinach)	PW132 (cDNA)	M58635	118(26)	8
Triticum durum (wheat)	pTd4.90 (cDNA)	X63669	113(23)	23
Vigna unguiculata	(cDNA)	X79604	99	59
Zea mays (maize)	9c2 (cDNA)	J04176	120(27)	109
	6B6 (cDNA)	M57249	99	2
Zinnia elegans	(cDNA)	U19266	95	127

[a]Other cDNA sequences or EST have been deposited in databases but not yet published. These sequences may be accessed online at URL:http:ncbi.nlm.nih.gov (78).

[b]The deduced number of amino acids are indicated as well as the signal peptide (between parentheses) when given by the authors.

the RNA coding for these proteins (3). Several cDNAs encoding LTP-like proteins can be identified in the same plant: castor bean (115, 120), barley (34, 71, 122), and broccoli (88). In *A. thaliana,* a cDNA and the corresponding gene were characterized (111), but several other different sequences are present in the databases.

This complexity was confirmed by Southern blot analysis, which suggested the presence of several LTP genes: at least two in cotton (63), maize (109), *Gerbera hybrida* (58), and tomato (113); three in rice (119); four in broccoli (88); five in *Sorghum* (84); seven in barley (122); and as many as 14 in loblolly pine (55). However, possibly because high stringency hybridization conditions were used, only one gene was detected in carrot (105) and spinach (8). In barley, levels of sequence homology and chromosome location divide the LTP genes into two families, comprising at least seven genes distributed among three chromosomes (3, 5, and 7) (122). Four of these genomic sequences have been determined in barley (60, 100, 122), as well as one additional gene coding for a shorter (7 kDa) LTP that was called *ltp2,* which introduced some confusion with the other LTP genes and which we call 7-kDa-LTP gene (54).

Until now, 14 genomic sequences have been determined for plant LTPs: one in *A. thaliana* (111), tobacco (30), and rice (119); two in *Sorghum* (84); four in broccoli (88); and five in barley (1, 60, 100, 122). Almost all of these genes have an intron placed in the region corresponding to the C-terminus of the protein, generally two codons before the stop codon. The length of the intron varies from one gene to another: 89 bp in rice (119), 114 bp in one of the two *Sorghum* genes (84), 115 bp in *A. thaliana* (111), 133 bp in barley (60, 100), 271 bp in broccoli (88), and as high as 980 bp in tobacco (30). However, no intron has been found in one of the barley genes (*gblt 4.9*) (122) or in one of the *Sorghum* genes (*ltp2*) (84).

Organ-Specific and Developmental Expression

In order to understand the in vivo role played by plant LTPs, it is important to determine when and where the LTP genes are expressed. This was done by using Northern blot analyses of mRNA from various tissues, by in situ hybridization, and more recently by following the expression for the reporter gene β-glucuronidase-LTP (*GUS*-LTP) constructs in transgenic plants.

The expression patterns reported for the different plant LTP genes were found to be complex and, in large part, temporally and spatially controlled. The studies, generally made by Northern blot analysis and, in one case (88), by reverse transcriptase polymerase chain reaction, led to several unexpected observations. When the vegetative organs were considered, one surprising finding was that no LTP gene transcript was detected in the roots of various plants (9, 30, 34, 63, 71, 88, 113, 120, 122). However, an LTP gene is expressed—albeit weakly—in rice seedlings (119). Depending on the LTP genes considered, the expression was found to be active in the aerial portions of the plants (leaves, stems, shoot meristems) (9, 30, 31, 63, 71, 88, 111, 119). It was also found that, in a given plant, each LTP gene displays its own peculiar pattern of expression. In barley, one of the two seedling-specific LTP genes is more expressed in leaves than in coleoptiles of seedlings, whereas the reverse is true for the other gene (34). In castor bean, two LTP genes (*ltpC1* and *ltpC2*) have been found exclusively expressed in the cotyledons (120), where another LTP gene (*nsltpD*) is expressed in the axis (115).

Other observations are that the expression of LTP genes is higher in young tissues than in old ones, as observed in tobacco leaves (30), and that the expression of an LTP gene was found to be the highest in the upper part of the tobacco plant and to decline toward the base, which indicates that these LTP genes are expressed in a developmental gradient in this plant. LTP genes were found to be highly expressed early in development in embryo cotyledons and leaf primordia of *A. thaliana* (111) and in somatic embryos of carrot (105).

In flower development, LTP genes are expressed at early stages. Very high expression in inflorescences was found for LTP genes in various plants (29, 30, 56, 88, 105, 111). Highest transcript levels were observed in young developing inflorescences of carrot (105) and *A. thaliana* (111), in the sepals of unopened flowers of tobacco (30), in flower buds of broccoli (88), in microspores of rapeseed (32), and in corolla and carpel of *G. hybrida* (58). The latter genes are strictly specific for the inflorescences, whereas the other LTP genes can be expressed in other organs.

Cell-Type Specificity

A surprising observation, provided by studies using in situ hybridization, is that the LTP gene expression is mainly restricted to defined cell layers, generally situated peripherally. In maize seedlings, for example, the LTP gene transcripts accumulate mainly in the epidermis of the coleoptiles as well as in the leaf veins (104). In carrot (105), the LTP gene is expressed in protoderm cells (which give rise to the epidermis) of somatic and zygotic embryos and in the shoot apical meristem. This cell type–specific expression was also observed in the shoot apical meristem of tobacco, particularly in the epidermis (105). The same pattern was observed in *A. thaliana* in which LTP gene transcripts were detected first in the protoderm cells of the embryo cotyledons, then in the leaf primordia of young seedlings, and at a later stage of development in the epidermal cells of meristem and leaves (111). In the succulent plant, *Pachyphytum,* a cDNA-encoding LTP-like protein has been isolated among epidermis-specific genes (18). In cotton, LTP genes specific for fibers, which are highly elongated trichomes that grow from epidermal cells of ovules, have been characterized (63). Studies by tissue-print hybridization carried out in barley seedlings also led to the conclusion that the LTP genes are expressed in the epidermis of both leaf and coleoptile (34).

An interesting model in cereal seeds is the aleurone layer, composed of specialized cells rich in lipid bodies and aleurone grains, surrounding the starchy endosperm (48). In barley, the transcripts of *ltp1* (100) or of the 7-kDa LTP genes were found to be specifically accumulated in the aleurone layer. This observation is consistent with studies by Northern blot analysis of RNAs extracted from the aleurone layer of barley seeds (75).

Another example of cell specificity of gene LTP expression is given by floral organs. Koltunow et al (56) found that an LTP gene, with a yet unpublished sequence, was specifically expressed in the tapetum layer of the tobacco anther. In the same plant, transcripts for a gene encoding an LTP showed a restricted distribution to the outermost cell layer in the floral apical meristem at the stage of transition to floral development (29). Another LTP gene, *E2*, was also found to be exclusively expressed in the tapetal cells of *B. napus* anthers (32). LTP gene expression was also detected in the epidermal cells of several floral organs, including *A. thaliana* (111) and *G. hybrida* (58).

It was of interest to extend these observations by the study of transgenic plants containing a fusion of a promoter region of LTP genes to the reporter gene *GUS*. This was carried out in barley using *ltp1* or the 7-kDa LTP genes (54, 100), in *A. thaliana* (111) using the *LTP1* gene, and in *B. oleracea* using the wax 9D gene (88). The data obtained by this approach are remarkably coherent, in some aspects, with the in situ hybridization observations. *GUS* activity was detected in the aleurone layer of barley seeds in the case of two genes (54, 100). When the 7-kDa LTP gene is considered, it was found that the corresponding promoter is able to direct aleurone-specific expression in immature barley grains or in transgenic rice (54). The expression of broccoli LTP promoter-*GUS* constructs was detected in the epidermal cells of various organs of transgenic tobacco (88). In the case of *A. thaliana* (111), the expression of the LTP-*GUS* activity was dependent on the stage of development. In young seedlings, *GUS* activity was detected in the cotyledons and in the hypocotyl. Then, as seedlings matured, *GUS* activity was observed in the shoot meristem, the vascular tissue, in the leaf primordia, and then in the tips of maturing leaves. In adult plants, high *GUS* staining was detected in leaf and stem epidermal cells, guard cells, and flowers. Although no activity was detected in flower buds, high activity was observed in the nectaries, in the stigma, and in pollen grains of opened flowers.

Some of these observations made by the *GUS* approach are not coherent with the in situ hybridization data, probably because of some artifactual expression of this reporter gene: In *A. thaliana*, no clear epidermis specificity of the *GUS* expression was observed in young seedlings, although in some cases (guard cells, floral nectaries, pollen grains, lateral roots, and stipules) the *GUS* activity was not confirmed by in situ hybridization (111).

Localization of LTPs

LTPs have been purified as soluble proteins after precipitation of cell membranes (50, 51, 124), although they were found to be partly bound to membranes (mitochondria, endoplasmic reticulum) (25). Several unexpected discoveries have changed this conclusion. One finding is that all known plant LTPs are synthesized as precursors with N-terminal extensions having the sequence characteristics of signal peptides. The length of this signal peptide varies from 21 to 27, depending on the LTP gene (Table 1). The longest leader sequence (35 amino acids) was postulated for LTPs from barley (54). As shown in the case of LTP genes from spinach (9) and barley (64, 65) but not in castor bean (126), this signal peptide is able to direct the cotranslational insertion of the polypeptide into the lumen of the endoplasmic reticulum in vitro. It has thus been suggested that LTPs participate in the lipid movements within the lumen of the endoplasmic reticulum (65). This hypothesis is interesting in reason of the presence of plasmodesmata, establishing communications implying membrane continuity and lipid movements between adjacent cells (38, 62). However, because no endoplasmic reticulum retention signal (KDEL) is present at the carboxyterminus of LTPs, they are expected to enter the secretory pathway. Several observations have indeed confirmed that LTPs are secreted proteins: In barley, an LTP first identified as a putative amylase protease inhibitor was found to be secreted into aleurone cell culture medium (75); in carrot, an LTP was detected among extracellular proteins in embryogenic cell cultures (105). A similar observation was made in grapevine because several isoforms of LTP have been purified from the extracellular medium of somatic embryo cultures (20). The extracellular location is consistent with the fact that an LTP from maize is synthesized on membrane-bound polysomes (117).

In addition, immunocytochemical studies carried out at the ultrastructural level using a polyclonal antibody against a fusion protein corresponding to the LTP1 gene of *A. thaliana* clearly indicated that in this plant the LTP was localized to the cell wall (112). In castor bean, a partial localization of LTP in the cell wall was also found (114). Immunocytochemistry at the light microscopy level also indicated some signals corresponding to LTP-immunogold complexes in the cell wall of epidermal cells of maize coleoptiles (104). In broccoli leaves, the wax 9 protein was present mainly in the cell wall of the epidermis and mesophyll tissues as well as in the phloem. The LTP is so abundant in young leaves that it constitutes more than 90% of the protein extracted from these leaves by a brief solvent wash (89). A similar preferential localization of LTP to the cell walls was found in barley, as indicated by immunological staining of tissue prints (71). This distribution of LTP was confirmed by the fact that these proteins could be extracted by a short dip of intact barley leaves in buffer (71). In conjunction with these data, it is worth noting that a high expression of LTP gene is observed in cotton fibers, particularly in the phase of primary cell wall synthesis (63). Similarly, an LTP gene is specifically expressed during the differentiation of tracheary elements of *Zinnia elegans* in relation to cell wall thickening (127).

It is easy to conclude, from all these observations, that plant LTPs are extracellular proteins. In agreement with this conclusion, no LTP was detected in the stroma of chloroplasts (95). However, it cannot be excluded that some LTP isoforms are addressed to cytosolic compartments. For example, on the basis of immunocytochemical evidence, LTP from castor bean appears to be partly located within the glyoxysomes (114, 126). However, it remains to be explained whether an LTP isoform is truly located in the cytosol and how it is addressed to this compartment in spite of the presence of a signal peptide.

PROPOSED BIOLOGICAL ROLES

On the basis of their in vitro properties of transferring lipids and binding acyl chains, the suggestion that LTPs could be involved in many aspects of cell function where movement of lipids is thought to be important, such as membrane biogenesis and turnover, was a logical one. This hypothesis has been presented in several reviews (2, 50, 52, 124). However, the fact that LTPs are located extracellularly made this hypothesis inconsistent. On the basis of this external location as well as on novel properties discovered in recent years, other roles have been suggested.

Cutin Formation and Embryogenesis

The hypothesis of the group of de Vries (105), which suggests that LTPs are involved in the secretion or deposition of extracellular lipophilic material, including cutin, has been argued convincingly in recent years. According to this theory, LTPs carry acyl monomers necessary for the biosynthesis of cutin. Different facts support this. 1. LTPs are mainly located in the cell wall and are secreted. 2. LTP gene expression and LTP gene products accumulation was detected in high levels in peripheral cell layers, including epidermis. 3. LTPs, particularly in young leaves where cutin deposition is active, are mainly concentrated in the surface wax (88). 4. LTPs are able to bind acyl chains (42, 69).

The "cutin theory" thus seems highly convincing. However, it remains to be validated, either with an antisense approach or by studying eceriferum mutants in *A. thaliana* affected in their epicuticular wax (49). In relation to this theory, LTPs were also supposed to participate in the deposition of lipophilic material in floral organs: corolla (58), stigma and nectaries (111), and anther (tapetum and pollen), where sporopollenin biosynthesis occurs (32).

The finding that LTP is secreted into the medium of embryogenic cell cultures suggested a new role for LTP in somatic embryogenesis (105). LTP could be involved in the early steps of embryogenesis by participating in the formation of a protecting layer around the young embryo (zygotic or somatic). The high LTP gene expression in young embryos (105, 111) is in agreement with this hypothesis.

Defense Reactions Against Pathogens

The unexpected antibiotic properties of LTP were discovered by screening plant proteins for their ability to inhibit the growth of fungal and bacterial pathogens (72, 110).

Several LTP-like proteins purified from barley leaves or an LTP isolated from maize leaves were shown to inhibit the growth of a bacterial pathogen and a fungus (72). This was confirmed by the isolation of LTP-like proteins from cell-wall preparations from the leaves of *A. thaliana* or spinach (96). A synergistic effect against the fungus occurred when the LTPs were combined with thionins (72). The cloning of three barley antipathogen proteins allowed the study of the induction of the corresponding genes. All LTP mRNA levels were significantly increased when the barley leaves were inoculated with a fungus (71). A similar antifungal activity has been described for a 9-kDa basic protein purified from radish seeds and exhibiting a high sequence homology with LTP in its N-terminal end (110). This protein is able to inhibit in vitro the growth of several fungi. In broccoli, the wax 9 protein was found to inhibit the growth of a fungus (89).

It can be concluded that the antifungal activity of LTPs varies between pathogens and that the pattern of expression of LTP genes in response to infection of barley plants by pathogens is rather complex (33). The idea of a defense-protein shield, suggested by Garcia-Olmedo et al (33), is supported by the fact that LTPs were found, at least for some isoforms, to be accumulated at the tissue surface at much higher concentrations than those required to inhibit many pathogens in vitro. It is interesting to note that an LTP-like protein is secreted into the medium of rice suspension cultures in response to a treatment by salicylic acid known as an elicitor to induce pathogen defense processes (67).

However, we do not yet know how LTPs inhibit the growth of the pathogens. Because of their high isoelectric point, LTPs may act as membrane permeabilizing agents. The fact that LTPs are induced could be due to the fact that cutin biosynthesis is generally stimulated by pathogen infection.

Symbiosis

Another relationship between bacteria and plants is the nodule initiation in legume roots following interaction with symbiotic bacteria such as *Rhizobium*. An early stage of the infection occurs in the root hairs to which *Rhizobium* binds and results in elongation and curling of the root hair that facilitates the penetration of the bacteria. A cDNA, coding for an LTP-like protein, has been isolated from a cDNA library constructed from RNA of *Vigna unguiculata* roots one and four days after inoculation with *Rhizobium* (59). The mRNA levels increased in root hairs after inoculation with *Rhizobium*. Because LTP-like transcripts are absent from differentiated nodules, it was concluded that the LTP gene was transiently expressed during nodule development.

Adaptation of Plants to Various Environmental Conditions

An important domain of research for plant biologists is the study of the ability of plants to modify their metabolism and development in response to changes in the environment, including changes involving temperature, drought, and salt stress. It is of interest to characterize genes involved in acclimation to low or high temperatures or to shortage of water. The search for stress-induced genes has led to the characterization of genes encoding LTP-like proteins. One example is the low-temperature response studied in barley (46, 122). Several genes induced by cold treatment code for LTP-like proteins and are also induced by a treatment of abscisic acid (ABA), which is common with genes induced by low temperature or drought (46). Although all these genes are induced by cold, there are varietal differences in the response of the barley LTP gene family to low temperature. For example, one gene (*blt 4.1*) was upregulated by low temperature in the winter cultivars but not in a spring cultivar (122). In other barley varieties, no induction was observed for the expression of three LTP genes in barley because of cold or other factors (drought, salicylate), and only a moderate increase was observed because of NaCl or ABA.

The isolation of several genomic clones in barley has led to a study of the regulatory elements. A putative ABA-responsive (ABRE) element (GTACGTGG) and a low-temperature responsive element (ACACGTCA) were found in the barley LTP genes (122). Another LTP gene was found to be regulated by auxin (naphtaleneacetic acid) during tracheary element formation in *Z. elegans* (127).

Other LTP genes respond to salt stress or drought stress. In tomato, a gene encoding an LTP-like protein is expressed, specifically in stems, only when plants are treated with NaCl, mannitol, ABA, or high temperature (46). Another gene, only partially homologous to LTP, was also induced by drought in barley

(86). Other genes were found to be expressed in the roots and stems of drought-stressed barley plants (122).

The response of LTP-genes to developmental and environmental signals is thus complex, and additional information is needed to determine the involvement of these proteins in the adaptation of plants to several stresses. However, all these conditions—drought, cold, and salt stresses—are related to desiccation or water stress. If LTPs are involved in cutin deposition, the induction of LTP genes by conditions leading to desiccation seems logical. The promoter regions of the *ltp1* gene from *A. thaliana* contain sequences homologous to putative regulatory elements of genes in the phenylpropanoid biosynthetic pathway and sequence elements that have been found in the promoters of stress-induced genes (111).

THE SEARCH FOR NOVEL LIPID-TRANSFER PROTEINS

The search for novel LTPs located in the cytoplasm is of major interest. Two categories of proteins have recently been considered.

The first category is acyl-CoA-binding proteins (ACBP). A cDNA encoding this protein was isolated by PCR, using the consensus sequences of ACBP, from a library constructed from mRNAs from oilseed rape (44). This cDNA encodes a protein 92 amino acids in length that is conserved when compared with other ACBP. In spite of its similar size, no homology was found with plant LTPs, including the absence of cysteine residues as well as the lack of signal peptide that indicates that the protein is cytosolic. To confirm this promising finding, functional tests on the product of this ACBP gene are needed. In addition to this gene, several expressed sequence tags, homologous to ACBP, have been detected in *A. thaliana* (45, 78; RS Pakovsky & JB Ohlrogge, unpublished manuscript).

Soluble LTPs worth studying are the specific LTPs, such as the PI-TP transferring preferentially PI and characterized in mammalian cells and in yeasts (37, 123). A growing interest in these proteins started with the discovery that a cytosolic PI-TP, a product of an *sec14* gene in *Saccharomyces cerevisiae,* is essential both for protein transport from a late yeast Golgi compartment and for yeast viability (5, 19). It was then proposed that the SEC14 protein might act as a sensor of Golgi membrane phospholipid composition, which allowed the regulation of PC biosynthesis needed for keeping a PC/PI ratio compatible with the secretory function of Golgi membranes (5, 19, 61, 68, 99). However, in another yeast, *Yarrowia lipolytica,* SEC14P was demonstrated to allow the dimorphic transition from the yeast to the mycelial form that characterizes this species (61). It was also recently suggested that, in mammalian cells, the inositol lipid kinases prefer as substrates the PI molecules bound to PI-TP (21).

It is thus of interest to search for plant genes homologous to *sec14.* This was recently initiated by following a complementation approach of yeast *sec14* mutant from cDNA from *A. thaliana,* which led to the isolation of cDNA clones homologous to the yeast gene and apparently encoding cytosolic protein, because no signal peptide is present. The properties of the products of these novel genes remain to be characterized (N Jouannic, M Lepetit & V Arondel, unpublished manuscript; R Dewey, unpublished manuscript). Other SEC14 homologous genes have been deposited in databases (78).

Also a phosphatidylinositol-3-kinase cDNA was identified in *A. thaliana* (121). The protein presents homology in its N-terminal part with a calcium-dependent phospholipid-binding domain, which suggests its involvement in vesicle transport of PI.

CONCLUDING REMARKS

It is clear that the main question concerning plant LTPs is in regard to their biological role. Although the plant LTPs, by reason of their in vitro functional properties, are ideal candidates to play a major role in intracellular lipid dynamics, their external location and their secretion make such a role unlikely (Figure 5). The term "lipid transfer protein" is still valid and, moreover, confirmed by structural studies that showed that these proteins are efficiently built for binding and releasing acyl chains. The problem remaining is to elucidate where lipids are transported in vivo by LTPs. The location of all LTP isoforms as well as their functional properties should be studied after their purification or expression of their corresponding cDNAs in heterologous organisms. It would also be worth studying all LTP isoforms of one plant in order to have a complete description of one LTP gene family by choosing, for example, *A. thaliana* or rice as models. In addition, it is worth studying the functional properties of the various isoforms of LTP, including the determination of the structure-function relationship and the elucidation of their mode of action.

In order to determine the biological roles of plant LTPs, several approaches are possible. The antisense strategy might provide a clear answer. The Somerville group has obtained transgenic *A. thaliana* plants in which the amount of the *LTP1* product has been strongly reduced (Thomas & Somerville, unpublished manuscript) (14) by expression of an antisense construct. This decrease in the LTP1 content has not provoked changes in the morphology of plants—except that the transgenic plants are very late flowering—and no defect in cutin and wax composition. However, the fact that several LTP genes are present in *A. thaliana* makes the antisense approach rather complex.

Another strategy would be to profit from the enormous progress in the genetics of *A. thaliana* (57) by mapping LTP genes. Also, the study of the

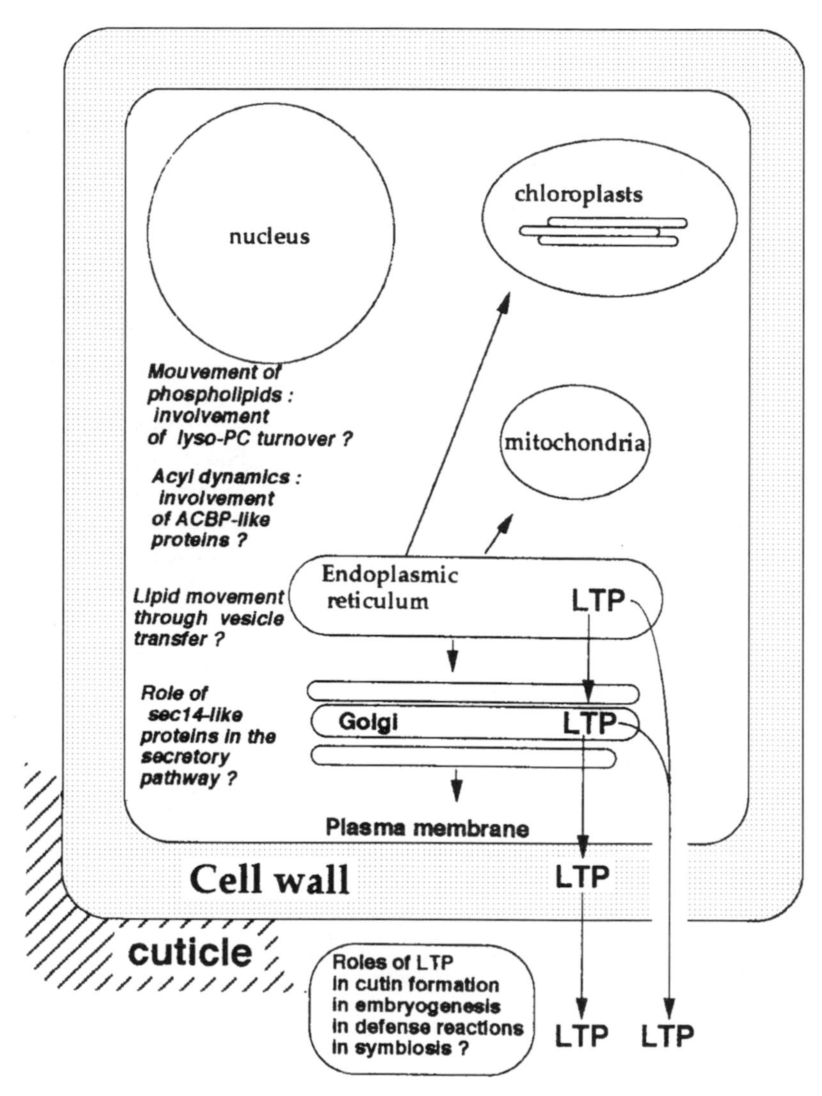

Figure 5 Proposed biological roles of lipid-transfer proteins. LTPs are synthesized as precursor proteins and entered in the secretory pathway after import into the endoplasmic reticulum. They were found to be located in the cell wall (89, 105, 111). They were proposed to be involved in several biological roles, mainly based on this external location, including the formation of protective barriers such as the cuticle. LTP might transport the acyl monomers needed for the synthesis of cutin (42). All things considered, the participation of LTPs in the intracellular lipid flow seems unlikely. Other proteins could be involved in this flow: ACBP in the acyl dynamics (44) and SEC14-like proteins in establishing a PI/PC ratio in Golgi membranes, which allows an efficient secretory process (68). Other pathways could be operative either through lyso-PC turnover or through vesicle transfer (10, 74).

genes involved in mutations in plants might reveal that some of these genes are encoding LTPs.

It can be easily predicted that in the future our knowledge of plant LTPs will increase considerably. I hope that the novel observations that will be made will help to elucidate the role of these enigmatic proteins.

ACKNOWLEDGMENTS

I am much indebted to the Ministère de L'Education Nationale, the Centre National de la Recherche Scientifique, and the Université Pierre et Marie Curie (Paris 6) for supporting this research. I am grateful to all the colleagues from my group as well as from the researchers at the Laboratoire Européen Associé (Perpignan and Barcelone) for helpful discussions. My thanks to several colleagues (W. J. Broughton, W. Broekaert, S. De Vries, R. Dewey, F. Garcia-Olmedo, M. A. Hughes, P. Kolattukudy, D. Marion, M. Ptak, C. R. Somerville, P. Schnable, S. W. Suh, T. H. Teeri, P. von Wettstein-Knowles, M. Yamada) who have communicated unpublished information and preprints. Dr. Mark Jacobs is acknowledged for his kind help in reviewing this paper. I also thank Marie-France Laforge for help in the preparation of the manuscript.

Literature Cited

1. Abad MS, Rosenberg CA, Shah DM. 1995. A genomic clone encoding a phospholipid transfer protein from barley. *Plant Physiol.* 108:871
2. Arondel V, Kader JC. 1990. Lipid transfer in plants. *Experientia* 46:579–85
3. Arondel V, Tchang F, Baillet B, Vignols F, Grellet F, et al. 1991. Multiple mRNA coding for phospholipid-transfer protein from *Zea mays* arise from alternative splicing. *Gene* 99:133–36
4. Arondel V, Vergnolle C, Tchang F, Kader JC. 1990. Bifunctional lipid-transfer: fatty acid-binding proteins in plants. *Mol. Cell. Biochem.* 98:49–56
5. Bankaitis VA, Aitken JR, Cleves AE, Dowhan W. 1990. An essential role for a phospholipid transfer protein in yeast Golgi function. *Nature* 347:561–62
6. Baud F, Pebay-Peyroula E, Cohen-Addad C, Odani S, Lehmann MS. 1993. Crystal structure of hydrophobic protein from soybean: a member of a new cysteine-rich family. *J. Mol. Biol.* 231:877–87
7. Bayon Y, Croset M, Daveloose D, Guerbette F, Chirouze V, et al. 1995. Effect of specific phospholipid molecular species incorporated in human platelet membranes on thromboxane A2/prostaglandin H2 receptors. *J. Lipid Res.* 36: 47–56
8. Bernhard WR, Somerville CR. 1989. Coidentity of putative amylase inhibitors from barley and finger millet with phospholipid transfer proteins inferred from amino acid sequence homology. *Arch. Biochem. Biophys.* 269:695–97
9. Bernhard WR, Thoma S, Botella J, Somerville CR. 1991. Isolation of a cDNA clone for spinach lipid transfer protein and evidence that the protein is synthesized by the secretory pathway. *Plant Physiol.* 95:164–70
10. Bessoule JJ, Testet E, Cassagne C. 1995. Synthesis of phosphatidylcholine in the chloroplast envelope after import of lysophosphatidylcholine from endoplasmic reticulum membranes. *Eur. J. Biochem.* 228:490–97

11. Bouillon P, Drischel C, Vergnolle C, Duranton H, Kader JC. 1987. The primary structure of spinach leaf phospholipid-transfer protein. *Eur. J. Biochem.* 166:387–91
12. Breu V, Guerbette F, Kader JC, Kannangara CG, Svensson B, von Wettstein-Knowles. 1989. A 1O-kD barley basic protein transfers phosphatidylcholine from liposomes to mitochondria. *Carlsberg Res. Commun.* 54: 81–84
13. Browse J, Somerville C. 1991. Glycerolipid synthesis: biochemistry and regulation. *Annu. Rev. Plant. Physiol. Plant Mol. Biol.* 42:467–506
14. Browse J, Somerville CR. 1994. Glycerolipids. In *Arabidopsis,* ed. EM Meyerowitz, CR Somerville, pp. 881–912. Plainview, NY: Cold Spring Harbor
15. Campos FAP, Richardson M. 1984. The complete amino acid sequence of the a amylase inibitor 1-2 from seeds of ragi (Indian finger millet, Eleusine coracana Gaertn.). *FEBS Lett.* 167:221–25
16. Castagnaro A, Garcia-Olmedo F. 1994. A fatty-acid-binding protein from wheat kernels. *FEBS Lett.* 349:117–19
17. Chasan R. 1991. Lipid transfer proteins: moving molecules? *Plant Cell* 3:842–43
18. Clark AM, Bohnert HJ. 1993. Epidermis-specific transcripts: nucleotide sequence of a full-length cDNA of EPI12, encoding a putative lipid transfer protein. *Plant Physiol.* 103:677–78
19. Cleves A, McGee T, Bankaitis VA. 1991. Phospholipid transfer proteins: a biological debut. *Trends Cell Biol.* 1:30–34
20. Coutos-Thevenot P, Jouenne T, Maes O, Guerbette F, Grosbois M, et al. 1993. Four 9-kDa proteins excreted by somatic embryos of grapevine are isofoms of lipid-transfer proteins. *Eur. J. Biochem.* 217:885–89
21. Cunningham E, Thomas GMH, Ball A, Hiles I, Cockcroft S. 1995. Phosphatidylinositol transfer protein dictates the rate of inositol trisphosphate production by promoting the synthesis of PIP2. *Curr. Biol.* 5:775–83
22. Désormeaux A, Blochet JE, Pézolet M, Marion D. 1992. Amino acid sequence of a non-specific wheat phospholipid transfer protein and its conformation as revealed by infrared and Raman spectroscopy: role of disulfide bridges and phospholipids in the stabilization of the α-helix structure. *Biochim. Biophys. Acta* 1121:137–52
23. Dieryck W, Gautier MF, Lullien V, Joudrier P. 1992. Nucleotide sequence of a cDNA encoding a lipid transfer protein from wheat. *Plant Mol. Biol.* 19:707–9
24. Dieryck W, Lullien-Pellerin V, Marion D, Joudrier P, Gautier MF. 1995. Purification and activity of a wheat 9-kDa lipid transfer protein expressed in *Escherichia coli* as a fusion with the maltose binding protein. *Protein Expr. Purif.* 6:597–603
25. Douady D, Grosbois M, Guerbette F, Kader JC. 1986. Phospholipid transfer protein from maize seedlings is partly membrane-bound. *Plant Sci.* 45:151–56
26. Douady D, Grosbois M, Guerbette F, Kader JC. 1982. Purification of a basic phospholipid transfer protein from maize seedlings. *Biochim. Biophys. Acta* 710:143–53
27. Douady D, Guerbette F, Kader JC. 1984. Purification of phospholipid transfer protein from maize seeds using a two-step chromatographic procedure. *Physiol. Vég.* 23:373–80
28. Fielding CJ. 1993. Lipid transfer proteins: catalysts, transmembrane carriers and signalling intermediates for intracellular and extracellular lipid reactions. *Curr. Opin. Lipidology* 4:218–22
29. Fleming AJ, Kuhlemeier C. 1994. Activation of basal cells of the apical meristem during sepal formation in tomato. *Plant Cell* 6:789–98
30. Fleming AJ, Mandel T, Hofmann S, Sterk P, de Vries SC, Kuhlemeier C. 1992. Expression pattern of a tobacco lipid transfer protein gene within the shoot apex. *Plant J.* 2:855–62
31. Fleming AJ, Mandel T, Roth I, Kuhlemeier C. 1993. The patterns of gene expression in the tomato shoot apical meristem. *Plant Cell* 5:297–309
32. Foster GD, Robinson SW, Blundell RP, Roberts MR, Hodge R, et al. 1992. A *Brassica napus* mRNA encoding a protein homologous to phospholipid transfer proteins, is expressed specifically in the tapetum and developing microspores. *Plant Sci.* 84:184–92
33. Garcia-Olmedo F, Molina A, Segura A, Moreno M. 1995. The defensive role of nonspecific lipid-transfer proteins in plants. *Trends Microbiol.* 3:72–74
34. Gausing K. 1994. Lipid transfer protein genes specifically expressed in barley leaves and coleoptiles. *Planta* 192:574–80
35. Geldwerth D, de Kermel A, Zachowski A, Guerbette F, Kader JC, et al. 1991. Use of spin-labeled and fluorescent lipids to study the activity of the phospholipid transfer protein from maize

seedlings. *Biochim. Biophys. Acta* 1082: 255–64
36. Gincel E, Simorre JP, Caille A, Marion D, Ptak M, Vovelle F. 1994. Three-dimensional structure in solution of a wheat lipid-transfer protein from multidimensional 1H-NMR data. *Eur. J. Biochem.* 226:413–22
37. Gnamusch E, Kalaus C, Hrastnik C, Paltauf F, Daum G. 1992. Transport of phospholipids between subcellular membranes of wild-type yeast cells and of the phosphatidylinositol transfer protein-deficient strain *Saccharomyces cerevisiae sec* 14. *Biochim. Biophys. Acta* 1121:120–26
38. Grabski S, De Feijter AW, Schindler M. 1993. Endoplasmic reticulum forms a dynamic continuum for lipid diffusion between contiguous soybean root cells. *Plant Cell* 5:25–38
39. Grosbois M, Guerbette F, Jolliot A, Quintin F, Kader JC. 1993. Control of maize lipid transfer protein activity by oxido-reducing conditions. *Biochim. Biophys. Acta* 1170:197–203
40. Grosbois M, Guerbette F, Kader JC. 1989. Changes in level and activity of phospholipid transfer protein during maturation and germination of maize seeds. *Plant Physiol.* 90:1560–64
41. Harryson P, Hmyene A, Guerbette F, Kader JC, Sandelius AS. 1994. The effects of a non-specific lipid transfer protein on a cell-free reconstitution of phosphatidylinositol transfer to plant plasma membranes. *Plant Sci.* 99:55–62
42. Hendriks T, Meijer EA, Thoma S, Kader JC, De Vries SC. 1994. The carrot extracellular lipid transfer protein EP2: quantitative aspects with respect to its putative role in cutin synthesis. In *Plant Molecular Biology,* ed. G Coruzzi, P Puigdomenech, pp. 85–94. Berlin: Springer-Verlag
43. Henrissat B, Popineau Y, Kader JC. 1988. Hydrophobic-cluster analysis of plant protein sequences. *Biochem. J.* 255:901–5
44. Hills MJ, Dann R, Lydiate D, Sharep A. 1994. Molecular cloning of a cDNA from *Brassica napus* L. homologous of acyl-CoA-binding protein. *Plant Mol. Biol.* 25:917–20
45. Hofte H, Desprez T, Amselem J, Chiapello H, Caboche M, et al. 1993. An inventory of 1152 expressed sequence tags obtained by partial sequencing of cDNAs from *Arabidopsis thaliana. Plant J.* 4:1051–61
46. Hughes MA, Dunn MA, Pearce RS, White AJ, Zhang L. 1992. An abscisic-acid-responsive, low temperature barley gene has homology with a maize phospholipid transfer protein. *Plant Cell Environ.* 15:861–65
47. Hwang KY, Kim KK, Min K, Eom SH, Yu YG, et al. 1993. Crystallization and preliminary X-ray crystallographic analysis of probable amylase/protease inhibitor-B from rice seeds. *J. Mol. Biol.* 229:255–57
48. Jakobsen K, Klemsdal SS, Aalen RB, Bosnes M, Alexander D, Olsen OA. 1989. Barley aleurone development: molecular cloning of aleurone-specific cDNAs from immature grains. *Plant Mol. Biol.* 12:285–93
49. Jenks MA, Tuttle HA, Eigenbrode SD, Feldman KA. 1995. Leaf epicuticular waxes of the eceriferum mutants in Arabidopsis. *Plant Physiol.* 108:369–77
50. Kader JC. 1993. Lipid transport in plants. In *Plant Lipid Metabolism,* ed. TS Moore, pp. 303–30. Boca Raton, FL: CRC Press
51. Kader JC. 1975. Proteins and the intracellular exchange of lipids: stimulation of phospholipid exchange between mitochondria and microsomal fractions by proteins isolated from potato tuber. *Biochim. Biophys. Acta* 380:31–44
52. Kader JC, Douady D, Mazliak P. 1982. Phospholipid transfer proteins. In *Phospholipids, A Comprehensive Treatise,* ed. JN Hawthorne, GB Ansell, pp. 279–311. Amsterdam: Elsevier
53. Kader JC, Julienne M, Vergnolle C. 1984. Purification and characterization of a spinach leaf protein capable of transferring phospholipids from liposomes to mitochondria or chloroplasts. *Eur. J. Biochem.* 139:411–16
54. Kalla R, Shimamoto K, Potter R, Nielsen PS, Linnestad C, Olsen OA. 1994. The promoter of the barley aleurone-specific gene encoding a putative 7-kDa lipid transfer protein confers aleurone cell-specific expression in transgenic rice. *Plant J.* 6:849–60
55. Kinlaw CS, Gerttulla SM, Carter MC. 1994. Lipid transfer protein genes of loblolly pine are members of a complex gene family. *Plant Mol. Biol.* 26:1213–16
56. Koltunow AM, Truettner J, Cox KH, Wallroth M, Goldberg RB. 1990. Different temporal and spatial gene expression patterns occur during anther development. *Plant Cell* 2:1201–24
57. Koornneef M. 1994. Arabidopsis genetics. In *Arabidopsis,* ed. EM Meyerowitz, CR Somerville, pp. 89–120. Plainview, NY: Cold Spring Harbor
58. Kotilainen M, Helariutta Y, Elomaa P,

Paulin L, Teeri TH. 1994. A corolla- and carpel-abundant, nonspecific lipid transfer protein gene is expressed in the epidermis and parenchyma of Gerbear hybrida var Regina (Compositae). *Plant Mol. Biol.* 26:971–78
59. Krause A, Sigrist CJA, Dehning I, Sommer H, Broughton WJ. 1994. Accumulation of transcripts encoding a lipid transfer-like protein during deformation of nodulation-competent *Vigna unguiculata* root hairs. *Mol. Plant-Microbe Interact.* 7:411–18
60. Linnestad C, Lönneborg A, Kalla E, Alsen OA. 1991. Promoter of a lipid transfer protein gene expressed in barley aleurone cells contains similar myb and myc recognition sites as the maize Bz-Mcc allele. *Plant Physiol.* 97:841–43
61. Lopez MC, Nicaud JM, Skinner HB, Vergnolle C, Kader JC, et al. 1994. A phosphatidylinositol/phosphatidylchol ine transfer protein is required for differentiation of the dimorphic yeast *Yarrowia lipolytica* from the yeast to the mycelial form. *J. Cell. Biol.* 124:113–27
62. Lucas WJ, Wolf S. 1993. Plasmodesmata: the intercellular organelles of green plants. *Trends Cell Biol.* 3:308–15
63. Ma DP, Tan H, Si Y, Creech RG, Jenkins JN. 1995. Differential expression of a lipid transfer protein gene in cotton fiber. *Biochim. Biophys. Acta* 1257:81–84
64. Madrid SM. 1991. The barley lipid transfer protein is targeted into the lumen of the endoplasmic reticulum. *Plant Physiol. Biochem.* 29:695–703
65. Madrid SM, Von Wettstein D. 1990. Reconciling contradictory notions on lipid transfer proteins in higher pants. *Plant Physiol. Biochem.* 29:705–11
66. Masuta C, Furuno M, Tanaka H, Yamada M, Koiwai A. 1992. Molecular cloning of a cDNA clone for tobacco lipid transfer protein and expression of the functional protein in *Escherichia coli. FEBS Lett.* 311:119–23
67. Masuta C, Van den Bulcke M, Bauw G, Van Montagu M, Caplan AB. 1991. Differential effects of elicitors on the viability of rice suspension cells. *Plant Physiol.* 97:619–29
68. McGee TP, Skinner HB, Whitters EA, Henry SA, Bankaitis VA. 1994. A phosphatidylinositol transfer protein controls the phosphatidylcholine content of yeast Golgi membranes. *J. Cell. Biol.* 124: 273–87
69. Meijer EA, De Vries SC, Sterk P, Gadella DWJ, Wirtz KWA, Hendriks T. 1993. Characterization of the non-specific lipid transfer protein EP2 from carrot (*Daucus carota* L.). *Mol. Cell. Biochem.* 123:159–66
70. Miquel M, Block MA, Joyard J, Dorne AJ, Dubacq JP, et al. 1987. Protein mediated transfer of phosphatidylcholine from liposomes to spinach chloroplast envelope membranes. *Biochim. Biophys. Acta* 937:219–28
71. Molina A, Garcia-Olmedo F. 1993. Developmental and pathogen induced expression of three barley genes encoding lipid transfer proteins. *Plant J.* 4:983–91
72. Molina A, Segura A, Garcia-Olmedo F. 1993. Lipid transfer proteins (nsLTPs) from barley and maize leaves are potent inhibitors of bacterial and fungal plant pathogens. *FEBS Lett.* 316:119–22
73. Moreau F, Davy de Virville J, Hoffekt M, Guerbette F, Kader JC. 1994. Use of a fluorimetric method to assay the binding and transfer of phospholipids by lipid transfer proteins from maize seedlings and Arabidopsis leaves. *Plant Cell Physiol.* 35:267–74
74. Moreau P, Cassagne C. 1994. Phospholipid trafficking and membrane biogenesis. *Biochim. Biophys. Acta* 1197: 257–90
75. Mundy J, Rogers JC. 1986. Selective expression of a probable amylase/protease inhibitor in barley aleurone cells: comparison to the barley amylase/subtilisin inhibitor. *Planta* 169:51–63
76. Neumann GM, Condron R, Svensson B, Polya GM. 1993. Phosphorylation of barley and wheat phospholipid transfer proteins by wheat calcium-dependent protein kinase. *Plant Sci.* 92:159–67
77. Neumann GM, Condron R, Thomas I, Polya GM. 1995. Purification, characterization and sequencing of a family of Petunia petal lipid transfer proteins phosphorylated by plant calcium-dependent protein kinase. *Plant Sci.* 107: 129–45
78. Newman T, de Bruijn FJ, Green P, Keegstra K, Kende H, et al. 1994. Genes galore: a summary of methods for accessing results from large-scale partial sequencing of anonymous Arabidopsis cDNA clones. *Plant Physiol.* 106:1241–55
79. Nishida I, Yamada M. 1985. Semisynthesis of a spin-labeled monogalactosyldiacylglycerol and its application in the assay for galactolipid transfer activity in spinach leaves. *Biochim. Biophys. Acta* 813:298–306
80. Ostergaard J, Hojrup P, Knudsen J. 1995. Amino acid sequences of three acyl-binding/lipid-transfer proteins from rape seedlings. *Biochim. Biophys. Acta* 1254:169–79

81. Ostergaard J, Vergnolle C, Schoentgen F, Kader JC. 1993. Acyl-binding/lipid-transfer proteins from rape seedlings, a novel category of proteins interacting with lipids. *Biochim. Biophys. Acta* 1170: 109–17
82. Oursel A, Escoffier A, Kader JC, Dubacq JP, Trémolières A. 1987. Last step in the cooperative pathway for galactolipid synthesis in spinach leaves: formation of monogalactosyldiacylglycerol with C18 polyunsaturated acyl groups at both carbon atoms of the glycerol. *FEBS Lett.* 219:393–99
83. Pebay-Peyroula E, Cohen-Addad C, Lehmann MS, Marion D. 1992. Crystallographic data for the 9000 dalton wheat non-specific phospholipid transfer protein. *J. Mol. Biol.* 226:563–64
84. Pelèse-Siebenbourg F, Caelles C, Kader JC, Delseny M, Puigdomenech P. 1994. A pair of genes coding for lipid-transfer proteins in *Sorghum vulgare*. *Gene* 148: 305–8
85. Petit MC, Sodano P, Marion D, Ptak M. 1994. Two-dimensional 1H-NMR studies of maize lipid-transfer protein: sequence-specific assignment and secondary structure. *Eur. J. Biochem.* 222: 1047–54
86. Plant AL, Cohen A, Moses MS, Bray EA. 1991. Nucleotide sequence and spatial expression pattern of a drought- and absissic acid-induced gene of tomato. *Plant Physiol.* 97:900–6
87. Polya GM, Chandra S, Chung R, Neumann GM, Hoj PB. 1992. Purification and characterization of wheat and pine small basic protein substrates for plant calcium-dependent protein kinase. *Biochim. Biophys. Acta* 1120:273–80
88. Pyee J, Kolattukudy PE. 1995. The gene for the major cuticular wax-associated protein and three homologous genes from broccoli (*Brassica oleracea*) and their expression patterns. *Plant J.* 7:49–59
89. Pyee J, Yu H, Kolattukudy PE. 1994. Identification of a lipid transfer protein as the major protein in the surface wax of broccoli (*Brassica oleracea*) leaves. *Arch. Biochem. Biophys.* 311:460–68
90. Qu R, Huang AHC. 1990. Oleosin KD18 on the surface of oil bodies in maize: genomic and cDNA sequences and the deduced protein structure. *J. Biol. Chem.* 265:2238–43
91. Raynal P, Pollard HB. 1994. Annexins: the problem of assessing the biological role of the gene family of multifunctional calcium- and phospholipid-binding proteins. *Biochim. Biophys. Acta* 1197:63–93
92. Record E, Asther M, Marion D, Asther M. 1995. Purification and characterization of a novel specific phosphatidylglycerol-phosphatidylinositol transfer protein with high activity from *Aspergillus oryzae*. *Biochim. Biophys. Acta* 1256:18–24
93. Rickers J, Spener F, Kader JC. 1985. A phospholipid transfer protein that binds fatty acids. *FEBS Lett.* 180:29–32
94. Schrier SL, Zachowski A, Hervé P, Kader JC, Devaux PF. 1992. Transmembrane redistribution of phospholipids of the human red cell membrane during hypotonic hemolysis. *Biochim. Biophys. Acta* 1105:170–76
95. Schwitzguebel JP, Siegenthaler PA. 1985. Evidence for a lack of phospholipid transfer protein in the stomata of Spinach chloroplasts. *Plant Sci.* 40: 167–71
96. Segura A, Moreno M, Garcia-Olmedo F. 1993. Purification and antipathogenic activity of lipid transfer proteins from the leaves of Arabidopsis and spinach. *FEBS Lett.* 3:243–46
97. Shin DH, Lee JY, Hwang KY, Kim KK, Su SW. 1995. High-resolution crystal structure of the non-specific lipid-transfer protein from maize seedlings. *Structure* 3:189–99
98. Simorre JP, Caille A, Marion D, Marion D, Ptak M. 1991. Two- and three-dimensional 1H NMR studies of a wheat phospholipid transfer protein: sequential resonance assignments and secondary structure. *Biochemistry* 30:11600–8
99. Skinner HB, Alb JG Jr, Whitters EA, Helmkamp GM Jr, Bankaitis VA. 1993. Phospholipid transfer activity is relevant to but not sufficient for the essential function of the yeast SEC 14 gene product. *EMBO J.* 12:4775–84
100. Skriver K, Leah R, Müller-Uri F, Olsen FL, Mundy J. 1992. Structure and expression of the barley lipid transfer protein gene Ltp1. *Plant Mol. Biol.* 18: 585–89
101. Slabas AR, Fawcett T. 1992. The biochemistry and molecular biology of plant lipid biosynthesis. *Plant Mol. Biol.* 19:169–91
102. Somerville CR, Browse J. 1991. Plant lipids: metabolism, mutants, and membranes. *Science* 252:80–87
103. Sorensen SB, Bech LM, Muldbjerg M, Beenfeldt T, Breddam K. 1993. Barley lipid transfer protein 1 is involved in beer foam formation. *MBAA Tech. Q.* 30:136–45
104. Sossountzov L, Ruiz-Avila L, Vignols F, Jolliot A, Arondel V, et al. 1991. Spatial and temporal expression of a

maize lipid tranfer protein gene. *Plant Cell* 3:923–33

105. Sterk P, Booij H, Scheleekens GA, Van Kammen A, de Vries SC. 1991. Cell-specific expression of the carrot EP2 lipid transfer protein gene. *Plant Cell* 3:907–21
106. Takishima K, Watanabe S, Yamada M, Mamiya G. 1986. The amino-acid sequence of the nonspecific lipid transfer protein from germinated castor bean endosperms. *Biochim. Biophys. Acta* 870: 248–55
107. Takishima K, Watanabe S, Yamada M, Suga T, Mamiya G. 1988. Amino acid sequences of two nonspecific lipid-transfer proteins from germinated castor bean. *Eur. J. Biochem.* 177:241–49
108. Tan H, Okazaki K, Kubota I, Kamiryo T, Utiyama H. 1990. A novel peroxisomal nonspecific lipid-transfer protein from *Candida tropicalis:* gene structure, purification and possible role in β-oxidation. *Eur. J. Biochem.* 190:107–12
109. Tchang F, This P, Stiefel V, Arondel V, Morch MD, et al. 1988. Phospholipid transfer protein: full-length cDNA and amino-acid sequence in maize. Amino-acid sequence homologies between plant phospholipid transfer proteins. *J. Biol. Chem.* 263:16489–855
110. Terras FRG, Schofs HME, de Bolle MFC, Van Leuven F, Rees SB, et al. 1992. In vitro antifungal activity of a radish (*Raphanus sativus* L.) seed protein homologous to nonspecific lipid transfer proteins. *Plant Physiol.* 100: 1055–58
111. Thoma S, Hecht U, Kippers A, Botella J, De Vries S, Somerville CR. 1994. Tissue-specific expression of a gene encoding a cell wall-localized lipid transfer protein from Arabidopsis. *Plant Physiol.* 105:35–45
112. Thoma S, Kaneto Y, Somerville CR. 1993. A nonspecific lipid transfer protein from Arabidopsis is a cell wall protein. *Plant J.* 3:427–36
113. Torres-Schumann S, Godoy JA, Pintor-Toro JA. 1992. A probable lipid transfer protein gene is induced by NaCl in stems of tomato plants. *Plant Mol. Biol.* 18: 749–57
114. Tsuboi S, Osafune T, Tsugeki R, Nishimura M, Yamada M. 1992. Nonspecific lipid transfer protein in castor bean cotyledon cells: subcellular localization and a possible role in lipid metabolism. *J. Biochem.* 111:500–8
115. Tsuboi S, Suga T, Takishima K, Mamiya G, Matsui K, et al. 1991. Organ-specific occurrence and expression of the isoforms of nonspecific lipid transfer protein in castor bean seedlings and molecular cloning of a full-length cDNA for a cotyledon-specific isoform. *J. Biochem.* 110:823–31
116. Tsuboi S, Watanabe SI, Ozeki Y, Yamada M. 1989. Biosynthesis of nonpecific lipid transfer proteins in germinating castor bean seeds. *Plant Physiol.* 90:841–45
117. Vergnolle C, Arondel V, Grosbois M, Guerbette F, Jolliot A, Kader JC. 1988. Synthesis of phospholipid transfer proteins from maize seedlings. *Biochem. Biophys. Res. Commun.* 157: 37–41
118. Vergnolle C, Arondel V, Jolliot A, Kader JC. 1992. Phospholipid transfer proteins from higher plants. *Methods Enzymol.* 209:522–30
119. Vignols F, Lund G, Pammi S, Trémousaygue D, Grellet F, et al. 1994. Characterization of a rice gene coding for a lipid transfer protein. *Gene* 142: 265–70
120. Weig A, Komor E. 1992. The lipid-transfer protein C of *Ricinus communis* L.: isolation of two cDNA sequences which are strongly and exclusively expressed in cotyledons after germination. *Planta* 187:367–71
121. Welters P, Takegawa K, Emr SD, Chrispeels MJ. 1994. AtVPS34, a phosphatidylinositol 3-kinase of *Arabidopsis thaliana,* is an essential protein with homology to a calcium-dependent lipid binding domain. *Proc. Natl. Acad. Sci. USA* 91:11398–402
122. White A, Dunn MA, Brown K, Hughes MA. 1994. Comparative analysis of genomic sequence and expression of a lipid transfer protein gene family in winter barley. *J. Exp. Bot.* 45:1885–92
123. Wirtz KWA. 1991. Phospholipid transfer proteins. *Annu. Rev. Biochem.* 60: 73–99
124. Yamada M. 1992. Lipid transfer proteins in plants and microorganisms. *Plant Cell Physiol.* 33:1–6
125. Yamada M, Tanaka T, Kader JC, Mazliak P. 1978. Transfer of phospholipids from microsomes to mitochondria in germinating castor bean endosperm. *Plant Cell Physiol.* 19:173–76
126. Yamada M, Tsuboi S, Koso EM, Osafune T, Ehara T, et al. 1995. Approach to in vivo function of nonspecific lipid transfer proteins in higher plants. In *Plant Lipid Metabolism,* ed. JC Kader, P Mazliak, pp. 206–9. Dordrecht: Kluwer

127. Ye ZH, Varner JE. 1993. Gene expression patterns associated with in vitro tracheary element formation in isolated single mesophyll cells of Zinnia elegans. *Plant Physiol.* 103:805–13

128. Yu YG, Chung CH, Fowler A, Suh SW. 1988. Amino acid sequence of a probable amylase/protease inhibitor from rice seeds. *Arch. Biochem. Biophys.* 265: 466–75

Annu. Rev. Plant Physiol. Plant Mol. Biol. 1996. 47:655–84

REGULATION OF LIGHT HARVESTING IN GREEN PLANTS

P. Horton, A. V. Ruban, and R. G. Walters
Robert Hill Institute, Department of Molecular Biology and Biotechnology, University of Sheffield, Sheffield S10 2TN, United Kingdom

KEY WORDS: photosynthesis, chlorophyll fluorescence, light-harvesting complex, thylakoid membrane, photoinhibition

ABSTRACT

When plants are exposed to light intensities in excess of those that can be utilized in photosynthetic electron transport, nonphotochemical dissipation of excitation energy is induced as a mechanism for photoprotection of photosystem II. The features of this process are reviewed, particularly with respect to the molecular mechanisms involved. It is shown how the dynamic properties of the proteins and pigments of the chlorophyll *a/b* light-harvesting complexes of photosystem II first enable the level of excitation energy to be sensed via the thylakoid proton gradient and subsequently allow excess energy to be dissipated as heat by formation of a nonphotochemical quencher. The nature of this quencher is discussed, together with a consideration of how the variation in capacity for energy dissipation depends on specific features of the composition of the light-harvesting system. Finally, the prospects for future progress in understanding the regulation of light harvesting are assessed.

CONTENTS

1040-2519/96/0601-0655$08.00

INTRODUCTION

The molecular basis of the high quantum efficiency of light harvesting and photochemistry in higher plant photosystem II (PSII) is now understood with high spatial and temporal resolution. Under many conditions in the field, quantum efficiency is actively downregulated by physiological control mechanisms that optimize plant performance and provide protection from the damaging effects of excess light (47). We review the state of knowledge of how PSII is regulated, in the context of the structural and photophysical details of the pigments and proteins of the photosynthetic membrane. This review complements recent Annual Review articles dealing with physiological aspects (21) and the main method of investigation, measurement of chlorophyll (chl) fluorescence (66).

THE MOLECULAR ENVIRONMENT FOR REGULATION OF LIGHT HARVESTING

Photosystem II Light-Harvesting System

Light harvesting refers to the process of absorption of light and the subsequent transfer of energy to the photosynthetic reaction center (RC). 200–300 chl molecules are structurally and functionally associated with an RC to form a functional PSII unit (Figure 1). The light-harvesting system consists of two distinct types of pigment-protein complex. First, two related chl proteins, CP47 and CP43, which bind ~50 chl *a* molecules, are linked to the RC forming the PSII core. Second, bound to the cores are the light-harvesting complexes (LHCII), consisting of polypeptides of 20–30 kDa that bind chl *a,* chl *b,* and xanthophylls (reviewed in 57, 92; see Green & Durnford, this volume). LHCIIb is the main complex. It is trimeric and binds ~60% PSII chl. LHCIIb is composed of polypeptides of ~28, ~27, and ~25 kDa—encoded by *Lhcb1, Lhcb2,* and *Lhcb3,* respectively—organized into proximal (25 and 28 kDa) and distal (27 and 28 kDa) populations. The minor complexes, LHCIIa (CP29), LHCIIc (CP26), and LHCIId (CP24) are monomeric, and each bind only ~5% PSII chl. They have been suggested to link the PSII core to LHCIIb. All LHCII types have a similar structure and possess three-membrane spanning α-helices.

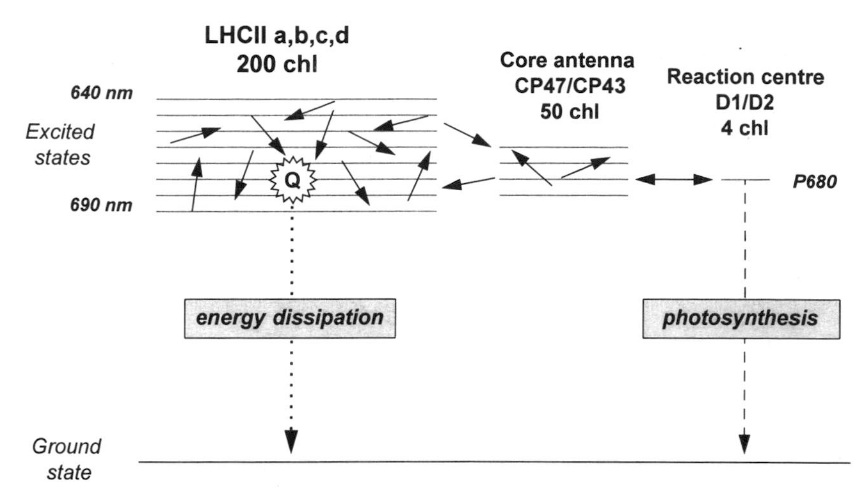

Figure 1 Schematic representation of the organization and energetics of PSII. The absence of a significant energy gradient between LHCII, the core antenna, and the reaction center core, together with the reversibility of charge separation, results in equilibration of energy between them. As a consequence of the large number of chlorophyll molecules in LHCII, there is a high probability that an excited state will be found in this complex. Hence, a quencher (Q) within LHCII provides effective energy dissipation. The arrows represent the randomness of energy transfer events and indicate the relative numbers of chl molecules in each set of complexes. The wavelengths correspond with approximate ranges of excited state energy levels of the chls.

The chls are arranged perpendicularly to the plane of the membrane in two layers close to the stromal and lumen surface. Many are exposed on the periphery possibly with the neaxanthin and xanthophyll cycle carotenoids (70). Two lutein molecules are embedded within the complex. The boundary lipids associated with LHCII complete a specific membrane environment for each complex (122).

A typical LHCII system contains five LHCIIb and three or four minor complexes, which together form a large oligomeric antenna in the thylakoid membrane (6, 57). The pigment-protein complexes confer new features on the pigments. Studies on isolated LHCIIb have shown that pigment properties are drastically altered when macroaggregates of trimers are formed (48), in particular a reduction in fluorescence yield by as much as 20-fold (78, 102). Examination of LHCIIb aggregates by CD suggests that they can self-organize to give ordered chiral arrays that confer specific features (30), including promotion of long-range energy migration (5). Spectral indicators of macroaggregation have been found in thylakoid membranes and in leaves (31, 63, 103), which suggests that the higher-order associations between pigments may be of great significance in vivo. Two other features of the membrane environment of LHCII are important. First, the stromal surface is dominated by the influence of membrane appression. Divalent cations probably allow cross-bridges between adjacent surfaces of LHCII, which stabilizes grana formation. It has been suggested that this allows additional chl-chl interactions such that the macrostructure of the LHCII system is three dimensional rather than two dimensional (30). Second, there are changes in the local environment when the leaf is illuminated: The stroma becomes more alkaline (rising to pH 8); there are increases in [Mg^{2+}]; and the lumen-facing domains are also exposed to changes in pH, from an estimated 7 in the dark to a minimum of about 4.5 in the light. These conditions appear to be important signals in the regulation of light harvesting.

Energy Transfer

Energy transfer in PSII is governed by two principles. First, the PSII RC is a "shallow trap" with a small energy gradient from the antenna chl to the RC. Second, primary charge separation is reversible. Therefore, energy transfer and trapping is an equilibrium involving the exciton and the radical pair; the position of equilibrium is set by the number of chl in the antenna and the presence of an electron of Q_A, which controls the rate of charge separation (115). This exciton-radical pair model for the PSII antenna is fundamental to understanding the regulation of light harvesting (Figure 1). The absorption spectra of the PSII pigment proteins and their relative concentrations show that there is a 60% probability of an excited state populating LHCIIb pigments, so that energy dissipation could occur efficiently via even relatively weak traps in these complexes (58, 59). Energy transfer within the PSII antenna is rapid and efficient (124). PSII functions as a so-called connected package comprising a small group of units that form a "lake" of pigments in which an excitation on any chl can be trapped by any of the RCs in the group (123). Connectivity is mediated by LHCII and promotes efficient light capture; it may also be important in regulation by similarly increasing the efficiency of energy dissipation by a nonphotochemical quencher (15).

Concluding Remarks

The light-harvesting system contains large numbers of chl concentrated into a small volume, held in place to give efficient, ultrafast energy transfer. Nonradiative and radiative energy losses are minimal; almost 100% quantum efficiency of trapping can occur. This represents a remarkable molecular design in which pigment-pigment interactions are finely controlled to prevent, for example, energy losses from "concentration" quenching (8). Similarly, the arrangement of pigment-protein complexes in the membrane must be controlled by recognition between complexes to confer a light-harvesting function.

Although the complexity of the light-harvesting system is usually considered in terms of the requirement for high quantum efficiency of photosynthesis, its real significance may reside in the flexibility it confers, which allows regulation of light harvesting (48). It is easy to imagine how structural alterations can occur in this system to reduce efficiency as a means of photoprotection in conditions of excess light.

DOWNREGULATION OF PHOTOSYNTHETIC EFFICIENCY

The Requirement for Regulation of Light Harvesting

The extremely high oxidizing potential needed for H_2O oxidation is the unique feature of PSII, one that requires regulation of light harvesting (4). The pigments and proteins (mainly D1) of the RC can be photochemically damaged by formation of the triplet state of P680 and consequent singlet O_2 production or because of oxidation by $P680^+$ (4, 16). Furthermore, unlike PSI, where $P700^+$ can be an effective energy quencher, PSII cannot accumulate $P680^+$.

Damage to the RC occurs under all conditions, and it appears to be an inevitable consequence of PSII function, with an RC being inactivated after 10^6–10^7 turnovers (90). The rate of damage is high in excess light and is exacerbated in the presence of other stress factors. Repair involves disassembly, proteolysis, and introduction of newly synthesized D1 polypeptide (16). This is not energetically costly, but it has slow kinetics. When the rate of damage exceeds the rate of repair, inactive centers accumulate, which leads to a decline in photosynthesis. Furthermore, a damaged RC may generate oxidized chl and free radicals and cause widespread damage to the thylakoid. Finally, because the repair of accumulated damaged centers is slow, further losses in photosynthesis would result even after the excess light is removed.

It is inferred that there has been strong selective pressure to reduce the increase in excitation density in excess light (12). Reductions in excitation density result from physiological responses, such as alteration in leaf angle or chloroplast movement to reduce light absorption. At the molecular level, excitation density is decreased by disrupting the efficient functioning (downregulation) of the light-harvesting system by the dissipation of the unwanted energy as heat (16, 21, 47).

Evidence for the Regulation of PSII Efficiency

In leaves, the redox state of PSII, determined by the level of photochemical quenching of chl fluorescence (qP), is more oxidized than expected from the quantum efficiency of photosynthetic gas exchange (130). The expected linear relationship between qP and quantum efficiency was found in uncoupled thylakoids but not coupled thylakoids, which behaved like leaves (131); therefore, the thylakoid ΔpH exerts control over PSII. In isolated chloroplasts, this results from two effects: first, a nonproductive PSII charge separation arising from electron cycling in PSII and second, a process that correlated with nonphotochemical quenching of chl fluorescence (qN) (97). In leaves, all of the effects on Q_A with increased light intensity were accounted for by qN. Later, it was shown that the product of qN (measured as the resulting lower value for F_v/F_m, F_v'/F_m') and Q_A reduction (measured as qP) yields the quantum efficiency of PSII (ϕ_{PSII}) (32). At limiting light intensity, ϕ_{PSII} is constant: Near to the inflexion point on the irradiance curve, the decline in ϕ_{PSII} arose mainly from the fall in F_v'/F_m', with qP maintained near 1. This is the state of perfect regulation, i.e. RCs remaining open despite increasing light intensity and increasing limitation by electron transfer capacity. At higher intensity, further declines in ϕ_{PSII} are associated with continued decreases in F_v'/F_m' but also now with a fall in qP. The onset of light stress is defined as a significant decrease of qP—empirically determined as <0.6 (84). Thus, dissection of ϕ_{PSII} into qN-related and qP components allows assessment of the extent to which a leaf is optimally acclimated to environmental conditions or whether it is under stress. In a leaf acclimated to its light environment, excess excitation is minimized either by photosynthetic electron transport or by the nonradiative dissipation responsible for qN (23, 89). Conversely, a leaf not acclimated to environmental conditions, e.g. a shade leaf exposed to full sunlight, shows low levels of total quenching (a large excess excitation) because of low rates of photosynthesis and of nonradiative decay.

Physiological Requirements of Regulation

Dissipation of excess absorbed excitation energy requires that the level of excess can be accurately sensed and transduced to the light-harvesting system. It is widely accepted that the level of thylakoid energization (the ΔpH) is the light sensor. This is complicated because ΔpH has multiple effects: It is the driving force for ATP synthesis, high ΔpH restricts electron transport at the cytochrome b_6f complex, and high ΔpH inhibits electron donation to PSII. Based on transgenic plants whose steady state ΔpH is lower because the amount of cytochrome b_6f complex is low, high ΔpH is proposed to be a primary cause of photoinhibition (51). In fact, the optimum value of ΔpH has to simultaneously allow high rates of photosynthesis and energy dissipation in high light. Yet, in low light, quenching needs to be switched off at ΔpH levels that are still sufficient for ATP synthesis. It is therefore probable that ΔpH is regulated within narrow limits and that qN is controlled by small ΔpH changes resulting from highly cooperative control of qN by ΔpH (116).

Nonphotochemical Quenching

Nonphotochemical fluorescence quenching arises from a number of processes in the thylakoid membrane (for reviews, see 46, 66), but the major fraction depends on the ΔpH and is called energy-dependent quenching (qE). It forms within minutes of exposure to light and relaxes with a $t_{\frac{1}{2}}$ of 1–2 min in darkness. A second component (qT) arising from the fluorescence decrease associated with a state transition is usually small and unlikely to be a significant factor in photoprotection. The third type of qN has been called qI because of its association with stronger degrees of light stress when quenching increasingly becomes either irreversible or slowly reversible. A part of this quenching occurs because the photodamaged RC quenches fluorescence. However, a significant proportion arises even in the absence of any loss of PSII activity and has been referred to as a "sustained quenching" and given, like qE, a role in photoprotection (20). In fact, this form of qN strongly resembles qE in that they are both related to thylakoid energization (104, 125). It has also been suggested that in leaves the ΔpH may sometimes persist in darkness, which suggests that qI may often be exactly the same process as qE (34, 35). Therefore, whereas the rapidly relaxing phase of qN can be unequivocally attributed to qE, assigning these slower phases to a particular process is much more problematic (125).

Because qN decreases the quantum yield of PSII, it is in a sense photoinhibitory (85). Upon transition from high to low light, the qN-related decrease in quantum yield could lead to "lost" photosynthesis, particularly in the case of long-lived qI states. It has been suggested that qI is an important reason for the differences between observed and predicted rates of photosynthesis in the field (83). Similarly, the relaxation of qI in chilled maize correlated with the restoration of CO_2 fixation upon return to ambient temperature (28). It has been argued that, except under the most extreme conditions, "photoinhibition" detected as a decrease in F_v/F_m does not arise from RC damage but from protective energy dissipation (22–24). However, the observed correlation between the decline in F_v/F_m and the decrease in the number of active PSII centers (84, 90) suggests that sustained downregulation of light harvesting and inactivation of the RC may proceed in a concerted manner or even arise from the same mechanism.

MECHANISM OF qE

Site of Quenching

A number of studies have attempted to discover whether qE occurs in the RC or in the light-harvesting system. Low pH, similar to that occurring in the thylakoid lumen in the light, causes quenching of PSII in vitro because of inactivation of electron donation to P680 (18, 67) caused by release of Ca^{2+} (68). It has been proposed that quenching arises from rapid recombination between $P680^{+}$ and the acceptor side of PSII (67). A thermoluminescence signal attributed to recombination between Q_A^- and the Tyr D^+ is known to change when the donor side is inhibited in this way. This signal occurs in plants grown under intermittent light (IML) but not in fully green plants (61), which argues that this recombination is not the mechanism of qE. In fact, heat evolution associated with qE occurs within 1.4 μs, faster than rates of charge recombination (79). Spectral analysis of qE and qP shows that different pigment populations are quenched: qP preferentially quenches the PSII core and qE quenches emitters near 680 and 700 nm, which suggests that excitation in LHCII is selectively quenched (103). A mathematical treatment incorporating the exciton-radical pair model gave best fit of steady state fluorescence data to a "two-state" antenna in which light-harvesting units are switched on or off (126). The appearance of a new short lifetime component in thylakoids displaying qE suggested the formation of a static quenching complex in the antenna (36). Other approaches have also suggested that qE occurs in the antenna and include the effect on quantum yield (32), the inhibition of qE by DCCD binding to LHCII (109, 128), the role of the xanthophyll cycle (20), alterations in qE in plants lacking LHCII components (15, 45, 54, 56, 72, 73), and observations made of LHCII in vitro (95, 102, 106, 108, 112).

Effect of ΔpH

An outline of the features of the transduction pathway by which light-dependent H^+ translocation triggers the process of quenching is shown in Figure 2. Formation and relaxation of ΔpH and qE occur with different kinetics, which suggests that ΔpH causes a conformational change in a thylakoid membrane constituent resulting in formation of a quencher. This conclusion was supported by the observation that the steady state relationship between ΔpH and qE depends on experimental conditions and plant material (65, 81, 82). In the presence of antimycin A, there is a strong shift to a lower pH requirement (65, 80), so that this reagent can be observed to almost completely inhibit qE (88). Dibucaine shifts the titration to a much higher pH (65). As a result, qE can be observed without lumen acidification (82), and complex interactions exist such as dibucaine altering the I_{50} for antimycin inhibition (82). These data can be explained if the qE site is treated as an allosteric enzyme, with H^+ as "substrate" and qE as "velocity of reaction," regulated by binding of various effectors. As discussed below, the xanthophyll cycle carotenoids are the physiological regulators in vivo.

It has been suggested that qE was not caused directly by the lumen pH but by a localized dibucaine-insensitive H^+ domain (76, 77, 82), possibly within

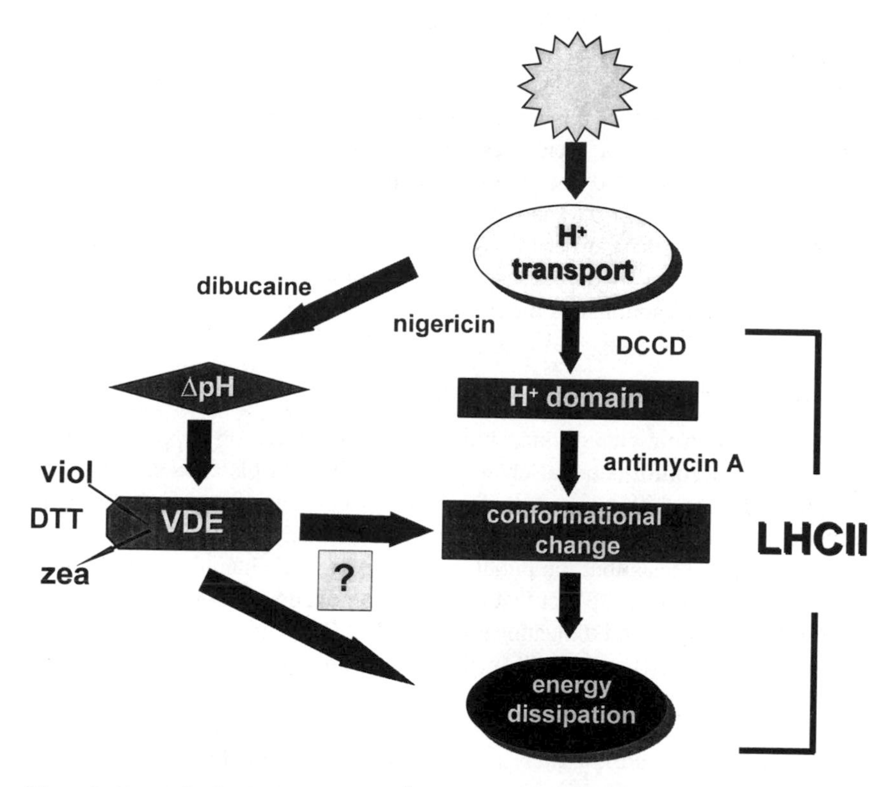

Figure 2 Events linking light-dependent H^+ translocation to the induction of energy dissipation. The sites of action of various inhibitors are shown. One point of controversy is whether the formation of zeaxanthin controls LHCII conformation or acts directly as an excitation energy quencher. ΔpH (as measured by 9-aminoacridine or neutral red) and the putative H^+ domain are distinguished by the effects of dibucaine. The conformational change is that monitored as ΔA_{535}. VDE, violaxanthin deepoxidation.

LHCIIb. There is independent evidence for a H^+ domain in LHCIIb that can be eliminated by high salt treatment of thylakoids (100). The idea that this domain and qE are linked is strengthened by the fact that high salt treatment renders qE sensitive to inhibition by dibucaine (76, 77).

Direct evidence for LHCII as the site of protonation in qE has been obtained using the carboxyl-modifying agent DCCD, which binds to asp and glu residues in relatively hydrophobic domains of membrane proteins. DCCD is a partial inhibitor of qE (109), and this inhibition correlates with labeling of LHCIIa and LHCIIc (128). Two of the amino acid residues bound by DCCD are glutamates on the lumen-facing region of the LHCIIc. Examination of the homologous LHCIIb structure shows that one of these residues is located on the D helix lying parallel to the membrane surface, whereas the other is close by just off the end of the B helix (127). Protonation of these residues may be the primary trigger for qE and may induce a conformational change in the lumen-facing domain of the complex, which affects the photophysical properties of neighboring pigments. These two residues cannot form the whole proton domain discussed above, and their direct role in qE is not proven by these observations. An alternative idea is that they may constitute a proton channel (53, 127) to the qE locus, which is elsewhere in LHCII. There is evidence that LHCIIb can sequester large numbers of H^+ upon aggregation (91; AV Ruban, unpublished data).

pH-Dependent Conformational Changes

The light-scattering changes accompanying ΔpH formation support the conclusion that a conformational change was involved in qE. The apparent absorption change with a λ_{max} at 530–540 nm correlates well with qE both kinetically and in magnitude in leaves and in isolated chloroplasts (9, 10, 82, 110). The hypothesis that the putative conformational change is the cause of qE is strengthened by the fact that antimycin blocked both ΔA_{535} and qE with an unchanged ΔpH, and dibucaine eliminated ΔpH but accelerated the rates of formation of ΔA_{535} and qE (82). Similarly, inhibition of qE in leaves by dithiothreitol also blocked ΔA_{535} (10). The origin of ΔA_{535} remains uncertain. It has been suggested that it arises from ΔpH-dependent macroscopic changes in membrane structure that may be linked to changes in chloroplast volume, but there may be a more specific origin because similar absorption changes occur upon aggregation of LHCII in vitro (108). It has been noted that the λ_{max} is variable from 515 nm to 540 nm, and this correlated in one study with the zeaxanthin (zea) content of the thylakoids (107). Because aggregation of purified carotenoids can give rise to a band varying from 510 nm for violaxanthin (viol) to 540 nm for zea, it has been speculated that the ΔA_{535} may include a contribution from changes in xanthophyll conformation (107).

Absorption changes in the Soret and Q_y bands of chl *a* and chl *b* and in the carotenoid region have also been reported to accompany qE (48, 105, 108). These changes are again similar to those observed when quenching is induced in isolated LHCII. Aggregates of purified chl and zea undergo similar changes (105), which suggests that the conformational change involved in qE is associated with changes in the environment of the pigments bound to LHCII. However, in other studies, these absorbance changes were not detected (10), and their direct involvement in the formation of the "quencher" has been questioned (36). It has also been suggested that ΔA_{535} is not a direct indicator of the "quencher" because the divalent cation ionophore A23187 has been reported to eliminate qE without reducing the extent of light scattering (76).

Independent evidence for conformational changes in LHCII has also been obtained from CD in isolated thylakoids. The large CD originates from the macroorganization of LHCII (31), and the light-dependent, uncoupler-sensitive decrease in the signal amplitude, ΔCD, is therefore strong evidence for a change in LHCII structure (30). The mechanism of the ΔCD and its relationship to qE are unclear. Although requiring a ΔpH, the light intensity dependence of ΔCD shows that it does not depend on a photosynthetic process, unlike qE (30). At present, we conclude that the change in LHCII conformation associated with qE potentiates the light-dependent ΔCD and is not directly linked to it.

Further work is required to assess how and if these changes in conformation and pigment photophysics are related to the mechanism of quenching. Models for qE propose that a protonation-dependent conformational change in LHCII results in a quenching pigment interaction (48–50), or that protonation alters the LHCII structure to make it sensitive to light-dependent quenching (30), or that protonation of a glu liganded to a chl weakens the chl binding, which allows quenching by a neighboring pigment (17). The key principle common to each of these models is that LHCII offers many opportunities for quenching and that structural change is envisaged to remove "antiquenching" features required for efficient light harvesting (48, 50).

Xanthophyll Cycle

There is now broad agreement that a close relationship exists between nonphotochemical energy dissipation and the xanthophyll cycle, the reversible light-dependent deepoxidation of viol to zea. The essential features of the xanthophyll cycle have been extensively reviewed (20, 21, 24, 94), and therefore only a few key issues are addressed here.

LOCATION OF XANTHOPHYLL CYCLE CAROTENOIDS Xanthophyll cycle carotenoids are associated with all LHC components, including LHCI (71, 121). For LHCII, there is enrichment in the minor complexes with LHCIIa, LHCIIc, and LHCIId probably containing 1 mol of viol per monomer (7, 93, 96, 113). The amount binding to LHCIIb may be variable; a stoichiometry of 1 per trimer has been reported (96, 113), although negligible viol was found in one study (7). Analysis of LHCII prepared from light-treated leaves showed a deepoxidation state approximately equal to the thylakoid as a whole in LHCIIb, LHCIIc, and LHCIId, with a much-reduced level in LHCIIa (113). Slightly different results were obtained by Phillip & Young (96). In contrast, other studies have suggested that the viol bound to LHCIIb is not available for deepoxidation (7). The tendency for the loss of pigments seems to be greater for zea than for viol. LHCIIb prepared from light-treated leaves showed much smaller amounts of zea than the amount of viol found in samples from dark-adapted leaves (43, 71), although this was not observed in other reports (96, 113). This variability in the carotenoid content of LHCIIb suggests binding to the periphery of the complex, a conclusion consistent with the location of the epoxidase and deepoxidase on opposite surfaces of the thylakoid.

In most studies only ~60% of viol is available for conversion to zea. It has been suggested that the unconverted viol pool is bound to just one of the LHCII components, such as LHCIIb (94). Circumstantial evidence that supports this view can be found in the much higher conversion ceiling in plants lacking LHCIIb (54, 72). However, viol is present but unavailable in all LHCII complexes (96, 113), which indicates that the organization of the complexes in the membrane may control viol availability. Recently, it has been shown that viol availability is increased if isolated thylakoids are unstacked by removal of Mg^{2+} (25). Light-dependent changes in membrane structure may make some viol unavailable because the changes occur more rapidly than deepoxidation. It is important to note that the viol availability may approach 100% when plants are grown in high light (21, 23, 75), which again may be due to altered membrane organization. The kinetics of epoxidation also depend on the LHCII content of the plant and on the illumination conditions (52, 55). Under prolonged exposure to excess light, there is a decrease in the rate of epoxidation: It is hypothesized that zea is redistributed toward the PSII core (25).

When plants are grown in excess light there is an increase in the xanthophyll cycle pool size. It has not been determined whether more xanthophyll is bound to each LHCII or whether the additional carotenoid is free in the membrane or bound to other proteins (e.g. ELIPS). In LHCII-depleted thylakoids, xanthophyll is proposed to bind other thylakoid proteins (52, 69).

Role of the Xanthophyll Cycle in qE

Zea differs from viol in the absence of two epoxide groups and the resultant lengthening of the conjugated chain length from 9 in viol to 11 in zea. These changes confer both photophysical and physicochemical differences. The important question is, Which of these properties explain their role in qE?

PHYSICOCHEMICAL EFFECTS A study of the six xanthophylls in LHCII indicated a linear relationship between the conjugated chain length and the aggregation index, an indicator of apparent polarity. Viol was the most polar and zea the least (107). This was partly attributed to that fact that zea adopts a planar conformation, whereas the end groups are twisted out of plane in viol. Modeling of the van der Waals surfaces of viol and zea reveal important differences that again appear to be determined by the number of carbon double bonds (AV Ruban, unpublished data). These effects do not automatically explain the role of the xanthophyll cycle in qE, but as a general principle viol and zea are predicted to differ markedly in their interactions with LHCII. Direct evidence for the dependence of LHCII properties on the zea:viol ratio has been obtained: Although LHCIIb prepared from zea-enriched light-treated leaves had the same fluorescence yield as that isolated from dark-adapted leaves, there were subtle differences. First, there was a general increase in the tendency for aggregation (49), including a shift to a higher pK, in the zea-enriched LHCII (91). Second, the potential for light-induced fluorescence quenching in LHCII was increased (43).

Similar results were obtained when exogenous carotenoids were added to LHCII. Viol inhibited whereas zea stimulated the quenching brought about by dilution of detergent-solubilized LHCII (LHCIIa, LHCIIb, and LHCIIc) into a low-detergent solution at low pH (111, 112). The action of viol resembled that of a specific "detergent," which seemed to antagonize the protein-protein and/or chl-chl interactions associated with quenching (48). It was found that only the conjugated chain length was significant in determining the inhibition/stimulation of quenching in LHCIIb (95). The effect of viol could be duplicated by any carotenoid with $n < 9$, and any with $n \geq 11$ were able to stimulate quenching. However, some specificity was found in that only viol could bring about reversal of quenching.

PHOTOPHYSICAL EFFECTS There has been a radical development of ideas concerning the role of the S_1 ($2'A_g$) states of carotenoids in energy transfer (16, 26, 27, 33, 86, 87). These states are hidden (i.e. they can not be populated directly from the ground state and therefore are not observed in an absorption spectrum), and it has required application of the energy gap law and fast spectroscopy before their energy levels could be determined (26). It was found that the S_1 of viol was 150120 cm^{-1} (equivalent to 661 nm), whereas for zea it was 13935 cm^{-1} (equivalent to 718 nm). Because the main emission from chl *a* in LHCII is ~680 nm, there is a possibility for energy transfer from viol to chl as expected for a light-harvesting function but not from zea to chl. More importantly, there could be energy transfer from chl to zea and, because of the high probability of nonradiative decay from the S_1 state, this provides the basis of a direct quenching of chl fluorescence (26), in agreement with earlier predictions (87). This is an attractive hypothesis for which experimental support is being sought, e.g. quenching of chl fluorescence by β-carotene in benzene has been investigated (27). It should be pointed out that interactions between chl and carotenoid are likely to be complex: Increases of chl fluorescence yield (antiquenching) have been reported for many carotenoids including zea (117). Moreover, because the S_1 state cannot be populated by inductive resonance, the Dexter electron exchange process must be invoked. This exchange requires orbital contact between donor and acceptor molecules. It remains to be established whether such association exists in LHCII.

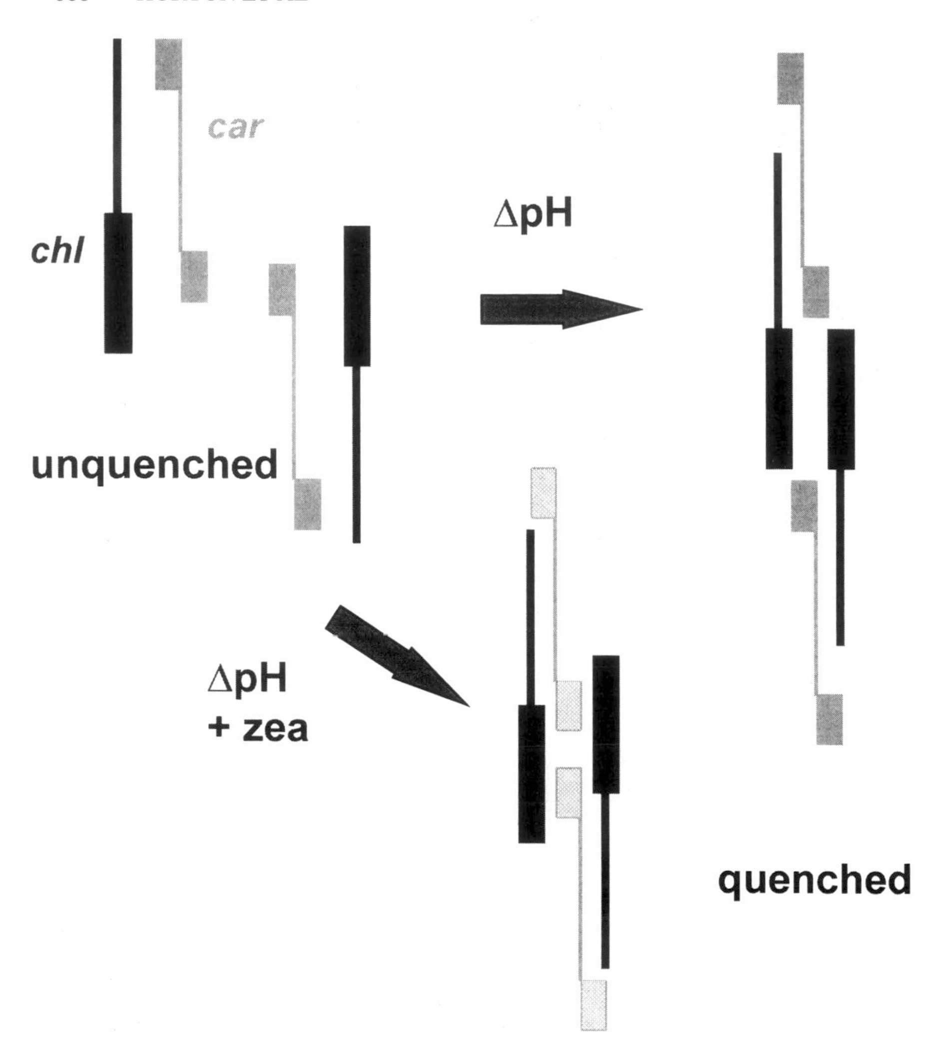

Figure 3 Quenching and antiquenching interactions between chlorophyll and carotenoid. Shown are two chl and two car (viol, open symbol; zea, filled symbol) exposed at the periphery of LHCII. In the unquenched state, viol prevents close chl-chl interaction (antiquenching). In the presence of high ΔpH, antiquenching is overcome under the influence of protonation to give chl-chl quenching. Alternatively, if viol is deepoxidized to zea and a moderate ΔpH is present, the antiquenching effect is overridden by its ability to accept energy from chl, with quenching arising in a chl-zea complex. This scheme explains the opposite effects of viol and zea on qE and LHCII in vitro and embodies their physicochemical and photophysical differences.

The effects of exogenous carotenoids on LHCIIb fluorescence in vitro provide at least partial support for this mechanism (95). The switch between those carotenoids that inhibit quenching and those that stimulate it occurs at an S_1 energy level that coincides with the chl Q_y emission. Therefore, it has been proposed that two opposing effects of carotenoids take place, both of which depend on chl-carotenoid association: the tendency for antiquenching and the ability of carotenoids to quench chl fluorescence (Figure 3).

INVESTIGATIONS OF THE ROLE OF THE XANTHOPHYLL CYCLE IN FLUORESCENCE QUENCHING Strong correlations have been found between the level of zea in leaves and qN (20–22, 24), although such data do not allow a distinction between direct and indirect roles of the xanthophyll cycle in qE. In isolated chloroplasts, qE and zea were also correlated (37–39), and a linear relationship was found between qE and [zea + ant] × lumen acidity under a variety of conditions in lettuce and pea chloroplasts (40). Different results have been obtained with chloroplasts isolated from dark-adapted leaves containing only trace amounts of zea and from leaves light-treated to induce maximum conversion of viol to zea. Both chloroplast types were found to show the same saturating qE (81, 99), which was antimycin-sensitive (81), correlated identically to ΔA_{535} (82), and showed the same relationship of quenching of F_o and F_m (81). We conclude that there is a common mechanism involved with and without zea. An important difference is that the qE-vs-ΔpH titration curve was shifted to smaller ΔpH in the chloroplasts containing zea (81, 99). This "light activation" was also observed for pH-dependent quenching in thylakoids (98). The fact that high levels of quenching can be observed in thylakoids lacking zea provides unequivocal evidence that there is no obligatory requirement for the presence of this carotenoid for qE under the conditions used in these experiments.

At present, there is no obvious explanation for the difference between these results and those of Gilmore & Yamamoto (37–40). In the former, it should be noted that the qE-vs-ΔpH plot reached an endpoint beyond which a further increase in ΔpH caused no increase in qE. In the Gilmore & Yamamoto studies, there is a linear relationship up to the maximum ΔpH attained. In this case, it is possible that ΔpH saturating for qE was not reached because of an intrinsically lower qE sensitivity to pH that rendered qE essentially completely dependent on zea.

The activating role of the xanthophyll cycle in qE can be explained mechanistically by the allosteric interaction between H^+ binding (by LHCII) and the association of viol or zea with the complex (50, 105). In this model, which is discussed in more detail later, there is preferential binding of zea to the protonated quenched form and/or preferential binding of viol to the deprotonated unquenched form. The correlations between [zea] and qE in vivo are consistent with this model if it is assumed that the ΔpH is not saturating for qE. However, some constraint needs to be put on such an indirect role for the xanthophyll cycle—a stoichiometric increase in H^+-binding affinity is required to maintain the linear dependency of qE on $[H^+]$[zea] (36).

Studies of Plants Deficient in LHCII

qE has been studied in the chl *b*–deficient mutants of barley; these plants are deficient in LHCIIb but possess varying amounts of the minor LHCII (93). Except in one case (3), the magnitude of qE is smaller than the wild type (45, 72, 73), which shows the importance of LHCIIb. Despite the reduced qE, both the content of xanthophyll cycle carotenoid and the light-induced deepoxidation state are higher in the chl *b*–deficient mutants. IML plants that have no LHCIIb and only LHCIIc show a reduced but significant level of quenching (15, 54). If the chl *b*–less mutant is grown under IML conditions, no LHCII is present, and there is no DTT-sensitive qN (45, 56). The results suggest that only the minor complexes are involved in qE, but conversely, the decreased amplitude of qE suggests that efficient quenching requires a complete LHCII system. A simple interpretation of these data is risky, because complex interactions may occur. Quantitative analysis shows that a larger ΔpH is required for quenching in the chl *b*–deficient barley mutant compared with the wild type (116). One explanation of the inefficiency of qE in plants lacking LHCIIb is that this complex is necessary to allow connectivity among PSII units (15).

Studies on Isolated LHCII

Trimeric LHCIIb has a fluorescence yield close to that of free chl. When detergent concentration falls below the critical micelle concentration, LHCIIb forms aggregates; this process is promoted by a variety of factors including Mg^{2+}, low pH, and glycerol (48). In the aggregated state, the fluorescence yield is quenched by up to 90% (78, 102). Examination of aggregates shows them to be either two- or three-dimensional depending on conditions, and they may be small (~10 trimers) or very large (hundreds of trimers).

Aggregation of LHCII has profound effects on the bound pigments. 1. Polarized light spectroscopy indicates increased exciton coupling between chl *b* and between chl *a* as the reason for red-shifted chls seen in the absorption spectrum (91; AV Ruban, SLS Kwa, R van Grondelle, P Horton & JP Dekker, unpublished data). 2. Raman spectroscopy shows new H bonds formed to the formyl group of a chl *b* and the keto group of a chl *a* and a twisting of a carotenoid molecule (106). 3. Examination of LHCII down to 4 K shows the presence of at least six red-shifted emitters in the aggregated complex (101). 5. Changes in Z-band thermoluminescence indicate increased local chl concentration in the aggregated LHCII (44, 91). 6. In the aggregated state, light-induced fluorescence quenching and ΔCD have been observed (30). Despite these data, the photophysical properties of aggregated LHCII are still not completely understood, and the mechanism of quenching has not yet been established.

The strong quenching in LHCIIb aggregates led to the suggestion that qE might arise through a similar process (49); conformational changes and macroscopic alterations in membrane structure were consistent with such a suggestion, as were the indications of the occurrence of qE in LHCII. A detailed examination of quenching of LHCIIb in vitro revealed a number of features similar to qE. 1. Quenching was promoted by lowering the pH with a pK ≈ 6.0 (91, 112), similar to those estimated for the thylakoid lumen in the light and for pH-dependent quenching of thylakoids (98). 2. Quenching was inhibited in the presence of antimycin but enhanced by dibucaine (49, 111, 108), similar to their effects on qE. 3. Absorbance changes in the Soret region and in the Q_y band of chl *a* and chl *b* were observed for LHCII aggregation and qE (48, , 108, 111). 4. Xanthophyll cycle carotenoids exert control of quenching in all the LHCII components, with the contrasting effects of either exogenous or endogenous viol and zea exactly the same as for qE (see above) (91, 95, 111, 112). LHCIIa and LHCIIc show stronger quenching and larger absorbance changes than LHCIIb (112). Significantly, only quenching in these complexes was inhibited by DCCD, consistent with its specificity of binding to these complexes.

The transitions involved in LHCII aggregation have been explored using 77 K fluorescence. LHCIIb trimers have a single emission band at ~680 nm. During the first stage in aggregation achieved upon dilution into low detergent, this band is quenched, and also new red-shifted emitters appear. Induction of deeper quenching by lowering the pH or adding dibucaine is accompanied by alteration in the shape of the spectrum with loss of 680 nm and 700 nm emission (AV Ruban, unpublished data). Examination by sucrose gradient ultracentrifugation confirmed that the first stages were the formation of rather small aggregates, whereas large macroaggregates were formed later (111). Other experiments have shown that aggregation occurs in at least two stages, only one of which is inhibited by antimycin (108). The same analysis of 77 K fluorescence gave clear evidence for changes in LHCII aggregation state upon illumination of leaves (110). During the first period of qN formation, the quenching spectrum is very similar to that found for the initial phase of LHCII aggregation, whereas in the final transition to maximum qN the spectrum contains a strong F700 component. Moreover the spectrum for reversible qN (i.e. qE) is that of aggregated LHCII, whereas the irreversible component is a spectrum of unaggregated LHCII. Thus, qI is very similar to the initial process of aggregation, whereas qE is the quenching within the aggregated state.

Such studies suggest a continuum of LHCII aggregation states in the thylakoid. LHCII is already in a partially aggregated state in darkness (30, 31), and light activation and qE formation may both arise from an increase in aggregation state. Aggregation is an imprecise term, and it is currently unclear exactly what different features of LHCII/LHCII interaction relate to quenching. Moreover, not only is LHCII heterogeneous in vivo, but there are many other membrane constituents interacting with it. Further, the change from an unquenched to a quenched state can take place within a few seconds, too short a time for large-scale macroaggregation to occur (111, 112). We consider it likely that interactions between only two complexes may be required or that changes within a complex are sufficient for quenching. The process of aggregation may be an unavoidable consequence of establishing the conditions (e.g. accessibility of water to the complex) necessary for quenching.

It is interesting that in pine needles during winter LHCII appears to exist in a strongly aggregated state (85a). Similarly, in zero CO_2–stressed leaves, LHCII aggregation has been reported (118). In both cases, this quenching appears when PSII reaction centers are lost, and possibly this allows unrestrained aggregation of LHCII as observed in vitro.

A Molecular Mechanism for qE: The Principle of Allosteric Regulation

The mechanism of qE involves the following steps: light-induced H^+ transfer, H^+ binding to one or more LHCII polypeptides, and conformational transition in LHCII resulting in formation of a quencher. The conversion of viol into zea is an integral part of the process of forming a quenching complex involving chl bound to LHCII. This complex may either be chl-zea or chl-chl. In the former, quenching may arise by direct energy transfer to zea, but in the latter the role of the xanthophyll cycle is in the modulation of the formation of the complex, with the mechanism of quenching the same as for "concentration quenching" (8). In vivo data and the effects of viol and zea on isolated LHCII suggest that both types of quenching may contribute to qE.

The dual effects of ΔpH and the xanthophyll cycle on LHCII give rise to the model presented in Figure 4, which postulates four states of the LHCII system. In addition to differing in their H^+ and viol/zea binding, these states differ in the extent and even mechanism of quenching. Each can be observed in isolated LHCII and inferred from examination of intact systems. As discussed previously (47), the molecular processes underlying the observed behavior of a leaf depend on the particular blend of transition from I to III, I to II, and II to IV. A sudden strong illumination of a dark-adapted leaf would favor formation of III, whereas the more gradual changes may favor I to II to IV. It is interesting in this regard that it has been shown that a higher level of reversible qN is found if the light intensity is raised in a number of small steps, compared with a single transition from dark to saturating light (2).

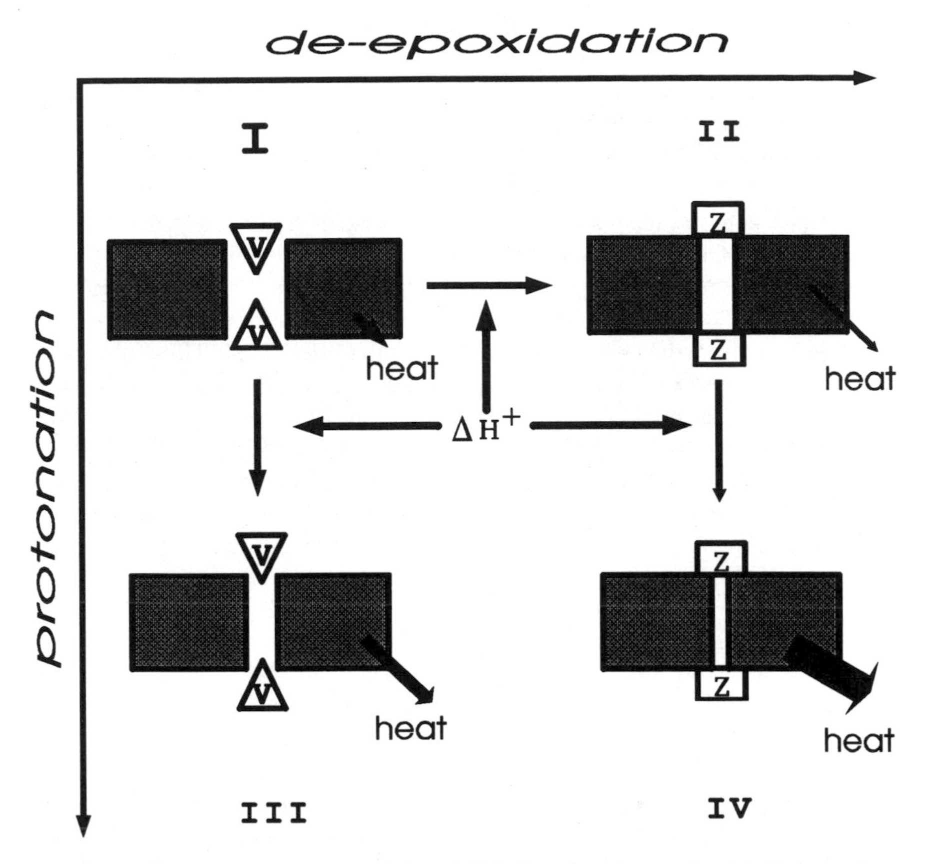

Figure 4 The allosteric model for regulation of qE. It shows how the rate of energy dissipation in LHCII is controlled by structural changes brought about allosterically by protonation and deepoxidation. Four states of LHCII are shown. I, unprotonated, binds viol, unquenched. II, unprotonated, binds zea, slightly quenched perhaps by chl-zea, the light-activated state. III, protonated, viol displaced from its binding site, quenched by chl-chl. IV, protonated, binds zea, strongly quenched by chl-zea. The equilibrium between I and II will depend on the zea:viol ratio. The pK for I → III is <5 and for II → IV is >5.5, which explains the sensitivity of qE to ΔpH. For convenience, this model shows two units, but there may be several interacting units giving rise to cooperative H^+ binding. The biochemical nature of the unit is unspecified—it may be a particular LHCII monomer, an LHCIIb trimer, the complete LHCII oligomer, or the whole LHCII/RC. The transition between III and IV is not allowed, which explains control of viol availability. In vivo, the composition of these states will depend on the exact nature of the exposure to light, the composition of the thylakoid membrane, and other processes affecting ΔpH such as metabolism and temperature.

Slowly Relaxing Nonphotochemical Quenching

At low temperature the slowly relaxing component of qN (qI) appears to result from a persistent ΔpH (34, 35), but this does not appear to be the case at higher temperatures. A component of qI, stable in the dark for over 15 min, can be eliminated if leaves are infiltrated with nigericin (104). However, this stable quenched state was maintained in chloroplasts isolated from these leaves, and it was shown that a bulk phase ΔpH was not responsible. This energy-dependent qI accounted for all of the slowly reversible qN under some conditions, although as the length of exposure or the magnitude of the excess light increased the proportion of qI sensitive to nigericin decreased. These observations suggest a second condition of a "qE-like" qI in which a stable protonated LHCII state can exist. As with qE, spectral analysis suggests that this qI selectively quenches energy in LHCII (60, 104, 105). In Figure 3, this qI state could be state III, which would have an increased stability in the presence of a high zea:viol ratio. Alternatively, if the formation of states III and IV involves increased protein-protein interaction, a high level of qE may result in more stable LHCII aggregation.

Another qI state occurs when the LHCII contains a high ratio of zea:viol but is unprotonated (state II in Figure 3). This is a characteristic of the light-activated thylakoid and probably accounts for a part of the sustained qN first described by Demmig et al (19) and the qI remaining after nigericin infiltration of leaves.

The physiological significance of qI has not been adequately explained. There is no convincing argument that possession of a sustained rather than rapidly reversible qN confers any advantage. To the contrary, slow relaxation may cause lost photosynthesis in fluctuating light environments. It has been suggested that qI is necessary if the qE state is to reach a maximum value or if qE is to be formed rapidly (110). In extreme conditions of light excess, the overnight retention of zea, and therefore some qI (1, 75), has been explained by the need to respond to light early in the morning in the absence of significant levels of photosynthetic electron transport. In such extreme conditions, maximum quantum yield is sacrificed to attain high levels of energy dissipation. In molecular terms, such tradeoffs are explained by the constraints on LHCII organization: The transition from an unquenched state to one in which over 80% of absorbed energy is dissipated as heat requires a large change in LHCII macroorganization that cannot be quickly formed or reversed and which cannot be brought about by protonation alone at physiological ΔpH—hence light activation is required.

The heterogeneity of LHCII should also be considered. The greater strength and sensitivities of quenching observed for isolated minor LHCII and the enrichment of xanthophyll cycle carotenoids in these complexes suggest that they may be the main qE sites. However, the deepoxidation state is very low in LHCIIa, perhaps modulating the strength of quenching in this complex. Perhaps qI is associated with the achievement of higher deepoxidation states in the minor complexes located closer to the RC. The consistent association between at least a part of qI with a decrease in the number of functional PSII indicates that changes are occurring in, or very close to, the RC (84, 89, 90). It has been suggested that these inactive centers protect functional centers by acting as energy quenchers (16).

Control over Quenching Capacity

The maximum extent of qE is variable among species of contrasting ecology (21, 62). For some species, growth in low light has been observed to result in a lower qE capacity than growth in high light (23, 129). Changes were also observed in grown plants at low temperature (2) and when photosynthetic capacity was genetically reduced (11). These responses illustrate the physiological role of qE in removing excess absorbed energy and in the integration of the light-absorbing and energy-utilizing sides of photosynthesis. Two features have been shown to accompany the acclimation of plants to excess light: the increase in size of the xanthophyll cycle pool (21, 23, 42, 64, 75, 110) and the decrease in the amount of LHCIIb (74). The increase in xanthophyll cycle pool size is accompanied by an increase in the maximum deepoxidation state, which together result in higher zea contents. If zea is a direct quencher of excitation energy, then this alone could explain the higher qE capacity. This explanation seems insufficient given the observations made on mutants that have high zea:chl ratios but low qE.

An answer might therefore lie in the composition and organization of the thylakoid membranes of plants acclimated to excess light. In some cases, there is a loss of LHCIIb, mainly the peripheral trimers containing the Lhcb2 gene product; this protein is the 27-kDa polypeptide, although the actual electrophoretic mobility is often 25 kDa (74). It has been shown that this pool of LHCIIb has a reduced tendency for aggregation in vitro compared with the inner pool enriched in the 28-kDa polypeptide (119). Perhaps the loss of the peripheral pool favors high qE because of an increased tendency for aggregation. In addition, the inner LHCIIb population might be more sensitive to the xanthophyll cycle carotenoids. In other studies, no significant changes in the chl *a*:chl *b* ratio were observed upon exposure to excess light (75). There was a decrease in thylakoid content per chloroplast and a reduction in the number of thylakoids per grana (C Maxwell, unpublished data). These changes might confer more flexibility to light-induced structural change and thereby increase the dynamic range of LHCII function. At present, these ideas remain highly speculative because there has been no systematic investigation of the relationship between LHCII content, xanthophyll cycle activity, membrane structure, and the quantitative aspects of qE.

Recent work has provided another possible explanation for the quenching characteristics of plants exposed to excess light. ELIPS are proteins enriched in xanthophyll cycle carotenoids. They share sequence homology with LHCII, and they are selectively induced during extended exposure to light stress (57; Green & Durnford, this volume). The synthesis of the Cbr protein (an algal ELIP) correlated with a decrease in uncoupler sensitivity of qN and a sustained decrease in F_v/F_m (14). Another LHCII homologue, the *PsbS* gene product, CP22 (29), may also be involved in the high light response. This protein has been found in large amounts in the highly quenched thylakoids isolated from pine needles during winter (85a). Perhaps ELIPS and CP22 are synthesised to specifically provide strong quenching centers in the PSII antenna.

CONCLUDING REMARKS—THE NEED FOR NEW APPROACHES

Experiments and their interpretation in the study of the regulation of light harvesting have adopted a "black box" approach in which the macromolecular features of PSII have been largely ignored. This has often led to futile arguments and unnecessary polarization of ideas. Future work needs to take into account the complexity that such features can cause in the fluorescence characteristics of intact systems and to recognize that new methods and approaches are required.

High-resolution electron microscopy of PSII reveals a compact unit containing LHCII and the RC core (13), within which there are likely to be strong protein-protein interactions. Furthermore, long-range interactions may extend between membranes in the granal stacks. Protonation of LHCII could then trigger concerted allosteric structural and functional alterations in the RC and the antenna. The minor LHCII exert control over both qE and H^+ release from the donor side of PSII, which almost certainly indicates interaction between these proteins and the extrinsic lumen polypeptides involved in H_2O oxidation. Ca^{2+}-binding sites are also present on both these groups of proteins, and pH-dependent Ca^{2+} release could be a common feature in regulation of PSII (47). Data suggest that such considerations are applicable at least for the qI state: Quenching in the antenna and inactivation of PSII reaction centers are closely correlated (84), and recovery from the qI state can involve simultaneous zea deepoxidation and D1 synthesis (55).

There are several proton targets within PSII: the minor LHCII, LHCIIb, the PSII donor side, and the violaxanthin deepoxidase (Figure 5). These appear to have different apparent pK values and to differentially respond to the lumen

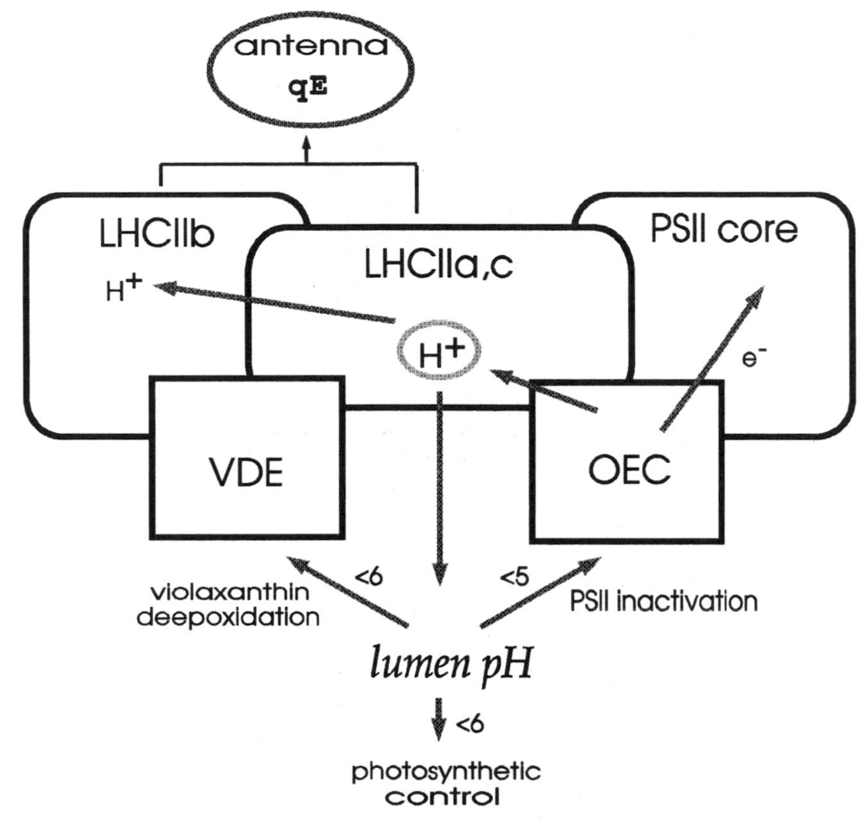

Figure 5 Control of PSII by protonation. There are four H^+ targets: LHCII, violaxanthin deepoxidation (VDE), the donor side of PSII (OEC), and the cytochrome b_6f complex (photosynthetic control). Each has a different sensitivity to the concentration and location of H^+. The balance of these effects determines the extent, site, and stability of quenching and the degree of PSII inactivation. The model proposes lumen pH control over VDE (76, 77) and PSII (18, 67), whereas qE responds to sequestered H^+ in an LHCII membrane domain that may be directly connected to the site of H_2O oxidation.

pH and to a localized H^+ domain, so that the distribution between downregulation of the antenna and of the RC may depend upon very subtle features of thylakoid structure. Furthermore, the pH-dependent control of LHCII structure may have other functions in addition to the control of energy dissipation. The putative H^+ channels in the minor LHCII could provide a mechanism for the regulation of ΔpH. Similarly, the regulation of light harvesting should be viewed in the context of the consequences of light-dependent ΔpH formation: Inhibition of the PSII donor side by low lumen pH can apparently give rise to a stimulation of cyclic electron transport around PSII, which could be photoprotective (97). However, strong inhibition at this site promotes photodegradation of D1 (4, 120). Therefore, effective energy dissipation in the antenna could be particularly important under conditions of high ΔpH. In a sense, qE and its underlying changes in LHCII function could be viewed as mechanisms for controlling ΔpH in addition to or instead of the redox state of the acceptor side of PSII. The observation that decrease in ΔpH in transgenic plants does not result in an increased sensitivity to photoinhibition despite the elimination of qN (51) is consistent with these notions.

For soluble enzymes, it is established from structural and kinetic information that monomers in an oligomer can interact so that information is transmitted throughout the whole protein through cooperative allosteric effects. There are several examples of the involvement of H^+ in these events. For membrane complexes, such behavior is not established so clearly. There is a shortage of methods to explore aspects of macroscopic processes in membrane protein complexes such as those that appear to be central to the regulation of light harvesting. Totally new approaches may be needed to describe the network of interactions within the heterogeneous LHCII matrix in which a variety of external stimuli control the energy levels and spatial relationships of the bound pigments allowing control over excitation transfer to the RC.

Using genetic engineering to control the content of individual proteins will yield new insights. Development of methods for the reconstitution of LHCII from purified pigments and LHCII polypeptides synthesized from Lhcb genes expressed in *Escherichia coli* now offers exciting possibilities to explore both pigment-protein interaction in LHCII monomers and also the process of oligomerization. For example, the amino acid residues needed for LHCII trimerization have recently been identified (92), and the chl *a*:chl *b* ratio of LHCIIa decreased by substitution of a single glutamate residue on the C helix for a glutamine (41). In the future, such approaches may allow discovery of key structural features controlling protein-protein interaction within the entire LHCII system. Combined with state-of-the-art spectroscopic methods, structural alterations may explain how the properties and composition of the pigments and proteins are optimized to allow for both efficient light harvesting and effective energy dissipation. These approaches need to be aligned with detailed analysis of all the events occurring upon exposure of plants to excess light (e.g. 89, 114) so that it can be elucidated which of the responses revealed in vitro are significant in vivo.

ACKNOWLEDGMENTS

Our work in this area is supported by the UK Biotechnology and Biological Sciences Research Council. We wish to acknowledge the contributions of many colleagues to the development of the ideas expressed in this review, and in particular those made by Dr. Andrew Young concerning the xanthophyll cycle.

Literature cited

1. Adams WW III, Demmig-Adams B. 1995. The xanthophyll cycle and sustained thermal energy dissipation activity in *Vinca minor* and *Euonymus kiautschovicus* in winter. *Plant Cell Environ.* 18:117–27
2. Adams WW III, Hoehn A, Demmig-Adams B. 1995. Chilling temperatures and the xanthophyll cycle: a comparison of warm-grown and overwintering spinach. *Aust. J. Plant Physiol.* 22:75–85
3. Andrews JR, Fryer MJ, Baker NR. 1995. Consequences of LHCII deficiency for photosynthetic regulation in *chlorina* mutants. *Photosynth. Res.* 44:81–91

3a. Baker NR, Bowyer JRG, eds. 1994. *Photoinhibition of Photosynthesis: From Molecular Mechanisms to the Field.* Oxford: Bios Sci.

4. Barber J. 1995. Molecular basis of the vulnerability of photosystem II to damage by light. *Aust. J. Plant Physiol.* 22:201–8
5. Barzda V, Garab G, Gulbinas V, Valkunas L. 1996. Energy migration in LHCII-aggregates *Biochim. Biophys. Acta.* In press
6. Bassi R, Dianese P. 1992. A supramolecular light-harvesting complex from chloroplast photosystem II membranes. *Eur. J. Biochem.* 204:317–26
7. Bassi R, Pineau B, Dainese P, Marquardt J. 1993. Carotenoid-binding proteins of photosystem II. *Eur. J. Biochem.* 212:297–303
8. Beddard GS, Porter G. 1976. Concentration quenching in chlorophyll. *Nature* 260:366–67
9. Bilger W, Björkman O. 1990. Role of the xanthophyll cycle in photoprotection elucidated by measurements of absorbance changes, fluorescence and photosynthesis in leaves of *Hedera canariensis. Photosynth. Res.* 25: 173–85
10. Bilger W, Björkman O. 1994. Relationships among violaxanthin deepoxidation, thylakoid membrane conformation, and nonphotochemical chlorophyll fluorescence quenching in leaves of cotton *Gossypium hirsutum* L. *Planta* 193:238–46
11. Bilger W, Fisahn J, Brummet W, Kossmann J, Willmitzer L. 1995. Violaxanthin pigment contents in potato and tobacco plants with genetically reduced photosynthetic capacity. *Plant Physiol.* 108:1479–86
12. Björkman O, Demmig-Adams B. 1995. Regulation of photosynthetic light energy capture, conversion, and dissipation in leaves of higher plants. In *Ecophysiology of Photosynthesis: Ecological Studies,* ed. ED Schulze, MM Caldwell, 100:14–47. Berlin: Springer-Verlag
13. Boekema EJ, Hankamer B, Bald D, Kruip J, Nield J, Boonstra AF, et al. 1995. Supramolecular organization of the photosystem II complex from green plants and cyanobacteria. *Proc. Natl. Acad. Sci. USA* 92:175–79
14. Braun P, Malkin S, Zamir A. 1996. Possible role of Cbr, an early light induced protein, in nonphotochemical quenching in *Dunaliella bardawil. Plant Physiol.* In press
15. Briantais J-M. 1994. Light harvesting chlorophyll a-b complex requirement for regulation of photosystem II photochemistry by nonphotochemical quenching. *Photosynth. Res.* 40:287–94
16. Chow WS. 1994. Photoprotection and photoinhibitory damage. *Adv. Mol. Cell Biol.* 10:151–96
17. Crofts AR, Yerkes CT. 1994. A molecular mechanism for qE quenching. *FEBS Lett.* 352:265–70
18. Crofts J, Horton P. 1991. Dissipation of excitation energy by photosystem II particles at low pH. *Biochim. Biophys. Acta* 1058:187–93
19. Demmig B, Winter K, Kruger A, Czygan FC. 1987. Photoinhibition and zeaxanthin formation in intact leaves: a possible role of the xanthophyll cycle in the dissipation of excess light. *Plant Physiol.* 84:218–24
20. Demmig-Adams B. 1990. Carotenoids and photoprotection: a role for the xanthophyll zeaxanthin. *Biochim. Biophys. Acta* 1020:1–24
21. Demmig-Adams B, Adams WW III. 1992. Photoprotection and other responses of plants to high light stress. *Annu. Rev. Plant. Physiol. Plant Mol. Biol.* 43:599–626
22. Demmig-Adams B, Adams WW III. 1996. Xanthophyll cycle in nature: uniform response to excess direct sunlight

among higher plant species. *Planta.* In press

23. Demmig-Adams B, Adams WW III, Logan BA, Verhoevan AS. 1995. Xanthophyll cycle-dependent energy dissipation and flexible photosystem II efficiency in plants acclimated to light stress. *Aust. J. Plant Physiol.* 22:249–60
24. Demmig-Adams B, Gilmore AM, Adams WW III. 1996. Changing views of in vivo functions of carotenoids in higher plants. *FASEB J.* In press
25. Färber A, Jahns P. 1995. Xanthophyll cycle pigments in intermittent light grown pea plants. See Ref. 74a, in press
26. Frank HA, Cua A, Chynwat V, Young AJ, Goztola D, Wasielewski MR. 1994. Photophysics of the carotenoids associated with the xanthophyll cycle in photosynthesis. *Photosynth. Res.* 41:389–95
27. Frank HA, Cua A, Chynwat V, Young AJ, Zhu Y, Blankenship RE. 1995. Quenching of chlorophyll excited states by carotenoids. See Ref. 74a, in press
28. Fryer MJ, Oxborough K, Martin B, Ort DR, Baker NR. 1995. Factors associated with depression of photosynthetic quantum efficiency in maize at low growth temperature. *Plant Physiol.* 108: 761–67
29. Funk C, Schroder WP, Green BR, Renger G, Andersson B. 1994. The intrinsic 22 kDa protein is a chlorophyll binding sub-unit of photosystem II. *FEBS Lett.* 342:261–66
30. Garab G. 1996. Chirally organised macrodomains in thylakoid membranes. In *Light as an Energy Source and Information Carrier in Plant Physiology,* ed. RC Jennings, G Zucchelli, F Ghetti, G Colombetti. New York: Plenum. In press
31. Garab G, Kieleczawa J, Sutherland JC, Bustamante C, Hind C. 1991. Organisation of pigment-protein complexes into macrodomains in the thylakoid membranes of wild-type and chlorophyll b–less mutant of barley as revealed by circular dichroism. *Photochem. Photobiol.* 54:273–81
32. Genty B, Briantais J-M, Baker NR. 1989. The relationship between the quantum yield of photosynthetic electron transport and quenching of chlorophyll fluorescence. *Biochim. Biophys. Acta* 990:87–92
33. Gillbro T, Andersson PO, Liu RSH, Asato AE, Takaishi S, Cogdell RJ. 1993. Location of the carotenoid $2A_g$-state and its role in photosynthesis. *Photochem. Photobiol.* 57:44–48
34. Gilmore AM, Björkman O. 1994. Adenine nucleotides and the xanthophyll cycle in leaves. II. Comparison of the effects of CO_2- and temperature-limited photosynthesis on photosystem II fluorescence quenching, the adenylate energy charge and violaxanthin de-epoxidation in cotton. *Planta* 192: 537–44
35. Gilmore AM, Björkman O. 1995. Temperature-sensitive coupling and uncoupling of ATPase-mediated, nonradiative energy dissipation: similarities between chloroplasts and leaves. *Planta* 197: 646–54
36. Gilmore AM, Hazlett TL, Govindjee. 1995. Xanthophyll cycle–dependent quenching of photosystem II chlorophyll a fluorescence: formation of a quenching complex with a short fluorescence lifetime. *Proc. Natl. Acad. Sci. USA* 92:2273–77
37. Gilmore AM, Mohanty N, Yamamoto HY. 1994. Epoxidation of zeaxanthin and antheraxanthin reverses non-photochemical quenching of photosystem II chlorophyll a fluorescence in the presence of trans-thylakoid ΔpH. *FEBS Lett.* 350:271–74
38. Gilmore AM, Yamamoto HY. 1991. Zeaxanthin formation and energy-dependent fluorescence quenching in pea chloroplasts under artificially mediated linear and cyclic electron transport. *Plant Physiol.* 96:635–43
39. Gilmore AM, Yamamoto HY. 1992. Dark induction of zeaxanthin-dependent nonphotochemical fluorescence quenching mediated by ATP. *Proc. Natl. Acad. Sci. USA* 89:1899–903
40. Gilmore AM, Yamamoto HY. 1992. Linear models relating xanthophylls and lumen acidity to non-photochemical fluorescence quenching: evidence that antheraxanthin explains zeaxanthin-independent quenching. *Photosynth. Res.* 35:67–78
41. Giuffra E, Cugini D, Croce R, Bassi R. 1995. In vitro reconstitution with pigments of maize photosystem II antenna CP29. See Ref. 74a, in press
42. Gray GR, Savitch LV, Ivanov AG, Huner NPA. 1996. Photosystem II excitation pressure and development of resistance to photoinhibition. II. Adjustment of photosynthetic capacity in *Triticum aestivum* and *Secale cereale. Plant Physiol.* In press
43. Gruszecki WI, Kernen P, Krupa Z, Strasser RJ. 1994. Involvement of xanthophyll pigments in regulation of light-driven excitation quenching in light-harvesting complex of photosystem II. *Biochim. Biophys. Acta* 1183: 235–42

44. Hagen H, Pascal AA, Horton P, Inoue Y. 1995. Influence of changes in the photon protective energy dissipation on red light induced detrapping of the thermoluminescence Z-band. *Photochem. Photobiol.* 62:514–21
45. Härtel H, Lokstein H, Rank B. 1996. Kinetic studies on the xanthophyll cycle in barley leaves: influence of antenna size and relations to nonphotochemical fluorescence quenching. *Plant Physiol.* In press
46. Horton P. 1995. Nonphotochemical quenching of chlorophyll fluorescence. In *Light as an Energy Source and Information Carrier in Plant Physiology,* ed. RC Jennings, G Zucchelli, F Ghetti, G Colombetti. New York: Plenum. In press
47. Horton P, Ruban AV. 1992. Regulation of photosystem II. *Photosynth. Res.* 34: 375–85
48. Horton P, Ruban AV. 1994. The role of LHCII in energy quenching. See Ref. 3a, pp. 111–28
49. Horton P, Ruban AV, Rees D, Pascal AA, Noctor G, Young AJ. 1991. Control of the light-harvesting function of chloroplast membranes by aggregation of the LHCII chlorophyll-protein complex. *FEBS Lett.* 292:1–4
50. Horton P, Ruban AV, Walters RG. 1994. Regulation of light harvesting in green plants: indication by nonphotochemical quenching of chlorophyll fluorescence. *Plant Physiol.* 106:415–20
51. Hurry V. 1995. Nonphotochemical quenching in xanthophyll cycle mutants of *Arabidopsis* and tobacco deficient in cytochrome b_6f and ATPase activity. See Ref. 74a, in press
52. Jahns P. 1995. The xanthophyll cycle in intermittent light-grown pea plants. *Plant Physiol.* 108:149–56
53. Jahns P, Junge W. 1990. Dicyclohexylcarbodiimide-binding proteins related to the short circuit of the proton pumping activity of photosystem II. *Eur. J. Biochem.* 193:731–36
54. Jahns P, Krause GH. 1994. Xanthophyll cycle and energy-dependent fluorescence quenching in leaves from pea plants grown under intermittent light. *Planta* 192:176–82
55. Jahns P, Miehe B. 1996. Kinetic correlation of recovery from photoinhibition and zeaxanthin epoxidation. *Planta.* In press
56. Jahns P, Schweig S. 1995. Energy-dependent fluorescence in thylakoids from intermittent-light grown pea plants: evidence for an interaction of zeaxanthin and the chrorophyll a/b binding protein CP26. *Plant Physiol. Biochem.* 33:683–87
57. Jansson S. 1994. The light harvesting chlorophyll a/b–binding proteins. *Biochim. Biophys. Acta* 1184:1–19
58. Jennings RC, Bassi R, Garlaschi FM, Dainese P. 1993. Distribution of the chlorophyll spectral forms in the chlorophyll-protein complexes of photosystem II antenna. *Biochemistry* 32: 3203–10
59. Jennings RC, Garlaschi FM, Bassi R, Zucchelli G, Vianelli A, Dainese P. 1993. A study of photosystem II fluorescence emission in terms of the antenna chlorophyll-protein complexes. *Biochim. Biophys. Acta* 1183: 194–200
60. Jennings RC, Zucchelli G, Garlaschi F, Vianelli A. 1992. A comparison of the light-induced, non-reversible fluorescence quenching in photosystem II with quenching due to open reaction centres in terms of the chlorophyll emission spectral forms. *Biochim. Biophys. Acta* 1101:79–83
61. Johnson G, Krieger A. 1994. Thermoluminescence as a probe of photosystem II in intact leaves: non-photochemical fluorescence quenching in peas grown in an intermittent light regime. *Photosynth. Res.* 413:371–79
62. Johnson GN, Young AJ, Scholes JD, Horton P. 1993. The dissipation of excess excitation energy in British plant species. *Plant Cell Environ.* 16: 673–79
63. Kolubayev T, Geacintov NE, Paillotin G, Breton J. 1986. Domain sizes in chloroplasts and chlorophyll-protein complexes probed by fluorescence yield quenching induced by singlet-triplet exciton annihilation. *Biochim. Biophys. Acta* 376:105–15
64. Korolova OY, Thiele A, Krause GH. 1995. Increased xanthophyll cycle activity as an important factor in acclimation of the photosynthetic apparatus to high-light stress at low temperature. See Ref. 74a, in press
65. Krause GH, Laasch H, Weis E. 1988. Regulation of thermal dissipation of absorbed light energy in chloroplasts indicated by energy-dependent fluorescence quenching. *Plant Physiol. Biochem.* 26:445–52
66. Krause GH, Weis E. 1991. Chlorophyll fluorescence and photosynthesis: the basics. *Annu. Rev. Plant Physiol. Plant Mol. Biol.* 42:313–49
67. Krieger A, Moya I, Weis E. 1992. Energy-dependent quenching of chlorophyll a fluorescence: effect of pH on stationary fluorescence and picosecond relaxation kinetics in thylakoid mem-

branes and photosystem II preparations. *Biochim. Biophys. Acta* 1102:167–76

68. Krieger A, Weis E. 1993. The role of calcium in the pH-dependent control of photosystem II. *Photosynth. Res.* 37: 117–30
69. Krol M, Spangfort MD, Huner NP, Öquist G, Gustafsson P Jansson S. 1995. Chlorophyll a/b–binding proteins, pigment conversions, and early light-induced proteins in chlorophyll b–less barley mutant. *Plant Physiol.* 107:873–83
70. Kühlbrandt W, Wang DN, Fujiyoshi Y. 1994. Atomic model of plant light-harvesting complex by electron crystallography. *Nature* 367:614–21
71. Lee AI, Thornber JP. 1995. Analysis of the pigment stoichiometry of pigment-protein complexes from barley: the xanthophyll cycle intermediates occur mainly in the light harvesting complexes of PSI and PSII. *Plant Physiol.* 107:565–74
72. Leverenz JW, Öquist G, Wingle G. 1992. Photosynthesis and photoinhibition in leaves of chlorophyll b–less barley in relation to absorbed light. *Physiol. Plant.* 85:495–502
73. Lokstein H, Härtel H, Hoffmann P, Woitke P, Renger G. 1994. The role of light-harvesting complex II in excess excitation energy dissipation: an in vivo fluorescence study on the origin of high-energy quenching. *J. Photochem. Photobiol. B* 26:174–85
74. Mäenpää P, Andersson B. 1989. Photosystem II heterogeneity and long-term acclimation of light harvesting. *Z. Naturforsch. Teil C* 44:403–6

74a. Mathis P, ed. 1996. *Photosynthesis: From Light to Biosphere.* Dordrecht: Kluwer. In press

75. Maxwell C, Griffiths H, Young AJ. 1994. Photosynthetic acclimation to light regime and water stress by C3-CM epiphyte *Guzmania monostachia:* gas exchange characteristics, photochemical efficiency and the xanthophyll cycle. *Funct. Ecol.* 8:746–54
76. Mohanty N, Gilmore AM, Yamamoto HY. 1995. Mechanism of nonphotochemical chlorophyll fluorescence quenching. I. Resolution of rapidly reversible absorbance changes at 530 nm and fluorescence quenching by the effects of antimycin, dibucaine, and cation exchanger A23187. *Aust. J. Plant Physiol.* 22:239–47
77. Mohanty N, Yamamoto HY. 1995. Mechanism of nonphotochemical chlorophyll fluorescence quenching. II. The role of de-epoxidised xanthophylls and sequestered thylakoid membrane protons as probed by dibucaine. *Aust. J. Plant Physiol.* 22:231–38
78. Mullineaux CW, Pascal AA, Horton P, Holzwarth AR. 1992. Excitation energy quenching in aggregates of the LHCII chlorophyll-protein complex: a time-resolved fluorescence study. *Biochim. Biophys. Acta* 1141:23–28
79. Mullineaux CW, Ruban AV, Horton P. 1994. Prompt heat release associated with ΔpH-dependent quenching in spinach thylakoid membranes. *Biochim. Biophys. Acta* 1185:119–23
80. Noctor G, Horton P. 1990. Uncoupler titration of energy-dependent chlorophyll fluorescence and photosystem II photochemical yield in intact pea chloroplasts. *Biochim. Biophys. Acta* 1016: 228–34
81. Noctor G, Rees D, Young A, Horton P. 1991. The relationship between zeaxanthin, energy-dependent quenching of chlorophyll fluorescence and the trans-thylakoid pH-gradient in isolated chloroplasts. *Biochim. Biophys. Acta* 1057: 320–30
82. Noctor G, Ruban AV, Horton P. 1993. Modulation of ΔpH-dependent nonphotochemical quenching of chlorophyll fluorescence in isolated chloroplasts. *Biochim. Biophys. Acta* 1183:339–44
83. Ogren E. 1994. The significance of photoinhibition for photosynthetic productivity. See Ref. 3a, pp. 433–47
84. Öquist G, Chow WS, Anderson JM. 1992. Photoinhibition of photosynthesis represents a mechanism for the long term regulation of photosystem II. *Planta* 186:450–60
85. Osmond CB. 1994. What is photoinhibition? Some insights from comparisons of shade and sun plants. See Ref. 3a, pp. 1–24

85a. Ottander C, Campbell D, Öquist G. 1995. Seasonal changes in photosystem II organisation and pigment composition in *Pinus sylvestris. Planta* 197:176–83

86. Owens TG. 1994. Excitation energy transfer between chlorophylls and carotenoids: a proposed molecular mechanism for nonphotochemical quenching. See Ref. 3a, pp. 95–109
87. Owens TG, Shreve AP, Albrecht AC. 1992. Dynamics and mechanism of singlet energy transfer between carotenoids and chlorophylls: light harvesting and nonphotochemical fluorescence quenching. In *Research in Photosynthesis,* ed. N Murata, 4:179–86. Dordrecht: Kluwer
88. Oxborough K, Horton P. 1987. Characterisation of the effects of antimycin A upon the high energy state quenching

of chlorophyll fluorescence qE in spinach and pea chloroplasts. *Photosynth. Res.* 12:119–28
89. Park YI, Chow WS, Anderson J. 1996. Differential susceptibility of photosystem II to photoinhibition in light-acclimated pea leaves depends on the capacity for photochemical and non-radiative dissipation of excess light. *Plant Sci.* In press
90. Park YI, Chow WS, Anderson J. 1995. Light activation of functional photosystem II in leaves of peas grown in moderate light depends on photon exposure. *Planta* 196:401–11
91. Pascal AA. 1995. *Spinach LHCII: spectral and biochemical changes associated with aggregation and with violaxanthin de-epoxidation.* PhD thesis. Univ. Sheffield. 157 pp.
92. Paulsen H. 1995. Chlorophyll a/b–binding proteins. *Photochem. Photobiol.* 62: 367–82
93. Peter GF, Thornber JP. 1991. Biochemical composition and organisation of higher plant photosystem II light harvesting pigment proteins. *J. Biol. Chem.* 266:16745–54
94. Pfündel E, Bilger W. 1994. Regulation and possible function of the violaxanthin cycle. *Photosynth. Res.* 42:89–110
95. Phillip D, Ruban AV, Horton P, Asato A, Young AJ. 1996. Quenching of chlorophyll fluorescence in the major light harvesting complex of photosystem II. *Proc. Natl. Acad. Sci. USA* In press
96. Phillip D, Young AJ. 1995. Occurrence of the carotenoid lactucaxanthin in higher plant LHCII. *Photosynth. Res.* 43:273–82
97. Rees D, Horton P. 1990. The mechanisms of changes in photosystem II efficiency in spinach thylakoids. *Biochim. Biophys. Acta* 1016:219–27
98. Rees D, Noctor G, Ruban AV, Crofts J, Young A, Horton P. 1992. pH dependent chlorophyll fluorescence quenching in spinach thylakoids from light treated or dark adapted leaves. *Photosynth. Res.* 31:11–19
99. Rees D, Young AJ, Noctor G, Britton G, Horton P. 1989. Enhancement of the ΔpH-dependent dissipation of excitation energy in spinach chloroplasts by light activation: correlation with the synthesis of zeaxanthin. *FEBS Lett.* 256:85–90
100. Renganathan M, Dilley RA. 1994. Evidence that the intrinsic membrane protein LHCII in thylakoids is necessary for maintaining localised Δ_{H+} energy coupling. *J. Bioenerg. Biomembr.* 26: 101–9
101. Ruban AV, Dekker JP, Horton P, Van Grondelle R. 1995. Temperature dependence of chlorophyll fluorescence from the light harvesting complex of higher plants. *Photochem. Photobiol.* 61: 216–21
102. Ruban AV, Horton P. 1992. Mechanism of ΔpH-dependent dissipation of absorbed excitation energy by photosynthetic membranes I: spectroscopic analysis of isolated light-harvesting complexes. *Biochim. Biophys. Acta* 1102: 30–38
103. Ruban AV, Horton P. 1994. Spectroscopy of non-photochemical and photochemical quenching of chlorophyll fluorescence in leaves; evidence for a role of the light harvesting complex of photosystem II in the regulation of energy dissipation. *Photosynth. Res.* 40: 181–90
104. Ruban AV, Horton P. 1995. An investigation of the sustained component of nonphotochemical quenching of chlorophyll fluorescence in isolated chloroplasts and leaves of spinach. *Plant Physiol.* 108:721–26
105. Ruban AV, Horton P. 1995. Regulation of non-photochemical quenching of chlorophyll fluorescence in plants. *Aust. J. Plant Physiol.* 22:221–30
106. Ruban AV, Horton P, Robert B. 1995. Resonance Raman spectroscopy of the photosystem II light harvesting complex of green plants: a comparison of the trimeric and aggregated states. *Biochemistry* 34:2333–37
107. Ruban AV, Horton P, Young AJ. 1993. Aggregation of higher plant xanthophylls: differences in absorption spectra and in the dependency on solvent polarity. *J. Photobiol. Photochem. B* 21: 229–34
108. Ruban AV, Rees D, Pascal AA, Horton P. 1992. Mechanism of ΔpH-dependent dissipation of absorbed excitation energy by photosynthetic membranes II: the relationships between LHCII aggregation in vitro and qE in isolated thylakoids. *Biochim. Biophys. Acta* 1102: 39–44
109. Ruban AV, Walters RG, Horton P. 1992. The molecular mechanism of the control of excitation energy dissipation in chloroplast membranes; inhibition of ΔpH-dependent quenching of chlorophyll fluorescence by dicyclohexylcarbodiimide. *FEBS Lett.* 309:175–79
110. Ruban AV, Young AJ, Horton P. 1993. Induction on nonphotochemical energy dissipation and absorbance changes in leaves: evidence for changes in the state of the light harvesting system of pho-

tosystem II in vivo. *Plant Physiol.* 102: 741–50

111. Ruban AV, Young AJ, Horton P. 1994. Modulation of chlorophyll fluorescence quenching in isolated light harvesting complex of photosystem II. *Biochim. Biophys. Acta* 1186:123–27
112. Ruban AV, Young AJ, Horton P. 1996. Dynamic properties of the minor chlorophyll a/b binding proteins of photosystem II: an in vitro model for photoprotective energy dissipation in the photosynthetic membrane of green plants. *Biochemistry.* In press
113. Ruban AV, Young AJ, Pascal AA, Horton P. 1994. The effects of illumination on the xanthophyll composition of the photosystem II light harvesting complexes of spinach thylakoid membranes. *Plant Physiol.* 104:227–34
114. Russell AW, Critchley C, Robinson SA, Franklin LA, Seaton GCR, et al. 1995. Photosystem II regulation and dynamics of the chloroplast D1 protein in *Arabidopsis* leaves during photosynthesis and photoinhibition. *Plant Physiol.* 107: 943–52
115. Schatz GH, Brock H, Holzwarth AR. 1988. Kinetic and energetic model for the primary processes of photosystem II. *Biophys. J.* 54:397–405
116. Schonknecht G, Neimanis S, Gerst U, Heber U. 1996. The pH-dependent regulation of photosynthetic electron transport. See Ref. 74a, in press
117. Searle G, Brody S, van Hoek A. 1990. Evidence for the formation of a chlorophyll a/zeaxanthin complex in lecithin liposomes from fluorescence decay kinetics. *Photochem. Photobiol.* 52: 401–7
118. Siffel P, Vacha F. 1996. LHC aggregation in intact leaves of tobacco plants stressed by CO_2 starvation. See Ref. 74a, in press
119. Spangfort M, Andersson B. 1989. Subpopulations of the main chlorophyll a/b light harvesting complex of photosystem II- isolation and biochemical characterisation. *Biochim. Biophys. Acta* 977: 163–70
120. Spetea C, Hideg E, Vass I. 1996. Both acceptor and donor side mechanisms of photoinhibition of photosystem II are involved in the light-induced degradation of the reaction centre II D1 protein at low pH. *Plant Sci.* In press
121. Thayer SS, Björkman O. 1992. Carotenoid distribution and de-epoxidation in thylakoid pigment protein complexes from cotton leaves and bundle sheath cells of maize. *Photosynth. Res.* 33:213–25
122. Tremolieres A, Dainese P, Bassi R. 1994. Heterogeneous lipid distribution among chlorophyll-binding proteins of photosytem II in maize mesophyll chloroplasts. *Eur. J. Biochem.* 2212:721–30
123. Trissl HW, Lavergne J. 1995. Fluorescence induction from photosystem II: analytical equations for the yields of photochemistry and fluorescence derived from analysis of a model including exciton pair equilibrium and restricted energy transfer between photosynthetic units. *Aust. J. Plant Physiol.* 22:183–93
124. van Grondelle R, Dekker JP, Gillbro T, Sundstrom V. 1994. Energy transfer in photosynthesis. *Biochim. Biophys. Acta* 1187:1–65
125. Walters RG, Horton P. 1991. Resolution of components of nonphotochemical fluorescence quenching in barley leaves. *Photosynth. Res.* 27:121–33
126. Walters RG, Horton P. 1993. Theoretical assessment of alternative mechanisms for non-photochemical quenching of PSII fluorescence in barley leaves. *Photosynth. Res.* 36:119–39
127. Walters RG, Horton P. 1995. DCCD binds to lumen exposed glutamate residues in LHCIIc. See Ref. 74a, in press
128. Walters RG, Ruban AV, Horton P. 1994. Light-harvesting complexes bound by dicyclohexylcarbodiimide during inhibition of protective energy dissipation. *Eur. J. Biochem.* 226:1063–69
129. Webster JI, Young AJ, Horton P. 1996. Carotenoid composition of *Digitalis purpurea* in relation to non-photochemical quenching. See Ref. 74a, in press
130. Weis E, Berry J. 1987. Quantum efficiency of photosystem II in relation to energy dependent quenching of chlorophyll fluorescence. *Biochim. Biophys. Acta* 894:198–208
131. Weis E, Lechtenberg D. 1989. Fluorescence analysis during steady state photosynthesis. *Philos. Trans. R. Soc. London Ser. B* 323:253–68

Annu. Rev. Plant Physiol. Plant Mol. Biol. 1996. 47:685–714

THE CHLOROPHYLL-CAROTENOID PROTEINS OF OXYGENIC PHOTOSYNTHESIS

B. R. Green

Department of Botany, University of British Columbia, Vancouver, British Columbia, Canada V6T 1Z4

D. G. Durnford

Department of Applied Science, Brookhaven National Laboratory, Upton, Long Island, New York 11973

KEY WORDS: Chl *a/b* (CAB) proteins, fucoxanthin–Chl *a/c* proteins (FCP), Chl *a*–peridinen proteins, light-harvesting antenna, algae

ABSTRACT

The chlorophyll-carotenoid binding proteins responsible for absorption and conversion of light energy in oxygen-evolving photosynthetic organisms belong to two extended families: the Chl *a* binding core complexes common to cyanobacteria and all chloroplasts, and the nuclear-encoded light-harvesting antenna complexes of eukaryotic photosynthesizers (Chl *a/b,* Chl *a/c,* and Chl *a* proteins). There is a general consensus on polypeptide and pigment composition for higher plant pigment proteins. These are reviewed and compared with pigment proteins of chlorophyte, rhodophyte, and chromophyte algae. Major advances have been the determination of the structures of LHCII (major Chl *a/b* complex of higher plants), cyanobacterial Photosystem I, and the peridinen–Chl *a* protein of dinoflagellates to atomic resolution. Better isolation methods, improved transformation procedures, and the availability of molecular structure models are starting to provide insights into the pathways of energy transfer and the macromolecular organization of thylakoid membranes.

CONTENTS

1040-2519/96/0601-0685$08.00

INTRODUCTION

The pigment proteins of photosynthesis are responsible for the absorption of light energy and the primary steps in its conversion to other forms of energy. All of the pigment proteins bind both chlorophylls (Chl) and carotenoids. The Chls do most of the light harvesting whereas carotenoids protect against excess light energy (185). In some marine organisms, the latter also contribute significantly to light harvesting (78). Both Chls and carotenoids are essential for correct folding of the proteins that bind them (135). Although the pigment proteins tend to be called simply "Chl proteins" or "Chl protein complexes" when discussed as holoproteins, they are more correctly called "Chl-carotenoid proteins." This review considers the two major types of Chl-carotenoid proteins: the Chl *a* proteins of cyanobacteria and eukaryotic chloroplasts, and the nuclear-encoded light-harvesting antenna proteins of eukaryotes, which bind a variety of other Chls as well as Chl *a*.

Considerable progress in understanding the molecular makeup of eukaryotic and cyanobacterial photosynthetic membranes has been made since the publication of several comprehensive reviews on chlorophyll-carotenoid proteins (14, 27, 35, 64, 163). Both chloroplast-encoded (Chl *a* binding) and nuclear-encoded (Chl *a*/*b* binding) polypeptides of the two photosystems have now been identified by peptide sequencing, and their genes have been cloned. It is now clear that the peripheral light-harvesting antennae (Chl *a*/*b* and Chl *a*/*c* proteins) of all eukaryotes are encoded by a large gene family that also includes a group of eukaryotic stress-response genes and related prokaryotic genes (42, 69). The structures of two pigment-protein complexes have been solved to atomic resolution: the major Chl *a*/*b* light-harvesting antenna complex (LHCII) of higher plants (105) and the entire Photosystem I complex of cyanobacteria (100).

This review focuses on the structural and evolutionary aspects of photosynthetic pigment proteins. Complementary viewpoints and additional information are found in other recent reviews (70, 84, 90, 119, 125, 135) and in a book dedicated to developmental aspects of Chl proteins (161). The roles of Chl *a*/*b* proteins in regulation of light harvesting are considered in this volume by Horton et al (85).

PIGMENT PROTEINS OF THE CORE COMPLEXES

The core complexes of the two photosystems include all the cofactors necessary for charge separation and electron transfer, as well as proximal (Chl *a*) light-harvesting antennae. The pigment proteins bind only Chl *a* and β-carotene and are chloroplast encoded in all eukaryotes. The organization and composition of the core complexes are highly conserved among green plants, cyanobacteria, and all classes of eukaryotic algae (Table 1). The degree of protein sequence conservation is on the order of 70–75% identity between cyanobacteria and chloroplasts, or higher if conservative substitutions are considered, which provides further evidence that all chloroplasts originated from a cyanobacterium-like ancestor that established an endosymbiotic relationship with a nonphotosynthetic eukaryote host (33, 63).

Photosystem II

The core complex of PSII includes all the proteins, pigments, and cofactors necessary for light-driven movement of electrons from water to reduced plastoquinone (57). It contains three Chl-protein complexes: the reaction center (RC), where the initial charge separation occurs, and two internal Chl *a* light-harvesting antennae, CP47 and CP43 (Table 1).

PSII REACTION CENTER (PSII-RC) The reaction center, strictly defined, contains four to six Chl *a*'s, two pheophytins, and two quinones involved in the initial charge separation, bound by a pair of hydrophobic polypeptides, D1 (*psb*A) and D2 (*psb*D) (149, 155). Whether there are four or six Chl *a* molecules per Reacton Center (RC) is the subject of lively debate. Spectroscopic similarities and conservation of chromophore-binding residues suggest that PSII-RC and the purple bacterial reaction center have a common ancestor (120). A number of hypothetical PSII models built on the bacterial template (e.g. 147, 174) have been useful in designing site-directed mutants in D1 and D2 (40, 173, 177). In addition to D1 and D2, PSII-RC preparations contain three small proteins:

Table 1 Chlorophyll *a* complexes of cyanobacteria and chloroplasts. Figures not cited in the text are consensus values from a number of sources. Polypeptides are listed under their most commonly used names

		Pigments/cofactors	Polypeptides (encoded by)
Photosystem I	Core complex (CPI)	75–100 Chl *a*	PsaA or PSI-A (*psa*A)
		12–15 β-carotene	PsaB or PSI-B (*psa*B)
		phylloquinone	
		4Fe-4S cluster	
Photosystem II	Reaction center (RC)	4–6 Chl *a*	D1 or PsbA (*psb*A)
		2 pheophytin *a*	D2 or PsbD (*psb*D)
		2 plastoquinone	cytochrome b_{559} (*psb*E,*psb*F)
		1 nonheme Fe	PsbI protein (*psb*I)
		1–2 β-carotene	PsbW protein (*psb*W)
	CP47 (CPa-1)	20–22 Chl *a*	PsbB (*psb*B)
		2–4 β-carotene	
	CP43 (CPa-2)	20 Chl *a*	PsbC (*psb*C)
		5 β-carotene	
Other	CP43′ (CPVI-4)	Chl *a*	IsiA (isiA)
		unidentified carotenoid	(cyanobacteria only)

cytochrome b_{559}, PsbI, and PsbW (89, 149, 155), none of which is thought to bind Chl.

CP47 AND CP43 CP47 and CP43 (also known as CPa-1 and CPa-2) are the internal light-harvesting proteins of PSII (22). The polypeptides of 52 and 48 kDa, encoded by *psb*B and *psb*C genes, are very hydrophobic and are predicted to form six transmembrane helices, with both amino and carboxyl termini on the stromal surface (22, 150). The best available data (Table 1) show about 20 Chl *a* and four to five β-carotene molecules per polypeptide chain, determined using amino acid analysis (3), in agreement with earlier work based on radiolabeling (41). CP47 and CP43 were reported to contain a small amount of lutein in addition to β-carotene (13).

In addition to their role as antennae, these polypeptides may also contribute to the protein environment of the water-splitting apparatus. Both CP47 and CP43 polypeptides have a large loop between the fifth and sixth helices, exposed on the lumenal (interior) surface (22). In CP47, this loop is in contact with the 33-kDa polypeptide (*psb*O gene product) and may be involved in stabilizing the tetramanganese cluster (22, 52, 76). Mutants lacking CP43 cannot evolve O_2, but this could be an indirect effect because all PSII polypeptides are depleted (173, 177).

Combined Reaction Center and Internal Antenna in PSI

In PSI, the reaction center and internal antenna are combined in a single Chl-protein complex that accepts electrons from plastocyanin or a cytochrome and delivers them to a small ferredoxin-like molecule, encoded by the *psaC* gene (23, 36, 60, 61). This Chl-carotenoid protein consists of two hydrophobic polypeptides of about 82 kDa (*psa*A and *psa*B gene products) that bind the Chl *a* dimer (P700) and the initial electron acceptors A_0 (Chl *a*), A_1 (phylloquinone), and A_2 (a 4Fe-4S center). The two polypeptides together bind 75–100 additional Chl *a* and 12–15 β-carotene molecules. These additional Chl molecules are the "built-in" Chl *a* antenna, filling the same role as the separate antenna complexes CP47 and CP43 in PSII.

The structure of a trimeric cyanobacterial PSI complex has recently been determined to 4.0–4.5 Å resolution by X-ray crystallography (100, 154). The central part of the structure has some striking similarities to the bacterial RC, which had been predicted earlier on the basis of spectroscopic similarities (61, 126). The RC chlorophylls and quinones are organized in pairs around a twofold axis, with the single 4Fe-4S cluster (A_2) above them, surrounded by five transmembrane helices from each of the PsaA and PsaB polypeptides. There are a total of 11 transmembrane helices in each of PsaA and PsaB (100, 154), plus two large surface helices lying parallel to the membrane plane (154). The 65 antenna Chl *a* molecules resolved in the X-ray structure appear to be hung on the outside of the central protein mass at various levels in the (presumed) lipid bilayer, but they are all oriented approximately perpendicular to the membrane axis (154).

CP43′ (CPVI-4 or CPIIIb) and the Prokaryotic Chl a/b Antennae

A novel Chl *a* protein related to CP43 (referred to as CP43′, CPVI-4, or CPIIIb) is made by certain cyanobacteria in response to iron starvation (158). Its 35-kDa polypeptide, encoded by the *isi*A gene (158), differs from the CP43 polypeptide in having a much shorter lumenal loop between (predicted) helices V and VI. Because phycobilisomes are rapidly degraded under nutrient deprivation (74), CP43′ could maintain some of the cell's light-harvesting capacity or act as a Chl reservoir to speed recovery from iron limitation (28).

A new light has been cast on this story with the discovery that the Chl *a/b* antenna polypeptides of prochlorophytes are encoded by homologues of the *isi*A genes (109, 171). Prochlorophytes are cyanobacteria-like prokaryotes that have both Chl *a* and Chl *b* and lack phycobilisomes (117, 139). When they were first discovered, it was thought that they could be directly related to the ancestral endosymbiont from which green chloroplasts are descended (117). However, molecular phylogeny unequivocally groups prochlorophytes with

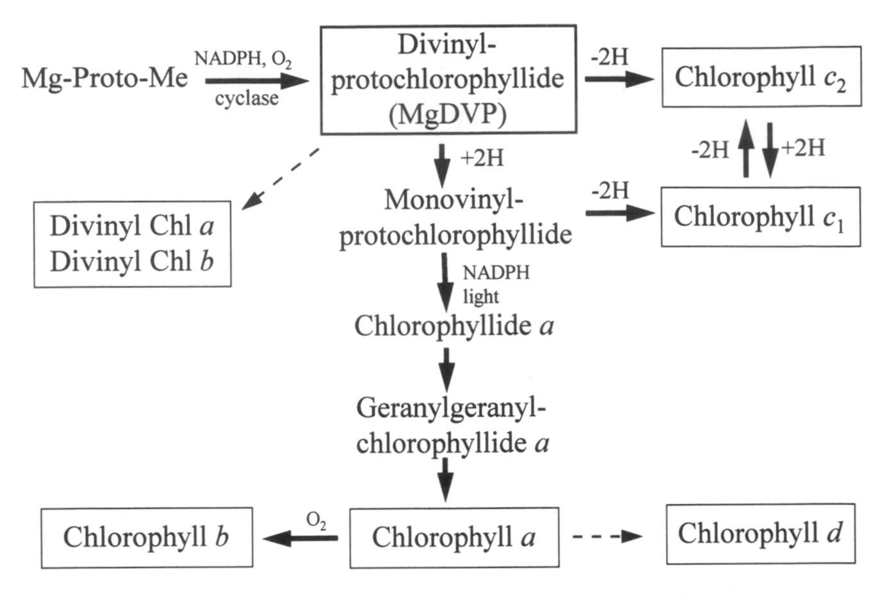

Figure 1 Biosynthetic pathways connecting the chlorophylls, adapted from Reference 92a. Pigments known to be involved in light harvesting are boxed. Dotted lines indicate plausible biosynthetic relationships. Conversion of Chl *a* to Chl *b* requires oxidation of a methyl to a formyl group (176), Chl *a* to Chl *d* the oxidation of a vinyl to a formyl group (150).

cyanobacteria rather than chloroplasts and furthermore shows that the three known genera (*Prochloron, Prochlorothrix,* and *Prochlorococcus*) are not closely related to one another (62, 129, 139, 167).

The Chl *a*/*b* protein sequences from all three prochlorophytes are remarkably similar, both to one another and to the CP43′ (*isi*A) proteins of several cyanobacteria (109, 171). The fact that very closely related proteins can bind either Chl *a* and Chl *b* or Chl *a* alone shows that it is possible for proteins to evolve the capacity to bind different accessory pigments with fairly minor sequence changes. In fact, the three prochlorophytes have rather different pigment compositions (117). *Prochlorococcus* has divinyl Chls *a* and *b,* and the *Prochloron* antenna contains the Chl *c*-like pigment MgDVP (Mg-2,4-divinyl phaeoporphyrin a_5 monomethyl ester or divinyl protochlorophyllide) as well as Chls *a* and *b* (Figure 1; 107, 117). The addition of chlorophylls other than Chl *a* to the ancestral CP43′ protein may have improved its absorption characteristics and light-harvesting efficiency compared with that of the modern cyanobacterial CP43′, which is not an efficient antenna (28).

One possible evolutionary scenario is that the common ancestor of all cyanobacteria may have been able to make and use a whole range of chlorophylls as light-harvesting pigments in addition to phycobiliproteins, and that different descendants lost the ability to use one or more of them (24, 68). Alternatively, the ancestor could have made only Chl *a,* and the ability to synthesize Chl *b* might have arisen independently several times. Conversion of Chl *a* to Chl *b* requires only the conversion of a methyl to a formyl group, which could be accomplished by a monoxygenase recruited from another metabolic pathway. MgDVP is a precursor of all the Chls and is only one step removed from Chl *c* (Figure 1). Both scenarios are consistent with the variety of pigments found in the three known prochlorophytes, and both predict that photosynthetic prokaryotes with other pigment complements will be discovered.

Chl *a/b* PROTEINS OF HIGHER PLANTS

"What Green Band That Is..."

In the past few years, a general consensus has been reached about the polypeptide and pigment compositions of the higher plant Chl *a/b*–carotenoid complexes (92). The DNA sequences of the genes are available and have been matched to the proteins they encode by peptide sequencing (reviewed in 68, 71, 88, 90). This has resolved most of the confusion in protein identification that resulted from different methods of separation and the fact that pigment proteins, by their very nature, are labile in the presence of detergents but cannot be isolated without them. We can now answer the question "Just what green band is that?"—the subtitle of an earlier review (64). A standard nomenclature for the genes (and thus the polypeptides) has been codified (92), although different names for the pigment-protein complexes themselves are still in use (Table 2).

To summarize, there are 10 Chl *a/b* (CAB) proteins, which with their associated carotenoids make up the six main separable pigment-protein complexes (Table 2). The major light-harvesting complex LHCII is probably organized into trimeric particles that transfer much of their excitation energy to PSII but that under certain conditions dissociate from it and migrate independently between stacked and unstacked regions of the thylakoid membrane. CP29 and CP26 are closely associated with the PSII core (6, 29, 65, 136), although they can be removed from it without affecting PSII activity. CP24 is removed from PSII along with LHCII (29). The two LHCI complexes associated with PSI are called LHCI-680 and LHCI-730 after their fluorescence emission maxima (15, 131). The former can be split into two pigmented subcomplexes, LHCI-680A and LHCI-680B, each enriched for a single polypeptide (99, 164). The 730-nm fluorescence may be associated with the *Lhca4* polypeptide (164).

It is generally agreed that the Chl *a/b* ratio of LHCII is 1.3–1.4 (105, 128), but there is a wide range of reported values for the other Chl *a/b* proteins

Table 2 Chlorophyll *a/b* proteins of green plants

Pigment-protein complex				
Name in text	Other names	Chl *a/b* ratio (Ref.)	Proteins encoded by	LHCI/LHCII polypeptide type
LHCI-680A	LHCIa	1.4 (140), 1.9 (164), 3.1 (131)	*Lhca3*	III
LHCI-680B			*Lhca2*	II
LHCI-730	LHCIb	2.3 (140), 2.9 (164), 3.2 (131)	*Lhca1*	I
			Lhca4	IV
LHCII	LHCIIb	1.4 (38), 1.33 (137)	*Lhcb1*	I
			Lhcb2	II
			Lhcb3	III
CP29	LHCIIa,CP29-Type II	2.3 (137), 2.8 (38), 3.1 (77)	*Lhcb4*	
CP26	LHCIIc, CP29-Type I	1.8 (137), 2.2 (38), 3.3 (170)	*Lhcb5*	
CP24	LHCIId	0.9 (137), 1.6 (38)	*Lhcb6*	
CP22	intrinsic 22-kDa protein of PSII	6 (53)	*psbS*	

(Table 2). In general, the values approach 3 for the LHCI and minor PSII complexes, with the exception of CP24, which always has a very low Chl *a/b* ratio. The PSII-associated complexes are reported to differ in their Chl-to-protein ratios, with values of 12, 8, 9, and 5 for LHCII, CP29, CP26, and CP24, respectively (38). This means that the different Chl *a/b* ratios could result from all the proteins binding the same amount of Chl *a* but different amounts of Chl *b*. All the CABs bind lutein, as well as the xanthophylls neoxanthin and violaxanthin. The latter are enriched in CP29, CP26, and CP24 compared with LHCII (13, 110, 146). The role of the Chl *a/b* proteins in the xanthophyll cycle and nonphotochemical quenching is discussed in Horton et al (85).

CP22 is a recent addition to the Chl *a/b* family (54, 55). Its 22-kDa polypeptide is encoded by the *psb*S gene, and it is predicted to form four transmembrane helices rather than three (98, 178). It has a Chl *a/b* ratio of about 6 and appears to have about half as much Chl/polypeptide as LHCII (55). It is stable in the absence of Chl, which suggests that it may have a role as a Chl reservoir or carrier rather than as a light-harvesting complex (53). Alternatively, it may act as a linker between LHCII and the PSII core (97).

The Chl a/b (CAB) Polypeptides of PSII

LHCII preparations contain three closely related polypeptides (Types I, II, and III encoded by *Lhcb1, Lhcb2,* and *Lhcb3,* respectively) in ratios that vary from 10:3:1 to 20:3:1 depending on plant growth conditions and the preparative method (116). All three of them can only be resolved simultaneously on certain gel systems (71, 116, 157) or by HPLC (39). All plants have multiple *Lhcb1* genes encoding almost identical polypeptides (27, 35). Type II and Type III polypeptides are encoded by smaller numbers of genes (35, 71). Single amino acid differences among the population of Type I polypeptides or posttranslational modifications such as phosphorylation (4, 5) could explain the resolution of multiple Type I bands on the novel denaturing gel systems used by Staehelin and coworkers (6, 49, 157). Type I and II but not Type III LHCII polypeptides are phosphorylated (5, 6, 8, 85). The Type III polypeptide is present early in development and under conditions where Chl *b* is limiting (45, 75, 180), which leads to the suggestion that it might act as a linker between LHCII trimers and PSII core (75, 90).

CP29 as originally defined had two polypeptides (29), which were shown to be the products of two different genes by peptide sequencing (138). Unfortunately, when improved isolation methods allowed the separation of the original CP29 complex into two Chl *a*/*b* complexes each containing a single polypeptide (38, 49, 77, 136), one of them was called CP29 and the other CP26. Although this is a violation of standard nomenclature rules [the "new" CP29 could have been called CP28 (66)], it does satisfy the "easy-recognition" criterion (64) and has been generally adopted (Table 2; 49, 92).

The controversies about the number and composition of Chl *a*/*b* proteins (64, 90, 162) result from the fact that the PSII antenna proteins are all related in sequence and associated with one another in vivo, which makes them difficult to distinguish by all available methods. The only way to be sure a new polypeptide is unique is by partial peptide sequencing, preferably accompanied by the sequencing of the corresponding gene. For example, a second polypeptide in CP26 has been identified by protein sequencing (121). When the gene sequence is available it will be possible to tell whether it is a variant of the same Type or should be called *Lhcb7*. At present, there is no evidence for allelic variants at CAB gene loci. The evidence for and against several putative CAB polypeptides in plants is discussed by Jansson (90). More information about individual CAB polypeptides and genes can be found in References 68, 88, 99, 135, and 163.

THE ATOMIC STRUCTURE OF LHCII

Probably the most significant recent achievement in the field of Chl-carotenoid proteins is the determination of the structure of purified pea LHCII to 3.4-Å resolution by electron crystallography (105). This structure is not only important because of the dominant role of LHCII in light harvesting but also because it provides a general model for the overall folding of all the Chl *a*/*b* proteins, as all Chl *a*/*b* proteins have substantial regions of sequence conservation (69,

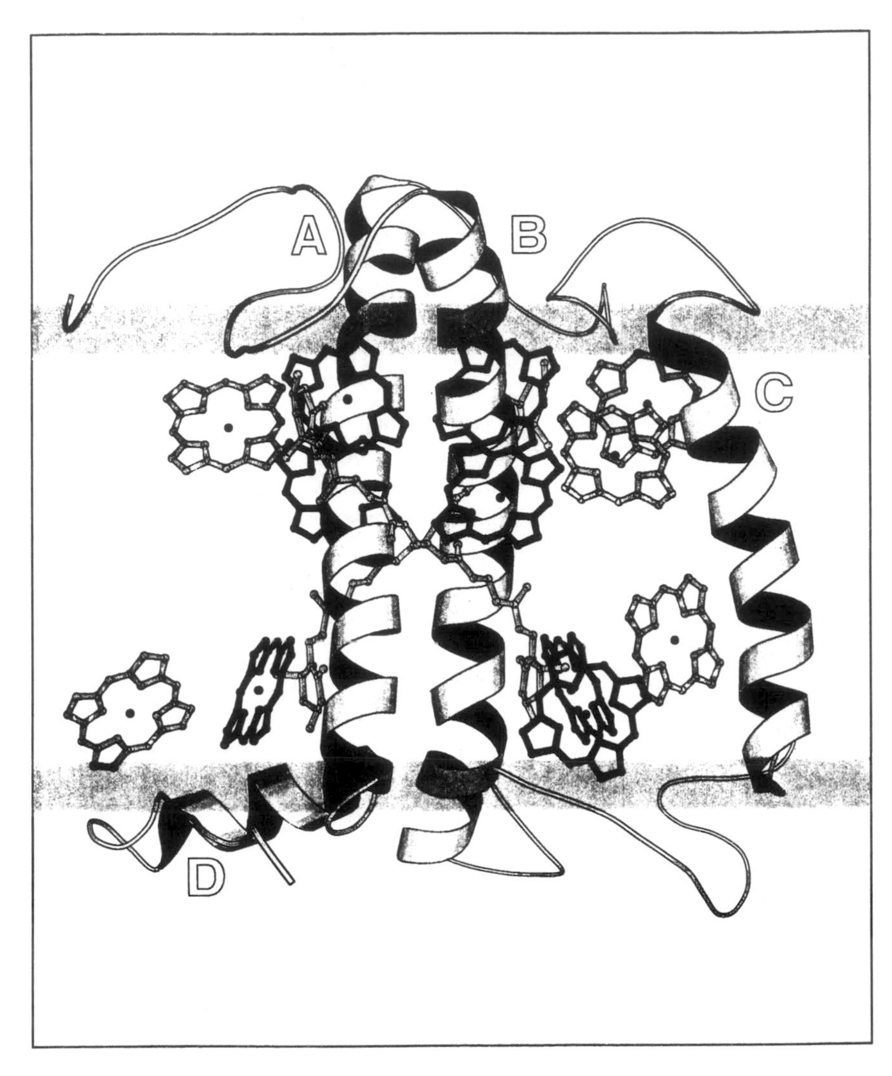

Figure 2 Structure of pea LHCII as determined by electron crystallography (105). α-helices are represented by ribbons, and the edge of the lipid bilayer is represented by shading. Chl molecules are represented by generic porphyrin rings, with proposed Chl *a* molecules shaded darker. Helices are denoted by letters as in Reference 105; B, C, and A are the first, second, and third transmembrane helices, respectively (cf Figure 3). Drawing courtesy of W Kühlbrandt.

104, 105, 163). In fact, all members of the extended family of proteins, which also includes the fucoxanthin–Chl *a/c* proteins (FCP) and early light-inducible proteins (ELIP), are predicted to have the same overall fold (67, 105).

The LHCII polypeptide folds into three membrane-spanning helices, with an additional small amphipathic helix near the C-terminal end (Figure 2; 105). The first (B) and third (A) helices (already known from sequence analysis to be related to each other) cross each other at an angle of about 30° to the membrane normal and are held together by reciprocal ion pairs involving an Arg on one helix and a Glu on the other. If the flattened helices in Figure 3 could be imagined in three dimensions, the side-chain of Arg 70 on helix 1 would be oriented almost straight up and as forming an ion pair with Glu 180 on Helix 3, while Arg 185 on Helix 3 would also be pointing straight up and bonding with Glu 65 on Helix 1. The two helices and the loops at their N-terminal ends are related by a twofold symmetry axis, along with two carotenoid molecules that connect the loops on either end of each helix, and four pairs of Chl molecules (Figure 2).

Seven of the 12 Chl molecules visible in the structure were provisionally identified as Chl *a*s because of their close proximity to one of the carotenoid molecules, which allows efficient energy transfer; the other five Chls, including three bound by the middle helix, were assigned to Chl *b* (105). Three of the Chl Mg atoms are ligated in a novel way, by the carbonyl groups of Glu residues involved in Glu-Arg ion pairs. Two of the symmetrical Chls (a1 and a4) are bound to the carbonyls of the Glu residues cross-linking Helices 1 and 3, and Chl *b5* is ligated by the carbonyl of another Glu-Arg pair formed between adjacent turns of the second transmembrane helix (Figures 2 and 3). Another symmetrical pair of Chl *a* molecules (a2 and a5) are ligated by His and Asn side-chains in Helix 1 and Helix 3, respectively (Figure 3); Chl *a3* is bound by a Gln in Helix 1; and Chl *b3* by His in Helix 4(D). No ligands could be assigned to the other Chls. The residues binding the four symmetrical Chls are conserved in all members of the extended family (67).

In terms of protein structure, the CAB proteins differ from one another in the surface-exposed loops that connect the helices and probably mediate protein-protein interactions, and to a lesser extent in the middle transmembrane helix that binds Chl *b* (Figure 3). There are several conserved motifs in these exposed regions, however, that appear to shield the carotenoid head groups and Chl *a* molecules (67). Only one of these motifs is universally conserved in Chl *a/c* proteins (67). Now that it is possible to reconstitute LHCII from proteins expressed in vitro and to obtain trimers and two-dimensional crystals from them (81, 82), it should be possible to determine more precisely the roles of specific motifs and individual residues in the CAB polypeptides.

The structure of LHCII differs markedly from the light-harvesting antennae of purple bacteria (96, 118). The latter are oligomeric proteins forming large

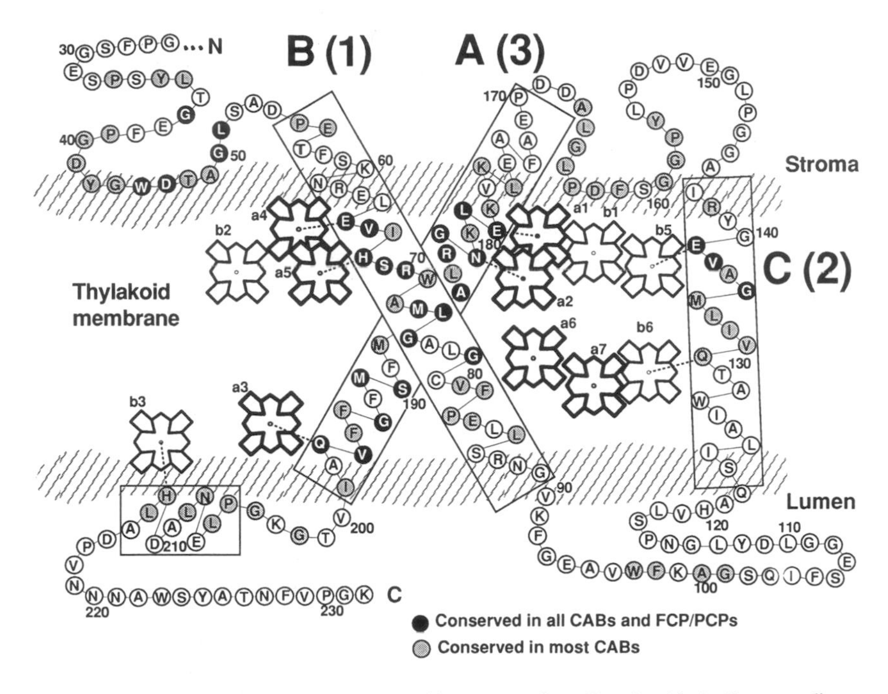

Figure 3 Model of LHCII showing amino acid sequence of pea Type I, with shading according to degree of conservation among CABs and FCPs/iPCPs (see legend). Helices are lettered (105) and numbered (68); hatching indicates edge of lipid bilayer.

rings of overlapping bacteriochlorophyll molecules. By analogy with an accelerator storage ring, the excitation energy could be very rapidly delocalized over the ring, enabling transfer of energy to any neighboring reaction center or antenna complex regardless of its orientation (118). The bacterial and eukaryotic antenna structures are clearly two independently derived structural solutions to the light-harvesting problem. Other unique three-dimensional solutions are found in the soluble membrane-extrinsic antennas: the phycobilisomes (51), the *Prosthecochloris* Chl *a* (FMO) protein (165), and the dinoflagellate peridinen Chl *a* protein (E Hofmann, F Sharples, P Wrench, R Hiller, W Welte & K Diederichs, personal communication).

THE LIGHT-HARVESTING PROTEINS OF ALGAE

Photosynthetic eukaryotes are traditionally divided into three major groups largely on the basis of their light-harvesting pigments (142). The Chlorophytes (green algae and higher plants) have Chl *a*/*b* antennae, the Chromophytes have Chl *a*/*c* antennas, and the Rhodophytes (red algae) have only Chl *a* and use phycobilisomes (extrinsic phycobilin-containing structures) as the major PSII antenna. A taxonomic difficulty arises in some groups whose pigment compositions do not put them into the same taxa as classification based on the number of membranes around the chloroplast, flagellar structure, or other anatomical features (Table 3). The Eustigmatophytes have only Chl *a* but do not have phycobilisomes, the Prasinophytes have the Chl *c*-like pigment MgDVP as well as Chls *a* and *b,* and the Cryptophytes have Chls *a* and *c* as well as phycobilins located in the lumen of the thylakoid. The Dinophyta (dinoflagellates) and Euglenophyta are the only taxa with three chloroplast envelope membranes, but their antenna pigments and the sequences of their antenna proteins group them with the Chromophytes and Chlorophytes, respectively. In spite of different pigment compositions, immunological and sequence data (where available) show that all the antenna polypeptides are structurally related to one another (68). Structures of all the Chls known to be bound by members of this extended family are boxed in Figure 1. Information on algal Chl proteins up to 1990 was thoroughly reviewed by Hiller et al (78).

Chlorophytes

CHLOROPHYTA The green algae have about the same number and complexity of Chl proteins as the higher plants (7, 16, 17), but only a few chlorophyte CAB gene sequences have been reported (68, 90). These genes can be assigned to LHCI or LHCII but not to Types within them (Table 3, Figure 4). There may be additional members of the CAB family in *Chlamydomonas* that are not found in higher plants (16, 17, 56). Some marine siphonous green algae contain the carotenoids siphonoxanthin and siphonein, which increase blue light–harvesting capacity (78).

PRASINOPHYTA The LHC complexes from the prasinophyte *Mantoniella* have at least two polypeptides of 20–21 kDa (78, 94, 144, 152) arranged into larger oligomeric complexes (144). The presence of a Chl *c*-like pigment suggested they might be related to the Chl *a*/*c* proteins, but gene and protein sequencing showed them to be more related to Chl *a*/*b* proteins (Figure 4; 94, 145, 152). There is some evidence that a unique PSI-associated antenna may not exist in *Mantoniella* and that the same protein complex excites both photosystems (153).

EUGLENOPHYTA *Euglena* is generally included with the chlorophytes because of its Chl *a*/*b* antenna. Its LHCI and LHCII polypeptides are definitely related to those of green algae (Figure 4; 86, 123). The presence of a third membrane surrounding the *Euglena* chloroplast suggests that it may have evolved from a eukaryotic rather than a prokaryotic endosymbiont (58). The only remaining

Table 3 Chlorophyll-carotenoid proteins of eukaryotic algae

		# Memb	Most studied genera	Chlorophylls[a]	Carotenoids[a]	Polypeptides Number	Polypeptides Mol. wt. (kDa)	References Proteins	References Genes
Chlorophytes	Chlorophyta	2	*Chlamydomonas, Dunaliella*	*a, b*	lut, neo, zea, vio	10–11	20–30	16, 17, 78	56, 68
	Prasinophyta	2	*Mantoniella*	*a, b,* MgDVP	lut, neo, zea, ant, pras	2–3	20–22, 23–25	144, 152	94, 145
	Euglenophyta	3	*Euglena*	*a, b*	dia, dit, neo	6–8	26–28 (polyprotein)	37	86, 123
Rhodophytes		2	Porphyridium	*a* only	lut, neo, zea	6	18–24	181, 182	
Chromophytes	Phaeophyta (brown algae)	4	*Fucus, Dictyota, Laminaria*	*a, c1, c2*	fuco	1–3	17–21	32, 43, 78	30
	Bacilliarophyta (diatoms)	4	*Phaeodactylum,*	*a, c1, c2*	fuco	1–3	17–20	32, 78	73
			Odontella					102	102
	Chrysophyta	4	*Giraudyopsis,*	*a, c*	fuco, zea,	1	20	134	132
			Ochromonas	*a, c* (?)	ant	2	21, 26	72	
	Raphidophyta	4	*Heterosigma*	*a, c1, c2*	fuco, dia, dino	8–10	16–28	47	47
	Xanthophyta	4	*Pleurochloris*	*a, c*	dia, vau, het	3	17–22	25, 78	
	Haptophyta	4	*Pavlova,*	*a, c1, c2*	fuco	1–4	17–21	78, 79	
			Isochrysis			3	18, 20, 24	109	109
	Eustigmatophyta	4	*Nannochloropsis,*	*a* only	vio, vau		26	78, 114, 159	
			Monodus				23	11	
	Cryptophyta	4	*Chroomonas,*	*a, c2*	allo		20, 24	19	
			Cryptomonas				18, 19, 22	20	
Dinoflagellates	Chl *a/c* (iPCP)	3	*Amphidinium, Gonyaulax, Symbiodinium*	*a, c2*	peridinin	1–3	19–24 (poly protein)	79, 87, 95	80
	Chl *a* (sPCP)			*a*	peridinin	1–2	15–17, 35	127	127

[a]MgDVP, divinylprotochlorophyllide; lut, lutein; neo, neoxanthin; zea, zeaxanthin; vio, violaxanthin; pras, prasinoxanthin; dia, diadinoxanthin; dit, diatoxanthin; fuco, fucoxanthin; ant, antheraxanthin; dino, dinoxanthin; vau, vaucheriaxanthin; het, heteroxanthin; allo, alloxanthin

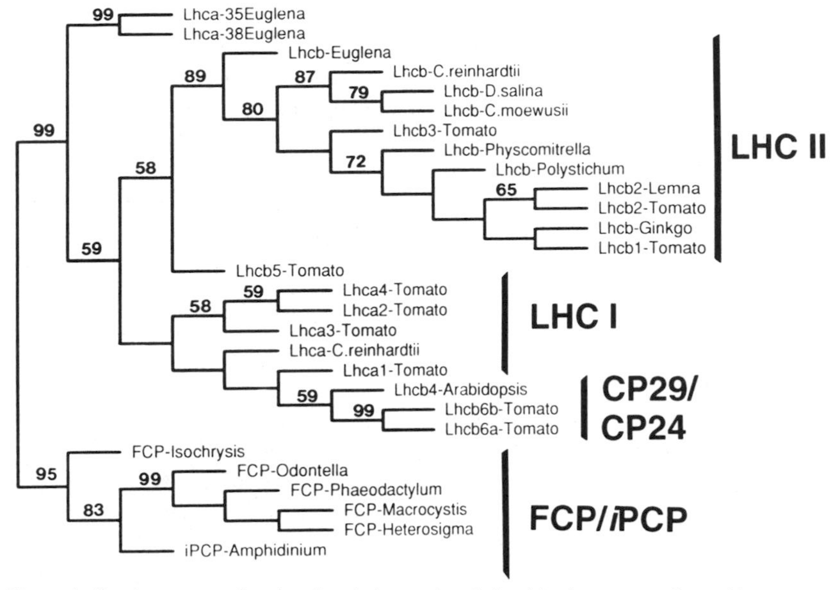

Figure 4 Parsimony tree showing the phylogenetic relationships between amino acid sequences of the main types of light-harvesting proteins, calculated with the PHYLIP package (50). Proteins labeled by gene type and organism name (see Tables 2 and 3). Numbers on branches are bootstrap values (100 replicates) and give an estimate of branch stability. Variable regions that could not be aligned across all sequences were omitted from the analysis.

traces of the endosymbiont would be the chloroplast surrounded by a membrane originating from the phagocytic host (58, 59). In support of this idea, phylogenetic studies on 18S rRNA show that the "host" is unrelated to the green algae or any other photosynthetic eukaryote (34). The LHCII and LHCI proteins are synthesized as polyproteins that are cleaved after import into the chloroplast (86, 123). It has now been shown convincingly that the polyprotein precursor is cotranslationally inserted into the ER membrane and passed to the Golgi before arriving in the chloroplast by an unknown mechanism (160).

Rhodophytes

The rhodophyte plastid genome carries many more genes than the higher plant plastid genome, including most of the genes encoding photosynthetic proteins (143). Along with the presence of phycobilisomes, this suggests the red algal chloroplast is more recently evolved from an endosymbiotic cyanobacterial ancestor. It has now been discovered that the red algae have a PSI-associated antenna complex with four to five polypeptides immunologically related to Chl *a*/*b* and Chl *a*/*c* antenna proteins, even though the polypeptides bind only Chl *a* (181, 182). These polypeptides are similar enough to higher plant CABs that they can be assigned to one of the four LHCI Types. The *Porphyridium cruentum Lhca*1 gene encodes a polypeptide having about 30% identity with higher plant *Lhca*1 (E Gantt & S Tan, personal communication). These findings are very significant, because they suggest that all the eukaryotic nuclear-encoded light-harvesting antenna proteins are descended from a common ancestral protein (181) and call into question classification schemes that put the red algae on a completely separate branch from chlorophyte and chromophyte algae.

Chromophytes

The chromophytes have a smaller number (1–3) of fucoxanthin–Chl *a*/*c* polypeptides (FCP) with a typical size range of 17–22 kDa compared with 20–30 kDa for the chlorophyte CABs (Table 3; 78). As biochemical methods improve, the number of FCP polypeptides increases: The raphidophyte *Heterosigma* now has at least eight immunologically distinguishable polypeptides (47). The Chl *a*/*c* ratios of the major chromophyte LHCs are varied, ranging from 1 to 5.6 (78, 79). In general, they have a higher carotenoid:Chl ratio than the higher plant LHCs, and those carotenoids, especially fucoxanthin, make a significant contribution to spectral coverage in the green region of the spectrum (78). The xanthophyte *Pleurochloris* is the only chromophyte where a PSI-associated antenna with a fluoresence emission different from the main LHC has been reported (25). There is no evidence for specific antenna polypeptides associated with PSI or PSII in other chromophytes.

Gene sequences for the major Chl *a*/*c* polypeptides of two diatoms (73, 102), two phaeophytes (10, 30), a raphidophyte (48), a chrysophyte (132), and a haptophyte (108) are now available. Sequence comparisons demonstrate that the Chl *a*/*c* proteins are indeed related to the Chl *a*/*b* proteins (Figure 3), although the relationship is distant (Figure 4). Southern blots indicate the presence of multigene families in several groups (10, 19, 48). The FCPs are all nuclear encoded and synthesized on cytoplasmic ribosomes. The presence of four membranes around the chromophyte chloroplast suggests that chromophytes originated from a secondary endosymbiosis between a photosynthetic eukaryote and a nonphotosynthetic host (33, 44, 59). Very little is known about how the Chl *a*/*c* polypeptides make their way from cytoplasm to thylakoid across four membranes, but there is evidence that they are cotranslationally inserted into the endoplasmic reticulum (18), which provides a possible explanation for how they cross the outermost membrane.

Dinoflagellates

The dinoflagellates are not considered true chromophytes because they have peridinin rather than fucoxanthin as their major carotenoid and three membranes rather than four around the chloroplast. They have two types of light-harvesting complex: a membrane-intrinsic peridinin–Chl *a*/*c* complex (*i*PCP) and a water-soluble peridinin–Chl *a* complex (*s*PCP) (78). The sequences of the iPCP polypeptides are related to those of chromophyte FCPs (79, 87) and clearly belong to the FCP branch of the CAB/FCP family (Figure 4). The *Amphidinium i*PCP polypeptides are translated as a polyprotein (80) that is posttranslationally cleaved to as many as 10 individual but closely related polypeptides (79, 80). Because the dinoflagellate chloroplast is similar to the *Euglena* chloroplast in being surrounded by three membranes and importing light-harvesting complex proteins as polyproteins, a similar transport mechanism (160) may be involved.

The polypeptide(s) of the *s*PCPs are also nuclear encoded but are not synthesized as polyproteins, although some species have a 35-kDa polypeptide that contains two repeats of the same sequence whereas others have a homodimer of two 15-kDa polypeptides (127). The gene sequence shows no similarity to any other pigment protein (127). The three-dimensional structure of the *Amphidinium s*PCP has been solved by X-ray crystallography (E Hofmann, F Sharples, P Wrench, R Hiller, W Welte & K Diederich, personal communication). It is quite unlike any other light-harvesting protein: The eight peridinens and two Chl *a*s float in a "boat" made of 16 α-helices. This is the first solved structure of an antenna protein having carotenoids as the major light-absorbing pigment.

EARLY LIGHT-INDUCIBLE PROTEINS AND THEIR PROKARYOTIC RELATIVES

The early light-inducible proteins (ELIP) are included with the light-harvesting complexes because their proteins are predicted to have three transmembrane helices and because they share conserved sequences with the Chl *a*/*b* proteins (68, 69). They are part of a complex gene family, like the CAB gene family. They are induced primarily by high light stress (111, 112). It has been suggested that ELIPs act as Chl scavengers after light-induced breakdown of Chl proteins (2). Their association with massive β-carotene accumulation in *Dunaliella* led to the suggestion that this ELIP could have a carotenoid-binding role (111). ELIPs have been proposed to bind xanthophyll cycle pigments in intermittent light-grown plants, which do not accumulate Chl *a*/*b* proteins, and to act as sinks for excitation energy under high light (101). The isolation of a single ELIP polypeptide binding lutein and a small amount of Chl has recently been reported (1).

ELIPs have not been reported in chromophyte or rhodophyte algae. However, a gene for a one-helix protein (HLIP) expressed under high light stress conditions and related to the ELIPs has been found in cyanobacteria (42).

HLIP-like genes have also been found on the plastid genomes of two rhodophyte algae (42; M Reith, personal communication) and on the cyanelle genome of the glaucophyte *Cyanophora paradoxa* (VL Stirewalt & DA Bryant, personal communication). Because these one-helix proteins are related to both first and third helices of the higher plant ELIPs, and because two transmembrane helices are required to make the reciprocal ion pairs and bind Chl, it has been suggested that they form homodimers (42, 67). To date, there is no evidence that HLIPs of cyanobacteria or red algal chloroplasts bind Chl.

EVOLUTIONARY RELATIONSHIPS IN THE CAB/FCP/ELIP/HLIP FAMILY

All the Chl *a/b,* Chl *a/c,* and Chl *a/a* light-harvesting antenna proteins are part of an extended gene family that also includes the ELIPs (68, 69, 90). Conserved residues in the first and third helices of all these proteins include those ligating the four core Chl *a* molecules, the ionic cross-bridges, and the pattern of small residues (G,A,S,C) that allow the helices to pack together efficiently (67, 105). The discovery of the cyanobacterial HLIP gene, clearly sharing the same conserved residues, supports the idea that prokaryotic HLIPs and eukaryotic antenna proteins had a common ancestor.

Because the CAB family includes a four-helix member (CP22 or PsbS) that clearly originated from the duplication of a two-helix protein (98, 178), it has been proposed that a similar gene duplication followed by deletion of the fourth helix gave rise to the three-helix CABs/FCPs/ELIPs (68). The second helix of the two-helix ancestor would have been acquired by a previous fusion of an HLIP-like gene with the gene for another one-helix protein (67). However, there is no obvious relatedness between the second helix of the eukaryotic ELIPs and that of the CABs, so it is possible the middle helices were acquired independently.

A phylogenetic tree (Figure 4) generated using the parsimony method (50) shows the FCPs on a separate branch from the CABs. LHCII and LHCI CABs diverged before the green alga/higher plant split, because the *Chlamydomonas* sequences group with one or the other, but the three LHCII Types (Lhcb1, Lhcb2, and Lhcb3) diverged later. In all the trees, Lhcb5 groups with LHCII, and Lhcb4 and Lhcb6 group together, sometimes closer to the LHCII group and sometimes to the LHCI group (90; DG Durnford, unpublished data). Three of the LHCI sequences cluster together, but all four of them are more divergent from one another than the three LHCII types (DG Durnford, unpublished manuscript). Considering that rhodophyte algae have LHCI but not LHCII and that the LHCI genes appear to have been diverging from one another longer than the LHCII genes, it is possible that LHCII originated from LHCI.

MACROMOLECULAR ORGANIZATION IN THE PHOTOSYNTHETIC MEMBRANE

Lateral Heterogeneity of the Thylakoid Membrane System

Although the thylakoid membrane system is continuous, two distinct regions can be differentiated in higher plants: single unappressed thylakoids (stroma lamellae) and grana stacks consisting of a number of appressed thylakoids. Most of PSII is segregated in the granal regions, and most of PSI is segregated in the stromal regions (8). Furthermore, the PSII units found in stromal regions ($PSII_\beta$) have smaller antenna sizes than the $PSII_\alpha$ units of the grana regions, and at least part of the population is photochemically inactive, i.e. unable to reduce Qb (8, 119). PSI is also heterogeneous: PSI_α found in the margins of the granal stacks has a larger antenna size than PSI_β in the stromal regions because of association with LHCII (9, 183). Although detergent fractionation is used in many cases to separate granal and stromal fractions, work from the laboratory of Albertsson (9, 183) involves mechanical fragmentation followed by two-phase separation, which argues against fortuitous associations provoked by detergent solubilization.

LHCII is found in both regions of the thylakoid membrane system and can transfer excitation energy to PSI as well as PSII. A subpopulation of LHCII can move from granal to stromal regions depending on light quality, redox potential, and phosphorylation (reviewed in 4, 5, 8). This implies that LHCII trimers are not strongly attached to PSII and are able to move independently in the thylakoid membrane system. There are different ratios of the three LHCII polypeptides in stromal, granal, and granal margin fractions (9).

There is little or no evidence for lateral heterogeneity of the thylakoid membrane in chromophytes (113, 141, 175). The one possible exception is found in the xanthophyte *Pleurochloris,* where examination of freeze-fracture specimens by electron microscopy does show some evidence for lateral heterogeneity within the thylakoid membranes, and there appear to be small but significant changes in particle distribution on dark-light adaptation (26).

LHCII Trimers

There is good evidence that LHCII exists as a trimer in vivo (reviewed in 90, 104). The LHCII trimers used for molecular structure determination by X-ray crystallography contained both Type I and Type II polypeptides, which have only minor differences in amino acid sequence outside of the first 20 residues (104, 128). It is not known whether LHCII in the plant is present as homo- or heterotrimers, although homotrimers of Type I LHCII have been isolated by chromatography (128). Developmental studies suggest that LHCII is first organized in monomers and only becomes trimeric when the thylakoid membrane structure is approaching its mature state (45). A WYGPDR motif conserved in LHCII and CP26 but not in CP29, CP24, or LHCI (68) is essential for the formation of trimers in vitro (81). Phosphatidylglycerol is also involved in trimer stabilization (128).

Photosystem I

Both monomeric and trimeric forms of the PSI holocomplex can be isolated from cyanobacteria; the trimer is probably the native form (reviewed in 103). It is generally agreed that higher plant PSI is monomeric and that LHCI remains associated with the PSI core under all physiological conditions (e.g. 9, 21, 183). It has been estimated that there are two copies of each of the four LHCI Chl proteins in each PSI holocomplex, which are possibly arranged in a single layer around the PSI core (21, 90). Cross-linking studies suggest that *Lhca*2 and *Lhca*3 form homodimers, but *Lhca*1 and *Lhca*4 may form heterodimers (91). However, developmental studies have been interpreted in favor of trimers (46). Several models for the arrangement of Chl proteins in higher plant PSI have been proposed (46, 90, 91, 99).

Photosystem II

There is an ongoing debate about whether PSII is a dimer (12, 20, 38, 115, 137, 148) or a monomer (83), with the weight of evidence currently favoring the dimer. The clearest picture comes from negatively stained cyanobacterial PSII particles and spinach PSII cores (without Chl *a*/*b* antennas), subjected to sophisticated image analysis (20). These particles clearly have an axis of twofold symmetry bisecting their largest dimension (Figure 5). Molecular weight estimates and the presence of two separate 33-kDa (PsbO) polypeptides per particle support the idea that they are dimeric. Images supporting the dimer concept have also been obtained by atomic force microscopy (156) and from two-dimensional crystals of PSII depleted in LHCII (115). On the other hand, images of ordered arrays of particles in PSII membranes containing the full complement of CAB proteins have been interpreted as having a monomeric PSII core (83).

Spinach PSII particles containing the Chl *a*/*b* antenna complexes have two symmetrical appendages peripheral to the core dimer (Figure 5; 20). A trimer of LHCII fits snugly into each corner, leaving several small regions of electron density unaccounted for. There is not yet enough evidence to localize the minor CAB proteins (CP29, CP26, CP24, CP22). A number of imaginative PSII models have been proposed on the basis of Chl proteins co-isolated after mild detergent solubilization and electrophoresis (90).

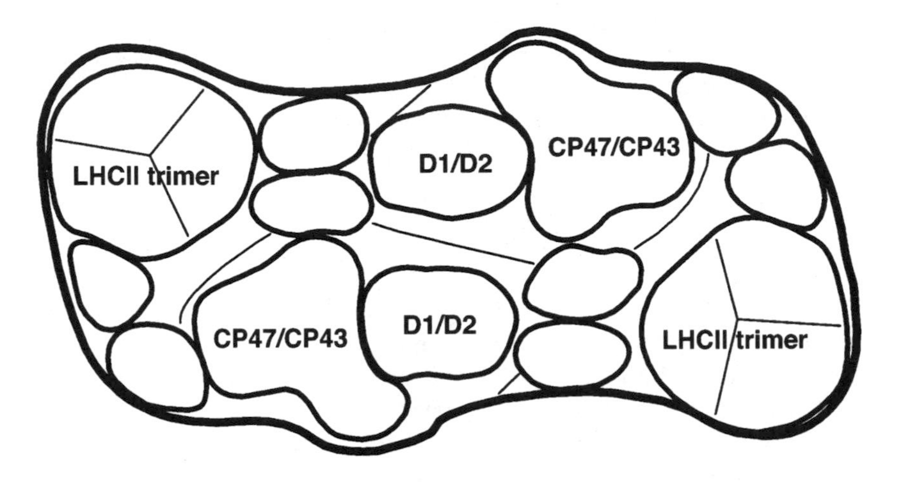

Figure 5 Model of PSII dimer (top view). Drawn from Reference 20. The unlabeled ovals are regions of electron density that cannot be assigned but probably represent the minor CAB complexes CP29, CP26, CP24, and CP22.

ENERGY TRANSFER AND THE PIGMENT-PROTEIN COMPLEXES

There are now many well-defined active preparations of PSII and PSI holo-complexes, core complexes, reaction centers, and antenna complexes. These improved biochemical preparations have yielded higher quality biophysical data for studying the detailed mechanisms of exciton and electron transfer (172). These studies (briefly summarized) have shown that:

1. Excitation energy is rapidly equilibrated throughout the PSI and PSII pigment beds within a few tens of picoseconds (166, 172). All types of particles, including isolated Chl *a/b* antennae (106, 124, 130, 131, 166), have several fast fluorescence decay components, probably resulting from competing energy transfer processes.
2. Within isolated trimeric LHCII, excitation energy equilibrates in less than 10 ps, but the dominant fluorescence lifetime is several ns, giving maximum opportunity for excitation to be transferred to reaction centers (106, 130).
3. The absorption and fluorescence maxima of all the PSII complexes, both Chl *a* and Chl *a/b*–binding, are rather similar, i.e. there are no differences in energy that would funnel excitation energy downhill toward the reaction center (93). The absorption of each isolated Chl protein or complex can be resolved into a number of Gaussian curves (93, 186). These may represent

Chl molecules in different protein environments, but they could also reflect collective excitation of groups of pigments close enough together for excitonic coupling (170).

4. Long wavelength fluorescence is not the sole property of LHCI-730: PSI cores from cyanobacteria and higher plants have several "red" Chls close to the reaction center (168, 179, 184). These low-energy Chls are believed to act as a "sink" for accumulation of excitation energy prior to its transfer to the reaction center.

Considering that there are Chls in many different orientations and environments within each Chl *a/b* protein (see the LHCII structure) to say nothing of the Chls on PsaA/B, CP47, and CP43, it is not surprising that we are far from understanding the contributions of any one Chl protein or its interactions with the others. Readers interested in pursuing these questions will find an entry to the literature in the cited references.

FOR THE FUTURE

The protein-binding proteins of the oxygenic photosynthetic apparatus and the genes encoding them are close to being sorted out. In contrast, our knowledge of how these components are assembled into an exquisitely engineered energy transfer apparatus is still primitive. It is a mystery why there should be such a variety of different light-harvesting proteins, especially in higher plants that are more likely to suffer from too much rather than too little light absorption. Developing techniques for targeted mutagenesis of nuclear genes in *Chlamydomonas reinhardtii* may soon make it possible to study the effects of deleting one or more CAB proteins. In vitro reconstitution and electron/X-ray crystallography should give a better idea of how the pigments are bound and the degree of adaptability among pigment-binding proteins. Better three-dimensional structures will provide a framework for understanding the mechanisms of energy transfer, which are far from being rigorously described at the quantum mechanical level.

ACKNOWLEDGMENTS

We thank all our colleagues who sent reprints and unpublished information, supplied missing references, and provided clarification. We particularly thank W. Kühlbrandt for Figure 2, M. Rögner for help with Figure 5, M. Roobal-Boza for reading the manuscript, and L. Balakshin for manuscript preparation. B. R. Green thanks the Natural Sciences and Engineering Research Council of Canada for continuous support of her research on pigment proteins.

Literature Cited

1. Adamska I, Kloppstech K. 1995. Light stress proteins (ELIPs); the intriguing relatives of *cab* gene family. See Ref. 116a, 3:887–92
2. Adamska I, Ohad I, Kloppstech K. 1992. Synthesis of the early light-inducible protein is controlled by blue light and related to light stress. *Biochim. Biophys. Acta* 89:2610–13
3. Alfonso M, Montoya G, Cases R, Rodriguez R, Picorel R. 1994. Core antenna complexes, CP43 and CP47, of higher plant photosystem II: spectral properties, pigment stoichiometry, and amino acid composition. *Biochemistry* 33:10494–500
4. Allen JF. 1992. Protein phosphorylation in regulation of photosynthesis. *Biochim. Biophys. Acta* 1098:275–335
5. Allen JF. 1995. Thylakoid protein phosphorylation, state 1 state 2 transitions, and photosystem stoichiometry adjustment: redox control at multiple levels of gene expression: minireview. *Physiol. Planta.* 93:196–205
6. Allen KD, Staehelin LA. 1992. Biochemical characterization of photosystem-II antenna polypeptides in grana and stroma membranes of spinach. *Plant Physiol.* 100:1517–26
7. Allen KD, Staehelin LA. 1994. Polypeptide composition, assembly and phosphorylation patterns of the photosystem II antenna system of *Chlamydomonas reinhardtii. Planta* 194:42–54
8. Anderson JM, Andersson B. 1988. The dynamic photosynthetic membrane and regulation of solar energy conversion. *Trends Biochem. Sci.* 13:351–55
9. Andreasson E, Albertsson PA. 1993. Heterogeneity in photosystem-I: the larger antenna of photosystem-I alpha is due to functional connection to a special pool of LHCII. *Biochim. Biophys. Acta* 1141:175–82
10. Apt KE, Clendennen SK, Powers DA, Grossman AR. 1995. The gene family encoding the fucoxanthin chlorophyll proteins from the brown alga *Macrocystis pyrifera. Mol. Gen. Genet.* 246: 455–64
11. Arsalane W, Rousseau B, Thomas JC. 1992. Isolation and characterization of native pigment-protein complexes from two *Eustigmatophyceae. J. Phycol.* 28: 32–36
12. Bassi R, Dainese P. 1992. A supramolecular light-harvesting complex from chloroplast photosystem-II membranes. *Eur. J. Biochem.* 204:317–26
13. Bassi R, Pineau B, Dainese P, Marquardt J. 1993. Carotenoid-binding proteins of photosystem II. *Eur. J. Biochem.* 212:297–303
14. Bassi R, Rigoni F, Giacometti GM. 1990. Chlorophyll binding proteins with antenna function in higher plants and green algae. *Photochem. Photobiol.* 52: 1187–206
15. Bassi R, Simpson D. 1987. Chlorophyll-protein complexes of barley photosystem I. *Eur. J. Biochem.* 163:221–30
16. Bassi R, Soen SY, Frank G, Zuber H, Rochaix JD. 1992. Characterization of chlorophyll *a/b* proteins of photosystem I from *Chlamydomonas reinhardtii. J. Biol. Chem.* 267:25714–21
17. Bassi R, Wollman F-A. 1991. The chlorophyll-*a/b* proteins of photosystem-II in *Chlamydomonas reinhardtii*: isolation, characterization and immunological cross-reactivity to higher-plant polypeptides. *Planta* 183:423–33
18. Bhaya D, Grossman AR. 1991. Targeting proteins to diatom plastids involves transport through an endoplasmic reticulum. *Mol. Gen. Genet.* 229:400–4
19. Bhaya D, Grossman AR. 1993. Characterization of gene clusters encoding the fucoxanthin chlorophyll proteins of the diatom *Phaeodactylum tricornutum. Nucleic Acids Res.* 21:4458–66
20. Boekema EJ, Hankamer B, Bald D, Kruip J, Nield J, et al. 1995. Supramolecular structure of the photosystem II complex from green plants and cyanobacteria. *Proc. Natl. Acad. Sci. USA* 92:175–79
21. Boekema EJ, Wynn RM, Malkin R. 1992. The structure of spinach photosystem I studied by electron microscopy. *Biochim. Biophys. Acta* 1017:49–56
22. Bricker TM. 1990. The structure and function of CPa-1 and CPa-2 in photosystem II. *Photosynth. Res.* 24:1–13
23. Bryant DA. 1992. Molecular biology of photosystem I. In *The Photosystems: Structure, Function and Molecular Bi-*

ology, ed. J Barber, pp. 501–41. Amsterdam: Elsevier

24. Bryant DA. 1992. Puzzles of chloroplast ancestry. *Curr. Biol.* 2:240–42

24a. Bryant DA, ed. 1994. *The Molecular Biology of Cyanobacteria.* Dordrecht: Kluwer

25. Büchel C, Wilhelm C. 1993. Isolation and characterization of a photosystem-I–associated antenna (LHC-I) and a photosystem-I core complex from the chlorophyll-*c*-containing alga *Pleurochloris meiringensis* (Xanthophyceae). *J. Photochem. Photobiol. B* 20:87–93

26. Büchel C, Wilhelm C, Hauswirth N, Wild A. 1992. Evidence for a lateral heterogeneity by patch-work like areas enriched with photosystem I complexes in the three thylakoid lamellae of *Pleurochloris meiringensis* (Xanthophyceae). *Cryptogam. Bot.* 2:375–86

27. Buetow DE, Chen H, Erdös G, Yi LSH. 1988. Regulation and expression of the multigene family coding light-harvesting chlorophyll *a*/*b*-binding proteins of photosystem II. *Photosynth. Res.* 18:61–97

28. Burnap RL, Troyan T, Sherman LA. 1993. The highly abundant chlorophyll-protein complex of iron-deficient *Synechococcus* sp PCC7942 (Cp43′) is encoded by the *isi*A gene. *Plant Physiol.* 103:893–902

29. Camm EL, Green BR. 1989. The chlorophyll *ab* complex, CP29, is associated with the photosystem-II reaction center core. *Biochim. Biophys. Acta* 974:180–84

30. Caron L, Douady D, Rousseau B, Quinet-Szely M, Berkaloff C. 1995. Light-harvesting complexes from a brown alga: biochemical and molecular study. See Ref. 116a, 3:223–26

31. Deleted in proof

32. Caron L, Remy R, Berkaloff C. 1988. Polypeptide composition of light-harvesting complexes from some brown algae and diatoms. *FEBS Lett.* 229:11–15

33. Cavalier-Smith T. 1993. The origin, losses and gains of chloroplasts. See Ref. 112a, pp. 291–348

34. Cavalier-Smith T, Allsopp MTEP, Chao EE. 1994. Chimeric conundra: Are nucleomorphs and chromists monophyletic or polyphyletic? *Proc. Natl. Acad. Sci. USA* 91:11368–72

35. Chitnis PR, Thornber JP. 1988. The major light-harvesting complex of photosystem II: aspects of its molecular and cell biology. *Photosynth. Res.* 16:41–63

36. Chitnis PR, Xu Q, Chitnis VP, Nechushtai R. 1995. Function and organization of photosystem I polypeptides. *Photosynth. Res.* 44:23–40

37. Cunningham FX, Schiff JA. 1986. Chlorophyll-protein complexes from *Euglena gracilis* and mutants deficient in chlorophyll *b*. II. Polypeptide composition. *Plant Physiol.* 80:231–38

38. Dainese P, Bassi R. 1991. Subunit stoichiometry of the chloroplast photosystem-II antenna system and aggregation state of the component chlorophyll-a/b binding proteins. *J. Biol. Chem.* 266:8136–42

39. Damm I, Green BR. 1994. Separation of closely related intrinsic membrane polypeptides of the photosystem-II light-harvesting complex (LHC-II) by reversed-phase high-performance liquid chromatography on a poly (styrene-divinylbenzene) column. *J. Chromatogr.* 664:33–38

40. Debus RJ. 1992. The manganese and calcium ions of photosynthetic oxygen evolution. *Biochim. Biophys. Acta* 1102:269–352

41. deVitry C, Wollman F-A, Delepelaire P. 1984. Function of the polypeptides of the photosystem II reaction center in *Chlamydomonas reinhardtii. Biochim. Biophys. Acta* 767:415–22

42. Dolganov NAM, Bhaya D, Grossman AR. 1995. Cyanobacterial protein with similarity to the chlorophyll *a/b* binding proteins of higher plants: evolution and regulation. *Proc. Natl. Acad. Sci. USA.* 92:636–40

43. Douady D, Rousseau B, Caron L. 1994. Fucoxanthin-chlorophyll ***a/c*** light-harvesting complexes of *Laminaria saccharina:* partial amino acid sequences and arrangement in thylakoid membranes. *Biochemistry* 33:3165–70

44. Douglas SE. 1994. Chloroplast origins and evolution. See Ref. 24a, pp. 91–118

45. Dreyfuss BW, Thornber JP. 1994. Assembly of the light-harvesting complexes (LHCs) of photosystem II: monomeric LHC IIb complexes are intermediates in the formation of oligomeric LHC IIb complexes. *Plant Physiol.* 106:829–39

46. Dreyfuss BW, Thornber JP. 1994. Organization of the light-harvesting complex of photosystem I and its assembly during plastid development. *Plant Physiol.* 106:841–48

47. Durnford DG, Green BR. 1994. Characterization of the light harvesting proteins of the chromophytic alga, *Olisthodiscus luteus* (*Heterosigma carterae*). *Biochim. Biophys. Acta* 1184:118–26

48. Durnford DG, Green BR. 1995. Characterization of a gene encoding a fu-

coxanthin-chlorophyll protein from the chromophytic alga, *Heterosigma carterae*. See Ref. 116a, 1:963–66
49. Falbel TG, Staehelin LA. 1992. Species-related differences in the electrophoretic behavior of CP29 and CP26: an immunochemical analysis. *Photosynth. Res.* 34:249–62
50. Felsenstein J. 1992. PHYLIP (phylogeny inference package). Univ. Wash., Seattle
51. Ficner R, Lobeck K, Schmidt G, Huber R. 1992. Isolation, crystallization, crystal structure analysis and refinement of β-phycoerythrin from the red alga *Porphyridium sordidum* at 2.2 Å resolution. *J. Mol. Biol.* 228:935–50
52. Frankel LK, Bricker TM. 1995. Interaction of the 33-kDa extrinsic protein with photosystem II: identification of domains on the 33-kDa protein that are shielded from NHS-biotinylation by photosystem II. *Biochemistry* 34:7492–97
53. Funk C, Adamska I, Green BR, Andersson B, Renger G. 1995. The nuclear-encoded chlorophyll-binding PSII-S protein is stable in the absence of pigments. *J. Biol. Chem.* 270:30141–47
54. Funk C, Schröder WP, Green BR, Renger G, Andersson B. 1994. The intrinisic 22-kDa protein is a chlorophyll-binding subunit of photosystem II. *FEBS Lett.* 342:261–66
55. Funk C, Schröder WP, Napiwotzki A, Tjus SE, Renger G, Andersson B. 1995. The PSII-S protein of higher plants: a new type of pigment-binding protein. *Biochemistry* 34:11133–41
56. Gagné G, Guertin M. 1992. The early genetic response to light in the green unicellular alga *Chlamydomonas eugametos* grown under light/dark cycles involves genes that represent direct responses to light and photosynthesis. *Plant Mol. Biol.* 18:429–45
57. Ghanotakis DF, Yocum CF. 1990. Photosystem II and the oxygen-evolving complex. *Annu. Rev. Plant Physiol. Plant Mol. Biol.* 41:255–76
58. Gibbs SP. 1978. The chloroplasts of *Euglena* may have evolved from symbiotic green algae. *Can. J. Bot.* 56:2883–89
59. Gibbs SP. 1981. The chloroplast endoplasmic reticulum: structure, function and evolutionary significance. *Int. Rev. Cytol.* 72:49–99
60. Golbeck JH. 1992. Structure and function of photosystem I. *Annu. Rev. Plant. Physiol. Plant Mol. Biol.* 43:293–324
61. Golbeck JH. 1993. Shared thematic elements in photochemical reaction centers. *Proc. Natl. Acad. Sci. USA* 90:1642–46
62. Golden SS, Morden CW, Greer KL. 1993. Comparison of sequences and organization of photosynthesis genes among the prochlorophyte *Prochlorothrix hollandica,* cyanobacteria, and chloroplasts. See Ref. 112a, pp. 141–58
63. Gray MW, Doolittle WF. 1982. Has the endosymbiont hypothesis been proven? *Microbiol. Rev.* 46:1–42
64. Green BR. 1988. The chlorophyll-protein complexes of higher plant photosynthetic membranes, or Just what green band is that? *Photosynth. Res.* 15:3–32
65. Green BR, Camm EL. 1990. Relationship of Chl *a*/*b*-binding and related polypeptides in PSII core particles. In *Current Research in Photosynthesis,* ed. M Baltscheffsky, 1:659–62. Dordrecht: Kluwer
66. Green BR, Durnford DG, Aebersold R, Pichersky E. 1992. Evolution of structure and function in the chlorophyll *a*/*b* antenna protein families. In *Research in Photosynthesis,* ed. N Murata, pp. 195–201. Dordrecht: Kluwer
67. Green BR, Kühlbrandt W. 1995. Sequence conservation of light-harvesting and stress-response proteins in relation to the three-dimensional molecular structure of LHCII. *Photosynth. Res.* 44:139–48
68. Green BR, Pichersky E. 1994. Hypothesis for the evolution of three-helix Chl *a*/*b* and Chl *a*/*c* light-harvesting antenna proteins from two-helix and four-helix ancestors. *Photosynth. Res.* 39:149–62
69. Green BR, Pichersky E, Kloppstech K. 1991. The chlorophyll *a*/*b*-binding light-harvesting antennas of green plants: the story of an extended gene family. *Trends Biochem. Sci.* 16:181–86
70. Green BR, Salter AH. 1995. Light regulation of nuclear-encoded thylakoid proteins. In *Molecular Genetics of Photosynthesis,* ed. B Anderson, AH Salter, J Barber, pp. 75–103. Oxford: Oxford Univ. Press
71. Green BR, Shen D, Aebersold R, Pichersky E. 1992. Identification of the polypeptides of the major light-harvesting complex of photosystem II (LHC II) with their genes in tomato. *FEBS Lett.* 305:18–22
72. Grevby C, Sundqvist C. 1992. Characterization of light-harvesting complex in *Ochromonas danica* (Chrysophyceae). *J. Plant Physiol.* 140:414–20
73. Grossman AR, Manodori A, Snyder D. 1990. Light-harvesting proteins of diatoms: the relationship to the chlorophyll *a*/*b* binding proteins of higher plants

and their mode of transport into plastids. *Mol. Gen. Genet.* 224:91–100

74. Grossman AR, Schaefer MR, Chiang GG, Collier JL. 1993. The phycobilisome, a light-harvesting complex responsive to environmental conditions. *Microbiol. Rev.* 57:725–49
75. Harrison MA, Nemson JA, Melis A. 1993. Assembly and composition of the chlorophyll *a-b* light-harvesting complex of barley (*Hordeum vulgare* L.): immunochemical analysis of chlorophyll *b*-less and chlorophyll *b*-deficient mutants. *Photosynth. Res.* 38:141–51
76. Hayashi H, Fujimura Y, Mohanty PS, Murata N. 1993. The role of CP-47 in the evolution of oxygen and the binding of the extrinsic 33-kDa protein to the core complex of photosystem-II as determined by limited proteolysis. *Photosynth. Res.* 36:35–42
77. Henrysson T, Schröder WP, Spangfort M, Akerlund H-E. 1989. Isolation and characterization of the chlorophyll *a/b* protein complex CP29 from spinach. *Biochim. Biophys. Acta* 977:301–8
78. Hiller RG, Anderson JM, Larkum AWD. 1991. The chlorophyll-protein complexes of algae. See Ref. 150a, pp. 529–47
79. Hiller RG, Wrench PM, Gooley AP, Shoebridge G, Breton J. 1993. The major intrinsic light-harvesting protein of *Amphidinium:* characterization and relation to other light-harvesting proteins. *Photochem. Photobiol.* 57:125–31
80. Hiller RG, Wrench PM, Sharples FP. 1995. The light-harvesting chlorophyll *a-c*-binding protein of dinoflagellates: a putative polyprotein. *FEBS Lett.* 363: 175–78
81. Hobe S, Foster R, Klingler J, Paulsen H. 1995. N-proximal sequence motif in light-harvesting chlorophyll *a/b*-binding protein is essential for the trimerization of light-harvesting chlorophyll *a/b* complex. *Biochemistry* 34:10224–28
82. Hobe S, Prytulla S, Kühlbrandt W, Paulsen H. 1994. Trimerization and crystallization of reconstituted light-harvesting chlorophyll *a/b* complex. *EMBO J.* 13: 3423–29
83. Holzenburg A, Bewley MC, Wilson FH, Nicholson WV, Ford RC. 1993. Three-dimensional structure of photosystem II. *Nature* 363:470–74
84. Hoober JK, White RA, Marks DB, Gabriel JL. 1994. Biogenesis of thylakoid membranes with emphasis on the process in *Chlamydomonas. Photosynth. Res.* 39:15–31
85. Horton P, Ruban AV, Walters RG. 1996. Regulation of light-harvesting in green plants. *Annu. Rev. Plant Physiol. Plant Mol. Biol.* 47:655–84
86. Houlné G, Schantz R. 1988. Characterization of cDNA sequences for LHC I apoproteins in *Euglena gracilis:* the mRNA encodes a large precursor containing several consecutive divergent polypeptides. *Mol. Gen. Genet.* 213: 479–86
87. Iglesias-Prieto R, Govind NS, Trench RK. 1993. Isolation and characterization of 3 membrane-bound chlorophyll protein complexes from 4 dinoflagellate species. *Philos. Trans. R. Soc. London Ser. B* 340:381–92
88. Ikeuchi M. 1992. Subunit proteins of photosystem-I: mini review. *Plant Cell Physiol.* 33:669–76
89. Irrgang KD, Shi LX, Funk C, Schröder WP. 1995. A nuclear-encoded subunit of the photosystem II reaction center. *J. Biol. Chem.* 270:17588–93
90. Jansson S. 1994. The light-harvesting chlorophyll *a/b* binding proteins. *Biochim. Biophys. Acta* 1184:1–19
91. Jansson S, Andersen B, Scheller HV. 1995. Subunit organization of the higher plant photosystem I (PSI) holocomplex. See Ref. 116a, 3:385–88
92. Jansson S, Pichersky E, Bassi R, Green BR, Ikeuchi M, et al. 1992. A nomenclature for the genes encoding the chlorophyll *a/b*-binding proteins of higher plants. *Plant Mol. Biol. Rep.* 10:242–53

92a. Jeffrey SW. 1989. Chlorophyll ***c*** pigments and their distribution in the chromophyte algae. In *The Chromophyte Algae: Problems and Perspectives,* ed. JC Green, BSC Leadbeater, WL Diver, pp. 13–36. Oxford: Clarendon

93. Jennings RC, Garlaschi FM, Bassi R, Zucchelli G, Vianelli A, Dainese P. 1993. A study of photosystem-II fluorescence emission in terms of the antenna chlorophyll-protein complexes. *Biochim. Biophys. Acta* 1183:194–200
94. Jiao S, Fawley MW. 1994. A cDNA clone encoding a light-harvesting protein from *Mantoniella squamata. Plant Physiol.* 104:797–98
95. Jovine RVM, Johnsen G, Prezelin BB. 1995. Isolation of membrane bound light-harvesting-complexes from the dinoflagellates *Heterocapsa pygmaea* and *Prorocentrum minimum. Photosynth. Res.* 44:127–38
96. Karrasch S, Bullough PA, Ghosh R. 1995. The 8.5– angstrom projection map of the light-harvesting complex I from *Rhodospirillum rubrum* reveals a ring composed of 16 subunits. *EMBO J.* 14: 631–38
97. Kim S, Pichersky E, Yocum CF. 1994.

Topological studies of spinach 22-kDa protein of photosystem II. *Biochim. Biophys. Acta* 1188:339–48
98. Kim S, Sandusky P, Bowlby NR, Aebersold R, Green BR, et al. 1992. Characterization of a spinach *psb*S cDNA encoding the 22 kDa protein of photosystem II. *FEBS Lett.* 314:67–71
99. Knoetzel J, Svendsen I, Simpson DJ. 1992. Identification of the photosystem-I antenna polypeptides in barley: isolation of 3 pigment-binding antenna complexes. *Eur. J. Biochem.* 206:209–15
100. Krauss N, Hinrichs W, Witt I, Fromme P, Pritzkow W, et al. 1993. 3-Dimensional structure of system-I of photosynthesis at 6 angstrom resolution. *Nature* 361:326–31
101. Krol M, Spangfort MD, Huner NPA, Oquist G, Gustafsson P, Jansson S. 1995. Chlorophyll *a/b*-binding proteins, pigment conversions, and early light-induced proteins in a chlorophyll *b*-less barley mutant. *Plant Physiol.* 107:873–83
102. Kroth-Pancic PG. 1995. Nucleotide sequence of two cDNAs encoding fucoxanthin chlorophyll *a/c* proteins in the diatom *Odontella sinensis. Plant Mol. Biol.* 27:825–28
103. Kruip J, Bald D, Boekema EJ, Rögner M. 1994. Evidence for the existence of trimeric and monomeric Photosystem I complexes in thylakoid membranes from cyanobacteria. *Photosynth. Res.* 40:279–86
104. Kühlbrandt W. 1994. Structure and function of the plant light-harvesting complex, LHC-II. *Curr. Opin. Struct. Biol.* 4:519–28
105. Kühlbrandt W, Wang DN, Fujiyoshi Y. 1994. Atomic model of plant light-harvesting complex by electron crystallography. *Nature* 367:614–21
106. Kwa SLS, Groeneveld FG, Dekker JP, Van Grondelle R, Van Amerongen H, et al. 1992. Steady-state and time-resolved polarized light spectroscopy of the green plant light-harvesting complex-II. *Biochim. Biophys. Acta* 1101: 143–46
107. Larkum AWD, Scaramuzzi C, Cox GC, Hiller RG, Turner AG. 1994. Light-harvesting chlorophyll *c*-like pigment in *Prochloron. Proc. Natl. Acad. Sci. USA.* 91:679–83
108. La Roche J, Henry D, Wyman K, Sukenik A, Falkowski P. 1994. Cloning and nucleotide sequence of a cDNA encoding a major fucoxanthin-, chlorophyll *a/c*–containing protein from the chrysophyte *Isochrysis galbana:* implications for evolution of the *cab* gene family. *Plant Mol. Biol.* 25:355–68
109. La Roche J, Partensky F, Falkowski P. 1995. The major light-harvesting chl binding protein of *Prochlorococcus marinus* is similar to CP43′, a chl binding protein induced by iron-depletion in cyanobacteria. See Ref. 116a, 1:171–74
110. Lee ALC, Thornber JP. 1995. Analysis of the pigment stoichiometry of pigment-protein complexes from barley (*Hordeum vulgare*): the xanthophyll cycle intermediates occur mainly in the light-harvesting complexes of photosystem I and photosystem II. *Plant Physiol.* 107:565–74
111. Lers A, Levy H, Zamir A. 1991. Co-regulation of a gene homologous to early light-induced genes in higher plants and beta-carotene biosynthesis in the alga *Dunaliella bardawail. J. Biol. Chem.* 266:13698–705
112. Levy H, Gokhman I, Zamir A. 1992. Regulation and light-harvesting complex II association of a *Dunaliella* protein homologous to early light-induced proteins in higher plants. *J. Biol. Chem.* 267:18831–36
112a. Lewin RA, ed. 1993. *Origins of Plastids: Symbiogenesis, Prochlorophytes, and the Origins of Chloroplasts.* New York: Chapman Hall
113. Lichtlé C, McKay RML, Gibbs SP. 1992. Immunogold localization of photosystem I and photosystem II light-harvesting complexes in cryptomonad thylakoids. *Biol. Cell* 74:187–94
114. Livne A, Katcoff D, Yacobi YZ, Sukenik A. 1992. Pigment-protein complexes of *Nannochloropsis* sp. (Eustigmatophyceae): an alga lacking chlorophylls *b* and *c*. In *Research in Photosynthesis,* ed. N Murata, pp. 203–6. Boston: Kluwer
115. Lyon MK, Marr KM, Furcinitti PS. 1993. Formation and characterization of two-dimensional crystals of photosystem II. *J. Struct. Biol.* 110:133–40
116a. Machold O. 1991. The structure of light-harvesting complex II as deduced from its polypeptide composition and stoichiometry I: studies with *Vicia faba. J. Plant Physiol.* 38:678–84
116a. Mathis P, ed. 1996. *Photosynthesis: From Light to Biosphere.* Dordrecht: Kluwer
117. Matthijs HCP, van der Staay GWM, Mur LR. 1994. Prochlorophytes: the 'other' cyanobacteria? See Ref. 24a, pp. 49–64
118. McDermott G, Prince SM, Freer AA, Hawthornthwaite-Lawless AM, Papiz MZ, et al. 1995. Crystal structure of an

integral membrane light-harvesting complex from photosynthetic bacteria. *Nature* 374:517–21

119. Melis A. 1991. Dynamics of photosynthetic membrane composition and function. *Biochim. Biophys. Acta* 1058: 87–106
120. Michel H, Deisenhofer J. 1988. Relevance of the photosynthetic reaction center from purple bacteria to the structure of photosystem II. *Biochemistry* 27:1–7
121. Morishige DT, Thornber JP. 1994. Identification of a novel light-harvesting complex II protein (LHC IIc′). *Photosynth. Res.* 39:33–38
122. Morishige DT, Thornber JP. 1992. Identification and analysis of a barley cDNA clone encoding the 31-kilodalton LHC IIa (CP29) apoprotein of the light-harvesting antenna complex of photosystem II. *Plant Physiol.* 98:238–45
123. Muchhal US, Schwarzbach SD. 1992. Characterization of a *Euglena* gene encoding a polyprotein precursor to the light-harvesting chlorophyll-*a/b*-binding protein of photosystem II. *Plant Mol. Biol.* 18:287–99
124. Mukerji I, Sauer K. 1993. Energy transfer dynamics of an isolated light harvesting complex of photosystem-I from spinach: time-resolved fluorescence measurements at 295-K and 77-K. *Biochim. Biophys. Acta* 1142:311–20
125. Nechushtai R, Cohen Y, Chitnis PR. 1995. Assembly of the chlorophyll-protein complexes. *Photosynth. Res.* 44: 165–81
126. Nitschke N, Rutherford AW. 1991. Photosynthetic reaction centres: variations on a common theme? *Trends Biochem. Sci.* 16:241–45
127. Norris BJ, Miller DJ. 1994. Nucleotide sequence of a cDNA clone encoding the precursor of the peridinin-chlorophyll *a*-binding protein from the dinoflagellate *Symbiodinium* sp. *Plant Mol. Biol.* 24:673–77
128. Nussberger S, Dörr K, Wang DN, Kühlbrandt W. 1993. Lipid-protein interactions in crystals of plant light-harvesting complex. *J. Mol. Biol.* 234: 347–56

128a. Ort DT, Yocum CF, eds. 1996. *Oxygenic Photosynthesis: The Light Reactions.* Dordrecht: Kluwer

129. Palenik B, Haselkorn R. 1992. Multiple evolutionary origins of prochlorophytes: the Chl *b* containing prokaryotes. *Nature* 355:265–67
130. Pålsson LO, Spangfort MD, Gulbinas V, Gillbro T. 1994. Ultrafast chlorophyll *b* chlorophyll *a* excitation energy transfer in the isolated light harvesting complex, LHC II, of green plants: implications for the organisation of chlorophylls. *FEBS Lett.* 339:134–38
131. Pålsson LO, Tjus SE, Andersson B, Gillbro T. 1995. Ultrafast energy transfer dynamics resolved in isolated spinach light-harvesting complex I and the LHC I-730 subpopulation. *Biochim. Biophys. Acta* 1230:1–9
132. Passaquet C, Lichtlé C. 1995. Molecular study of a light-harvesting apoprotein of *Giraudyopsis stellifer* (Chrysophyceae). *Plant Mol. Biol.* 29:135–48
133. Deleted in proof
134. Passaquet C, Thomas JC, Caron L, Hauswirth N, Puel F, Berkaloff C. 1991. Light-harvesting complexes of brown algae: biochemical characterization and immunological relationships. *FEBS Lett.* 280:21–26
135. Paulsen H. 1995. Chlorophyll *a/b*-binding proteins. *Photochem. Photobiol.* 62: 367–82
136. Peter GF, Thornber JP. 1991. Biochemical composition and organization of higher plant photosystem-II light-harvesting pigment-proteins. *J. Biol. Chem.* 266:16745 54
137. Peter GF, Thornber JP. 1991. Biochemical evidence that the higher plant photosystem II core complex is organized as a dimer. *Plant Cell Physiol.* 32:1237–50
138. Pichersky E, Subramaniam R, White MJ, Reid J, Aebersold R, Green BR. 1991. Chlorophyll *a/b* binding (CAB) polypeptides of CP29, the internal chlorophyll *a/b* complex of PSII: characterization of the tomato gene encoding the 26-kDA (type I) polypeptide, and evidence for a second CP29 polypeptide. *Mol. Gen. Genet.* 227:277–84
139. Post AF, Bullerjahn GS. 1994. The photosynthetic machinery in prochlorophytes: structural properties and ecological significance. *FEMS Microbiol. Rev.* 13:393–413
140. Preiss S, Peter GF, Anandan S, Thornber JP. 1993. The multiple pigment-proteins of the photosystem-I antenna. *Photochem. Photobiol.* 57:152–57
141. Pyszniak AM, Gibbs SP. 1992. Immunocytochemical localization of photosystem-I and the fucoxanthin-chlorophyll a/c light-harvesting complex in the diatom *Phaeodactylum tricornutum. Protoplasma* 166:208–17
142. Raven PH. 1970. A multiple origin for plastids and mitochondria. *Science* 169: 641–46
143. Reith M. 1995. Molecular biology of rhodophyte and chromophyte plastids.

Annu. Rev. Plant Physiol. Plant Mol. Biol. 46:549–75
144. Rhiel E, Lange W, Mörschel E. 1993. The unusual light-harvesting complex of *Mantoniella squamata*: supramolecular composition and assembly. *Biochim. Biophys. Acta* 1143:163–72
145. Rhiel E, Mörschel E. 1993. The atypical chlorophyll *a/b/c* light-harvesting complex of *Mantoniella squamata*: molecular cloning and sequence analysis. *Mol. Gen. Genet.* 240:403–13
146. Ruban AV, Horton P. 1992. Mechanism of delta pH-dependent dissipation of absorbed excitation energy by photosynthetic membranes. 1. Spectroscopic analysis of isolated light-harvesting complexes. *Biochim. Biophys. Acta* 1102:30–38
147. Ruffle SV, Donnelly D, Blundell TL, Nugent JHA. 1992. A three-dimensional model of the photosystem II reaction centre of *Pisum sativum. Photosynth. Res.* 34:287–300
148. Santini C, Tidu V, Tognon G, Magaldi AG, Bassi R. 1994. Three-dimensional structure of the higher-plant photosystem II reaction centre and evidence for its dimeric organization in vivo. *Eur. J. Biochem.* 221:307–15
149. Satoh K. 1993. Isolation and properties of the photosystem II reaction center. In *The Photosynthetic Reaction Center,* ed. J Deisenhofer, JR Norris, 1:289–318. San Diego: Academic
150. Sayre RT, Wrobelboerner EA. 1994. Molecular topology of the photosystem II chlorophyll alpha binding protein, CP 43: topology of a thylakoid membrane protein. *Photosynth. Res.* 40:11–19
150a. Scheer H, ed. 1991. *Chlorophylls*. Baton Rouge, LA: CRC Press
151. Schelvis JPM, Germano M, Aartsma TJ, Van Gorkom HJ. 1995. Energy transfer and trapping in photosystem II core particles with closed reaction centers. *Biochim. Biophys. Acta* 1230:165–69
152. Schmitt A, Frank G, James P, Staudenmann W, Zuber H, Wilhelm C. 1994. Polypeptide sequence of the chlorophyll *a/b/c*-binding protein of the prasinophycean alga *Mantoniella squamata. Photosynth. Res.* 40:269–77
153. Schmitt A, Herold A, Welte C, Wild A, Wilhelm C. 1993. The light-harvesting system of the unicellular alga *Mantoniella squamata* (Prasinophyceae): evidence for the lack of a photosystem I–specific antenna complex. *Photochem. Photobiol.* 57:132–38
154. Schubert WD, Klukas O, Krauss N, Saenger W, Fromme P, Witt HT. 1996. Present state of the crystal structure analysis of photosystem I. See Ref. 116a, 2:3–10
155. Seibert M. 1993. Biochemical, biophysical, and structural characterization of the isolated photosystem II reaction center complex. In *The Photosynthetic Reaction Center,* ed. J Deisenhofer, JR Norris, 1:319–56. San Diego: Academic
156. Seibert M. 1995. Reflections on the nature and function of the photosystem II reaction centre. *Aust. J. Plant Physiol.* 22:161–66
157. Sigrist M, Staehelin LA. 1994. Appearance of Type 1, 2, and 3 light-harvesting complex II and light-harvesting complex I proteins during light-induced greening of barley (*Hordeum vulgare*) etioplasts. *Plant Physiol.* 104:135–45
158. Straus NA. 1994. Iron deprivation: physiology and gene regulation. See Ref. 24a, pp. 731–50
159. Sukenik A, Livne A, Neori A, Yacobi YZ, Katcoff D. 1992. Purification and characterization of a light-harvesting chlorophyll-protein complex from the marine Eustigmatophyte *Nannochloropsis* sp. *Plant Cell Physiol.* 33:1041–48
160. Sulli C, Schwartzbach SD. 1995. The polyprotein precursor to the Euglena light-harvesting chlorophyll *a/b*-binding protein is transported to the Golgi apparatus prior to chloroplast import and polyprotein processing. *J. Biol. Chem.* 270:13084–90
161. Sundqvist C, Ryberg M. 1993. *Pigment-Protein Complexes in Plastids: Synthesis and Assembly.* New York: Academic
162. Thornber JP, Morishige DT, Anandan S, Peter GF. 1991. Chlorophyll-carotenoid protein of higher plant thylakoids. See Ref. 150a, pp. 549–85
163. Thornber JP, Peter GF, Morishige DT, Gomez S, Anandan S, et al. 1993. Light harvesting in photosystem-I and photosystem-II. *Biochem. Soc. Trans.* 21:15–18
164. Tjus SE, Roobol-Boza M, Pålsson LO, Andersson B. 1995. Rapid isolation of photosystem I chlorophyll-binding proteins by anion exchange perfusion chromatography. *Photosynth. Res.* 45:41–49
165. Tronrud DE, Schmid MF, Matthews BW. 1986. Structure and x-ray amino acid sequence of a bacteriochlorophyll *a* protein from *Prosthecochloris aestuarii* refined at 1.9 Å resolution. *J. Mol. Biol.* 188:443–53
166. Turconi S, Weber N, Schweitzer G, Strotmann H, Holzwarth AR. 1994. Energy transfer and charge separation kinetics in Photosystem I. 2. Picosecond fluorescence study of various PS I particles and light-harvesting complex iso-

lated from higher plants. *Biochim. Biophys. Acta* 1187:324–34

167. Urbach E, Robertson DL, Chisholm S. 1992. Multiple evolutionary origins of prochlorophytes within the cyanobacterial radiation. *Nature* 355:267–70
168. Valkunas L, Liuolia V, Dekker JP, Van Grondelle R. 1995. Description of energy migration and trapping in Photosystem I by a model with two distance scaling parameters. *Photosynth. Res.* 43: 149–54
169. Deleted in proof
170. Van Amerongen H, Van Bolhuis BM, Betts S, Mei R, Van Grondelle R, et al. 1994. Spectroscopic characterization of CP26, a chlorophyll *a/b* binding protein of the higher plant photosystem II complex. *Biochim. Biophys. Acta* 1188:227–34
171. Van der Staay GWM, Ducret A, Aebersold R, Li R, Golden SS, et al. 1995. The Chl *a/b* antenna from prochlorophytes is related to the iron stress-induced Chl *a* antenna (IsiA) from cyanobacteria. See Ref. 116a, 1:175–78
172. Van Grondelle R, Dekker JP, Gillbro T, Sundstrom V. 1994. Energy transfer and trapping in photosynthesis. *Biochim. Biophys. Acta* 1187:1–655
173. Vermaas WFJ. 1993. Molecular-biological approaches to analyze photosystem II structure and function. *Annu. Rev. Plant Physiol. Plant Mol. Biol.* 44:457–81
174. Vermaas WFJ, Styring S, Schröder WP, Andersson B. 1993. Photosynthetic water oxidation: the protein framework. *Photosynth. Res.* 38:249–63
175. Vesk M, Dwarte D, Fowler S, Hiller RG. 1992. Freeze fracture immunocytochemistry of light-harvesting pigment complexes in a Cryptophyte. *Protoplasma* 170:166–76
176. von Wettstein D, Gough S, Kannangara CG. 1995. Chlorophyll biosynthesis. *Plant Cell* 7:1039-57
177. Webber AN, Bingham SE, Lee H. 1995. Genetic engineering of thylakoid protein complexes by chloroplast transformation in *Chlamydomonas reinhardtii*. *Photosynth. Res.* 44:191–205
178. Wedel N, Klein R, Ljungberg U, Andersson B, Herrmann RG. 1992. The single-copy gene *psb*S codes for a phylogenetically intriguing 22 kDa polypeptide of photosystem II. *FEBS Lett.* 314:61–66
179. Werst M, Jia YW, Mets L, Fleming GR. 1992. Energy transfer and trapping in the photosystem-I core antenna: a temperature study. *Biophys. J.* 61:868–78
180. White MJ, Green BR. 1988. Intermittent-light chloroplasts are not developmentally equivalent to chlorina f2 chloroplasts in barley. *Photosynth. Res.* 15:195–203
181. Wolfe GR, Cunningham FX, Durnford D, Green BR, Gantt E. 1994. Evidence for a common origin of chloroplasts with light-harvesting complexes of different pigmentation. *Nature* 367:566–68
182. Wolfe GR, Cunningham FX, Grabowski B, Gantt E. 1994. Isolation and characterization of photosystems I and II from the red alga *Porphyridium cruentum. Biochim. Biophys. Acta* 1188:357–66
183. Wollenberger L, Weibull C, Albertsson PA. 1995. Further characterization of the chloroplast grana margins: the non-detergent preparation of granal photosystem I cannot reduce ferredoxin in the absence of NADP(+) reduction. *Biochim. Biophys. Acta* 1230:10–22
184. Woolf VM, Wittmershaus BP, Vermaas WFJ, Tran TD. 1994. Resolution of low-energy chlorophylls in photosystem I of *Synechocystis* sp. PCC 6803 at 77 and 295 K through fluorescence excitation anisotropy. *Photosynth. Res.* 40:21–34
185. Yamamoto HY, Bassi R. 1996. Carotenoids: localization and function. See Ref. 128a, in press
186. Zucchelli G, Dainese P, Jennings RC, Breton J, Garlaschi FM, Bassi R. 1994. Gaussian decomposition of absorption and linear dichroism spectra of outer antenna complexes of photosystem II. *Biochemistry* 33:8982–90

AUTHOR INDEX

A

B

D

E

F

H

K

L

N

O

P

Q

R

S

T

U

V

W

X

Y

Z

SUBJECT INDEX

B

C

D

G

H

I

M

N

O

S

T

U

V

W

X

Y

Z

CUMULATIVE INDEXES

CONTRIBUTING AUTHORS, VOLUMES 37–47

CHAPTER TITLES, VOLUMES 37–47

Intracellular Communication

Cell Development

TISSUE, ORGAN, AND WHOLE PLANT EVENTS

Signal Transduction in the Plant/Hormonal Regulation

ACCLIMATION AND ADAPTATION

Economic Botany

Physiological Ecology

Plant Genetics/Evolution